MODERN FOOD MICRO- BIOLOGY

MODERN FOOD MICRO- BIOLOGY

Third Edition

JAMES M. JAY
Wayne State University

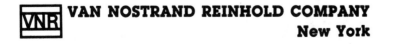 **VAN NOSTRAND REINHOLD COMPANY**
New York

Printed in the United States of America

Designed by Anna Kurz

Van Nostrand Reinhold Company Inc.
115 Fifth Avenue
New York, New York 10003

Van Nostrand Reinhold Company Limited
Molly Millars Lane
Wokingham, Berkshire RG11 2PY, England

Van Nostrand Reinhold
480 La Trobe Street
Melbourne, Victoria 3000, Australia

Macmillan of Canada
Division of Canada Publishing Corporation
164 Commander Boulevard
Agincourt, Ontario M1S 3C7, Canada

16 15 14 13 12 11 10 9 8 7 6 5 4 3 2 1

Library of Congress Cataloging-in-Publication Data

Jay, James M. (James Monroe), 1927–
    Modern food microbiology.

    Bibliography: p.
    Includes index.
    1. Food—Microbiology. I. Title.
QR115.J3   1986   576'.163   85-26623
ISBN 0-442-24445-2

CONTENTS

Preface   xiii

I. Historical Background   1

1. History of Microorganisms in Food   3
   Historical Developments   5
   References   9

II. Sources, Types, Incidence, and Behavior of Microorganisms in Foods   11

2. The Role and Significance of Microorganisms   13
   Primary Sources of Microorganisms Found in Foods   15
   Synopsis of Common Food-borne Bacteria   18
   Synopsis of Genera of Molds Common to Foods   25
   Synopsis of Genera of Yeasts Common to Foods   30
   References   32

3. Intrinsic and Extrinsic Parameters of Foods That
   Affect Microbial Growth   33
   Intrinsic Parameters   33
   Extrinsic Parameters   49
   References   56

4. Incidence and Types of Microorganisms in Foods   61
   Meats   62
   Poultry   73
   Seafood   75
   Vegetables   78
   Dairy Products   80
   Delicatessen and Related Foods   81
   Frozen Meat Pies   84
   Dehydrated Foods   84
   References   85

III. Determining Microorganisms and Their Products in Foods — 95

5. **Culture, Microscopic, and Sampling Methods** 97
   Conventional SPC 97
   Membrane Filters 101
   Microscope Colony Counts 102
   Agar Droplets 103
   Dry Films 103
   Most Probable Numbers (MPN) 103
   Dye-reduction 104
   Roll Tubes 105
   Direct Microscopic Count (DMC) 105
   Microbiological Examination of Surfaces 106
   Air Sampling 109
   Metabolically Injured Organisms 110
   Enumeration and Detection of Food-borne Organisms 117
   References 120

6. **Physical, Chemical, and Immunologic Methods** 128
   Physical Methods 128
   Chemical Methods 135
   Immunologic Methods 147
   References 159

7. **Bioassay and Related Methods** 172
   Whole-animal Assays 172
   Animal Models Requiring Surgical Procedures 178
   Cell Culture Systems 180
   References 185

IV. Food Spoilage — 191

8. **Spoilage of Fruits and Vegetables** 193
   Microbial Spoilage of Vegetables 194
   Spoilage of Fruits 203
   References 204

9. **Spoilage of Fresh and Processed Meats, Poultry, and Seafood 205**
Spoilage of Fresh Beef, Pork, and Related Meats 209
Detection and Mechanism of Meat Spoilage 212
Spoilage of Fresh Livers 216
Spoilage of Vacuum-packaged Meats 218
Spoilage of Frankfurters, Bologna, Sausage, and Luncheon
  Meats 222
Spoilage of Bacon and Cured Hams 224
Spoilage of Poultry 225
Spoilage of Fish and Shellfish 227
References 232

10. **Spoilage of Miscellaneous Foods 239**
Eggs 239
Cereals, Flour, and Dough Products 241
Bakery Products 242
Dairy Products 242
Sugars, Candies, and Spices 244
Nutmeats 245
Beers, Wines, and Fermented Foods 246
Mayonnaise and Salad Dressings 248
Canned Foods 249
References 253

V. Food Preservation   257

11. **Food Preservation with Chemicals 259**
Benzoic Acid and the Parabens 259
Sorbic Acid 262
The Propionates 264
Sulfur Dioxide and Sulfites 264
Nitrites and Nitrates 265
NaCl and Sugars 272
Indirect Antimicrobials 274
Acetic and Lactic Acids 278
Antibiotics 278
Antifungal Agents for Fruits 284
Ethylene and Propylene Oxides 285

Miscellaneous Chemical Preservatives 286
References 289

12. Food Preservation Using Irradiation 297
Characteristics of Radiations of Interest in Food
   Preservation 299
Principles Underlying the Destruction of Microorganisms by
   Irradiation 300
Processing of Foods for Irradiation 302
Application of Radiation 303
Radappertization, Radicidation, and Radurization of Foods 304
Legal Status of Food Irradiation 309
Effect of Irradiation on Food Constituents 311
Storage Stability of Irradiated Foods 313
References 313

13. Food Preservation with Low Temperatures 317
Temperature Growth Minima of Food-borne Microorganisms 320
Preparation of Foods for Freezing 320
Freezing of Foods and Freezing Effects 321
Storage Stability of Frozen Foods 322
Effect of Freezing upon Microorganisms 325
References 329

14. Food Preservation with High Temperatures 331
Factors That Affect Heat Resistance in Microorganisms 332
Relative Heat Resistance of Microorganisms 337
Thermal Destruction of Microorganisms 338
Aseptic Packaging 343
References 344

15. Preservation of Foods by Drying 346
Preparation and Drying of Low-moisture Foods 346
Effects of Drying upon Microorganisms 348
Storage Stability of Dried Foods 350
Intermediate-moisture Foods (IMF) 351
References 359

16. Fermented Foods and Related Products of Fermentation 362
Fermentation—Defined and Characterized 363
Lactic Acid Bacteria 364
Products of Fermentation 369
Single-cell Protein 391
Lactic Antagonism 395
Apparent Health Benefits of Fermented Foods 396
References 399

VI. Indicator and Food-borne Pathogens   407

17. Indices of Food Sanitary Quality; Microbiological Standards and Criteria 409
Coliform Bacteria as Indicators of Food Sanitary Quality 410
Enterococci as Indicators of Food Sanitary Quality 414
Other Indicators 420
Total Counts as Indicators of Food Sanitary Quality 421
Microbiological Standards and Criteria 423
Sampling Plans 430
Microbiological Guidelines 432
References 433

18. Staphylococcal Gastroenteritis 437
Potentially Pathogenic Species and Strains 437
Habitat and Distribution 439
Incidence in Foods 439
Nutritional Requirements for Growth 439
Temperature Growth Range 440
Effect of Salts and Other Chemicals 440
Effect of pH, a_w and Other Parameters 440
Staphylococcal Enterotoxins—Types and Incidence 442
The Gastroenteritis Syndrome 450
Incidence and Vehicle Foods 451
Ecology of *S. aureus* Growth 451
Prevention of Staphylococcal and Other Food Poisoning
    Syndromes 453
References 453

19. Food Poisoning Caused by Gram-positive Sporeforming Bacteria 459
Clostridium perfringens Food Poisoning 459
Botulism 469
Bacillus cereus Gastroenteritis 480
References 482

20. Food-borne Gastroenteritis Caused by *Salmonella* and *Escherichia* 489
Salmonellosis 489
E. coli Gastroenteritis Syndromes 502
References 508

21. Food-borne Gastroenteritis Caused by *Vibrio*, *Yersinia*, and *Campylo-bacter* Species 515
Vibrio parahaemolyticus 515
Other Vibrios 520
Yersinia enterocolitica 522
Campylobacter jejuni 527
Prevention 533
References 533

22. Other Proven and Suspected Food-borne Pathogens 541
Mycotoxins 541
Viruses 551
Aeromonas hydrophila, Plesiomonas shigelloides, and Other
    Gram-negative Bacteria 557
Listeriosis 559
Histamine-associated (Scombroid) Poisoning 560
Cestodes 562
Nematodes 563
Trematodes 565
Protozoa 566
Paralytic Shellfish Poisoning 567
Ciguatera Poisoning 567
Other Poisonous Fishes 568
References 568

VII. Psychrotrophs, Thermophiles and Radiation-resistant Microorganisms · 577

23. Characteristics and Growth of Psychrotrophic Microorganisms 579
    Temperature-induced Changes 580
    The Effect of Low Temperatures on Microbial Physiologic
      Mechanisms 582
    Nature of the Low Heat Resistance of Psychrotrophs 587
    References 589

24. Characteristics and Growth of Thermophilic Microorganisms 593
    Thermostability 594
    Other Aspects of Thermophilic Microorganisms 598
    References 601

25. Nature of Radiation Resistance in Microorganisms 603
    The Microbiology of *M. radiodurans* Strains 604
    The General Nature of Radioresistance in Other
      Microorganisms 611
    References 613

Appendix 617

Index 623

PREFACE

This third edition of *Modern Food Microbiology* is designed primarily for a second or subsequent course in microbiology in liberal arts, food science, nutrition, or related course programs. Organic chemistry and biochemistry are desirable prerequisites, but students with a good grasp of general biology and chemistry should be able to handle most of the material.

The twenty-five chapters are organized in seven sections. Section I (Chapter 1) consists of an outline of the history of food microbiology. In section II (Chapters 2, 3, and 4), the sources and types of microorganisms in foods are presented along with discussions of the parameters that affect the growth and activity of the food flora. Three new chapters (5, 6, and 7) form section III, and these chapters are devoted to methods of determining microorganisms and/or their products in foods. Essentially all information in the text on methodology has been placed in these three chapters. Section IV consists of three chapters (8, 9, and 10) devoted to microbial food spoilage, with emphasis on mechanisms to the extent known. Methods of preserving foods are presented in the six chapters (11 through 16) that constitute section V, and section VI comprises six chapters (17 through 22) devoted to indicator and food-borne pathogens. The final section, VII, consists of three chapters (23, 24, and 25) that deal with psychrotrophs, thermophiles, and radiation-resistant microorganisms. Some of the material in these three chapters may be combined with Chapters 12, 13, and 14 for textbook use or may be used in an advanced food microbiology course. The relationship between the common genera of food-borne bacteria is depicted schematically in the appendix section.

Numerous references have been consulted in the preparation of this edition, and I thank the many investigators whose findings I have drawn upon so heavily. I have been assisted in this revision by a number of individuals who critiqued various sections of the draft, and they are listed below in alphabetical order.

R. W. Bennett	M. R. Grula	J. L. McDonel
L. R. Beuchat	M. K. Hamdy	J. W. Peterson
L. B. Bullerman	P. A. Hartman	J. A. Seiter
M. P. Doyle	C. W. Houston	L. A. Shelef
L. A. Dreyfus	R. T. Jones	W. H. Sperber
K. E. Fagerman	J. Kim	

Those who assisted me with the previous editions are acknowledged in those editions. Finally, I thank R. Firstenberg-Eden and J. L. McDonel for allowing me to use some of their unpublished illustrations, and P. A. Hartman for suggesting and initiating the section on air sampling.

HISTORICAL

BACKGROUND

1.

HISTORY OF

MICROORGANISMS IN FOOD

Although it is extremely difficult to pinpoint the precise beginnings of human awareness of the presence and role of microorganisms in foods, the available evidence indicates that this knowledge preceded the establishment of bacteriology or microbiology as a science. The era prior to the establishment of bacteriology as a science may be designated the prescientific era. This era may be further divided into what has been called the **food-gathering period** and the **food-producing period.** The former covers the time from human origin over one million years ago up to eight thousand years ago. During this period, humans were presumably carnivorous, with plant foods coming into their diet later in this period. It is also during this period that foods were first cooked.

The food-producing period dates from about eight thousand to ten thousand years ago and, of course, includes the present time. It is presumed that the problems of spoilage and food poisoning were encountered early in this period. With the advent of prepared foods, the problems of disease transmission by foods and of faster spoilage caused by improper storage made their appearance. Spoilage of prepared foods apparently dates from around 6000 B.C. The practice of making pottery was brought to Western Europe about 5000 B.C. from the Near East. The first boiler pots are thought to have originated in the Near East about eight thousand years ago. The arts of cereal cookery, brewing, and food storage were either started at about this time or stimulated by this new development. The first evidence of beer manufacture has been traced to ancient Babylonia, as far back as 7000 B.C. (7). The Sumerians of about 3000 B.C. are believed to have been the first great livestock breeders and dairymen and were among the first to make butter. Salted meats, fish, fat, dried skins, wheat, and barley are also known to have been associated with this culture. Milk, butter, and cheese were used by the Egyptians as early as 3000 B.C. Between 3000 B.C. and 1200 B.C., the Jews used salt from the Dead Sea in the preservation of various foods. The Chinese and Greeks used salted fish in their diet, and the Greeks are credited with passing this practice on to the Romans, whose diet included pickled meats. Mummification and preservation of foods were related technologies that seem to have influenced each other's development. Wines are known to have been prepared by the Assyrians by 3500 B.C. Fermented sausages were prepared and consumed by the ancient Babylonians and the people of ancient China as far back as 1500 B.C. (7).

Another method of food preservation that apparently arose during this time was the use of oils such as olive and sesame. Jensen (6) has pointed out that the use of oils leads to high incidences of staphylococcal food poisoning. The Romans excelled in the preservation of meats other than beef by around 1000 B.C. and are known to have used snow to pack prawns and other perishables, according to Seneca. The practice of smoking meats as a form of preservation is presumed to have emerged sometime during this period, as did the making of cheese and wines. It is doubtful whether man at this time understood the nature of these newly found preservation techniques. It is also doubtful whether the role of foods in the transmission of disease or the danger of eating meat from infected animals was recognized.

Few advances were apparently made toward understanding the nature of food poisoning and food spoilage between the time of the birth of Christ and A.D. 1100 Ergot poisoning (caused by *Claviceps purpurea,* a fungus that grows on rye and other grains) caused many deaths during the Middle Ages. Over forty thousand deaths due to ergot poisoning were recorded in France alone in A.D. 943, but it was not known that the toxin of this disease was produced by a fungus. Meat butchers are mentioned for the first time in 1156, and by 1248 the Swiss were concerned with marketable and non-marketable meats. In 1276 a compulsory slaughter and inspection order was issued for public abattoirs in Augsburg. Although people were aware of quality attributes in meats by the thirteenth century, it is doubtful that there was any knowledge of the causal relationship between meat quality and microorganisms.

Perhaps the first man to suggest the role of microorganisms in spoiling foods was A. Kircher, a monk, who as early as 1658 examined decaying bodies, meat, milk, and other substances and saw what he referred to as "worms" invisible to the naked eye. Kircher's descriptions lacked precision, however, and his observations did not receive wide acceptance. In 1765, L. Spallanzani showed that beef broth that had been boiled for an hour and sealed remained sterile and did not spoil. Spallanzani performed this experiment to disprove the doctrine of the spontaneous generation of life. However, he did not convince the proponents of the theory since they believed that his treatment excluded oxygen, which they felt was vital to spontaneous generation. In 1837 Schwann showed that heated infusions remained sterile in the presence of air, which he supplied by passing it through heated coils into the infusion. While both of these men demonstrated the idea of the heat preservation of foods, neither took advantage of his findings with respect to application. The same may be said of D. Papin and G. Leibniz, who hinted at the heat preservation of foods at the turn of the eighteenth century.

The event that led to the discovery of canning had its beginnings in 1795, when the French government offered a prize of 12,000 francs for the discovery of a practical method of food preservation. In 1809, a Parisian confectioner, François (Nicholas) Appert, succeeded in preserving meats in glass bottles that had been kept in boiling water for varying periods of time. This discovery was made public in 1810, when Appert was issued a patent for his process. Not being a scientist, Appert was probably unaware of the long-range significance of his discovery or why it worked. This, of course, was the beginning of canning as it is known and practiced today. This event occurred some fifty years before L. Pasteur demonstrated the role of microorganisms in the spoilage of French wines, a development that gave rise to the rediscovery of bacteria. A. Leeuwenhoek in the Netherlands had examined bacteria through a microscope and described them in 1683, but it is unlikely that Appert was aware of this development since he was not a scientist, and Leeuwenhoek's report was not available in French.

The first man to appreciate and understand the presence and role of microorganisms in food was Pasteur. In 1837 he showed that the souring of milk was caused by microorganisms, and in about 1860 he used heat for the first time to destroy undesirable organisms in wine and beer. This process is, of course, now known as pasteurization.

HISTORICAL DEVELOPMENTS

Some of the more significant dates and events in the history of food preservation, food spoilage, food poisoning, and food legislation are listed below.

Food preservation

1782—Canning of vinegar was introduced by a Swedish chemist.

1810—Preservation of food by canning was patented by Appert in France.

1810—Peter Durand was issued a British patent to preserve food in "glass, pottery, tin or other metals or fit materials." Patent was later acquired by Hall, Gamble, and Donkin, possibly from Appert.

1813—Donkin, Hall, and Gamble introduced the practice of post-processing incubation of canned foods.

1813—Use of SO_2 as a meat preservative is thought to have originated around this time.

1825—T. Kensett and E. Daggett were granted U.S. patent for preserving food in tin cans.

1835—A patent was granted to Newton in England for making condensed milk.

1837—Winslow was the first to can corn from the cob.

1839—Tin cans came into wide use in the United States.

1839—L. A. Fastier was given a French patent for the use of brine bath to raise the boiling temperature of water.

1840—Fish and fruit were first canned.

1841—S. Goldner and J. Wertheimer were issued British patents for brine baths based on Fastier's method.

1842—A patent was issued to H. Benjamin in England for freezing foods by immersion in an ice and salt brine.

1843—Sterilization by steam was first attempted by I. Winslow in Maine.

1845—S. Elliott introduced canning to Australia.

1853—R. Chevallier-Appert obtained a patent for sterilization of food by autoclaving.

1854—Pasteur began wine investigations. Heating to remove undesirable organisms was introduced commercially in 1867–68.

1855—Grimwade in England was the first to produce powdered milk.

1856—A patent for the manufacture of unsweetened condensed milk was granted to Gail Borden in the United States.

1861—I. Solomon introduced the use of brine baths to the United States.

1865—The artificial freezing of fish on a commercial scale was begun in the United States. Eggs followed in 1889.

1874—The first extensive use of ice in transporting meat at sea was begun.

1874—Steam pressure cookers or retorts were introduced.

1878—The first successful cargo of frozen meat went from Australia to England. The first from New Zealand to England was sent in 1882.

1880—The pasteurization of milk was begun in Germany.

1882—Krukowitsch was the first to note the destructive effects of ozone on spoilage bacteria.

1886—A mechanical process of drying fruits and vegetables was carried out by an American, A. F. Spawn.

1887—Malted milk first appeared.

1890—The commercial pasteurization of milk was begun in the United States.

1890—Mechanical refrigeration for fruit storage was begun in Chicago.

1893—The Certified Milk movement was begun by H. L. Coit in New Jersey.

1895—The first bacteriological study of canning was made by Russell.

1907—E. Metchnikoff and coworkers isolated and named one of the yogurt bacteria, *Lactobacillus bulgaricus*.

1907—The role of acetic acid bacteria in cider production was noted by B. T. P. Barker.

1908—Sodium benzoate was given official sanction by the United States as a preservative in certain foods.

1916—The quick freezing of foods was achieved in Germany by R. Plank, E. Ehrenbaum, and K. Reuter.

1917—Clarence Birdseye in the United States began work on the freezing of foods for the retail trade.

1917—Franks was issued a patent for preserving fruits and vegetables under CO_2.

1928—Heat-process calculations were completed for the canning industry.

1928—The first commercial use of controlled-atmosphere storage of apples was made in Europe (first used in New York in 1940).

1929—A patent issued in France proposed the use of high-energy radiation for the processing of foods.

1929—Birdseye frozen foods were placed in retail markets.

1943—B. E. Proctor in the United States was the first to employ the use of ionizing radiation to preserve hamburger meat.

1954—The antibiotic nisin was patented in England for use in certain processed cheese to control clostridial defects.

1955—Sorbic acid was approved for use as a food preservative.

1955—The antibiotic chlortetracycline was approved for use in fresh poultry (oxytetracycline followed a year later). Approval was rescinded in 1966.

1967—The first commercial facility designed to irradiate foods was planned and designed in the United States.

Food spoilage

1659—Kircher demonstrated the occurrence of bacteria in milk; Bondeau did the same in 1847.

1680—Leeuwenhoek was the first to observe yeast cells.

1780—Scheele identified lactic acid as the principal acid in sour milk.

1836—Latour discovered the existence of yeasts.

1839—Kircher examined slimy beet juice and found organisms that formed slime when grown in sucrose solutions.

1857—Pasteur showed that the souring of milk was caused by the growth of organisms in it.

1866—L. Pasteur's *Étude sur le Vin* was published.

1867—Martin advanced the theory that cheese ripening was similar to alcoholic, lactic, and butyric fermentations.

1873—The first reported study on the microbial deterioration of eggs was carried out by Gayon.

1873—Lister was first to isolate *Streptococcus lactis* in pure culture.

1876—Tyndall observed that bacteria in decomposing substances were always traceable to air, substances, or containers.

1878—Cienkowski reported the first microbiological study of sugar slimes and isolated *Leuconostoc mesenteroides* from them.

1887—Forster was the first to demonstrate the ability of pure cultures of bacteria to grow at 0°C.

1888—Miquel was the first to study thermophilic bacteria.

1895—The first records on the determination of numbers of bacteria in milk were those of Von Geuns in Amsterdam.

1895—S. C. Prescott and W. Underwood traced the spoilage of canned corn to improper heat processing for the first time.

1902—The term *psychrophile* was first used by Schmidt-Nielsen for microorganisms that grow at 0°C.

1912—The term *osmophilic* was coined by Richter to describe yeasts that grow well in an environment of high osmotic pressure.

1915—*Bacillus coagulans* was first isolated from coagulated milk by B. W. Hammer.

1917—*Bacillus stearothermophilus* was first isolated from cream-style corn by P. J. Donk.

1933—Olliver and Smith in England observed spoilage by *Byssochlamys fulva;* first described in the United States in 1964 by D. Maunder.

Food poisoning

1820—The German poet Justinus Kerner described "sausage poisoning" (which in all probability was botulism) and its high fatality rate.

1857—Milk was incriminated as transmitter of typhoid fever by W. Taylor of Penrith, England.

1870—Francesco Selmi advanced his theory of ptomaine poisoning to explain illness contracted by eating certain foods.

1888—Gaertner first isolated *Salmonella enteritidis* from meat that had caused fifty-seven cases of food poisoning.

1894—T. Denys was the first to associate staphylococci with food poisoning.

1896—Van Ermengen first discovered *Clostridium botulinum.*

1926—The first report of food poisoning by streptococci was made by Linden, Turner, and Thom.

1938—Outbreaks of *Campylobacter* enteritis were traced to milk in Illinois.

1939—Gastroenteritis caused by *Yersinia enterocolitica* was first recognized by Schleifstein and Coleman.

1945—McClung was the first to prove the etiologic status of *Clostridium perfringens (welchii)* in food poisoning.

1951—*Vibrio parahaemolyticus* was shown to be an agent of food poisoning by T. Fujino of Japan.

1955—Similarities between cholera and *Escherichia coli* gastroenteritis in infants were first noted by S. Thompson.

1960—The production of aflatoxins by *Aspergillus flavus* was first reported.

1963—The salmonellae surveillance program in the United States was started.

1969—*C. perfringens* enterotoxin was demonstrated by C. L. Duncan and D. H. Strong.

1969—*C. botulinum* type G was first isolated in Argentina by Gimenez and Ciccarelli.

1971—First U.S. food-borne outbreak of *Vibrio parahaemolyticus* gastroenteritis occurred in Maryland.

1975—*Salmonella* enterotoxin was demonstrated by L. R. Koupal and R. H. Deibel.

1976—First U.S. food-borne outbreak of *Yersinia enterocolitica* gastroenteritis occurred in New York.

1978—Documented food-borne outbreak of gastroenteritis caused by the Norwalk virus occurred in Australia.

1979—Food-borne gastroenteritis caused by non-01 *Vibrio cholerae* occurred in Florida. Earlier outbreaks occurred in Czechoslovakia (1965) and Australia (1973).

1983—*Campylobacter jejuni* enterotoxin described by Ruiz-Palacios et al.

Food legislation

1890—The first national meat inspection law was enacted. It required the inspection of meats for export only.

1895—The previous meat inspection act was amended to strengthen its provisions.

1906—The U.S. Federal Food and Drug Act was passed by Congress.

1910—New York City Board of Health issued an order requiring the pasteurization of milk.

1939—The New Food, Drug, and Cosmetic Act became law.

1954—The Miller Pesticide Chemicals Amendment to the Food, Drug, and Cosmetic Act was passed by Congress.

1957—The U.S. Compulsory Poultry and Poultry Products law was enacted.

1958—The Food Additives Amendment to the Food, Drug, and Cosmetics Act was passed.

1962—The Talmadge-Aiken Act (allowing for federal meat inspection by states) was enacted into law.

1963—The U.S. Food and Drug Administration approved the use of irradiation for the preservation of bacon.

1967—The U.S. Wholesome Meat Act was passed by Congress and enacted into law on December 15.

1968—The Food and Drug Administration withdrew its 1963 approval of irradiated bacon.

1968—The Poultry Inspection Bill was signed into law.

1969—The U.S. Food and Drug Administration established an allowable level of 20 ppb of aflatoxin for edible grains and nuts.

1973—The state of Oregon adopted microbial standards for fresh and processed retail meats. They were repealed in 1977.

1984—A bill was introduced in the U.S. House of Representatives that would promote the irradiation of foods by, among other things, defining irradiation as a process rather than as a food additive.

REFERENCES

1. Bishop, P. W. 1978. Who introduced the tin can? Nicolas Appert? Peter Durand? Bryan Donkin? *Food Technol.* 32(4):60–67.

2. Brandly, P. J., G. Migaki, and K. E. Taylor. 1966. *Meat hygiene*. 3d ed., ch. 1. Philadelphia: Lea & Febiger.
3. Farrer, K. T. H. 1979. Who invented the brine bath?—The Isaac Solomon myth. *Food Technol*. 33(2):75–77.
4. Goldblith, S. A. 1971. A condensed history of the science and technology of thermal processing. *Food Technol*. 25(12):44–50.
5. Goldblith, S. A., M. A. Joslyn, and J. T. R. Nickerson. 1961. *Introduction to thermal processing of foods*. Vol. 1. Westport, Conn.: AVI.
6. Jensen, L. B. 1953. *Man's foods*. Chs. 1, 4, 12. Champaign, Ill.: Garrard Press.
7. Pederson, C. S. 1971. *Microbiology of food fermentations*. Westport, Conn.: AVI.
8. Schormüller, J. 1966. *Die Erhaltung der Lebensmittel*. Stuttgart. Ferdinand Enke Verlag.
9. Stewart, G. F., and M. A. Amerine. 1973. *Introduction to food science and technology*. Ch. 1. New York: Academic Press.
10. Tanner, F. W. 1944. *The microbiology of foods*. 2d ed. Champaign, Ill.: Garrard Press.
11. Tanner, F. W., and L. P. Tanner. 1953. *Food-borne infections and intoxications*. 2d ed. Champaign, Ill.: Garrard Press.

II

SOURCES, TYPES, INCIDENCE, AND BEHAVIOR OF MICROORGANISMS IN FOODS

2.

THE ROLE AND

SIGNIFICANCE OF MICROORGANISMS

Since human food sources are of plant and animal origin, it is important to understand the biological principles of the microbial flora associated with plants and animals in their natural habitats and respective roles. While it sometimes appears that microorganisms are trying to ruin our food sources by infecting and destroying plants and animals, including man, this is by no means their primary role in nature. In our present view of life on this planet, the primary function of microorganisms in nature is self-perpetuation. During this process, the heterotrophs carry out the following general reaction:

All organic matter
(carbohydrates, proteins, lipids, etc.)
↓
Energy + Inorganic compounds
(nitrates, sulfates, etc.)

This, of course, is essentially nothing more than the operation of the nitrogen cycle and the cycle of other elements (see Fig. 2-1). As will be discussed in a later chapter, the microbial spoilage of foods may be viewed simply as an attempt by the food flora to carry out what appears to be their primary role in nature. This should not be taken in the teleological sense. In spite of their simplicity when compared to higher forms, microorganisms are capable of carrying out many complex chemical reactions essential to their perpetuation. To do this, they must obtain nutrients from organic matter, some of which constitutes our food supply.

If one considers the types of microorganisms associated with plant and

13

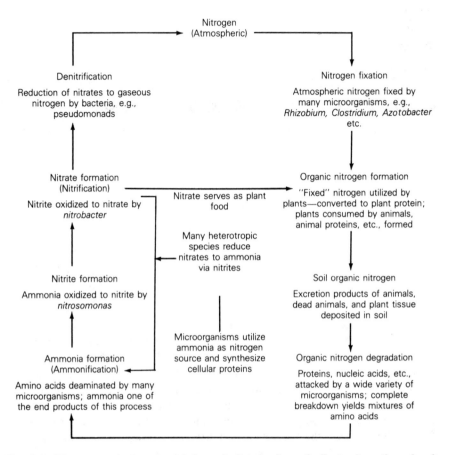

Fig. 2-1. Nitrogen cycle in nature is here depicted schematically to show the role of microorganisms. (From *Microbiology* by M. J. Pelczar and R. Reid, 6; copyright © 1965 by McGraw-Hill Book Company, used with permission of the publisher)

animal foods in their natural states, one can then predict the general types of microorganisms to be expected on this particular food product at some later stage in its history. Results from many laboratories show that untreated foods may be expected to contain varying numbers of bacteria, molds, or yeasts, and the question often arises as to the safety of a given food product based upon total microbial numbers. The question should be twofold: What is the total number of microorganisms present per g or ml? and what *types* of organisms are represented in this number? It is necessary to know which organisms are associated with a particular food in its natural state, and which of the organisms present are not normal for that particular food. It is, therefore, of value to know the general distribution of bacteria in nature and the general types of organisms normally present under given conditions where foods are grown and handled.

BACTERIA. The most important genera of bacteria known to occur in foods are listed below in alphabetical order. Some of these are highly desirable in certain foods, while others bring about spoilage or cause food poisoning or gastroenteritis in humans. Over one-third belong to the family Enterobacteriaceae.

Acetobacter	*Corynebacterium*	*Proteus*
Acinetobacter	*Enterobacter*	*Pseudomonas*
Aeromonas	*Erwinia*	*Salmonella*
Alcaligenes	*Escherichia*	*Serratia*
Alteromonas	*Flavobacterium*	*Shigella*
Bacillus	*Lactobacillus*	*Staphylococcus*
Brochothrix	*Leuconostoc*	*Streptococcus*
Campylobacter	*Micrococcus*	*Vibrio*
Citrobacter	*Moraxella*	*Yersinia*
Clostridium	*Pediococcus*	

MOLDS. Some of the most common genera of molds associated with foods are:

Alternaria	*Colletotrichum*	*Mucor*
Aspergillus	*Fusarium*	*Penicillium*
Botrytis	*Geotrichum*	*Rhizopus*
Byssochlamys	*Gloeosporium*	*Sporotrichum*
Cephalosporium	*Helminthosporium*	*Thamnidium*
Cladosporium	*Monilia (Neurospora)*	*Trichothecium*

YEASTS. The most common genera of yeasts encountered in and on foods are:

Brettanomyces	*Kloeckera*	*Saccharomycopsis*
Candida	*Kluyveromyces*	*Schizosaccharomyces*
Debaryomyces	*Mycoderma*	*Torulopsis (Torula)*
Endomycopsis	*Rhodotorula*	*Trichosporon*
Hansenula	*Saccharomyces*	

PRIMARY SOURCES OF MICROORGANISMS FOUND IN FOODS

The genera listed above represent perhaps the most important organisms normally found in foods. Each genus has its own particular functions of nutrition and consequent degradative processes. These organisms may be generally associated with the particular aspects of the food environment presented below.

SOIL AND WATER. It may be assumed that at one time all microorganisms existed in water. The drying of surface soils gives rise to dust that, when disseminated by winds, carries adhering microorganisms to many places including other areas of the soil, rivers, and oceans. The formation of clouds over large bodies of water and the subsequent rainfall over land as well as other waters has the same effect. It is not surprising, then, that soil and water microorganisms are often one and the same.

The following genera of food-borne bacteria generally found in soils and waters may be expected in foods: *Acinetobacter, Alcaligenes, Bacillus, Citrobacter, Clostridium, Corynebacterium, Enterobacter, Micrococcus, Proteus, Pseudomonas, Serratia,* and *Streptomyces,* among others.

Many of the molds listed above occur in soils and also in waters. Molds are in general very widespread in nature, where they participate in the degradation of both plant and animal matter as well as cause many diseases of plants and animals (see Chapter 8). Among those that are nearly always present in soils are *Aspergillus, Rhizopus, Penicillium, Trichothecium, Botrytis, Fusarium,* and others.

A large number of yeast genera are associated with plants and may therefore be expected to be found in soils. Their numbers in water are generally low.

PLANTS AND PLANT PRODUCTS. Most of the organisms discussed above for soil and water are found also on plants, since soil and water constitute the primary sources of microorganisms to plants. On the other hand, there are some bacteria that are associated more with plants than with soil. Among these genera are *Acetobacter, Erwinia, Flavobacterium, Kurthia, Lactobacillus, Leuconostoc, Listeria, Pediococcus,* and *Streptococcus.* Any or all of the other genera may at times be found on plants and in plant products along with genera not listed.

Among the molds, the most important plant-borne genera are those that cause the spoilage of vegetables and fruits (so-called market diseases). These are presented in more detail in Chapter 8. The genus *Saccharomyces* is the most notable of the yeasts that may be found on many plant products, especially fruits. Others include the genera *Rhodotorula* and *Torula.*

FOOD UTENSILS. The genera of microorganisms to be found on food utensils depend upon the types of foods handled, the care of these utensils, their storage, and other factors. If vegetables are handled in a given set of utensils, one would expect to find some or all of the organisms associated with vegetables. When utensils are cleaned with hot or boiling water, the remaining flora would normally be those best able to withstand the effects of this treatment. Utensils that are stored in the open where dust might collect should be expected to have airborne bacteria, yeasts, and molds.

INTESTINAL TRACT OF MAN AND ANIMALS. There are several genera of bacteria that are more commonly found in this environment than in soils, water, or other places. Among these are: *Bacteroides, Escherichia, Lactobacillus, Proteus, Salmonella, Shigella, Staphylococcus,* and *Streptococcus.* The most notable of these is the genus *Escherichia,* which has as its natural habitat the intestinal tract of man and other mammals. Species of other genera common to the intestinal tract include *Clostridium, Citrobacter, Enterobacter,* and *Pseudomonas.* From the intestinal tracts of animals, intestinal microorganisms find their way directly to the soil and water. And from the soil they may find their way onto plants, dust, utensils, and so on. Molds are not thought to be transmitted by fecal sources, though the yeast genus *Candida* is very often found in the intestinal tract of man. (See Chapter 22 for discussion of the enteroviruses.)

FOOD HANDLERS. The microflora on the hands and outer garments of food handlers generally reflects the environment and habits of the individuals. This flora would normally consist of organisms found on any object handled by the individual as well as some of those picked up from dust, water, soil, and the like. In addition, there are several genera of bacteria that are specifically associated with the hands, nasal cavities, and mouth. Among these are the genera *Micrococcus* and *Staphylococcus,* the most notable of which are the staphylococci, which are found on hands, arms, in nasal cavities, the mouth, and other parts of the body. While the genera *Salmonella* and *Shigella* are basically intestinal forms, they may be deposited onto foods and utensils by food handlers if sanitary practices are not followed by each individual. Any number of molds and yeasts may be found on the hands and garments of food handlers depending upon the immediate history of each individual.

ANIMAL FEEDS. Any one or all of the genera of bacteria, yeasts, and molds cited earlier in this chapter may be found in animal feeds. The types of organisms to be found would depend, of course, on the source of the feeds, the treatment given them to destroy microorganisms, the containers in which they are stored, and the like. As discussed further in Chapter 20, animal feeds are of great importance in the spread of food-poisoning *Salmonella.* Organisms from this source have been shown to be rapidly disseminated throughout processing plants where feeds are handled.

ANIMAL HIDES. Just about any or all of the microorganisms associated with soils, water, animal feeds, dust, and fecal matter may be found on the hides of animals. From animal hides, these organisms may again be deposited in the air, onto the hands of workers, and directly into foods. Some members of the hide flora find their way into the lymphatic system of slaughter animals from which they migrate after slaughter into the muscle tissue proper.

AIR AND DUST. The types of organisms to be found in air and dust, with the exception of some of the pathogens, include the genera of bacteria, the eighteen genera of molds, and many of the yeasts. Although *Staphylococcus* and *Salmonella* spp. may at times be found in air and dust, these are not the major sources of these organisms to foods. Notable among the bacterial genera in air and dust are *Bacillus* and *Micrococcus* spp., all of which are able to endure dryness to varying degrees. Notable among the yeasts is the genus *Torulopsis*, and many mold genera may be found from time to time.

THE PRIMARY SOURCES OF FOOD-POISONING BACTERIA. The most important food-poisoning and gastroenteritis-causing bacteria belong to the genera *Staphylococcus, Salmonella, Clostridium,* and *Campylobacter.* The staphylococci are associated with the nasal cavities of man and animals as well as with other parts of the body. Salmonellae are indigenous to the intestinal tract of man and animals but may enter foods from other sources contaminated from fecal matter. The clostridia are basically soil forms while *Campylobacter* spp. are animal associated. In addition to the above, *Bacillus cereus, Vibrio parahaemolyticus, Yersinia enterocolitica,* and *Vibrio cholerae* all cause gastroenteritis, and these syndromes are discussed in Chapters 18, 19, 20, 21, and 22.

SYNOPSIS OF COMMON FOOD-BORNE BACTERIA

The following are brief descriptions of each of the most common bacterial genera. These synopses are not meant to be employed alone in the identification of these organisms but are designed to characterize these important genera with respect to their roles in foods. For the purpose of generic and species identification, *Bergey's Manual* should be consulted (2, 3).

Acetobacter. This genus contains three species and nine subspecies. They are gram-negative, rod-shaped cells that are strict aerobes. They are commonly found in fermented grain mash, mother of vinegar, beer, wines, and souring fruits and vegetables. Some species such as *A. aceti* subsp. *aceti* oxidize ethanol to acetic acid and thereby give rise to vinegar, perhaps their greatest industrial use. The G + C content of DNA = 55–64 moles %.

Acinetobacter. This genus of gram-negative rods shows some affinity to the family Neisseriaceae and only two species are recognized (*A. calcoaceticus* and *A. lwoffii*). Some organisms formerly classified as *Achromobacter* have been placed in this genus. They are strict aerobes that do not reduce nitrates.

While they are related to the genus *Moraxella,* they differ in being oxidase negative. Although rod-shaped cells are formed in young cultures, old cultures contain many coccoid-shaped cells. While generally regarded as being important in meats, recent evidence suggests that their incidence is quite low in these products. They are common in soils and water, and the G + C content of DNA = 39–47 moles % (5).

Aeromonas. This genus is represented by three species and eight subspecies and belongs to the Vibrionaceae. They are gram-negative rods that are facultatively anaerobic and display respiratory and fermentative metabolism. They are proteolytic and produce copious quantities of gases from those sugars fermented. Growth is over the pH range 5.5–9.0. They are found in the aquatic environment, where they are normal inhabitants of the intestines of fish. Some are fish pathogens, and some *A. hydrophila* strains are human pathogens. The G + C content of their DNA is 57–65 moles %.

Alcaligenes. This genus consists of nine species. They are gram-negative rods that sometimes stain variable or gram positive. They do not ferment sugars, but produce alkaline reactions that are also produced in litmus milk. They do not produce pigments. They are widely distributed in nature in decomposing matter of all types, in raw milk, poultry products, the intestinal tract, and so on. The G + C content of DNA = 58–70 moles %.

Alteromonas. These organisms are common on marine fish and are closely related to *Alcaligenes* and *Pseudomonas.* While the G + C content of DNA of *Pseudomonas* spp. is 57–70%, the range for *Alteromonas* is 43.2–48%. Most require Na$^+$ for growth and they are nonfermentative. They are gram-negative, straight or curved rods, motile by means of a single polar flagellum, strict aerobes, lack arginine dihydrolase, and do not accumulate polyhydroxybutyrate. Some, such as *A. putrefaciens* and *A. nigrifaciens,* were formerly classified as pseudomonads.

Bacillus. This genus belongs to the family Bacillaceae. Some thirty-one species are recognized. Most are aerobic, gram-positive rods that produce endospores. They often exist in long chains on cultural media. Most are mesophiles with some being psychrotropic and some thermophilic in nature. The thermophilic members are of great importance in the canning industry due to the extreme heat resistance of their spores. This genus contains one pathogen for man and other vertebrates—*B. anthracis,* which causes anthrax, and one food-poisoning species, *B. cereus.* Some species are insect pathogens. The nonpathogens are widely distributed and may be found in air, dust, soil, water, and on utensils and in various foods. They are important in the spoilage of many foods held above refrigerator temperatures. The G + C content of DNA = 32–62 moles %.

Brochothrix. Only one species (*B. thermosphacta*) is recognized in this genus and it was placed orginally in the genus *Microbacterium*. The genus is placed tentatively in the family Lactobacteriaceae but differs from typical lactobacilli in being catalase positive (10). Exponential-phase cells are rods while stationary-phase cells are coccoids, typical of coryneform bacteria. Glucose metabolism is fermentative, and the G + C content of DNA is 35 moles %. *B. thermosphacta* is found most on processed meats stored in gas-impermeable packages, where it is sometimes the dominant organism.

Campylobacter. Four species and four subspecies are represented by this genus of slender, spirally curved rods which are 1.5–3.5 μm in length and 0.2–0.4 μm in width. They are motile by a corkscrewlike motion and possess polar flagellae at one or both ends. They are inactive on carbohydrates, are microaerophilic to anaerobic, and have a G + C content of DNA of 30–35 moles %. Once classified as *Vibrio* spp., they can be found in the oral cavity and intestinal canal, and some are human and animal pathogens (9).

Citrobacter. This genus belongs to the family Enterobacteriaceae, and three species are recognized. The slow, lactose-fermenting, gram-negative rods previously designated paracolons and the Bethesda-Ballerup group have been placed in this genus. All members can use citrate as their sole carbon source. They are common in fecal matter, water, and some foods.

Clostridium. This is an important group of anaerobic bacteria that belong to the family Bacillaceae. Some seventy-six species are now recognized. They are gram-positive, sporeforming rods. The genus contains some thermophilic species that are of great importance in the canning industry due to the extreme heat resistance of their endospores. The etiologic agents of tetanus, gas gangrene, perfringens food poisoning, and botulism are members of this genus. Some species are of commercial importance in the production of certain solvents. They are very widely distributed in nature, soils, water, the intestinal tract of man and animals, and other places. They may be found in many foods where they may or may not grow. Many species are strongly proteolytic as are many of the *Bacillus* spp. in the same family. The G + C content of DNA = 23–43 moles %.

Corynebacterium. This genus belongs to the coryneform group and contains some thirty-two species. They are aerobic, gram-positive rods that often show granules and club-shaped swellings. They are nonsporeformers. This group contains the etiologic agent of diphtheria and other species common on the body. Most are mesophiles with some being psychrotrophs. These organisms are widely distributed among certain plants such as wheat, beans, and tomatoes. They are also found in the intestinal tract of man and animals and have been isolated from spoiling foods of various types. The G + C content of DNA = 57–60 moles %.

Enterobacter. These organisms belong to the family Enterobacteriaceae and are represented by nine species. They are short, gram-negative, nonpigment-forming rods that grow well on many cultural media and ferment glucose and lactose with the production of acid and gas. In their sugar fermentations, these organisms produce two or more times as much CO_2 as hydrogen gas. The most distinctive taxonomic features are: MR− and VP+ (MR = methyl red; VP = Voges-Proskauer reaction). They are widely distributed in nature, being especially present on plants and grain, and in water and the intestinal tract. This genus is one of the coliform genera along with *Escherichia*. Formerly *Aerobacter*. The G + C content of DNA = 52–59 moles %.

Erwinia. This genus belongs to the family Enterobacteriaceae. Some twenty-one species are now recognized. They are gram-negative rods that are motile. Many species grow within the psychrotrophic range and many produce pigments of various shades of red when growing on culture media and certain foods. They are characteristically associated with plants where they cause dry necroses, galls, wilts, and soft rots. These are the most important bacteria in the cause of market diseases of fruits and vegetables (see Chapter 6). The G + C content of DNA = 53.6–54.1 moles %.

Escherichia. As stated above, this group belongs to the family Enterobacteriaceae and contains three species. They are short, gram-negative rods that are indistinguishable from *Enterobacter* on culture media and under the microscope. They differ from *Enterobacter* spp. in being MR+ and VP−. Their main habitat appears to be the intestinal tract of man and other animals from which they may be found in soils, waters, and many other places in nature. *E. coli* is the most important member of the coliform group along with *E. aerogenes*. Their presence in large numbers in foods is generally taken to indicate fecal contamination. The G + C content of DNA = 50–51 moles %.

Flavobacterium. This genus contains eighteen species. They are gram-negative rods that generally produce yellow to red pigments on agar. They are widely distributed in soils and water and on fish and plants. They may be isolated from decaying plant materials. Many are psychrophilic while most are meso-philes. For motile species, the G + C content of DNA = 63–70 moles %, while for some nonmotile species 30–42 moles %.

Lactobacillus. This group belongs to the family Lactobacillaceae and contains some thirty-five species. They are long, gram-positive, nonsporing rods that are catalase negative. They often occur in long chains when viewed under the microscope. Most are microaerophilic or anaerobic while both homofermentative and heterofermentative species exist among them. They are widely distributed among plants and in dairy products. Some are employed in the production of fermented milks such as acidophilus and bulgaricus milks. Some are important in cheese making. Many are used in the microbiological assay of B vitamins and amino acids due to their exacting growth requirements. One species, *L. thermophilus*, survives milk pasteurization temperatures. They are common in and on cured and processed meat products. G + C content = 35–53%.

Leuconostoc. This genus belongs to the family Streptococcaceae. It contains at least six species. They are gram-positive, spherical to oval, catalase-negative, heterofermentative organisms. They are widely distributed among plants, from which they find their way into milk and dairy products. Some cause problems in sugar refineries, where they form slime in sugar lines. Some species are employed in dairy starter cultures while others are often found in cured meat products. Some synthesize the medically important polymer dextran. The G + C content of DNA of most = 38–42 moles %.

Micrococcus. This group belongs to the family Micrococcaceae and contains at least nine species. They are gram-positive cocci that are catalase positive in contrast to the streptococci. Some produce nonpigmented colonies while others produce pink to orange-red to red pigments on culture media. Most or all can tolerate high levels of salt. They are widely distributed in nature, on the skin of man, and the hides of animals, as well as in dust, soil, water, and many foods. Most are mesophiles while some are capable of growth in the psychrophilic range. Several species are associated with dairy products, through which they enter processed meats such as frankfurters. The formerly recognized genera *Gaffkya* and *Sarcina* have been reduced to species of this genus. The G + C content of DNA = 66–75 moles %.

Moraxella. This genus is a member of the family Neisseriaceae as is the genus *Acinetobacter*. They are very short and plump gram-negative rods that differ from *Acinetobacter* in being sensitive to penicillin, oxidase positive, and having a G + C content of DNA of 40–46 moles %. Their metabolism is oxidative and they do not form acid from glucose. Because of the similarity of the two groups, some investigators use the designation "*Acinetobacter-Moraxella* group."

Pediococcus. These homofermentative cocci belong to the family Streptococcaceae and differ microscopically from streptococci and leuconostocs by their arrangement in pairs and tetrads resulting from cell division in two planes. Like the other lactic acid genera, they are widespread in nature, especially on plants. At least six species are recognized. Most are important as starters in certain fermented foods. The G + C content of DNA = 34–44 moles %.

Proteus. These organisms belong to the family Enterobacteriaceae and consist of six species. Like the other members of this family, they are gram-negative rods that are aerobic and often show pleomorphism. All are motile while all but one species hydrolyze urea. They are found in the intestinal tract of man and animals and on decaying materials in general. They may be isolated from spoiled eggs and meats, especially those allowed to spoil above refrigerator temperatures. For most species, G + C content is 38–42 moles %.

Pseudomonas. These important bacteria belong to the family Pseudomonadaceae and while eighty-seven species are recognized, over two hundred others have been reported. They are short, gram-negative, aerobic rods that usually produce a single polar flagellum. Many psychrotrophic species and strains as well as mesophiles exist in this genus. They are widely distributed in nature in soils and water, on plants, and in the intestinal canal of man and other animals. These are by far the most important bacteria in the low-temperature spoilage of foods such as meats, poultry, eggs, and seafoods. Some strains produce fluorescent pigments (pyocyanin and/or fluorescin). Most are capable of oxidizing glucose to gluconic acid, 2-keto-gluconic acid, or other intermediates. Many are plant pathogens, causing leaf spot, leaf stripe, and related diseases. Many of those that cause food spoilage do not produce the water-soluble pigment but may fluoresce under ultraviolet light. The G + C content of DNA ranges from 58 to 70 moles %.

Salmonella. This important genus of bacteria belongs to the family Enterobacteriaceae and while only five species are recognized, over two thousand serovars (serotypes) are known, with new ones added yearly. They are gram-negative, short rods that are aerobic and do not produce pigments on culture media. Most ferment glucose and other simple sugars with the production of acid and gas. They generally do not ferment lactose although some do. These are basically intestinal forms, as are most of the Enterobacteriaceae genera. They may be found widely distributed in nature. The etiologic agents of typhoid and paratyphoid fevers belong to this group as well as those that cause food-borne salmonellosis in man. All species and strains of this genus are undesirable in foods. The G + C content of DNA = 50–53 moles %.

Serratia. This genus belongs to the family Enterobacteriaceae and contains nine species. They are gram-negative, aerobic, proteolytic, and mesophilic rods that generally produce red pigments on culture media and on certain foods. Most species are rather widespread in nature in water, soil, and

ɣing plant and animal matter. The G + C content of DNA ranges from
ₒ 59 moles %.

Shigella. These organisms also belong to the family Enterobacteriaceae
and contain at least four species. They are short, gram-negative rods that
are nonmotile. They are aerobic and mesophilic in nature and occur in
polluted waters and in the human intestinal canal, where they cause bacillary
dysentery and other intestinal disorders. Their primary sources to foods
are polluted water and human carriers. These organisms are undesirable in
foods.

Staphylococcus. These organisms belong to the family Micrococcaceae and
consist of thirteen species and four subspecies (8). They are gram-positive
cocci that divide irregularly as do the micrococci. They are catalase positive
in contrast to the streptococci. Most *S. aureus* strains produce a golden
pigment and coagulate blood plasma, while *S. epidermidis* is nonpigmented
and does not produce coagulase. Both are common in the nasal cavities of
man and certain other animals as well as on the skin and other parts of
the body. *S. aureus* produces boils, carbuncles, and an important food-
poisoning syndrome in man. Their presence in foods in large numbers is
undesirable. The G + C content of DNA ranges from 30 to 40 moles %.

Streptococcus. This genus belongs to the family Streptococcaceae and consists
of twenty-seven species. They are gram-positive, catalase-negative cocci
that often appear as spherical to ovoid forms. All produce small colonies
when growing on culture media, as do the lactobacilli. They are nonpigmented
and microaerophilic in nature. Some are associated with the upper respiratory
tract of man and other animals, where they may cause diseases such as
scarlet fever and septic sore throat. Others are found in the intestinal tract
of man and animals and tend to be rather widespread on plants and plant
parts, and in dairy products. While most are mesophilic, some grow within
the psychrotrophic range. Some cause mastitis in cattle while others are
important in dairy starter cultures. One species, *S. lactis,* is the most
common cause of sour milk, and is important in the manufacture of cheese.
The presence of some species in foods in large numbers may indicate fecal
contamination. The G + C content of DNA = 33–42 moles %.

It has been proposed that *S. faecalis* and *S. faecium* be placed in a new
genus, *Enterococcus.* If this proposal is adopted, the new genus would be
characterized by a G + C content of DNA of 37–45 moles %, with *E.
faecalis* being the type species (7). DNA hybridization studies reveal a close
relationship between these two species but not to other members of the
genus *Streptococcus* (see Chapter 17).

Vibrio. This genus belongs to the family Vibrionaceae and consists of straight
or curved rods that are motile by sheathed polar flagella when cultured in
liquid media. They are facultative anaerobes capable of respiratory and

fermentative metabolism, and all utilize glucose. Most species require Na^+. The G + C content of DNA = 38–51 moles %. About twenty species are recognized, including *V. cholerae* and *V. parahaemolyticus*. The former is endemic in coastal waters of Texas, Louisiana, and Florida, while the latter is widely distributed in marine waters and is a common contaminant of shellfish. A detailed discussion of vibrios in the environment has been presented by Colwell et al. (4).

Yersinia. Five species are recognized in this genus including the agent of plague, *Y. pestis*. They are gram-negative rods, 1–2 μm in length, nonmotile at 37°C but motile when grown < 30°C, produce acid from glucose, and belong to the family Enterobacteriaceae. The G + C content of their DNA is 45.8–46.8 moles %. They are widely distributed in nature and some are adapted to specific animal hosts. *Y. enterocolitica* causes gastroenteritis in man.

SYNOPSIS OF GENERA OF MOLDS COMMON TO FOODS

Unlike the true bacteria and most yeasts, molds grow in the form of a tangled mass which spreads rapidly and may cover several inches of area in two to three days. The total of the mass or any large single portion of it is referred to as **mycelium.** The mycelium is composed of branches or filaments referred to as **hyphae.** At the time of asexual reproduction, **sporangiophores** or **conidiophores** are sent up which bear, usually at their tips, sporangia or conidia (see Figs. 2-2*K* and 2-3*A*). In those molds that produce sporangia, the spores are borne by these structures and are responsible for the various colors displayed by molds. Conidia represent unprotected spores. In addition to these characteristic and common asexual spores, some molds produce other asexual spores. **Chlamydospores** result when a thick wall develops around any cell of the mycelium. These structures are somewhat resistant to adverse environmental conditions. **Arthrospores** or **oidia** result from fragmentation by some molds that produce a septate mycelium. Chlamydospores and arthrospores are a bit more difficult to destroy than other parts of the mold mycelium, and they sometimes cause concern in the food industry. Molds also reproduce by sexual means where they form either ascospores, oospores, or zygospores.

Most molds of importance in foods are now placed in the *Fungi imperfecti* group. These are molds whose sex cycles are not known. Most or all are thought to be related to the *Ascomycetes* group.

The descriptions below are not meant to be detailed enough to permit generic identifications of these organisms except for the most common ones. For this purpose, the reader should consult a more detailed reference such as Barnett (1).

Alternaria. These molds produce septate mycelia with dark conidiophores and dark conidia. The conidia have both cross and longitudinal septa and

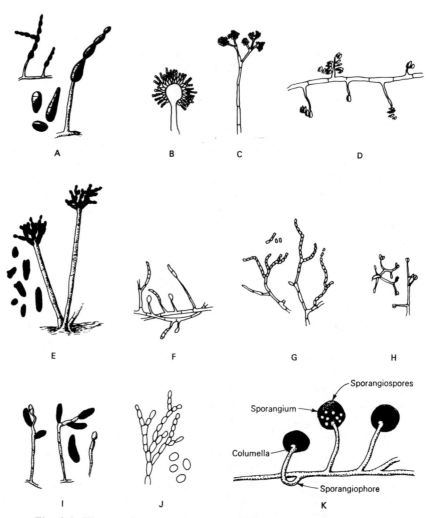

Fig. 2-2. Illustrated genera of common food-borne molds (see text for identification).

are variously shaped. They are active in the spoilage of many plant products (Fig. 2-2A). Some species produce mycotoxins.

Aspergillus. These molds produce upright conidiophores that are simple and terminate in a globose or clavate swelling. The conidia are one-celled, globose, and variously colored in mass. These molds appear yellow to green to black on a large number of foods. They produce septate mycelia. Some species produce aflatoxins and ochratoxins while others are employed as commercial sources of proteases and citric acid. Some species are field and some are storage fungi. Aspergilli may be found on cakes, fruits, vegetables,

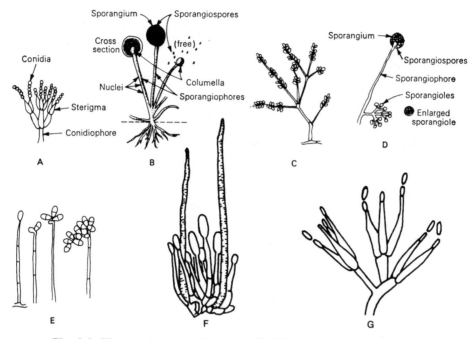

Fig. 2-3. Illustrated genera of common food-borne molds (see text for identification).

meats, and other products. The *A. glaucus* group is the second most common group of molds on hams and sausage (Fig. 2-2*B*).

Botrytis (Sclerotinia).   These organisms produce long, slender, and often pigmented conidiophores. The mycelium is septate and the conidia are borne on apical cells. The conidia are gray in mass and one-celled. Black irregular sclerotia are frequently produced. They cause a "gray mold" condition on many plants and plant foods. They are important causes of market diseases of fruits and vegetables. *B. cinerea* is presented in Fig. 2-2*C*, showing conidiophores and conidia.

Cephalosporium.   These molds produce a septate mycelium with simple and slender or swollen conidiophores. The microspores of certain species of *Fusarium* are similar in many ways to these. A water-mount of *Cephalosporium* sp. is presented in Fig. 2-2*D*.

Cladosporium.   This genus is characterized by the production of septate mycelia with dark conidiophores variously branched near the apex of the middle portion. Conidia are dark, one- or two-celled, and some are lemon shaped. One species, *C. herbarum*, produces black spots on beef (Fig. 2-2*E*).

Fusarium (Gibberella). These molds produce an extensive mycelium that is cottony in culture with tinges of pink, purple, or yellow. The conidia are canoe shaped and borne singly or in chains. They are important in the spoilage of many vegetables and fruits including the "neck rot" of bananas. Some produce T-2 toxin, vomitoxin, and zearalenone (Fig. 2-2*F*).

Geotrichum (Oidium). These are yeastlike fungi that are usually but not always white. The mycelium is septate and reproduction occurs by fragmentation of mycelium into arthrospores. These organisms are sometimes referred to as the "dairy mold" since they impart flavor and aroma to many types of cheese. They are also referred to as "machinery molds," since they build up on food-contact equipment in food-processing plants, especially tomato canning plants. They display a dimorphic form and a feathery appearance and are capable of infecting ripe fruits. *G. albidum* (*Oospora* or *Oidium lactis*) is shown in Fig. 2-2*G*.

Gloeosporium. These molds produce simple and variable-length conidiophores. The conidia are hyaline, one-celled, and sometimes curved. They cause anthracnoses in plants. *G. fructigenum* is shown in Fig. 2-2*H*.

Helminthosporium. These molds produce a light to dark mycelium in culture. Their conidiophores are short or long, septate, simple or branched. The conidia are borne successively on new growing tips; they are dark and typically contain more than three cells. This genus contains both plant pathogens and saprophytes. *H. satiuum* is shown in Fig. 2-2*I*.

Monilia. These molds produce white or gray mycelia that bear branched conidiophores. The conidia are pink or tan in mass. Some species are the imperfect stages of *Neurospora*. Some of those whose perfect states are *Monilinia* or *Sclerotinia* spp. cause brown rot of fruits. *N. sitophila* is often referred to as the "red bread mold." *M. americana* is shown in Fig. 2-2*J*.

Mucor. These molds produce nonseptate mycelia that give rise to conidiophores that bear columella and a sporangium at the apex. The spores are smooth, regular, and borne within the sporangium. They may be found growing on a large number of foods (Fig. 2-2*K*).

Penicillium. These organisms produce septate mycelia that bear conidiophores arising singularly or sometimes in synnemata with branches near the apex to form a brushlike, conidia-bearing apparatus. The conidia form by pinching off from phialides. Typical colors on foods are blue to blue-green. These molds are important in the making of some cheeses. Some are important in the production of antibiotics (penicillin, e.g.). They are widespread in soil, air, dust, and many other places and may be found on foods such as breads, cakes, fruits, and preserves. They are the most common molds on hams and sausages, and some cause soft rots of fruits. The perfect state of some is either *Talaromyces* or *Eupenicillium* (Fig. 2-3A).

Rhizopus. These molds produce nonseptate mycelia that give rise to stolons and rhizoids. The sporangiophores arise at nodes and bear columella and sporangia at the apex. The spores are borne within the sporangia and are usually black in color. Like the penicillia, they are very widespread in nature and may be found growing on foods such as fruits, cakes, preserves, and bread. One species, *R. stolonifer,* is often called the "bread mold." Some are employed in the fermentation of starch to alcohol (Fig. 2-3B).

Sporotrichum. These molds produce septate mycelia that give rise to conidiophores that bear spores near the apex. The conidia are hyaline, one-celled, globose, or ovoid, attached apically and laterally. They have been reported to grow at and below 0°C. Some produce "white spot" on refrigerated beef (Fig. 2-3C).

Thamnidium. These organisms produce nonseptate mycelia that bear sporangiophores with large sporangia at the tip and lateral sporangioles near the base. They are sometimes found on refrigerated meats, especially hindquarters held for long periods of time, where they cause a condition often referred to as "whiskers." They may be found in a large number of decaying foods such as eggs (Fig. 2-3D).

Trichothecium (Cephalothecium). These forms produce septate mycelia that bear long, slender, and simple conidiophores. The conidia appear singly, apically, and sometimes in groups of chains. Some are pink on foods such as fruits and vegetables. *T. roseum* is shown in Fig. 2-3E.

Byssochlamys. Members of this genus of *Ascomycetes* produce clusters of asci each of which contains eight ascospores. The asci appear to be without a covering wall or ascocarp. The ascospores of these organisms are heat resistant, resulting in spoilage of some high-acid canned foods. They can grow under conditions of low Eh. They exist in soils and can be recovered from ripening fruits, especially grapes. They are the most heat resistant of all spoilage molds. The conidial structures of *B. fulva* are shown in Fig. 2-3F.

Colletotrichum. This genus belongs to the order Melanconiales. They produce simple but elongate conidiophores and hyaline conidia that are one-celled, ovoid, or oblong. The acervuli produced are disc or cushion shaped, waxy, and generally dark in color. They are common contaminants of fruits and vegetables. The conidiophores, conidia, and spines of *C. lindemutheanum* are illustrated in Fig. 2-3G.

SYNOPSIS OF GENERA OF YEASTS COMMON TO FOODS

Yeasts are microscopic organisms that may be differentiated from the common bacteria by their larger cell size, their oval, elongate, elliptical, or spherical cell shapes, and by their production of buds during the process of division. The sizes of yeast cells vary, some ranging from 5 to 8 μm in diameter, while others may be as large as 100 μm in length. In general, older yeast cells tend to be smaller in size than young growing cells. Yeasts can grow over rather wide ranges of pH, alcohol, and sugar concentrations. Some have been reported to grow at a pH as low as 1.5 and in up to 18% ethanol. Many grow in the presence of 55% or more sucrose. These organisms produce pigments of many colors, with red and black pigment producers being common. A microscopic slide of growing yeast cells generally reveals cells in varying stages of budding with some showing several buds. The true yeasts or **ascosporogenous** yeasts reproduce by sexual reproduction involving copulation to varying degrees. True yeasts also produce asexual spores and **chlamydospores**. The latter are durable and tend to be produced when a culture encounters unfavorable conditions of growth. The ascosporogenous yeasts presented below all belong to the family Endomycetaceae.

The **asporogenous** yeasts (false yeasts, wild yeasts) do not display sexual reproduction and are all placed in the family Cryptococcaceae. Some of these are yeast-like organisms which are sometimes placed in the *Fungi imperfecti* group along with some of the more common molds. The wild yeasts are difficult to distinguish morphologically and culturally but may be identified serologically. While some are strains of *Saccharomyces cerevisiae* or *S. carlsbergensis*, some are representatives of many other genera including *Candida*, *Debaryomyces*, and *Pichia*.

Yeasts are often placed into groupings based upon some particular function or activity. For example, **film yeasts** are those that grow at the surface of certain acid products such as sauerkraut and pickles. Some film yeasts are species and strains of the genera *Candida* and *Hansenula*. These organisms are capable of oxidizing acids and alcohols as sources of energy. **Top yeasts** are those that carry out the conversion of sugars to alcohol at the top of a vessel, while **bottom yeasts** are those that can do the same but from the bottom of the vessel. **Apiculate** or lemon-shaped yeasts are undesirable in wine fermentations, where they produce off-flavors.

Brettanomyces. These are acid-producing asporogenous yeasts that produce oval, elongate, spherical, or olive-shaped cells. Reproduction is by multipolar budding that often leads to the formation of chains of cells. Some species carry out an after-fermentation in certain European beers and ales while others have been isolated from spoiled pickles.

Candida. These are yeastlike organisms that are sometimes placed among *Fungi imperfecti* in the family Moniliaceae along with the genera *Trichothecium* and *Geotrichum*. They reproduce by either fragmentation of mycelium into blastospores, or by budding. These are asporogenous yeasts that produce a pseudomycelium. A few species are of both industrial and medical importance. They are common on many foods such as fresh and cured meats. One species causes rancidity of margarine.

Debaryomyces. These are ascosporogenous yeasts that sometimes produce a pseudomycelium. They generally reproduce by multipolar budding and also by sexual means. This genus is often found on the surface of spoiling foods such as fresh and cured meats, sausage, pickle brine, and wines.

Hansenula. These are ascosporogenous yeasts that produce spherical, elongate, or oval cells and often a pseudomycelium. They reproduce by multipolar budding and by sexual means. When the latter occurs, hat-shaped spores are produced inside of the asci. They are common on citrus fruits, grapes, grape products, and in olive brine and fruit juice concentrates.

Kloeckera. These are nonsporeforming yeasts common on fruits, where they are disseminated and also consumed by fruit flies. They possess both fermentative and oxidative abilities. Some cause off-flavor and turbidity in wines.

Kluyveromyces (Fabospora). These are ascosporeforming yeasts. The ascospores are reniform to oblong with asci containing one to many spores, which are released at maturity. Vegetative reproduction is by multilateral budding. They are vigorous fermenters of sugars including lactose. *K. marxianus* produces ethanol directly from D-xylose, and *K. fragilis* and *K. lactis* produce β-galactosidase (lactase).

Mycoderma. These are asporogenous yeasts that usually grow on the surface of beers, pickle brines, fruit juices, vinegar, and other related products and produce a heavy film or pellicle. One species, *M. vini*, is associated with the "wine flower" condition of wines, vinegar, and related products.

Rhodotorula. These are asporogenous yeasts that sometimes produce a primitive type of pseudomycelium. Reproduction is by multipolar budding. Many produce red pigments both on cultural media and on various foods. They are widespread in nature and are often found in the air and dust.

Saccharomyces. These are ascosporogenous yeasts that produce ovoid, spherical, or elongate cells. Reproduction is by multipolar budding and ascus formation where from one to four spores are formed. This group represents the yeasts of greatest industrial importance, especially *S. cerevisiae*, which is employed in the brewing, baking, and distilling industries. These organisms are very widespread on fruits (especially grapes) and vegetables, where they cause a fermentation of sugars leading to CO_2 and ethanol. Most osmophilic yeasts belong to this genus, including those formerly classified as *Zygosaccharomyces*.

Saccharomycopsis (Endomycopsis). These are ascosporeforming or true yeasts that are oxidative. They are common on stored cereal grains and commonly form films on fermenting sauerkraut and cucumbers. At least one species, *E. fibuliger,* possesses strong alpha and beta-amylase activities.

Schizosaccharomyces. These are ascosporogenous yeasts that reproduce either by fission, arthrospores, or sexual means. When the latter occurs, the asci contain four to eight spores that are oval, spherical, or kidney shaped. They occur in sugar and other related products.

Torulopsis (Torula). These are asporogenous yeasts that produce spherical or oval cells. They reproduce by multipolar budding and sometimes produce a primitive type of pseudomycelium. They are widespread in nature and may be seen growing on refrigerated foods of many types.

Trichosporon. These are nonascosporeforming oxidative yeasts found in a variety of foods including fermenting maple sap, meats, and beers. At least one species is lipolytic (*T. pullulans*). They belong to the family Cryptococcaceae.

REFERENCES

1. Barnett, R.L. 1960. *Illustrated genera of imperfect fungi.* 2d ed. Minneapolis: Burgess.
2. *Bergey's manual of determinative bacteriology.* 1974. 8th ed., ed. R. E. Buchanan and N. E. Gibbons. Baltimore: Williams & Wilkins.
3. *Bergey's manual of systematic bacteriology.* 1984. Vol. 1. Ed. N. R. Kreig. Baltimore: Williams & Wilkins.
4. Colwell, R. R., ed. 1984. *Vibrios in the environment.* New York: Wiley.
5. Henriksen, S. D. 1973. *Moraxella, Acinetobacter,* and the *Mimae. Bacteriol. Rev.* 37:522–61.
6. Pelczar, M. J., Jr., and R. D. Reid. 1965. *Microbiology.* 2d ed. Ch. 36 and app. B. New York: McGraw-Hill.
7. Schleifer, K. H., and R. Kilpper-Balz. 1984. Transfer of *Streptococcus faecalis* and *Streptococcus faecium* to the genus *Enterococcus* nom. rev. as *Enterococcus faecalis* comb. nov. and *Enterococcus faecium* comb. nov. *Int. J. Syst. Bacteriol.* 34:31–34.
8. Skerman, V. B. D., V. McGowan, and P. H. A. Sneath. 1980. Approved lists of bacterial names. *Int. J. Syst. Bacteriol.* 30:225–420.
9. Smibert, R. M. 1978. The genus *Campylobacter. Ann. Rev. Microbiol.* 32:673–709.
10. Sneath, P. H. A., and D. Jones. 1976. *Brochothrix,* a new genus tentatively placed in the family *Lactobacillaceae. Int. J. Syst. Bacteriol.* 26:102–4.

3.

INTRINSIC

AND EXTRINSIC PARAMETERS

OF FOODS THAT AFFECT

MICROBIAL GROWTH

Since our foods are of plant and animal origin, it is worthwhile to consider those characteristics of plant and animal tissues that affect the growth of microorganisms. The plants and animals that serve as food sources have all evolved mechanisms of defense against the invasion and proliferation of microorganisms, and some of these remain in effect in fresh foods. By taking these natural phenomena into account, one can make effective use of each or all in preventing or retarding the microbial spoilage of the products that are derived from them.

INTRINSIC PARAMETERS

Those parameters of plant and animal tissues that are an inherent part of the tissues are referred to as **intrinsic parameters** (52). These parameters are:

1. pH
2. moisture content
3. oxidation-reduction potential (Eh)
4. nutrient content
5. antimicrobial constituents
6. biological structures

Each of these is discussed below with emphasis placed upon their effects on microorganisms in foods.

pH

It has been well established that most microorganisms grow best at pH values around 7.0 (6.6–7.5), while few grow below 4.0 (see Fig. 3-1). Bacteria tend to be more fastidious in their relationships to pH than molds and yeasts, with the pathogenic bacteria being the most fastidious. With respect to pH minima and maxima of microorganisms, those represented in Fig. 3-1 should not be taken to be precise boundaries, since the actual values are known to be dependent upon other growth parameters. For example, the pH minima of certain lactobacilli have been shown to be dependent upon the type of acid used, with citric, hydrochloric, phosphoric, and tartaric acids permitting growth at lower pH than acetic or lactic acids (46).

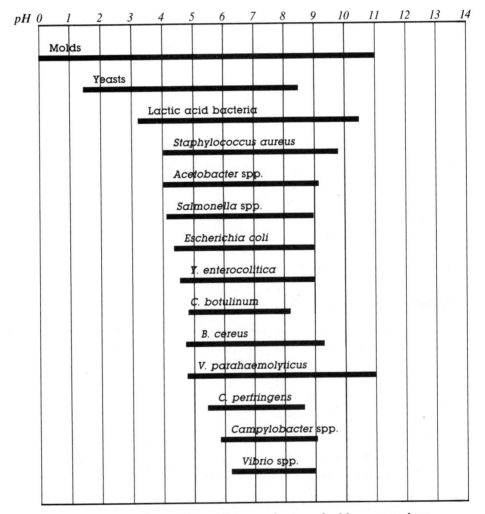

Fig. 3-1. Approximate pH growth ranges for some food-borne organisms.

In the presence of 0.2 M NaCl, *Alcaligenes faecalis* has been shown to grow over a wider pH range than in the absence of NaCl or in the presence of 0.2 M sodium citrate (Fig. 3-2). Of the foods presented in Table 3-1, it can be seen that fruits, soft drinks, vinegar, and wines all fall below the point at which bacteria normally grow. The excellent keeping quality of these products is due in great part to pH. It is a common observation that fruits generally undergo mold and yeast spoilage, and this is due to the capacity of these organisms to grow at pH values < 3.5, which is considerably

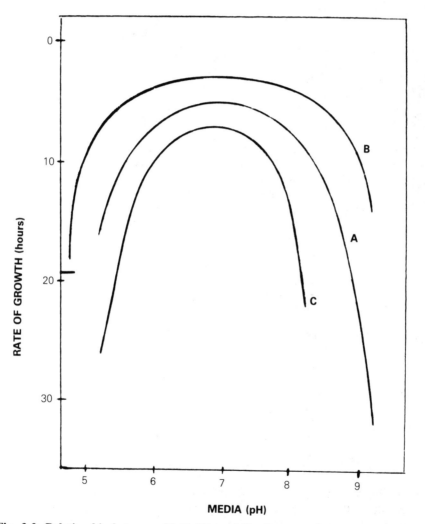

Fig. 3-2. Relationship between pH, NaC1, and Na citrate on the rate of growth of *Alcaligenes faecalis* in 1% peptone: A = 1% peptone; B = 1% peptone + 0.2 M NaC1; C = 1% peptone + 0.2 M Na citrate. (Redrawn from 70: used with permission of publisher)

Table 3-1. Approximate pH values of some fresh fruits and vegetables.

Product	pH
VEGETABLES	
Asparagus (buds and stalks)	5.7–6.1
Beans (string and Lima)	4.6 & 6.5
Beets (sugar)	4.2–4.4
Broccoli	6.5
Brussels sprouts	6.3
Cabbage (green)	5.4–6.0
Carrots	4.9–5.2; 6.0
Cauliflower	5.6
Celery	5.7–6.0
Corn (sweet)	7.3
Eggplant	4.5
Lettuce	6.0
Olives	3.6–3.8
Onions (red)	5.3–5.8
Parsley	5.7–6.0
Parsnip	5.3
Potatoes (tubers & sweet)	5.3–5.6
Pumpkin	4.8–5.2
Rhubarb	3.1–3.4
Spinach	5.5–6.0
Squash	5.0–5.4
Tomatoes (whole)	4.2–4.3
Turnips	5.2–5.5
FRUITS	
Apples	2.9–3.3
Bananas	4.5–4.7
Figs	4.6
Grapefruit (juice)	3.0
Limes	1.8–2.0
Melons (honey dew)	6.3–6.7
Oranges (juice)	3.6–4.3
Plums	2.8–4.6
Watermelons	5.2–5.6
Grapes	3.4–4.5

below the minima for most food spoilage and all food poisoning bacteria. It may be further noted from Table 3-2 that most of the meats and seafoods have a final ultimate pH of about 5.6 and above. This makes these products susceptible to bacterial as well as to mold and yeast spoilage. Likewise, most vegetables have higher pH values than fruits, and consequently vegetables should be subject more to bacterial than fungal spoilage (see below).

Table 3-2. Approximate pH values of dairy, meat, poultry, and fish products.

Product	pH
MEAT AND POULTRY	
Beef (ground)	5.1–6.2
Ham	5.9–6.1
Veal	6.0
Chicken	6.2–6.4
FISH AND SHELLFISH	
Fish (most species)[a]	6.6–6.8
Clams	6.5
Crabs	7.0
Oysters	4.8–6.3
Tuna fish	5.2–6.1
Shrimp	6.8–7.0
Salmon	6.1–6.3
White fish	5.5
DAIRY PRODUCTS	
Butter	6.1–6.4
Buttermilk	4.5
Milk	6.3–6.5
Cream	6.5
Cheese (American mild and Cheddar)	4.9, 5.9

[a] Just after death.

With respect to the keeping quality of meats, it is well known that meat from fatigued animals spoils faster than that from rested animals, and that this is a direct consequence of final pH attained upon completion of rigor mortis. Upon the death of a well-rested meat animal, the usual 1% glycogen is converted into lactic acid, which directly causes a depression in pH values from about 7.4 to about 5.6, depending upon the type of animal. Callow (16) found the lowest pH values for beef to be 5.1 and the highest 6.2 after rigor mortis. The usual pH value attained upon completion of rigor mortis of beef is around 5.6 (5). The lowest and highest values for lamb and pork, respectively, were found by Callow to be 5.4 and 6.7, and 5.3 and 6.9. Briskey (8) reported that the ultimate pH of pork may be as low as approximately 5.0 under certain conditions. The effect of pH of this magnitude upon microorganisms, especially bacteria, is obvious. With respect to fish, it has been known for some time that halibut, which usually attains an ultimate pH of about 5.6, has better keeping qualities than most other fish, whose ultimate pH values range between 6.2 and 6.6 (62).

While some foods are characterized by what Weiser (79) referred to as inherent acidity, some foods owe their acidity or pH to the actions of certain microorganisms. Weiser referred to the latter type as biological acidity, which is displayed by products such as fermented milks, sauerkraut, and pickles. Regardless of the source of acidity, the effect upon keeping quality appears to be the same.

Some foods are better able to resist changes in pH than others. Those that tend to resist changes in pH are said to be **buffered**. In general, meats are more highly buffered than vegetables. Contributing to the buffering capacity of meats are their various proteins. Vegetables are generally low in proteins and consequently lack the buffering capacity to resist changes in their pH by the growth of microorganisms (see Table 8-1 for the general chemical composition of vegetables).

The natural or inherent acidity of foods may be thought of as nature's way of protecting the respective plant or animal tissues from destruction by microorganisms. It is of interest that fruits should have pH values below those required by many spoilage organisms. The biological function of the fruit is the protection of the plant's reproductive body, the seed. This one fact alone has no doubt been quite important in the evolution of present-day fruits. While the pH of a living animal favors the growth of most spoilage organisms, other intrinsic parameters come into play to permit the survival and growth of the animal organism.

pH EFFECTS. Adverse pH affects at least two aspects of a respiring microbial cell: the functioning of its enzymes and the transport of nutrients into the cell. The cytoplasmic membrane of microorganisms is relatively impermeable to H^+ and ^-OH ions. Their concentration in the cytoplasm therefore probably remains reasonably constant despite wide variations that may occur in the pH of the surrounding medium (64). The intracellular pH of resting baker's yeast cells was found by Conway and Downey (24) to be 5.8. While the outer region of the cells during glucose fermentation was found to be more acidic, the inner cell remained more alkaline. On the other hand, Peña et al. (57) did not support the notion that the pH of yeast cells remains constant with variations in pH of medium. It appears that the internal pH of most all cells is near neutrality. Bacteria such as *Sulfolobus* and *Methanococcus* may be exceptions, however. When microorganisms are placed in environments below or above neutrality, their ability to proliferate depends upon their ability to bring the environmental pH to a more optimum value or range. When placed in acid environments, the cells must either keep H^+ from entering or expel H^+ ions as rapidly as they enter. Such key cellular compounds as DNA and ATP require neutrality (9). When most microorganisms grow in acid media, their metabolic activity results in the medium or substrate becoming less acidic, while those that grow in high pH environments tend to effect a lowering of pH. The amino acid decarboxylases that have optimum activity at around pH 4.0 and almost no activity at pH 5.5 cause a spontaneous adjustment of pH toward neutrality when cells are grown in the acid range. Bacteria such as *Clostridium ace-*

tobutylicum raise substrate pH by reducing butyric acid to butanol, while *Enterobacter aerogenes* produces acetoin from pyruvic acid to raise the pH of its growth environment. When amino acids are decarboxylated, the increase in pH occurs from the resulting amines. When grown in the alkaline range, a group of amino acid deaminases that have optimum activity at about pH 8.0 cause the spontaneous adjustment of pH toward neutrality as a result of the organic acids that accumulate.

With respect to the transport of nutrients, the bacterial cell tends to have a residual negative charge. Nonionized compounds can, therefore, enter cells while ionized cannot. At neutral or alkaline pH, organic acids do not enter, while at acid pH values these compounds are nonionized and can enter the negatively charged cells. Also, the ionic character of side chain ionizable groups is affected on either side of neutrality, resulting in increasing denaturation of membrane and transport enzymes.

Among the other effects that are exerted on microorganisms by adverse pH is that of the interaction between H^+ and the enzymes in the cytoplasmic membrane. The morphology of some microorganisms can be affected by pH. The length of the hyphae of *Penicillium chrysogenum* has been reported to decrease when grown in continuous culture where pH values increased above 6.0. Pellets of mycelium rather than free hyphae were formed at about pH 6.7 (64). Extracellular H^+ and K^+ may be in competition where the latter stimulates fermentation, for example, while the former represses it. The metabolism of glucose by yeast cells in an acid medium was markedly stimulated by K^+ (65). Glucose was consumed 83% more rapidly in the presence of K^+ under anaerobic conditions and 69% more under aerobic conditions.

Other environmental factors interact with pH as noted above. With respect to temperature, pH of substrate becomes more acid as temperature increases. Concentration of salt has definite effect on pH-growth rate curves as illustrated in Fig. 3-2, where it can be seen that the addition of 0.2 M NaCl broadened the pH growth range of *Alcaligenes faecalis*. A similar result was noted for *Escherichia coli* by these investigators. When salt content exceeds this optimal level, the pH growth range is narrowed. An adverse pH makes cells much more sensitive to toxic agents of a wide variety, and young cells are more susceptible to pH changes than older or resting cells.

When microorganisms are grown on either side of their optimum pH range, an increased lag phase results. The increased lag would be expected to be of longer duration if the substrate is a highly buffered one in contrast to one that has poor buffering capacity. In other words, the length of the lag phase may be expected to reflect the time necessary for the organisms to bring the external environment within their optimum pH growth range. Analysis of the substances that are responsible for the adverse pH is of value in determining not only speed of subsequent growth but also minimum/maximum pH at which an organism may grow. In a study of the minimum pH at which salmonellae would initiate growth, Chung and Goepfert

(22) found the minimum pH to be 4.05 when hydrochloric and citric acids were used but 5.4 and 5.5 when acetic and propionic acids were used, respectively. This is undoubtedly a reflection of the ability of the organisms to alter their external environment to a more favorable range in the case of hydrochloric and citric acids than with the other acids tested. It is also possible that factors other than pH come into play in the varying abilities of organic acids as growth inhibitors. For more information on pH and acidity, see Corlett and Brown (25).

Moisture content

One of man's oldest methods of preserving foods is drying or desiccation, and precisely how this method came to be used is not known. The preservation of foods by drying is a direct consequence of removal or binding of moisture without which microorganisms do not grow. It is now generally accepted that the water requirements of microorganisms should be defined in terms of the **water activity (a_w)** in the environment. This parameter is defined by the ratio of the water vapor pressure of food substrate to the vapor pressure of pure water at the same temperature, i.e., $a_w = p/p_0$, where p = vapor pressure of solution and p_0 = vapor pressure of solvent (usually water). This concept is related to relative humidity (R.H.) in the following way: R.H. = $100 \times a_w$ (19). Pure water has an a_w of 1.00, a 22% NaCl solution (w/v) has a_w of 0.86, while a saturated solution of NaCl is 0.75 (Table 3-3).

The a_w of most fresh foods is above 0.99. The minimum values reported for the growth of some microorganisms in foods are presented in Table 3-4 (see also Chapter 15). In general, bacteria require higher values of a_w for growth than fungi, with gram-negative bacteria having higher requirements than gram positives. Most spoilage bacteria do not grow below a_w 0.91,

Table 3-3.   Relationship between water activity and concentration of salt solutions.

Water activity	Sodium chloride concentration	
	Molal	Percent, w/v
0.995	0.15	0.9
0.99	0.30	1.7
0.98	0.61	3.5
0.96	1.20	7
0.94	1.77	10
0.92	2.31	13
0.90	2.83	16
0.88	3.33	19
0.86	3.81	22

From *The Science of Meat and Meat Products*, by the American Meat Institute Foundation. W. H. Freeman and Company, San Francisco, Copyright © 1960 (33).

Table 3-4. Approximate minimum a_w values for the growth of microorganisms of importance in foods.

Organisms	a_w	Organisms	a_w
GROUPS		GROUPS	
Most spoilage bacteria	0.9	Halophilic bacteria	0.75
Most spoilage yeasts	0.88	Xerophilic molds	0.61
Most spoilage molds	0.80	Osmophilic yeasts	0.60
SPECIFIC ORGANISMS		SPECIFIC ORGANISMS	
Clostridium botulinum, Type E	0.97	*Candida scottii*	0.92
Pseudomonas spp.	0.97	*Trichosporon pullulans*	0.91
Acinetobacter spp.	0.96	*Candida zeylanoides*	0.90
Escherichia coli	0.96	*Endomyces vernalis*	0.89
Enterobacter aerogenes	0.95	*Staphylococcus aureus*	0.86
Bacillus subtilis	0.95	*Alternaria citri*	0.84
Clostridium botulinum,		*Penicillium patulum*	0.81
Types A and B	0.94	*Aspergillus glaucus*[a]	0.70
Candida utilis	0.94	*Aspergillus conicus*	0.70
Vibrio parahaemolyticus	0.94	*Aspergillus echinulatus*	0.64
Botrytis cinerea	0.93	*Saccharomyces rouxii*	0.62
Rhizopus stolonifer	0.93	*Monascus bisporus*	
Mucor spinosus	0.93	(*Xeromyces bisporus*)	0.61

[a] Perfect stages of the *A. glaucus* group are found in the genus *Eurotium*.

while spoilage molds can grow as low as 0.80. With respect to food-poisoning bacteria, *Staphylococcus aureus* has been found to grow as low as 0.86, while *Clostridium botulinum* does not grow below 0.94. Just as yeasts and molds grow over a wider pH range than bacteria, the same is true for a_w. The lowest reported values for bacteria of any type is 0.75 for halophilic (literally, "salt-loving") bacteria, while xerophilic ("dry-loving") molds and osmophilic (preferring high osmotic pressures) yeasts have been reported to grow at a_w values of 0.65 and 0.60, respectively. When salt is employed to control a_w, it can be seen from Table 3-3 that an extremely high level is necessary to achieve a_w values below 0.80.

Certain relationships have been shown to exist between a_w, temperature, and nutrition. First, at any temperature, the ability of microorganisms to grow is reduced as the a_w is lowered. Second, the range of a_w over which growth occurs is greatest at the optimum temperature for growth; and third, the presence of nutrients increases the range of a_w over which the organisms can survive (51). The specific values given above, then, should be taken only as reference points, since a change in temperature or nutrient content might permit growth at lower values of a_w.

EFFECTS OF LOW a_w. The general effect of lowering a_w below optimum is to increase the length of the lag phase of growth and to decrease the growth rate and size of final population. This effect may be expected to result from adverse influences of lowered water on all metabolic activities since all chemical reactions of cells require an aqueous environment. It must be kept in mind, however, that a_w is influenced by other environmental parameters such as pH, temperature of growth, and Eh. In their study of the effect of a_w on the growth of *Enterobacter aerogenes* in culture media, Wodzinski and Frazier (80) found that the lag phase and generation time were progressively lengthened until no growth occurred with a lowering of a_w. The minimum a_w was raised, however, when incubation temperature was decreased. When both the pH and temperature of incubation were made unfavorable, the minimum a_w for growth was higher. The interaction between a_w, pH, and temperature on the growth of molds on jam was shown by Horner and Anagnostopoulos (42). The interaction between a_w and temperature was the most significant.

With regard to specific compounds used to lower water activity, results akin to those seen with adsorption and desorption systems (see Chapter 15) have been reported. In a study on the minimum a_w for the growth and germination of *Clostridium perfringens*, Kang et al. (47) found the value to be between 0.97 and 0.95 in complex media when sucrose or NaCl were used to adjust a_w but 0.93 or below when glycerol was used. In another study, glycerol was found to be more inhibitory than NaCl to relatively salt-tolerant bacteria but less inhibitory than NaCl to salt-sensitive species when compared at similar levels of a_w in complex media (49). In their studies on the germination of *Bacillus* and *Clostridium* spores, Jakobsen and Murrell (44) observed strong inhibition of spore germination when a_w was controlled by NaCl and $CaCl_2$, but less inhibition when glucose and sorbitol were used, and very little inhibition when glycerol, ethylene glycol, acetamide, or urea was used. The germination of clostridial spores was completely inhibited at a_w 0.95 with NaCl, but no inhibition occurred at the same a_w when urea, glycerol, or glucose was employed. In another study, the limiting a_w for the formation of mature spores by *B. cereus* strain T was shown to be about 0.95 for glucose, sorbitol, and NaCl but about 0.91 for glycerol (45). Both yeasts and molds have been found to be more tolerant to glycerol than to sucrose (42). Using a glucose minimal medium and *Pseudomonas fluorescens*, Prior (59) found that glycerol permitted growth at lower a_w values than either sucrose or NaCl. It was further shown by this worker that the catabolism of glucose, sodium lactate, and DL-arginine was completely inhibited by a_w values greater than the minimum for growth when a_w was controlled with NaCl. The control of a_w with glycerol allowed catabolism to continue at a_w values below that for growth on glucose. In all cases where NaCl was used by this investigator to adjust a_w, substrate catabolism ceased at an a_w greater than the minimum for growth, while glycerol permitted catabolism at lower a_w values than the minimum for growth. In spite of some reports to the contrary, it appears that glycerol is clearly less inhibitory to respiring organisms than agents such as sucrose and NaCl.

There are some definite effects of lowered a_w on microorganisms. Some bacteria accumulate proline as a response to low a_w, and increases in some "pool" amino acids have been reported to occur in some salt-tolerant *Staphylococcus aureus* strains (12). Using a defined medium, Christian (18) found that *Salmonella oranienburg* at a_w of < 0.97 had the added requirement for the amino acid proline, and Christian and Waltho (21) showed that proline stimulated respiration at reduced values of a_w. With *S. aureus* MF-31 in 10% NaCl, proline was shown to accumulate by transport while glutamine accumulated via synthesis (1). As to the trigger for proline synthesis, K^+ accumulates inside of cells as a_w is lowered and catalyzes the formation of proline precursors (76). However, proline synthesis is not stimulated by osmotic stress among enteric bacteria but instead by transport from the culture medium. Osmophilic yeasts accumulate polyhydric alcohols to a concentration commensurate with their extracellular a_w. According to Pitt (58), the xerophilic fungi accumulate compatible solutes or osmoregulators as a consequence of the need for high internal solutes if growth at low a_w is to be possible. In a comparative study of xero-tolerant and nonxero-tolerant yeasts to water stress, Edgley and Brown (27) found that *Saccharomyces rouxii* responded to low a_w controlled by polyethylene glycol by retaining within the cells increasing proportions of glycerol. However, the amount did not change greatly, nor did the level of arabitol change appreciably by a_w. On the other hand, a nontolerant *S. cerevisiae* responded to a lowering of a_w by synthesizing more glycerol but retaining less. The *S. rouxii* response to low a_w was at the level of glycerol permeation/transport, while that for *S. cerevisiae* was metabolic. It appears from this study that a low a_w forces *S. cerevisiae* to divert a greater proportion of its metabolic activity to glycerol production accompanied by an increase in the amount of glucose consumed during growth.

While yeasts concentrate polyols as "osmoregulators" and enzyme protectors (10), halophilic bacteria operate under low a_w conditions by virtue of their ability to accumulate KCl in the same general manner. In the case of halophilic bacteria, KCl is a requirement while osmophilic yeasts have a high tolerance for high solute concentrations (11).

It is known that the growth of at least some cells may occur in high numbers at reduced a_w values, while at the same time certain extracellular products are not produced. For example, reduced a_w results in the cessation of enterotoxin B production by *S. aureus* even though high numbers of cells are produced at the same time (74). In the case of *Neurospora crassa,* low a_w resulted in nonlethal alterations of permeability of the cell membrane, leading to loss of several essential molecules (17). Similar results were observed with electrolytes or nonelectrolytes.

Overall, the effect of lowered a_w on the nutrition of microorganisms appears to be of a general nature where cell requirements that must be mediated through an aqueous milieu are progressively shut off. In addition to the effect on nutrients, lowered a_w undoubtedly has adverse effects on the functioning of the cell membrane, which must be kept in a fluid state. The drying of internal parts of cells would be expected to occur upon placing cells in a medium of lowered a_w to a point where equilibrium of water between cells and substrate occurs. While the mechanisms are not entirely clear at this time, all microbial cells may require the same effective internal a_w. Those that can grow under extreme conditions of low a_w apparently do so by virtue of their ability to concentrate salts, polyols, and amino acids (and possibly other types of compounds) to internal levels not only sufficient to prevent the cells from losing water but which may allow the cell to extract water from the water-depressed external environment. For more information, see Christian (20).

Oxidation-reduction potential (O/R,Eh)

It has been known for many years that microorganisms display varying degrees of sensitivity to the oxidation-reduction potential of their growth medium (41). The O/R potential of a substrate may be defined generally as the ease with which the substrate loses or gains electrons. When an element or compound loses electrons, the substrate is said to be oxidized while a substrate that gains electrons becomes reduced:

$$Cu \underset{\text{reduction}}{\overset{\text{oxidation}}{\rightleftarrows}} Cu + e.$$

Oxidation may also be achieved by the addition of oxygen as illustrated in the following reaction:

$$2\,Cu + O_2 \rightarrow 2\,CuO$$

Therefore, a substance that readily gives up electrons is a good reducing agent, while one that readily takes up electrons is a good oxidizing agent. When electrons are transferred from one compound to another, a potential difference is created between the two compounds. This difference may be measured by use of an appropriate instrument and expressed as millivolts (mv). The more highly oxidized a substance, the more positive will be its electrical potential, and the more highly reduced a substance, the more negative will be its electrical potential. When the concentration of oxidant and reductant is equal, a zero electrical potential exists. The O/R potential of a system is expressed by the symbol Eh. Aerobic microorganisms require positive Eh values (oxidized) for growth while anaerobes require negative Eh values (reduced). (See Fig. 3-3.) Among the substances in foods that

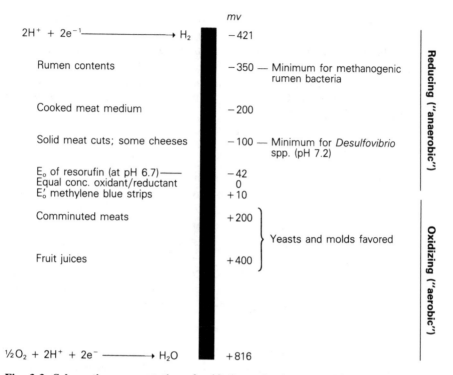

Fig. 3-3. Schematic representation of oxidation-reduction potentials relative to the growth of certain microorganisms.

help to maintain reducing conditions are —SH groups in meats and ascorbic acid and reducing sugars in fruits and vegetables.

According to Frazier (37), the O/R potential of a food is determined by: (1) the characteristic O/R potential of the original food; (2) the **poising capacity,** that is, the resistance to change in potential of the food; (3) the oxygen tension of the atmosphere about the food; and (4) the access which the atmosphere has to the food.

With respect to Eh requirements of microorganisms, some bacteria require reduced conditions for growth initiation (Eh of about -200 mv) while others require a positive Eh for growth. In the former category are the anaerobic bacteria such as the genus *Clostridium,* while in the latter belong aerobic bacteria such as the genus *Bacillus.* Some aerobic bacteria actually grow better under slightly reduced conditions, and these organisms are often referred to as **microaerophiles.** Examples of microaerophilic bacteria are lactobacilli and streptococci. Some bacteria have the capacity to grow under either aerobic or anaerobic conditions. Such types are referred to as **facultative anaerobes.** Most molds and yeasts encountered in and on foods are aerobic though a few tend to be facultative anaerobes.

In regard to the Eh of foods, plant foods, especially plant juices, tend to have Eh values of from $+300$ to $+400$. It is not surprising to find that aerobic bacteria and molds are the common cause of spoilage of products of this type. Solid meats have Eh values of around -200 mv while in minced meats the Eh is generally around $+200$ mv. Cheeses of various types have been reported to have Eh values on the negative side from -20 to around -200 mv.

With respect to the Eh of prerigor as opposed to postrigor muscles, Barnes and Ingram (3, 4) undertook a study of the measurement of Eh in muscle over periods of up to 30 h postmortem and its effect upon the growth of anaerobic bacteria. These authors found that the Eh of the sternocephalicus muscle of the horse immediately after death was $+250$ mv, at which time clostridia failed to multiply. At 30 h postmortem, the Eh had fallen to about -130 mv in the absence of bacterial growth. When bacterial growth was allowed to occur, the Eh fell to about -250 mv. Growth of clostridia was observed at Eh values of -36 mv and below. These authors confirmed for horse meat the finding of Robinson et al. (63) for whale meat: that anaerobic bacteria do not multiply until the onset of rigor mortis, because of the high Eh in prerigor meat. The same is undoubtedly true for beef, pork, and other meats of this type.

Eh EFFECTS. Microorganisms affect the Eh of their environments during growth just as they do pH. This is true especially of aerobes, which can lower the Eh of their environment while anaerobes cannot. As aerobes grow, O_2 in the medium is depleted, resulting in the lowering of Eh. Growth is not slowed, however, as much as might be expected due to the ability of cells to make use of O_2-donating or hydrogen-accepting substances in the medium. The result of this is that the medium becomes poorer in oxidizing and richer in reducing substances (51). The Eh of a medium can be reduced by microorganisms by their production of certain metabolic by-products such as H_2S, which has the capacity to lower Eh to -300 mv. Since H_2S reacts readily with O_2, it will accumulate only in anaerobic environments (9).

As noted above, Eh is dependent upon the pH of the substrate and the direct relationship between these two factors is the rH value defined in the following way:

$$Eh = 2.303 \frac{RT}{F} (rH - 2pH),$$

where $R = 8.315$ joules, $F = 96{,}500$ coulombs, and $T =$ absolute temperature (54). Therefore, the pH of a substrate should be stated when Eh is given. Normally Eh is taken at pH 7.0 (expressed Eh'). When taken at pH 7.0, 25°C, and with all concentrations at 1.0 M, Eh $=$ Eh$_0'$ (simplified Nernst equation). In nature, Eh tends to be more negative under progressively alkaline conditions.

Among naturally occurring nutrients, ascorbic acid and reducing sugars in plants and fruits and —SH groups in meats are of primary importance. The presence or absence of appropriate quantities of oxidizing/reducing agents in a medium is, therefore, of obvious value to the growth and activity of all microorganisms.

While the growth of anaerobes is normally believed to occur at reduced values of Eh, the exclusion of O_2 may be necessary for some anaerobes. When *Clostridium perfringens, Bacteroides fragilis,* and *Peptococcus magnus* were cultured in the presence of O_2, inhibition of growth occurred even when the medium was at a negative Eh of -50 mv (74). These investigators found that growth occurred in media with an Eh as high as $+325$ mv when no O_2 was present.

With regard to the effect of Eh on lipid production by *Saccharomyces cerevisiae,* it has been shown that anaerobically grown cells produce a lower total level, a highly variable glyceride fraction, and decreased phospholipid and sterol components than aerobically grown cells (61). The lipid produced by anaerobically grown cells was characterized by a high content (up to 50% of total acid) of 8:0 to 14:0 acids and a low level of unsaturated fatty acid in the phospholipid fraction. In aerobically grown cells, 80–90% of the fatty acid component was associated with glyceride, and the phospholipid was found to be 16:1 and 18:1 acids. Unlike aerobically grown cells, anaerobically grown *S. cerevisiae* cells were found to have a lipid and sterol requirement. For more on Eh, see Brown and Emberger (13).

Nutrient content

In order to grow and function normally, the microorganisms of importance in foods require the following:

1. water
2. source of energy
3. source of nitrogen
4. vitamins and related growth factors
5. minerals

The importance of water to the growth and welfare of microorganisms was presented earlier in this chapter. With respect to the other four groups of substances, molds have the lowest requirement, followed by yeasts, gram-negative bacteria, and gram-positive bacteria.

As sources of energy, food-borne microorganisms may utilize sugars, alcohols, and amino acids. Some few microorganisms are able to utilize complex carbohydrates such as starches and cellulose as sources of energy by first degrading these compounds to simple sugars. Fats are used also by microorganisms as sources of energy, but these compounds are attacked by a relatively small number of microbes in foods.

The primary nitrogen sources utilized by heterotrophic microorganisms are amino acids. A large number of other nitrogenous compounds may serve this function for various types of organisms. Some microbes, for example, are able to utilize nucleotides and free amino acids, while others are able to utilize peptides and proteins. In general, simple compounds such as amino acids will be utilized by most all organisms before any attack is made upon the more complex compounds such as high molecular weight proteins. The same is true of polysaccharides and fats.

Microorganisms may require B vitamins in low quantities and most all natural foods tend to have an abundant quantity for those organisms that are unable to synthesize their essential requirements. In general, gram-positive bacteria are the least synthetic and must, therefore, be supplied with one or more of these compounds before they will grow. The gram-negative bacteria and molds are able to synthesize most or all of their requirements. Consequently, these two groups of organisms may be found growing on foods low in B vitamins. Fruits tend to be lower in B vitamins than meats and this fact along with the usual low pH and positive Eh of fruits all help to explain the usual spoilage of these products by molds rather than bacteria.

Antimicrobial constituents

The stability of some foods against attack by microorganisms is due to the presence of certain naturally occurring substances that have been shown to have antimicrobial activity. Some spices are known to contain essential oils that possess antimicrobial activity. Among these are eugenol in cloves, allicin in garlic, cinnamic aldehyde and eugenol in cinnamon, allyl isothiocyanate in mustard, eugenol and thymol in sage, and carvacrol (isothymol) and thymol in oregano (69). Cows' milk contains several antimicrobial substances including lactoferrin, conglutinin, and the lactoperoxidase system (see Chapter 11). Casein as well as some free fatty acids that occur in milk have been shown to be antimicrobial. Eggs contain lysozyme as does milk, and this enzyme along with conalbumin provides fresh eggs with a fairly efficient antimicrobial system. The hydroxycinnamic acid derivatives (*p*-coumaric, ferulic, caffeic, and chlorogenic acids) found in fruits, vegetables, tea, molasses, and other plant sources all show antibacterial and some antifungal activity. (See Chapter 11 for more specific information on the above and other related antimicrobials.)

Biological structures

The natural covering of some foods provides excellent protection against the entry and subsequent damage by spoilage organisms. In this category are such structures as the testa of seeds, the outer covering of fruits, the shell of nuts, the hide of animals, and the shells of eggs. In the case of nuts such as pecans and walnuts, the shell or covering is sufficient to prevent the entry of all organisms. Once cracked, of course, nutmeats are subject to spoilage by molds. The outer shell and membranes of eggs, if intact, prevent the entry of nearly all microorganisms when stored under the proper conditions of humidity and temperature. Fruits and vegetables with damaged covering undergo spoilage much faster than those not damaged. The skin covering of fish and meats such as beef and pork prevents the contamination and spoilage of these foods partly because it tends to dry out faster than freshly cut surfaces.

Taken together, these six intrinsic parameters represent nature's way of preserving plant and animal tissues from microorganisms. By determining the extent to which each exists in a given food, one can predict the general types of microorganisms that are likely to grow and consequently the overall stability of this particular food. Their determination may also aid one in determining age and possibly the handling history of a given food.

EXTRINSIC PARAMETERS

The extrinsic parameters of foods are those properties of the storage environment that affect both the foods and their microorganisms. Those of greatest importance to the welfare of food-borne organisms are: (1) temperature of storage, (2) relative humidity of environment, and (3) presence and concentration of gases in the environment.

Temperature of storage

Microorganisms grow over a very wide range of temperatures. Therefore, it would be well to consider at this point the temperature growth ranges for organisms of importance in foods as an aid in selecting the proper temperature for the storage of different types of foods (see Fig. 3-4).

The lowest temperature at which a microorganism has been reported to grow is $-34°C$ while the highest is somewhere in excess of $90°C$. It is customary to place microorganisms into three groups based upon their temperature requirements for growth. Those organisms that grow well below $20°C$ and have their optimum between $20°$ and $30°C$ are referred to as **psychrophiles** or **psychrotrophs** (see Chapter 13). Those that grow well between $20°$ and $45°C$ with optima between $30°$ and $40°C$ are referred to as **mesophiles,** while those that grow well at and above $45°C$ with optima between $55°$–$65°C$ are referred to as **thermophiles.** Physiological properties of these groups are treated in Chapters 23 and 24.

With regard to bacteria, psychrotrophic species and strains are found among the following genera of those presented in the previous chapter: *Alcaligenes, Alteromonas, Brochothrix, Corynebacterium, Flavobacterium, Lactobacillus, Micrococcus, Pseudomonas, Streptococcus, Streptomyces,* and others. The psychrotrophs found most commonly on foods are those that belong to the genera *Alcaligenes, Pseudomonas,* and *Streptococcus.*

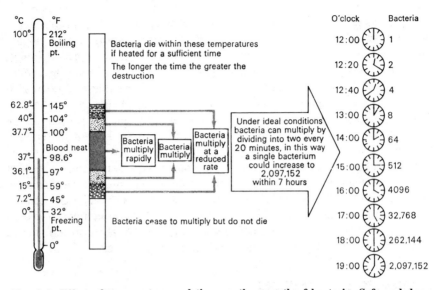

Fig. 3-4. Effect of temperature and time on the growth of bacteria. Safe and dangerous temperatures for foodstuffs. (From Hobbs, 41, reproduced with permission of the publisher)

These organisms grow well at refrigerator temperatures and cause spoilage of meats, fish, poultry, eggs, and other foods normally held at this temperature. Standard plate counts of viable organisms on such foods are generally higher when the plates are incubated at about 7°C for at least 7 days than when incubated at 30°C and above. Mesophilic species and strains are known among all genera presented in the previous chapter and may be found on foods held at refrigerator temperature. They apparently do not grow at this temperature but do grow at temperatures within the mesophilic range if other conditions are suitable. It should be pointed out that some organisms can grow over a range from 0° and 30°C or above. One such organism is *Streptococcus faecalis.*

Most thermophilic bacteria of importance in foods belong to the genera *Bacillus* and *Clostridium.* While only a few species of these genera are thermophilic, they are of great interest to the food microbiologist and food technologist in the canning industry.

Just as molds are able to grow over wider ranges of pH, osmotic pressure, and nutrient content, they are also able to grow over wider ranges of temperature than bacteria. Many molds are able to grow at refrigerator temperatures, notably some strains of *Aspergillus, Cladosporium,* and *Thamnidium,* which may be found growing on eggs, sides of beef, and fruits. Yeasts grow over the psychrophilic and mesophilic temperature ranges but generally not within the thermophilic range.

The quality of the food product must also be taken into account in selecting a storage temperature. While it would seem desirable to store all foods at refrigerator temperatures or below, this is not always best for the maintenance of desirable quality in some foods. For example, bananas keep better if stored at 13°–17°C than at 5°–7°C. A large number of vegetables are favored by temperatures of about 10°C, including potatoes, celery, cabbage, and many others. In every case, the success of storage temperature depends to a great extent upon the R.H. of the storage environment and the presence or absence of gases such as CO_2 and O_3.

Temperature of storage is the single most important parameter that affects the spoilage of highly perishable foods, and this fact has been emphasized by the work of Olley and Ratkowsky and their co-workers. According to these investigators, spoilage can be predicted by a spoilage rate curve (54). The general spoilage curve has been incorporated into the circuitry of a temperature function integrator which reads out the equivalent days of storage at 0°C and thus makes it possible to predict the remaining shelf life at 0°C. It has been shown that the rate of spoilage of fresh poultry at 10°C is about twice that at 5°C, and that at 15°C is about three times that at 5°C (27). Instead of using the Arrhenius law equation, the following was developed to describe the relationship between temperature and growth rate of microorganisms between the minimum and optimum temperatures (60):

$$\sqrt{r} = b(T - T_0)$$

where r = the growth rate; b = the slope of the regression line; and T_0 = a conceptual temperature of no metabolic significance. The linear relationship has been shown to apply to spoilage bacteria and fungi when growing in foods or when utilizing amino acids (60).

Relative humidity of environment
The relative humidity (R.H.) of the storage environment is important both from the standpoint of a_w within foods and the growth of microorganisms at the surfaces. When the a_w of a food is set at 0.60, it is important that this food be stored under conditions of R.H. that do not allow the food to pick up moisture from the air and thereby increase its own surface and subsurface a_w to a point where microbial growth can occur. When foods with low a_w values are placed in environments of high R.H., the foods pick up moisture until equilibrium has been established. Likewise, foods with a high a_w lose moisture when placed in an environment of low R.H. There is a relationship between R.H. and temperature which should be borne in mind in selecting proper storage environments for the storage of foods. In general, the higher the temperature, the lower the R.H., and vice versa.

Foods that undergo surface spoilage from molds, yeasts, and certain bacteria should be stored under conditions of low R.H. Improperly wrapped meats such as whole chickens and beef cuts tend to suffer surface spoilage in the refrigerator much before deep spoilage occurs, due to the generally high R.H. of the refrigerator and the fact that the meat spoilage flora is essentially aerobic in nature. While it is possible to lessen the chances of surface spoilage in certain foods by storing under low conditions of R.H., it should be remembered that the food itself will lose moisture to the atmosphere under such conditions and thereby become undesirable. In selecting the proper environmental conditions of R.H., consideration must be given to both the possibility of surface growth and the desirable quality to be maintained in the foods in question. By altering the gaseous atmosphere, it is possible to retard surface spoilage without lowering R.H.

Presence and concentration of gases in the environment

The storage of food in atmospheres containing increased amounts of CO_2 up to about 10% is referred to as "controlled atmosphere" or c-a storage. The effect of c-a storage on plant organs has been known since 1917 (72), and was first put into commercial use in 1928. The use of c-a storage for fruits is employed in a number of countries, with apples and pears being the fruits most commonly treated. The concentration of CO_2 generally does not exceed 10% and is applied either from mechanical sources or by use of dry ice (solid CO_2). Carbon dioxide has been shown to retard fungal rotting of fruits caused by a large variety of fungi. While the precise mechanism of action of CO_2 in retarding fruit spoilage is not known, it is probable that it acts as a competitive inhibitor of ethylene action. Ethylene seems to act as a senescence factor in fruits (66), and its inhibition would have the effect of maintaining a fruit in a better state of natural resistance to fungal invasion (see below). For a more detailed treatment of the c-a storage of fruits and vegetables, see Smith (72).

It has also been known for many years that ozone added to food storage environments has a preservative effect upon certain foods. At levels of several ppm, this gas has been tried with several foods and found to be effective against spoilage microorganisms. It is effective against a variety of microorganisms (15). Since it is a strong oxidizing agent, it should not be used on high-lipid-content foods, since it would cause an increase in rancidity. Both CO_2 and O_3 are effective in retarding the surface spoilage of beef quarters under long-term storage.

EFFECT OF CO_2 AND O_2. The use of CO_2 atmospheres to extend the storage life of meats has received a lot of attention during the past decade. During the 1930s, the meat storage lockers of ships transporting meat carcasses were enriched with CO_2 as a means of increasing storage life (36, 68). New interest in this practice seems to have developed along with the increasing use of vacuum packaging of meats. A large number of workers have found that hyperbaric CO_2 increases the shelf life of a variety of meats, and some of these findings are summarized below. For recent reviews, see Daniels et al. (26) and Genigeorgis (36). Further discussions of vacuum packaging are presented in the next chapter.

In general, the inhibitory effects of CO_2 increase with decreasing temperature due primarily to the increased solubility of CO_2 at lower temperatures, and the pH of meats stored in high CO_2 environments tends to be slightly lower than that of air-stored controls due to carbonic acid formation. Gram-negative bacteria are more sensitive to CO_2 than gram positives (71, 74), with pseudomonads being among the most sensitive and the lactic acid bacteria and anaerobes being among the most resistant.

With steaks stored in 100% CO_2, significantly lower counts were seen 16–27 days after slaughter, compared to steaks stored in 100% N_2 or 100% O_2 or in air (43). Steaks stored at 1°C in 15% CO_2 + 85% air, or in 15% CO_2 + 85% O_2 for up to 20 days had consistently lower counts (about one-third) than those stored in air (2). The CO_2 + O_2 mixture was slightly more effective than the CO_2 + air mixture. When lamb chops were stored at −1°C in 80% O_2 + 20% N_2, or 80% air or O_2 + 20% CO_2, or 80% N_2 or H_2 + 20% CO_2, psychrotrophic organisms decreased successively when compared to those stored in air (53). Using slices of lean beef inoculated with pseudomonads and *Moraxella-Acinetobacter* spp. and stored at 5°C in 70–85% O_2 + 15% CO_2, shelf life was increased by 9 days compared to storage in air (23). In a study of pork stored at 4° or 14°C, the time needed for the APC at 4°C to reach 5×10^6 organisms/cm^2 was about three times longer when stored with 5 atm CO_2, and about fifteen times longer than in air (6). With poultry stored in air, APC of drip after 16 days at 10°C was log 9.40, while in 20% CO_2 log APC was 6.14 (77). Carbon dioxide was less effective against coliforms in the latter study with log numbers of 3.28 and 3.21 for air and 20% CO_2 storage respectively.

Carbon dioxide atmosphere storage is effective with fish. Using 80% CO_2 + air, log numbers after 14 days at 35°C were approximately 6.00/cm^2 compared to air controls with numbers > 10.5 cm^2. The pH of CO_2-stored products after 14 days decreased from around 6.75 to around 6.30, while controls increased to around 7.45 (56). The shelf life of rockfish and salmon at 4.5°C was extended by 20–80% CO_2 (14). At least 1 log difference in bacterial counts over controls was obtained when trout and croaker were stored in CO_2 environments at 4°C (39). When fresh shrimp or prawns were packed in ice in an atmosphere of 100% CO_2, they were edible for up to 2 weeks, and bacterial counts after 14 days were lower than air-packed controls after 7 days (50). When cod fillets were stored at 2°C, air-stored samples spoiled in 6 days with aerobic plate counts (APC) of $\log_{10}$ 7.7, while samples stored in 50% CO_2 + 50% O_2 or 50% CO_2 + 50% N_2 or 100% CO_2 did not show bacterial spoilage until, respectively, 26, 34, and 34 days, with respective APCs of 7.2, 6.6, and 5.5/g (73). It was suggested that the use of 50% CO_2 + 50% O_2 is technically more feasible than the use of 100% CO_2. While the practical upper limit of CO_2 for red meats is around 20%, higher concentrations can be used with fish since they contain lower levels of myoglobin.

The overall effect of high concentrations of CO_2 in meat packs is to shift the flora from a heterogeneous one consisting of gram-negative bacteria to one consisting primarily of lactobacilli and other lactic acid bacteria. In a study by Blickstad et al. (6), > 90% of the flora of pork consisted of *Pseudomonas* spp. after aerobic storage for 8 days at 4°C and 3 days at 14°C, but after 5 atm CO_2 storage, the lactobacilli dominated at both temperatures. This effect can be seen from Table 3-5, where smoked pork loins and frankfurter sausage were held from 48 to 140 days at 4°C in the presence of 100% CO_2, 100% N_2, and in vacuum. While pH did not decrease with the pork loins, decrease did occur with frankfurters in CO_2 and in vacuum, reflective of the domination by the lactobacilli. The shelf life of both products increased in the order, vacuum < N_2 < CO_2 (7). Several investigators have found the shelf life of raw meats to increase in the order N_2 < vacuum < CO_2. In another study, the initial flora of normal, low-pH beef consisted

Table 3-5. The effect of storage in vacuum, pure CO_2, or pure N_2 on the microflora of smoked pork loins and frankfurter sausage held from 48 to 140 days at 4°C.

		Smoked pork loins		
	0 day	*Vacuum 48 d*	*CO_2 48 d*	*N_2 48 d*
Log APC/g	2.5	7.6	6.9	7.2
pH	5.8	5.8	5.9	5.9
Dominant flora (%)	Flavo (20)	Lactos (52)[a]		
	Arthro (20)		Lactos (74)[b]	
	Yeasts (20)			Lactos (67)[c]
	Pseudo (11)			
	Coryne (10)			

		Frankfurter sausage		
	0 day	*Vacuum 98 d*	*CO_2 140 d*	*N_2 140 d*
Log APC/g	1.7	9.0	2.4	4.8
pH	5.9	5.4	5.6	5.9
Dominant flora (%)	Bac (34)	Lactos (38)	Lactos (88)[d]	Lactos (88)[e]
	Coryne (34)			
	Flavo (8)			
	Broch (8)			

Percent flora represented by *Lactobacillus viridescens:* [a]40; [b]72; [c]50; [d]22; [e]35.
Flavo = *Flavobacterium;*
Arthro = *Arthrobacter;*
Pseudo = *Pseudomonas;*
Coryne = *Corynebacterium;*
Bac = *Bacillus;*
Broch = *Brochothrix.*
Adapted from (7).

of 13% *Pseudomonas fluorescens* and 87% nonfluorescent *Pseudomonas* spp. After vacuum storage for 21 days at 4°C, 96% of the flora consisted of homofermentative lactic acid bacteria, while after 51 days in 100% CO_2, 100% of the flora consisted of these organisms (32). In a study employing inoculated, high-pH slices of meat stored in a mixture of 20% CO_2 + 80% air, *Enterobacter* and *Brochothrix thermosphacta* were unaffected at 3°C compared to air-stored controls (38).

With regard to the relative sensitivity of microorganisms to CO_2 inhibition, the fungus *Sclerotium rolfsii* was inhibited by 0.12 atm CO_2 while *Streptococcus cremoris* required ~ 11 atm (30). *Proteus* and *Micrococcus* spp. are less sensitive than the pseudomonads but not as resistant as *S. cremoris*. From their study of CO_2 atmospheres on the flora of normal, low-pH beef, Erichsen and Molin (32) found resistance to CO_2 as follows: *Pseudomonas* spp. < *B. thermosphacta* < lactic acid bacteria. Storage of meat in 100% CO_2 extended shelf life the greatest, followed by vacuum packaging. To effect a 50% reduction in growth rate compared to controls, 0.5 atm of CO_2 was required for *Pseudomonas fragi*, 1.3 atm for *Bacillus cereus*, and 0.6 atm for *S. cremoris* (30).

With regard to the mechanism of CO_2 inhibition of microorganisms, two explanations have been offered. King and Nagel (48) found that CO_2 blocked the metabolism of *Pseudomonas aeruginosa* and appeared to effect a mass action on enzymatic decarboxylations. Sears and Eisenberg (67) found that CO_2 affected the permeability of cell membranes, and Enfors and Molin (29) found support for the latter hypothesis in their studies on the germination of *Clostridium sporogenes* and *C. perfringens* endospores. At 1 atm CO_2, spore germination of these two species was stimulated while *B. cereus* spore germination was inhibited. As was shown by others, CO_2 is more stimulatory at low pH than high. With 55 atm CO_2, only 4% germination of *C. sporogenes* spores occurred, while with *C. perfringens* 50 atm reduced germination to 4% (29). These authors suggested that CO_2 inhibition was due to its accumulation in the membrane lipid bilayer such that increased fluidity results. In an earlier study, the generation time for eight strains of *C. perfringens* in 100% CO_2 and 100% N_2 grown in thioglycollate medium at 1 atm was about the same—12.9–17.2 and 12.9–16.9, respectively (55). With *P. fluorescens* in a minimal medium, CO_2 pressure of 100 mm Hg stimulated growth but at higher concentrations, the growth rate declined (38). In a complex medium at 30°C, maximum inhibition of this organism was achieved at 250 mm Hg. These investigators along with Enfors and Molin (31) found that CO_2 inhibition increases as temperature is lowered.

Since extremely high pressures of CO_2 are required to prevent the germination of *C. sporogenes* and *C. perfringens* spores, the storage of meats at temperatures of growth of *C. botulinum* in 1 atm CO_2 may present concerns relative to the possible germination of botulinal spores. In a culture medium flushed with CO_2, germination of types A and B spores of *C.*

botulinum was enhanced (34). In the case of salmonellae and staphylococci, the use of up to 60% CO_2 had no stimulatory effect although *Yersinia enterocolitica* may be of concern (71).

The effect of vacuum packaging and gas atmospheres on microorganisms is discussed further in Chapter 4.

REFERENCES

1. Anderson, C. B., and L. D. Witter. 1982. Glutamine and proline accumulation by *Staphylococcus aureus* with reduction in water activity. *Appl. Environ. Microbiol.* 43:1501–3.
2. Bala, K., W. C. Stringer, and H. D. Naumann. 1977. Effect of spray sanitation treatment and gaseous atmospheres on the stability of prepackaged fresh beef. *J. Food Sci.* 42:743–46.
3. Barnes, E. M., and M. Ingram. 1955. Changes in the oxidation-reduction potential of the sterno-cephalicus muscle of the horse after death in relation to the development of bacteria. *J. Sci. Food Agr.* 6:448–55.
4. ———. 1956. The effect of redox potential on the growth of *Clostridium welchii* strains isolated from horse muscle. *J. Appl. Bacteriol.* 19:117–28.
5. Bate-Smith, E. C. 1948. The physiology and chemistry of rigor mortis, with special reference to the aging of beef. *Adv. Food Res.* 1:1–38.
6. Blickstad, E., S.-O. Enfors, and G. Molin. 1981. Effect of hyperbaric carbon dioxide pressure on the microbial flora of pork stored at 4° or 14°C. *J. Appl. Bacteriol.* 50:493–504.
7. Blickstad, E., and G. Molin. 1983. The microbial flora of smoked pork loin and frankfurter sausage stored in different gas atmospheres at 4°C. *J. Appl. Bacteriol.* 54:45–56.
8. Briskey, E. J. 1964. Etiological status and associated studies of pale, soft, exudative porcine musculature. *Adv. Food Res.* 13:89–178.
9. Brock, T. D., D. W. Smith, and M. T. Madigan. 1984. *Biology of microorganisms.* 257–60. Englewood Cliffs, N.J.: Prentice-Hall.
10. Brown, A. D. 1964. Aspects of bacterial response to the ionic environment. *Bacteriol. Rev.* 28:296–329.
11. ———. 1974. Microbial water relations: Features of the intracellular composition of sugar-tolerant yeasts. *J. Bacteriol.* 118:769–77.
12. ———. 1976. Microbial water stress. *Bacteriol. Rev.* 40:803–46.
13. Brown, M. H., and O. Emberger. 1980. Oxidation-reduction potential. In *Microbial ecology of foods,* vol. 1, ICMSF, 112–15. New York: Academic Press.
14. Brown, W. D., M. Albright, D. A. Watts, B. Heyer, B. Spruce, and R. J. Price. 1980. Modified atmosphere storage of rockfish (*Sebastes miniatus*) and silver salmon (*Oncorhynchus kisutch*). *J. Food Sci.* 45:93–96.
15. Burleson, G. R., T. M. Murray, and M. Pollard. 1975. Inactivation of viruses and bacteria by ozone, with and without sonication. *Appl. Microbiol.* 29:340–44.
16. Callow, E. H. 1949. Science in the imported meat industry. *J. Roy. Sanitary Inst.* 69:35–39.
17. Charlang, G., and N. H. Horowitz. 1974. Membrane permeability and the loss

of germination factor from *Neurospora crassa* at low water activities. *J. Bacteriol.* 117:261–64.

18. Christian, J. H. B. 1955. The water relations of growth and respiration of *Salmonella oranienburg* at 30°C. *Aust. J. Biol. Sci.* 8:490–97.

19. ———. 1963. Water activity and the growth of microorganisms. In *Recent advances in food science*, vol. 3., ed. J. M. Leitch and D. N. Rhodes, 248–55. London: Butterworths.

20. ———. 1980. Reduced water activity. In *Microbial ecology of foods*, vol. 1, ICMSF, 70–91. New York: Academic Press.

21. Christian, J. H. B., and J. A. Waltho. 1966. Water relations of *Salmonella oranienburg*; stimulation of respiration by amino acids. *J. Gen. Microbiol.* 43:345–55.

22. Chung, K. C., and J. M. Goepfert. 1970. Growth of *Salmonella* at low pH. *J. Food Sci.* 35:326–28.

23. Clark, D. S., and C. P. Lentz. 1973. Use of mixtures of carbon dioxide and oxygen for extending shelf-life of prepackaged fresh beef. *Can. Inst. Food Sci. Technol. J.* 6:194–96.

24. Conway, E. J., and M. Downey. 1950. pH values of the yeast cell. *Biochem. J.* 47:355–60.

25. Corlett, D. A., Jr., and M. H. Brown. 1980. pH and acidity. In *Microbial ecology of foods*, vol. 1, ICMSF, 92–111. New York: Academic Press.

26. Daniels, J. A., R. Krishnamurthi, and S. S. H. Rizvi. 1985. A review of effects of carbon dioxide on microbial growth and food quality. *J. Food Protect.* 48:532–37.

27. Daud, H. B., T. A. McMeekin, and J. Olley. 1978. Temperature function integration and the development and metabolism of poultry spoilage bacteria. *Appl. Environ. Microbiol.* 36:650–54.

28. Edgley, M., and A. D. Brown. 1978. Response of xerotolerant and nontolerant yeasts to water stress. *J. Gen. Microbiol.* 104:343–45.

29. Enfors, S.-O., and G. Molin. 1978. The influence of high concentrations of carbon dioxide on the germination of bacterial spores. *J. Appl. Bacteriol.* 45:279–85.

30. ———. 1980. Effect of high concentrations of carbon dioxide on growth rate of *Pseudomonas fragi*, *Bacillus cereus* and *Streptococcus cremoris*. *J. Appl. Bacteriol.* 48:409–16.

31. ———. 1981. The influence of temperature on the growth inhibitory effect of carbon dioxide on *Pseudomonas fragi* and *Bacillus cereus*. *Can. J. Microbiol.* 27:15–19.

32. Erichsen, I., and G. Molin. 1981. Microbial flora of normal and high pH beef stored at 4°C in different gas environments. *J. Food Protect.* 44:866–69.

33. Evans, J. B., and C. F. Niven, Jr. 1960. Microbiology of meat: Bacteriology. In *The science of meat and meat products*. San Francisco: Freeman.

34. Foegeding, P. M., and F. F. Busta. 1983. Effect of carbon dioxide, nitrogen and hydrogen gases on germination of *Clostridium botulinum* spores. *J. Food Protect.* 46:987–89.

35. Frazier, W. C. 1968. *Food microbiology*. 2d ed., 171. New York: McGraw-Hill.

36. Genigeorgis, C. A. 1985. Microbial and safety implications of the use of modified atmospheres to extend the storage life of fresh meat and fish. *Int. J. Food Microbiol.* 1:237–51.
37. Gill, C. O., and K. H. Tan. 1979. Effect of carbon dioxide on growth of *Pseudomonas fluorescens. Appl. Environ. Microbiol.* 38:237–40.
38. ———. 1980. Effect of carbon dioxide on growth of meat spoilage bacteria. *Appl. Environ. Microbiol.* 39:317–19.
39. Hanks, H., R. Nickelson, II, and G. Finne. 1980. Shelf-life studies on carbon dioxide packaged finfish from the Gulf of Mexico. *J. Food Sci.* 45:157–62.
40. Hewitt, L. F. 1950. *Oxidation-reduction potentials in bacteriology and bio-chemistry.* 6th ed. Edinburgh: Livingston.
41. Hobbs, B. C. 1968. *Food poisoning and food hygiene.* 2d ed. London: Edward Arnold.
42. Horner, K. J., and G. D. Anagnostopoulos. 1973. Combined effects of water activity, pH and temperature on the growth and spoilage potential of fungi. *J. Appl. Bacteriol.* 36:427–36.
43. Huffman, D. L. 1974. Effect of gas atmospheres on microbial quality of pork. *J. Food Sci.* 39:723–25.
44. Jakobsen, M., and W. G. Murrell. 1977. The effect of water activity and the a_w-controlling solute on germination of bacterial spores. *Spore Res.* 2:819–34.
45. ———. 1977. The effect of water activity and a_w-controlling solute on sporulation of *Bacillus cereus* T. *J. Appl. Bacteriol.* 43:239–45.
46. Juven, B. J. 1976. Bacterial spoilage of citrus products at pH lower than 3.5. *J. Milk Food Technol.* 39:819–22.
47. Kang, C. K., M. Woodburn, A. Pagenkopf, and R. Cheney. 1969. Growth, sporulation, and germination of *Clostridium perfringens* in media of controlled water activity. *Appl. Microbiol.* 18:798–805.
48. King, A. D., Jr., and C. W. Nagel. 1975. Influence of carbon dioxide upon the metabolism of *Pseudomonas aeruginosa. J. Food Sci.* 40:362–66.
49. Marshall, B. J., F. Ohye, and J. H. B. Christian. 1971. Tolerance of bacteria to high concentrations of NaCl and glycerol in the growth medium. *Appl. Microbiol.* 21:363–64.
50. Matches, J. R., and M. E. Lavrisse. 1985. Controlled atmosphere storage of spotted shrimp (*Pandalus platyceros*). *J. Food Protect.* 48:709–11.
51. Morris, E. O. 1962. Effect of environment on micro-organisms. In *Recent advances in food science,* vol. 1, ed. J. Hawthorn and J. M. Leitch, 24–36. London: Butterworths.
52. Mossel, D. A. A., and M. Ingram. 1955. The physiology of the microbial spoilage of foods. *J. Appl. Bacteriol.* 18:232–68.
53. Newton, K. G., J. C. L. Harrison, and K. M. Smith. 1977. The effect of storage in various gaseous atmospheres on the microflora of lamb chops held at −1°C. *J. Appl. Bacteriol.* 43:53–59.
54. Olley, J., and D. A. Ratkowsky. 1973. The role of temperature function integration in monitoring fish spoilage. *Food Technol. New Zealand* 8:13–17.
55. Parekh, K. G., and M. Solberg. 1970. Comparative growth of *Clostridium perfringens* in carbon dioxide and nitrogen atmospheres. *J. Food Sci.* 35:156–59.
56. Parkin, K. L., M. J. Wells, and W. D. Brown. 1981. Modified atmosphere storage of rockfish fillets. *J. Food Sci.* 47:181–84.

57. Peña, A., G. Cinco, A. Gómez-Puyou, and M. Tuena. 1972. Effect of pH of the incubation medium on glycolysis and respiration in *Saccharomyces cerevisiae*. *Arch. Biochem. Biophys.* 153:413–25.
58. Pitt, J. I. 1975. Xerophilic fungi and the spoilage of foods of plant origin. In *Water relations of foods,* ed. R. B. Duckworth, 273–307. London: Academic Press.
59. Prior, B. A. 1978. The effect of water activity on the growth and respiration of *Pseudomonas fluorescens. J. Appl. Bacteriol.* 44:97–106.
60. Ratkowsky, D. A., J. Olley, T. A. McMeekin, and A. Ball. 1982. Relationship between temperature and growth rate of bacterial cultures. *J. Bacteriol.* 149:1–5.
61. Rattray, J. B. M., A. Schibeci, and D. K. Kidby. 1975. Lipids of yeasts. *Bacteriol. Rev.* 39:197–231.
62. Reay, G. A., and J. M. Shewan. 1949. The spoilage of fish and its preservation by chilling. *Adv. Food Res.* 2:343–98.
63. Robinson, R. H. M., M. Ingram, R. A. M. Case, and J. G. Benstead. 1952. Whale-meat: bacteriology and hygiene. *Special Report of the Service of Food Investigation Board,* no. 59. London.
64. Rose, A. H. 1965. *Chemical microbiology,* ch. 3. London: Butterworths.
65. Rothstein, A., and G. Demis. 1953. The relationship of the cell surface to metabolism. The stimulation of fermentation by extracellular potassium. *Arch. Biochem. Biophys.* 44:18–29.
66. Salisbury, F. B., and C. Ross. 1969. *Plant physiology.* 467. Belmont, Cal.: Wadsworth.
67. Sears, D. F., and R. M. Eisenberg. 1961. A model representing a physiological role of CO_2 at the cell membrane. *J. Gen. Physiol.* 44:869–87.
68. Seideman, S. C., and P. R. Durland. 1984. The utilization of modified gas atmosphere packaging for fresh meat: A review. *J. Food Qual.* 6:239–52.
69. Shelef, L. A. 1983. Antimicrobial effects of spices. *J. Food Safety* 6:29–44.
70. Sherman, J. M., and G. E. Holm. 1922. Salt effects in bacterial growth. II. The growth of *Bacterium coli* in relation to H-ion concentration. *J. Bacteriol.* 7:465–70.
71. Silliker, J. H., and S. K. Wolfe. 1980. Microbiological safety considerations in controlled-atmosphere storage of meats. *Food Technol.* 34(3):59–63.
72. Smith, W. H. 1963. The use of carbon dioxide in the transport and storage of fruits and vegetables. *Adv. Food Res.* 12:95–146.
73. Stenstrom, I.-M. 1985. Microbial flora of cod fillets (*Gadus morhua*) stored at 2°C in different mixtures of carbon dioxide and nitrogen/oxygen. *J. Food Protect.* 48:585–89.
74. Stier, R. F., L. Bell, K. A. Ito, B. D. Shafer, L. A. Brown, M. L. Seeger, B. H. Allen, M. N. Porcuna, and P. A. Lerke. 1981. Effect of modified atmosphere storage on *C. botulinum* toxigenesis and the spoilage microflora of salmon fillets. *J. Food Sci.* 46:1639–42.
75. Troller, J. A. 1971. Effect of water activity on enterotoxin B production and growth of *Staphylococcus aureus. Appl. Microbiol.* 21:435–39.
76. ———. 1984. Effect of low moisture environments on the microbial stability of foods. In *Food microbiology,* ed. A. H. Rose, 173–98. New York: Academic Press.
77. Wabeck, C. J., C. E. Parmalee, and W. J. Stadelman. 1968. Carbon dioxide preservation of fresh poultry. *Poultry Sci.* 47:468–74.

78. Walden, W. C., and D. J. Hentges. 1975. Differential effects of oxygen and oxidation-reduction potential on the multiplication of three species of anaerobic intestinal bacteria. *Appl. Microbiol.* 30:781–85.
79. Weiser, H. H. 1962. *Practical food microbiology and technology.* Ch. 8. Conn.: AVI.
80. Wodzinski, R. J., and W. C. Frazier. 1961. Moisture requirements of bacteria. II. Influence of temperature, pH, and maleate concentration on requirements of *Aerobacter aerogenes. J. Bacteriol.* 81:353–58.

4.

INCIDENCE AND TYPES
OF MICROORGANISMS IN FOODS

In general, the numbers and types of microorganisms present in a finished food product are influenced by the following: (1) the general environment from which the food was originally obtained, (2) the microbiological quality of the food in its raw or unprocessed state, (3) the sanitary conditions under which the product is handled and processed, and (4) the adequacy of subsequent packaging, handling, and storage conditions in maintaining the flora at a low level. In producing good-quality market foods, it is important to keep microorganisms at a low level for reasons of aesthetics, public health, and product shelf life. Other than those foods that have been made sterile, all foods should be expected to contain a certain number of microorganisms of one type or another. Ideally, the numbers of organisms should be as low as is possible under good conditions of production. Excessively high numbers of microorganisms in fresh foods present cause for alarm. It should be kept in mind that the inner parts of healthy plant and animal tissues are generally sterile and that it is theoretically possible to produce many foods free of microorganisms. This objective becomes impractical, however, when mass production and other economic considerations are realized. The number of microorganisms in a fresh food product, then, may by taken to reflect the overall conditions of raw product quality, processing, handling, storage, and so forth. This is one of the most important uses made of the standard plate count in food microbiology. The question immediately arises as to the attainability of low numbers under the best and most economic production conditions known. With few exceptions, it is difficult to know what is the lowest number of microorganisms attainable under good production conditions because of the many variables that must be considered. With advances in modern technology, it has been possible to reduce the microbial load in a large number of foods over what was possible even 10 years ago.

The microbiology of various groups of foods is treated below. With each group, the sources of organisms are listed and discussed, along with their incidence as reported by various workers. It should be noted that the numbers reported for a particular product by different investigators do not always reflect the microbiology of that product under ideal conditions of production, handling, and storage. The chapters on food spoilage and food poisoning should be consulted for additional information on the types of organisms associated with the foods discussed.

MEATS
Fresh red meats

The reported incidence of microorganisms in red meats is presented in Table 4-1. Comminuted meats such as ground beef invariably have higher numbers of microorganisms than noncomminuted meats such as steaks. This has been reported rather consistently for over 70 years, and there are several reasons for the generally higher counts on these types of products than one finds on whole meats. First, commercial ground meats generally consist of trimmings from various cuts. These pieces have been handled excessively and consequently normally contain more microorganisms than meat cuts such as steaks. Second, ground meat provides a greater surface area, which itself accounts in part for the increased flora. It should be recalled that as particle size is reduced, the total surface area increases with a consequent increase in surface energy. Third, this greater surface area of ground meat favors the growth of aerobic bacteria, the usual low-temperature spoilage flora. Fourth, in some commercial establishments, the meat grinders, cutting knives, and storage utensils are rarely cleaned as often and as thoroughly as is necessary to prevent the successive buildup of microbial numbers. This may be illustrated by data obtained by the author from a study of the bacteriology of several areas in the meat department of a large grocery store. The blade of the meat saw and the cutting block were swabbed immediately after they were cleaned on three different occasions with the following mean results. The saw blade had a total $\log/in^2$ count of 5.28, with 2.3 coliforms, 3.64 enterococci, 1.60 staphylococci, and 3.69 micrococci. The cutting block had a mean $\log/in^2$ count of 5.69, with 2.04 coliforms, 3.77 enterococci, < 1.00 staphylococci, and 3.79 micrococci. These are among the sources of the high total bacterial count to comminuted meats. Fifth, one heavily contaminated piece of meat is sufficient to contaminate others as well as the entire lot as they pass through the grinder. This heavily contaminated portion is often in the form of lymph nodes which are generally imbedded in fat. These organs have been shown to contain high numbers of microorganisms, and account in part for hamburger meat having a generally higher total count than ground beef. In some states,

Table 4-1.   Relative percentage of organisms in red meats that meet specified target numbers.

Products	Number of samples	Microbial group/Target	% Samples meeting target	Reference
Raw	735	APC: $\log_{10}$ 6.00 or less/g	76	144
beef	735	Coliforms: log 2.00 or less/g	84	144
patties	735	*E. coli:* log 2.00 or less/g	92	144
	735	*S. aureus:* 2.00 or less/g	85	144
	735	Presence of salmonellae	0.4	144
Fresh	1,830	APC: 6.70 or less/g	89	13
ground	1,830	*S. aureus:* 3.00 or less/g	92	13
beef[a]	1,830	*E. coli:* 1.70 or less/g	84	13
	1,830	Presence of salmonellae	2	13
	1,830	Presence of *C. perfringens*	20	13
Fresh	1,090	APC: $\geqslant$ 7.00 or less/g at 35°C	88	110
ground	1,090	Fecal coliforms: $\leqslant$2.00/g	76	110
beef	1,090	*S. aureus:* <2.00/g	91	110
Frozen	604	APC: 6.00 or less/g	67	47
ground beef	604	*E. coli:* <2.70/g	85	47
patties	604	*E. coli:* >3.00/g MPN	9	47
Fried	107	APC at 21°C: 72 h, <3.00/g	76	25
hamburger	107	Absence of enterococci, coliforms, *S. aureus*, salmonellae	100	25
Comminuted	113	Coliforms: 2.00 or less/g	42	129
big game	113	*E. coli:* 2.00 or less/g	75	129
meats	113	*S. aureus:* 2.00 or less/g	96	129
Fresh	67	APC: 5.70 or less/g	75	145
pork	67	*E. coli:* 2.00 or less/g	88	145
sausage	67	*S. aureus:* 2.00 or less/g	75	145
	560	Presence of salmonellae	28	145
Pork trimmings for above sausage	528	Presence of salmonellae	28	145
Vacuum-packaged	180	APC: <3.00/g	97	143
sliced imported	180	Absence of *E. coli* or *S. aureus* in 0.1 g; or salmonellae in 25 g	100	143
canned ham samples[b]	180	Coliforms in 0.1 g portions	2	143

[a] Under Oregon law that was in effect at the time (see Chapter 17).
[b] The machine-sliced and vacuum-packed samples were obtained from 16- to 21-lb. canned, refrigerated, imported hams.

the former may contain up to 30% beef fat while the latter should not contain more than 20% fat.

Both bacilli and clostridia may be found in meats of all types. In a study of the incidence of putrefactive anaerobe (P.A.) spores in fresh and cured pork trimmings and canned pork luncheon meat, Steinkraus and Ayres (137) found these organisms to occur at very low levels, generally less than 1/g. In a study of the incidence of clostridial spores in meats, Greenberg et al. (48) found a mean P.A. spore count/g of 2.8 from 2,358 meat samples. Of the 19,727 P.A. spores isolated, only 1 was a *C. botulinum* spore, and it was recovered from chicken. The large number of meat samples studied by these investigators consisted of beef, pork, and chicken, obtained from all parts of the United States and Canada. The significance of P.A. spores in meats is due to the problems encountered in the heat destruction of these forms in the canning industry (see Chapter 14).

The incidence of *C. perfringens* in a variety of American foods was studied by Strong et al. (140). They recovered the organism from 16.4% of raw meats, poultry, and fish tested; from 5% of spices; from 3.8% of fruits and vegetables; from 2.7% of commercially prepared frozen foods; and from 1.8% of home-prepared foods. Others have found low numbers of this organism in both fresh and processed meats. In ground beef, *C. perfringens* at 100 or less/g was found in 87% of ninety-five samples, while forty-five of the ninety-five (47%) samples contained this organism at levels < 1,000/g (80). One group was unable to recover *C. perfringens* from pork carcasses, hearts, and spleens, but 21.4% of livers were positive (7). Commercial pork sausage was found to have a prevalence of 38.9%. The significance of this organism in foods is discussed in Chapter 19.

Some members of the family Enterobacteriaceae have been found to be common in fresh and frozen beef, pork, and related meats. Of 442 meat samples examined by Stiles and Ng (139), 86% yielded enteric bacteria with all 127 ground beef samples being positive. The most frequently found were *Escherichia coli* biotype I (29%), *Serratia liquefaciens* (17%), and *Enterobacter agglomerans* (12%). A total of 721 isolates (32%) were represented by *Citrobacter freundii*, *Klebsiella pneumoniae*, *Enterobacter cloacae*, and *E. hafniae*. In an examination of 702 foods for fecal coliforms by MPN representing ten food categories, the highest number was found in the 119 ground beef samples, with geometric mean by AOAC procedure being 59/g (3). Mean number for 94 pork sausage samples was 7.9/g. From 32 samples of minced goat meat, mean coliform count was log 2.88, mean Enterobacteriaceae was 3.07, while log APC mean was 6.57 (96). Information on the incidence of coliforms, enterococci, and other indicator organisms is presented in Chapter 17.

Soy-extended ground meats

The addition of soy protein (soybean flour, soy flakes, texturized soy protein) at levels of 10–30% to ground meat patties is fairly widespread in the fast-food industry, at least in the United States, and the microbiology of these soy blends has been investigated. The earliest, most detailed study is that of Craven and Mercuri (20), who found that when ground beef or chicken was extended with 10 or 30% soy, APCs of these products increased over unextended controls when both were stored at 4°C for up to 8–10 days. While coliforms were also higher in beef-soy mixtures than in controls, this was not true for the chicken-soy blends. In general, APCs were higher at the 30% level of soy than at 10%. These findings have been supported by several investigators (8, 57, 72, 149). In one study in which 25% soy was used with ground beef, the mean time to spoilage at 4°C for the beef-soy blend was 5.3 days compared to 7.5 days for the unextended ground beef (8). In another study using 10, 20, and 30% soy, APC increased significantly both with time and concentration of soy in the blend (72).

With regard to the microbiological quality of soy products, the geometric mean APC of 1,226 sample units of seasoned product was found to be 1,500/g, with fungi, coliforms, *E. coli,* and *S. aureus* counts of 25, 3, 3, and 10/g, respectively (148).

Just why bacteria grow faster in the meat-soy blends than in nonsoy controls is not clear. The soy itself does not alter the initial flora and the general spoilage pattern of meat-soy blends is not unlike that of all-meat controls. One notable difference is a slightly higher pH (0.3–0.4 unit) in soy-extended products, and this alone could account for the faster growth rate. This was assessed by Harrison et al. (56) by using organic acids to lower the pH of soy blends to that of beef. By adding small amounts of a 5% solution of acetic acid to 20% blends, spoilage was delayed about 2 days over controls, but not all of the inhibitory activity was due to pH depression alone. With 25% fat in the ground meat, bacterial counts did not increase proportionally to those of soy-extended beef (72). It is possible that soy protein increases the surface area of soy-meat mixtures so that aerobic bacteria of the type that predominate on meats at refrigerator temperatures are favored, but data along these lines are wanting. The spoilage of soy-meat blends is discussed further in Chapter 9, and the subject has been reviewed by Draughon (23).

Mechanically deboned meat, poultry, and fish

When meat animals are slaughtered for human consumption, meat from the carcasses is removed by meat cutters. However, the most economical way to salvage the small bits and pieces of lean meat left on carcass bones is by mechanical means (mechanical deboning). Mechanically deboned meat (MDM) is removed from bones by machines. The production of MDM began in the 1970s, preceded by chicken meat in the late 1950s and fish in the late 1940s (32, 39). During the deboning process, small quantities of bone powder become part of the finished product, and the 1978 USDA regulation limits the amount of bone (based on calcium content) to no more than 0.75% (the calcium content of meat is 0.01%). MDM must contain a minimum of 14% protein and no more than 30% fat. The most significant parametrical difference between MDM and conventionally processed meat relative to microbial growth is the higher pH of the former, typically 6.0–7.0 (32, 33). The increased pH is due to the incorporation of marrow in MDM.

While most studies on the microbiology of MDM have shown these products to be not unlike those produced by conventional methods, some have found higher counts. The microbiological quality of deboned poultry was compared to other raw poultry products, and while the counts were comparable, MPN coliform counts of the commercial MDM products ranged from 460 to > 1,100/g. Six of fifty-four samples contained salmonellae, four contained *C. perfringens,* but none contained *S. aureus* (106). The APC of hand-boned lamb breasts was found to be 680,000 while for mechanically deboned lamb allowed to age for 1 week, APC was 650,000/g (34). Commercial samples of mechanically deboned fish were found to contain tenfold higher numbers of organisms than conventionally processed fish but different methods were used to perform the counts on fish frames and the mechanically deboned flesh (116). These investigators did not find *S. aureus* and concluded that the spoilage of MDF was similar to that for the traditionally processed products. In a more recent study, MDM was found to support the more rapid growth of psychrotrophic bacteria than lean ground beef (118).

Several studies have revealed the absence of *S. aureus* in MDM, reflecting perhaps the fact that these products are less handled by meat cutters. In general, the mesophilic flora count is a bit higher than that for psychotrophs, and fewer gram negatives tend to be found. Field (32) concluded that with good manufacturing practices, MDM should present no microbiological problems, and a similar conclusion was reached by Froning (39) relative to deboned poultry and fish.

Hot-boned meats

In the conventional processing of meats (cold boning), carcasses are chilled after slaughter for 24 + h and processed in the chilled state (postrigor). Hot boning (hot processing) involves the processing of meats generally within 1–2 h after slaughter (prerigor) while the carcass is still "hot."

In general, the microbiology of hot-boned meats is comparable to that of cold-boned meats, but some differences have been reported. One of the earliest studies on hot-boned hams evaluated the microbiological quality of cured hams made from hot-boned meat (hot-processed hams). These hams were found to contain a significantly higher APC (at 37°C) than cold-boned hams, and 67% of the former yielded staphylococci to 47% of the latter (115). Mesophiles counted at 35°C were significantly higher on hot-boned prime cuts than comparable cold-boned cuts, both before and after vacuum-packaged storage at 2°C for 20 days (77). Coliforms, however, were apparently not affected by hot boning. Another early study is that of Barbe et al. (5), who evaluated nineteen paired hams (hot- and cold-boned) and found that the former contained 200 bacteria/g, while 220/g were found in the latter. In a study of hot-boned carcasses held at 16°C and cold-boned bovine carcasses held at 2°C for up to 16 h postmortem, no significant differences in mesophilic and psychrotrophic counts were found (69). Both hot-boned and cold-boned beef initially contained low bacterial counts, but after a 14-day storage period the hot-boned meats contained higher numbers than the cold-boned (40). These investigators found that the temperature control of hot-boned meat during the early hours of chilling is critical, and in a later study found that chilling to 21°C within 3–9 h was satisfactory (41).

In a study of sausage made from hot-boned pork, significantly higher counts of mesophiles and lipolytics were found in the product made from hot-boned pork than in the cold-boned product, but no significant differences in psychrotrophs were found (87).

The effect that delayed chilling might have on the flora of hot-boned beef taken about 1 h after slaughter was examined by McMillin et al. (90). Portions were chilled for 1, 2, 4, and 8 h after slaughter, and subsequently ground, formed into patties, frozen, and examined. No significant differences were found between this product and a cold-boned product relative to coliforms, staphylococci, psychrotrophs, and mesophiles. A numerical taxonomy study of the flora from hot-boned and cold-boned beef both at the time of processing and after 14 days of vacuum storage at 2°C revealed no statistically significant differences in the flora (82). The predominant organisms, after storage, for both products were streptococci and lactobacilli, while in the freshly prepared hot-boned product (before storage) more staphylococci and bacilli were found. Overall, though, the two products were comparable.

Restructured lamb roast made from 10% and 30% MDM and hot-boned meat was examined for microorganisms, and overall the two uncooked products were of good quality (117). The uncooked products had counts < 3.0 × 10⁴/g, with generally higher numbers in products containing the higher amounts of MDM. Coliforms and fecal coliforms especially were higher in products with 30% MDM, and this was thought to be caused by contamination of shanks and pelvic regions during slaughtering and evisceration. Not detected in either uncooked product (in 0.1 g) were *Staphylococcus aureus* and *Clostridium perfringens,* while no salmonellae, *Yersinia enterocolitica,* or *Campylobacter jejuni* were found in 25 g samples. Cooking reduced cell counts in all products to < 30/g.

A summary of the work of ten groups of investigators made by Kotula (76) on the effect of hot-boning on the microbiology of meats revealed that six found no effect, three found only limited effects, while only one found higher counts. Kotula concluded that hot-boning per se has no effect on microbial counts. Hot-boning is often accompanied by prerigor pressurization consisting of the application of around 15,000 psi for 2 min. This process improves muscle color, overall shelf appearance, and increases tenderization. It appears not to have any effect on the microbial flora.

EFFECT OF ELECTRICAL STIMULATION. Before hot-boning, carcasses are electrically stimulated to speed the conversion of glycogen to lactic acid. The resulting rapid drop in carcass pH eliminates the toughening associated with cutting up prerigor meat. By this method, an electric stunner is attached to a carcass and repeated pulses of 0.5 to 1.0 or more seconds are administered to the product at 400+ volt potential differences between the electrodes. A summary of the findings of ten groups of researchers on what effect, if any, electrical stimulation had on the microbial flora revealed that six found no effect, two found a slight effect, and two found some effect (76). The meats studied included beef, lamb, and pork.

Among investigators who found a reduction of APC by electrical stimulation were Ockerman and Szczawinski (105), who found that the process significantly reduced APC of samples of beef inoculated before electrical stimulation but when samples were inoculated immediately after the treatment, no significant reductions occurred. The latter finding suggests that the disruption of lysosomal membranes and the consequent release of catheptic enzymes, which has been shown to accompany electrical stimulation (26), should not affect microorganisms. The tenderization associated with electrical stimulation of meats is presumed to be at least in part the result of lysosomal destruction (26). In one study, no significant reduction in surface flora was observed, whereas significant reduction was found to occur on the muscle above the aitch bone of beef carcasses (86). These workers exposed meatborne bacteria to electrical stimulation on culture media and found that gram-positive bacteria were the most sensitive to electrical stimulation, followed by gram negatives and sporeformers. When exposed to a 30-volt, 5-min treatment in saline or phosphate-buffered saline, a 5 log-cycle reduction occurred with *E. coli, Pseudomonas putrefaciens,* and *P. fragi,* while in 0.1% peptone or 2.5 M sucrose solutions, essentially no changes occurred.

It appears that electrical stimulation per se does not exert measurable effects on the microbial flora of hot-boned meats.

Organ and variety meats

The meats discussed in this section are livers, kidneys, hearts, and tongues of bovine, porcine, and ovine origins. They differ from the skeletal muscle parts of the respective animals in having both higher pH and glycogen levels, especially in the case of liver. The pH of fresh beef and pork liver ranges from 6.1 to 6.5, and that of kidneys from 6.5 to 7.0. Most investigators have found generally low numbers of microorganisms on these products, with surface numbers ranging from $\log_{10}$ 1.69 to $4.20/cm^2$ for fresh livers, kidneys, hearts, and tongues (52, 53, 104, 122). The initial flora has been reported to consist largely of gram-positive cocci, coryneforms, aerobic sporeformers, *Moraxella-Acinetobacter,* and *Pseudomonas* spp. (52, 53, 104, 122, 126). In a detailed study by Hanna et al. (53), micrococci, streptococci, and coryneforms were clearly the three most dominant groups on fresh livers, kidneys, and hearts. In one study, coagulase-positive staphylococci, coliforms, and *C. perfringens* counts ranged from log_{10} 0.19–1.37/cm² but no salmonellae were found (122).

Vacuum-packaged beef and pork livers and beef kidneys had lower counts after 7- and 14-day storage at 2°C, or 7 days at 5°C, than those stored in air or PVC film (44, 53). As is the case for vacuum-packaged nonorgan meats, pH values of livers decreased during storage for up to 28 days concomitant with increases in homo- and heterofermentative lactic acid bacteria in livers and kidneys (53).

Vacuum-packaged meats
It is estimated that at least 80% of fresh beef in the United States in the mid-1980s leaves the packing plant in vacuum packages, and because of the longer shelf life of this product, its microbiological quality is of great interest. Vacuum packaging is achieved by placing meats into plastic bags or pouches followed by the removal of air using a vacuum packaging machine and the closure of bag or pouch, often by a heat sealer. From the microbiological standpoint, the most dramatic effect of this process is the change in the gaseous environment of the product. While not all O_2 is removed before sealing, some of what remains is consumed by the aerobic flora and the meat itself, resulting in an increased level of CO_2, which is inhibitory to the flora (see Chapter 3). The relative quantities of O_2 and CO_2 in the headspace of stored vacuum packages is controlled largely by the degree to which the plastic package impedes the flow of these gases.

A large number of investigators have shown that when O_2-permeable packaging is used, refrigerated fresh meats undergo gram-negative spoilage accompanied by increased pH and foul odors, with *Pseudomonas* spp. being the predominant organisms (see Chapter 9).

It has been shown that the storage life of vacuum-packaged beef is inversely related to film permeability, with the longest shelf life (> 15 weeks) obtained with a "zero" O_2 film and the shortest (2–4 weeks) with a highly permeable film (920 cc $O_2/m^2/24$ h/atm at 25°C, R.H. 100%). Growth of *Pseudomonas* spp. increases with increasing film permeability (100). On the other hand, if O_2-impermeable packaging is used, the growth of lactic acid bacteria and sometimes that of *Brochothrix thermosphacta* is favored because of increased levels of CO_2 and a lowered Eh (123, 124, 152). These organisms typically effect a decrease in pH and create an unfavorable environment for most food-borne pathogens and gram-negative bacteria. With an initial pH of 5.45–5.65, vacuum-packaged beef inoculated with lactic acid bacteria showed pH of 5.30–5.45 after 24–35 days at 5°C (28). With vacuum-packaged cooked luncheon meats held for 16 days at 5°C, plate counts on LBS agar essentially equaled the APC on APT agar, indicating that the flora was composed essentially of lactics (73). After 15 weeks, 0.6–0.8% lactic acid was produced and the pH of some products decreased to < 5.0.

The findings by Henry et al. (60) for vacuum-packaged lamb chops and steaks are typical of those with other vacuum-packaged fresh meats. The initial count averaged $\log_{10}$ 2.53/cm^2 and the flora consisted of streptococci, micrococci, coagulase-negative staphylococci, *Moraxella-Acinetobacter* types, and *Pseudomonas*. After a 4-day storage at 2°C in PVC film, the pseudomonads constituted 86% of the flora. However, after 5–15 days at 2°C in high O$_2$-barrier film, *Lactobacillus* spp. constituted 80–90% of the flora, with *L. cellobiosus* being the most prominent. The $\log_{10}$ APC after the 15-day storage was 6.26/cm^2. In another study of vacuum-packaged lamb stored up to 21 days at 1–3°C, the initial flora was dominated by groups II and III *Coryne-bacterium* spp. and *B. thermosphacta* while after 21 days lactobacilli and *Moraxella-Acinetobacter* types dominated (54). For vacuum-packaged beef, *B. thermosphacta* constituted about 5% of the flora (146).

In a study by Patterson and Gibbs (109), pH 6.6 beef was vacuum packaged 1 day after slaughter. The initial surface count was log 2.48/cm^2 and after 0.2°C storage for 8 weeks, the count increased to log 7.69/cm^2. With regard to the initial flora, 85% consisted of *Aeromonas* spp. and 15% of gram-positive, catalase-positive bacteria. After 8 weeks, *B. thermosphacta* constituted 39% of the flora, homofermentative lactobacilli 22%, and psychrotrophic Enterobacteriaceae 39%. Most of the latter types (93%) were similar to *Serratia liquefaciens*. While vacuum packaging has been shown to extend the shelf life of both whole and cut-up fresh poultry (79), its effectiveness for poultry is highly temperature dependent, with temperatures at or below 0°C showing the greatest effect (91).

Vacuum-packaged cooked meats have been studied in the same way as raw meats. Of 113 samples of sliced bologna, 63% contained < 10^2 and 93% < 10^4/g of *B. thermosphacta*, while *C. perfringens, S. aureus,* and salmonellae were generally absent (108). Only 3 of 153 isolates from vacuum-packaged sliced cooked meat products were *B. thermosphacta*, while most of the others were unclassified lactic cocci and rods (93). With meat loaves packaged in vacuum or nitrogen, no significant differences in numbers of organisms were found after 49-day storage at temperatures from −4 to 7°C, but the initial flora of *Pseudomonas* (32–34%), *Brochothrix* (24–38%), micrococci (9–22%), and lactobacilli (7–20%) was changed to predominantly lactobacilli (62–72%) among the psychrotrophic flora by day 49 for both treatments (81). In another study, beef roasts were stored in a combination of atmospheres ranging from 100% O$_2$ to no O$_2$, and controls were vacuum packaged; but after 35 days at 1–3°C, counts of psychrotrophic flora and lactobacilli were not statistically significant (18). The possible reason for this was the high initial counts of about 10^6/in^2.

When nitrites are present in vacuum-packaged meats, domination of the flora by lactic acid bacteria is even more pronounced, since these organisms are insensitive to nitrite and since the microaerophilic environment is more favorable to them than to gram negatives (102). *B. thermosphacta* and the Enterobacteriaceae were more inhibited by nitrites under these conditions. Even though lactics outgrow the more aerobic types in the vacuum package environment, their lag phase is increased under these conditions, resulting in increased product shelf life.

The relative dominance of the flora of vacuum-packaged meats by lactobacilli versus *B. thermosphacta* is an interesting one. As noted above, the latter organism is more sensitive to nitrites than the former so that its relative absence in nitrite-treated products would be expected. Lamb chops with a storage life of 2 weeks at $-1°C$ in air showed a storage life increase of 8 weeks in O_2-free atmospheres, and *B. thermosphacta* was the major organism in each of a combination of gas atmospheres including nitrogen, O_2, and CO_2 (99). It has been reported that lactobacilli markedly inhibit the growth of *B. thermosphacta* in vacuum-packaged beef when the two organisms are present in approximately equal numbers but not on beef incubated in air (121). In the latter study, up to 75% CO_2 did not inhibit the ultimate growth of *B. thermosphacta*. However, when the initial flora of vacuum-packaged meats contains significant numbers of *B. thermosphacta*, this organism grows faster than the homo- or heterofermentative lactics and becomes dominant (27).

Reports vary on the relative dominance by *B. thermosphacta* and the lactic acid bacteria in vacuum-packaged meats. In one study, vacuum-packaged luncheon meats with lactobacilli as the dominant flora were considered acceptable up to 21 days with counts of $10^8/g$ (27). These investigators found that more homofermentative lactics were required to produce product offness than heterolactics because of the end products produced by the latter. While about 9 days were required for *B. thermosphacta* and heterofermentative lactics to reach a count of $10^8/g$, homofermentative lactics required 12–20 days. With N_2- and CO_2-packaged frankfurters stored at $-4°$ to $7°C$ for 49 days, *B. thermosphacta* decreased from about 48% to about 5%, while lactobacilli increased from around 6% to 94–96% under CO_2 (128). With raw pork stored at $2°C$ for 14 days, about 30% of the flora consisted of lactobacilli, about 18% of *B. thermosphacta,* and the remainder of pseudomonad types (45). The initial counts were approximately $10^3/g$, and the CO_2 content in packages was around 12.3%.

In a study of the populations of lactic acid bacteria on vacuum-packaged beef, 18 of 177 psychrotrophic lactic acid bacteria were *Leuconostoc mesenteroides* and the remainder were lactobacilli, atypical streptobacteria, or atypical betabacteria (60). From vacuum-packaged beef loins stored at $4°C$, fewer pseudomonads and more lactobacilli were found after 3 weeks than in controls (98). While some of the reported variations may result from product differences, it is not easy to classify correctly the lactic rods to genus and species; and until a common taxonomic scheme is followed by all investigators, variations in the lactic flora of vacuum-packaged meats as reported by different investigators may be expected to continue.

With regard to safety, raw beef was inoculated with types A and B spores of *Clostridium botulinum* and stored for up to 15 days at $25°C$ in one study.

Toxin was first detected after 6 days, always accompanied by significant organoleptic changes indicating that the vacuum-packaged toxic samples should be rejected before consumption (59). In another study, the storage of vacuum-packaged, N_2-flushed, CO_2-flushed packages of fresh cod, whiting, or flounder fillets that were inoculated with five strains of type E *C. botulinum* spores resulted in toxin production in all products after 1–3 days when they were stored at 26°C, at which time the products were generally unacceptable organoleptically (111). Continuous storage of cod fillets at 4°C resulted in toxin being detected prior to sensory rejection. The vacuum packaging of cooked bologna-type sausage resulted in the restriction by the normal flora of growth of *Yersinia enterocolitica* and salmonellae but not *Staphylococcus aureus* (103). *C. perfringens* was completely inhibited by the normal flora, and all pathogens were inhibited by the lactic flora with greater inhibition occurring as the storage temperature was decreased.

Overall, vacuum-packaged meats of all types may be regarded as being safe and generally free of food-borne pathogens with the possible exception of *Y. enterocolitica* and *S. aureus*. The generally present lactic flora represses the growth of most nonlactics and this inhibitory effect cannot be explained by the reduced pH alone. The spoilage of vacuum-packaged meats is discussed in Chapter 9.

Sausage, Bacon, Bologna, and Related Meat Products

In addition to the meat components, sausages and frankfurters have additional sources of organisms in the seasoning and formulation ingredients that are usually added in their production. Many spices and condiments have high microbial counts. The lactic acid bacteria and yeasts in some composition products are usually contributed by milk solids. In the case of pork sausage, natural casings have been shown to contain high numbers of bacteria. In their study of salt-packed casings, Riha and Solberg (120) found counts to range from log 4.48 to 7.77, and from 5.26 to 7.36 for wet-packed casings. Over 60% of the isolates from these natural casings consisted of *Bacillus* spp. followed by clostridia and pseudomonads. Of the individual ingredients of fresh pork sausage, casings have been shown to contribute the largest number of bacteria (144).

Processed meats such as bologna, salami, and others may be expected to reflect the sum of their ingredient makeup with regard to microbial numbers and types. The microflora of frankfurters has been shown to consist largely of gram-positive organisms with micrococci, bacilli, lactobacilli, microbacteria, streptococci, and leuconostocs along with yeasts (22). In a study of slime from frankfurters, these investigators found that 275 of 353 isolates were bacteria while 78 were yeasts. *B. thermosphacta* was the most conspicuous single isolate. With regard to the incidence of *C. botulinum* spores in liver sausage, 3 of 276 heated (75°C for 20 min) and 2 of 276 unheated commercial preparations contained type A botulinal toxin (58).

The most probable number of botulinal spores in this product was estimated to be 0.15/kg.

Wiltshire bacon has been reported to have a total count generally in the range of log 5–6/g (64), while high-salt vacuum-packaged bacon has been reported to have a generally lower count—about log 4/g. The flora of vacuum-packaged sliced bacon consists largely of catalase-positive cocci, such as micrococci and coagulase-negative staphylococci, as well as catalase-negative bacteria of the lactic acid types, such as lactobacilli, leuconostocs, pediococci, and Group D streptococci (2, 15, 75). The flora in cooked salami has been found to consist mostly of lactobacilli.

The so-called soul foods may be expected to contain high numbers of organisms since they consist of offal parts that are in direct contact with the intestinal-tract flora, as well as other parts such as pig feet and pig ears that do not receive much care during slaughtering and processing. This was confirmed in a study by Stewart (138), who found the geometric mean APC of log 7.92/g for chitterlings (pig intestines), 7.51/g for maws, and 7.32/g for liver pudding. For *S. aureus,* log 5.18, 5.70, and 5.15/g respectively were found for chitterlings, maws, and liver pudding.

POULTRY

Whole poultry tends to have a lower microbial count than cut-up poultry. Most of the organisms on such products are at the surface so that surface counts/cm^2 are generally more valid than counts on surface and deep tissues. May (89) has shown how the surface counts of chickens build up through successive stages of processing. In a study of whole chickens from six commercial processing plants, the initial mean total surface count was log 3.30/cm^2. After being cut up, the mean total count increased to log 3.81, and further increased to log 4.08 after packaging. The conveyor over which these birds moved showed a count of log 4.76/cm^2. When the above procedures were repeated for five retail grocery stores, May found that the mean count before cutting was log 3.18, which increased to log 4.06 after cutting and packaging. The cutting block was shown to have a total count of log 4.68/cm^2.

The changes in enteric bacteria during various stages of poultry chilling were studied by Cox et al. (19). Carcass counts before chilling were log 3.17/cm^2 for APC, and log 2.27/cm^2 for Enterobacteriaceae. However, after chilling, the latter organisms were reduced more than the APC. *Escherichia* spp. constituted 85% of enterics at day 0, but after 10 days at 4°C, they were reduced to 14%, while *Enterobacter* spp. increased from 6% to 88% during the same time.

Poultry represents the single most important food source of salmonellae to man. Of 50 frozen comminuted turkey meat samples examined, 38% yielded salmonellae (51). Their incidence in dressed broiler-fryer chickens was investigated by Woodburn (156). It was found that 72 of 264 birds (27%) harbored salmonellae representing thirteen serovars. Among the serovars, *S. infantis*, *S. reading*, and *S. blockley* were the most common. Salmonellae were isolated from the surfaces of 24 of 208 (11.5%) turkey carcasses before further processing (11). After processing into uncooked rolls, 90 of 336 (26.8%) yielded salmonellae. From the processing plants, 24% of processing equipment yielded salmonellae. Almost one-third of the workers had the organisms on their hands and gloves. Of twenty-three serovars recovered, *S. sandiego* and *S. anatum* were recovered most frequently. In fresh-ground turkey meats, salmonellae were found in 28% of 75 samples by another group of workers (50). Almost one-half of the samples had total counts above log 7.00/g. Ninety-nine percent harbored coliforms, 41% *E. coli*, 52% *C. perfringens*, and 69% *S. aureus*. About 14% of 101 chicken samples were positive for fecal coliforms by MPN (3).

Of the various cooked poultry products, precooked turkey rolls have been found to have considerably lower microbial numbers of all types (Table 4-2). In an examination of 118 samples of cooked broiler products, *C. perfringens* was found in 2.6% (85).

Table 4-2. General microbiological quality of turkey meat products.

Products	Number of samples	Microbial group/Target	% Samples meeting target	Reference
Precooked	6	APC: log 3.00/g	100	92
turkey	6	Coliforms: log 2.00 or less/g	67	92
rolls	6	Enterococci: log 2.00 or less/g	83	92
	48	Presence of salmonellae	4	92
	48	Presence of *C. perfringens*	0	92
Precooked	30	APC: $<$log 2.00/g	20	157
turkey rolls/	29	Presence of coliforms	21	157
sliced turkey meat	29	Presence of *E. coli* or salmonellae	0	157
Ground fresh	74	APC: log 7.00 or less/g	51	50
turkey meat	75	Presence of coliforms	99	50
	75	Presence of *E. coli*	41	50
	75	Presence of fecal streptococci	95	50
	75	Presence of *S. aureus*	69	50
	75	Presence of salmonellae	28	50
Frozen ground	50	APC 32°C: $<10^6$/g	54	51
turkey meat	50	Psychrotrophs: $<10^6$/g	32	51
	50	MPN *E. coli:* <10/g	80	51
	50	MPN *S. aureus:* <10/g	94	51
	50	MPN fecal streptococci: <10/g	54	51

The microbial flora of fresh poultry consists largely of pseudomonads and other closely related gram-negative bacteria as well as coryneforms, yeasts, and other organisms (78). Other microorganisms normally present on poultry products are discussed in Chapter 9.

SEAFOOD

The incidence of microorganisms in seafood such as shrimp, oysters, and clams would depend greatly upon the quality of water from which these animals are harvested. Assuming good-quality waters, most of the organisms are picked up during the various stages of processing. In the case of breaded raw shrimp, the breading process would be expected to add organisms if not properly done or if the ingredients are of poor microbiological quality. In their study of 91 samples of shrimp of various types, Silverman et al. (127) found that all precooked samples except one had total counts of < log 4.00/g. Of the raw samples, 59% had total counts below log 5.88 while 31% were below log 5.69/g. In a study of 204 samples of frozen, cooked, and peeled shrimp, 52% had total counts < log 4.70/g, and 71% had counts of log 5.30 or less/g (84). The general microbiological quality of a variety of seafood is presented in Table 4-3.

In a study of haddock fillets, most microbial contamination was found to occur during filleting and subsequent handling prior to packaging (101). These investigators showed that total count increased from log 5.61 in the morning to log 5.65 at noon, and to log 5.94/g in the evening for one particular processor. According to their study, results obtained in other companies were generally similar if the nighttime cleanup was good. In the case of shucked, soft-shell clams, the same general pattern of buildup was demonstrated from morning to evening. The mean clostridial count for both haddock fillets and soft-shell clams was less than 2/g, with clams being slightly higher than haddock fillets for these organisms, although both were low. Total counts on fresh perch fillets produced under commercial conditions were found to average log 5.54/g with yeast and mold counts of about log 2.69/g (71).

Clams may be expected to contain the organisms that inhabit the waters from which they are obtained. Of sixty clam samples from the coast of Florida, 43% contained salmonellae, which were also found in oysters at a level of 2.2/100 g oyster meats (38). Hard-shell clams have been shown to retain *S. typhimurium* more efficiently than *E. coli* (150).

In a study of the flora of raw Pacific shrimp taken from docks, *Moraxella* spp. constituted 30–60% followed by types I and II pseudomonads (8–22%), *Acinetobacter* (4–24%), *Flavobacterium-Cytophaga* (7–16%), and *Pseudomonas* types III and IV (8–22%); but following blanching and machine peeling, *Acinetobacter* spp. represented 16–35%, *Pseudomonas* types III and IV 2–76%, and *Flavobacterium-Cytophaga* 3–37% (83). The initial flora

Table 4-3. General microbiological quality of various seafood products.

Products	Number of samples	Microbial group/Target	% Samples meeting target	References
Frozen catfish	41	APC 32°C: 10^5/g or less	100	35
fillet	41	MPN coliforms: <3/g	100	35
	41	MPN *S. aureus:* <3/g	100	35
Frozen salmon	43	APC 32°C: 10^5/g or less	98	35
steaks	43	MPN coliforms: <3/g	93	35
	43	MPN *S. aureus:* <3/g	98	35
Fresh clams	53	APC 32°C: 10^5/g or less	53	35
	53	MPN coliforms: <3/g	51	35
	53	MPN *S. aureus:* <3/g	91	35
Fresh oysters	59	APC 32°C: 10^7/g or less	49	35
	59	MPN coliforms: 1,100 or less/g	22	35
	59	MPN *S. aureus:* <3/g	90	35
Shucked oysters	1,337	APC 30°C: 10^6/g or less	51	154
(retail)	1,337	MPN coliforms: 460 or less/g	94	154
	1,337	MPN fecal coliforms: 460 or less/g	96	154
Blue crabmeat	896	APC 30°C: 10^6/g or less	61	154
(retail)	896	MPN coliforms: 1,100/g or less	93	154
	896	MPN *E. coli:* <3/g	97	154
	896	MPN *S. aureus:* 1,100/g or less	94	154
Hard-shell clams	1,124	APC 30°C: 10^6/g or less	99.8	154
(wholesale)	1,130	MPN coliforms: 460/g or less	96	154
	161	MPN fecal coliforms: <3/g	91	154
Soft-shell clams	351	APC 30°C: 10^6/g or less	96	154
(wholesale)	363	MPN coliforms: 460/g or less	98	154
	75	MPN fecal coliforms: <3/g	72	154
Peeled shrimp	1,468	APC 30°C: 10^7/g or less	94	147
(raw)	1,468	MPN coliforms: 64/g or less	97	147
	1,468	MPN *E. coli:* <3/g	97	147
	1,468	MPN *S. aureus:* 64/g or less	97	147
Peeled shrimp	1,464	APC 30°C: 10^5/g or less	81	147
(cooked)	1,464	MPN coliforms: <3/g	86	147
	1,464	MPN *E. coli:* <3/g	99	147
	1,464	MPN *S. aureus:* <3/g	99	147
Lobster tail	1,315	APC 30°C: 10^6/g or less	74	147
(frozen, raw)	1,315	MPN coliforms: 64 or less/g	91	147
	1,315	MPN *E. coli:* <3/g	95	147
	1,315	MPN *S. aureus:* <3/g	76	147

Table 4-3, *continued.*

Product	Number of samples	Microbial group/Target	% Samples meeting target	References
Retail frozen,	27	APC: 6.00 or less/g	52	152
breaded,	27	Coliforms: 3.00 or less/g	100	152
raw shrimp	27	Presence of *E. coli*	4	152
	27	Presence of *S. aureus*	59	152
Fresh	335	APC: ≤7.00/g	93	4
channel	335	Fecal coliforms: 2.60/g	70.7	4
catfish	335	Presence of salmonellae	4.5	4
Frozen	342	APC: ≤7.00/g	94.5	4
channel	342	Fecal coliforms: 2.60/g	92.4	4
catfish	342	Presence of salmonellae	1.5	4
Frozen,	204	APC: <4.70/g	52	84
cooked,	204	APC: 5.30 or less/g	71	84
peeled	204	Coliforms: none or <0.3/g	52.4	84
shrimp	204	Coliforms: <3/g	75.2	84

of herring fillets has been found to be dominated by *Alteromonas putrefaciens* and *Pseudomonas* spp., with the latter dominating at 2°C and *A. putrefaciens* more predominant at 2–15°C (94). In a *C. perfringens* survey of 287 retail samples of fresh fish and shellfish, 10% were positive for this organism (1).

In general, frozen seafood as well as other frozen products have lower microbial counts than the comparable fresh products. In a study of 597 fresh and frozen seafood from retail stores, APC geometric means for the 240 frozen products ranged from log 3.54–4.97/g and from 4.89–8.43/g for the 357 fresh products (35). For coliforms, geometric mean MPN counts ranged from 1 to 7.7 cells/g for frozen and from 7.8 to 4,800/g for fresh. By MPN, only 4.7% of the 597 were positive for *E. coli,* 7.9% were positive for *S. aureus,* and 2% were positive for *C. perfringens.* All were negative for salmonellae and *Vibrio parahaemolyticus* (see Table 4-3).

Plate counts are generally higher on seafood when incubated at 30°C than at 35°C, and this is reflected in results from fresh crabmeat, clams, and oysters evaluated by Wentz et al. (154). APC geometric means for 896 crabmeat samples at 35°C was log 5.15 and 5.72 at 30°C; for 1,337 shucked oysters 5.59 at 35° and 5.95 at 30°; and for 358 soft-shell clams, log APC was 2.83 at 35° and 4.43 at 30°. This was also seen in raw in-shell shrimp and frozen raw lobster tails, where geometric mean APC for shrimp at 35°C was log 5.48 and 5.90/g at 30°C, while for lobster tail, 4.62 at 35° and 5.15/g at 30° (147).

VEGETABLES

The incidence of microorganisms in vegetables may be expected to reflect the sanitary quality of the processing steps and the microbiological condition of the raw product at the time of processing. In a study of green beans before blanching, Splittstoesser et al. (135) showed that the total counts ranged from log 5.60 to over 6.00 in two production plants. After blanching, the total numbers were reduced to log 3.00–3.60/g. After passing through the various processing stages and packaging, the counts ranged from log 4.72 to 5.94/g. In the case of French-style beans, one of the greatest buildups in numbers of organisms occurred immediately after slicing. This same general pattern was shown for peas and corn. Preblanch green peas from three factories showed total counts/g between log 4.94 and 5.95. These numbers were reduced by blanching and again increased successively with each processing step. In the case of whole-kernel corn, the postblanch counts rose both after cutting and at the end of the conveyor belt to the washer. Whereas the immediate postblanch count was about log 3.48, the product had total counts of about log 5.94/g after packaging. Between 40 and 75% of the bacterial flora of peas, snap beans, and corn was shown to consist of leuconostocs and streptococci, while many of the gram-positive, catalase-positive rods resembled corynebacteria (133, 134).

Lactic acid cocci have been associated with many raw and processed vegetables (95). These cocci have been shown to constitute from 41–75% of the APC flora of frozen peas, snap beans, and corn (130). It has been shown that fresh peas, green beans, and corn all contained coagulase-positive staphylococci after processing (133). Peas were found to have the highest count (log 0.86/g), while 64% of corn samples contained this organism. These authors found that a general buildup of staphylococci occurred as the vegetables underwent successive stages of processing, with the main source of organisms coming from the hands of employees. Although staphylococci may be found on vegetables during processing, they are generally unable to proliferate in the presence of the more normal lactic flora. Both coliforms (but not *E. coli*) and enterococci have been found at most stages during raw vegetable processing, but they appear to present no public health hazard (130).

In a study of the incidence of *C. botulinum* in 100 commercially available frozen vacuum pouch-pack vegetables, the organism was not found in fifty samples of string beans, but types A and B spores were found in six of fifty samples of spinach (65). The general microbiological quality of some vegetables is presented in Table 4-4.

In a study of 575 packages of frozen vegetables processed by twenty-four factories in twelve states, Splittstoesser and Corlett (131) found that peas yielded some of the lowest counts (mean of approximately log 1.93/g) while chopped broccoli yielded the highest mean APCs—log 3.26/g. Using the three-class sampling plan of the International Commission on Micro-

Table 4-4. General microbiological quality of frozen vegetables.

Products	Number of samples	Microbial group/Target	% Samples meeting target	References
Cauliflower	1,556	APC at 35°C: 10^5/g or less	75	6
	1,556	MPN coliforms: <20/g	79	6
	1,556	MPN *E. coli:* <3/g	98	6
Corn	1,542	APC at 35°C: 10^5/g or less	94	6
	1,542	MPN coliforms: <20/g	71	6
	1,542	MPN *E. coli:* <3/g	99	6
Peas	1,564	APC at 35°C: 10^5/g or less	95	6
	1,564	MPN coliforms: <20/g	78	6
	1,564	MPN *E. coli:* <3/g	99	6
Blanched vegetables (17 different)	575	Absence of fecal coliforms	63	134
	575	$n = 5, c = 3, m = 10, M = 10^3$	33	134
	575	$n = 5, c = 3, m = 50, M = 10^3$	70	134
Cut green beans, leaf spinach, peas	144	Mean APC range for group: log 4.73–4.93/g	—	131
Lima beans, corn, broccoli spears, brussels sprouts	170	Mean APC range for group: 5.30–5.36/g	—	131
French-style green beans, chopped greens, squash	135	Mean APC range: log 5.48–5.51/g	—	131
Chopped spinach, cauliflower	80	Mean APC range: log 5.54–5.65/g	—	131
Chopped broccoli	45	Mean APC: 6.26/g	—	131
Freshly formed Cheddar cheese	236	<500 coliforms/g	95	10
	237	<1,000 *S. aureus*/g	100	10
	250	Presence of salmonellae	0	10

biological Specifications for Foods (ICMSF), the acceptance rate for the 115 lots would have been 74% for the *m* specification of 10^5/g and 84% for *M* of 10^6/g. In a study of seventeen different frozen blanched vegetables, 63% were negative for fecal coliforms and 33% of the 575 examined were acceptable when $n = 5, c = 3, m = 10$, and $M = 10^3$, while 70% were acceptable if $n = 5, c = 3, m = 50$, and $M = 10^3$ (134). In another study, mean APC at 30°C for 1,556 frozen retail cauliflower samples was log 4.65/g; for 1,542 sample units of frozen corn, log 3.93/g; and for 1,564 units of frozen peas, log 3.83/g with 5/g or less of coliforms and < 3/g of *E. coli* for all samples (6). Based on APC, 97.2–99.6% of the latter foods were acceptable by ICMSF's sampling plan $n = 5$, $c = 3$, $m = 10^5$, and $M = 10^6$.

DAIRY PRODUCTS

The microbial flora of raw milk consists of those organisms that may be present on the cow's udder and hide and on milking utensils or lines. Under proper handling and storage conditions, the predominant flora is gram positive. While yeasts, molds, and gram-negative bacteria may be found along with lactic acid bacteria, most or all of these types are more heat sensitive than gram positives and are more likely to be destroyed during pasteurization. Studies over the past decade have revealed the presence of psychrotrophic sporeformers and mycobacteria in raw milks. Psychrotrophic *Bacillus* spp. were found in 25–35% of ninety-seven raw milk samples (125). These organisms were shown to grow at or below 7°C. Psychrotrophic clostridia were isolated from four of forty-eight raw milk samples (9). Sporeformers can survive pasteurization temperatures in their spore state and cause problems in the refrigerated products because of their proteolytic abilities. *Mycobacterium* and *Nocardia* spp. have been isolated from about 69% of fifty-one raw milks (63), but forty-three samples of pasteurized milk were negative. The presence of these organisms in raw milk is not surprising since they are quite abundant in soils. Also abundant in soils are psychrotrophic pseudomonads, and they are not uncommon in raw milks.

The incidence of microorganisms in commercial cream cheese was studied by Fanelli et al. (31), who found that the total count/g was about log 2.69. In commercial sour cream, the total count/g ranged from log 4.79–8.58, with most being lactic acid bacteria. Commercial onion dips with sour cream bases had total counts of log 7.76–8.28 and log 4.34/g of yeasts and molds. The majority of bacteria in this product were found to be lactic acid types also. The microbial flora of fermented dairy products is discussed in Chapter 16.

The potential botulism hazard posed by imitation cheeses was investigated by Kautter et al. (70). Spores of *C. botulinum* types A and B were inoculated into fifty samples of eleven imitation cheeses with a_w ranging from 0.942–0.973 and pH from 5.53–6.14, with examination for toxin after 289 days. Only one sample became toxic. From other tests, toxic samples were all organoleptically unacceptable, indicating that these products do not present a botulism hazard.

Milk continues to serve as a vehicle for certain diseases, and an excellent review on the subject has been prepared by Bryan (11). Outbreaks over the past decade have generally involved the consumption of raw milk, certified raw milk, homemade ice cream containing fresh eggs, and dried and pasteurized milks contaminated after heat processing. For 1980–81, over 538 cases of salmonellosis in the United States were traced to dairy products with Cheddar cheese, raw milk, and certified raw milk as vehicles. During 1980–82, over 172 cases of campylobacteriosis were traced to raw milk and certified raw milk in the United States, and milk-borne outbreaks have been reported in Canada and several European countries. For nine different microbial diseases traced to dairy products for the period 1970–79, ice cream (often contaminated by eggs) was the leading vehicle, followed by cheese and unspecified types of milk (11).

DELICATESSEN AND RELATED FOODS

Delicatessen foods, such as salads and sandwiches, are sometimes involved in food-poisoning outbreaks. These foods are often prepared by hand, and this direct contact may lead to an increased incidence of food-poisoning agents such as *S. aureus*. Once organisms such as these enter meat salads or sandwiches, they may grow well because of the reduction in numbers of the normal food flora by the prior cooking of salad ingredients.

In a study of retail salads and sandwiches, 36% of fifty-three salads were found to have total counts > log 6.00/g, but only 16% of the sixty sandwiches had counts as high (17). With respect to coliforms, 57% of sandwiches were found to harbor < log 2.00/g. *S. aureus* was present in 60% of sandwiches and 39% of salads. Yeasts and molds were found in high numbers, with six samples containing > log 6.00/g.

In a study of 517 salads from around 170 establishments, 71–96% were found to have APC < log 5.00/g (108). Ninety-six to 100% of salads contained coagulase-positive *S. aureus* at levels < log 2.00/g. Salads included chicken, egg, macaroni, shrimp, and others. *S. aureus* was recovered in low numbers from 6 of 64 salads in another study (36). The twelve different salads examined by these investigators had total counts between log 2.08 and 6.76, with egg, shrimp, and some of the macaroni salads having the highest counts. Neither salmonellae nor *C. perfringens* were found in any product. A study of forty-two salads by Harris et al. (56) revealed the products to be of generally good microbial quality. Mean APC was log 5.54/g and mean coliform counts were log 2.66/g for the six different products. Staphylococci were found in some products, especially ham salad.

Fresh green salads (green, mixed green, and coleslaw) were found to contain mean total counts of log 6.67 for coleslaw to log 7.28 for green salad (37). Fecal coliforms were found in 26% of mixed, 28% of green, and 29% of coleslaw, while the respective percentage findings for *S. aureus* were 8, 14, and 3. With respect to parsley, *E. coli* was found on eleven of sixty-four samples of fresh and unwashed products and on over 50% of frozen samples (67). Mean APC of fresh washed parsley was log 7.28/g. Neither salmonellae nor *S. aureus* were found in any samples.

In a study of the microbiological quality of imitation-cream pies from plants operated under poor sanitary conditions, Surkiewicz (141) found that the microbial load increased successively as the products were carried through the various processing steps. For example, in one instance the final mixture of the synthetic pie base contained fewer than log 2.00 bacteria/g after final heating to 160°F. After overnight storage, however, the count rose to log 4.15. The pie topping ingredients to be mixed with pie base had a rather low count—log 2.78/g. After being deposited on the pies, the pie topping showed a total count of log 7.00/g. In a study of the microbiological quality of French fries, Surkiewicz et al. (142) demonstrated the same pattern—that is, the successive buildup of microorganisms as the fries underwent processing. Since these products are cooked late in their processing, the incidence of organisms in the finished state does not properly reflect the actual state of sanitation during processing.

Geometric mean APC of 1,187 sample units of refrigerated biscuit dough was found to be 34,000/g, while for fungi, coliforms, E. coli, and S. aureus, the mean counts were 46, 11, < 3, and < 3/g, respectively (148). In the same study, geometric mean APC of 1,396 units of snack cake was 910/g, with < 3/g of coliforms, E. coli, and S. aureus (see Table 4-5).

A bacteriological study of 580 frozen cream-type pies (lemon, coconut, chocolate, and banana) showed them to be of excellent quality, with 98% having APC of log 4.70 or less/g (84). The overall microbiological quality of other related products is presented in Table 4-5.

Table 4-5. General microbiological quality of miscellaneous food products.

Products	Number of samples	Microbial group/Target	% samples meeting target	Reference
Frozen cream-	465	APC: $\leq 10^4$/g	96	151
type pies	465	Fungi: 10^3/g or less	98	151
	465	Coliforms: <10/g	89	151
	465	E. coli: 10/g or less	99	151
	465	S. aureus: <25/g	99	151
	465	O salmonellae	100	151
Frozen breaded	1,590	APC 30°C: 10^5/g or less	99	155
onion rings	1,590	MPN coliforms: <3/g	89	155
(pre- or	1,590	MPN E. coli: <3/g	99	155
partially cooked)	1,590	MPN S. aureus: <10/g	99.6	155
Frozen tuna	1,290	APC 30°C: 10^5/g or less	97.6	155
pot pies	1,290	MPN coliforms: 64/g or less	93	155
	1,290	MPN E. coli: <3/g	97	155
	1,290	MPN S. aureus: <10/g	98	155
Tofu	60	APC: $>10^6$/g	83	119
(commercial)	60	Psychrotrophs: $<10^4$/g	83	119
	60	Coliforms: $\sim 10^3$/g	67	119
	60	S. aureus: <10/g	100	119

Table 4–5, *continued*.

Products	Number of samples	Microbial group/Target	% samples meeting target	Reference
Dry food-grade gelatin	185	APC: 3.00 or less/g	74	84
Delicatessen salads	764	Within AAFES microbial limits[a]	44	35
	764	APC: 5.00 or less/g	84	35
	764	Coliforms: 1.00 or less/g	78	35
	764	Yeasts and molds: 1.30 or less/g	55	35
	764	Fecal streptococci: 1.00/g	77	35
	764	Presence of *S. aureus*	9	35
	764	Pres. of *C. perfringens;* salmonellae	0	35
	517	APC: 5.00 or less/g	26–85	107
	517	Coliforms: 2.00 or less/g	36–79	107
	517	*S. aureus:* 2.00 or less/g	96–100	107
Retail trade salads	53	APC: >6.00/g	36	17
	53	Coliforms: 2.00 or less/g	57	17
	53	Presence of *S. aureus*	39	17
Retail trade sandwiches	62	APC: >6.00/g	16	17
	62	Coliforms: >3.00/g	12	17
	62	Presence of *S. aureus*	60	17
Imported spices and herbs	113	APC: 6.00 or less/g	73	66
	114	Spores: 6.00 or less/g	75	66
	113	Yeasts and molds: 5.00 or less/g	97	66
	114	TA spores: 3.00 or less/g	70	66
	114	Pres. of *E. coli, S. aureus,* salmonellae	0	66
Processed spices	114	APC: 5.00 or less/g	70	114
	114	APC: 6.00 or less/g	91	114
	114	Coliforms: 2.00 or less/g	97	114
	114	Yeasts and molds: 4.00 or less/g	96	114
	114	*C. perfringens:* <2.00/g	89	114
	110	Presence of *B. cereus*	53	113
Dehydrated space foods	129	APC: <4.00/g	93	112
	129	Coliforms: <1/g	98	112
	129	*E. coli:* negative in 1 g	99	112
	102	Fecal streptococci: 1.30/g	88	112
	104	*S. aureus:* negative in 5 g	100	112
	104	Salmonellae: negative in 10 g	98	112

[a] Army and Air Force Exchange Service.

FROZEN MEAT PIES

The microbiological quality of frozen meat pies has steadily improved since these products were first marketed. Any and all of the ingredients added may increase the total number of organisms, and the total count of the finished product may be taken to reflect the overall quality of ingredients, handling, and storage. Many investigators have suggested that these products should be produced with total counts not to exceed log 5.00/g. In a study of 48 meat pies, 84% had APC < log 5 (87) while in another study of 188 meat pies, 93% had counts less than log 5.00 (74). Accordingly, a microbiological criterion of log 5.00 seems attainable for such products (see Chapter 17 for further information on microbiological standards and criteria).

In a recent study of 1,290 frozen tuna pot pies, geometric mean APC at 35°C was log 3.20 while at 30°C it was log 3.38/g (155). Coliforms averaged 5/g, *E. coli* < 3/g, and *S. aureus* < 10/g (Table 4-5).

DEHYDRATED FOODS

In a detailed study of the microbiology of dehydrated soups, Fanelli et al. (30) showed that approximately seventeen different kinds of dried soups from nine different processors had total counts of less than log 5.00/g. These soups included chicken noodle, chicken rice, beef noodle, vegetable, mushroom, pea, onion, tomato, and others. Some of these products had total counts as high as log 7.30/g, while some had counts as low as around log 2.00. These investigators further found that reconstituted dehydrated onion soup showed a mean total count of log 5.11/ml, with log 3.00 coliforms, log 4.00 aerobic sporeformers, and log 1.08/ml of yeast and molds. Upon cooking, the total counts were reduced to a mean of log 2.15, while coliforms were reduced to < log 0.26, sporeformers to log 1.64, and yeasts and molds to < log 1.00/ml. In a study of dehydrated sauce and gravy mixes, soup mixes, spaghetti sauce mixes, and cheese sauce mixes, *C. perfringens* was isolated from ten of fifty-five samples (97). The facultative anaerobe counts ranged from log 3.00 to > log 6.00/g.

In a study of 185 samples of food-grade dry gelatin, no samples exceeded an APC of log 3.70/g (84). Of 129 dehydrated space food samples examined, 93% contained total counts < log 4.00/g (112).

Powdered eggs and milk often contain high numbers of microorganisms on the order of log 6–8/g. One reason for the generally high numbers in dried products is that the organisms have been concentrated on a per gram basis along with product concentration. The same is generally true for fruit juice concentrates, which tend to have higher numbers of microorganisms than the fresh, nonconcentrated products.

The author has investigated the incidence and types of organisms on raw squash seeds to be roasted for food use. Some twelve samples of this product showed a mean total count of log 7.99 and the presence of log 4.72

coliforms. Most of the latter were of the nonfecal type. By adding flour batter and salt to these seeds and roasting, the total count was reduced to less than log 2.00/g. The microbiology of desiccated foods is discussed further in Chapter 15.

Enteral Nutrient Solutions (Enteral Foods)

Enteral nutrient solutions (ENS) are liquid foods that are administered by tube. They are available either as powdered products requiring reconstitution or as liquids. They are generally administered to certain patients in hospitals or other patient care facilities but may be administered in the home. Administration is by continuous drip from enteral feeding bags, and the process may go on for 8 h or longer with the ENS at room temperature. Enteral foods are made by several commercial companies either as complete diets that only require reconstituting with water before use, or as incomplete meals that require supplementation with milk, eggs, or the like prior to use. ENS-use preparations are nutritionally complete with varying concentrations of proteins, peptides, carbohydrates, and so forth, depending upon patient need.

The microbiology of ENS has been addressed by some hospital researchers, who have found the products to contain varying numbers and types of bacteria and to be the source of patient infections. Numbers as high as 10^8/ml have been found in some ENS at time of infusion (42). In a study of one reconstituted commercial ENS, the initial count of 9×10^3/ml increased to 7×10^4/ml after 8 h at room temperature (62). Numbers as high as 1.2×10^5/ml were found in another sample of the same preparation, and the most frequently isolated organism was *Staphylococcus epidermidis* with *Corynebacterium, Citrobacter,* and *Acinetobacter* spp. among the other isolates. From a British study, enteral feeds yielded 10^4–10^6 organisms/ml with coliforms and *Pseudomonas aeruginosa* as the predominant types (46).

The capacity of five different commercial ENS to support the growth of *Enterobacter cloacae* under use conditions has been demonstrated (29), and the addition of 0.2% potassium sorbate was shown to reduce numbers of this organism by 3 log cycles over controls. Patients are known to have contracted *E. cloacae* and *Salmonella enteritidis* infections from ENS (14, 46). Procedures that should be employed in the preparation/handling of ENS to minimize microbial problems have been noted (49). For more information, see Fagerman et al. (29).

REFERENCES

1. Abeyta, C., Jr. 1983. Comparison of iron milk and official AOAC methods for enumeration of *Clostridium perfringens* from fresh seafoods. *J. Assoc. Off. Anal. Chem.* 66:1175–77.
2. Allen, J. R., and E. M. Foster. 1960. Spoilage of vacuum-packed sliced processed meats during refrigerated storage. *Food Res.* 25:1–7.

—

3. Andrews, W. H., A. P. Duran, F. D. McClure, and D. E. Gentile. 1979. Use of two rapid A-1 methods for the recovery of fecal coliforms and *Escherichia coli* from selected food types. *J. Food Sci.* 44:289–91, 293.
4. Andrews, W. H., C. R. Wilson, P. L. Poelma, and A. Romero. 1977. Bacteriological survey of the channel catfish (*Ictalurus punctalus*) at the retail level. *J. Food Sci.* 42:359–63.
5. Barbe, C. D., R. W. Mandigo, and R. L. Henrickson. 1966. Bacterial flora associated with rapid-processed ham. *J. Food Sci.* 31:988–93.
6. Barnard, R. J., A. P. Duran, A. Swartzentruber, A. H. Schwab, B. A. Wentz, and R. B. Read, Jr. 1982. Microbiological quality of frozen cauliflower, corn, and peas obtained at retail markets. *Appl. Environ. Microbiol.* 44:54–58.
7. Bauer, F. T., J. A. Carpenter, and J. O. Reagan. 1981. Prevalence of *Clostridium perfringens* in pork during processing. *J. Food Protect.* 44:279–83.
8. Bell, W. N., and L. A. Shelef. 1978. Availability and microbial stability of retail beef-soy blends. *J. Food Sci.* 43:315–18, 333.
9. Bhadsavle, C. H., T. E. Shehata, and E. B. Collins. 1972. Isolation and identification of psychrophilic species of *Clostridium* from milk. *Appl. Microbiol.* 24:699–702.
10. Brodsky, M. H. 1984. Bacteriological survey of freshly formed Cheddar cheese. *J. Food Protect.* 47:546–48.
11. Bryan, F. L. 1983. Epidemiology of milk-borne diseases. *J. Food. Protect.* 46:637–49.
12. Bryan, F. L., J. C. Ayres, and A. A. Kraft. 1968. Salmonellae associated with further-processed turkey products. *Appl. Microbiol.* 16:1–9.
13. Carl, K. E. 1975. Oregon's experience with microbiological standards for meat. *J. Milk Food Technol.* 38:483–86.
14. Casewell, M. W., J. E. Cooper, and M. Webster. 1981. Enteral feeds contaminated with *Enterobacter cloacae* as a cause of septicaemia. *Brit. Med. J.* 282:973.
15. Cavett, J. J. 1962. The microbiology of vacuum packed sliced bacon. *J. Appl. Bacteriol.* 25:282–89.
16. Chipley, J. R., and E. K. Heaton. 1971. Microbial flora of pecan meat. *Appl. Microbiol.* 22:252–53.
17. Christiansen, L. N., and N. S. King. 1971. The microbial content of some salads and sandwiches at retail outlets. *J. Milk Food Technol.* 34:289–93.
18. Christopher, F. M., S. C. Seideman, Z. L. Carpenter, G. C. Smith, and C. Vanderzant. 1979. Microbiology of beef packaged in various gas atmospheres. *J. Food Protect.* 42:240–44.
19. Cox, N. A., A. J. Mercuri, B. J. Juven, and J. E. Thomson. 1975. *Enterobacteriaceae* at various stages of poultry chilling. *J. Food Sci.* 40:44–46.
20. Craven, S. E., and A. J. Mercuri. 1977. Total aerobic and coliform counts in beef-soy and chicken-soy patties during refrigerated storage. *J. Food Protect.* 40:112-15.
21. DeBoer, E., and E. M. Boot. 1983. Comparison of methods for isolation and confirmation of *Clostridium perfringens* from spices and herbs. *J. Food Protect.* 46:533–36.
22. Drake, S. D., J. B. Evans, and C. F. Niven, Jr. 1958. Microbial flora of packaged frankfurters and their radiation resistance. *Food Res.* 23:291–96.
23. Draughon, F. A. 1980. Effect of plant-derived extenders on microbiological stability of foods. *Food Technol.* 34(10):69–74.

24. Duitschaever, C. L., D. R. Arnott, and D. H. Bullock. 1973. Bacteriological quality of raw refrigerated ground beef. *J. Milk Food Technol.* 36:375–77.
25. Duitschaever, C. L., D. H. Bullock, and D. R. Arnott. 1977. Bacteriological evaluation of retail ground beef, frozen beef patties, and cooked hamburger. *J. Food Protect.* 40:378–81.
26. Dutson, T. R., G. C. Smith, and Z. L. Carpenter. 1980. Lysosomal enzyme distribution in electrically stimulated ovine muscle. *J. Food Sci.* 45:1097–98.
27. Egan, A. F., A. L. Ford, and B. J. Shay. 1980. A comparison of *Microbacterium thermosphactum* and lactobacilli as spoilage organisms of vacuum-packaged sliced luncheon meats. *J. Food Sci.* 45:1745–48.
28. Egan, A. F., and B. J. Shay. 1982. Significance of lactobacilli and film permeability in the spoilage of vacuum-packaged beef. *J. Food Sci.* 47:1119–22, 1126.
29. Fagerman, K. E., J. D. Paauw, M. A. McCamish, and R. E. Dean. 1984. Effects of time, temperature, and preservative on bacterial growth in enteral nutrient solutions. *Amer. J. Hosp. Pharm.* 41:1122–26.
30. Fanelli, M. J., A. C. Peterson, and M. F. Gunderson. 1965. Microbiology of dehydrated soups. I. A survey. *Food Technol.* 19:83–86.
31. ———. 1965. Microbiology of dehydrated soups. III. Bacteriological examination of rehydrated dry soup mixes. *Food Technol.* 19:90–94.
32. Field, R. A. 1976. Mechanically deboned red meat. *Food Technol.* 30(9):38–48.
33. ———. 1981. Mechanically deboned red meat. *Adv. Food Res.* 27:23–107.
34. Field, R. A., and M. L. Riley. 1974. Characteristics of meat from mechanically deboned lamb breasts. *J. Food Sci.* 39:851–52.
35. Foster, J. F., J. L. Fowler, and J. Dacey. 1977. A microbial survey of various fresh and frozen seafood products. *J. Food Protect.* 40:300–303.
36. Fowler, J. L., and W. S. Clark, Jr. 1975. Microbiology of delicatessen salads. *J. Milk Food Technol.* 38:146–49.
37. Fowler, J. L., and J. F. Foster. 1976. A microbiological survey of three fresh green salads: Can guidelines be recommended for these foods? *J. Milk Food Technol.* 39:111–13.
38. Fraiser, M. B., and J. A. Koburger. 1984. Incidence of salmonellae in clams, oysters, crabs and mullet. *J. Food Protect.* 47:343–45.
39. Froning, G. W. 1981. Mechanical deboning of poultry and fish. *Adv. Food Res.* 27:109–47.
40. Fung, D. Y. C., C. L. Kastner, M. C. Hunt, M. E. Dikeman, and D. H. Kropf. 1980. Mesophilic and psychrotrophic bacterial populations on hot-boned and conventionally processed beef. *J. Food Protect.* 43:547–50.
41. Fung, D. Y. C., C. L. Kastner, C.-Y. Lee, M. C. Hunt, M. E. Dikeman, and D. H. Kropf. 1981. Initial chilling rate effects of bacterial growth on hot-boned beef. *J. Food Protect.* 44:539–44.
42. Furtado, D., A. Parrish, and P. Beyer. 1980. Enteral nutrient solutions (ENS): In vitro growth supporting properties of ENS for bacteria. *J. Paren. Ent. Nutri.* 4:594.
43. Gabis, D. A., B. E. Langlois, and A. W. Rudnick. 1970. Microbiological examination of cocoa powder. *Appl. Microbiol.* 20:644–45.
44. Gardner, G. A. 1971. A note on the aerobic microflora of fresh and frozen porcine liver stored at 5°C. *J. Food Technol.* 6:225–31.
45. Gardner, G. A., A. W. Carson, and J. Patton. 1967. Bacteriology of prepacked

pork with reference to the gas composition within the pack. *J. Appl. Bacteriol.* 30:321–33.

46. Gill, K. J., and P. Gill. 1981. Contaminated enteral feeds. *Brit. Med. J.* 282:1971.
47. Goepfert, J. M. 1977. Aerobic plate count and *Escherichia coli* determination on frozen ground-beef patties. *Appl. Environ. Microbiol.* 34:458–60.
48. Greenberg, R. A., R. B. Tompkin, B. O. Bladel, R. S. Kittaka, and A. Anellis. 1966. Incidence of mesophilic *Clostridium* spores in raw pork, beef, and chicken in processing plants in the United States and Canada. *Appl. Microbiol.* 14:789–93.
49. Gröschel, D. H. M. 1983. Infection control considerations in enteral feeding. *Nutri. Supp. Serv.* 3(6):48–49.
50. Guthertz, L. S., J. T. Fruin, R. L. Okoluk, and J. L. Fowler. 1977. Microbial quality of frozen comminuted turkey meat. *J. Food Sci.* 42:1344–47.
51. Guthertz, L. S., J. T. Fruin, D. Spicer, and J. L. Fowler. 1976. Microbiology of fresh comminuted turkey meat. *J. Milk Food Technol.* 39:823–29.
52. Hanna, M. O., G. C. Smith, J. W. Savell, F. K. McKeith, and C. Vanderzant. 1982. Microbial flora of livers, kidneys and hearts from beef, pork and lamb: Effects of refrigeration, freezing and thawing. *J. Food Protect.* 45:63–73.
53. ———. 1982. Effects of packaging methods on the microbial flora of livers and kidneys from beef or pork. *J. Food Protect.* 45:74–81.
54. Hanna, M. O., C. Vanderzant, Z. L. Carpenter, and G. C. Smith. 1977. Microbial flora of vacuum-packaged lamb with special reference to psychrotrophic, gram-positive, catalase-positive pleomorphic rods. *J. Food Protect.* 40:98–100.
55. Harris, N. D., S. R. Martin, and L. Ellias. 1975. Bacteriological quality of selected delicatessen foods. *J. Milk Food Technol.* 38:759–61.
56. Harrison, M. A., F. A. Draughton, and C. C. Melton. 1983. Inhibition of spoilage bacteria by acidification of soy extended ground beef. *J. Food Sci.* 48:825–28.
57. Harrison, M. A., C. C. Melton, and F. A. Draughon. 1981. Bacterial flora of ground beef and soy extended ground beef during storage. *J. Food Sci.* 46:1088–90.
58. Hauschild, A. H. W., and R. Hilsheimer. 1983. Prevalence of *Clostridium botulinum* in commercial liver sausage. *J. Food Protect.* 46:242–44.
59. Hauschild, A. H. W., L. M. Poste, and R. Hilsheimer. 1985. Toxin production by *Clostridium botulinum* and organoleptic changes in vacuum-packaged raw beef. *J. Food Protect.* 48:712–16.
60. Henry, K. G., J. W. Savell, G. C. Smith, J. G. Ehlers, and C. Vanderzant. 1983. Physical, sensory and microbiological characteristics of lamb retail cuts vacuum packaged in high oxygen-barrier film. *J. Food Sci.* 48:1735–40, 1749.
61. Hitchener, B. J., A. F. Egan, and P. J. Rogers. 1982. Characteristics of lactic acid bacteria isolated from vacuum-packaged beef. *J. Appl. Bacteriol.* 52:31–37.
62. Hostetler, C., T. O. Lipman, M. Geraghty, and R. H. Parker. 1982. Bacterial safety of reconstituted continuous drip tube feeding. *J. Paren. Ent. Nutri.* 6:232–35.
63. Hosty, T. S., and C. I. McDurmont. 1975. Isolation of acid-fast organisms from milk and oysters. *Hlth Lab. Sci.* 12:16–19.
64. Ingram, M. 1960. Bacterial multiplication in packed Wiltshire bacon. *J. Appl. Bacteriol.* 23:206–15.
65. Insalata, N. F., J. S. Witzeman, J. H. Berman, and E. Borker. 1968. A study

of the incidence of the spores of *Clostridium botulinum* in frozen vacuum pouch-pack vegetables. *Proc., 96th Ann. Meet., Am. Pub. Health Assoc.*, 124.

66. Julseth, R. M., and R. H. Deibel. 1974. Microbial profile of selected spices and herbs at import. *J. Milk Food Technol.* 37:414–19.

67. Käferstein, F. K. 1976. The microflora of parsley. *J. Milk Food Technol.* 39:837–40.

68. Kajs, T. M., R. Hagenmaier, C. Vanderzant, and K. F. Mattil. 1976. Microbiological evaluation of coconut and coconut products. *J. Food Sci.* 41:352–56.

69. Kastner, C. L., L. O. Leudecke, and T. S. Russell. 1976. A comparison of microbial counts on conventionally and hot-boned bovine carcasses. *J. Milk Food Technol.* 39:684–85.

70. Kautter, D. A., R. K. Lynt, T. Lilly, Jr., and H. M. Solomon. 1981. Evaluation of the botulism hazard from imitation cheeses. *J. Food Sci.* 46:749–50, 764.

71. Kazanas, N., J. A. Emerson, H. L. Seagram, and L. L. Kempe. 1966. Effect of γ-irradiation on the microflora of fresh-water fish. I. Microbial load, lag period, and rate of growth on yellow perch (*Perca flavescens*) fillets. *Appl. Microbiol.* 14:261–66.

72. Keeton, J. T., and C. C. Melton. 1978. Factors associated with microbial growth in ground beef extended with varying levels of textured soy protein. *J. Food Sci.* 43:1125–29.

73. Kempton, A. G., and S. R. Bobier. 1979. Bacterial growth in refrigerated, vacuum-packed luncheon meats. *Can. J. Microbiol.* 16:287–97.

74. Kereluk, K., and M. F. Gunderson. 1959. Studies on the bacteriological quality of frozen meat pies. I. Bacteriological survey of some commercially frozen meat pies. *Appl. Microbiol.* 7:320–23.

75. Kitchell, A. G. 1962. Micrococci and coagulase negative staphylococci in cured meats and meat products. *J. Appl. Bacteriol.* 25:416–31.

76. Kotula, A. W. 1981. Microbiology of hot-boned and electrostimulated meat. *J. Food Protect.* 44:545–49.

77. Kotula, A. W., and B. S. Emswiler-Rose. 1981. Bacteriological quality of hot-boned primal cuts from electrically stimulated beef carcasses. *J. Food Sci.* 46:471–74.

78. Kraft, A. A., J. C. Ayres, G. S. Torrey, R. H. Salzer, and G. A. N. DaSilva. 1966. Coryneform bacteria in poultry, eggs and meat. *J. Appl. Bacteriol.* 29:161–66.

79. Kraft, A. A., V. Reddy, R. J. Hasiak, K. D. Lind, and D. E. Galloway. 1982. Microbiological quality of vacuum packaged poultry with or without chlorine treatment. *J. Food Sci.* 47:380–85.

80. Ladiges, W. C., J. F. Foster, and W. M. Ganz. 1974. Incidence and viability of *Clostridium perfringens* in ground beef. *J. Milk Food Technol.* 37:622–23.

81. Lee, B. H., R. E. Simard, C. L. Laleye, and R. A. Holley. 1984. Shelf-life of meat loaves packaged in vacuum or nitrogen gas. I. Effect of storage temperature, light and time on the micro-flora change. *J. Food Protect.* 47:128–33.

82. Lee, C. Y., D. Y. C. Fung, and E. L. Kastner. 1982. Computer-assisted identification of bacteria on hot-boned and conventionally processed beef. *J. Food Sci.* 47:363–67, 373.

83. Lee, J. S., and D. K. Pfeifer. 1977. Microbiological characteristics of Pacific shrimp (*Pandalus jordani*). *Appl. Environ. Microbiol.* 33:853–59.

84. Leininger, H. V., L. R. Shelton, and K. H. Lewis. 1971. Microbiology of frozen cream-type pies, frozen cooked-peeled shrimp, and dry food-grade gelatin. *Food Technol.* 25:224–29.

85. Lillard, H. S. 1971. Occurrence of *Clostridium perfringens* in broiler processing and further processing operations. *J. Food Sci.* 36:1008–10.

86. Lin, C. K., W. H. Kennick, W. E. Sandine, and M. Koohmaraie. 1984. Effect of electrical stimulation on meat microflora: Observations on agar media, in suspensions and on beef carcasses. *J. Food Protect.* 47:279–83.

87. Lin, H.-S., D. G. Topel, and H. W. Walker. 1979. Influence of prerigor and postrigor muscle on the bacteriological and quality characteristics of pork sausage. *J. Food Sci.* 44:1055–57.

88. Litsky, W., I. S. Fagerson, and C. R. Fellers. 1957. A bacteriological survey of commercially frozen beef, poultry and tuna pies. *J. Milk Food Technol.* 20:216–19.

89. May, K. N. 1962. Bacterial contamination during cutting and packaging chicken in processing plants and retail stores. *Food Technol.* 16:89–91.

90. McMillin, D. J., J. G. Sebranek, and A. A. Kraft. 1981. Microbial quality of hot-processed frozen ground beef patties processed after various holding times. *J. Food Sci.* 46:488–90.

91. Mead, G. C. 1983. Effect of packaging and gaseous environment on the microbiology and shelf life of processed poultry products. In *Food microbiology: Advances and prospects,* ed. T. A. Roberts and F. A. Skinner, 203–16. New York and London: Academic Press.

92. Mercuri, A. J., G. J. Banwart, J. A. Kinner, and A. R. Sessoms. 1970. Bacteriological examination of commercial precooked Eastern-type turkey rolls. *Appl. Microbiol.* 19:768–71.

93. Mol, J. H. H., J. E. A. Hietbrink, H. W. M. Mollen, and J. van Tinteren. 1971. Observations on the microflora of vacuum packed sliced cooked meat products. *J. Appl. Bacteriol.* 34:377–97.

94. Molin, G., and I.-M. Stenstrom. 1984. Effect of temperature on the microbial flora of herring fillets stored in air or carbon dioxide. *J. Appl. Bacteriol.* 56:275–82.

95. Mundt, J. O., W. F. Graham, and I. E. McCarty. 1967. Spherical lactic acid-producing bacteria of Southern-grown raw and processed vegetables. *Appl. Microbiol.* 15:1303–08.

96. Murthy, T. R. K. 1984. Relative numbers of coliforms, *Enterobacteriaceae* (by two methods), and total aerobic bacteria counts as determined from minced goat meat. *J. Food Protect.* 47:142–44.

97. Nakamura, M., and K. D. Kelly. 1968. *Clostridium perfringens* in dehydrated soups and sauces. *J. Food Sci.* 33:424–26.

98. Newsome, R. L., B. E. Langlois, W. G. Moody, N. Gay, and J. D. Fox. 1984. Effect of time and method of aging on the composition of the microflora of beef loins and corresponding steaks. *J. Food Protect.* 47:114–18.

99. Newton, K. G., J. C. L. Harrison, and K. M. Smith. 1977. The effect of storage in various gaseous atmospheres on the microflora of lamb chops held at −1° C. *J. Appl. Bacteriol.* 43:53–59.

100. Newton, K. G., and W. J. Rigg. 1979. The effect of film permeability on the storage life and microbiology of vacuum-packed meat. *J. Appl. Bacteriol.* 47:433–41.

101. Nickerson, J. T. R., and S. A. Goldblith. 1964. A study of the microbiological quality of haddock fillets and shucked, soft-shelled clams processed and marketed in the greater Boston area. *J. Milk Food Technol.* 27:7–12.
102. Nielsen, H.-J. S. 1983. Influence of nitrite addition and gas permeability of packaging film on the microflora in a sliced vacuum-packed whole meat product under refrigerated storage. *J. Food Technol.* 18:573–85.
103. Nielsen, H.-J. S. and P. Zeuthen. 1985. Influence of lactic acid bacteria and the overall flora on development of pathogenic bacteria in vacuum-packed, cooked emulsion-style sausage. *J. Food Protect.* 48:28–34.
104. Oblinger, J. L., J. E. Kennedy, Jr., C. A. Rothenberg, B. W. Berry, and N. J. Stern. 1982. Identification of bacteria isolated from fresh and temperature abused variety meats. *J. Food Protect.* 45:650–54.
105. Ockerman, H. W., and J. Szczawinski. 1983. Effect of electrical stimulation on the microflora of meat. *J. Food Sci.* 48:1004–1005, 1007.
106. Ostovar, K., J. H. MacNeil, and K. O'Donnell. 1971. Poultry product quality. 5. Microbiological evaluation of mechanically deboned poultry meat. *J. Food Sci.* 36:1005–7.
107. Pace, P. J. 1975. Bacteriological quality of delicatessen foods: Are standards needed? *J. Milk Food Technol.* 38:347–53.
108. Paradis, D. C., and M. E. Stiles. 1978. A study of microbial quality of vacuum packaged, sliced bologna. *J. Food Protect.* 41:811–15.
109. Patterson, J. T., and P. A. Gibbs. 1977. Incidence and spoilage potential of isolates from vacuum-packaged meat of high pH value. *J. Appl. Bacteriol.* 43:25–38.
110. Pivnick, H., I. E. Erdman, D. Collins-Thompson, G. Roberts, M. A. Johnston, D. R. Conley, G. Lachapelle, U. T. Purvis, R. Foster, and M. Milling. 1976. Proposed microbiological standards for ground beef based on a Canadian survey. *J. Milk Food Technol.* 39:408–12.
111. Post, L. S., D. A. Lee, M. Solberg, D. Furgang, J. Specchio, and C. Graham. 1985. Development of botulinal toxin and sensory deterioration during storage of vacuum and modified atmosphere packaged fish fillets. *J. Food Sci.* 50:990–96.
112. Powers, E. M., C. Ay, H. M. El-Bisi, and D. B. Rowley. 1971. Bacteriology of dehydrated space foods. *Appl. Microbiol.* 22:441–45.
113. Powers, E. M., T. G. Latt, and T. Brown. 1976. Incidence and levels of *Bacillus cereus* in processed spices. *J. Milk Food Technol.* 39:668–70.
114. Powers, E. M., R. Lawyer, and Y. Masuoka. 1975. Microbiology of processed spices. *J. Milk Food Technol.* 38:683–87.
115. Pulliam, J. D., and D. C. Kelley. 1965. Bacteriological comparisons of hot processed and normally processed hams. *J. Milk Food Technol.* 28:285–86.
116. Raccach, M., and R. C. Baker. 1978. Microbial properties of mechanically deboned fish flesh. *J. Food Sci.* 43:1675–77.
117. Ray, B., and R. A. Field. 1983. Bacteriology of restructured lamb roasts made with mechanically deboned meat. *J. Food Protect.* 46:26–28.
118. Ray, B., C. Johnson, and R. A. Field. 1984. Growth of indicator, pathogenic and psychrotrophic bacteria in mechanically separated beef, lean ground beef and beef bone marrow. *J. Food Protect.* 47:672–77.
119. Rehberger, T. G., L. A. Wilson, and B. A. Glatz. 1984. Microbiological quality of commercial tofu. *J. Food Protect.* 47:177–81.

120. Riha, W. E., and M. Solberg. 1970. Microflora of fresh pork sausage casings. 2. Natural casings. *J. Food Sci.* 35:860–63.
121. Roth, L. A., and D. S. Clark. 1975. Effect of lactobacilli and carbon dioxide on the growth of *Microbacterium thermosphactum* on fresh beef. *Can. J. Microbiol.* 21:629–32.
122. Rothenberg, C. A., B. W. Berry, and J. L. Oblinger. 1982. Microbiological characteristics of beef tongues and livers as affected by temperature-abuse and packaging systems. *J. Food Protect.* 45:527–32.
123. Seideman, S. C., C. Vanderzant, G. C. Smith, M. O. Hanna, and Z. L. Carpenter. 1976. Effect of degree of vacuum and length of storage on the microflora of vacuum packaged beef wholesale cuts. *J. Food Sci.* 41:738–42.
124. Seideman, S. C., and P. R. Durland. 1983. Vacuum packaging of fresh beef: A review. *J. Food Qual.* 6:29–47.
125. Shehata, T. E., and E. B. Collins. 1971. Isolation and identification of psychrophilic species of *Bacillus* from milk. *Appl. Microbiol.* 21:466–69.
126. Shelef, L. A. 1975. Microbial spoilage of fresh refrigerated beef liver. *J. Appl. Bacteriol.* 39:273–80.
127. Silverman, G. J., J. T. R. Nickerson, D. W. Duncan, N. S. Davis, J. S. Schachter, and M. M. Joselow. 1961. Microbial analysis of frozen raw and cooked shrimp. I. General results. *Food Technol.* 15:455–58.
128. Simard, R. E., B. H. Lee, C. L. Laleye, and R. A. Holley. 1983. Effects of temperature, light and storage time on the microflora of vacuum- or nitrogen-packed frankfurters. *J. Food Protect.* 46:199–205.
129. Smith, F. C., R. A. Field, and J. C. Adams. 1974. Microbiology of Wyoming big game meat. *J. Milk Food Technol.* 37:129–31.
130. Splittstoesser, D. F. 1973. The microbiology of frozen vegetables. How they get contaminated and which organisms predominate. *Food Technol.* 27:54–56.
131. Splittstoesser, D. F., and D. A. Corlett, Jr. 1980. Aerobic plate counts of frozen blanched vegetables processed in the United States. *J. Food Protect.* 43:717–19.
132. Splittstoesser, D. F., and I. Gadjo. 1966. The groups of micro-organisms composing the "total" count population in frozen vegetables. *J. Food Sci.* 31:234–39.
133. Splittstoesser, D. F., G. E. R. Hervey, II, and W. P. Wettergreen. 1965. Contamination of frozen vegetables by coagulase-positive staphylococci. *J. Milk Food Technol.* 28:149–51.
134. Splittstoesser, D. F., D. T. Queale, J. L. Bowers, and M. Wilkison. 1980. Coliform content of frozen blanched vegetables packed in the United States. *J. Food Safety* 2:1–11.
135. Splittstoesser, D. F., W. P. Wettergreen, and C. S. Pederson. 1961. Control of microorganisms during preparation of vegetables for freezing. I. Green beans. *Food Technol.* 15:329–31.
136. Splittstoesser, D. F., M. Wexler, J. White, and R. R. Colwell. 1967. Numerical taxonomy of gram-positive and catalase-positive rods isolated from frozen vegetables. *Appl. Microbiol.* 15:158–62.
137. Steinkraus, K. H., and J. C. Ayres. 1964. Incidence of putrefactive anaerobic spores in meat. *J. Food Sci.* 29:87–93.
138. Stewart, A. W. 1983. Effect of cooking on bacteriological population of "soul foods." *J. Food Protect.* 46:19–20.

139. Stiles, M. E., and L.-K. Ng. 1981. Biochemical characteristics and identification of *Enterobacteriaceae* isolated from meats. *Appl. Environ. Microbiol.* 41:639–45.

140. Strong, D. H., J. C. Canada, and B. B. Griffiths. 1963. Incidence of *Clostridium perfringens* in American foods. *Appl. Microbiol.* 11:42–44.

141. Surkiewicz, B. F. 1966. Bacteriological survey of the frozen prepared foods industry. *Appl. Microbiol.* 14:21–26.

142. Surkiewicz, B. F., R. J. Groomes, and A. P. Padron. 1967. Bacteriological survey of the frozen prepared foods industry. III. Potato products. *Appl. Microbiol.* 15:1324–31.

143. Surkiewicz, B. F., M. E. Harris, and J. M. Carosella. 1977. Bacteriological survey and refrigerated storage test of vacuum-packed sliced imported canned ham. *J. Food Protect.* 40:109–11.

144. Surkiewicz, B. F., M. E. Harris, R. P. Elliott, J. F. Macaluso, and M. M. Strand. 1975. Bacteriological survey of raw beef patties produced at establishments under federal inspection. *Appl. Microbiol.* 29:331–34.

145. Surkiewicz, B. F., R. W. Johnston, R. P. Elliott, and E. R. Simmons. 1972. Bacteriological survey of fresh pork sausage produced at establishments under federal inspection. *Appl. Microbiol.* 23:515–20.

146. Sutherland, J. P., J. T. Patterson, and J. G. Murray. 1975. Changes in the microbiology of vacuum-packaged beef. *J. Appl. Bacteriol.* 39:227–37.

147. Swartzentruber, A., A. H. Schwab, A. P. Duran, B. A. Wentz, and R. B. Read, Jr. 1980. Microbiological quality of frozen shrimp and lobster tail in the retail market. *Appl. Environ. Microbiol.* 40:765–69.

148. Swartzentruber, A., A. H. Schwab, B. A. Wentz, A. P. Duran, and R. B. Read, Jr. 1984. Microbiological quality of biscuit dough, snack cakes and soy protein meat extender. *J. Food Protect.* 47:467–70.

149. Thompson, S. G., H. W. Ockerman, V. R. Cahill, and R. F. Plimpton. 1978. Effect of soy protein flakes and added water on microbial growth (total counts, coliforms, proteolytics, staphylococci) and rancidity in fresh ground beef. *J. Food Sci.* 43:289–91.

150. Timoney, J. F., and A. Abston. 1984. Accumulation and elimination of *Escherichia coli* and *Salmonella typhimurium* by hard clams in an in vitro system. *Appl. Environ. Microbiol.* 47:986–88.

151. Todd, E. C. D., G. A. Jarvis, K. F. Weiss, G. W. Riedel, and S. Charbonneau. 1983. Microbiological quality of frozen cream-type pies sold in Canada. *J. Food Protect.* 46:34–40.

152. Vanderzant, C., M. O. Hanna, J. G. Ehlers, J. W. Savell, G. C. Smith, D. B. Griffin, R. N. Terrell, K. D. Lind, and D. E. Galloway. 1982. Centralized packaging of beef loin steaks with different oxygen-barrier films: Microbiological characteristics. *J. Food Sci.* 47:1070–79.

153. Vanderzant, C., A. W. Matthys, and B. F. Cobb, III. 1973. Microbiological, chemical, and organoleptic characteristics of frozen breaded raw shrimp. *J. Milk Food Technol.* 36:253–61.

154. Wentz, B. A., A. P. Duran, A. Swartzentruber, A. H. Schwab, and R. B. Read, Jr. 1983. Microbiological quality of fresh blue crabmeat, clams and oysters. *J. Food Protect.* 46:978–81.

155. Wentz, B. A., A. P. Duran, A. Swartzentruber, A. S. Schwab, and R. B.

Read, Jr. 1984. Microbiological quality of frozen breaded onion rings and tuna pot pies. *J. Food Protect.* 47:58–60.

156. Woodburn, M. 1964. Incidence of Salmonellae in dressed broiler-fryer chickens. *Appl. Microbiol.* 12:492–95.

157. Zottola, E. A., and F. F. Busta. 1971. Microbiological quality of further-processed turkey products. *J. Food Sci.* 36:1001–4.

III

DETERMINING

MICROORGANISMS AND

THEIR PRODUCTS IN FOODS

5.

CULTURE, MICROSCOPIC, AND SAMPLING METHODS

The examination of foods for the presence, types, and numbers of microorganisms and/or their products is basic to food microbiology. In spite of the importance of this, none of the methods in common use permits the determination of exact numbers of microorganisms in a food product. Although some methods of analysis are better than others, every method has certain inherent limitations associated with its use.

The four basic methods employed for "total" numbers are: (1) standard plate counts (SPC) for viable cells, (2) the most-probable numbers (MPN) method as a statistical determination of viable cells, (3) dye-reduction techniques to estimate numbers of viable cells that possess reducing capacities, and (4) direct microscopic counts (DMC) for both viable and nonviable cells. All of these are discussed in this chapter along with their uses in determining microorganisms from various sources. Detailed procedures for their use can be obtained from references in Table 5-1. In addition, variations of these basic methods for examining the microbiology of surfaces and of air are presented along with a summary of methods and attempts to improve their overall efficiency.

CONVENTIONAL SPC

By this method, portions of food samples are blended or homogenized, serially diluted in an appropriate diluent, plated in or onto a suitable agar medium, incubated at an appropriate temperature for a given time, after which all visible colonies are counted by use of a Quebec or electronic counter.

97

Table 5-1. Some standard references for methods of microbiological analysis of foods.

	Reference							
	16	117	81	76	121	82	11	9
Direct microscopic counts				×	×	×	×	×
Standard plate counts			×	×	×		×	×
Most probable numbers			×	×	×		×	×
Dye reductions				×				
Coliforms			×	×	×		×	×
Fungi	×				×		×	×
Fluorescent antibodies					×		×	×
Sampling plans				×		×	×	
Food commodities		×						

The SPC is by far the most widely used method for determining the numbers of viable cells or colony-forming units (cfu) in a food product. When total viable counts are reported for a product, the counts should be viewed as a function of at least some of the following factors:

1. sampling methods employed
2. distribution of the organisms in the food sample
3. nature of the food flora
4. nature of the food material
5. the preexamination history of the food product
6. nutritional adequacy of the plating medium employed
7. incubation temperature and time used
8. pH, a_w, and Eh of the plating medium
9. type of diluent used
10. relative number of organisms in food sample
11. existence of other competing or antagonistic organisms

In addition to the limitations noted, plating procedures for selected groups are further limited by the degree of inhibition and effectiveness of the selective and/or differential agents employed.

While the SPC is more often determined by pour plating, essentially comparable results can be obtained by surface plating. By the latter method, prepoured and hardened agar plates with dry surfaces are employed. The diluted specimens are planted onto the surface of replicate plates, and with the aid of bent glass rods ("hockey sticks") the 0.1-ml inoculum/plate is carefully and evenly distributed over the entire surface. Surface plating offers advantages in determining the numbers of heat-sensitive psychrotrophs in a food product because the organisms do not come in contact with melted agar. It is the method of choice when the colonial features of a colony are important to its presumptive identification, and for most selective media.

Strict aerobes are obviously favored by surface plating, but microaerophilic organisms tend to grow slower. Among the disadvantages of surface plating is the problem of spreaders (especially when the agar surface is not adequately dry prior to plating), and the crowding of colonies, which makes enumeration more difficult.

Homogenization of food samples

Prior to the mid to late 1970s, microorganisms were extracted from food specimens for plating almost universally by use of mechanical blenders (Waring type). Around 1971, the Colwell Stomacher was developed in England by Sharpe and Jackson (115), and this device is now the method of choice in many laboratories for homogenizing foods for counts. The Stomacher is a relatively simple device which homogenizes specimens in a special plastic bag by the vigorous pounding of two paddles. The pounding effects the shearing of food specimens, and microorganisms are released into the diluent. Three models of the machine are available, but the model 400 is most widely used in food microbiology laboratories. It can handle samples (diluent and specimen) of 40 to 400 ml.

The Stomacher has been compared to high-speed blenders for food analysis by a large number of investigators. Plate counts from Stomacher-treated samples are similar to those treated by blender (14, 34, 112, 114). The instrument is generally preferred over blending for the following reasons: (1) the need to clean and store blender containers is obviated, (2) heat buildup does not occur during normal operational times (usually 2 min), (3) the homogenates can be stored in the Stomacher bags in a freezer for further use, and (4) the noise level is not as unpleasant as that of mechanical blenders. In a study by Sharpe and Harshman (114), the Stomacher was shown to be less lethal than a blender to *Staphylococcus aureus, Streptococcus faecalis,* and *Escherichia coli.* One investigator reported that counts by using a Stomacher were significantly higher than when a blender was used (129), while other investigators obtained higher overall counts by blender than by Stomacher (5). The latter investigators showed that the Stomacher is food-specific, being better than high-speed blending for some types of foods but not for others. In another study, SPC determinations made by Stomacher, blender, and shaking were not significantly different, although significantly higher counts of gram-negative bacteria were obtained by Stomacher than by either of the other two methods (66).

Another advantage of the Stomacher over blending is the homogenization of meats for dye-reduction tests. Holley et al. (55) showed that the extraction of bacteria from meat by using a Stomacher does not cause extensive disruption of meat tissue and consequently fewer reductive compounds were present to interfere with resazurin reduction, while with blending the level of reductiv' compounds released made resazurin reduction results meaningless.

The spiral plater

The spiral plater is a mechanical device that distributes the liquid inoculum on the surface of a rotating plate containing a suitable poured and hardened agar medium. The dispensing arm moves from the near center of the plate toward the outside depositing the sample in an Archimedes spiral. The attached special syringe dispenses a continuously decreasing volume of sample so that a concentration range of up to 10,000:1 is effected on a single plate. Following incubation at an appropriate temperature, colony development reveals a higher density of deposited cells near the center of the plate with progressively fewer toward the edge.

The enumeration of colonies on plates prepared with a spiral plater is achieved by use of a special counting grid (Fig. 5-1A). Depending upon the relative density of colonies, colonies that appear in one or more specific area(s) of the superimposed grid are counted. An agar plate prepared by a spiral plater is shown in Fig. 5-1B, and the corresponding grid area counted is shown in Fig. 5-1C. In this particular example, a total sample volume of 0.0018 ml was deposited and the two grid areas counted contained forty-four and sixty-one colonies, respectively, resulting in a total count of 6.1 $\times$ 10^4 bacteria/ml.

The spiral plating device herein described was devised by Gilchrist et al.

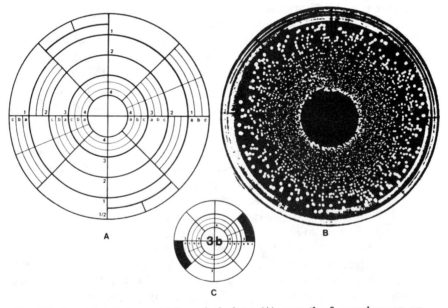

Fig. 5-1. Special counting grid for spiral plater (A); growth of organisms on an inoculated spiral plate (B); and areas of plate enumerated (C). In this example, the inoculum volume was 0.0010 ml, counts for the two areas shown were 44 and 63, and the averaged count was 6.1 $\times$ 10^4 bacteria/ml. (Courtesy of Spiral System Instruments, Bethesda, Maryland)

(43) although some of its principles were presented by earlier investigators among whom were Reyniers (108) and Trotman (128). The method has been studied by a rather large number of investigators and compared to other methods of enumerating viable organisms. It was compared to the SPC method by using 201 samples of raw and pasteurized milk; overall good agreement was obtained (31). A collaborative study from six analysts on milk samples showed that the spiral plater compared favorably with the SPC. A standard deviation of 0.109 was obtained by using the spiral plater compared to 0.110 for the SPC (96). In another study, the spiral plater was compared with three other methods (pour, surface plating, and drop count) and no difference was found between the methods at the 5% level of significance (65). In yet another study, the spiral plate maker yielded counts as good as those by the droplette method (52). Spiral plating is an official AOAC method (9).

Among the advantages of the spiral plater over standard plating are: less agar is used; fewer plates, dilution blanks, and pipettes are required; and three to four times more samples/h can be examined (70). Also, 50–60 plates/h can be prepared, and little training is required for its operation (65). Among the disadvantages is the problem that food particles may cause in blocking the dispensing stylus. It is more suited for use with liquid foods such as milk. A laser-beam counter has been developed for use with the plater. Because of the expense of the device, it is not likely to be available in laboratories that do not analyze large numbers of plates. The method is further described by AOAC (9) and in the *Bacteriological Analytical Manual* (11).

MEMBRANE FILTERS

Membranes with a pore size that will retain bacteria (generally 0.45 μm) but allow water or diluent to pass are used. Following the collection of bacteria upon filtering a given volume, the membrane is placed on an agar plate or an absorbent pad saturated with the culture medium of choice, and incubated appropriately. Following growth, the colonies are enumerated. Alternatively, a direct microscopic count (DMC) can be made. In this case, the organisms collected on the membrane are viewed and counted microscopically following appropriate staining, washing, and treatment of the membrane to render it transparent. These methods are especially suited for samples that contain low numbers of bacteria. While relatively large volumes of water can be passed through a membrane without clogging it, only small samples of dilute homogenates from certain foods can be used for a single membrane.

The overall efficiency of membrane filter methods for determining microbial numbers by the DMC has been improved by the introduction of fluorescent dyes. The use of fluorescent dyes and epifluorescent microscopes to enumerate bacteria in waters has been employed rather widely since the early 1970s.

Cellulose filters were among the earliest used; however, polycarbonate Nucleopore filters offer the advantage of retaining all bacteria on top of the filter. When lake and ocean waters were examined by using the two kinds of membranes, counts were twice as high with Nucleopore membranes as with cellulose membranes (54). A membrane filter–epifluorescent technique for enumerating bacteria in raw milk has been proposed (99). In order to lyse somatic cells in the milk to prevent clogging of the filter, 2-ml samples were treated with trypsin and Triton X-100. About 80% of the bacteria remained intact and were concentrated on a membrane. Following staining with acridine orange, the orange-red and green-fluorescing clumps were counted as viewed with a microscope fitted with an epifluorescent illuminator. The results were compared to those obtained by SPC and by standard Breed DMC, and the method was found suitable for samples of raw milk that contained between 5×10^3 and 5×10^8 bacteria/ml. The SPC and epifluorescent counts showed good overall correlation. The latter counts were in close agreement with Breed counts, and results by epifluorescent microscopy could be obtained in about 25 min. In another study (100), a direct epifluorescent filter technique (DEFT) was employed to enumerate bacteria in heat-treated milk and milk products, and by the technique as few as 5,700 bacteria/ml could be detected in about 20 min. The method was in agreement with plate counts for heat-treated milk, pasteurized cream, whey, and sweet cream butter but was not effective on evaporated and condensed milks and some reconstituted skim milk powders, pasteurized whey, and ripened cream butter. Results of a collaborative study among six laboratories comparing DEFT and SPC yielded correlations generally above 0.9, but the repeatability of DEFT was 1.5 times worse than plate counts and reproducibility only three times that for plate counts (98). The latter was due primarily to counting errors. To prepare solid foods for DEFT analysis, homogenates from a variety of foods were subjected to filtration through 5-μm nylon filters and as few as < 60,000 organisms/g could be detected (101). For more detailed information on this method, see Pettipher (97).

MICROSCOPE COLONY COUNTS

Microscope colony count methods involve the counting of microcolonies that develop in agar layered over microscope slides. The first was that of Frost, which consisted of spreading 0.1 ml of a milk-agar mixture over a 4-cm^2 area on a glass slide. Following incubation, drying, and staining, microcolonies are counted with the aid of a microscope.

In another method, 2 ml of melted agar are mixed with 2 ml of warmed milk and after mixing, 0.1 ml of the inoculated agar is spread over a 4-cm^2 area. Following staining with thionin blue, the slide is viewed with the 16-mm objective of a wide-field microscope (68).

AGAR DROPLETS

In the agar droplet method of Sharpe and Kilsby (116), the food homogenate is diluted in tubes of melted agar (at 45°C). For each food sample, three tubes of agar are used, the first tube being inoculated with 1 ml of food homogenate. After mixing, a sterile capillary pipette (ideally delivering 0.033 ml/drop) is used to transfer a line of 5 × 0.1-ml droplets to the bottom of an empty petri dish. With the same capillary pipette, three drops (0.1 ml) from the first 9-ml tube are transferred to the second tube and after mixing, another line of 5 × 0.1-ml droplets is placed next to the first. This step is repeated for the third tube of agar. Petri plates containing the agar droplets are incubated for 24 h and colonies are enumerated with the aid of a 10× viewer. Results using this method from pure cultures, meats, and vegetables compared favorably to those obtained by conventional plate counts; droplet counts from ground meat were slightly higher than plate counts. The method was about three times faster, and 24-h incubations gave counts equal to those obtained after 48 h by the conventional plate count. Dilution blanks are not required, and only one petri dish/sample is needed.

DRY FILMS

A dry film method consisting of two plastic films attached together on one side and coated with plate count agar ingredients and a cold-water-soluble jelling agent has been developed by the 3M Company (Petrifilm). If found successful on further testing, this method will simplify greatly the SPC. For use, 1 ml of diluent is placed between the two films and spread over the nutrient area by pressing. Following incubation, microcolonies appear red because of the presence of a tetrazolium dye in the nutrient phase. The method was compared to conventional SPC on 196 samples of raw milk and the correlation coefficient was 0.96 (79). A Petrifilm-VRB has been compared to violet red bile agar (VRBA) and MPN for coliform enumeration on 120 samples of raw milk. Plate counts by Petrifilm-VRB compared favorably to VRBA counts, and both were comparable to MPN results. A plot of coliform counts by Petrifilm-VRB against VRBA yielded a regression line with slope of 0.959 (89). In another study employing 108 milk samples, the dry film method was compared with SPC and a standard plate loop method. The correlation coefficient between dry film and SPC was 0.946 and the mean log difference was −0.177 (44). These investigators found the film method to be a suitable alternative to SPC.

MOST PROBABLE NUMBERS (MPN)

In this method, dilutions of food samples are prepared as for the SPC. Three serial aliquots or dilutions are then planted into nine or fifteen tubes of appropriate medium for the three- or five-tube method, respectively. Numbers of organisms in the original sample are determined by use of standard MPN tables. The method is statistical in nature, and MPN results are generally higher than SPC.

This method was introduced by McCrady in 1915. It is not a precise method of analysis—the 95% confidence intervals for a three-tube test range from 21 to 395! When the three-tube test is used, 20 of the 62 possible test combinations account for 99% of all results, while with the five-tube test, 49 of the possible 214 combinations account for 99% of all results (131). In a collaborative study on coliform densities in foods, a three-tube MPN value of 10 was found to be as high as 34, while in another phase of the study, the upper limit could be as high as 60 (118). Although Woodward (131) concluded that many MPN values are improbable, this method of analysis has gained popularity over the past decade. Among the advantages it offers are the following: (1) it is relatively simple, (2) results from one laboratory are more likely than SPC results to agree with those from another laboratory, (3) specific groups of organisms can be determined by use of appropriate selective and differential media, and (4) it is the method of choice for determining fecal coliform densities. Among the drawbacks to its use is the large volume of glassware required (especially for the five-tube method), the lack of opportunity to observe the colonial morphology of the organisms, and its lack of precision.

DYE-REDUCTION

Two dyes are commonly employed in this procedure to estimate the number of viable organisms in suitable products—methylene blue and resazurin. To conduct a dye-reduction test, properly prepared supernatants of foods are added to standard solutions of either dye for reduction from blue to white for methylene blue, and from slate blue to pink or white for resazurin. The time for dye reduction to occur is inversely proportional to the number of organisms in the sample.

Methylene blue and resazurin reduction by 100 cultures was studied in milk; with two exceptions, a good agreement was found between numbers of bacteria and time needed for reduction of the two dyes (42). In a study of resazurin reduction as a rapid method for assessing ground beef spoilage, reduction to the colorless state, odor scores, and SPC correlated significantly (111). One of the problems of using dye reduction for some foods is the existence of inherent reductive substances. This is true of raw meats, and Austin and Thomas (10) reported that resazurin reduction was less useful than with cooked meats. For the latter, approximately 600 samples were successfully evaluated by resazurin reduction by adding 20 ml of a 0.0001% resazurin solution to 100 g of sliced meat in a plastic pouch. Another way of getting around the reductive compounds in fresh meats is to homogenize samples by Stomacher rather than by Waring blender. Using Stomacher homogenates, raw meat was successfully evaluated by resazurin reduction when Stomacher homogenates were added to a solution of resazurin in 10% skim milk (55). Stomacher homogenates contained less disrupted tissue and consequently lower concentrations of reductive compounds. The method of Holley et al. (55) was evaluated further by Dodsworth and Kempton (30) who found that raw meat with an SPC $> 10^7$ bacteria/g could be detected within 2 h.

Dye-reduction tests have a long history of use in the dairy industry for assessing the overall microbial quality of raw milk. Among their advantages are: they are simple, rapid, and inexpensive; and only viable cells actively reduce the dyes. Disadvantages are: not all organisms reduce the dyes equally; and they are not applicable to food specimens that contain reductive enzymes unless special steps are employed as noted above.

ROLL TUBES

Screw-capped tubes or bottles of varying sizes are used in this method. Briefly, predetermined amounts of the melted and inoculated agar are added to the tube and the agar is made to solidify as a thin layer on the inside of the vessel. Following appropriate incubation, colonies are counted by rotating the vessel. It has been found to be an excellent method for enumerating fastidious anaerobes. For a review of the method, see Anderson and Fung (4).

DIRECT MICROSCOPIC COUNT (DMC)

In its simplest form, the DMC consists of making smears of food specimens or cultures onto a microscope slide, staining with an appropriate dye, and viewing and counting cells with the aid of a microscope (oil immersion objective). DMCs are most widely used in the dairy industry for assessing the microbial quality of raw milk and other dairy products, and the specific method employed is that originally developed by R. S. Breed (Breed count). Briefly, the method consists of adding 0.01 ml of a sample to a 1-cm^2 area on a microscope slide and following fixing, defatting of sample, and staining, the organisms or clumps of organisms are enumerated. The latter involves the use of a calibrated microscope (for further details, see 76). The method lends itself to the rapid microbiological examination of other food products such as dried and frozen foods.

Among the advantages of DMC are: it is rapid and simple; cell morphology can be assessed; and it lends itself to fluorescent probes for improved efficiency. Among its disadvantages are: it is a microscopic method and therefore fatiguing to the analyst; both viable and nonviable cells are enumerated; food particles are not always distinguishable from microorganisms; microbial cells are not uniformly distributed relative to single cells and clumps; some cells do not take the stain well and may not be counted; and

DMC counts are invariably higher than counts by SPC and other methods. In spite of its drawbacks, it remains the fastest way to make an assessment of microbial cells in a food product.

Howard mold counts

This is a microscope slide method developed by B. J. Howard in 1911 primarily for the purpose of monitoring tomato products. The method requires the use of a special chamber (slide) designed to enumerate mold mycelia. It is not valid on tomato products that have been comminuted. Similar to the Howard mold count is a method for quantifying *Geotrichum candidum* in canned beverages and fruits, and this method as well as the Howard mold count method is fully described by AOAC (9). The DEFT method described above has been shown to correlate well with the Howard mold count method on autoclaved and unautoclaved tomato concentrate, and it consequently could be used as an alternative to the Howard mold count (102).

MICROBIOLOGICAL EXAMINATION OF SURFACES

The need to maintain food contact surfaces in a hygienic state is of obvious importance. The primary problem that has to be overcome when examining surfaces or utensils for microorganisms is the removal of a significant percentage of the resident flora. Although a given method may not recover all organisms, its consistent use in specified areas of a food processing plant can still provide valuable information as long as it is realized that not all organisms are being recovered. The most commonly used methods for surface assessment in food operations are presented below.

Swab or swab-rinse methods

Swabbing is the oldest and most widely used method for the microbiological examination of surfaces not only in the food and dairy industries but also in hospitals and restaurants. The swab-rinse method was developed in 1917 by W. A. Manheimer and T. Ybanez. Either cotton or calcium alginate swabs are used. If one wishes to examine given areas of a surface, templates may be prepared with openings corresponding to the size of the area to be swabbed, for example, 1 in^2 or cm^2. The sterile template is placed over the surface and the exposed area is rubbed thoroughly with a moistened swab. The exposed swab is returned to its holder (test tube) containing a suitable diluent and stored at refrigerator temperatures until plated. The diluent should contain a neutralizer, if necessary. When cotton swabs are used, the organisms must be dislodged from the fibers. When calcium alginate swabs are used, the organisms are released into the diluent upon dissolution of the alginate by sodium hexametaphosphate. The organisms in the diluent are enumerated by a suitable method such as SPC, but any of the culture methods in this chapter may be used. Also, a battery of selective and differential culture media may be used to test specifically for given groups of organisms. In a recent innovation in the swab rinse method presented by Koller (69), 1.5 ml of fluid is added to a flat surface, swabbed for 15 sec over a 3-cm area, and volumes of 0.1 and 0.5 ml collected in microliter pipettes. The fluid may be surface or pour plated using plate count agar or selective media.

Concerning the relative efficacy of cotton and calcium alginate swabs, most agree that higher numbers of organisms are obtained by use of the latter. Using swabs, some researchers recovered as little as 10% of organisms from bovine carcasses (91); 47% of *Bacillus subtilis* spores from stainless steel surfaces (7); and up to 79% from meat surfaces (24, 95). Swab results from bovine carcasses were on the average 100 times higher than by contact plate method, and the deviation was considerably lower (91). The latter investigators found the swab method to be best suited for flexible, uneven, and heavily contaminated surfaces. The ease of removal of organisms depends upon the texture of the surface and the nature and types of flora. Even with its limitations, the swab-rinse method remains a rapid, simple, and inexpensive way to assess the microbiological flora of food surfaces and utensils.

Contact (RODAC) plate

The replicate organism direct agar contact (rodac) method employs special petri plates which are poured with 15.5–16.5 ml of an appropriate plating medium resulting in a raised agar surface. When the plate is inverted, the hardened agar makes direct contact with the surface. Originated by Gunderson and Gunderson in 1945, it was further developed in 1964 by Hall and Hartnett. When surfaces are examined that have been cleaned with certain detergents, it is necessary to include a neutralizer (lecithin, Tween 80, and so on) in the medium. Once exposed, plates are covered, incubated, and the colonies enumerated.

Perhaps the most serious drawbacks to this method are the covering of the agar surface by spreading colonies, and its ineffectiveness for heavily contaminated surfaces. These can be minimized by using plates with dried agar surfaces, and by using selective media (29). The rodac plate has been shown to be the method of choice when the surfaces to be examined are smooth, firm, and nonporous (7, 91). While it is not suitable for heavily contaminated surfaces, it has been estimated that a solution which contaminates a surface needs to contain at least 10 cells/ml before results can be achieved either by contact or by swabs (91). The latter investigators found that the contact plate removed only about 0.1% of surface flora. This suggests that 10 cfu/cm^2 detected by this method are referable to a surface that actually contains about 10^4 cfu/cm^2. When stainless steel surfaces were contaminated by *B. subtilis* endospores, 41% were recovered by the rodac plate compared to 47% by the swab method (7). In another study, swabs were better than contact plates when the contamination level was 100 or more organisms/21–25 cm^2 (113). On the other hand, contact plates give better results where low numbers exist. In terms of ranking of surface contamination, the two methods correlated well.

Agar syringe/"agar sausage" methods

The agar syringe method was proposed by W. Litsky in 1955 and subsequently modified (6). By this method, a 100-ml syringe is modified by removing the needle end to create a hollow cylinder that is filled with agar. A layer of agar is pushed beyond the end of the barrel by means of the plunger and pressed against the surface to be examined. The exposed layer is cut off and placed in a petri dish followed by incubation and colony enumeration. The "agar sausage" method proposed by ten Cate (126) is similar to the above method but employs plastic tubing rather than a modified syringe. The latter method has been used largely by European workers for assessing the surfaces of meat carcasses as well as for food plant surfaces. Both methods can be viewed as variations of the rodac plate and both have the same disadvantages—that is, spreading colonies and applicability limited to low levels of surface contaminants. Because clumps or chains of organisms on surfaces may yield single colonies, the counts obtained by these methods are lower than those obtained by methods that allow for the breaking up of chains or clumps.

For the examination of meat carcasses, Nortje et al. (92) compared three methods: a double swab, excision, and agar sausage. While the excision method was found to be the most reliable of the three, the modified agar sausage method correlated more closely with it than double swab, and the investigators recommended the agar sausage method because of its simplicity, speed, and accuracy.

Other surface methods

DIRECT SURFACE. A number of workers have employed direct surface agar plating methods, in which melted agar is poured onto the surface or utensil to be assessed. Upon hardening, the agar mold is placed in a petri dish and incubated. Angelotti and Foter (6) proposed this as a reference method for assessing surface contamination, and it is excellent for enumerating particulates containing viable microorganisms (35). It was used successfully to determine the survival of *Clostridium sporogenes* endospores on stainless steel surfaces (88). Although effective as a research tool, the method does not lend itself to routine use for food plant surfaces.

STICKY FILM. The sticky film method of Thomas has been used with some success by Mossel et al. (85). The method consists of pressing sticky film or tape against the surface to be examined and pressing the exposed side on an agar plate. It was shown to be less effective than swabs in recovering bacteria from wooden surfaces (85). An adhesive tape method has been employed successfully to assess microorganisms on meat surfaces (40). In a recent study, the swab, rodac, and adhesive tape (Mylar) methods were compared for the examination of pork carcasses, and the correlation between adhesive tape and rodac was better than that between adhesive tape and swab or between rodac and swab (27). Plastic strips attached to pads containing culture media have been used to monitor microorganisms on bottles (28).

SWAB/AGAR SLANT. The swab/agar slant method described in 1962 by N.-H. Hansen has been used with success by some European workers. The method involves sampling with cotton swabs that are transferred directly to slants. Following incubation, slants are grouped into one-half $\log_{10}$ units based upon estimated numbers of developed colonies. The average number of colonies is determined by plotting the distribution on probability paper. A somewhat similar method, the swab/agar plate, was proposed by Ølgaard (93). It requires a template, a comparator disc, and a reference table, making it a bit more complicated than the other methods noted.

ULTRASONIC DEVICES. Ultrasonic devices have been used to assess the microbiological contamination of surfaces, but the surfaces to be examined must be small in size and removable so that they can be placed inside a container immersed in diluent. Once the container is placed in an ultrasonic apparatus, the energy generated effects the release of microorganisms into the diluent. A more practical use of ultrasonic energy may be the removal of bacteria from cotton swabs in the swab-rinse method (105).

SPRAY GUN. A spray gun method was devised by Clark (24, 25) based on the impingement of a spray of washing solution against a circumscribed area of surface and the subsequent plating of the washing solution. Although the device is portable, a source of air pressure is necessary. It was shown to be much more effective than the swab method in removing bacteria from meat surfaces.

AIR SAMPLING

A variety of methods and devices exist for sampling air in food plants for the presence and relative numbers of microorganisms. They include impingement in liquids, impaction on solid surfaces, filtration, sedimentation, centrifugation, electrostatic precipitation, and thermal precipitation (121). Most commonly used are sedimentation, impaction, and impingement. These and other methods are more fully described in a manual that is periodically updated (75).

One of the simplest methods of air sampling is to open prepoured petri dishes for specific periods of time in the area to be assessed. While this method gives only an approximation of microbial numbers, it is effective in certain situations (53). Results are influenced by size of particles and by the speed and direction of air flow. If exposed for too long, drying of the agar surface will affect growth of deposited organisms. The presence and relative numbers of different types of organisms can be assessed by use of appropriate selective media.

The two most popular types of air samplers are the all-glass impinger and the Andersen (3) sieve sampler. They offer the advantage that a specific volume of air can be sampled. With the all-glass impinger, a specific quantity of diluent is placed in the impinger and air is pulled by vacuum through a capillary orifice with the air impinging on the diluent, where microorganisms are trapped in the liquid. The sample may then be plated as in the SPC or by use of selective media. This method does not work well when numbers of organisms are too small, and some gram-negative bacteria may be destroyed by high impingement velocity.

In the Andersen sampler, air is drawn by vacuum through one or more sieves, and the microorganisms are trapped on the surface of one or more agar plates. Optimum flow rate is 28.3 liters/min. Single- and multiple-stage sieve samplers have been compared (71), and the ranges and coefficients of variation for the all-glass and Andersen samplers have been determined (73). Of somewhat similar design are the Fort Detrick Slit Sampler and the Casella type single-stage slit sampler, both of which are described by D. A. Gabis et al. in Speck (121).

METABOLICALLY INJURED ORGANISMS

When microorganisms are subjected to environmental stresses such as sublethal heat and freezing, many of the individual cells undergo metabolic injury resulting in their inability to form colonies on selective media that uninjured cells can tolerate. Whether or not a culture has suffered metabolic injury can be determined by plating aliquots separately on a nonselective and a selective medium and enumerating the colonies that develop after suitable incubation. The colonies that develop on the nonselective medium represent both injured and uninjured cells, while only the uninjured cells develop on the selective medium. The difference between the number of colonies on the two media is a measure of the number of injured cells in the original culture or population. This principle is illustrated in Fig. 5-2 by data from Tomlins et al. (127) on sublethal heat injury of *Staphylococcus aureus*. These investigators subjected the organism to 52°C for 15 min in a phosphate buffer at pH 7.2 to inflict cell injury. The plating of cells at zero time and up to 15 min of heating on nonselective trypticase soy agar (TSA) and selective TSA + 7.0% NaCl (stress medium; TSAS) reveals only a slight reduction in numbers on TSA, while the numbers on TSAS were reduced considerably, indicating a high degree of injury relative to a level of salt that uninjured *S. aureus* can withstand. To allow the heat-injured cells to repair, the cells were placed in nutrient broth (recovery medium) followed by incubation at 37°C for 4 h. With hourly plating of

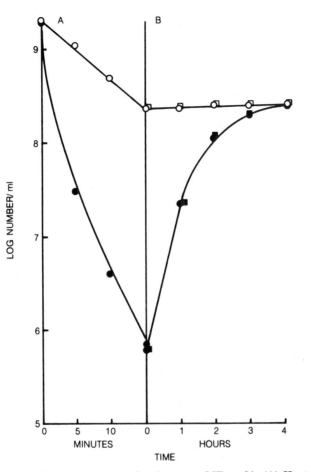

Fig. 5-2. Survival and recovery curve for *S. aureus* MF = 31. (A) Heat injury at 52°C for fifteen minutes in 100mM potassium phosphate buffer, pH 7.2. (B) Recovery from heat injury in NB at 37°C. Symbols: ○, samples plated on TSA to give a total viable count; ●, samples plated on TSAS to give an estimate of the uninjured population—cells recovered in NB containing 100 µg/ml of chloramphenicol; □, samples plated on TSA; ■, samples plated on TSAS. (Tomlins et al., 127; reproduced by permission of Nat'l Res. Coun. of Canada from *Can. J. Microbiol.* 17:759–65, 1971)

aliquots from the recovery medium onto TSAS, it can be seen that the injured cells regained their capacity to withstand the 7.0% NaCl in TSAS after the 4-h incubation.

The existence of metabolically injured cells in foods and their recovery during culturing procedures is obviously of great importance not only from the standpoint of pathogenic organisms but for spoilage organisms as well. The data cited above suggest that if a high-salt medium had been employed to examine a heat-pasteurized product for *S. aureus,* the number of viable cells found would have been lower than the actual number by a factor of 3 log cycles. Injury of food-borne microorganisms has been shown by a large number of investigators to be induced not only by sublethal heat and freezing but also by freeze drying, drying, irradiation, aerosolization, dyes, sodium azide, salts, heavy metals, antibiotics, essential oils, and other chemicals such as EDTA and sanitizing compounds.

The recognition of sublethal stresses on food-borne microorganisms and their effect upon growth under varying conditions dates back to the turn of the century. However, a full appreciation of this phenomenon did not come until the late 1960s. During the early part of the 1960s, it was observed that an initial rapid decrease in numbers of a metabolically injured organism was followed by only a limited recovery during the resuscitation process ("Phoenix phenomenon"). The increased nutritional requirement of bacteria that had undergone heat treatment was noted by Nelson (90) in 1943 (Nelson also reviewed the work of others up to that time). Gunderson and Rose (47) noted the progressive decrease in numbers of coliforms from frozen chicken products that grew on violet red bile agar (VRBA) with increasing storage time of products. Hartsell (51) inoculated foods with salmonellae, froze the inoculated foods, and then studied the fate of the organisms during freezer storage. More organisms could be recovered on highly nutritive nonselective media than on selective media such as MacConkey, deoxycholate, or VRB agars. The importance of the isolation medium in recovering stressed cells was noted also by Postgate and Hunter (104), and by Harris (49). In addition to the more exacting nutritional requirements of food-borne organisms that undergo environmental stresses, these organisms may be expected to manifest their injury via increased lag phases of growth, increased sensitivity to a variety of selective media agents, damage to cell membranes and TCA-cycle enzymes, breakdown of ribosomes, and DNA damage (22, 23, 46, 61, 62, 63, 64, 84, 107, 124, 127). While damage to ribosomes and cell membranes appear to be common consequences of sublethal heat injury, not all harmful agents produce identifiable injuries.

Recovery/repair
Metabolically injured cells can recover, at least in *S. aureus*, in no-growth media (60), and at a temperature of 15°C but not 10°C (41). In some instances at least, the recovery process is not instantaneous, for it has been shown that stressed coliforms do not all recover to the same degree but that the process takes place in a step-wise manner (78). Not all cells in a population suffer the same degree of injury. Hurst et al. (57) found dry-injured *S. aureus* cells that failed to develop on the nonselective recovery medium (TSA) but did recover when pyruvate was added to this medium. These cells were said to be severely injured in contrast to injured and uninjured cells. It has been found that sublethally heated *S. aureus* cells may recover their NaCl tolerance before certain membrane functions are restored (59). It is well established that injury repair occurs in the general absence of cell

wall and protein synthesis. It can be seen from Fig. 5-2 that the presence of chloramphenicol in the recovery medium had no effect upon the recovery of *S. aureus* from sublethal heat injury. The repair of cell ribosomes and membrane appears to be essential for recovery at least from sublethal heat, freezing, drying, and irradiation injuries.

The protection of cells from heat and freeze injury is favored by complex media and menstra or certain specific components thereof. Milk provides more protection than saline or mixtures of amino acids (83, 87), and the milk components that are most influential appear to be phosphate, lactose, and casein (64). Sucrose appears to be protective against heat injury (2, 72), while glucose has been reported to decrease heat protection for *S. aureus* (83). Nonmetabolizable sugars and polyols such as arabinose, xylose, and sorbitol have been found to protect *S. aureus* against sublethal heat injury but the mechanism of this action is unclear (120).

A summary of some of the many recovery or resuscitation methods employed to induce recovery of injured cells is presented in Table 5-2. The consequences of not employing a recovery step have been reviewed by Busta (19). The use of TSB with incubations ranging from 1 to 24 h at temperatures from 20° to 37°C is widely used for various organisms. The enumeration of sublethally heated *S. aureus* strains on various media has been studied (18, 36, 57). In one of these studies, seven staphylococcal media were compared on their capacity to recover nineteen strains of sublethally heated *S. aureus,* and the Baird-Parker medium was found to be clearly the best of those studied including nonselective TSA. Similar findings by others led to the adoption of this medium in the official methods of AOAC for the direct determination of *S. aureus* in foods that contain ≥ 10 cells/g (9). The greater efficacy of the Baird-Parker medium has been shown to result from its content of pyruvate. The use of this medium following recovery in an antibiotic-containing, nonselective medium has been suggested (57). While this approach may be suitable for *S. aureus* recovery, some problems may be expected to occur with the widespread use of antibiotics in recovery media to prevent cell growth. It has been shown that heat-injured spores of *C. perfringens* are actually sensitized to polymyxin and neomycin (13), and it is well established that the antibiotics that affect cell wall synthesis are known to induce L-phase variations in many bacteria.

Pyruvate is well established as an injury repair agent not only for injured *S. aureus* cells but for other organisms such as *E. coli.* Higher counts are obtained on media containing this compound when injured by a variety of agents. When added to TSB containing 10% NaCl, higher numbers of both stressed and nonstressed *S. aureus* were achieved (18), and the repair-detection of freeze- or heat-injured *E. coli* was significantly improved by pyruvate (80).

Catalase is another agent that increases recovery of injured aerobic organisms. First reported by Martin et al. (77), it has been found effective by many other investigators. It is effective for sublethally heated *S. aureus,*

Table 5-2. Summary of some methods employed to recover food-borne microorganisms subjected to various stresses.

Organisms	Method of injury	Stress medium	Recovery medium and method	Reference
Clostridium perfringens	0.1 N NaOH	TSN agar	TSN agar + lysozyme	33
C. perfringens spores	Ultra-high temp.	Polymyxin and neomycin	TSN agar + lysozyme; no antibiotics	13
C. botulinum Type E	Sublethal heat	PYE + 0.07% bile salts	PY agar	103
C. botulinum Type E	Irradiation	Polymyxin and neomycin	TP-EY medium + polymyxin and neomycin	110
C. botulinum spores	Irradiation	TYT + 5% NaCl	TYT thioglycolate medium	21
Coliforms	Freezing	VRBA	TSA surface plate; VRBA overlay	123
Enterobacteriaceae	Dried foods	E. E. medium	TSPB, 1-6 hr at 19–25°C	86
Escherichia coli	Freeze drying	Antibiotics and Nadeoxycholate	TSYA and minimal salts	119
E. coli	Organic and in-organic acids	VRBA	Nutrient agar at 37°C	109
E. coli	Frozen foods	VRBA and DCA	TSB for 1 hr at 25°C	122
E. coli	Freezing and sublethal heat	VRBA	3,3'-thiodipropionic acid supplemented media	80
Pseudomonas spp.	Freezing	Minimal agar	Trypticase soy agar	125
Salmonella typhimurium	Sublethal heat	EMBA + 2% NaCl	TSA at 37°C for 24 hr	22
S. anatum	Freeze drying	XLPA	XLPA + 0.25% Nadeoxycholate	106
S. anatum	Freezing	XLPA	XLPA + 0.2% Nadeoxycholate	64
S. typhimurium	Sublethal heat	TSA + yeast extract	M-9 agar at 37°C	45
S. senftenberg	Sublethal heat	TSYA	M-9 medium at 37°C for 2 hr	130
Salmonella spp.	Irradiation	SS and DCA	BGA; NB + TT + BG (MPN)	74
Shigella sonnei	Freezing	Synthetic medium	Blood heart inf., NA plates	87
Staphylococcus aureus	Sublethal heat	SM #110	Plate count agar	20
S. aureus	Sublethal heat	TSA + 7.5% NaCl	5% glucose or galactose	124
S. aureus	Sublethal heat	SM #110	Sodium pyruvate medium	12
S. aureus	Sublethal heat	TSA + 7.5% NaCl	Trypticase soy broth, 4 hr	62
S. aureus	Freeze drying	PCA + 7.5% NaCl	Plate count agar at 20–50°C	41

S. aureus	Sublethal heat	TSA + 7.0% NaCl	Catalase-treated media	77
Streptococcus faecalis	Sublethal heat	High salts in APT	All-purpose tween broth	17
S. faecalis	Sublethal heat	TSA + 6% NaCl	Synthetic nongrowth	23
S. faecium	Sublethal heat	TYGA + 2.5% NaCl	TYG medium at 37°C for 6 hr	32
Vibrio parahaemolyticus	Sublethal heat	TCBS + 6% NaCl	TCBS medium	15
Aspergillus parasiticus conidia	Sublethal heat	Reduced a_w medium; YEA + 10% NaCl	High a_w media; YEA	1
Yeasts	Sublethal heat	PDA at pH 3.5	Potato dextrose agar at pH ca. 8.0	90
Yeasts (8 strains)	Sublethal heat	Essential plant oils	YMPG broth, pH 5.5	26

BGA = brilliant green agar; DCA = deoxycholate agar; EE = Enterobacteriaceae medium; EMBA = eosin methylene blue agar; M-9 = a glucose salts medium; NA = nutrient agar; NB = nutrient broth; PDA = potato dextrose agar; PYE = peptone-yeast extract; SM #110 = Staphylococcus medium #110; SS = Salmonellae-Shigella medium; TCBS = thiosulphate-citrate-bile salts-sucrose medium; TSA = trypticase soy agar; TSB = trypticase soy broth; TSN = tryptone-sulfite-neomycin; TSPB = tryptone-soya-peptone broth; TSYA = tryptone-soya-yeast extract agar; TT = tetrathionate broth; TYGA = tryptone-yeast extract-glucose-phosphate medium; TYT = thiotone-yeast extract-trypticase-sodium thioglycollate; VRBA = violet red bile agar; XLPA = xylose-lysine-peptone agar; YEA = yeast extract agar.

Pseudomonas fluorescens, Salmonella typhimurium, and *E. coli* (77). It is effective also for *S. aureus* in the presence of 10% NaCl (18) and for water-stressed *S. aureus* (36). Another compound, shown to be as effective as pyruvate for heat-injured *E. coli,* is 3,3'-thiodipropionic acid (80).

Radiation injury of *Clostridium botulinum* type E spores by 4 kGy resulted in the inability to grow at 10°C in the presence of polymyxin and neomycin (110). The injured cells had a damaged post-germination system and formed aseptate filaments during outgrowth, but the germination lytic system was not damaged. The radiation injury was repaired at 30°C in about 15 h on TPEY agar without antibiotics. When *C. botulinum* spores are injured with hypochlorite, the L-alanine germination sites are modified resulting in the need for higher concentrations of alanine for repair (38). The L-alanine germination sites could be activated by lactate, and hypochlorite-treated spores could be germinated by lysozyme, indicating that the chloride removed spore coat proteins (39). More detailed information on spore injury has been provided by Foegeding and Busta (37).

Sublethally heat-stressed yeasts are inhibited by some essential oils (spices) at concentrations as low as 25 ppm (26). The spice oils affected colony size and pigment production.

Special plating procedures have been found by Speck et al. (123) and Hartman et al. (50) to allow for recovery from injury and subsequent enumeration in essentially one step. The procedures consist of using the agar overlay plating technique with one layer consisting of TSA, onto which are plated the stressed organisms. Following a 1- to 2-h incubation at 25°C for recovery, the TSA layer is overlaid with VRBA and incubated at 35°C for 24 h. The overlay method of Hartman et al. involved the use of a modified VRBA. The principle involved in the overlay technique could be extended to other selective media, of course. An overlay technique has been recommended for the recovery of coliforms. By this method, coliforms are plated with TSA and incubated at 35°C for 2 h followed by an overlay of VRBA (94).

In his comparison of eighteen plating media and seven enrichment broths to recover heat-stressed *Vibrio parahaemolyticus,* Beuchat (17) found that the two most efficient plating media were water blue-alizarin yellow agar and arabinose-ammonium-sulphate-cholate agar; arabinose-ethyl violet broth was the most suitable enrichment broth.

Mechanism
Pyruvate and catalase both act to degrade peroxides, suggesting that metabolically injured cells lack this capacity. The inability of heat-damaged *E. coli* cells to grow as well when surface-plated as when pour-plated with the same medium (48) may be explained by the loss of peroxides.

A large number of investigators have found that metabolic injury is accompanied by damage to cell membranes, ribosomes, DNA, or enzymes, as previously noted. The cell membrane appears to be the most universally affected (56). The lipid components of the membrane are the most likely targets, especially for sublethal heat injuries. Ribosomal damage is believed to result from the loss of Mg^{2+} and not to heat effects per se (58). On the other hand, ribosome-free areas have been observed by electron microscopy in heat-injured *S. aureus* cells (67). Following prolonged heating at 50°C, virtually no ribosomes were detected, and in addition, the cells were characterized by the appearance of surface blebs and exaggerated internal membranes (67). When *S. aureus* was subjected to acid injury by exposure to acetic, hydrochloric, and lactic acids at 37°C, coagulase and thermostable nuclease activities were reduced in injured cells (132). While acid injury did not affect cell membranes, RNA synthesis was affected.

ENUMERATION AND DETECTION OF FOOD-BORNE ORGANISMS

The synopses below are designed to illustrate the use of culture methods to enumerate or detect certain species or groups of organisms of interest in foods. For actual determinations, the appropriate reference in Table 5-1 should be consulted and followed.

ANAEROBES. SPC methods are employed with special attention to diluents and media prereduced to an appropriate negative Eh. When incubated under anaerobic conditions, it may be necessary to differentiate between facultative anaerobes and strict anaerobes.

BACILLUS CEREUS. SPC methods are followed with plating onto selective media such as KG or phenol red-egg yolk-polymyxin agars with enumeration of lecithinase-producing colonies followed by biochemical confirmations.

CAMPYLOBACTER JEJUNI. Following enrichment in Preston medium or nutrient broth containing vancomycin, trimethoprim lactate, and polymyxin B, sample is surface plated onto Campylobacter agar, Skirrow agar, or BHI containing the above antibiotics.

CLOSTRIDIUM BOTULINUM. Food aliquots are enriched in reduced media such as cooked meat or other nonselective media containing trypsin, followed by isolation on media such as veal liver–egg yolk agar. Suspect colonies are characterized biochemically. Toxigenesis is determined by injection of food homogenates or culture filtrates into protected and unprotected mice.

CLOSTRIDIUM PERFRINGENS. SPC methods are used with plating onto selective media such as tryptose-sulfite-cycloserine egg yolk or sulfite-polymyxin-sulfadiazine agars, followed by enumeration of suspect colonies and biochemical characterizations; or, by determining the presence of *C. perfringens* α-toxin in food products.

COLIFORMS. Presumptive and confirmed tests are carried out by inoculation of food dilutions into lauryl sulfate tryptose (LST) broth; gas-positive tubes of LST are inoculated into brilliant green lactose bile broth (BGLB), both at 35°C. Positive BGLB tubes are streaked into Eosin Methylene blue (EMB) agar and incubated at 35°C. LST may be inoculated and incubated at 44°C, followed by streaking onto EMB agar plates. For **fecal coliforms,** inoculate EC broth from positive LST tubes and incubate at 45.5°C. For **enteropathogenic *E. coli,*** the method is to examine LST-positive cultures with polyvalent EEC antisera and confirm, by biochemical means, serologically positive cultures taken from EMB plates. Alternatively, a rapid confirmatory test employing MUG (see Chapter 6) may be incorporated into the medium.

ENTEROBACTERIACEAE. Food homogenates are enriched in buffered glucose or brilliant green medium (EE broth), followed by plating onto selective media such as VRBA. Alternatively, 1% glucose is added to VRBA during reconstitution to make VRBG agar.

ENTEROCOCCI. SPC or MPN methods are employed, with direct plating onto selective media such as KF agar, or the use of broths such as KF or azide dextrose for MPN determinations, with confirmation of the latter medium on media such as Pfizer selective enterococcus agar. A fluorogenic selective and differential medium containing dyed starch and 4-methylumbelliferone-α-D-galactoside has been described (see Chapter 6).

GRAM-NEGATIVE BACTERIA. SPC methods are used, but with plating on or in selective media such as VRBA, EMB, MacConkey, or other appropriate media including peptone-bile-amphotericin-cycloheximide agar (PBAC) or plate count-monensin-KCl (PMK) agar (see Chapter 6).

LACTIC ACID BACTERIA. SPC methods are used, but with plating on selective media such as MRS, Eugon, or Elliker's lactic agars followed by biochemical characterization.

L-FORMS AND MYCOPLASMAS. Food homogenates or diluents are plated on L-form media such as BHI or TSA containing 1% agar, 10–20% horse serum, and 10–20% sucrose along with antibiotics such as methicillin or penicillin. The sample is incubated at 30°C for up to several weeks. Periodic examinations are made for the presence of "fried egg" or diffuse-spreading colonies characteristic of bacterial L-forms or mycoplasmas. Mycoplasmas have G + C values of 23–40%, while the values for bacterial L-forms are higher.

PSYCHROTROPHS. SPC methods are used and plates or broth are incubated at 5–7°C for up to 10 days.

SALMONELLAE AND SHIGELLAE. Food aliquots or dilutions thereof are pre-enriched in a nonselective broth and either plated directly onto selective media (such as brilliant green, XLD, and/or Hektoen agars), or planted into selective enrichment media such as selenite cystine or tetrathionate broths. The selective enrichment tubes are streaked onto selective media plates such as those above. Suspect colonies are characterized biochemically, by use of appropriate O and H antisera, and/or by fluorescent antibody procedures (see Chapter 6) in the case of salmonellae. Alternatively, commercial kits for the rapid detection of salmonellae by ELISA or DNA-DNA hybridization techniques may be used (see Chapter 6).

SPOREFORMERS. Food samples or diluents are heat treated at 70–80°C for 10 min followed by plating and incubation under aerobic and/or anaerobic conditions. **Flat-sour** spores are determined by steaming or boiling aliquots of food for around 5 min, plating with medium containing glucose and acid-base indicator, and incubating aerobically in the thermophilic range. **Sulfide-spoilers** are determined as above but with culturing in sulfide-containing media.

STAPHYLOCOCCUS AUREUS. SPC methods are used with direct plating onto media such as Baird-Parker agar, or onto similar media following enrichment in broth containing 7.5–10% NaCl. MPN methods may be used also with appropriate media. Suspect colonies should be confirmed for coagulase and/or thermostable nuclease (see Chapter 6).

THERMODURIC ORGANISMS. Following exposure of food products to specified temperature and time, use SPC methods and nonselective plating media. For **thermophiles,** use SPC or MPN methods and appropriate media with incubation at or above 55°C.

VIBRIO PARAHAEMOLYTICUS. Following enrichment in glucose salt teepol broth (GSTB) or alkaline peptone salt water, surface-plate onto thiosulfate-citrate-bile salts-sucrose (TCBS) agar, and confirm suspect colonies by biochemical means, including Kanagawa reaction.

VIRUSES. A slurry of food is prepared and clarified with appropriate agents, and the clarified effluent is concentrated by use of polyethylene glycol or other agents. Viruses are concentrated by ultracentrifugation and enumerated by plaque count following inoculation and incubation of suitable tissue culture system.

YEASTS AND MOLDS. Using SPC methods, the procedure is to plate onto media at pH 3.5 (such as acidified potato dextrose agar), or onto media containing antibacterial agents. Morphological characteristics of molds may be determined by culturing specimens directly on microscope slides, or by making preparations of mycelia from colonies on plates, followed by staining and viewing.

YERSINIA ENTEROCOLITICA. Following enrichment of the food homogenate at 4°C for up to 21 days, the samples are surface plated onto MacConkey or SS agar plates. The plates should be observed for lactose-negative colonies, which must be further characterized biochemically. Alternatively, food specimens may be plated on/in cefsulodin-irgasan-novobiocin (CIN) agar or broth along with bismuth sulfite agar. Strains that contain the 40–50 Mdal plasmid can be identified by use of Congo red dye in the medium.

REFERENCES

1. Adams, G. H., and Z. J. Ordal. 1976. Effects of thermal stress and reduced water activity on conidia of *Aspergillus parasiticus. J. Food Sci.* 41:547–50.
2. Allwood, M. C., and A. D. Russell. 1967. Mechanism of thermal injury in *Staphylococcus aureus.* I. Relationship between viability and leakage. *Appl. Microbiol.* 15:1266–69.
3. Andersen, A. A. 1958. New sampler for the collection, sizing, and enumeration of viable airborne particles. *J. Bacteriol.* 76:471–84.
4. Anderson, K. L., and D. Y. C. Fung. 1983. Anaerobic methods, techniques and principles for food bacteriology: A review. *J. Food Protect.* 46:811–22.
5. Andrews, W. H., C. R. Wilson, P. L. Poelma, A. Romero, R. A. Rude, A. P. Duran, F. D. McClure, and D. E. Gentile. 1978. Usefulness of the Stomacher in a microbiological regulatory laboratory. *Appl. Environ. Microbiol.* 35:89–93.
6. Angelotti, R., and M. J. Foter. 1958. A direct surface agar plate laboratory method for quantitatively detecting bacterial contamination on nonporous surfaces. *Food Res.* 23:170–74.
7. Angelotti, R., J. L. Wilson, W. Litsky, and W. G. Walter. 1964. Comparative evaluation of the cotton swab and rodac methods for the recovery of *Bacillus subtilis* spore contamination from stainless steel surfaces. *Hlth Lab. Sci.* 1:289–96.
8. Association of Official Analytical Chemists. 1977. Spiral plate method for bacterial count: Official first action. *J. Assoc. Off. Anal. Chem.* 60:493–94.
9. ———. 1984. *Official methods of analysis,* 14th ed. Washington, D.C.: AOAC.
10. Austin, B. L., and B. Thomas. 1972. Dye reduction tests on meat products. *J. Sci. Food Agric.* 23:542.
11. *Bacteriological analytical manual.* 6th ed. 1984. Washington, D.C.: AOAC.
12. Baird-Parker, A. C., and E. Davenport. 1965. The effect of recovery medium on the isolation of *Staphylococcus aureus* after heat treatment and after the storage of frozen or dried cells. *J. Appl. Bacteriol.* 28:390–402.
13. Barach, J. T., R. S. Flowers, and D. M. Adams. 1975. Repair of heat-injured *Clostridium perfringens* spores during outgrowth. *Appl. Microbiol.* 30:873–75.
14. Baumgart, J. 1973. Der "Stomacher"—ein neues Zerkleinerungsgerat zur Herstellung von Lebensmittelsuspensionen für die Keimzahlbestimmung. *Fleischwirtschaft* 53:1600.

15. Beuchat, L. R. 1976. Survey of media for the resuscitation of heat-stressed *Vibrio parahaemolyticus*. *J. Appl. Bacteriol.* 40:53–60.
16. ———, ed. 1978. *Food and beverage mycology.* Westport, Conn.: AVI.
17. Beuchat, L. R., and R. V. Lechowich. 1968. Effect of salt concentration in the recovery medium on heat-injured *Streptococcus faecalis. Appl. Microbiol.* 16:772–76.
18. Brewer, D. G., S. E. Martin, and Z. J. Ordal. 1977. Beneficial effects of catalase or pyruvate in a most-probable-number technique for the detection of *Staphylococcus aureus. Appl. Environ. Microbiol.* 34:797–800.
19. Busta, F. F. 1976. Practical implications of injured microorganisms in food. *J. Milk Food Technol.* 39:138–45.
20. Busta, F. F., and J. J. Jezeski. 1963. Effect of sodium chloride concentration in an agar medium on growth of heat-shocked *Staphylococcus aureus. Appl. Microbiol.* 11:404–7.
21. Chowdhury, M. S. U., D. B. Rowley, A. Anellis, and H. S. Levinson. 1976. Influence of postirradiation incubation temperature on recovery of radiation-injured *Clostridium botulinum* 62A spores. *Appl. Env. Microbiol.* 32:172–78.
22. Clark, C. W., and Z. J. Ordal. 1969. Thermal injury and recovery of *Salmonella typhimurium* and its effect on enumeration procedures. *Appl. Microbiol.* 18:332–36.
23. Clark, C. W., L. D. Witter, and Z. J. Ordal. 1968. Thermal injury and recovery of *Streptococcus faecalis. Appl. Microbiol.* 16:1764–69.
24. Clark, D. S. 1965. Method of estimating the bacterial population of surfaces. *Can. J. Microbiol.* 11:407–13.
25. ———. 1965. Improvement of spray gun method of estimating bacterial populations on surfaces. *Can. J. Microbiol.* 11:1021–22.
26. Conner, D. E., and L. R. Beuchat. 1984. Sensitivity of heat-stressed yeasts to essential oils of plants. *Appl. Environ. Microbiol.* 47:229–33.
27. Cordray, J. C., and D. L. Huffman. 1985. Comparison of three methods for estimating surface bacteria on pork carcasses. *J. Food Protect.* 48:582–84.
28. Cousin, M. A. 1982. Evaluation of a test strip used to monitor food processing sanitation. *J. Food Protect.* 45:615–19, 623.
29. deFigueiredo, M. P., and J. M. Jay. 1976. Coliforms, enterococci, and other microbial indicators. In *Food microbiology: Public health and spoilage aspects,* ed. M. P. deFigueiredo and D. F. Splittstoesser, 271–97. Westport, Conn.: AVI.
30. Dodsworth, P. J., and A. G. Kempton. 1977. Rapid measurement of meat quality by resazurin reduction. II. Industrial application. *Can. Inst. Food Sci. Technol. J.* 10:158–60.
31. Donnelly, C. B., J. E. Gilchrist, J. T. Peeler, and J. E. Campbell. 1976. Spiral plate count method for the examination of raw and pasteurized milk. *Appl. Environ. Microbiol.* 32:21–27.
32. Duitschaever, C. L., and D. C. Jordan. 1974. Development of resistance to heat and sodium chloride in *Streptococcus faecium* recovering from thermal injury. *J. Milk Food Technol.* 37:382–86.
33. Duncan, C. L., R. G. Labbe, and R. R. Reich. 1972. Germination of heat- and alkali-altered spores of *Clostridium perfringens* Type A by lysozyme and an initiation protein. *J. Bacteriol.* 109:550–59.

34. Emswiler, B. S., C. J. Pierson, and A. W. Kotula. 1977. Stomaching vs. blending. A comparison of two techniques for the homogenization of meat samples for microbiological analysis. *Food Technol.* 31(10):40–42.

35. Favero, M. S., J. J. McDade, J. A. Robertsen, R. K. Hoffman, and R. W. Edwards. 1968. Microbiological sampling of surfaces. *J. Appl. Bacteriol.* 31:336–43.

36. Flowers, R. S., S. E. Martin, D. G. Brewer, and Z. J. Ordal. 1977. Catalase and enumeration of stressed *Staphylococcus aureus* cells. *Appl. Environ. Microbiol.* 33:1112–17.

37. Foegeding, P. M., and F. F. Busta. 1981. Bacterial spore injury—an update. *J. Food Protect.* 44:776–86.

38. ———. 1983. Proposed role of lactate in germination of hypochlorite-treated *Clostridium botulinum* spores. *Appl. Environ. Microbiol.* 45:1369–73.

39. ———. 1983. Proposed mechanism for sensitization by hypochlorite treatment of *Clostridium botulinum* spores. *Appl. Environ. Microbiol.* 45:1374–79.

40. Fung, D. Y. C., C.-Y. Lee, and C. L. Kastner. 1980. Adhesive tape method for estimating microbial load on meat surfaces. *J. Food Protect.* 43:295–97.

41. Fung, D. Y. C., and L. L. VandenBosch. 1975. Repair, growth, and enterotoxigenesis of *Staphylococcus aureus* S-6 injured by freeze-drying. *J. Milk Food Technol.* 38:212–18.

42. Garvie, E. I., and A. Rowlands. 1952. The role of micro-organisms in dye-reduction and keeping-quality tests. II. The effect of micro-organisms when added to milk in pure and mixed culture. *J. Dairy Res.* 19:263–74.

43. Gilchrist, J. E., J. E. Campbell, C. B. Donnelly, J. T. Peeler, and J. M. Delaney. 1973. Spiral plate method for bacterial determinaton. *Appl. Microbiol.* 25:244–52.

44. Ginn, R. E., V. S. Packard, and T. L. Fox. 1984. Evaluation of the 3M dry medium culture plate (Petrifilm™ SM) method for determining numbers of bacteria in raw milk. *J. Food Protect.* 47:753–55.

45. Gomez, R. F., A. J. Sinskey, R. Davies, and T. P. Labuza. 1973. Minimal medium recovery of heated *Salmonella typhimurium* LT 2. *J. Gen. Microbiol.* 74:267–74.

46. Gray, R. J. H., L. D. Witter, and Z. J. Ordal. 1973. Characterization of milk thermal stress in *Pseudomonas fluorescens* and its repair. *Appl. Microbiol.* 26:78–85.

47. Gunderson, M. F., and K. D. Rose. 1948. Survival of bacteria in a precooked, fresh-frozen food. *Food Res.* 13:254–63.

48. Harries, D., and A. D. Russell. 1966. Revival of heat-damaged *Escherichia coli*. *Experientia* 22:803–4.

49. Harris, N. D. 1963. The influence of the recovery medium and the incubation temperature on the survival of damaged bacteria. *J. Appl. Bacteriol.* 26:387–97.

50. Hartman, P. A., P. S. Hartman, and W. W. Lanz. 1975. Violet red bile 2 agar for stressed coliforms. *Appl. Microbiol.* 29:537–39.

51. Hartsell, S. E. 1951. The longevity and behavior of pathogenic bacteria in frozen foods: The influence of plating media. *Amer. J. Pub. Hlth.* 41:1072–77.

52. Hedges, A. J., R. Shannon, and R. P. Hobbs. 1978. Comparison of the precision obtained in counting viable bacteria by the spiral plate maker, the droplette and the Miles & Misra methods. *J. Appl. Bacteriol.* 45:57–65.

53. Hill, R. A., D. M. Wilson, W. R. Burg, and O. L. Shotwell. 1984. Viable fungi in corn dust. *Appl. Environ. Microbiol.* 47:84–87.

54. Hobbie, J. E., R. J. Daley, and S. Jasper. 1977. Use of nucleopore filters for counting bacteria by fluorescence microscopy. *Appl. Environ. Microbiol.* 33:1225–28.

55. Holley, R. A., S. M. Smith, and A. G. Kempton. 1977. Rapid measurement of meat quality by resazurin reduction. I. Factors affecting test validity. *Can. Inst. Food Sci. Technol. J.* 10:153–57.

56. Hurst, A. 1977. Bacterial injury: A review. *Can. J. Microbiol.* 23:935–44.

57. Hurst, A., G. S. Hendry, A. Hughes, and B. Paley. 1976. Enumeration of sublethally heated staphylococci in some dried foods. *Can. J. Microbiol.* 22:677–83.

58. Hurst, A., and A. Hughes. 1978. Stability of ribosomes of *Staphylococcus aureus* S-6 sublethally heated in different buffers. *J. Bacteriol.* 133:564–68.

59. Hurst, A., A. Hughes, J. L. Beare-Rogers, and D. L. Collins-Thompson. 1973. Physiological studies on the recovery of salt tolerance by *Staphylococcus aureus* after sublethal heating. *J. Bacteriol.* 116:901–7.

60. Hurst, A., A. Hughes, D. L. Collins-Thompson, and B. G. Shah. 1974. Relationship between loss of magnesium and loss of salt tolerance after sublethal heating of *Staphylococcus aureus. Can. J. Microbiol.* 20:1153–58.

61. Hurst, A., A. Hughes, M. Duckworth, and J. Baddiley. 1975. Loss of D-alanine during sublethal heating of *Staphylococcus aureus* S-6 and magnesium binding during repair. *J. Gen. Microbiol.* 89:277–84.

62. Iandolo, J. J., and Z. J. Ordal. 1966. Repair of thermal injury of *Staphylococcus aureus. J. Bacteriol.* 91:134–42.

63. Jackson, H., and M. Woodbine. 1963. The effect of sublethal heat treatment on the growth of *Staphylococcus aureus. J. Appl. Bacteriol.* 26:152–58.

64. Janssen, D. W., and F. F. Busta. 1973. Influence of milk components on the injury, repair of injury, and death of *Salmonella anatum* cells subjected to freezing and thawing. *Appl. Microbiol.* 26:725–32.

65. Jarvis, B., V. H. Lach, and J. M. Wood. 1977. Evaluation of the spiral plate maker for the enumeration of micro-organisms in foods. *J. Appl. Bacteriol.* 43:149–57.

66. Jay, J. M., and S. Margitic. 1979. Comparison of homogenizing, shaking, and blending on the recovery of microorganisms and endotoxins from fresh and frozen ground beef as assessed by plate counts and the *Limulus* amoebocyte lysate test. *Appl. Environ. Microbiol.* 38:879–84.

67. Jones, S. B., S. A. Palumbo, and J. L. Smith. 1983. Electron microscopy of heat-injured and repaired *Staphylococcus aureus. J. Food Safety* 5:145–57.

68. Juffs, H. S., and F. J. Babel. 1975. Rapid enumeration of psychrotrophic bacteria in raw milk by the microscopic colony count. *J. Milk Food Technol.* 38:333–36.

69. Koller, W. 1984. Recovery of test bacteria from surfaces with a simple new swab-rinse technique: A contribution to methods for evaluation of surface disinfectants. *Zent. Bakteriol. Hyg. I. Orig. B.* 179:112–124.

70. Konuma, H., A. Suzuki, and H. Kurata. 1982. Improved Stomacher 400 bag applicable to the spiral plate system for counting bacteria. *Appl. Environ. Microbiol.* 44:765–69.

71. Kotula, A. W., J. R. Guilfoyle, B. S. Emswiler, and M. D. Pierson. 1978.

Comparison of single and multiple stage sieve samplers for airborne microorganisms. *J. Food Protect.* 41:447–49.

72. Lee, A. C., and J. M. Goepfert. 1975. Influence of selected solutes on thermally induced death and injury of *Salmonella typhimurium. J. Milk Food Technol.* 38:195–200.

73. Lembke, L. L., R. N. Kniseley, R. C. Van Nostrand, and M. D. Hale. 1981. Precision of the all-glass impinger and the Andersen microbial impactor for air sampling in solid-waste handling facilities. *Appl. Environ. Microbiol.* 42:222–25.

74. Licciardello, J. J., J. T. R. Nickerson, and S. A. Goldblith. 1970. Recovery of salmonellae from irradiated and unirradiated foods. *J. Food Sci.* 35:620–24.

75. Lioy, P. J., and M. J. Y. Lioy, eds. 1983. *Air sampling instruments for evaluation of atmospheric contamination.* 6th ed. Cincinnati: American Conference of Governmental Industrial Hygienists.

76. Marth, E. H., ed. 1978. *Standard methods for the examination of dairy products.* 14th ed. Washington, D.C.: American Public Health Association.

77. Martin, S. E., R. S. Flowers, and Z. J. Ordal. 1976. Catalase: Its effect on microbial enumeration. *Appl. Environ. Microbiol.* 32:731–34.

78. Maxcy, R. B. 1973. Condition of coliform organisms influencing recovery of subcultures on selective media. *J. Milk Food Technol.* 36:414–16.

79. McAllister, J. S., R. L. Nelson, P. E. Hanson, and K. F. McGoldrick. 1984. A dry media film for use as a replacement for the aerobic pour plate for enumeration of bacteria. In *Bacteriological proceedings,* 14.

80. McDonald, L. C., C. R. Hackney, and B. Ray. 1983. Enhanced recovery of injured *Escherichia coli* by compounds that degrade hydrogen peroxide or block its formation. *Appl. Environ. Microbiol.* 45:360–65.

81. *Microorganisms in foods.* 1982. Vol. 1, *Their significance and methods of enumeration,* 2d ed. ICMSF. Toronto: Univ. of Toronto Press.

82. *Microorganisms in foods.* 1982. Vol. 2, *Sampling for microbiological analysis: Principles and specific applications.* ICMSF. Toronto: Univ. of Toronto Press.

83. Moats, W. A., R. Dabbah, and V. M. Edwards. 1971. Survival of *Salmonella anatum* heated in various media. *Appl. Microbiol.* 21:476–81.

84. Moss, C. W., and M. L. Speck. 1966. Identification of nutritional components in trypticase responsible for recovery of *Escherichia coli* injured by freezing. *J. Bacteriol.* 91:1098–1104.

85. Mossel, D. A. A., E. H. Kampelmacher, and L. M. Van Noorle Jansen. 1966. Verification of adequate sanitation of wooden surfaces used in meat and poultry processing. *Zent. Bakteriol. Parasiten., Infek. Hyg. Abt. I.* 201:91–104.

86. Mossel, D. A. A., and M. A. Ratto. 1970. Rapid detection of sublethally impaired cells of Enterobacteriaceae in dried foods. *Appl. Microbiol.* 20:273–75.

87. Nakamura, M., and D. A. Dawson. 1962. Role of suspending and recovery media in the survival of frozen *Shigella sonnei. Appl. Microbiol.* 10:40–43.

88. Neal, N. D., and H. W. Walker. 1977. Recovery of bacterial endospores from a metal surface after treatment with hydrogen peroxide. *J. Food Sci.* 42:1600–1602.

89. Nelson, C. L., T. L. Fox, and F. F. Busta. 1984. Evaluation of dry medium film (Petrifilm VRB) for coliform enumeration. *J. Food Protect.* 47:520–25.

90. Nelson, F. E. 1943. Factors which influence the growth of heat-treated bacteria. I. A comparison of four agar media. *J. Bacteriol.* 45:395–403.

91. Niskanen, A., and M. S. Pohja. 1977. Comparative studies on the sampling and investigation of microbial contamination of surfaces by the contact plate and swab methods. *J. Appl. Bacteriol.* 42:53–63.
92. Nortje, G. L., E. Swanepoel, R. T. Naude, W. H. Holzapfel, and P. L. Steyn. 1982. Evaluation of three carcass surface microbial sampling techniques. *J. Food Protect.* 45:1016–17, 1021.
93. Ølgaard, K. 1977. Determination of relative bacterial levels on carcasses and meats—A new quick method. *J. Appl. Bacteriol.* 42:321–29.
94. Ordal, Z. J., J. J. Iandolo, B. Ray, and A. G. Sinskey. 1976. Detection and enumeration of injured microorganisms. In *Compendium of methods for the microbiological examination of foods,* ed. M. L. Speck, 163–69. Washington, D.C.: American Public Health Association.
95. Patterson, J. T. 1971. Microbiological assessment of surfaces. *J. Food Technol.* 6:63–72.
96. Peeler, J. T., J. E. Gilchrist, C. B. Donnelly, and J. E. Campbell. 1977. A collaborative study of the spiral plate method for examining milk samples. *J. Food Protect.* 40:462–64.
97. Pettipher, G. L. 1983. *The direct epifluorescent filter technique for the rapid enumeration of microorganisms.* New York: Wiley.
98. Pettipher, G. L., R. J. Fulford, and L. A. Mabbitt. 1983. Collaborative trial of the direct epifluorescent filter technique (DEFT), a rapid method for counting bacteria in milk. *J. Appl. Bacteriol.* 54:177–82.
99. Pettipher, G. L., R. Mansell, C. H. McKinnon, and C. M. Cousins. 1980. Rapid membrane filtration-epifluorescent microscopy technique for direct enumeration of bacteria in raw milk. *Appl. Environ. Microbiol.* 39:423–29.
100. Pettipher, G. L., and U. M. Rodrigues. 1981. Rapid enumeration of bacteria in heat-treated milk and milk products using a membrane filtration-epifluorescent microscopy technique. *J. Appl. Bacteriol.,* 50:157–66.
101. ———. 1982. Rapid enumeration of microorganisms in foods by the direct epifluorescent filter technique. *Appl. Environ. Microbiol.* 44:809–13.
102. Pettipher, G. L., R. A. Williams, and C. S. Gutteridge. 1985. An evaluation of possible alternative methods to the Howard mould count. *Lett. Appl. Microbiol.* 1:49–51.
103. Pierson, M. D., S. L. Payne, and G. L. Ades. 1974. Heat injury and recovery of vegetative cells of *Clostridium botulinum* type E. *Appl. Microbiol.* 27:425–26.
104. Postgate, J. R., and J. R. Hunter. 1963. Metabolic injury in frozen bacteria. *J. Appl. Bacteriol.* 26:405–14.
105. Puleo, J. R., M. S. Favero, and N. J. Petersen. 1967. Use of ultrasonic energy in assessing microbial contamination on surfaces. *Appl. Microbiol.* 15:1345–51.
106. Ray, B., J. J. Jezeski, and F. F. Busta. 1971. Repair of injury in freeze-dried *Salmonella anatum. Appl. Microbiol.* 22:401–7.
107. Ray, B., and M. L. Speck. 1973. Freeze-injury in bacteria. *CRC Crit. Rev. Clin. Lab. Sci.* 4:161–213.
108. Reyniers, J. A. 1935. Mechanising the viable count. *J. Pathol. Bacteriol.* 40:437–54.
109. Roth, L. A., and D. Kenna. 1971. Acid injury of *Escherichia coli. Can. J. Microbiol.* 17:1005–8.

110. Rowley, D. B., R. Firstenberg-Eden, and G. E. Shattuck. 1983. Radiation-injured *Clostridium botulinum* type E spores: Outgrowth and repair. *J. Food Sci.* 48:1829–31, 1848.
111. Saffle, R. L., K. N. May, H. A. Hamid, and J. D. Irby. 1961. Comparing three rapid methods of detecting spoilage in meat. *Food Technol.* 15:465–67.
112. Schiemann, D. A. 1977. Evaluation of the Stomacher for preparation of food homogenates. *J. Food Protect.* 40:445–48.
113. Scott, E., S. F. Bloomfield, and C. G. Barlow. 1984. A comparison of contact plate and calcium alginate swab techniques of environmental surfaces. *J. Appl. Bacteriol.* 56:317–20.
114. Sharpe, A. N., and G. C. Harshman. 1976. Recovery of *Clostridium perfringens*, *Staphylococcus aureus,* and molds from foods by the Stomacher: Effect of fat content, surfactant concentration, and blending time. *Can. Inst. Food Sci. Technol. J.* 9:30–34.
115. Sharpe, A. N., and A. K. Jackson. 1972. Stomaching: A new concept in bacteriological sample preparation. *Appl. Microbiol.* 24:175–78.
116. Sharpe, A. N., and D. C. Kilsby. 1971. A rapid, inexpensive bacterial count technique using agar droplets. *J. Appl. Bacteriol.* 34:435–40.
117. Silliker, J. H., ed. 1980. *Microbial ecology of foods.* Vol. 2. ICMSF. New York: Academic Press.
118. Silliker, J. H., D. A. Gabis, and A. May. 1979. ICMSF methods studies. XI. Collaborative/comparative studies on determination of coliforms using the most probable number procedure. *J. Food Protect.* 42:638–44.
119. Sinskey, T. J., and G. J. Silverman. 1970. Characterization of injury incurred by *Escherichia coli* upon freeze-drying. *J. Bacteriol.* 101:429–37.
120. Smith, J. L., R. C. Benedict, M. Haas, and S. A. Palumbo. 1983. Heat injury in *Staphylococcus aureus* 196E: Protection by metabolizable and non-metabolizable sugars and polyols. *Appl. Environ. Microbiol.* 46:1417–19.
121. Speck, M. L., ed. 1984. *Compendium of methods for the microbiological examination of foods.* Washington, D.C.: American Public Health Association.
122. Speck, M. L., and B. Ray. 1973. Recovery of *Escherichia coli* after injury from freezing. *Bull. Inst. Intern. Froid, Annexe* 5:37–46.
123. Speck, M. L., B. Ray, and R. B. Read, Jr. 1975. Repair and enumeration of injured coliforms by a plating procedure. *Appl. Microbiol.* 29:549–50.
124. Stiles, M. E., and L. D. Witter. 1965. Thermal inactivation, heat injury, and recovery of *Staphylococcus aureus. J. Dairy Sci.* 48:677–81.
125. Straka, R. P., and J. L. Stokes. 1959. Metabolic injury to bacteria at low temperatures. *J. Bacteriol.* 78:181–85.
126. ten Cate, L. 1963. An easy and rapid bacteriological control method in meat processing industries using agar sausage techniques in Rilsan artificial casing. *Fleischwarts.* 15:483–86.
127. Tomlins, R. I., M. D. Pierson, and Z. J. Ordal. 1971. Effect of thermal injury on the TCA cycle enzymes of *Staphylococcus aureus* MF 31 and *Salmonella typhimurium* 7136. *Can. J. Microbiol.* 17:759–65.
128. Trotman, R. E. 1971. The automatic spreading of bacterial culture over a solid agar plate. *J. Appl. Bacteriol.* 34:615–16.
129. Tuttlebee, J. W. 1975. The Stomacher—Its use for homogenization in food microbiology. *J. Food Technol.* 10:113–22.

130. Wilson, J. M., and R. Davies. 1976. Minimal medium recovery of thermally injured *Salmonella senftenberg* 4969. *J. Appl. Bacteriol.* 40:365–74.
131. Woodward, R. L. 1957. How probable is the most probable number? *J. Amer. Water Works Assoc.* 49:1060–68.
132. Zayaitz, A. E. K., and R. A. Ledford. 1985. Characteristics of acid-injury and recovery of *Staphylococcus aureus* in a model system. *J. Food Protect.* 48:616–20.

6.

PHYSICAL, CHEMICAL, AND
IMMUNOLOGIC METHODS

The methods of detecting microorganisms and their products covered in this chapter were developed since 1960. Most may be used to estimate numbers of organisms. Unlike direct microscopic counts, most of the enumeration methods that follow are based upon either the metabolic activity of the organisms on given substrates, measurements of growth response, or the measurement of some part of cells.

PHYSICAL METHODS

Impedance
Although the concept of electrical impedance measurement of microbial growth was advanced by G. N. Stewart in 1899, it was not until the 1970s that the method was employed for this purpose.

Impedance is the apparent resistance in an electric circuit to the flow of alternating current corresponding to the actual electrical resistance to a direct current. When microorganisms grow in culture media, they metabolize substrates of low conductivity into products of higher conductivity and thereby decrease the impedance of the media. When the impedance of broth cultures is measured, the curves are reproducible for species and strains, and mixed cultures can be identified by use of specific growth inhibitors. The technique has been shown capable of detecting as few as ten to one hundred cells (see Table 6-1). Cell populations of 10^5-10^6/ml can be detected in 3–5 h; and 10^4-10^5/ml in 5–7 h (218). The times noted are required for the organisms in question to attain a threshold of 10^6-10^7 cells/ml.

Impedance has been evaluated by a large number of investigators as a means of monitoring the overall microbial quality of various foods (24, 207). Two hundred samples of puréed vegetables were assessed and a 90–95% agreement was found between impedance measurements and plate count results relative to unacceptable levels of bacteria (85). Impedance analyses required 5 h and the method was reported to be applicable to cream pies, ground meat, and other foods. The microbiological quality of pasteurized milk was assessed by using the impedance detection time (IDT) of 7 h or less, which was equivalent to an aerobic plate count (APC) of 10,000/ml or more bacteria. Of 380 samples evaluated, 323 (85%) were correctly assessed by impedance (25). Using the same criterion for 27 samples of raw milk, 10 h were required for assessment. From a collaborative study of raw milk involving six laboratories, impedance results varied less than SPC results between laboratories (57). Impedance results have been compared to plate counts and Moseley test results to predict shelf life of pasteurized whole milk. IDTs at 18° and 21°C correlated better with Moseley results than plate counts as shelf-life predictors, and results were obtained by impedimetry in 25–38 h, compared to 7–9 days for Moseley test results (19). The findings of these investigators suggested that if the IDT was ≤ 6.1 h, the potential shelf life would be ≤ 9 days while an IDT of ≥ 12.4 h indicated a potential shelf life of ≥ 9 days. From another study with raw milk, impedance was found useful when a 7-h cut-off time (10^5 cfu/ml) was used to screen samples (76). Findings from this study were compared to SPC data. As a monitor of the postpasteurization contamination of cream, an IDT of > 20 h was found to denote cream of acceptable microbiological quality (83). A scattergram relating IDT to APC on 132 raw milk samples is presented in Fig. 6-1.

By use of impedance, the brewing industry test for detecting spoilage organisms in beer was shortened from 3 weeks or more to only 2–4 days. Yeasts growing in wort caused an increase in impedance while bacteria caused a decrease (49). For raw beef, IDTs for forty-eight samples were plotted against log bacterial counts and a regression coefficient of 0.97 was found (56). The IDT for meats was found to be about 9 h.

Impedance has been used to classify frozen orange juice concentrate as acceptable (< 10^4 cfu/ml) or unacceptable (> 10^4 cfu/ml). By using cut-off times of 10.2 h for bacteria, and 15.8 h for yeasts, 96% of 468 retail samples could be correctly classified (216). The method has been employed to detect starter culture failure within 2 h. In this application, complete starter failure occurred with approximately 10^5 phages/ml, resulting in the inhibition of normal impedance decrease caused by growth of normal lactic starter organisms (214). The relative level of contamination of meat surfaces by impedance has been assessed. With 10^7 cells and above/cm^2, detection could be made accurately within 2 h (22).

A relatively large number of studies have been conducted with impedance as a rapid method to detect coliforms. Fecal coliforms could be detected

Table 6-1. Reported minimum detectable levels of toxins or organisms by physical, chemical, and immunologic methods of analysis.

Methods	Toxin or organism	Sensitivity	Reference
Impedance	Coliforms in meats	10^3/g in 6.5 h	59
	Coliforms in culture media	10 in 3.8 h	56
	Fecal coliforms at 44.5° C	100 in 6.5–7.7 h	181
Microcalorimetry	S. aureus cells	2 cells in 12–13 h	125
	S. aureus	Minimum HPR[a] $\sim 10^4$ cells/ml	125
ATP measurement	Total flora of meats	10^4 cells/g in 20–25 min.	185
Radiometry	Frozen orange juice flora	10^4 cells/g in 6–10 h	86
	Salmonellae in foods; at 37° C with poly H antiserum	10^4–10^5 cells/g in 7 h	187
Glutamate decarboxylase	Coliforms in water	1–10 cells in 6 h	9
	E. coli	50,000 cells/ml in <10 h	140
ONPG	E. coli	1 cell in <20 h	215
Felix-01 phages	Salmonellae	<5 cells/ml in 24 h	91
Fluorescent antibody	Salmonellae	10^6 cells/ml	99
Thermostable nuclease	Staph. enterotoxin B	ca. 50 ng/ml	70
	From S. aureus	10 ng/g	118
Limulus lysate test	From S. aureus	2.5–5 ng	45
	Gram-negative endotoxins	2–6 pg of E. coli LPS	206
Radioimmunoassay	Staph. enterotoxins A, B, C, D, and E in foods	0.5–1.0 ng/g	17,137
	Staph. enterotoxin B in nonfat dry milk	2.2 ng/ml	149
Radioimmunoassay	Staph. enterotoxins A and B	0.1 ng/ml for A; 0.5 ng/ml for B	6
	Staph. enterotoxin C_2	100 pg	164
	E. coli ST_a enterotoxin	50–500 pg/tube	74
	Aflatoxin M_1 in milk	5–50 ng/assay	159
	Aflatoxin B_1	0.5–5.0 ng	43

Method	Analyte	Detection limit	Ref.
Electroimmunodiffusion	Ochratoxin A	20 ppb	1
	Bacterial cells	500–1,000 cells in 8–10 min.	190
	C. perfringens enterotoxin	10 ng	41
Micro-Ouchterlony	Botulinal toxins	3.7–5.6 mouse LD_{50}/0.1 ml	210
	S. aureus enterotoxins A and B	10–100 ng/ml	15,28
	C. perfringens type A toxin	500 ng/ml	71
Passive immune hemolysis	E. coli LT enterotoxin	<100 ng	48
Aggregate-hemagglutination	B. cereus enterotoxin	4 ng/ml	80
Latex agglutination	E. coli LT enterotoxin	32 ng/ml	55
Single radial immunodiffusion	S. aureus enterotoxins	0.3 μg/ml	136
Hemagglutination-inhibition	Staph. enterotoxin B	1.3 ng/ml	109
Reverse passive hemagglutination	Staph. enterotoxin B	1.5 ng/ml	182
ELISA	C. perfringens type A toxin	1 ng/ml	71
	Staph. enterotoxin A in wieners	0.4 ng	171
	Staph. enterotoxins A, B, and C in foods	0.1 ng/ml	188
	Staph. enterotoxins A, B, C, D, and E in foods	$\geq$ 1 ng/g	67
	Botulinal toxin type A	about 9 mouse LD_{50}/ml with monoclonal antibody	179
	Botulinal toxin type A	50–100 mouse i.p. LD_{50}	151
	Botulinal toxin type E	100 mouse LD_{50}	150
	Aflatoxin B_1	25 pg/assay	158
	Aflatoxin B_1	10 pg/ml	127
	Aflatoxin M_1 in milk	0.25 ng/ml in 3 h	159
	Aflatoxin B_1	<1 pg/assay	18
	Ochratoxin A	25 pg/assay	160
	Salmonellae	10^5 cells/ml	138
	Salmonellae	10^4–10^5 cells/ml	120

[a] Exothermic heat production rate

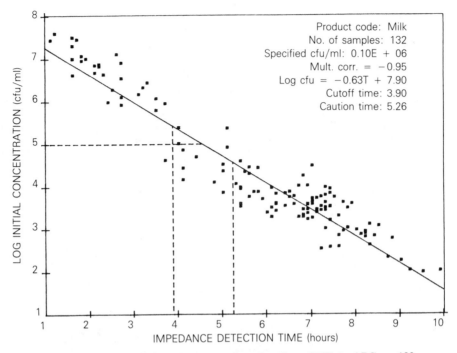

Fig. 6-1. Scattergram relating impedance detection time (IDT) to APC on 132 samples of raw milk. Samples containing >10⁵ mesophiles/ml were detected within 4 h. (Courtesy of Ruth Firstenberg-Eden, Bactomatic, Inc., Princeton, N.J.)

in 6.5–7.7 h when 100 log- or exponential-phase cells were inoculated into a selective medium and incubated at 44.5°C (181). Employing a new selective medium for coliforms, impedance was evaluated on seventy samples of ground beef, and 79% of the impedimetric results fell within the 95% confidence limits of the three-tube MPN procedure for coliforms, and fewer than 100 to 21,000 cells/g could be detected with results obtained within 24 h (131). In another study of coliforms in meats, the investigators found it necessary to develop a new selective medium that yielded impedance signals more consistent with cfu results. From an inoculum of ten coliforms into the new medium, the average IDT was 3.8 h. Of ninety-six meat samples, a correlation coefficient of 0.90 was found between impedance and corrected coliforms counts on violet red bile agar (59). An IDT of 6.5 h was required for meat samples with 10^3 coliforms, and the authors suggested that an impedance signal in 5.5 h or less denoted meat samples with coliforms $> 10^3$/g while the inability to detect by impedance in 7.6 h denoted coliform levels $< 10^3$/g (59). For coliforms in raw and pasteurized milk and two other dairy products, an IDT of < 9 h indicated coliforms were > 10/ml while an IDT of > 12 h indicated < 10 cells/ml (61).

Attention must be paid to culture media for impedimetry since not all media sustain smooth growth of given organisms. For example, it has been found that in brain heart infusion (BHI) broth *Pseudomonas* spp. exhibited a triphasic-type growth curve, making it difficult to determine the true IDT (60). On the other hand, smooth curves resulted when the organisms were grown in another medium—double plate count broth. Similarly, several standard selective media for coliforms were shown to yield less than ideal growth curves relative to impedance changes (59). The latter authors used impedance to determine total, mesophilic, and psychrotrophic numbers in raw milk. Total numbers (samples containing 1×10^5 cfu/ml) could be detected within 16 h; mesophiles within 4 h; and psychrotrophs within 21 h. With each group, correlation with APC was statistically significant (60).

In general, impedance methods measure changes with the conductance of the growth medium. The effect of microbial growth on changes in polarization capacitance and medium conductivity has been evaluated by use of an electrobacteriological impedance model. This model consists of a pair of electrodes inserted in a growth medium containing microorganisms (42). Upon passage through the medium of a small-amplitude AC current, total conductance (G_T) is measured. Conductance generally increases from 10 to 100% during growth, apparently from the increase in ion pairs as a result of metabolic activity. The parameter G_{pol} refers to capacitance polarization effect (electrode capacitance), and G_{sol} to medium conductance. The former is more useful in high-conductive media while the latter is useful in low-conductive media where changes are correlated with bacterial growth. Employing two media, yeast growth resulted in 20–100% changes in G_{pol} while G_{sol} changes were only 1–4% (62). Changes in pH were shown to affect G_{pol} more than G_{sol}. G_{pol} is believed by these investigators to be quite suitable for monitoring yeast growth and microbial growth in general.

Impedance has been compared to radiometry for the recovery of heat-stressed fecal coliforms from meat loaf. Radiometry detected in about 17 h and impedance in about 18 h. Either method was found suitable for determining, in cooked food meats, the criterion of 0 fecal coliforms/g (167). For more information on microbiological applications of impedance, see (58).

Microcalorimetry

This is the study of small heat changes: the measurement of the enthalpy change involved in the breakdown of growth substrates. The heat production that is measured is closely related to the cell's catabolic activities (66).

There are two types of calorimeters—batch and flow. Most of the early work was done with batch-type instruments. The thermal events measured by microcalorimetry are those from catabolic activities as noted above. One of the most widely used microcalorimeters for microbiological work is the Calvet instrument, which is sensitive to a heat flow of 0.01 cal/h from a 10-ml sample (66).

With respect to its use as a rapid method, most attention has been devoted to the identification and characterization of food-borne organisms. Microcalorimetric results vary according to the history of the organism, inoculum size, fermentable substrates, and the like. One group of investigators (155) found the variations such that the identification of microorganisms by this method was questioned, but in a later study in which a synthetic medium was used, Perry et al. (156) successfully characterized commercial yeast strains. The utility of the method to identify yeasts has been questioned (14) but by use of flow microcalorimetry yeasts could be characterized. The latter method is one in which a microcalorimeter is fitted with a flow-through calorimetric vessel. By use of a chemically defined medium containing seven sugars, thermograms were produced by nine lactic acid bacteria (belonging to the genera *Streptococcus, Leuconostoc,* and *Lactobacillus*) distinctive enough to recommend the method for their identification (68). All cultures were run at 37°C except *S. cremoris,* which was run at 30°C, and results were obtained within 24 h.

Microcalorimetry has been applied to the study of spoilage in canned foods (170), to differentiate between the Enterobacteriaceae, to detect the presence of *Staphylococcus aureus,* and to estimate bacteria in ground meat (81). Boling et al. (20) were able to differentiate between seventeen species of ten genera of Enterobacteriaceae by inoculating approximately 500 cells into BHI broth and recording the temperature changes over an 8–14-h incubation period. The thermograms produced were distinctive for each organism. In detecting *S. aureus,* Lampi et al. (125) achieved results in 2 h using an initial number of 10^7–10^8 cells/ml, and in 12–13 h when only 2 cells/ml were used. Russell et al. (169) examined over 250 cultures representing twenty-four genera and found that most of the organisms typically produced maximum outputs of 40–60 μcal/(sec) (ml) and returned to baseline in 5–7 h thereafter. Some profiles developed within 3 h while others required up to 14 h. A monitoring use of microcalorimetry is suggested by the work of Beezer et al. (13), who used flow microcalorimetry to determine the viability of recovered frozen cells of *Saccharomyces cerevisiae* within 3 h after thawing. When applied to comminuted meat, the peak exothermic heat production rate (HPR) could be recorded within 24 h for meats containing 10^5 to 10^8 cfu/g, and results by microcalorimetry correlated well with plate count results (81). With 10^2 cfu/ml, a measurable HPR was produced after 6 h, with peak HPR at 10 h.

Flow cytometry
Flow cytometry is the science of measuring components (cells) and the properties of individual cells in liquid suspension. In essence, suspended cells one by one are brought to a detector by means of a flow channel. Fluidic devices under laminar flow define the trajectories and velocities that cells traverse the detector, and among the cell properties that can be detected are fluorescence, absorbance, and light scatter. By use of flow sorting, individual cells are selected on the basis of their measured properties,

and from 1–3 or more global properties of the cell can be measured (134). Flow cytometers and cell sorters make use of one or more excitation sources such as argon, krypton, or helium-neon ion lasers, and one or two fluorescent dyes to measure and characterize several thousand cells/sec. When a dye is used, its excitation spectrum must match the light wavelengths of the excitation source (37). Two dyes may be used in combination to measure, for example, total protein and DNA content. In these instances, both dyes must excite at the same wavelength and emit at different wavelengths so that the light emitted by each dye is measured separately. The early history of flow cytometry has been reviewed by Horan and Wheeless (95).

While most studies have been conducted on mammalian cells, both DNA and protein have been measured in yeast cells. Typically, yeast cells are grown, fixed, and incubated in an RNAse solution for 1 h. Cell protein may be stained with fluorescein isothiocyanate and DNA with propidium iodide. Following necessary washing, the stained cells are suspended in a suitable buffer and are now ready for application to a flow cytometer. The one used by Hutter et al. (98) was equipped with a 50-mW argon laser. Yeast cells were excited at different wavelengths with the aid of special optical filters. By this method, baker's yeast was found to contain 4.6×10^{-14} g of DNA/cell, and the protein content/cell was found to be 1.1×10^{-11} g.

The possible application of flow cytometry to the analysis of foods for their content of microorganisms is suggested by the study of Van Dilla et al. (208), in which bacteria were stained with fluorescent dyes one of which bound preferentially to DNA rich in guanine-cytosine (G-C) while the other bound to adenine-thymine (A-T) rich DNA. Thus, the rapid identification of bacteria in food based upon their specific A-T/G-C ratios is possible.

In addition to the simultaneous measurement of protein and DNA in cells and the enumeration of cells based upon G-C and A-T content, flow cytometry has been used to distinguish between living and dead cells by dual staining; to determine the ploidy of yeast cells (97); to differentiate between spores and vegetative cells in *Bacillus* spp.; and to separate pathogenic and non-pathogenic amoebae (145).

CHEMICAL METHODS

Thermostable nuclease

The presence of *S. aureus* in significant numbers in a food can be determined by examining the food for the presence of thermostable nuclease (DNAse). This is possible because of the high correlation between the production of coagulase and thermostable nuclease by *S. aureus* strains, especially enterotoxin producers. For example, in one study 232 of 250 (93%) enterotoxigenic strains produced coagulase while 242 or 95% produced thermostable nuclease (124).

The examination of foods for this enzyme as an indirect test for *S. aureus* was first carried out by Chesbro and Auborn (31) employing a spectrophotometric method for nuclease determination. They showed that as the numbers of cells increased in ham sandwiches there was an increase in the amount of extractable thermostable nuclease of staphylococcal origin. They suggested that the presence of 0.34 unit of nuclease indicated certain staphylococcal growth and that at this level it was unlikely that enough enterotoxin was present to cause food poisoning. The 0.34 unit was shown to correspond to 9.5×10^{-3} μg enterotoxin by *S. aureus* strain 234. The reliability of the thermostable nuclease assay as an indicator of *S. aureus* growth has been shown by many others. It has been found to be as good as coagulase in testing for enterotoxigenic strains (147), and in another study all foods that contained enterotoxin contained thermostable nuclease, which was present in most foods with 1×10^6 *S. aureus* cells/g (153). On the other hand, thermostable nuclease is produced by some enterococci. Of 728 enterococci from milk and milk products, about 30% produced nuclease, with 4.3% of the latter (31 of the 728) being positive for thermostable nuclease (12).

The mean quantity of thermostable nuclease produced by enterotoxigenic strains is less than that for nonenterotoxigenic strains with 19.4 and 25.5 μg/ml, respectively, as determined in one study (147). For detectable levels of nuclease, 10^5–10^6 cells are needed, while for detectable enterotoxin, $> 10^6$ cells/ml are needed (148). During the recovery of heat-injured cells in trypticase soy broth (TSB), nuclease was found to increase during recovery but later decreased (220). The reason for the decrease was found to be proteolytic enzymes, and the decrease was reversed by the addition of protease inhibitors.

Several methods have been described for measuring thermostable nuclease in food products. By one, the assay medium containing DNA is flooded with 4 *N* HCl after samples have been incubated at 50°C for 1 h. This method produces results in 2.5 h and has been found sensitive to as little as 10 ng nuclease/g of food (118). Another similar method is that described by Lachica et al. (122, 123). It consists of combining DNA and toluidine blue 0 in a buffered salts solution with 1 percent agar. To a microscope slide, 3 ml of the molten DNA-dye preparation are layered. Small wells are cut in the agar layer and particles of food (about 5 mg) are added to the wells. The inoculated wells are covered, incubated at 37°C for 3 h, and read for the appearance of a bright pink halo around the food particles indicating reaction of DNAse with DNA. Heat-stable nuclease is detected by heating food samples at 97°C for 15 min before adding them to slide wells. These investigators found the technique to be reliable on inoculated beef and pork samples, and Tatini et al. (198) found the technique to be reliable on naturally contaminated products. The latter authors assessed heat-stable nuclease and *S. aureus* growth and enterotoxin production in broth, milk products, ground beef, and bologna. They also followed the production of nuclease in Genoa sausage and during the curing and smoking of sausage relative to the efficacy of heat-stable nuclease to assess the safety of these products and found the indirect test to be a reliable product indicator. A high statistical correlation has been found between heat-stable nuclease and *S. aureus* growth in Cheddar, Colby, and brick cheeses (35). Optimal assay conditions for thermostable nuclease include 50°C incubation at pH 10, which produce better results than 37°C and pH 9 (110).

To test for nuclease in casings of fermented sausage, 0.5-inch disks were removed from casings and placed on assay plates. The presence or absence of nuclease in steamed disks was determined by developing the plates. The method was shown capable of detecting about 2.5–5 ng of nuclease on salami casing disks (45).

The nuclease assay has been combined with a plating method for detecting *S. aureus*. By this method, *S. aureus* is plated on Baird-Parker agar plates and incubated for 24 h followed by a 2-h incubation at 60°C to inactivate heat-labile nuclease. Plates are finally covered with a layer of toluidine blue-DNA agar and observed for color change (121).

The assay of foods containing starter cultures for thermostable nuclease will not necessarily correlate with enterotoxin production. The enzyme was found in dry sausage containing enterotoxin-producing staphylococci but no toxin was found (148). On the other hand, it has been reported that the enzyme is produced under all conditions that permit cell growth (31).

While *S. epidermidis* and some micrococci produce nuclease, it is not as stable to heating as is that produced by *S. aureus* (124). Thermostable nuclease will withstand boiling for 15 min. It has been found to have a D value (D_{130}) of 16.6 min in BHI broth at pH 8.2, and a z value of 51 (46).

Among the advantages of testing for heat-stable nuclease as an indicator of *S. aureus* growth and activity are the following: (1) because of its heat-stable nature, the enzyme will persist even if the bacterial cells are destroyed by heat, chemicals, or bacteriophage, or if they are induced to L-forms; (2) the heat-stable nuclease can be detected faster than enterotoxin (about 3 h versus several days); (3) the nuclease appears to be produced by enterotoxigenic cells before enterotoxins appear (see Fig. 6-2); (4) the nuclease is detectable in unconcentrated cultures of food specimen while enterotoxin detection requires concentrated samples; and (5) the nuclease of concern is stable to heat as are the enterotoxins.

Limulus lysate for endotoxins

Gram-negative bacteria are characterized by their production of endotoxins, which consist of a lipopolysaccharide (LPS) layer (outer membrane) of the cell envelope. The LPS is pyrogenic and responsible for some of the symptoms that accompany infections caused by gram-negative bacteria.

The *Limulus* amoebocyte lysate (LAL) test employs a lysate protein obtained from the blood (actually haemolymph) cells (amoebocytes) of the horseshoe crab (*Limulus polyphemous*). The lysate protein is the most sensitive substance known for endotoxins. Of six different LAL preparations tested from five commercial companies, they were found to be from 3–300 times more sensitive to endotoxins than the U.S. Pharmacopeia rabbit pyrogen test (212). The LAL test is performed by adding aliquots of food suspensions or other test materials to small quantities of a lysate preparation followed by incubation at 37°C for 1 h. The presence of endotoxins causes gel formation of the lysate material. While most applications of this method have been made in clinical microbiology and in the pharmaceutical industry, food applications have been made.

The first food application was the use of LAL to detect the microbial spoilage of ground beef (103, 104). Endotoxin titers increase in proportion to viable counts of gram-negative bacteria (107). Since the normal spoilage of refrigerated fresh meats is caused by gram-negative bacteria, the LAL test is a good rapid indicator of the total numbers of gram-negative bacteria. The method has been found to be suitable for the rapid evaluation of the hygienic quality of milk relative to the detection of coliforms before and after pasteurization (199). For raw and pasteurized milk, it represents a method that can be used to determine the history of a milk product relative to its content of gram-negative bacteria. Since both viable and nonviable gram-negative bacteria are detected by LAL, a simultaneous plating is necessary to determine the numbers of cfu's. The method has been applied successfully to monitor milk and milk products (100, 219); microbial quality of raw fish (191); and cooked turkey rolls. In the latter, LAL titers and numbers of Enterobacteriaceae in vacuum-packaged rolls were found to have a statistically significant linear relationship (39). These investigators suggested that the test could be used to predict spoilage in such products.

LAL titers for foods can be determined either by direct serial dilutions or by MPN with results by the two methods being essentially similar (175). To extract endotoxins from foods, the Stomacher has been found to be generally better than the use of Waring blenders or the shaking of dilution bottles (106).

In this test, the proclotting enzyme of the *Limulus* reagent has been purified (196). It is a serine protease with a molecular weight of about 150,000 daltons. When activated with Ca^{2+} and endotoxin, gelation of the natural clottable protein occurs. The *Limulus* coagulogen has a molecular weight of 24,500. When it is acted upon by the *Limulus* clotting enzyme, the coagulogen releases a soluble peptide of about 45 amino acid residues and an insoluble coagulin of about 170 amino acids. The latter interacts with itself to form the clot which involves the cleavage of -arg-lys- or -arg-gly- linkages (197).

Commercial substrates are available that contain amino acid sequences similar to coagulogen. The chromogenic substrates used for endotoxin consist of these linked to *p*-nitroaniline. When the endotoxin-activated enzyme attacks the chromogenic substrate, free *p*-nitroaniline results and can be read at 405 nm. The amount of the chromogenic compound liberated is proportional to the quantity of endotoxin in the sample. Employing a chromogenic substrate, Tsuji et al. (205) have devised an automated method for endotoxin assay, and the method was shown sensitive to as little as 30 pg of endotoxin/ml.

Assuming that the quantity of endotoxin/gram-negative bacterial cells is fairly constant, and assuming further that cells of all genera contain the same given quantity, it is possible to calculate the number of cells (viable and nonviable) from which the experimentally determined endotoxin was derived. With a further assumption that the ratio of gram-negative to gram-positive bacteria is more or less constant for given products, one can make a 1-h estimate of the total numbers of bacteria in food products such as fresh ground beef (105). Low values by this procedure are more meaningful than high values, and as noted above, the latter need to be confirmed by other methods.

Overall, the value of the LAL test lies in the speed at which results can be obtained. Foods that have high LAL titers may be candidates for further testing by other methods, while those that have low titers may be placed immediately into categories of lower risk relative to numbers of gram-negative bacteria.

DNA-DNA hybridization

On the assumption that all members of a genus of bacteria share a given DNA sequence, and on the further assumption that this sequence is not shared by other organisms that might be associated with the first group, it is possible to identify unknown bacteria by using radiolabeled DNA probes consisting of fragments of the common DNA sequence to effect hybridization with DNA fragments from an unknown culture. The degree to which hybridization occurs is assessed by tests for the radiolabel, typically by autoradiography. One may also use DNA bound to membranes and measure the hybridization of complementary DNA. DNA-DNA hybridizations appear to offer great potential for microorganisms of interest in foods, especially pathogens.

Hybridizations are carried out by first preparing DNA fragments by treatment with restriction endonucleases. Following separation of the fragments on agarose gels, they are transferred to cellulose nitrate filters. DNA is hybridized on the filters to radioactive RNA or DNA, and the fragments in the DNA that contain transcribed sequences are detected by radioautography. Fragments of 500 nucleotide pairs or less may be under-represented or not detected (183). Alternatively, the inoculated hybridization filter may be placed on an agar plate and incubated for 24 h or more. Colonies that develop are lysed and the recovered DNA is subjected to autoradiography (90).

In their study of salmonellae in foods, Fitts et al. (64) used salmonella-specific cloned DNA fragments with *S. typhimurium* nick-translated with $(\alpha\text{-}^{32}P)dCTP$. Bacteria were collected on a series of filters including nitrocellulose. Following lysis of bacteria on the filters, their DNA was denatured (strands separated). The nick-translated probe was mixed with filters and incubated. Following appropriate washings, the filters were assessed with a radiocounter for hybridization. A commercially available salmonellae probe has been developed by preparing plasmid clones specific to salmonellae (see 63). Of the many clones tested, only ten did not react with other Enterobacteriaceae, and two of these hybridized to all salmonellae tested. The ten clones vary in size from 1.4 to 6.0 kilobases, and the use of combinations of all ensure that no salmonellae strains are missed (63).

DNA-DNA hybridization has been employed by Moseley and co-workers to study enterotoxigenic strains of *E. coli* (ETEC). Genes encoding for *E. coli* ST and LT (see Chapters 7 and 20) were detected among ETEC isolates by employing radiolabeled fragments of DNA as hybridization probes for homologous DNA sequences in colonies grown and lysed on nitrocellulose filters (144). The LT probe was prepared from plasmid DNA and consisted of a 0.5-Mdal fragment encoding a protein of the LT molecule. The ST probe was prepared from another plasmid and comprised a 157-base pair encoding a portion of the ST molecule. The technique was used to examine clinical specimens for ETEC strains that produced both LT and ST, or either alone, and was found to differentiate between the two heterologous genes encoding for ST (142). The DNA sequence homology between *E. coli* LT gene and *Vibrio cholerae* DNA was demonstrated by a hybridization technique (143). DNA-DNA hybridization has been employed to differentiate between lactic streptococcal phages (102), and to study the degree of similarity between strains of lactic streptococci. *S. thermophilus* strains showed little relationship to *S. cremoris, S. lactis,* and *S. diacetylactis* strains examined by this method (103). It has been used to qualitatively identify rotaviruses recovered from gastroenteritis (65).

More detailed information on this method can be found in Denhardt (38), Southern (183), Hill (89), and Fitts (63).

ATP measurement
Adenosine-triphosphate (ATP) is the primary source of energy in all living organisms. It disappears within 2 h after cell death (154), and the amount per cell is generally constant (203), with values of 10^{-18} to 10^{-17} mole/bacterial cell which corresponds to around 4×10^4 M ATP/10^5 cfu of bacteria (203). Among procaryotes, ATP in exponentially growing cells is regularly around 2–6 nmole ATP/mg dry weight regardless of mode of nutrition (111). In the case of rumen bacteria, the average cellular content was found to be 0.3 fg/cell, with higher levels found in rumen protozoal cells (152). The complete extraction and accurate measurement of cellular ATP can be equated to individual groups of microorganisms in the same general way as endotoxins for gram-negative bacteria.

One of the simplest ways to measure ATP is by use of the firefly luciferin-luciferase system. In the presence of ATP, luciferase emits light, which is measured with a liquid scintillation spectrometer or a luminometer. The amount of light produced by firefly luciferase is directly proportional to the amount of ATP added (154). Cellular ATP is extracted generally by boiling, and other details of its assay have been presented by Kimmich et al. (116) and Karl (111).

The application of ATP measurement as a rapid method for estimating microbial numbers has been developed and used more in clinical microbiology than other areas. In the clinical laboratory, it has been employed to screen urine specimens. The numbers of bacteria in urine are considered to be of etiologic importance if they occur at a level of 10^5 cfu/ml urine. In one study of 348 urine specimens, only an 89.4% agreement with culture methods was found (2). Using 10^5 organisms/ml or above to indicate bacteriuria and $< 10^5$ as negative, the ATP assay yielded 7% false positive and 27% false negative results. Thore et al. (204) assayed 2,018 urine specimens for ATP by the luciferase method setup to distinguish between samples containing $> 10^5$ cfu/ml and those containing $< 10^5$ cfu/ml. Ninety-two percent of those with $> 10^5$ cfu and 88% of those with $< 10^5$ were correctly classified by the ATP assay. The assay was a 15-min test, and samples were set at 13.5 nM ATP to define the limit between negative and positive results. These investigators and others cited by them found ATP analysis for bacteriuria to agree well with plate count results. In a comparative study of plate counts, the LAL test, and ATP measurement to rapidly assess contamination of intravenous fluids, the LAL test was found to be more sensitive than ATP, but the authors suggested that ATP measurements could be used (5).

The successful use of the method for bacteriuria and for assessing biomass in activated sludge (154) suggests that it should be of value for foods. It lends itself to automation and represents an excellent potential method for the rapid estimation of microorganisms in foods. The major problem that has to be overcome for food use is the removal of nonmicrobial ATP. The method was suggested to be of value for food use by Sharpe et al. (178) provided that microbial ATP levels are as great or greater than the intrinsic ATP in the food itself. They found that intrinsic levels decreased with food storage (Fig. 6-2). The utility of employing ATP analysis for foods was questioned by Williams (217), who called attention to the following: (1) one yeast strain was found to contain 300 times more ATP than the average for bacterial cells (some strains may contain as much as 1,000 times more); (2) sterile 0.1% peptone diluent gave readout values equivalent to about 200,000 organisms/ml; (3) heat-killed bacteria contain ATP; and (4) ATP is found in food itself, and there are interfering substances in foods such as potato starch and milk. The utility of the method for measuring rumen biomass has been questioned because of the discrepancy between the bacterial and protozoal content of ATP (152).

The problems noted above have been overcome to a large extent. Thore et al. (203) used Triton X-100 and apyrase to selectively destroy nonbacterial ATP in urine specimens and found that the resultant ATP levels were close to values observed in laboratory cultures with detection at 10^5 bacteria/ml.

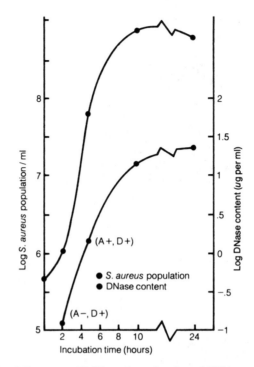

Fig. 6-2. Growth of *S. aureus* (196E) and production of DNAse and enterotoxins in Brain Heart Infusion Broth at 37°C. DNAse and enterotoxin D were detectable within 2 h at a population of 2 × 10⁶, whereas enterotoxin A was detected after 4 h at higher cell populations. DNAse was detectable in unconcentrated cultures, enterotoxins in fifty-fold concentrates. (Tatini et al., 198; copyright © 1975, Institute of Food Technologists)

In meats, the problem of nonmicrobial ATP was addressed by Stannard and Wood (185) by use of a three-stage process consisting of centrifugation, use of cation exchange resin, and filtration to get rid of food particles and collect bacteria on 0.22-μm filters. ATP analyses were carried out on bacteria eluted from the filter membranes, and 70–80% of most microorganisms were recovered on the filters. Linear relationship was shown between microbial ATP and bacterial numbers over the range 10^6–10^9 cfu/g. By the methods employed, results on ground beef were obtained in 20–25 min. In another study, seventy-five samples of ground beef were evaluated and a high correlation was found between $\log_{10}$ APC and $\log_{10}$ ATP when samples were incubated at 20°C (113). In this study, the amount of ATP/cfu ranged from 0.6 to 17.1 fg with fifty-one of the seventy-five samples containing ≤ 5.0 fg of ATP.

ATP analysis was combined with benzalkon-crystal violet (BC-ATP) to predict keeping quality of pasteurized milk, and results were compared to those from two commonly used shelf-life tests (213). The BC-ATP results

Table 6-2. Relation between total viable count and ATP during incubation at 37°C (178).

	ATP and mean viable count/g after:					
	0 h		*6 h*		*24 h*	
Sample	*ATP*[a]	*Viable count*	*ATP*[a]	*Viable count*	*ATP*[a]	*Viable count*
Crinkle-cut chip (frozen)	4.57×10^6 4.21×10^6	4.9×10^2	7.11×10^6 7.69×10^6	1.2×10^3	3.23×10^8 2.62×10^8	5.9×10^8
Frozen peas	5.11×10^7 5.43×10^7	4.1×10^4	3.18×10^8 3.54×10^8	1.3×10^7	3.90×10^8 5.23×10^8	7.7×10^8
Comminuted meat	1.51×10^7 1.60×10^7	1.3×10^3	3.62×10^6 4.94×10^6	6.4×10^3	7.88×10^5 8.34×10^5	2.2×10^3
Beef steaklet (frozen)	2.75×10^7 4.21×10^7 3.30×10^7	4.3×10^3	6.93×10^6 6.61×10^6	4.1×10^3	6.68×10^6 5.71×10^6	1.3×10^6
Beefburger (frozen)	3.92×10^7 4.89×10^7	3.1×10^4	5.65×10^6 4.39×10^6	3.2×10^3	3.76×10^6 3.70×10^6	2.2×10^5
Plaice fillet (frozen)	8.99×10^6 9.33×10^6	1.7×10^4	1.15×10^7 1.28×10^7	4.4×10^5	4.28×10^7 1.89×10^7	3.4×10^8
Fish finger (frozen)	1.96×10^9 1.79×10^9	6.0×10^5	6.08×10^7 4.26×10^7	2.2×10^6	1.38×10^9 1.56×10^9	6.4×10^8

[a] fg

obtained within 24 h were as good as those obtained from the shelf-life tests requiring 5 and 10 days for results. These investigators found that at an ATP content of $< 1,000$ relative light units it was always accompanied by bacterial counts of $< 10^6$/ml.

At the present time, ground beef can be extracted and read in 20 min and as few as 10^4 organisms/g can be detected by currently available instrumentation. However, these methods may not be useful when applied to seafoods that may contain luminescent bacteria such as *Photobacterium fisheri* and *Vibrio harveye,* for these organisms produce luciferase.

Radiometry

The radiometric detection of microorganisms is based upon the incorporation of a ^{14}C-labeled metabolite in a growth medium so that when the organisms utilize this metabolite, $^{14}CO_2$ is released and measured by use of a radioactivity counter. For organisms that utilize glucose, ^{14}C-glucose is usually employed. For those that cannot utilize this compound, others such as ^{14}C-formate or ^{14}C-glutamate are used. The overall procedure consists of using capped 50-ml serum vials to which are added anywhere from 12 to 36 ml of medium containing the labeled metabolite. The vials are made either aerobic or anaerobic by sparging with appropriate gases, and are then inoculated. Following incubation, the headspace is tested periodically for the presence of $^{14}CO_2$. The time required to detect the labeled CO_2 is inversely related to the number of organisms in a product.

The use of radiometry to detect the presence of microorganisms was first suggested by Levin et al. (128). It is confined largely to clinical microbiology but some applications have been made to foods and water. One of the earliest nonclinical uses was the detection of coliforms in water and sewage (174). These investigators showed that a direct relationship existed between coliform numbers and the amount of $^{14}CO_2$ produced by the organisms. The experimental detection of *S. aureus, S. typhimurium,* and spores of putrefactive anaerobe 3679 and *Clostridium botulinum* in beef loaf was studied by Previte (162). The inocula employed ranged from about 10^4-10^6/ml of medium, and the detection time ranged from 2 hr for *S. typhimurium* to 5–6 hr for *C. botulinum* spores. For these studies, 0.0139 μCi of ^{14}C-glucose/ml of tryptic soy broth was employed. In another study, Lampi et al. (125) found that 1 cell/ml of *S. typhimurium* or *S. aureus* could be detected by a radiometric method in 9 h. For 10^4 cells, 3 to 4 h were required. With respect to spores, a level of 90 of P.A. 3679 was detected in 11 h while 10^4 were detectable within 7 h. These and other investigators have shown that spores required 3–4 h longer for detection than vegetative cells. From the findings of Lampi et al., the radiometric detection procedure could be employed as a screening procedure for foods containing high numbers of organisms, for such foods produced results by this method within 5–6 h, while those with lower numbers required longer times.

The detection of nonfermenters of glucose by this method is possible when metabolites such as labeled formate and/or glutamate are used (163). These investigators further showed that a large number of food-borne organisms can be detected by this method in 1 to 6 h. The radiometric detection of 1 to 10 coliforms in water within 6 h was achieved by Bachrach and Bachrach (9) by employing ^{14}C-lactose with incubation at 37°C in a liquid medium. It is conceivable that a differentiation can be made between fecal *E. coli* and total coliforms by employing 45.5°C incubation along with 37°C incubation.

Radiometry has been used to detect organisms in frozen orange juice concentrate (86). The investigators used ^{14}C-glucose, 4 yeasts, and 4 lactic acid bacteria, and at an organism concentration of 10^4 cells, detection was achieved in 6–10 h. Of 600 juice samples examined, 44 with counts of 10^4/ml were detected in 12 h, and 41 of these in 8 h. No false negatives occurred and only two false positives were noted. The method was used for cooked foods to determine if counts were $< 10^5$ cfu/ml, and the results were compared to APC. Of 404 samples consisting of seven types of foods, around 75% were correctly classified as acceptable or unacceptable within 6 h (167). No more than five were incorrectly classified. The study employed ^{14}C-glucose, glutamic acid, and sodium formate.

A method for the detection of salmonellae in foods based on salmonellae poly H antibodies preventing salmonellae from producing $^{14}CO_2$ from (^{14}C)dulcitol has been developed (187). It was shown capable of detecting the organisms in 27 h compared to 4-5 days by culture methods. The agreement with culture methods on fifty-eight samples of food was 91%. The radiometric assay flask contained poly H serum and a control flask without poly H. Positive tests were determined by the difference in radioactivity of the two flasks following incubation at 37°C for 7 h. Results were achieved with an inoculum of 10^4–10^5 cells. The potential use of radiometry for field and surface samples is suggested by a procedure employed by Schrot et al. (172). By this method, membrane filter samples are collected and moistened with a small quantity of labeled medium in a closed container. The evolved $^{14}CO_2$ is trapped by $Ba(OH)_2$-moistened filter pads which are assayed later by a radioactivity counter.

Glucuronidase assay for *E. coli*

The enzyme β-glucuronidase (GUD) has been shown to be produced by 97% of *E. coli* strains tested (115). While up to 50% of salmonellae and shigellae produce the enzyme, with the exception of some *Bacteroides* spp. (36), other gram negatives apparently do not. In the presence of the substrate 4-methylumbelliferone-β-D-glucuronide (MUG), GUD produces a fluorogenic end product which is visible under black light.

As a rapid and more efficient way to detect *E. coli* in foods, Feng and Hartman (54) incorporated MUG into lauryl tryptose broth (LTB) and other coliform-selective media and found that in LTB-MUG, one *E. coli* cell could be detected in 20 h. While most positive reactions occurred in 4 h, some weak GUD-positive strains required up to 16 h for reaction. One important feature of this method is that fluorescence appears before gas production from lactose. The use of MUG in LTB was more effective than violet-red bile 2 agar in recovering heat- and chlorine-injured *E. coli* cells. The method lends itself to a microtitration assay plate method for MPN determinations. Of 120 *E. coli* strains tested by the above authors, 116 were GUD positive. That some salmonellae and shigellae are GUD positive does not invalidate the method since these organisms are of greater significance in foods than coliforms. Some *Corynebacterium* spp. have been shown to be GUD positive (36), but they normally do not develop on media selective for *E. coli*.

Employing the Feng and Hartman method using LTB-MUG, van Wart and Moberg (209) evaluated 1,020 specimens by a three-tube MPN and were able to detect more *E. coli*-positive samples than with a conventional MPN. The greater effectiveness of LTB-MUG resulted because some *E. coli* strains are anaerogenic. No false negative results were obtained. In another study, MPN enumeration of *E. coli* in seafoods was made by employing MUG in 3 media: lactose broth (LB-MUG), violet-red bile agar (VRB-MUG), and M-Endo broth (M-Endo-MUG). Specificities for the media were 96.5, 92.0, and 90%, respectively, for LB-MUG, VRB-MUG, and M-

Endo-MUG (3). This investigator indicated that either medium could be used for *E. coli* with results obtained in 24 h compared to 2 to 4 days by conventional coliform methods. With MUG added to lauryl sulfate broth (LSB-MUG) and compared to a conventional culture method for *E. coli* on 270 product samples, agreement between the two methods was 94.8% with 4.8% false positives but no false negatives with LSB-MUG (165). When fresh oysters were examined for *E. coli* by use of LSB-MUG, endogenous glucuronidase caused significant interference, which was eliminated when MUG was added to EC broth (117). By use of EC-MUG, *E. coli* was recovered from 102 of 103 fluorescing tubes. Of interest are the hemorrhagic colitis strains of *E. coli* that have been recovered from foods (see Chapter 20). An 0157:H7 isolate has been found to be nonfluorogenic in the MUG assay (40).

Employing the principle described above, a fluorogenic substrate (4-methylumbelliferone-α-D-galactoside) has been employed to differentiate species of fecal streptococci. The method employs dyed starch along with the substrate noted, both of which are added to a medium selective for fecal streptococci (130). By observing for starch hydrolysis and fluorescence, 86% of fecal streptococci from environmental samples were correctly differentiated.

Other chemical methods

Glutamate decarboxylase is found in *E. coli,* and on the assumption that this organism is the only bacterium containing this enzyme that is likely to be found in raw milk, its analysis forms the basis of a test for the organism in such products. Using an autoanalyzer, Moran and Witter (140) could detect as few as 50,000 *E. coli*/ml and the results were obtained in < 10 h.

β-galactosidase of fecal coliforms hydrolyzes *o*-nitrophenyl-β-D-galactoside (ONPG), and a rapid method for enumeration of these organisms in water based on this principle was proposed by Warren et al. (215). The organisms are collected on a 0.45-μm membrane, incubated in EC medium for 1 h, and filter-sterilized ONPG is added. Incubation is continued at 45.5°C until turbidity develops that can be read at 420 nm. The time for results depends upon the initial number of cells, with 1 cell being detected in < 20 h. Fecal coliforms other than *E. coli* are detected by the method. To detect a limit of 200 fecals or less/100 ml, no ONPG hydrolysis should be observed in < 11 h.

Felix-01 bacteriophages multiply in the presence of salmonellae. Based on this fact, a rapid method was developed for detecting < 5 salmonellae/ml of milk within 24 h of sample collection (92, 93). The cells are first removed by electropositive large-pore filters, exposing the filter eluate to Felix-01 bacteriophages, and analyzing the eluant-phage mixture for Felix-01 phages by HPLC (high performance liquid chromatography). The increase in phages was proportional to the number of salmonellae present.

When ^{35}S-labeled methionine is added to a methionine-free broth and inoculated with an isolated colony, the radiolabel is incorporated into certain proteins. Following separation of proteins on SDS-polyacrylamide gels (SDS-PAGE), the SDS-PAGE profiles are autoradiographed and used to characterize or identify species or strains. The ^{35}S-labeled protein bands can be scanned by a computer-assisted automated device and the determined patterns compared to known cultures. Results can be obtained in about 6 h following colony isolation, and the method has been employed successfully to characterize clinical strains of *Clostridium difficile* (195). This method may be viewed as a form of "fingerprinting" somewhat analogous to DNA-DNA hybridization and offers promise as a rapid method to identify food-borne pathogens.

A synopsis of other rapid methods for detecting microorganisms in foods has been presented by Goldschmidt and Fung (79).

IMMUNOLOGIC METHODS

Fluorescent antibody

This technique has had extensive use in clinical microbiology since its development in 1942, and from all indications it will continue to find application in food microbiology. In this technique, an antibody to a given antigen is made fluorescent by coupling it to a fluorescent compound, and when the antibody reacts with its antigen, the antigen-antibody complex emits fluorescence and can be detected by the use of a fluorescence microscope. The fluorescent markers used are rhodamine B, fluorescein isocyanate, and fluorescein isothiocyanate, with the latter being used most widely. The FA technique can be carried out by use of either of two basic methods. The direct method employs antigen and specific antibody to which is coupled the fluorescent compound (antigen coated by specific antibody with fluorescent label). With the indirect method, the homologous antibody is not coupled with the fluorescent label, but instead an antibody to this antibody is prepared and coupled (antigen coated by homologous antibody which is in turn coated by antibody to the homologous antibody bearing the fluorescent label). In the indirect method, the labeled compound detects the presence of the homologous antibody, while in the direct method it detects the presence of the antigen. The use of the indirect method eliminates the need to prepare FA for each organism of interest. The FA technique obviates the necessity of pure culture isolations of salmonellae if H antisera are employed. A commonly employed conjugate is polyvalent salmonellae OH globulin labeled with fluorescein isothiocyanate with somatic groups A to Z represented. Because of the cross-reactivity of salmonellae antisera with other closely related organisms (for example, *Arizona, Citrobacter, E. coli*), false positive results are to be expected when naturally contaminated foods are examined. The early history and development of the FA technique for clinical microbiology has been reviewed by Cherry and Moody (30), and for food applications by Ayres (8) and Goepfert and Insalata (77).

The application of FA to the examination of foods for organisms such as salmonellae is made after cultural enrichment procedures have been carried out. The general outline of a slide method is as follows. Smears are made on glass slides, air dried, fixed with an appropriate fixative such as Kirkpatrick's, rinsed with ethanol, and air dried. To the dried slides is added the conjugate (labeled antibody) followed by incubation in a moist chamber to allow antigen-antibody reaction to occur. The unbound conjugate is washed off, the slide rinsed in distilled water, air dried, and mounted with buffered saline (pH 7.5–9.5) containing glycerol. The smears are next examined under a fluorescence microscope equipped with appropriate exciter and barrier filters. Positive cells display a yellow-green fluorescence, and the degree of fluorescence is scored from 1+ to 4+ when manually done.

The first successful use of the FA technique for the detection of foodborne organisms was made by Russian workers who employed the technique to detect salmonellae in milk (7). The technique has been employed successfully to detect the presence of salmonellae in a large number of different types of foods and related products (Table 6-3). The procedure has been used also to monitor food-processing plant utensils and equipment. While most investigators have obtained some false positive results with FA, the incidence of false negatives has been lower. Of the false negatives obtained by Laramore and Moritz (126), 68% were from meatmeal samples. Eleven

Table 6-3. **The application of the fluorescent antibody (FA) technique to the detection of salmonellae in foods and related products.**

Foods examined	No. of samples	FA method used	% false negatives	% false positives	Reference
Meats	286	Indirect	0	14.7	72
Egg products	20	Indirect	0	0	84
Various foods (656 raw beef)	706	Direct	6.7	7	73
Dried foods and products	420	Indirect	0	0	180
Various inoculated foods	48	Direct	0	0	98
Various foods	3,991	Direct	<0.1	<0.2	173
Animal feed and ingredients	1,013	Direct	2.2	5.7	126
Various foods (39)	894	Direct	0	0	51
Various foods (7)	422	Indirect	0	7.3	78
Variety of human foods and animal feeds	65	Direct slide	0	4.6–14.8	99
Food products ca.	4,000	Direct	<1	7	52
Meatmeal	100	Auto direct	11.1	1.1	202
Powdered eggs and candy	201	Auto direct	0	5.3–6.6	202
Frog legs and dried products	283	Auto direct	0	4	146

percent of meatmeal samples examined by Thomason et al. (202) yielded false negative results. Just why this product is so conspicuous in this regard is not entirely clear at this time. With the exception of this product, the false negative findings have been extremely low, thus making FA an excellent screening test for salmonellae in foods. The adoption of the direct FA method in 1975 by AOAC as an official method was limited to its use as a screening method. The generally low percentage of false positive results is not significant in a screening test as long as false negative results can be avoided. The most obvious reason for false positive results is cross-reaction of salmonellae antibodies with related organisms as noted above.

The FA direct slide technique first described by Insalata et al. (99) has been adapted to an automated method (146, 202). By the automated method, slides are prepared by machine at the rate of 120/h and machine-read at the rate of 360/h. Both groups who have evaluated this procedure found that at least 10^6 salmonellae/ml of culture are necessary to obtain fluorescence readings within useful ranges. The automated procedure was found to correlate well with cultural methods on samples of milk powder, dried yeasts, and frog legs (146). While the method gave excellent results on samples of powdered eggs and chocolate candy, false negative results were obtained on meatmeal and a high rate of false positives on samples of poultry products and sausage (202). At least one group believes that the automated system shows potential for screening of samples for salmonellae and that all positives should be confirmed by manual methods (146).

A microcolony test was devised (200) and shown to possess several advantages over the direct FA test. Smears are easier to prepare, there is less fluorescence of background material, and sensitivity is greater. A semi-automated method based on the microcolony method was devised and shown to have an overall agreement of 89.6% with the standard culture techniques employing 144 samples of 67 foods, feeds, and environmental samples (11). However, 13.5% of raw meat products produced false negative results, which for all products was 3.5%. Only 1.3% false negatives and 5% false positive samples were found by Gibbs et al. (75) using the buffered peptone pre-enrichment broth proposed by the International Organization for Standardization. The FA method was compared with standard cultural methods on 546 samples including over 100 food samples.

Although the FA technique has had its widest food application in the detection of salmonellae, it has been used to trace bacteria associated with plants and to identify clostridia including *C. perfringens* and *C. botulinum*. It can be applied to any organism to which an antibody can be prepared. Among the many advantages of the procedure for detecting salmonellae in foods are the following: (1) FA is more sensitive than cultural methods; (2) FA is more rapid—time for results can be reduced from around 5 days to 18–24 hr; (3) reduction in time for negative tests on food products makes it possible to free foods earlier for distribution; (4) larger numbers of samples can be analyzed, making for increased sampling of food products; and (5) the qualitative determination of specific salmonellae serotypes is possible for FA if desired for epidemiologic reasons.

In spite of its drawbacks, the FA technique is official by AOAC and several other standard reference methods (see Chapter 5). It was designed to screen out salmonellae-negative samples, and all FA-positive samples should be confirmed by cultural and serological methods. It may be improved by use of IgG conjugates and by use of rhodamine-labeled normal rabbit serum as diluent for fluorescein-labeled salmonellae conjugates. Thomason (201) in a status report on FA noted that adequate supplies of reagents for the test are sometimes difficult to obtain.

Enrichment serology

The use of enrichment serology (ES) is a more rapid method for recovering salmonellae from foods than the conventional culture method (CCM). Originally developed by Sperber and Deibel (184), it is carried out in four steps: pre-enrichment in a nonselective medium for 18 h; selective enrichment in selenite-cystine and/or tetrathionate broth for 24 h; elective enrichment in M broth for either 6–8 h or 24 h; and agglutination with polyvalent H antisera at 50°C for 1 h. Results can be obtained in 50 h (depending upon elective enrichment time used) compared to 96–120 h by CCM. A modified ES method has been proposed involving a 6-h pre-enrichment, thus making it possible to obtain results in 32 h (192).

Employing the ES procedure on 105 samples of dried foods and feeds, all positive samples by CCM were positive by ES (184). Similar findings were achieved when 689 samples were examined by ES and CCM (53). The ES method was compared with three FA procedures for salmonellae detection in 347 food samples, and while 52 were positive by CCM, 51 were positive by ES (132). In the latter study, only one false negative was found by ES, and the authors preferred this method with a 6-h elective enrichment over CCM. In yet another comparison of ES with CCM using 2,208 samples, 95% of positive samples by CCM were positive by ES when elective enrichment was carried out for 24 h (21). ES was found to be as good as FA. The 6-h elective enrichment was unsuitable for comminuted meat products. When the ES and FA methods were combined, a greater accuracy was achieved than by use of either alone (88). Employing 126 samples, 66 were salmonellae-positive by CCM and FA, and 64 by ES. The latter exhibited a relatively low incidence of false positives while FA showed a higher incidence. The ES method was later shown capable of detecting 97–99% of salmonellae-positive samples (87). Using a modified ES as noted above on 3,486 samples, 96.7% were correctly analyzed (192). Overall, the ES method provides results in 32–50 h compared to 92–120 for CCM; results are comparable to both CCM and FA; and no specialized equipment or training is needed. Possible disadvantages to its use is the need for a minimum of about 10^7 cells/ml, and its lack of response to

nonmotile salmonellae. The latter can be overcome by use of a slide agglutination test from the elective enrichment broth employing polyvalent O
antiserum (184).

Radioimmunoassay

This technique consists of adding a radioactive label to an antigen, allowing
the labeled antigen to react with its specific antibody, and measuring the
amount of antigen that combined with the antibody by use of a counter to
measure radioactivity. Solid-phase RIA refers to methods that employ solid
materials or surfaces onto which a monolayer of antibody molecules binds
electrostatically. The solid materials used include polypropylene, polystyrene,
bromacetylcellulose, and so on. The ability of antibody-coated polymers
to bind specifically with radioactive tracer antigens is essential to the basic
principle of solid-phase RIA (29). When the free-labeled antigen is washed
out, the radioactivity measurements are quantitative. The label used by
many workers is ^{125}I.

Johnson et al. (108) developed a solid-phase RIA procedure for the determination of *S. aureus* enterotoxin B and found the procedure to be five
to twenty times more sensitive than the immunodiffusion technique. These
investigators found the sensitivity of the test to be in the 1–5 ng range
employing polystyrene and counting radioactivity with an integral counter.
Collins et al. (34) employed RIA for enterotoxin B with the concentrated
antibody coupled to bromacetylcellulose. Their findings indicated the procedure to be 100-fold more sensitive than immunodiffusion and to be reliable
at an enterotoxin level of 0.01 μg/ml. Staphylococcal enterotoxin A was
extracted from a variety of foods including ham, milk products, crab meat,
and so on, by Collins et al. (33) and measured by RIA all within 3–4 h.
They agreed with earlier workers that the method was highly sensitive
and useful to 0.001 μg/ml and quantitatively reliable to 0.01 μg/ml of enterotoxin A.

By iodination of enterotoxins, solid-phase RIA can be used to detect as
little as 1 ng toxin/g (17). When protein A was used as immunoabsorbent
to separate antigen-antibody complex from unreacted toxin, a sensitivity
of < 1.0 ng/g for staphylococcal enterotoxin A (SEA), SEB, SEC, SED,
and SEE was achieved within one working day (17, 137). In another study,
0.1 ng/ml of SEA and 0.5 ng/ml of SEB could be detected when protein
A was used (6). A sensitivity of 100 pg for SEC_2 was achieved by use of
a double-antibody RIA (164).

An RIA method has been developed for *E. coli* ST_a that detected 50 to
500 pg toxin/tube (74). The method measured ST both from human and
porcine strains and no cross-reactivity with ST_b was noted. Using solid-
phase sandwich RIA, colonies of *E. coli* that produced LT and colonies of
V. cholerae that produced cholera toxin (CT) could be differentiated from
nontoxigenic colonies (177). The method detected 5 to 25 pg of pure toxin

and was more sensitive than ELISA. A microtiter solid-phase RIA for LT was developed and found to be comparable in sensitivity to Y-1 adrenal cells (82).

For mycotoxins, an RIA method has been developed for ochratoxin A with a sensitivity of 20 ppb (1). A method has been developed for aflatoxin M_1 (AFM_1) in milk with a sensitivity of 5–50 ng/assay, but it was subject to interference by whole milk (159). An ELISA method detected as little as 0.25 ng AFM_1 in 3 h. The action level of the U.S. Food and Drug Administration for AFM_1 in milk is 0.5 ng/ml.

The RIA technique lends itself to the examination of foods for other biological hazards such as endotoxins, paralytic shellfish toxins, and the like. The detection and identification of bacterial cells within 8–10 min has been achieved (190) by use of ^{125}I-labeled homologous antibody filtered and washed on a Millipore membrane. Multibacterial species have been detected in one operation when mixtures of homologous antibodies were used (189). Indirect and direct methods may be used with the former requiring only one labeled globulin, while with the latter a labeled antibody to each organism in a mixed population is needed.

ELISA

The enzyme-linked immunosorbent assay (ELISA, enzyme immunoassay, or EIA) is an immunological method similar to RIA but employs an enzyme coupled to either antigen or antibody rather than a radioactive isotope. Essentially synonymous with ELISA are enzyme-multiplied immunoassay technique (EMIT), and indirect enzyme-linked antibody technique (ELAT). A typical ELISA is performed with a solid-phase (polystyrene) coated with antigen and incubated with antiserum. Following incubation and washing, an enzyme-labeled preparation of anti-immunoglobulin is added. After gentle washing, the enzyme remaining in the tube or microtiter well is assayed to determine the amount of specific antibodies in the initial serum. A commonly used enzyme is horseradish peroxidase, and its presence is measured by the addition of peroxidase substrate. The amount of enzyme present is ascertained by the colorimetric determination of enzyme substrate. Variations of this basic ELISA consist of a "sandwich" ELISA in which the antigen is required to have at least two binding sites. The antigen reacts first with excess solid-phase antibody, and following incubation and washing, the bound antigen is treated with excess labeled antibody. The "double sandwich" ELISA is a variation of the latter method and it employs a third antibody.

Among the applications of the ELISA technique in food microbiology is its use to detect staphylococcal enterotoxins, botulinal toxins, salmonellae, shigellae, mycotoxins, viruses, toxoplasma antibodies, and antibodies to gram-negative endotoxins. These are discussed further below.

STAPHYLOCOCCAL ENTEROTOXINS. When foods were spiked with staphylococcal enterotoxin A (SEA), from 72 to 90% was recovered from wieners, and 84 to 94% from mayonnaise (171). Using a double-antibody EIA, these investigators detected 0.4 ng of SEA in 20 h from extracts of wieners, 3.2 ng/ml in 1-3 h from milk, and 1.6 ng/ml from mayonnaise. By use of β-amylase coupled with SEB, Morita and Woodburn (141) detected < 5 ng/ml in food extracts using homogeneous EIA. By using polystyrene balls coated individually with antibody against SEA, SEB, and SEC, Stiffler-Rosenberg and Fey (188) detected 0.1 ng or less/ml when 20 ml of food extract was incubated with three balls followed by testing for uptake of toxin by competitive ELISA. With SEA coupled to alkaline phosphatase and an EIA procedure, Kauffman (112) detected as little as 2 ng/ml of food extract. Although the RIA procedure is more sensitive than ELISA for some metabolic products, Kuo and Silverman (119) found the latter to be as sensitive as RIA and noted that the sensitivity of ELISA can be decreased by food materials. From vanilla custard, SEB was detected at levels of 0.1 μg in 100 g using ELISA (23). In a study by Notermans et al. (151), ELISA detected < 0.5 μg of SEA, SEB, SEC, and SEE/100 g of minced meat product. It was noted that the extraction of staphylococcal enterotoxins was low in heated products due to inactivation by components such as gelatin.

As a means of simplifying ELISA and obviating the need for highly purified enterotoxin, Freed et al. (67) employed a solid-phase "double-antibody sandwich" method in which horseradish peroxidase was coupled to specific enterotoxin antibody (IgG). The solid-phase supports were polystyrene balls or microtiter plates. By use of the method with SEA, SEB, SEC, SED, and SEE added to a variety of foods, levels of $\geq$ 1 ng of enterotoxin/g of food could be detected and results could be obtained in one working day. Monoclonal antibodies have been prepared with the capacity to interact with SEA, SEB, SEC, SED, and SEE, and by use of a double-sandwich ELISA it is proposed as a screening method for food samples (135).

An ELISA method in combination with membrane filters has been devised for enterotoxigenic colonies of *S. aureus* (161). Organisms are grown on cellulose nitrate filters for 24 h, followed by incubation of filters with a fluorescein isothiocyanate-conjugated horseradish peroxidase-protein A conjugate. The latter step is completed in 3 h, thus allowing confirmation of enterotoxigenic strains within 27 h. The method was found to be sensitive to 500 pg of SEB.

BOTULINAL TOXINS. Using a "double-sandwich" technique, 50–100 mouse i.p. LD_{50} of *C. botulinum* type A toxin could be detected (151). No cross-reactions with types B, C, and E toxins were noted. In a later study employing the same technique, < 100 mouse i.p. LD_{50} of type E toxin were detected and no cross-reactivity was found between types A and B toxins (150). *C. botulinum* type G toxin was detected using a "double-sandwich" technique with alkaline phosphatase and polystyrene plates at levels of 1 mouse i.p. median lethal dose (129). Time required was 5.5–6.5 h, and no cross-reactions with *C. botulinum* types B, C, D, E, or F were noted, but specific and nonspecific reactions with *C. subterminale* and *C. botulinum* type A, respectively, did occur. Using a monoclonal antibody to type A toxin, an ELISA was developed with a sensitivity of about 9 mouse LD_{50}/ml (179). The assay was shown to be about ten times more sensitive to this toxin than other reported in vitro assays.

An ELISA has been developed for the detection of *C. perfringens* enterotoxin in individual colonies by use of a nitrocellulose colony-blot method (186). This application suggests that the method may be adapted to a variety of assay needs.

MYCOTOXINS. ELISA was employed by Lawellin et al. (127) to detect aflatoxin B_1 (AFB_1), and levels as low as < 10 pg/ml were detected. In another study, AFB_1 was detected at levels as low as 25 pg/assay using polystyrene microtiter plates (158). These authors found ELISA to be as good as or better than RIA, thin-layer chromatography, or HPLC. The lowest concentration reported detectable by ELISA was < 10 pg/ml (127) using a tube ELISA. For AFB_1 in corn and peanut butter, an ELISA with a sensitivity of 0.1 μg/ml has been developed (50). By use of nylon beads or Terasaki plates, AFB_1 could be detected at levels as low as 0.1 ng/ml, AFM_1 as low as 0.05 ng/ml, and T-2 toxin as low as 0.1 ng/ml (157). Using a monoclonal antibody in an ELISA, as little as 0.2 ng/ml of AFB_1 could be detected, with a working range of up to 10 ng/ml (26). T-2 toxin in corn was detected at a level of 0.05 ng/ml (69). Using a direct ELISA, Hu et al. (96) detected 10–25 pg/ml of AFM_1 in samples of urine and milk following pretreatment of samples with reversed-phase Sep-Pak cartridges. In another study, as little as 0.25 ng/ml of AFM_1 could be detected in milk (159).

Ochratoxin A (OA) produced by some aspergilli and penicillia was detected at levels as low as 25 pg/assay by ELISA in a study by Pestka et al. (160). For a review of immunoassays for mycotoxins, see (32).

SALMONELLAE AND OTHER GRAM-NEGATIVE BACTERIA. An EIA method for salmonellae was developed by Swaminathan and Ayres (194) employing a conjugate of IgG fraction of salmonellae polyvalent flagellar antibodies and horseradish peroxidase. When reacted with salmonellae and then treated with enzyme substrate, the cell wall and flagella were stained. The stained salmonellae could be differentiated from nonsalmonellae by light microscopy. A comparison of the above method with a culture method on 142 food samples showed overall agreement of 92.2%. False positives by the EIA technique amounted to 6.4%. When compared to FA, 95.8% agreement was found between the two methods. Using a similar method in which samples from enrichment cultures were first reacted with H-specific IgG followed by the addition of goat anti-rabbit antibody conjugated to alkaline phosphatase, Minnich et al. (138) found the method sensitive to about 10^5

salmonellae/ml, and advantages over the immunofluorescence and tube agglutination methods were noted. A direct EIA for salmonellae using polyclonal antibodies was developed in which the bacterial cells are attached to the walls of polystyrene microtitration plates employing a capture-antibody technique (4). The substrate used was 4-methylumbelliferyl-β-D-galactoside, and positive results are indicated by fluorescence under long-wave UV light. Results were obtained in 3 working days, and the sensitivity threshold was found to be 10^7 cells/ml (4).

One of the newest variations in the ELISA technique is the introduction of monoclonal antibodies. Employing immunoglobulin A monoclonal antibodies from a myeloma in which the myeloma protein binds to a flagellar determinant of salmonellae, Robison et al. (166) studied 100 salmonellae and found that 95% were detectable with the monoclonal antibody method. The method could detect as few as 10^6 cells/ml and the investigators suggested that results could be achieved in 36 h after culture initiation. It was successful in detecting salmonellae added to infant formula. A method highly specific for salmonellae was further developed employing monoclonal IgA antibodies attached to polycarbonate-coated metal beads (132). Selective enrichment is not required, and $< 10^6$ cells/ml can be detected with results obtained in 2 days. By contrast, the salmonellae agglutination test requires about 10^7 cells/ml for results. A commercially produced kit encompassing the Mattingly-Gehle procedure was evaluated with a variety of meat and poultry products by comparing it with two conventional culture methods (44). All samples positive by the culture methods were positive by the ELISA kit method, and results were obtained in 2–3 days.

The ELAT technique used by Krysinski and Heimsch (120) to detect *S. typhimurium* cells employed the use of cellulose acetate filter membranes that were spotted with broth-enriched samples. After the membranes were fixed, they were treated with an appropriate enzyme-conjugated antibody followed by incubation and then the enzyme substrate. While about 5×10^7 cells/ml are needed for detection by enrichment serology or FA, these authors found that only 10^4–10^5 cells were needed for results by ELAT.

With regard to the relative sensitivity of ELISA to salmonellae antibody, Carlsson et al. (27) found the procedure capable of measuring at least ten times lower concentrations than passive hemagglutination or quantitative precipitation methods. The method employed by these investigators required 26 to 28 h for results. Vos et al. (211) compared ELISA with passive HA for quantitative detection of antibodies to endotoxin and found ELISA to be two times more sensitive for LPS antibody than HA. In a similar manner, Keren (114) found that ELISA produced an antibody titer of 1:100,000 to LPS antigens of *Shigella flexneri* while passive hemagglutination gave a titer of only 1:640. These findings suggest that ELISA methods are capable of measuring equally lower levels of antigens. For more information on ELISA methods for salmonellae in foods, see Mattingly et al. (133) and Swaminathan et al. (193).

Gel diffusion

Gel diffusion methods have been widely used for the detection and quantitation of bacterial toxins and enterotoxins. The four most used are single-diffusion tube (Oudin), microslide double diffusion, micro-Ouchterlony slide, and electroimmunodiffusion. They have been employed to measure enterotoxins of staphylococci and *C. perfringens;* and the toxins of *C. botulinum.* The relative sensitivity of the various methods is presented in Table 6-1. Although they should be usable for any soluble protein to which an antibody can be made, they require that the antigen be in precipitable form. Perhaps the most widely used is the Crowle modification of the Ouchterlony slide test as modified by Casman and Bennett (28) and Bennett and McClure (15). The procedure for determining enterotoxins in foods is illustrated in Fig. 6-3, and further details on its use are presented in the *Bacteriological Analytical Manual* (10). The micro-Ouchterlony method can detect 0.1– 0.01 μg of staphylococcal enterotoxin, which is the same limit for the Oudin test. The double-diffusion tube test can detect levels as low as 0.1 μg/ml, but the incubation period required for such low levels is 3–6 days. This immunodiffusion method requires that extracts from a 100-g sample be concentrated to 0.2 ml (67). While other methods such as RIA and RPH are more sensitive and rapid than the gel diffusion methods, the latter continue to be widely used. Their reliability within their range of sensitivity is unquestioned.

Hemagglutination

While gel diffusion methods generally require at least 24 h for results, two comparable serologic methods yield results in 2–4 h—hemagglutination-inhibition (HI) and reverse passive hemagglutination (RPH). Unlike the gel diffusion methods, antigens are not required to be in precipitable form for these two tests.

In the HI test, specific antibody is kept constant and enterotoxin (antigen) is diluted out. Following incubation for about 20 min, treated sheep red blood cells (SRBCs) are added. Hemagglutination (HA) occurs only where antibody is not bound by antigen. HA is prevented (inhibited) where toxin is present in optimal proportions with antibody. The sensitivity of HI in detecting enterotoxins is noted in Table 6-1.

In contrast to HI, antitoxin globulin in RPH is attached directly to SRBCs and used to detect toxin. When diluted toxin preparations are added, the test is read for HA after incubation for 2 h. HA occurs only where optimal antigen-antibody levels occur. No HA occurs if no toxin or enterotoxin is present. The levels of two enterotoxins detected by RPH are indicated in Table 6-1.

Other immunologic methods

While the methods presented in this section have received only limited use for the analysis of food-borne pathogens, they are presented as examples of the magnitude of immunologic procedures.

100 g food + 500 ml 0.2M NaCl → Adjust to pH 7.5 → Centrifuge 10 min 32,800 xg

Concentrate to 15–20 ml dialysis in 30% PEG → Adjust to pH 7.5 → Centrifuge 10 min 32,800 xg

$CHCl_3$ extract and centrifuge → Add 40 vols. 0.005M PO_4 pH 5.7 → Percolate thru CM22-column (equil. 0.005M PO_4, pH 5.7)

Desorb enterotoxin 0.05M PO_4 NaCl pH 6.5 → Concentrate eluate (30% PEG) → $CHCl_3$ extract

Lyophilize → Rehydrate 0.15–0.2 ml (saline) → Serological assay for enterotoxin (microslide test)

Fig. 6-3. Schematic diagram for the extraction and serological assay of *Staphylococcus aureus* enterotoxin in food (16). (Copyright © 1980, Association of Official Analytical Chemists)

LYSIS INHIBITION TEST (LIT). This test has been employed to detect *E. coli* strains that produce heat-labile enterotoxin (LT). The test employs sheep erythrocytes, complement, anti-LT antibodies, and LT. Enterotoxin-sensitized SRBCs are hemolyzed in the presence of anti-LT antisera and complement. Immune hemolysis is inhibited if the anti-LT antisera is pre-exposed to soluble LT before the addition of LT-sensitized SRBCs. In testing seventy-five *E. coli* isolates that were positive by the Y-1 adrenal assay (see Chapter 7), all were positive by LIT (47). The test is run in microtiter plates and purified LT is required.

PASSIVE IMMUNE HEMOLYSIS (PIH). PIH is similar to LIT and has been employed also to detect *E. coli* strains that produce LT. This technique employs complement-mediated lysis of LT-sensitized SRBCs by antitoxin. PIH is more sensitive than LIT and does not require purified LT. Like LIT, the test is run in microtiter plates and has been shown to be sensitive to < 100 ng LT (48). While the PIH test correlated well with the Y-1 adrenal assay for human strains of LT-producing *E. coli,* correlation was poor with strains of nonhuman origin (176).

AGGREGATE HEMAGGLUTINATION. This test has been developed and used to detect enterotoxin of *B. cereus.* Antiserum to *B. cereus* enterotoxin was aggregated by use of glutaraldehyde and used to sensitize pretreated erythrocytes. Hemagglutination was carried out in microtiter plates and the sensitivity of the method for *B. cereus* enterotoxin was 4 ng/ml (80) while by gel immunodiffusion sensitivity was 2–3 μg/ml.

LATEX PARTICLE AGGLUTINATION TEST (LPAT). LPAT has been employed to detect LT-producing colonies of *E. coli.* In this method, LT-producing colonies effect the agglutination of latex particles pretreated with specific anti-LT serum. The sensitized latex is mixed with colony supernatant and the mixture is shaken for 3 min and observed for agglutination. LPAT results are obtained in minutes and the method has detected 8 μg/ml of LT with whole serum antibody, or 0.032 μg/ml using purified antibody. The test was found by Finkelstein and Yang (55) to be 16 to 1,000 times more sensitive than the Biken test (see below).

BIKEN TEST. This test is essentially a precipitin agar plate employing in one well bacterial colonies (*E. coli,* for example) lysed with polymyxin B to release LT and cholera enterotoxin (which cross-reacts with *E. coli* LT), and LT in an adjacent well. LT-producing colonies thus produce precipitin lines while negative colonies do not. In a study of this method by Honda et al. (94), 161 of 164 strains of *E. coli* gave similar results in the Biken test, by CHO assay, and by the PIH test. About 3 days are required for results. In a later study, Honda et al. (93) compared the Biken test using anti-LT with the CHO assay on 2,229 isolates of *E. coli* from travelers' diarrhea and found a 99.0% correlation between the two.

SINGLE RADIAL IMMUNODIFFUSION (SRD). This method was developed to screen large numbers of foods or cultures for enterotoxin-producing *S. aureus* strains. The method consists of adding polyvalent antiserum (types A, B, C, D, and E) to melted agar gel, which is layered onto the surface of microslides previously coated with 0.2% agar. Test preparations are added to wells and incubated in a moist chamber. The presence of enterotoxin is detected by a precipitin ring around the well. As employed by Meyer and Palmieri (136), the method could detect 0.3 μg/ml of the respective enterotoxins.

REFERENCES

1. Aalund, G., K. Brunfeldt, B. Hald, P. Krogh, and K. Poulsen. 1975. A radioimmunoassay for ochratoxin A: A preliminary investigation. *Acta Path. Microbiol. Scand. Sect. C.* 83:390–92.
2. Alexander, D. N., G. M. Ederer, and J. M. Jatsen. 1976. Evaluation of an adenosine 5′-triphosphate assay as a screening method to detect significant bacteriuria. *J. Clin. Microbiol.* 3:42–46.
3. Alvarez, R. J. 1984. Use of fluorogenic assays for the enumeration of *Escherichia coli* from selected seafoods. *J. Food Sci.* 49:1186–87, 1232.
4. Anderson, J. M., and P. A. Hartman. 1985. Direct immunoassay for detection of salmonellae in foods and feeds. *Appl. Environ. Microbiol.* 49:1124–27.
5. Anderson, R. L., A. K. Highsmith, and B. W. Holland. 1981. Comparison of standard pour plate, adenosine triphosphate, and *Limulus* amoebocyte lysate procedures for the detection of microbial contamination in intravenous fluids. *Bacteriol. Proc.* C-272.
6. Areson, P. W. D., S. E. Charm, and B. L. Wong. 1980. Determination of staphylococcal enterotoxins A and B in various food extracts, using staphylococcal cells containing protein. *J. Food Sci.* 45:400–401.
7. Arkhangel'skii, I. I., and V. M. Kartoshova. 1962. Accelerated methods of detecting *Salmonella* in milk. *Veterinariya* 9:74–78.
8. Ayres, J. C. 1967. Use of fluorescent antibody for the rapid detection of enteric organisms in egg, poultry and meat products. *Food Technol.* 21:631–40.
9. Bachrach, U., and Z. Bachrach. 1974. Radiometric method for the detection of coliform organisms in water. *Appl. Microbiol.* 28:169–71.
10. *Bacteriological analytical manual.* 1984. 6th ed. Washington, D.C.: AOAC.
11. Barrell, R. A. E., and A. M. Paton. 1979. A semi-automatic method for the detection of salmonellae in food products. *J. Appl. Bacteriol.* 46:155–59.
12. Batish, V. K., H. Chander, and B. Ranganathan. 1984. Incidence of enterococcal thermonuclease in milk and milk products. *J. Food Sci.* 49:1610–11, 1615.
13. Beezer, A. E., D. Newell, and H. J. V. Tyrrell. 1976. Application of flow microcalorimetry to analytical problems: The preparation, storage and assay of frozen inocula of *Saccharomyces cerevisiae*. *J. Appl. Bacteriol.* 41:197–207.
14. ———. 1978. Characterisation and metabolic studies of *Saccharomyces cerevisiae* and *Kluyveromyces fragilis* by flow microcalorimetry. *Antonie van Leeuwenhoek* 45:55–63.
15. Bennett, R. W., and F. McClure. 1976. Collaborative study of the serological identification of staphylococcal enterotoxins by the microslide gel double diffusion test. *J. Assoc. Off. Anal. Chem.* 59:594–600.

16. ———. 1980. Extraction and separation of staphylococcal enterotoxin in foods: Collaborative study. *J. Assoc. Off. Anal. Chem.* 63:1205–10.
17. Bergdoll, M. S., and R. Reiser. 1980. Application of radioimmunoassay for detection of staphylococcal enterotoxins in foods. *J. Food Protect.* 43:68–72.
18. Biermann, V. A., and G. Terplan. 1980. Nachweis von Aflatoxin B_1 mittels ELISA. *Arch. Lebensmittelhyg.* 31:51–57.
19. Bishop, J. R., C. H. White, and R. Firstenberg-Eden. 1984. Rapid impedimetric method for determining the potential shelflife of pasteurized whole milk. *J. Food Protect.* 47:471–75.
20. Boling, E. A., G. C. Blanchard, and W. J. Russell. 1973. Bacterial identifications by microcalorimetry. *Nature* 241:472–73.
21. Boothroyd, M., and A. C. Baird-Parker. 1973. The use of enrichment serology for *Salmonella* detection in human foods and animal feeds. *J. Appl. Bacteriol.* 36:165–72.
22. Bulte, M., and G. Reuter. 1984. Impedance measurement as a rapid method for the determination of the microbial contamination of meat surfaces, testing two different instruments. *Intern. J. Food Microbiol.* 1:113–25.
23. Büning-Pfaue, H., P. Timmermans, and S. Notermans. 1980. Einfache Methods für den Nachweis von Staphylokokken-Enteroxin-B in Vanillepudding mittels ELISA-Test. *Lebensm. Unters. Forsch.* 173:351–55.
24. Cady, P. 1975. Rapid automated bacterial identification by impedance measurements. In *New approaches to the identification of microorganisms*, ed. C.-G. Heden and T. Illeni, 73–99. New York: Wiley.
25. Cady, P., D. Hardy, S. Martins, S. W. Dufour, and S. J. Kraeger. 1978. Automated impedance measurements for rapid screening of milk microbial content. *J. Food Protect.* 41:277–83.
26. Candlish, A. A. G., W. H. Stimson, and J. E. Smith. 1985. A monoclonal antibody to aflatoxin B_1: Detection of the mycotoxin by enzyme immunoassay. *Lett. Appl. Microbiol.* 1:57–61.
27. Carlsson, H. E., A. A. Lindberg, and S. Hammarstrom. 1972. Titration of antibodies to *Salmonella* O antigens by enzyme-linked immunosorbent assay. *Infect. Immun.* 6:703–8.
28. Casman, E. P., and R. W. Bennett. 1965. Detection of staphylococcal enterotoxin in food. *Appl. Microbiol.* 13:181–89.
29. Catt, K., and G. W. Tregear. 1967. Solid-phase radioimmunoassay in antibody-coated tubes. *Science* 158:1570–72.
30. Cherry, W. B., and M. D. Moody. 1965. Fluorescent-antibody techniques in diagnostic bacteriol. *Bacteriol. Rev.* 29:222–50.
31. Chesbro, W. R., and K. Auborn. 1967. Enzymatic detection of the growth of *Staphylococcus aureus* in foods. *Appl. Microbiol.* 15:1150–59.
32. Chu, F. S. 1984. Immunoassays for analysis of mycotoxins. *J. Food Protect.* 47:562–69.
33. Collins, W. S., II, A. D. Johnson, J. F. Metzger, and R. W. Bennett. 1973. Rapid solid-phase radioimmunoassay for staphylococcal enterotoxin A. *Appl. Microbiol.* 25:774–77.
34. Collins, W. S., II, J. F. Metzger, and A. D. Johnson. 1972. A rapid solid phase radioimmunoassay for staphylococcal B enterotoxin. *J. Immunol.* 108:852–56.
35. Cords, B. R., and S. R. Tatini. 1973. Applicability of heat-stable deoxyribonuclease assay for assessment of staphylococcal growth and the likely presence of enterotoxin in cheese. *J. Dairy Sci.* 56:1512–19.

36. Dahlén, G., and A. Linde. 1973. Screening plate method for detection of bacterial β-glucuronidase. *Appl. Microbiol.* 26:863–66.
37. Dean, P. N., and D. Pinkel. 1978. High resolution dual laser flow cytometry. *J. Histochem. Cytochem.* 26:622–27.
38. Denhardt, D. T. 1966. A membrane-filter technique for the detection of complementary DNA. *Biochem. Biophys. Res. Comm.* 23:641–46.
39. Dodds, K. L., R. A. Holley, and A. G. Kempton. 1983. Evaluation of the catalase and *Limulus* amoebocyte lysate tests for rapid determination of the microbial quality of vacuum-packed cooked turkey. *Can. Inst. Food Sci. Technol. J.* 16:167–72.
40. Doyle, M. P., and J. L. Schoeni. 1984. Survival and growth characteristics of *Escherichia coli* associated with hemorrhagic colitis. *Appl. Environ. Microbiol.* 48:855–56.
41. Duncan, C. L., and E. B. Somers. 1972. Quantitation of *Clostridium perfringens* Type A enterotoxin by electroimmunodiffusion. *Appl. Microbiol.* 24:801–4.
42. Eden, G., and R. Eden. 1984. Enumeration of microorganisms by their dynamic AC conductance patterns. *IEEE Trans. Biomed. Engin.* BME 31:193–98.
43. El-Nakib, O., J. J. Pestka, and F. S. Chu. 1981. Determination of aflatoxin B_1 in corn, wheat, and peanut butter by enzyme-linked immunosorbent assay and solid phase radioimmunoassay. *J. Assoc. Off. Anal. Chem.* 64:1077–82.
44. Emswiler-Rose, B. S., W. D. Genle, R. W. Johnston, A. Okrend, A. Moran, and B. Bennett. 1984. An enzyme immunoassay technique for detection of salmonellae in meat and poultry products. *J. Food Sci.* 49:1018–20.
45. Emswiler-Rose, B. S., R. W. Johnston, M. E. Harris, and W. L. Lee. 1980. Rapid detection of staphylococcal thermonuclease on casings of naturally contaminated fermented sausages. *Appl. Environ. Microbiol.* 40:13–18.
46. Erickson, A., and R. H. Deibel. 1973. Turbidimetric assay of staphylococcal nuclease. *Appl. Microbiol.* 25:337–41.
47. Evans, D. J., Jr., and D. G. Evans. 1977. Inhibition of immune hemolysis: Serological assay for the heat-labile enterotoxin of *Escherichia coli*. *J. Clin. Microbiol.* 5:100–105.
48. ———. 1977. Direct serological assay for the heat-labile enterotoxin of *Escherichia coli*, using passive immune hemolysis. *Infect. Immun.* 16:604–9.
49. Evans, H. A. V. 1982. A note on two uses for impedimetry in brewing microbiology. *J. Appl. Bacteriol.* 53:423–26.
50. Fan, T. S. L., and F. S. Chu. 1984. Indirect enzyme-linked immunosorbent assay for detecting aflatoxin B_1 in corn and peanut butter. *J. Food Protect.* 47:263–66.
51. Fantasia, L. D. 1969. Accelerated immunofluorescence procedure for the detection of *Salmonella* in foods and animal by-products. *Appl. Microbiol.* 18:708–13.
52. Fantasia, L. D., J. P. Schrade, J. F. Yager, and D. Debler. 1975. Fluorescent antibody method for the detection of *Salmonella:* Development, evaluation, and collaborative study. *J. Assoc. Off. Anal. Chem.* 58:828–44.
53. Fantasia, L. D., W. H. Sperber, and R. H. Deibel. 1969. Comparison of two procedures for detection of *Salmonella* in food, feed, and pharmaceutical products. *Appl. Microbiol.* 17:540–41.
54. Feng, P. C. S., and P. A. Hartman. 1982. Fluorogenic assays for immediate confirmation of *Escherichia coli*. *Appl. Environ. Microbiol.* 43:1320–29.
55. Finkelstein, R. A., and Z. Yang. 1983. Rapid test for identification of heat-

labile enterotoxin-producing *Escherichia coli* colonies. *J. Clin. Microbiol.* 18:23–28.

56. Firstenberg-Eden, R. 1983. Rapid estimation of the number of microorganisms in raw meat by impedance measurement. *Food Technol.* 37(1):64–70.

57. ———. 1984. Collaborative study of the impedance method for examining raw milk samples. *J. Food Protect.* 47:707–12.

58. Firstenberg-Eden, R., and G. Eden. 1984. *Impedance microbiology.* New York: Wiley.

59. Firstenberg-Eden, R., and C. S. Klein. 1983. Evaluation of a rapid impedimetric procedure for the quantitative estimation of coliforms. *J. Food Sci.* 48:1307–11.

60. Firstenberg-Eden, R., and M. K. Tricarico. 1983. Impedimetric determination of total, mesophilic and psychrotrophic counts in raw milk. *J. Food. Sci.* 48:1750–54.

61. Firstenberg-Eden, R., M. L. Van Sise, J. Zindulis, and P. Kahn. 1984. Impedimetric estimation of coliforms in dairy products. *J. Food Sci.* 49:1449–52.

62. Firstenberg-Eden, R., and J. Zindulis. 1984. Electrochemical changes in media due to microbial growth. *J. Microbiol. Meth.* 2:103–13.

63. Fitts, R. 1985. Development of a DNA-DNA hybridization test for the presence of *Salmonella* in foods. *Food Technol.* 39(3):95–102.

64. Fitts, R., M. L. Diamond, C. Hamilton, and M. Nori. 1983. DNA-DNA hybridization assay for detection of *Salmonella* spp. in foods. *Appl. Environ. Microbiol.* 46:1146–51.

65. Flores, J., I. Perez, L. White, M. Perez, A. R. Kalica, R. Marquina, R. G. Wyatt, A. Z. Kapikian, and R. M. Chanock. 1982. Genetic relatedness among human rotaviruses as determined by RNA hybridization. *Infect. Immun.* 37:648–55.

66. Forrest, W. W. 1972. Microcalorimetry. *Meth. in Microbiol.* 6B:285–318.

67. Freed, R. C., M. L. Evenson, R. F. Reiser, and M. S. Bergdoll. 1982. Enzyme-linked immunosorbent assay for detection of staphylococcal enterotoxins in foods. *Appl. Environ. Microbiol.* 44:1349–55.

68. Fujita, T., P. R. Monk, and I. Wadso. 1978. Calorimetric identification of several strains of lactic acid bacteria. *J. Dairy Res.* 45:457–63.

69. Gendloff, E. H., J. J. Pestka, S. P. Swanson, and L. P. Hart. 1984. Detection of T-2 toxin in *Fusarium sporotrichioides*-infected corn by enzyme-linked immunosorbent assay. *Appl. Environ. Microbiol.* 47:1161–63.

70. Genigeorgis, C., and W. W. Sadler. 1966. Immunofluorescent detection of staphylococcal enterotoxin B. II. Detection in foods. *J. Food Sci.* 31:605–9.

71. Genigeorgis, C., G. Sakaguchi, and H. Riemann. 1973. Assay methods for *Clostridium perfringens* Type A enterotoxin. *Appl. Microbiol.* 26:111–15.

72. Georgala, D. L., and M. Boothroyd. 1964. A rapid immunofluorescence technique for detecting salmonellae in raw meat. *J. Hyg.* 62:319–26.

73. Georgala, D. L., M. Boothroyd, and P. R. Hayes. 1965. Further evaluation of a rapid immunofluorescence technique for detecting salmonellae in meat and poultry. *J. Appl. Bacteriol.* 28:421–25.

74. Giannella, R. A., K. W. Drake, and M. Luttrell. 1981. Development of a radioimmunoassay for *Escherichia coli* heat-stable enterotoxin: Comparison with the suckling mouse bioassay. *Infect. Immun.* 33:186–92.

75. Gibbs, P. A., J. T. Patterson, and J. Early. 1979. A comparison of the fluorescent

antibody method and a standardized cultural method for the detection of salmonellae. *J. Appl. Bacteriol.* 46:501–5.

76. Gnan, S., and L. O. Luedecke. 1982. Impedance measurements in raw milk as an alternative to the standard plate count. *J. Food Protect.* 45:4–7.

77. Goepfert, J. M., and N. F. Insalata. 1969. Salmonellae and the fluorescent antibody technique: A current evaluation. *J. Milk Food Technol.* 32:465–73.

78. Goepfert, J. M., M. E. Mann, and R. Hicks. 1970. One-day fluorescent-antibody procedure for detecting salmonellae in frozen and dried foods. *Appl. Microbiol.* 20:977–83.

79. Goldschmidt, M. C., and D. Y. C. Fung. 1978. New methods for microbiological analysis of food. *J. Food Protect.* 41:201–19.

80. Gorina, L. G., F. S. Fluer, A. M. Olovnikov, and Yu. V. Ezepcuk. 1975. Use of the aggregate-agglutination technique for determining exo-enterotoxin of *Bacillus cereus. Appl. Environ. Microbiol.* 29:201–4.

81. Gram, L., and H. Søgaard. 1985. Microcalorimetry as a rapid method for estimation of bacterial levels in ground meat. *J. Food Protect.* 48:341–45.

82. Greenberg, H. B., D. A. Sack, W. Rodriguez, R. B. Sack, R. G. Wyatt, A. R. Kalica, R. L. Horswood, R. M. Chanock, and A. Z. Kapikian. 1977. Microtiter solid-phase radioimmunoassay for detection of *Escherichia coli* heat-labile enterotoxin. *Infect. Immun.* 17:541–45.

83. Griffiths, M. W., and J. D. Phillips. 1984. Detection of post-pasteurization contamination of cream by impedimetric methods. *J. Appl. Bacteriol.* 57:107–14.

84. Haglund, J. R., J. C. Ayres, A. M. Paton, A. A. Kraft, and L. Y. Quinn. 1964. Detection of *Salmonella* in eggs and egg products with fluorescent antibody. *Appl. Microbiol.* 12:447–50.

85. Hardy, D., S. W. Dufour, and S. J. Kraeger. 1975. Rapid detection of frozen food bacteria by automated impedance measurements. *Proceedings, Institute of Food Technologists.*

86. Hatcher, W. S., S. DiBenedetto, L. E. Taylor, and D. I. Murdock. 1977. Radiometric analysis of frozen concentrated orange juice for total viable microorganisms. *J. Food Sci.* 42:636–39.

87. Hilker, J. S. 1975. Enrichment serology and fluorescent antibody procedures to detect salmonellae in foods. *J. Milk Food Technol.* 38:227–31.

88. Hilker, J. S., and M. Solberg. 1973. Evaluation of a fluorescent antibody-enrichment serology combination procedure for the detection of salmonellae in condiments, food products, food byproducts, and animal feeds. *Appl. Microbiol.* 26:751–56.

89. Hill, W. E. 1981. DNA hybridization method for detecting enterotoxigenic *Escherichia coli* in human isolates and its possible application to food samples. *J. Food Safety* 3:233–47.

90. Hill, W. E., J. M. Madden, B. A. McCardell, D. B. Shah, J. A. Jagow, W. L. Payne, and B. K. Boutin. 1983. Foodborne enterotoxigenic *Escherichia coli:* Detection and enumeration by DNA colony hybridization. *Appl. Environ. Microbiol.* 45:1324–30.

91. Hirsh, D. C., and L. D. Martin. 1983. Detection of *Salmonella* spp. in milk by using Felix-01 bacteriophage and high-pressure liquid chromatography. *Appl. Environ. Microbiol.* 46:1243–45.

92. ———. 1984. Rapid detection of *Salmonella* in certified raw milk by using charge-modified filters and Felix-01 bacteriophage. *J. Food Protect.* 47:388–90.

93. Honda, T., M. Arita, Y. Takeda, and T. Miwatani. 1982. Further evaluation of the Biken test (modified Elek test) for detection of enterotoxigenic *Escherichia coli* producing heat-labile enterotoxin and application of the test to sampling of heat-stable enterotoxin. *J. Clin. Microbiol.* 16:60–62.

94. Honda, T., S. Taga, T. Takeda, and T. Miwatani. 1981. Modified Elek test for detection of heat-labile enterotoxin of enterotoxigenic *Escherichia coli*. *J. Clin. Microbiol.* 13:1–5.

95. Horan, P. K., and L. L. Wheeless, Jr. 1977. Quantitative single cell analysis and sorting. *Science* 198:149–57.

96. Hu, W. J., N. Woychik, and F. S. Chu. 1984. ELISA of picogram quantities of aflatoxin M_1 in urine and milk. *J. Food Protect.* 47:126–27.

97. Hutter, K.-J., and H. E. Eipel. 1979. Microbial determinations by flow cytometry. *J. Gen. Microbiol.* 113:369–75.

98. Hutter, K.-J., M. Stöhr, and H. E. Eipel. 1980. Simultaneous DNA and protein measurements of microorganisms. In *Flow cytometry,* Vol. 4, eds. O. D. Laerum, T. Lindmo, and E. Thorud, 100–102. Bergen: Universitetsforlaget.

99. Insalata, N. F., C. W. Mahnke, and W. G. Dunlap. 1972. Rapid, direct fluorescent-antibody method for the detection of salmonellae in food and feeds. *Appl. Microbiol.* 24:645–49.

100. Jaksch, V. P., K.-J. Zaadhof, and G. Terplan. 1982. Zur Bewertung der hygienischen Qualität von Milchprodukten mit dem *Limulus*-Test. *Molkerei-Zeitung Welt der Milch* 36:5–8.

101. Jarvis, A. W. 1984. Differentiation of lactic streptococcal phages into phage species by DNA-DNA homology. *Appl. Environ. Microbiol.* 47:343–49.

102. Jarvis, A. W., and B. D. W. Jarvis. 1981. Deoxyribonucleic acid homology among lactic streptococci. *Appl. Environ. Microbiol.* 41:77–83.

103. Jay, J. M. 1974. Use of the *Limulus* lysate endotoxin test to assess the microbial quality of ground beef. *Bacteriol. Proc.,* 13.

104. Jay, J. M. 1977. The *Limulus* lysate endotoxin assay as a test of microbial quality of ground beef. *J. Appl. Bacteriol.* 43:99–109.

105. Jay, J. M. 1981. Rapid estimation of microbial numbers in fresh ground beef by use of the *Limulus* test. *J. Food Protect.* 44:275–78.

106. Jay, J. M., and S. Margitic. 1979. Comparison of homogenizing, shaking, and blending on the recovery of microorganisms and endotoxins from fresh and frozen ground beef as assessed by plate counts and the *Limulus* amoebocyte lysate test. *Appl. Environ. Microbiol.* 38:879–84.

107. Jay, J. M., S. Margitic, A. L. Shereda, and H. V. Covington. 1979. Determining endotoxin content of ground beef by the *Limulus* amoebocyte lysate test as a rapid indicator of microbial quality. *Appl. Environ. Microbiol.* 38:885–90.

108. Johnson, H. M., J. A. Bukovic, P. E. Kauffman, and J. T. Peeler. 1971. Staphylococcal enterotoxin B: Solid-phase radioimmunoassay. *Appl. Microbiol.* 22:837–41.

109. Johnson, H. M., H. E. Hall, and M. Simon. 1967. Enterotoxin B: Serological assay in cultures by passive hemagglutination. *Appl. Microbiol.* 15:815–18.

110. Kamman, J. F., and S. R. Tatini. 1977. Optimal conditions for assay of staphylococcal nuclease. *J. Food Sci.* 42:421–24.

111. Karl, D. M. 1980. Cellular nucleotide measurements and applications in microbial ecology. *Microbiol. Rev.* 44:739–96.
112. Kauffman, P. E. 1980. Enzyme immunoassay for staphylococcal enterotoxin A. *J. Assoc. Off. Anal. Chem.* 63:1138–43.
113. Kennedy, J. E., Jr., and J. L. Oblinger. 1985. Application of bioluminescence to rapid determination of microbial levels in ground beef. *J. Food Protect.* 48:334–40.
114. Keren, D. F. 1979. Enzyme-linked immunosorbent assay for immunoglobulin G and immunoglobulin A antibodies to *Shigella flexneri* antigens. *Infect. Immun.* 24:441–48.
115. Killian, M., and P. Bülow. 1976. Rapid diagnosis of *Enterobacteriaceae*. I. Detection of bacterial glycosidases. *Acta Pathol. Microbiol. Scand. Sect. B* 84:245–51.
116. Kimmich, G. A., J. Randles, and J. S. Brand. 1975. Assay of picomole amounts of ATP, ADP, and AMP using the luciferase enzyme system. *Anal. Biochem.* 69:187–206.
117. Koburger, J. A., and M. L. Miller. 1985. Evaluation of a fluorogenic MPN procedure for determining *Escherichia coli* in oysters. *J. Food Protect.* 48:244–45.
118. Koupal, A., and R. H. Deibel. 1978. Rapid qualitative method for detecting staphylococcal nuclease in foods. *Appl. Environ. Microbiol.* 35:1193–97.
119. Kuo, J. K. S., and G. J. Silverman. 1980. Application of enzyme-linked immunosorbent assay for detection of staphylococcal enterotoxins in food. *J. Food Protect.* 43:404–7.
120. Krysinski, E. P., and R. C. Heimsch. 1977. Use of enzyme-labeled antibodies to detect *Salmonella* in foods. *Appl. Environ. Microbiol.* 33:947–54.
121. Lachica, R. V. 1980. Accelerated procedure for the enumeration and identification of food-borne *Staphylococcus aureus*. *Appl. Environ. Microbiol.* 39:17–19.
122. Lachica, R. V. F., C. Genigeorgis, and P. D. Hoeprich. 1971. Metachromatic agar-diffusion methods for detecting staphylococcal nuclease activity. *Appl. Microbiol.* 21:585–87.
123. Lachica, R. V. F., P. D. Hoeprich, and C. Genigeorgis. 1972. Metachromatic agar-diffusion microslide technique for detecting staphylococcal nuclease in foods. *Appl. Microbiol.* 23:168–69.
124. Lachica, R. V., K. F. Weiss, and R. H. Deibel. 1969. Relationships among coagulase, enterotoxin, and heat-stable deoxyribonuclease production by *Staphylococcus aureus*. *Appl. Microbiol.* 18:126–27.
125. Lampi, R. A., D. A. Mikelson, D. B. Rowley, J. J. Previte, and R. E. Wells. 1974. Radiometry and microcalorimetry—techniques for the rapid detection of foodborne microorganisms. *Food Technol.* 28(10):52–55.
126. Laramore, C. R., and C. W. Moritz. 1969. Fluorescent-antibody technique in detection of salmonellae in animal feed and feed ingredients. *Appl. Microbiol.* 17:352–54.
127. Lawellin, D. W., D. W. Grant, and B. K. Joyce. 1977. Enzyme-linked immunosorbent analysis for aflatoxin B$_1$. *Appl. Environ. Microbiol.* 34:94–96.
128. Levin, G. V., V. R. Harrison, and W. C. Hess. 1956. Preliminary report on a one-hour presumptive test for coliform organisms. *J. Amer. Water Works Assoc.* 48:75–80.

129. Lewis, G. E., Jr., S. S. Kulinski, D. W. Reichard, and J. F. Metzger. 1981. Detection of *Clostridium botulinum* Type G toxin by enzyme-linked immunosorbent assay. *Appl. Environ. Microbiol.* 42:1018–22.

130. Littel, K. J., and P. A. Hartman. 1983. Fluorogenic selective and differential medium for isolation of fecal streptococci. *Appl. Environ. Microbiol.* 45:622–27.

131. Martins, S. B., and M. J. Selby. 1980. Evaluation of a rapid method for the quantitative estimation of coliforms in meat by impedimetric procedures. *Appl. Environ. Microbiol.* 39:518–24.

132. Mattingly, J. A., and W. D. Gehle. 1984. An improved enzyme immunoassay for the detection of *Salmonella*. *J. Food Sci.* 49:807–9.

133. Mattingly, J. A., B. J. Robison, A. Boehm, and W. D. Gehle. 1985. Use of monoclonal antibodies for the detection of *Salmonella* in foods. *Food Technol.* 39(3):90–94.

134. Mendelsohn, M. L. 1980. The attributes and applications of flow cytometry. In *Flow cytometry*, Vol. 4, eds. O. D. Laerum, T. Lindmo, and E. Thorud, 15–27. Bergen: Universitetsforlaget.

135. Meyer, R. F., L. Miller, R. W. Bennett, and J. D. MacMillan. 1984. Development of a monoclonal antibody capable of interacting with five serotypes of *Staphylococcus aureus* enterotoxin. *Appl. Environ. Microbiol.* 47:283–87.

136. Meyer, R. F., and M. J. Palmieri. 1980. Single radial immunodiffusion method for screening staphylococcal isolates for enterotoxin. *Appl. Environ. Microbiol.* 40:1080–85.

137. Miller, B. A., R. F. Reiser, and M. S. Bergdoll. 1978. Detection of staphylococcal enterotoxins A, B, C, D, and E in foods by radioimmunoassay, using staphylococcal cells containing protein A as immunoadsorbent. *Appl. Environ. Microbiol.* 36:421–26.

138. Minnich, S. A., P. A. Hartman, and R. C. Heimsch. 1982. Enzyme immunoassay for detection of salmonellae in foods. *Appl. Environ. Microbiol.* 43:877–83.

139. Mohr, H. K., H. L. Trenk, and M. Yeterian. 1974. Comparison of fluorescent-antibody methods and enrichment serology for the detection of *Salmonella*. *Appl. Microbiol.* 27:324–28.

140. Moran, J. W., and L. D. Witter. 1976. An automated rapid test for *Escherichia coli* in milk. *J. Food Sci.* 41:165–67.

141. Morita, T. N., and M. J. Woodburn. 1978. Homogeneous enzyme immune assay for staphylococcal enterotoxin B. *Infect. Immun.* 21:666–68.

142. Moseley, S. L., P. Echeverria, J. Seriwatana, C. Tirapat, W. Chaicumpa, T. Sakuldaipeara, and S. Falkow. 1982. Identification of enterotoxigenic *Escherichia coli* by colony hybridization using three enterotoxin gene probes. *J. Infect. Dis.* 145:863–69.

143. Moseley, S. L., and S. Falkow. 1980. Nucleotide sequence homology between the heat-labile enterotoxin gene of *Escherichia coli* and *Vibrio cholerae* deoxyribonucleic acid. *J. Bacteriol.* 144:444–46.

144. Moseley, S. L., I. Huq, A. R. M. A. Alim, M. So, S. M. Samadpour-Motalebi, and S. Falkow. 1980. Detection of enterotoxigenic *Escherichia coli* by DNA colony hybridization. *J. Infect. Dis.* 142:892–98.

145. Muldrow, L. L., R. L. Tyndall, and C. B. Fliermans. 1982. Application of flow cytometry to studies of pathogenic free-living amoebas. *Appl. Environ. Microbiol.* 44:1258–69.

146. Munson, T. E., J. P. Schrade, N. B. Bisciello, Jr., L. D. Fantasia, W. H. Hartung, and J. J. O'Connor. 1976. Evaluation of an automated fluorescent antibody procedure for detection of *Salmonella* in foods and feeds. *Appl. Environ. Microbiol.* 31:514–21.

147. Niskanen, A., and L. Koiranen. 1977. Correlation of enterotoxin and thermonuclease production with some physiological and biochemical properties of staphylococcal strains isolated from different sources. *J. Food Protect.* 40:543–48.

148. Niskanen, A., and E. Nurmi. 1976. Effect of starter culture on staphylococcal enterotoxin and thermonuclease production in dry sausage. *Appl. Environ. Microbiol.* 31:11–20.

149. Niyomvit, N., K. E. Stevenson, and R. F. McFeeters. 1978. Detection of staphylococcal enterotoxin B by affinity radioimmunoassay. *J. Food Sci.* 43:735–39.

150. Notermans, S., J. Dufrenne, and S. Kozaki. 1979. Enzyme-linked immunosorbent assay for detection of *Clostridium botulinum* type E toxin. *Appl. Environ. Microbiol.* 37:1173–75.

151. Notermans, S., J. Dufrenne, and M. van Schothorst. 1978. Enzyme-linked immunosorbent assay for detection of *Clostridium botulinum* toxin type A. *Japan. J. Med. Sci. Biol.* 31:81–85.

152. Nuzback, D. E., E. E. Bartley, S. M. Dennis, T. G. Nagaraja, S. J. Galitzer, and A. D. Dayton. 1983. Relation of rumen ATP concentration to bacterial and protozoal numbers. *Appl. Environ. Microbiol.* 46:533–38.

153. Park, C. E., H. B. El Derea, and M. K. Rayman. 1978. Evaluation of staphylococcal thermonuclease (TNase) assay as a means of screening foods for growth of staphylococci and possible enterotoxin production. *Can. J. Microbiol.* 24:1135–39.

154. Patterson, J. W., P. L. Brezonik, and H. D. Putnam. 1970. Measurement and significance of adenosine triphosphate in activated sludge. *Environ. Sci. & Technol.* 4:569–75.

155. Perry, B. F., A. E. Beezer, and R. J. Miles. 1979. Flow microcalorimetric studies of yeast growth: Fundamental aspects. *J. Appl. Bacteriol.* 47:527–37.

156. ———. 1983. Characterization of commercial yeast strains by flow microcalorimetry. *J. Appl. Bacteriol.* 54:183–89.

157. Pestka, J. J., and F. S. Chu. 1984. Enzyme-linked immunosorbent assay of mycotoxins using nylon bead and Terasaki plate solid phases. *J. Food Protect.* 47:305–8.

158. Pestka, J. J., P. K. Gaur, and F. S. Chu. 1980. Quantitation of aflatoxin B_1 and aflatoxin B_1 antibody by an enzyme-linked immunosorbent microassay. *Appl. Environ. Microbiol.* 40:1027–31.

159. Pestka, J. J., V. Li, W. O. Harder, and F. S. Chu. 1981. Comparison of radioimmunoassay and enzyme-linked immunosorbent assay for determining aflatoxin M_1 in milk. *J. Assoc. Off. Anal. Chem.* 65:294–301.

160. Pestka, J. J., B. W. Steinert, and F. S. Chu. 1981. Enzyme-linked immunosorbent assay for detection of ochratoxin A. *Appl. Environ. Microbiol.* 41:1472–74.

161. Peterkin, P. I., and A. N. Sharpe. 1984. Rapid enumeration of *Staphylococcus aureus* in foods by direct demonstration of enterotoxigenic colonies on membrane filters by enzyme immunoassay. *Appl. Environ. Microbiol.* 47:1047–53.

162. Previte, J. J. 1972. Radiometric detection of some food-borne bacteria. *Appl. Microbiol.* 24:535–39.

163. Previte, J. J., D. B. Rowley, and R. Wells. 1975. Improvements in a non-proprietary radiometric medium to allow the detection of some *Pseudomonas* species and *Alcaligenes faecalis*. *Appl. Microbiol.* 30:339–40.

164. Robern, H., M. Dighton, Y. Yano, and N. Dickie. 1975. Double-antibody radioimmunoassay for staphylococcal enterotoxin C_2. *Appl. Microbiol.* 30:525–29.

165. Robison, B. J. 1984. Evaluation of a fluorogenic assay for detection of *Escherichia coli* in foods. *Appl. Environ. Microbiol.* 48:285–88.

166. Robison, B. J., C. I. Pretzman, and J. A. Mattingly. 1983. Enzyme immunoassay in which a myeloma protein is used for detection of salmonellae. *Appl. Environ. Microbiol.* 45:1816–21.

167. Rowley, D. B., J. J. Previte, and H. P. Srinivasa. 1978. A radiometric method for rapid screening of cooked foods for microbial acceptability. *J. Food Sci.* 43:1720–22.

168. Rowley, D. B., P. Vandemark, D. Johnson, and E. Shattuck. 1979. Resuscitation of stressed fecal coliforms and their subsequent detection by radiometric and impedance techniques. *J. Food Protect.* 42:335–41.

169. Russell, W. J., J. F. Zettler, G. C. Blanchard, and E. A. Boling. 1975. Bacterial identification by microcalorimetry. In *New approaches to the identification of microorganisms,* ed. C.-G. Hedén and T. Illéni, 101–21. New York: Wiley.

170. Sacks, L. E., and E. Menefee. 1972. Thermal detection of spoilage in canned foods. *J. Food Sci.* 37:928–31.

171. Saunders, G. C., and M. L. Bartlett. 1977. Double-antibody solid-phase enzyme immunoassay for the detection of staphylococcal enterotoxin A. *Appl. Environ. Microbiol.* 34:518–22.

172. Schrot, J. R., W. C. Hess, and G. V. Levin. 1973. Method for radiorespirometric detection of bacteria in pure culture and in blood. *Appl. Microbiol.* 26:867–73.

173. Schultz, S. J., J. S. Witzeman, and W. M. Hall. 1968. Immunofluorescent screening for *Salmonella* in foods: Comparison with cultural methods. *J. Assoc. Off. Anal. Chem.* 51:1334–38.

174. Scott, R. M., D. Seiz, and H. J. Shaughnessy. 1964. I. Rapid carbon[14] test for coliform bacteria in water. II. Rapid carbon[14] test for sewage bacteria. *Amer. J. Pub. Hlth* 54:827–44.

175. Seiter, J. A., and J. M. Jay. 1980. Comparison of direct serial dilution and most-probable-number methods for determining endotoxins in meats by the *Limulus* amoebocyte lysate test. *Appl. Environ. Microbiol.* 40:177–78.

176. Serafim, M. B., A. F. Pestana de Castro, M. H. Lemos Dos Reis, and L. R. Trabulsi. 1979. Passive immune hemolysis for detection of heat-labile enterotoxin produced by *Escherichia coli* isolated from different sources. *Infect. Immun.* 24:606–10.

177. Shah, D. B., P. E. Kauffman, B. K. Boutin, and C. H. Johnson. 1982. Detection of heat-labile-enterotoxin-producing colonies of *Escherichia coli* and *Vibrio cholerae* by solid-phase sandwich radioimmunoassays. *J. Clin. Microbiol.* 16:504–8.

178. Sharpe, A. N., M. N. Woodrow, and A. K. Jackson. 1970. Adenosinetriphosphate

(ATP) levels in foods contaminated by bacteria. *J. Appl. Bacteriol.* 33:758–67.

179. Shone, C., P. Wilton-Smith, N. Appleton, P. Hambleton, N. Modi, S. Gatley, and J. Melling. 1985. Monoclonal antibody-based immunoassay for type A *Clostridium botulinum* toxin is comparable to the mouse bioassay. *Appl. Environ. Microbiol.* 50:63–67.

180. Silliker, J. H., A. Schmall, and J. Y. Chiu. 1966. The fluorescent antibody technique as a means of detecting salmonellae in foods. *J. Food Sci.* 31:240–44.

181. Silverman, M. P., and E. F. Munoz. 1979. Automated electrical impedance technique for rapid enumeration of fecal coliforms in effluents from sewage treatment plants. *Appl. Environ. Microbiol.* 37:521–26.

182. Silverman, S. J., A. R. Knott, and M. Howard. 1968. Rapid, sensitive assay for staphylococcal enterotoxin and a comparison of serological methods. *Appl. Microbiol.* 16:1019–23.

183. Southern, E. M. 1975. Detection of specific sequences among DNA fragments separated by gel electrophoresis. *J. Mol. Biol.* 98:503–17.

184. Sperber, W. H., and R. H. Deibel. 1969. Accelerated procedure for *Salmonella* detection in dried foods and feeds involving only broth cultures and serological reactions. *Appl. Microbiol.* 17:533–39.

185. Stannard, C. J., and J. M. Wood. 1983. The rapid estimation of microbial contamination of raw meat by measurement of adenosine triphosphate (ATP). *J. Appl. Bacteriol.* 55:429–38.

186. Stelma, G. N., Jr., C. H. Johnson, and D. B. Shah. 1985. Detection of enterotoxin in colonies of *Clostridium perfringens* by a solid phase enzyme-linked immunosorbent assay. *J. Food Protect.* 48:227–31.

187. Stewart, B. J., M. J. Eyles, and W. G. Murrell. 1980. Rapid radiometric method for detection of *Salmonella* in foods. *Appl. Environ. Microbiol.* 40:223–30.

188. Stiffler-Rosenberg, G., and H. Fey. 1978. Simple assay for staphylococcal enterotoxins A, B, and D: Modification of enzyme-linked immunosorbent assay. *J. Clin. Microbiol.* 8:473–79.

189. Strange, R. E., and K. L. Martin. 1972. Rapid assays for the detection and determination of sparse populations of bacteria and bacteriophage T7 with radioactively labelled homologous antibodies. *J. Gen. Microbiol.* 72:127–41.

190. Strange, R. E., E. O. Powell, and T. W. Pearce. 1971. The rapid detection and determination of sparse bacterial populations with radioactively labelled homologous antibodies. *J. Gen. Microbiol.* 67:349–57.

191. Sullivan, J. D., Jr., P. C. Ellis, R. G. Lee, W. S. Combs, Jr., and S. W. Watson. 1983. Comparison of the *Limulus* amoebocyte lysate test with plate counts and chemical analyses for assessment of the quality of lean fish. *Appl. Environ. Microbiol.* 45:720–22.

192. Surdy, T. E., and S. G. Haas. 1981. Modified enrichment-serology procedure for detection of salmonellae in soy products. *Appl. Environ. Microbiol.* 42:704–7.

193. Swaminathan, B., J. A. G. Aleixo, and S. A. Minnich. 1985. Enzyme immunoassays for *Salmonella:* One-day testing is now a reality. *Food Technol.* 39(3):83–89.

194. Swaminathan, B., and J. C. Ayres. 1980. A direct immunoenzyme method for the detection of salmonellae in foods. *J. Food Sci.* 45:352–55, 361.

195. Tabaqchali, S., D. Holland, S. O'Farrell, and R. Silman. 1984. Typing scheme for *Clostridium difficile:* Its application in clinical and epidemiological studies. *Lancet* 1:935–37.

196. Tai, J. Y., and T.-Y. Liu. 1977. Studies on *Limulus* amoebocyte lysate. Isolation of pro-clotting enzyme. *J. Biol. Chem.* 252:2178–81.

197. Tai, J. Y., R. C. Seid, Jr., R. D. Hurn, and T.-Y. Liu. 1977. Studies on *Limulus* amoebocyte lysate. II. Purification of the coagulogen and the mechanism of clotting. *J. Biol. Chem.* 252:4773–76.

198. Tatini, S. R., H. M. Soo, B. R. Cords, and R. W. Bennett. 1975. Heat-stable nuclease for assessment of staphylococcal growth and likely presence of enterotoxins in foods. *J. Food Sci.* 40:352–56.

199. Terplan, V. G., K.-J. Zaadhof, and S. Buchholz-Berchtold. 1975. Zum nachweis von Endotoxinen gramnegativer Keime in Milch mit dem *Limulus*-test. *Arch. Lebensmittelhyg.* 26:217–21.

200. Thomason, B. M. 1971. Rapid detection of *Salmonella* microcolonies by fluorescent antibody. *Appl. Microbiol.* 22:1064–69.

201. Thomason, B. M. 1981. Current status of immunofluorescent methodology for salmonellae. *J. Food Protect.* 44:381–84.

202. Thomason, B. M., G. A. Hebert, and W. B. Cherry. 1975. Evaluation of a semiautomated system for direct fluorescent antibody detection of salmonellae. *Appl. Microbiol.* 30:557–64.

203. Thore, A., S. Ånséhn, A. Lundin, and S. Bergman. 1975. Detection of bacteriuria by luciferase assay of adenosine triphosphate. *J. Clin. Microbiol.* 1:1–8.

204. Thore, A., A. Lundin, and S. Ånséhn. 1983. Firefly luciferase ATP assay as a screening method for bacteriuria. *J. Clin. Microbiol.* 17:218–24.

205. Tsuji, K., P. A. Martin, and D. M. Bussey. 1984. Automation of chromogenic substrate *Limulus* amebocyte lysate assay method for endotoxin by robotic system. *Appl. Environ. Microbiol.* 48:550–55.

206. Tsuji, K., and K. A. Steindler. 1983. Use of magnesium to increase sensitivity of *Limulus* amoebocyte lysate for detection of endotoxin. *Appl. Environ. Microbiol.* 45:1342–50.

207. Ur, A., and D. Brown. 1975. Monitoring of bacterial activity by impedance measurements. In *New approaches to the identification of microorganisms,* ed. C.-G. Heden and T. Illeni, 61–71. New York: Wiley.

208. Van Dilla, M. A., R. G. Langlois, D. Pinkel, D. Yajko, and W. K. Hadley. 1983. Bacterial characterization by flow cytometry. *Science* 220:620–22.

209. Van Wart, M., and L. J. Moberg. 1984. Evaluation of a novel fluorogenic-based method for detection of *Escherichia coli. Bacteriol. Proc.,* 201.

210. Vermilyea, B. L., H. D. Walker, and J. C. Ayres. 1968. Detection of botulinal toxins by immunodiffusion. *Appl. Microbiol.* 16:21–24.

211. Vos, J. G., J. Buys, J. G. Hanstede, and A. M. Hagenaars. 1979. Comparison of enzyme-linked immunosorbent assay and passive hemagglutination method for quantification of antibodies to lipopolysaccharide and tetanus toxoid in rats. *Infect. Immun.* 24:798–803.

212. Wachtel, R. E., and K. Tsuji. 1977. Comparison of *Limulus* amebocyte lysates and correlation with the United States Pharmacopeial pyrogen test. *Appl. Environ. Microbiol.* 33:1265–69.

213. Waes, G. M., and R. G. Bossuyt. 1982. Usefulness of the benzalkon-crystal

violet-ATP method for predicting the keeping quality of pasteurized milk. *J. Food Protect.* 45:928–31.

214. ———. 1984. Impedance measurements to detect bacteriophage problems in cheddar cheesemaking. *J. Food Protect.* 47:349–51.

215. Warren, L. S., R. E. Benoit, and J. A. Jessee. 1978. Rapid enumeration of fecal coliforms in water by a colorimetric β-galactosidase assay. *Appl. Environ. Microbiol.* 35:136–41.

216. Weihe, J. L., S. L. Seist, and W. S. Hatcher, Jr. 1984. Estimation of microbial populations in frozen concentrated orange juice using automated impedance measurements. *J. Food Sci.* 49:243–45.

217. Williams, M. L. R. 1971. The limitations of the DuPont luminescence biometer in the microbiological analysis of foods. *Can. Inst. Food Technol. J.* 4:187–89.

218. Wood, J. M., V. Lach, and B. Jarvis. 1977. Detection of food-associated microbes using electrical impedance measurements. *J. Appl. Bacteriol.* 43:14–15.

219. Zaadhof, K.-J., and G. Terplan. 1981. Der *Limulus*-Test—ein Verfahren zur Beurteilung der mikrobiologischen Qualität von Milch und Milchprodukten. *Deut. Molkereizeitung* 34:1094–98.

220. Zayaitz, A. E. K., and R. A. Ledford. 1982. Proteolytic inactivation of thermonuclease activity of *Staphylococcus aureus* during recovery from thermal injury. *J. Food Protect.* 45:624–26.

7.

BIOASSAY AND RELATED METHODS

After establishing the presence of pathogens or toxins in foods or food products, the next important concern is whether or not the organisms/toxins are biologically active. For this purpose, experimental animals are employed where feasible. When it is not feasible to use whole animals or animal systems, a variety of tissue culture systems have been developed which by a variety of responses provide information on the biological activity of pathogens or their toxic products. These bioassay and related tests are the methods of choice for some food-borne pathogens, and some of the principal ones are described in Table 7-1.

WHOLE-ANIMAL ASSAYS
Mouse lethality
This method was first employed for food-borne pathogens around 1920 and continues to be an important bioassay method. To test for botulinal toxins in foods, appropriate extracts are made and portions are treated with trypsin (for toxins of nonproteolytic *Clostridium botulinum* strains). Pairs of mice are injected intraperitoneally (IP) with 0.5 ml of trypsin-treated and untreated preparations. Untreated preparations that have been heated for 10 min at 100°C are injected into a pair of mice. All injected mice are observed for 72 h for symptoms of botulism or death. Mice injected with the heated preparations should not die since the botulinal toxins are heat-labile. Specificity in this test can be achieved by protecting mice with known botulinal antitoxin and in a similar manner, the specific serologic type of botulinal toxin can be determined (see Chapter 19 for toxin types).

Table 7-1. Some bioassay models used to assess the biological activity of various food-borne pathogens and/or their products.

Organism	Toxin/product	Bioassay method	Sensitivity	Reference
A. hydrophila	Cytotoxic enterotoxin	Infant mouse intestines	~30 ng	1
B. cereus	Diarrheagenic toxin	Monkey feeding		90
	Diarrheagenic toxin	Rabbit ileal loop		85
	Diarrheagenic toxin	Rabbit skin		33
	Diarrheagenic toxin	Guinea pig skin		32
	Diarrheagenic toxin	Mouse lethality		62
	Emetic toxin	Rhesus monkey emesis		58
C. jejuni	Viable cells	Adult mice	10^4 cells	5
	Viable cells	Chickens	90 cells	74
	Viable cells	Chickens	10^3–10^6 cells	78
	Viable cells	Neonatal mice		26
	Culture supernatants	Adult rat jejunal loops		25
	Enterotoxin	Rat ileal loop		45, 75
C. botulinum	A, B, E, F, G toxins	Mouse lethality		3
C. perfringens A	Enterotoxin	Mouse lethality, LD_{50}	1.8 μg	27
	Enterotoxin	Mouse ileal loop, 90-min test	1.0 μg	98
	Enterotoxin	Rabbit ileal loop, 90-min test	6.25 μg	37
	Enterotoxin	Guinea pig skin (erythemal activity)	0.06–0.125 mg/ml	27, 87
Infant botulism	Endospores	7–12-day-old rats	1,500 spores	61
	Endospores	9-day-old mice	700 spores	88
	Endospores	Adult germ-free mice	10 spores	60
E. coli	LT	Rabbit ileal loop, 18-h test		76
	ST	Suckling mouse (fluid accumulation)		17, 28
	ST	Rabbit ileal loop, 6-h test		23
	ST_a	Suckling mouse		17
	ST_a	1–3-day-old piglets		9
	ST_b	Jejunal loop of pig		63
	ST_b	Weaned piglets, 7–9 weeks old		9

Table 7-1, *continued.*

Organism	Toxin/product	Bioassay method	Sensitivity	Reference
Salmonella spp.	Heat-labile cytotoxin	Rabbit ileal loop (protein synthesis inhibition)		47
S. aureus	SEB	Skin of specially sensitized guinea pigs	0.1–1.0 pg	79
	All enterotoxins	Emesis in rhesus monkeys	5μg/2–3 kg body wt.	59
	SEA, SEB	Emesis in suckling kittens	0.1, 0.5 μg/kg body wt.	4
V. parahaemolyticus	Broth cultures	Rabbit ileal loop; response in 50% animals	10^2 cells	94
	Viable cells	Adult rabbit ileal loop, invasiveness		7
	Thermostable direct toxin	Mouse lethality, death in 1 min	5 μg/mouse	38, 40
	Thermostable direct toxin	Mouse lethality, LD_{50} by IP route	1.5 μg	99
	Thermostable direct toxin	Rabbit ileal loop	250 μg	99
	Thermostable direct toxin	Guinea pig skin	2.5 μg/g	99
V. vulnificus	Culture filtrates	Rabbit skin permeability		6
V. cholerae (non-01)	Exterotoxin	Suckling mice		65
Y. enterocolitica	Heat-stable toxin	Sereny test		8
	Heat-stable toxin	Suckling mouse (oral)		8
	Heat-stable toxin	Suckling mouse (oral)	110 ng	67
	Enterotoxin	Rabbit ileal loop, 6- and 18-h tests		69
	Viable cells	Mouse diarrhea		80
	Viable cells	Rabbit diarrhea	50% infectious dose = 2.9 $\times$ 10^8	70
	Viable cells	Lethality in suckling mice by IP injection	14 cells	2
	Viable cells	Lethality of gerbils by IP injection	100 cells	81

LT = heat labile toxin; ST = heat stable toxin; SEA = Staphylococcal enterotoxin A.

Mouse lethality may be employed for other toxins. Stark and Duncan (87) used the method for *Clostridium perfringens* enterotoxin. Mice were injected IP with enterotoxin preparations and observed for up to 72 h for lethality. The mouse-lethal dose was expressed as the reciprocal of the highest dilution that was lethal to the mice within 72 h. Genigeorgis et al. (27) employed the method by use of intravenous (IV) injections. *C. perfringens* enterotoxin preparations were diluted in phosphate buffer, pH 6.7, to achieve a concentration of 5 to 12 $\mu g/ml$. From each dilution prepared, 0.25 ml was injected IV into six male mice weighing 12–20 g, the number of deaths were recorded, and the LD_{50} was calculated.

Suckling (infant) mouse

This animal model was introduced by Dean et al. (17) primarily for *Escherichia coli* enterotoxins and is now used for this and some other food-borne pathogens. Typically, mice are separated from their mothers and given oral doses of the test material consisting of 0.05–0.1 ml with the aid of a blunt 23-gauge hypodermic needle. A drop of 5% Evans blue dye/ml of test material may be used to determine the presence of the test material in the small intestine. The animals are usually held at 25°C for 2 h, after which they are sacrificed. The entire small intestine is removed and the relative activity of test material is determined by the ratio of gut weight to body weight (GW/BW). Giannella (28) found the following GW/BW ratios for *E. coli* enterotoxins: < 0.074 = negative test; 0.075 − 0.082 = intermediate (should be retested); and > 0.083 = positive test. The investigator found the day-to-day variability among various *E. coli* strains to range from 10.5 to 15.7% and about 9% for replicate tests with the same strain. A GW/BW of 0.060 was considered negative for *E. coli* ST_a by Mullan et al. (64). In studies with *E. coli* ST, Wood et al. (97) treated as positive GW/BW ratios that were > 0.087, while Boyce et al. (8) held mice at room temperature for 4 h for *Yersinia enterocolitica* heat-stable enterotoxin and considered GW/BW of 0.083 or greater to be positive. In studies with *Y. enterocolitica,* Okamoto et al. (68), keeping mice for 3 h at 25°C, considered a GW/BW of 0.083 to be positive.

In using the suckling mouse model, test material may also be injected percutaneously directly into the stomach through the mouse's translucent skin, or by administration either orogastrically or intraperitoneally. For the screening of large numbers of cultures, the intestines may be examined visually for dilation and fluid accumulation (71). Infant mice along with 1- to 3-day-old piglets are the animals of choice for *E. coli* enterotoxin ST_a, while ST_b is inactive in the suckling mouse but active in piglets and weaned pigs (9, 44). The infant mouse assay does not respond to choleragen or to the LT of *E. coli*. It correlates well with the 6-h rabbit ileal loop assay for the ST_a of *E. coli*.

Suckling mice have been used for lethality studies by employing IP injections. Aulisio et al. (2) used 1- to 3-day-old Swiss mice and injected 0.1

ml of diluted culture. The mice were observed for 7 days, and deaths that occurred within 24 h were considered nonspecific, while deaths occurring between days 2 and 7 were considered specific for *Y. enterocolitica*. By this method, an LD_{50} can be calculated relative to numbers of cells/inoculum. In the case of *Y. enterocolitica*, Aulisio et al. found the LD_{50} to be fourteen cells, and the average time for death of mice to be 3 days.

Rabbit and mouse diarrhea

Rabbits and mice have been employed to test for diarrheagenic activity of some food-borne pathogens. Employing young rabbits weighing 500–800 g, Pai et al. (70) inoculated orogastrically with approximately 10^{10} cells of *Y. enterocolitica* suspended in 10% sodium bicarbonate. Diarrhea developed in 87% of forty-seven rabbits after a mean time of 5.4 days. Bacterial colonization occurred in all animals regardless of dose of cells.

Mice deprived of water for 24 h were employed by Schiemann (80) to test for the diarrheagenic activity of *Y. enterocolitica*. The animals were given inocula of 10^9 cells/ml in peptone water, and fresh drinking water was allowed 24 h later. After 2 days, feces of mice were examined for signs of diarrhea.

Infant rabbits have been used by Smith (82) to assay enterotoxins of *E. coli* and *Vibrio cholerae*. Infant rabbits 6 to 9 days old are administered 1–5 ml of culture filtrate via stomach tube. Following return to their mothers, they are observed for diarrhea. Diarrhea after 6–8 h is a positive response. If death of animals occurs, a large volume of yellow fluid is found in the small and large intestines. The quantitation of enterotoxin is achieved by ascertaining the ratio of intestinal weight to total body weight. Young pigs have been used in a similar way to assay porcine strains of *E. coli* for enterotoxin activity.

Monkey feeding

The use of rhesus monkeys (*Macaca mulatta*) to assay staphylococcal enterotoxins was first developed in 1931 by Jordan and McBroom (43). Next to man, this is perhaps the animal most sensitive to staphylococcal enterotoxins. When enterotoxins are to be assayed by this method, young rhesus monkeys weighing 2–3 kg are selected. The food homogenate, usually in solution in 50-ml quantities, is administered via stomach tube. The animals are then observed continuously for 5 h. Vomiting in at least two of six animals denotes a positive response. Rhesus monkeys have been shown to respond to levels of enterotoxins A and B as low as approximately 5 µg/2–3 kg body weight (59).

Kitten (cat) test

This method was developed by Dolman et al. (20) as an assay for staphylococcal enterotoxins. The original test employed the injection of filtrates into the abdominal cavity of very young kittens (250–500 g). This procedure leads to false positive results. The most commonly used method consists of administering the filtrates IV and observing the animals continuously for emesis. When cats weighing 2 to 4 kg are used, positive responses occur in 2 to 6 h (14). Emesis has been reported to occur with 0.1 and 0.5 µg of staphylococcal enterotoxin A (SEA) and SEB/kg body weight (4). The test tends to lack the specificity of the monkey-feeding test since staphylococcal culture filtrates containing other by-products may also induce emesis. Kittens are much easier to obtain and maintain than rhesus monkeys and in this regard the test has value.

Rabbit and guinea pig skin tests

The skin of these two animals is used to assay toxins for at least two properties. The vascular permeability test is generally done by use of albino rabbits weighing 1.5 to 2.0 kg. Typically, 0.05 to 0.1 ml of culture filtrate is inoculated intradermally (ID) in a shaved area of the rabbit's back and sides. From 2 to 18 h later, a solution of Evans blue dye is administered IV and 1–2 h are allowed for permeation by the dye. The diameters of two zones of blueing are measured and the area approximated by squaring the average of the two values. Areas of 25 cm^2 are considered positive. *E. coli* LT gives a positive response in this assay (24). Employing this assay, permeability has been shown to be a function of the *E. coli* diarrheagenic enterotoxin.

Similar but not identical to the permeability factor test is a test of erythemal activity that employs guinea pigs. The method has been employed by Stark and Duncan (87) to test for erythemal activity of *C. perfringens* enterotoxin. Guinea pigs weighing 300–400 g are depilated (back and sides) and marked in 2.5-cm squares, and duplicate 0.05-ml samples of toxic preparations are injected ID in the center of the squares. Animals are observed after 18–24 h for erythema at the injection site. In the case of *C. perfringens* enterotoxin, a concentric area of erythema is produced without necrosis. A unit of erythemal activity is defined as the amount of enterotoxin producing an area of erythema 0.8 cm in diameter. The enterotoxin preparation used by Stark and Duncan contained 1,000 erythemal units/ml. To enhance readings, 1 ml of 0.5% Evans blue can be injected intracardially (IC) 10 min following the skin injections and the diameters read 80 min later (27). The specificity of the skin reactions can be determined by neutralizing the enterotoxin with specific antisera prior to injections. The erythema test was found to be 1,000 times more sensitive than the rabbit ileal loop technique for assaying the enterotoxin of *C. perfringens* (36).

Sereny test

This method is used to test for virulence of viable bacterial cultures. It was proposed by Sereny in 1955, and the guinea pig is the animal most often used. The test consists of administering with the aid of a loop a drop of cell suspension, containing 1.5×10^{10} to 2.3×10^{10}/ml in phosphate-buffered saline, into the conjunctivae of guinea pigs weighing about 400 g each. The animal's eyes are examined daily for 5 days for evidence of keratoconjunctivitis. When strains of unknown virulence are evaluated, it is important that known positive and negative strains are tested also.

A mouse Sereny test has been developed using Swiss mice and administering one-half the dose noted above.

ANIMAL MODELS REQUIRING SURGICAL PROCEDURES
Ligated loop techniques

These techniques are based on the fact that certain enterotoxins elicit fluid accumulation in the small intestines of susceptible animals. While they may be performed with a variety of animals, rabbits are most often employed. Young rabbits 7 to 20 weeks old and weighing 1.2–2.0 kg are kept off food and water for a period of 24 h, or off food for 48–72 h with water ad libitum prior to surgery. Under local anaesthesia, a midline incision about 2 in long is made just below the middle of the abdomen through the muscles and peritoneum in order to expose the small intestines (16). A section of the intestine midway between its upper and lower ends or just above the appendix is tied with silk or other suitable ligatures in 8–12-cm segments with intervening sections of at least 1 cm or more. Up to six sections may be prepared by single or double ties.

Meanwhile, the specimen or culture to be tested is prepared, suspended in sterile saline, and injected intraluminally into the ligated segments. A common inoculum size is 1 ml, although smaller and larger doses may be used. Different doses of test material may be injected into adjacent loops or into loops separated by a blank loop or by a sham (inoculated with saline). Following injection, the abdomen is closed with surgical thread and the animal is allowed to recover from anaesthesia. The recovered animal may be kept off food and water for an additional 18–24 h period, or water or feed or both may be allowed. With ligatures intact, the animals may not survive beyond 30–36 h (10).

To assess the effect of the materials previously injected into ligated loops, the animal is sacrificed and the loops are examined and measured for fluid accumulation. The fluid may be aspirated and measured. The reaction can be quantitated by measuring loop fluid volume to loop length ratios (10), or by determining the ratio of fluid volume secreted/mg dry weight intestine (57). The appearance of a ligated rabbit ileum 24 hr after injection of a *C. perfringens* culture is presented in Fig. 7-1. The minimum amount of *C. perfringens* enterotoxin necessary to produce a loop reaction has been reported variously to be 28 to 40 μg and as high as 125 μg of toxin by the standard loop technique. The 90-min loop technique has been found to respond to as little as 6.25 μg and the standard technique to 29 μg of toxin (27).

This technique was developed originally to study the mode of action of

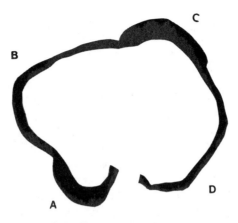

Fig. 7-1. Gross appearance of the ligated rabbit ileum 24 h after injection of 2 ml of cultures of *Clostridium perfringens* grown 4 h at 37°C in skim milk. Loop A, strain NCTC 8798, 8 ml of fluid in loop; loop B, 2 ml of sterile milk, negative loop; loop C, strain T-65, 10 ml of fluid; loop D, strain 6867, negative loop. (Redrawn from Duncan et al., 22; copyright © 1968, American Society for Microbiology)

the cholera organism in producing the disease (16). It has been employed widely in studies on the virulence and pathogenesis of food-borne pathogens including *Bacillus cereus, C. perfringens, E. coli, Vibrio parahaemolyticus,* and others.

Although the rabbit loop is the most widely used of ligated loop methods, other animal models have come into prominence in the past decade. The mouse intestinal loop may be used for *E. coli* enterotoxins. As used by Punyashthiti and Finkelstein (72), Swiss mice 18–22 g are deprived of food 8 h before use. The abdomen is opened under light anaesthesia, and two 6-cm loops separated by 1-cm interloops are prepared. The loops are inoculated with 0.2 ml of test material followed by closing of the abdomen. Animals are deprived of food and water and killed 8 h later. Fluid is measured and the length of the loops determined. Results are considered positive when the ratio of fluid to length is 50 or more mg/cm. In this study, positive loops generally had ratios between 50 and 100 but occasionally approached 200 or more. Alternatively, the net increase in weight of loops in mg can be used to measure the intensity of a toxic reaction (98). With the mouse loop, 1 μg of enterotoxin can be detected (98).

Ligated loop methods have been developed for other animals including 1- to 2-week-old calves and weaned and neonatal pigs.

The RITARD model
The removable intestinal tie-adult rabbit diarrhea (RITARD) method was developed by Spira et al. (86). Rabbits weighing 1.6–2.7 kg are kept off

food for 24 h but allowed water. Under local anaesthesia, the cecum is brought out and ligated close to the ileocecal junction. The small intestine is now brought out and a slip knot tied to close it in the area of the mesoappendix. Test material in 10 ml of phosphate-buffered saline is injected into the lumen of the anterior jejunum. After injection, intestine and cecum are returned to the peritoneal cavity and the incision closed. With the animal kept in a box, the temporary tie is removed 2–4 h after administration of the test dose, and the slip knot in the intestine is released. Sutures are applied as needed. The animal is now returned to its cage and provided with food and water. Animals are observed for diarrhea or death at 2-h intervals up to 124 h. At autopsy, small intestine and adjacent sections are tied and removed for fluid measurement. Enterotoxigenic strains of *E. coli* produce severe and watery diarrhea, and the susceptibility of animals to *V. cholerae* infections is similar in this system to that in the infant rabbit model.

The gist of the RITARD model is that the animals are not altered except that the cecum is ligated to prevent it from taking up fluid from the small intestine, and a temporary reversible obstruction is placed on the ileum long enough to allow the inoculated organism to initiate colonization of the small intestine. The method has been successfully used as an animal model for *Campylobacter jejuni* infection (12).

CELL CULTURE SYSTEMS

A variety of cell culture systems are employed to assess certain pathogenic properties of viable cells. The properties often assessed are invasiveness, permeability, cytotoxicity, adherence/adhesion/binding, and other more general biological activities. Some cell cultures are used to assess various properties of toxins and enterotoxins. Some examples of these models are summarized in Table 7-2, and brief descriptions are presented below.

Table 7-2. Summary of some tissue and cell culture systems employed to study biological activity of gastroenteritis-causing organisms or their products.

Culture system	Pathogen/toxin	Demonstration/use	Reference
CHO monolayer	*E. coli* LT; *V. cholerae* toxin	Biological activity	34
	V. parahaemolyticus	Biological activity	39
	Salmonella toxin	Biological activity	77
	C. jejuni enterotoxin	Biological activity	45, 75
CHO floating cell assay	*Salmonella* toxin	Biological activity	41
HeLa cells	*E. coli*	Invasiveness	3
	Y. enterocolitica	Invasiveness	3, 50, 80, 95
	V. parahaemolyticus	Adherence	13, 42
	C. jejuni	Invasiveness	51

Table 7-2, *continued.*

Culture system	Pathogen/toxin	Demonstration/use	Reference
Vero cells	*C. perfringens* enterotoxin	Mode of action	52
	E. coli LT	Biological activity, assay	83
	A. hydrophila toxin	Cytotoxicity	1
	C. perfringens enterotoxin	Binding	53, 54
	C. perfringens enterotoxin	Biological activity	53
	Salmonella cytotoxin	Protein synthesis inhib.	47
	V. vulnificus	Cytotoxicity	6
Y-1 adrenal cells	*E. coli* LT	Biological activity, assay	21, 76
	V. cholerae toxin	Biological activity, assay	21, 76
	V. mimicus	Biological activity	84
Rabbit int. epith. cells	*C. perfringens* enterotoxin	Binding	53
	Salmonella cytotoxin	Protein synthesis inhib.	47
Murine spleen cells	Staph. enterotoxins A, B, and E	Binding	11
Macrophages	*Y. enterocolitica*	Phagocytosis	95
Human peripheral lymphocytes	Staph. enterotoxin A	Biological effects	48
Human laryngeal carcinoma	*E. coli, Shigella*	Invasiveness	3
Henle 407 human intestine	*E. coli, Shigella*	Invasiveness	3
Human fetal intestinal cells	*V. parahaemolyticus*	Adherence	13, 35
	Enteropathogenic *E. coli*	Adherence	56
	B. cereus toxins	Biological activity	93
Human intestinal cells	*V. parahaemolyticus*	Adherence	31, 42
Human ileal cells	Enterotoxigenic *E. coli*	Adherence	18
Human mucosal cells	*E. coli*	Adherence	66, 91
	V. parahaemolyticus	Adherence	73
Human uroepithelial cells	*E. coli*	Adhesion	89
Viable human duodenal biopsies	*E. coli*	Adherence	46
Rat hepatocytes	*C. perfringens* enterotoxin	Amino acid transport	29, 30
	C. perfringens enterotoxin	Membrane permeability	15
Guinea pig intest. cells	*V. parahaemolyticus*	Adherence	42

Human mucosal cells

As employed by Ofek and Beachey (66), human buccal mucosa cells (about 2×10^5 in phosphate-buffered saline) are mixed with 0.5 ml of washed *E. coli* cells—2×10^8/ml. The mixture is rotated for 30 min at room temperature. Epithelial cells are separated from the bacteria by differential centrifugation followed by drying and staining with gentian violet. Adherence is determined by microscopic counting of bacteria/epithelial cell. As employed by Thorne et al. (91), *E. coli* cells are labeled with ^{3}H-amino acids (alanine and leucine), or fluorescein isothiocyanate. In their use of this method, Reyes et al. (73) mix *V. parahaemolyticus* cells with mucosal epithelial cells and incubate at 37°C for 5 min followed by filtering. The unbound cells are washed off and the culture is dried, fixed, and stained with Giemsa. Adherence is quantitated by counting the total number of *V. parahaemolyticus* adhering to fifty buccal cells as compared to controls. Best results are obtained when approximately 10^9 bacterial cells and 10^5 buccal cells are suspended together in phosphate-buffered saline at pH 7.2 for 5 min. All of twelve strains tested adhered. Adherence apparently bears no relationship to pathogenicity for *V. parahaemolyticus*.

Human fetal intestine (HFI)

By this adherence model, HFI cells are employed in monolayers. The monolayers are thoroughly washed, inoculated with a suspension of *V. parahaemolyticus,* and incubated at 37°C for up to 30 min. Adherence is determined by the microscopic examination of stained cells after washing away unattached bacteria. All strains of *V. parahaemolyticus* tested adhered, but those from food-poisoning cases have a higher adherence ability than those from foods (35). By use of this method, the adherence of an entero-pathogenic strain of *E. coli* of human origin has been found to be plasmid-mediated (96).

Human ileal and intestinal cells

To study adherence of enterotoxigenic *E. coli* (ETEC), Deneke et al. (18) used ileal cells from adult humans in a filtration-binding assay. The cells were mixed with bacteria grown in ^{3}H-alanine and leucine. The amount of binding was determined with a scintillation counter. ETEC strains of human origin bound to a greater extent than controls. Binding to human ileal cells was 10- to 100-fold greater than to human buccal cells.

Monolayers of human intestine cells were employed by Gingras and Howard (31) to study adherence of *V. parahaemolyticus*. The bacterium was grown in the presence of ^{14}C-labeled valine, and the labeled cells were added to monolayers and incubated for up to 60 min. Following incubation, unattached cells were removed and those adhering were counted by radioactive counts of monolayers. The adhered cells were also enumerated microscopically. The Kanagawa-positive and -negative organisms adhered similarly. No correlation was found between hemolysis production and adherence.

Guinea pig intestinal cells

To study adherence of *V. parahaemolyticus,* Iijima et al. (42) employed adult guinea pigs weighing about 300 g and fasted them for 2 days before use. Under anaesthesia, the abdomen was opened and the small intestine tied approximately 3 cm distal from the stomach. The intestine was injected with 1.0 ml of a suspension of 2×10^8 cells of adherence-positive and adherence-negative strains, followed by closing of the abdomen. Six h later, the animals were sacrificed and the small intestine removed and cut into four sections. Following homogenization with 3% NaCl, the number of cells in the homogenate was determined by plating. With adherence-positive cells, larger numbers were found in the homogenates, especially in the upper section of the intestine.

A recently reported adhesion model consists of immobilizing soluble mucosal glycoproteins from mouse intestines on polystyrene (49). Using this model, it was shown that two plasmid-bearing strains of *E. coli* (K88 and K99) adhered readily, as do other adhering strains of this organism.

HeLa cells

This cell line is widely used to test for the invasive potential of intestinal pathogens as well as for adherence. Although HeLa cells seem to be preferred, other cell lines such as human laryngeal carcinoma and Henle 407 human intestine may be employed. In general, monolayers of cells are prepared by standard culture techniques on a chamber slide, and inoculated with 0.2 ml of a properly prepared test culture suspension. Following incubation for 3 h at 35°C to allow for bacterial growth, monolayer cells are washed, fixed, and stained for viewing under the light microscope. In the case of invasive *E. coli,* cells will be present in the cytoplasm of monolayer cells but not in the nucleus. In addition, invasive strains are phagocytized to a greater extent than noninvasives and the number of bacteria/cell is > 5. According to *BAM* (3), at least 0.5% of the HeLa cells should contain no less than five bacteria. Positive responses to this test are generally confirmed by the Sereny test (see 3).

A modification of the above is used for invasive *Yersinia.* By this method, 0.2 ml of a properly prepared bacterial suspension is inoculated into chamber slides containing the HeLa cell monolayer. Following incubation for 1.5 h at 35°C, the cells are washed, fixed, and stained for microscopic examination. Invasive *Y. enterocolitica* are present in the cytoplasm—usually in the phagolysome. Infectivity rates are generally greater than 10%. While invasive *E. coli* are confirmed by the Sereny test, this is not done with *Y. enterocolitica,* even though invasive, since this organism may not yield a positive Sereny test.

HeLa cells have been used to test for adherence of *V. parahaemolyticus,* and to study the penetration of *Y. enterocolitica.* Strains of the latter that gave an index of 3.7 to 5.0 were considered penetrating (80). The infectivity of HeLa cells by *Y. enterocolitica* has been studied by use of cell monolayers in roller tubes. The number of infecting bacterial cells is counted at random in 100 stained HeLa cells for up to 24 h (19).

Chinese hamster ovary (CHO) cells

The CHO assay was developed by Guerrant et al. (34) for *E. coli* enterotoxins and employs CHO cells grown in a medium containing fetal calf serum. Upon establishment of a culture of cells, enterotoxin is added. Microscopic examinations are made 24–30 h later to determine whether cells have become bipolar and elongated at least three times their width, and whether or not their knoblike projections have been lost. The morphological changes in CHO cells caused both by cholera toxin and *E. coli* enterotoxin have been shown to parallel the elevation of cyclic AMP. It has been found to be 100 to 10,000 times more sensitive than skin permeability and ileal loop assays for *E. coli* enterotoxins. For the LT of *E. coli*, CHO has been found to be 5 to 100 times more sensitive than skin permeability and rabbit ileal loop assays (34).

Vero cells

This monolayer consists of a continuous cell line derived from African green monkey kidneys and it has been employed by Speirs et al. (83) to assay for *E. coli* LT. Vero cell results compare favorably with Y-1 adrenal cells (see below), and the test was found by these authors to be the simplest and most economical of the two to maintain in the laboratory. Toxigenic strains produce a morphological response to Vero cells similar to Y-1 cells.

A highly sensitive and reproducible biological assay for *C. perfringens* enterotoxin employing Vero cells has been developed by McDonel and McClane (55). The assay is based on the observation that the enterotoxin inhibits plating efficiency of Vero cells grown in culture. The inhibition of plating efficiency detected as little as 0.1 ng of enterotoxin, and a linear dose-response curve was obtained with 0.5 to 5 ng (5 to 50 ng/ml). The authors proposed a new unit of biological activity—the plating efficiency unit (PEU)—as that amount of enterotoxin that causes a 25% inhibition of the plating of 200 cells inoculated into 100 μl of medium.

Y-1 adrenal cell assay

In this widely used assay, mouse adrenal cells (Y-1) are grown in a monolayer using standard cell culture techniques. With monolayer cells in microtiter plate wells, test extracts or filtrates are added to the microtiter wells followed by incubation at 37°C. In testing *E. coli* LT, heated and unheated culture filtrates of known positive and negative LT-producing strains are added to monolayers in microtiter plates and results are determined by microscopic examinations. The presence of 50% or more rounded cells in monolayers of unheated filtrates and 10% or less for heated filtrates denotes a positive response. The specificity of the response can be determined by the use of specific antibodies in toxin-containing filtrates. Details of this method for food-borne pathogens are presented in *BAM* (3).

Other assays

An immunofluorescence method was employed by Boutin et al. (7) using 6-week-old rabbit ileal loops inoculated with *V. parahaemolyticus*. The loops were removed 12–18 h after infection, and placed in trays, cut into tissue sections, and cleaned by agitation. Tissue sections were fixed and stained with fluorescein isothiocyanate-stained agglutinins to *V. parahae-molyticus*. The reaction of the tagged antibody with *V. parahaemolyticus* cells in the tissue was assessed microscopically. By use of immunofluorescence, it was possible to demonstrate the penetration by this organism into the lamina propria of the ileum and thus the tissue invasiveness of the pathogen. Both Kanagawa-positive and -negative cells penetrated the lamina according to this method.

REFERENCES

1. Asao, T., Y. Kinoshita, S. Kozaki, T. Uemura, and G. Sakaguchi. 1984. Purification and some properties of *Aeromonas hydrophila* hemolysin. *Infect. Immun.* 46:122–27.
2. Aulisio, C. C. G., W. E. Hill, J. T. Stanfield, and J. A. Morris. 1983. Pathogenicity of *Yersinia enterocolitica* demonstrated in the suckling mouse. *J. Food Protect.* 46:856–60.
3. *Bacteriological analytical manual.* 1978. Washington, D.C.: U.S. Food and Drug Administration.
4. Bergdoll, M. S. 1972. The enterotoxins. In *The Staphylococci*, ed. J. O. Cohen, 301–31. New York: Wiley.
5. Blaser, M. J., D. J. Duncan, G. H. Warren, and W.-L. L. Wang. 1983. Experimental *Campylobacter jejuni* infection of adult mice. *Infect. Immun.* 39:908–16.
6. Boutin, B. K., A. L. Reyes, and R. M. Twedt. 1984. Toxicity testing of *Vibrio vulnificus* culture filtrates in the rabbit back permeability model and five cell culture lines. *Bacteriol. Proc.*, 200.
7. Boutin, B. K., S. F. Townsend, P. V. Scarpino, and R. M. Twedt. 1979. Demonstration of invasiveness of *Vibrio parahaemolyticus* in adult rabbits by immunofluorescence. *Appl. Environ. Microbiol.* 37:647–53.
8. Boyce, J. M., D. J. Evans, Jr., D. G. Evans, and H. L. DuPont. 1979. Production of heat-stable, methanol-soluble enterotoxin by *Yersinia enterocolitica*. *Infect. Immun.* 25:532–37.
9. Burgess, M. N., R. J. Bywater, C. M. Cowley, N. A. Mullan, and P. M. Newsome. 1978. Biological evaluation of a methanol-soluble, heat-stable *Escherichia coli* enterotoxin in infant mice, pigs, rabbits, and calves. *Infect. Immun.* 21:526–31.
10. Burrows, W., and G. M. Musteikis. 1966. Cholera infection and toxin in the rabbit ileal loop. *J. Infect. Dis.* 116:183–90.

11. Buxser, S., P. F. Bonventre, and D. L. Archer. 1981. Specific receptor binding of staphylococcal enterotoxins by murine splenic lymphocytes. *Infect. Immun.* 33:827–33.

12. Caldwell, M. B., R. I. Walker, S. D. Stewart, and J. E. Rogers. 1983. Simple adult rabbit model for *Campylobacter jejuni* enteritis. *Infect. Immun.* 42:1176–82.

13. Carruthers, M. M. 1977. In vitro adherence of Kanagawa-positive *Vibrio parahaemolyticus* to epithelial cells. *J. Infect. Dis.* 136:588–92.

14. Clark, W. G., and J. S. Page. 1968. Pyrogenic reponses to staphylococcal enterotoxins A and B in cats. *J. Bacteriol.* 96:1940–46.

15. Dasgupta, B. R., and M. W. Pariza. 1982. Purification of two *Clostridium perfringens* enterotoxin-like proteins and their effects on membrane permeability in primary cultures of adult rat hepatocytes. *Infect. Immun.* 38:592–97.

16. De, S. N., and D. N. Chatterje. 1953. An experimental study of the mechanism of action of *Vibrio cholerae* on the intestinal mucous membrane. *J. Path. Bacteriol.* 66:559–62.

17. Dean, A. G., Y.-C. Ching, R. G. Williams, and L. B. Harden. 1972. Test for *Escherichia coli* enterotoxin using infant mice: Application in a study of diarrhea in children in Honolulu. *J. Infect. Dis.* 125:407–11.

18. Deneke, C. F., K. McGowan, G. M. Thorne, and S. L. Gorbach. 1983. Attachment of enterotoxigenic *Escherichia coli* to human intestinal cells. *Infect. Immun.* 39:1102–6.

19. Devenish, J. A., and D. A. Schiemann. 1981. HeLa cell infection by *Yersinia enterocolitica:* Evidence for lack of intracellular multiplication and development of a new procedure for quantitative expression of infectivity. *Infect. Immun.* 32:48–55.

20. Dolman, C. E., R. J. Wilson, and W. H. Cockroft. 1936. A new method of detecting *Staphylococcus* enterotoxin. *Can. J. Pub. Hlth* 27:489–93.

21. Donta, S. T., H. W. Moon, and S. C. Whipp. 1974. Detection of heat-labile *Escherichia coli* enterotoxin with the use of adrenal cells in tissue culture. *Science* 183:334–36.

22. Duncan, C. L., H. Sugiyama, and D. H. Strong. 1968. Rabbit ileal loop response to strains of *Clostridium perfringens*. *J. Bacteriol.* 95:1560–66.

23. Evans, D. G., D. J. Evans, and N. F. Pierce. 1973. Differences in the response of rabbit small intestine to heat-labile and heat-stable enterotoxins of *Escherichia coli*. *Infect. Immun.* 7:873–80.

24. Evans, D. J., Jr., D. G. Evans, and S. L. Gorbach. 1973. Production of vascular permeability factor by enterotoxigenic *Escherichia coli* isolated from man. *Infect. Immun.* 8:725–30.

25. Fernandez, H., U. F. Neto, F. Fernandes, M. D. A. Pedra, and L. R. Trabulsi. 1983. Culture supernatants of *Campylobacter jejuni* induce a secretory response in jejunal segments of adult rats. *Infect. Immun.* 40:429–31.

26. Field, L. H., J. L. Underwood, L. M. Pope, and L. J. Berry. 1981. Intestinal colonization of neonatal animals by *Campylobacter fetus* subsp. *jejuni*. *Infect. Immun.* 33:884–92.

27. Genigeorgis, C., G. Sakaguchi, and H. Riemann. 1973. Assay methods for *Clostridium perfringens* type A enterotoxin. *Appl. Microbiol.* 26:111–15.

28. Giannella, R. A. 1976. Suckling mouse model for detection of heat-stable *Esch-*

erichia coli enterotoxin: Characteristics of the model. *Infect. Immun.* 14:95–99.

29. Giger, O., and M. W. Pariza. 1978. Depression of amino acid transport in cultured rat hepatocytes by purified enterotoxin from *Clostridium perfringens*. *Biochem. Biophys. Res. Comm.* 82:378–83.

30. ———. 1980. Mechanism of action of *Clostridium perfringens* enterotoxin. Effects on membrane permeability and amino acid transport in primary cultures of adult rat hepatocytes. *Biochim. Biophys. Acta* 595:264–76.

31. Gingras, S. P., and L. V. Howard. 1980. Adherence of *Vibrio parahaemolyticus* to human epithelial cell lines. *Appl. Environ. Microbiol.* 39:369–71.

32. Glatz, B. A., and J. M. Goepfert. 1973. Extracellular factor synthesized by *Bacillus cereus* which evokes a dermal reaction in guinea pigs. *Infect. Immun.* 8:25–29.

33. Glatz, B. A., W. M. Spira, and J. M. Goepfert. 1974. Alteration of vascular permeability in rabbits by culture filtrates of *Bacillus cereus* and related species. *Infect. Immun.* 10:299–303.

34. Guerrant, R. L., L. L. Brunton, T. C. Schaitman, L. L. Rebhun, and A. G. Gilman. 1974. Cyclic adenosine monophosphate and alteration of Chinese hamster ovary cell morphology: A rapid, sensitive in vitro assay for the enterotoxins of *Vibrio cholerae* and *Escherichia coli*. *Infect. Immun.* 10:320–27.

35. Hackney, C. R., E. G. Kleeman, B. Ray, and M. L. Speck. 1980. Adherence as a method for differentiating virulent and avirulent strains of *Vibrio parahaemolyticus*. *Appl. Environ. Microbiol.* 40:652–58.

36. Hauschild, A. H. W. 1970. Erythemal activity of the cellular enteropathogenic factor of *Clostridium perfringens* type A. *Can. J. Microbiol.* 16:651–54.

37. Hauschild, A. H. W., R. Hilsheimer, and C. G. Rogers. 1971. Rapid detection of *Clostridium perfringens* enterotoxin by a modified ligated intestinal loop technique in rabbits. *Can. J. Microbiol.* 17:1475–76.

38. Honda, T., K. Goshima, Y. Takeda, Y. Sugino, and T. Miwatani. 1976. Demonstration of the cardiotoxicity of the thermostable direct hemolysin (lethal toxin) produced by *Vibrio parahaemolyticus*. *Infect. Immun.* 13:163–71.

39. Honda, T., M. Shimizu, Y. Takeda, and T. Miwatani. 1976. Isolation of a factor causing morphological changes in Chinese hamster ovary cells from the culture filtrate of *Vibrio parahaemolyticus*. *Infect. Immun.* 14:1028–33.

40. Honda, T., S. Taga, T. Takeda, M. A. Hasibuan, Y. Takeda, and T. Miwatani. 1976. Identification of lethal toxin with the thermostable direct hemolysin produced by *Vibrio parahaemolyticus*, and some physicochemical properties of the purified toxin. *Infect. Immun.* 13:133–39.

41. Houston, C. W., F. C. W. Koo, and J. W. Peterson. 1981. Characterization of *Salmonella* toxin released by mitomycin C-treated cells. *Infect. Immun.* 32:916–26.

42. Iijima, Y., H. Yamada, and S. Shinoda. 1981. Adherence of *Vibrio parahaemolyticus* and its relation to pathogenicity. *Can. J. Microbiol.* 27:1252–59.

43. Jordan, E. O., and J. McBroom. 1931. Results of feeding *Staphylococcus* filtrates to monkeys. *Proc. Soc. Exp. Biol. Med.* 29:161–62.

44. Kennedy, D. J., R. N. Greenberg, J. A. Dunn, R. Abernathy, J. S. Ryerse, and R. L. Guerrant. 1984. Effects of *Escherichia coli* heat-stable enterotoxin ST_b on intestines of mice, rats, rabbits, and piglets. *Infect. Immun.* 46:639–41.

45. Klipstein, F. A., and R. F. Engert. 1984. Properties of crude *Campylobacter jejuni* heat-labile enterotoxin. *Infect. Immun.* 45:314–19.
46. Knutton, S., D. R. Lloyd, D. C. A. Candy, and A. S. McNeish. 1984. In vitro adhesion of enterotoxigenic *Escherichia coli* to human intestinal epithelial cells from mucosal biopsies. *Infect. Immun.* 44:514–18.
47. Koo, F. C. W., J. W. Peterson, C. W. Houston, and N. C. Molina. 1984. Pathogenesis of experimental salmonellosis: Inhibition of protein synthesis by cytotoxin. *Infect. Immun.* 43:93–100.
48. Langford, M. P., G. J. Stanton, and H. M. Johnson. 1978. Biological effects of staphylococcal enterotoxin A on human peripheral lymphocytes. *Infect. Immun.* 22:62–68.
49. Laux, D. C., E. F. McSweegan, and P. S. Cohen. 1984. Adhesion of enterotoxigenic *Escherichia coli* to immobilized intestinal mucosal preparations: A model for adhesion to mucosal surface components. *J. Microbiol. Meth.* 2:27–39.
50. Lee, W. H., P. McGrath, P. H. Carter, and E. L. Eide. 1977. The ability of some *Yersinia enterocolitica* strains to invade HeLa cells. *Can. J. Microbiol.* 23:1714–22.
51. Manninen, K. I., J. F. Prescott, and I. R. Dohoo. 1982. Pathogenicity of *Campylobacter jejuni* isolates from animals and humans. *Infect. Immun.* 38:46–52.
52. McClane, B. A., and J. L. McDonel. 1980. Characterization of membrane permeability alterations induced in Vero cells by *Clostridium perfringens* enterotoxin. *Biochim. Biophys. Acta* 600:974–85.
53. McDonel, J. L. 1980. Binding of *Clostridium perfringens* (^{125}I)enterotoxin to rabbit intestinal cells. *Biochem.* 19:4801–7.
54. McDonel, J. L., and B. A. McClane. 1979. Binding versus biological activity of *Clostridium perfringens. Biochem. Biophys. Res. Comm.* 87:497–504.
55. ———. 1981. Highly sensitive assay for *Clostridium perfringens* enterotoxin that uses inhibition of plating efficiency of Vero cells grown in culture. *J. Clin. Microbiol.* 13:940–46.
56. McNeish, A. S., P. Turner, J. Fleming, and N. Evans. 1975. Mucosal adherence of human enteropathogenic *Escherichia coli. Lancet* 2:946–48.
57. Mehlman, I. J., M. Fishbein, S. L. Gorbach, A. C. Sanders, E. L. Eide, and J. C. Olson, Jr. 1976. Pathogenicity of *Escherichia coli* recovered from food. *J. Assoc. Off. Anal. Chem.* 59:67–80.
58. Melling, J., B. J. Capel, P. C. B. Turnbull, and R. J. Gilbert. 1976. Identification of a novel enterotoxigenic activity associated with *Bacillus cereus. J. Clin. Pathol.* 29:938–40.
59. Minor, T. E., and E. H. Marth. 1976. *Staphylococci and their significance in foods.* 127–57. New York: Elsevier.
60. Moberg, L. J., and H. Sugiyama. 1979. Microbial ecological basis of infant botulism as studied with germfree mice. *Infect. Immun.* 25:653–57.
61. ———. 1980. The rat as an animal model for infant botulism. *Infect. Immun.* 29:819–21.
62. Molnar, D. M. 1962. Separation of the toxin of *Bacillus cereus* into two components and nonidentity of the toxin with phospholipase. *J. Bacteriol.* 84:147–53.
63. Moon, H. W., E. M. Kohler, R. A. Schneider, and S. C. Whipp. 1980. Prevalence of pilus antigens, enterotoxin types, and enteropathogenicity among K88-negative enterotoxigenic *Escherichia coli* from neonatal pigs. *Infect. Immun.* 27:222–30.

64. Mullan, N. A., M. N. Burgess, and P. M. Newsome. 1978. Characterization of a partially purified, methanol-soluble heat-stable *Escherichia coli* enterotoxin in infant mice. *Infect. Immun.* 19:779–84.

65. Nishibuchi, M., and R. J. Seidler. 1983. Medium-dependent production of extracellular enterotoxins by non-01 *Vibrio cholera, Vibrio mimicus,* and *Vibrio fluvialis. Appl. Environ. Microbiol.* 45:228–31.

66. Ofek, I., and E. H. Beachey. 1978. Mannose binding and epithelial cell adherence of *Escherichia coli. Infect. Immun.* 22:247–54.

67. Okamoto, K., T. Inoue, H. Ichikawa, Y. Kawamoto, and A. Miyama. 1981. Partial purification and characterization of heat-stable enterotoxin produced by *Yersinia enterocolitica. Infect. Immun.* 31:554–59.

68. Okamoto, K., T. Inoue, K. Shimizu, S. Hara, and A. Miyama. 1982. Further purification and characterization of heat-stable enterotoxin produced by *Yersinia enterocolitica. Infect. Immun.* 35:958–64.

69. Pai, C. H., and V. Mors. 1978. Production of enterotoxin by *Yersinia enterocolitica. Infect. Immun.* 19:908–11.

70. Pai, C. H., V. Mors, and T. A. Seemayer. 1980. Experimental *Yersinia enterocolitica* enteritis in rabbits. *Infect. Immun.* 28:238–44.

71. Pai, C. H., V. Mors, and S. Toma. 1978. Prevalence of enterotoxigenicity in human and nonhuman isolates of *Yersinia enterocolitica. Infect. Immun.* 22:334–38.

72. Punyashthiti, K., and R. A. Finkelstein. 1971. Enteropathogenicity of *Escherichia coli.* I. Evaluation of mouse intestinal loops. *Infect. Immun.* 4:473–78.

73. Reyes, A. L., R. G. Crawford, P. L. Spaulding, J. T. Peeler, and R. M. Twedt. 1983. Hemagglutination and adhesiveness of epidemiologically distinct strains of *Vibrio parahaemolyticus. Infect. Immun.* 39:721–25.

74. Ruiz-Palacios, G., E. Escamilla, and N. Torres. 1981. Experimental *Campylobacter* diarrhea in chickens. *Infect. Immun.* 34:250–55.

75. Ruiz-Palacios, G. M., J. Torres, E. Escamilla, B. R. Ruiz-Palacios, and T. Tamayo. 1983. Cholera-like enterotoxin produced by *Campylobacter jejuni. Lancet* 2:250–53.

76. Sack, D. A., and R. B. Sack. 1975. Test for enterotoxigenic *Escherichia coli* using Y-1 adrenal cells in miniculture. *Infect. Immun.* 11:334–36.

77. Sandefur, P. D., and J. W. Peterson. 1977. Neutralization of *Salmonella* toxin-induced elongation of Chinese hamster ovary cells by cholera antitoxin. *Infect. Immun.* 15:988–92.

78. Sanyal, S. C., K. M. N. Islam, P. K. B. Neogy, M. Islam, P. Speelman, and M. I. Huq. 1984. *Campylobacter jejuni* diarrhea model in infant chickens. *Infect. Immun.* 43:931–36.

79. Scheuber, P. H., H. Mossmann, G. Beck, and D. K. Hammer. 1983. Direct skin test in highly sensitized guinea pigs for rapid and sensitive determination of staphylococcal enterotoxin B. *Appl. Environ. Microbiol.* 46:1351–56.

80. Schiemann, D. A. 1981. An enterotoxin-negative strain of *Yersinia enterocolitica* serotype 0:3 is capable of producing diarrhea in mice. *Infect. Immun.* 32:571–74.

81. Schiemann, D. A., and J. A. Devenish. 1980. Virulence of *Yersinia enterocolitica* determined by lethality in Mongolian gerbils and by the Sereny test. *Infect. Immun.* 29:500–506.

82. Smith, H. W. 1972. The production of diarrhea in baby rabbits by the oral

administration of cell-free preparations of enteropathogenic *Escherichia coli* and *Vibrio cholerae:* The effect of antisera. *J. Med. Microbiol.* 5:299–303.

83. Speirs, J. I., S. Stavric, and J. Konowalchuk. 1977. Assay of *Escherichia coli* heat-labile enterotoxin with Vero cells. *Infect. Immun.* 16:617–22.

84. Spira, W. M., and P. J. Fedorka-Cray. 1983. Production of cholera toxin-like toxin by *Vibrio mimicus* and non-01 *Vibrio cholera:* Batch culture conditions for optimum yields and isolation of hypertoxigenic lincomycin-resistant mutants. *Infect. Immun.* 42:501–9.

85. Spira, W. M., and J. M. Goepfert. 1972. *Bacillus cereus*-induced fluid accumulation in rabbit ileal loops. *Appl. Microbiol.* 24:341–48.

86. Spira, W. M., R. B. Sack, and J. L. Froehlich. 1981. Simple adult rabbit model for *Vibrio cholerae* and enterotoxigenic *Escherichia coli* diarrhea. *Infect. Immun.* 32:739–47.

87. Stark, R. L., and C. L. Duncan. 1971. Biological characteristics of *Clostridium perfringens* type A enterotoxin. *Infect. Immun.* 4:89–96.

88. Sugiyama, H., and D. C. Mills. 1978. Intraintestinal toxin in infant mice challenged intragastrically with *Clostridium botulinum* spores. *Infect. Immun.* 21:59–63.

89. Svanborg Eden, G., R. Eriksson, and L. A. Hanson. 1977. Adhesion of *Escherichia coli* to human uroepithelial cells in vitro. *Infect. Immun.* 18:767–74.

90. Terranova, W., and P. A. Blake. 1978. Current concepts: *Bacillus cereus* food poisoning. *New Engl. J. Med.* 298:143–44.

91. Thorne, G. M., C. F. Deneke, and S. L. Gorbach. 1979. Hemagglutination and adhesiveness of toxigenic *Escherichia coli* isolated from humans. *Infect. Immun.* 23:690–99.

92. Tsuji, T., T. Honda, and T. Miwatani. 1984. Comparison of effects of nicked and unnicked *Escherichia coli* heat-labile enterotoxin on Chinese hamster ovary cells. *Infect. Immun.* 46:94–97.

93. Turnbull, P. C. B., J. M. Kramer, K. Jorgensen, R. J. Gilbert, and J. Melling. 1979. Properties and production characteristics of vomiting, diarrheal, and necrotizing toxins of *Bacillus cereus*. *Amer. J. Clin. Nutri.* 32:219–28.

94. Twedt, R. M., J. T. Peeler, and P. L. Spaulding. 1980. Effective ileal loop dose of Kanagawa-positive *Vibrio parahaemolyticus*. *Appl. Environ. Microbiol.* 40:1012–16.

95. Une, T. 1977. Studies on the pathogenicity of *Yersinia enterocolitica*. II. Interaction with cultured cells *in vitro*. *Microbiol. Immunol.* 21:365–77.

96. Williams, P. H., M. I. Sedgwick, N. Evans, P. J. Turner, R. H. George, and A. S. McNeish. 1978. Adherence of an enteropathogenic strain of *Escherichia coli* to human intestinal mucosa is mediated by a colicinogenic conjugative plasmid. *Infect. Immun.* 22:393–402.

97. Wood, L. V., W. H. Wolfe, G. Ruiz-Palacios, W. S. Foshee, L. I. Corman, F. McCleskey, J. A. Wright, and H. L. DuPont. 1983. An outbreak of gastroenteritis due to a heat-labile enterotoxin-producing strain of *Escherichia coli*. *Infect. Immun.* 41:931–34.

98. Yamamoto, K., I. Ohishi, and G. Sakaguchi. 1979. Fluid accumulation in mouse-ligated intestine inoculated with *Clostridium perfringens* enterotoxin. *Appl. Environ. Microbiol.* 37:181–86.

99. Zen-Yoji, H., Y. Kudoh, H. Igarashi, K. Ohta, and K. Fukai. 1975. Further studies on characterization and biological activities of an enteropathogenic toxin of *Vibrio parahaemolyticus*. *Toxicon* 13:134–35.

IV

FOOD

SPOILAGE

8.

SPOILAGE OF FRUITS AND VEGETABLES

Spoiled food may be defined as food that has been damaged or injured so as to make it undesirable for human use. Food spoilage may be caused by insect damage, physical injury of various kinds such as bruising and freezing, enzyme activity, or microorganisms. Only that caused by microorganisms will be treated here.

As pointed out in Chapter 2, the microbial spoilage of foods should not be viewed as a sinister plot on the part of microorganisms deliberately to destroy our foods, but as the normal function of these organisms in the total ecology of all living organisms. While plants and animals have both evolved intrinsic mechanisms that aid them in combating harmful microbes, many microorganisms can and do overcome these forces and bring about the destruction of the plant and animal organic matter by converting it into inorganic compounds. These primary activities of microorganisms on foods bring about the phenomenon we refer to as spoilage.

It has been estimated that 20% of all fruits and vegetables harvested for human consumption are lost through microbial spoilage by one or more of 250 market diseases (1). The primary causative agents of microbial spoilage are the bacteria, yeasts, and molds. While viruses have the capacity to damage both plant and animal tissues, these agents along with the *Rickettsia* are not generally regarded as being important in food spoilage as it is now recognized. With respect to yeasts, molds, and bacteria, the latter two are by far the most important etiologic agents of food spoilage in general. On the basis of their growth requirements (discussed in Chapter 3), each may be expected to occupy its own niche with respect to the various types of foods and even within the same food.

MICROBIAL SPOILAGE OF VEGETABLES

The general composition of higher plants is presented in Table 8-1, and the composition of twenty-one common vegetables is presented in Table 8-2. It can be seen that the average water content of vegetables is about 88% with an average content of 8.6% carbohydrates, 1.9% proteins, 0.3% fat, and 0.84% ash. While the percentage composition of vitamins, nucleic acids, and other plant constituents is not given, the total of these compounds is generally less than 1%. From the standpoint of nutrient content, vegetables are capable of supporting the growth of molds, yeasts, and bacteria, and consequently of being spoiled by any or all of these organisms. The higher water content of vegetables favors the growth of spoilage bacteria and the relatively low carbohydrate and fat contents suggest that much of this water is in available form. The pH range of most vegetables is within the growth range of a large number of bacteria, and it is not surprising, therefore, that bacteria are common agents of vegetable spoilage. The relatively high O/R potential of vegetables and their lack of high poising capacity suggest that the aerobic and facultative anaerobic types would be more important than the anaerobes. This is precisely the case, since some of the most ubiquitous etiologic agents in the bacterial spoilage of vegetables are species of the genus *Erwinia*, and are associated with plants and vegetables in their

Table 8-1. General chemical composition of higher plant material.

A. CARBOHYDRATES AND RELATED COMPOUNDS
1. Polysaccharides—pentosan (araban), hexosans (cellulose, starch, xylans, fructans, mannans, galactans, levans).
2. Oligosaccharides—tetrasaccharide (stachyose), trisaccharides (robinose, mannotriose, raffinose), disaccharides (maltose, sucrose, cellobiose, melibiose, trehalose).
3. Monosaccharides—hexoses (mannose, glucose, galactose, fructose, sorbose), pentoses (arabinose, xylose, ribose, L-rhamnose, L-fucose).
4. Sugar alcohols—glycerol, ribitol, mannitol, sorbitol, inositols.
5. Sugar acids—uronic acids, ascorbic acid.
6. Esters—tannins.
7. Organic acids—citric, shikimic, D-tartaric, oxalic, lactic, glycolic, malonic, etc.
B. PROTEINS—albumins, globulins, glutelins, prolamines, peptides, and amino acids.
C. LIPIDS—fatty acids, fatty acid esters, phospholipids, glycolipids, etc.
D. NUCLEIC ACIDS AND DERIVATIVES—purine and pyrimidine bases, nucleotides, etc.
E. VITAMINS—fat-soluble (A, D, E), water-soluble (thiamine, niacin, riboflavin, etc.).
F. MINERALS—Na, K, Ca, Mg, Mn, Fe, etc.
G. WATER
H. OTHERS—alkaloids, porphyrins, aromatics, etc.

Table 8-2. Approximate percentage chemical composition of 21 vegetable foods (7).

Vegetable	Water	Carbohydrates	Proteins	Fat	Ash
Beans, green	89.9	7.7	2.4	0.2	0.8
Beets	87.6	9.6	1.6	0.1	1.1
Broccoli	89.9	5.5	3.3	0.2	1.1
Brussels sprouts	84.9	8.9	4.4	0.5	1.3
Cabbage	92.4	5.3	1.4	0.2	0.8
Cantaloupe	94.0	4.6	0.2	0.2	0.6
Cauliflower	91.7	4.9	2.4	0.2	0.8
Celery	93.7	3.7	1.3	0.2	1.1
Corn	73.9	20.5	3.7	1.2	0.7
Cucumbers	96.1	2.7	0.7	0.1	0.4
Lettuce	94.8	2.9	1.2	0.2	0.9
Onions	87.5	10.3	1.4	0.2	0.6
Peas	74.3	17.7	6.7	0.4	0.9
Potatoes	77.8	19.1	2.0	0.1	1.0
Pumpkin	90.5	7.3	1.2	0.2	0.8
Radishes	93.6	4.2	1.2	0.1	1.0
Spinach	92.7	3.2	2.3	0.3	1.5
Squash, summer	95.0	3.9	0.6	0.1	0.4
Sweet potatoes	68.5	27.9	1.8	0.7	1.1
Tomatoes	94.1	4.0	1.0	0.3	0.6
Watermelon	92.1	6.9	0.5	0.2	0.3
Mean	88.3	8.6	2.0	0.3	0.8

natural growth environment. The common spoilage pattern displayed by these organisms is referred to as **bacterial soft rot.** It is described below, together with some of the vegetables most commonly affected.

Bacterial agents

BACTERIAL SOFT ROT. This type of spoilage is caused by *Erwinia carotovora* and pseudomonads such as *P. marginalis,* with the former being the more important. *Bacillus* and *Clostridium* spp. have been implicated, but their roles are probably secondary.

The causative organisms break down pectins, giving rise to a soft, mushy consistency, sometimes a bad odor, and a water-soaked appearance. Some of the vegetables affected by this disease are: asparagus, onions, garlic, beans (green, lima, and wax), carrots, parsnips, celery, parsley, beets, endives, globe artichokes, lettuce, rhubarb, spinach, potatoes, cabbage, brussels sprouts, cauliflower, broccoli, radishes, rutabagas, turnips, tomatoes, cucumbers, cantaloupes, peppers, and watermelons.

While the precise manner in which *Erwinia* spp. bring about soft rot is not yet well understood, it is very likely that these organisms, present on the susceptible vegetables at the time of harvest, subsist upon vegetable sap until the supply is exhausted. The cementing substance of the vegetable body then induces the formation of pectinases, which act by hydrolyzing pectin, thereby producing the mushy consistency. In potatoes, tissue maceration has been shown to be caused by an endopolygalacturonate trans-eliminase of *Erwinia* origin (6). Because of the early and relatively rapid growth of these organisms, molds, which tend to be crowded out, are of less consequence in the spoilage of vegetables that are susceptible to bacterial agents. Once the outer plant barrier has been destroyed by these pectinase producers, nonpectinase producers no doubt enter the plant tissues and help bring about fermentation of the simple carbohydrates that are present. The quantities of simple nitrogenous compounds present, the vitamins (especially the B-complex group), and minerals are adequate to sustain the growth of the invading organisms until the vegetables have been essentially consumed or destroyed. The malodors that are produced are probably the direct result of volatile compounds (such as NH_3, volatile acids, and the like) produced by the flora. When growing in acid media, microorganisms tend to decarboxylate amino acids, leaving amines which cause an elevation of pH toward the neutral range and beyond. Complex carbohydrates such as cellulose are generally the last to be degraded, and a varied flora consisting of molds and other soil organisms is usually responsible, since cellulose degradation by *Erwinia* spp. is doubtful. Aromatic constituents and porphyrins are probably not attacked until late in the spoilage process, and again by a varied flora of soil types.

The genus *Erwinia* belongs to the family Enterobacteriaceae. Of the twenty-one species listed in *Bergey's Manual*, all are associated with plants where they are known to cause plant diseases of the rot and wilt types. These are gram-negative rods that are related to the genera *Proteus, Serratia, Escherichia, Salmonella,* and others. *Erwinia* spp. normally do not require organic nitrogen compounds for growth, and the relatively low levels of proteins in vegetables make them suitable for the task of destroying plant materials of this type. The pectinase produced by these organisms is actually a **protopectinase,** since the cementing substance of plants as it actually exists in the plant is protopectin. Many *Erwinia* spp. such as *E. carotovora* are capable of fermenting many of the sugars and alcohols that exist in certain vegetables such as rhamnose, cellobiose, arabinose, mannitol, and so forth—compounds which are not utilized by many of the more common bacteria. While most *Erwinia* spp. grow well at about 37°C, most are also capable of good growth at refrigerator temperatures with some strains reported to grow at 1°C.

OTHER BACTERIAL SPOILAGE CONDITIONS. "Black leg" of potatoes, especially in Scotland, is caused mainly by *E. carotovora* var *atroseptica. E. carotovora* var. *carotovora* also causes black leg, generally at temperatures of around 25°C or above (4). With respect to the source of infecting organisms, some investigators have found soil to be the source while others have been unable to confirm this. A study by Jones and Paton (3) suggests that the black leg organisms may exist as L-phase variants in soils and infect potatoes in this state. Once healthy tissues have been invaded, the L-phase variants then revert to classical forms.

Although *Erwinia* spp. constitute the most important group of bacteria causing vegetable spoilage, members of the genus *Pseudomonas* are pathogens of edible plants. In addition to *P. marginalis* noted previously which causes soft rot, *P. glycinea* causes a disease of soybeans; *P. apii* causes **bacterial blight** of celery; *P. cichorii* causes **bacterial zonate spot** of cabbage and lettuce; *P. lachrymans* causes **angular leaf spot** of cucumbers; *P. maculicola* causes **bacterial leaf spot** of broccoli and cauliflower; *P. phaseolicola* causes **halo blight** of beans; *P. pisi* causes **bacterial blight** of peas; and *P. tomato* causes **bacterial speck** of tomatoes. **Black rot** of cabbage and cauliflower is caused by *Xanthomonas campestris;* **common blight** of beans by *X. phaseoli;* and **bacterial spot** of tomatoes and peppers by *X. vesicatoria*. **Bacterial wilt** of beans is caused by *Corynebacterium flaccumfaciens;* **bacterial canker** by *C. michiganense;* **ring rot** of potatoes by *C. sepedonicum* (4); and **side slime** of lettuce is caused by *P. marginalis*. In addition to cabbage and cauliflower, *X. campestris* causes diseases of a wide variety of other plants and food crops including **citrus canker.** The Florida citrus industry was devastated during the early part of the 1900s by this organism, and new problems arose in 1984. Numerous pathovars exist based primarily on plant hosts, and the organism is unique in that it is one of only a small number of bacteria that contain branched 3-hydroxyl fatty acids in the lipopolysaccharide of its cell envelope. It invades plants through leaf stomata.

The appearance of some market vegetables undergoing bacterial and fungal spoilage may be seen in Figs. 8-1 and 8-2.

While the genera *Erwinia* and *Pseudomonas* are the most important bacteria that cause vegetable spoilage, the molds are by far the most important group of organisms in vegetable spoilage. One of the most versatile groups belongs to the genus *Botrytis*, which causes **gray mold rot** on at least twenty-six common vegetables.

Fungal agents

A synopsis of some of the common spoilage conditions of vegetables and fruits is presented in Table 8-3. Among these spoilage conditions, some are initiated preharvest and others postharvest. Among the former are the following: *Botrytis* invades the flower of strawberries to cause gray mold rot; *Colletotrichum* invades the epidermis of bananas to initiate banana anthracnose; and *Gloeosporium* invades the lenticels of apples to initiate lenticel rot (2). The largest number of market fruit and vegetable spoilage conditions occur after harvesting, and while the fungi most often invade bruised and damaged products, some enter specific areas. For example, *Thielaviopsis* invades the fruit stem of pineapples to cause black rot of this

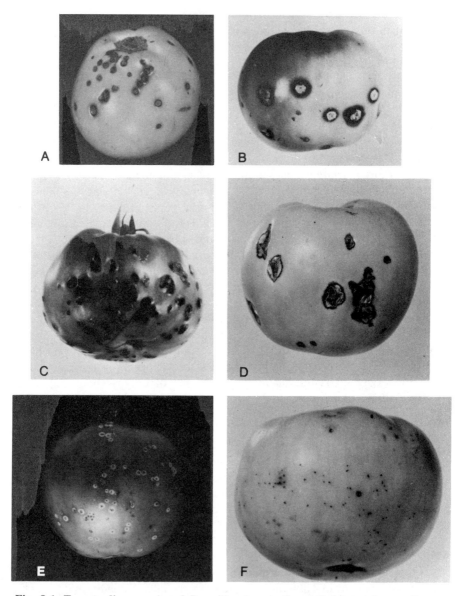

Fig. 8-1. Tomato diseases: *A* and *B*, nailhead spot; *C* and *D*, bacterial spot; *E*, bacterial canker; *F*, bacterial speck. (From *Agriculture Handbook 28*, USDA, **1968**, "Fungus and bacterial diseases of fresh tomatoes")

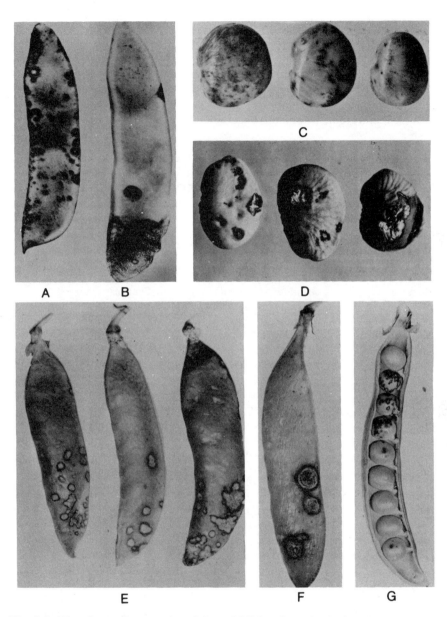

Fig. 8-2. Lima bean diseases: *A* and *B*, pod blight; *C*, seed spotting; *D*, yeast spot. Pea diseases: *E*, pod spot; *F*, anthracnose; *G*, scab. (From *Agriculture Handbook 303*, USDA, 1966, Chapter 5)

fruit, while *Colletotrichum* invades the crown cushion of bananas to cause banana crown rot (2). Some of the conditions listed in Table 8-3 are discussed further below.

GRAY MOLD ROT. This disease is caused by *Botrytis cinerea,* which produces a gray mycelium. This type of spoilage is favored by high humidity and warm temperatures. Among the vegetables affected are: asparagus, onions,

Table 8-3. Synopsis of some common fungal fruit and vegetable spoilage conditions (market diseases) along with etiologic agents and typical products affected.

Spoilage condition	Etiologic agent	Typical products affected
Alternaria rot	*A. tenuis*	Citrus fruits
Anthracnose (bitter rot)	*Colletotrichum musae*	Bananas
Black rot	*Ceratocystis fimbriata*	Sweet potatoes
Blue mold rot	*Penicillium digitatum*	Citrus fruits
Brown rot	*Monilinia fructicola* (*= Sclerotinia fructicola*)	Peaches, cherries
Brown rot	*Phytophora* spp.	Citrus fruits
Cladosporium rot	*C. herbarum*	Cherries, peaches
Crown rot	*Colletotrichum musae* (*= Gloeosporium musarum*), *Fusarium roseum, Verticillium theobromae, Ceratocystis paradoxa*	Bananas
Downy mildew	*Plasmapara viticole, Phytophora* spp., *Bremia* spp.	Grapes
Dry rot	*Fusarium* spp.	Potatoes
Gray mold rot	*Botrytis cinerea*	Grapes, many others
Green mold rot	*Penicillium digitatum*	Citrus fruits
Lenticel rot	*Cryptosporiopsis malicorticis* (*= Gloeosporium perennans*), *Phylctaena vagabunda*	Apples, pears
Pineapple black rot	*Ceratocystis paradoxa* (*= Thielaviopsis paradoxa*)	Pineapples
Phytophora rot	*Colletotrichum coccodes*	Vegetables
Pink mold rot	*Trichothecium roseum*	
Rhizopus soft rot	*Rhizopus stolonifer*	Sweet potatoes, tomatoes
Slimy brown rot	*Rhizoctonia* spp.	Vegetables
"Smut" (black mold rot)	*Aspergillus niger*	Peaches, apricots
Sour rot	*Geotrichum candidum*	Tomatoes, citrus fruits
Stem-end rot	*Phomopsis citri, Diploida natalensis, Alternaria citri*	Citrus fruits
Watery soft rot	*Sclerotinia sclerotiorum*	Carrots

garlic, beans (green, lima, and wax), carrots, parsnips, celery, tomatoes, endives, globe artichokes, lettuce, rhubarb, cabbage, brussels sprouts, cauliflower, broccoli, radishes, rutabagas, turnips, cucumbers, pumpkin, squash, peppers, sweet potatoes, and others. In this disease, the causal fungus grows on decayed areas in the form of a prominent gray mold. It can enter fruits and vegetables through the unbroken skin, or through cuts and cracks.

SOUR ROT (OOSPORA ROT, WATERY SOFT ROT). This market disease of vegetables is caused by *Geotrichum candidum* and other organisms. Among the vegetables affected are asparagus, onions, garlic, beans (green, lima, and wax), carrots, parsnips, parsley, endives, globe artichokes, lettuce, cabbage, brussels sprouts, cauliflower, broccoli, radishes, rutabagas, turnips, and tomatoes. The causal fungus of this disease is widely distributed in soils and on decaying fruits and vegetables. *Drosophila melanogaster* (fruit fly) carries spores and mycelial fragments on its body from decaying fruits and vegetables to growth cracks and wounds in healthy fruits and vegetables. Since the fungus cannot enter through the unbroken skin, infections usually start in openings of one type or another (5).

RHIZOPUS SOFT ROT. This market disease is caused by *R. stolonifer* and other species that make vegetables soft and mushy. Cottony growth of the mold with small black dots of sporangia often covers the vegetables. Among those affected are beans (green, lima, and wax), carrots, sweet potatoes, potatoes, cabbage, brussels sprouts, cauliflower, broccoli, radishes, rutabagas, turnips, cucumbers, cantaloupes, pumpkins, squash, watermelons, and tomatoes. This fungus is spread by *D. melanogaster,* which lays its eggs in the growth cracks on various fruits and vegetables. The fungus is widespread and is disseminated also by other means. Infections usually occur through wounds and other skin breaks.

PHYTOPHORA ROT. This market disease is caused by *Phytophora* spp. It occurs largely in the field as a blight and fruit rot of market vegetables. It appears to be more variable than some other market diseases and affects different plants in different ways. Among the vegetables affected are asparagus, onions, garlic, cantaloupes, watermelons, tomatoes, eggplants, and peppers.

ANTHRACNOSE. This plant disease is characterized by spotting of leaves, fruit, or seed pods. It is caused by *Colletotrichum coccodes* and other species. These fungi are considered weak plant pathogens. They live from season to season on plant debris in the soil and on the seed of various plants such as the tomato. Their spread is favored by warm, wet weather. Among the vegetables affected are beans, cucumbers, watermelons, pumpkins, squash, tomatoes, and peppers.

For further information on market diseases of fruits and vegetables, the

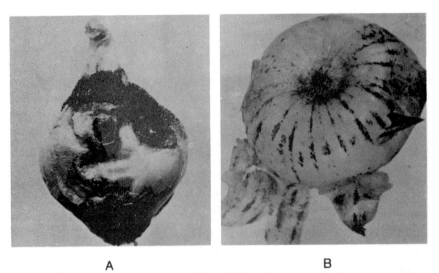

Fig. 8-3. Onion diseases: *A*, white rot; *B*, black mold rot; *C*, diplodia stain. (From *Agriculture Handbook 303*, USDA, 1966, Chapter 5)

monographs issued by the Agricultural Research Service of the U.S. Department of Agriculture should be consulted (see Fig. 8-3 for several fungal diseases of onions).

SPOILAGE OF FRUITS

The general composition of eighteen common fruits is presented in Table 8-4, which shows that the average water content is about 85% and the average carbohydrate content is about 13%. The fruits differ from vegetables in having somewhat less water but more carbohydrate. The mean protein, fat, and ash content of fruits are, respectively, 0.9, 0.5, and 0.5%—somewhat lower than vegetables except for ash content. Though not shown in the table, fruits contain vitamins and other organic compounds just as do vegetables. On the basis of nutrient content, these products would appear to be capable of supporting the growth of bacteria, yeasts, and molds. However, when the pH of fruits alone is considered, it is found to be below the level that generally favors bacterial growth. This one fact alone would seem to be sufficient to explain the general absence of bacteria in the incipient spoilage of fruits. The wider pH growth range of molds and yeasts suits them as spoilage agents of fruits. With the exception of pears, which sometimes undergo **Erwinia rot,** bacteria are of no known importance in the initiation of fruit spoilage. Just why pears with a reported pH range of 3.8–4.6 should

Table 8-4. **Approximate percentage composition of eighteen common fruits (7).**

Fruit	Water	Carbohydrate	Protein	Ash	Fat
Apples	84.1	14.9	0.3	0.3	0.4
Apricots	85.4	12.9	1.0	0.6	0.1
Bananas	74.8	23.0	1.2	0.8	0.2
Blackberries	84.8	12.5	1.2	0.5	1.0
Cherries, sweet and sour	83.0	14.8	1.1	0.6	0.5
Figs	78.0	19.6	1.4	0.6	0.4
Grapefruit	88.8	10.1	0.5	0.4	0.2
Grapes, American type	81.9	14.9	1.4	0.4	1.4
Lemons	89.3	8.7	0.9	0.5	0.6
Limes	86.0	12.3	0.8	0.8	0.1
Oranges	87.2	11.2	0.9	0.5	0.2
Peaches	86.9	12.0	0.5	0.5	0.1
Pears	82.7	15.8	0.7	0.4	0.4
Pineapples	85.3	13.7	0.4	0.4	0.2
Plums	85.7	12.9	0.7	0.5	0.2
Raspberries	80.6	15.7	1.5	0.6	1.6
Rhubarb	94.9	3.8	0.5	0.7	0.1
Strawberries	89.9	8.3	0.8	0.5	0.5
Mean	84.9	13.2	0.88	0.53	0.46

undergo bacterial spoilage is not clear. It is conceivable that *Erwinia* initiates its growth on the surface of this fruit where the pH is presumably higher than on the inside.

A variety of yeast genera can usually be found on fruits, and these organisms often bring about the spoilage of fruit products, especially in the field. Many yeasts are capable of attacking the sugars found in fruits and bringing about fermentation with the production of alcohol and CO_2. Due to their generally faster growth rate than molds, they often precede the latter organisms in the spoilage process of fruits in certain circumstances. It is not clear whether some molds are dependent upon the initial action of yeasts in the process of fruit and vegetable spoilage. The utilization or destruction of the high molecular weight constituents of fruits is brought about more by molds than yeasts. Many molds are capable of utilizing alcohols as sources of energy and when these and other simple compounds have been depleted, these organisms proceed to destroy the remaining parts of fruits such as the structural polysaccharides and rinds.

REFERENCES

1. Beraha, L., M. A. Smith, and W. R. Wright. 1961. Control of decay of fruits and vegetables during marketing. *Dev. Indust. Microbiol.* 2:73–77.
2. Eckert, J. W. 1979. Fungicidal and fungistatic agents: Control of pathogenic microorganisms on fresh fruits and vegetables after harvest. In *Food mycology,* ed. M. E. Rhodes, 164–99. Boston: Hall.
3. Jones, S. M., and A. M. Paton. 1973. The L-phase of *Erwinia carotovora* var. *atroseptica* and its possible association with plant tissue. *J. Appl. Bacteriol.* 36:729–37.
4. Lund, B. M. 1971. Bacterial spoilage of vegetables and certain fruits. *J. Appl. Bacteriol.* 34:9–20.
5. McColloch, L. P., H. T. Cook, and W. R. Wright. 1968. Market diseases of tomatoes, peppers, and eggplants. *Agricultural Handbook no. 28.* Washington, D.C.: Agricultural Research Service, U.S.D.A.
6. Mount, M. S., D. F. Bateman, and H. G. Basham. 1970. Induction of electrolyte loss, tissue maceration, and cellular death of potato tissue by an endopolygalacturonate trans-eliminase. *Phytopathol.* 60:924–1000.
7. Watt, B. K., and A. L. Merrill. 1950. Composition of foods—raw, processed, prepared. *Agricultural Handbook no. 8,* Washington, D.C.: Agricultural Research Service, U.S.D.A.

9.

SPOILAGE OF FRESH AND PROCESSED MEATS, POULTRY, AND SEAFOOD

Meats are the most perishable of all important foods, and the reasons for this may be seen in Table 9-1, in which the chemical composition of a typical adult mammalian muscle postmortem is presented. Meats contain an abundance of all nutrients required for the growth of bacteria, yeasts, and molds, and an adequate quantity of these constituents exist in fresh meats in available form. The general chemical composition of a variety of meats is presented in Table 9-2.

The genera of bacteria most often found on fresh and spoiled meats, poultry, and seafood are listed in Table 9-3. Not all of the genera indicated for a given product would be found at all times, of course. Those that are more often found during spoilage are indicated under the various products. In Table 9-4 are listed the genera of molds most often identified from meats and related products, while the identified yeasts are listed in Table 9-5. When spoiled meat products are examined, only a few of the many genera of bacteria, molds, or yeasts are found, and in almost all cases one or more genera are found to be characteristic of the spoilage of a given type of meat product. The presence of the more varied flora on nonspoiled meats, then, may be taken to represent the organisms that exist in the original environment of the product in question, or contaminants picked up during processing, handling, packaging, and storage.

The question arises, then, as to why only a few types predominate in spoiled meats. It is helpful here to return to the intrinsic and extrinsic parameters that affect the growth of spoilage microorganisms. Fresh meats such as beef, pork, and lamb as well as fresh poultry, seafood, and processed meats all have pH values within the growth range of most of the organisms

Table 9-1. Chemical composition of typical adult mammalian muscle after rigor mortis but before degradative changes post-mortem (percent wet weight) (67).

Water	75.5%
Protein	18.0
Myofibrillar—myosin, tropomyosin, X protein ...7.5%	
—actin ...2.5	
Sarcoplasmic—myogen, globulins ...5.6	
—myoglobin ...0.36	
—haemoglobin ...0.04	
Mitochondrial—cytochrome C ... ca. 0.002	
Sarcoplasmic reticulum, collagen, elastin, "reticulin," insoluble enzymes, connective tissue ...2.0	
Fat	3.0
Soluble nonprotein substances	3.5
Nitrogenous—creatine ...0.55	
—inosine monophosphate ...0.30	
—di- and tri-phosphopyridine nucleotides ...0.07	
—amino acids ...0.35	
—carnosine, anserine ...0.30	
Carbohydrate—lactic acid ...0.90	
—glucose-6-phosphate ...0.17	
—glycogen ...0.10	
—glucose ...0.01	
Inorganic—total soluble phosphorous ...0.20	
—potassium ...0.35	
—sodium ...0.05	
—magnesium ...0.02	
—calcium ...0.007	
—zinc ...0.005	
Traces of glycolytic intermediates, trace metals, vitamins, etc. ... ca. 0.10	

Reprinted with permission from R. A. Lawrie, *Meat Science,* copyright 1966, Pergamon Press.

Table 9-2. Approximate percentage chemical composition of nine meats and meat products (118).

Meats	Water	Carbohydrates	Proteins	Fat	Ash
Beef, hamburger	55.0	0	16.0	28.0	0.8
Beef, round	69.0	0	19.5	11.0	1.0
Bologna	62.4	3.6	14.8	15.9	3.3
Chicken (broiler)	71.2	0	20.2	7.2	1.1
Frankfurters	60.0	2.7	14.2	20.5	2.7
Lamb	66.3	0	17.1	14.8	0.9
Liver (beef)	69.7	6.0	19.7	3.2	1.4
Pork, medium	42.0	0	11.9	45.0	0.6
Turkey, medium fat	58.3	0	20.1	20.2	1.0

Table 9-3.   The genera of bacteria most frequently found on meats, poultry, and seafood.

Genus	Gram reaction	Fresh meats	Fresh livers	Processed meats	Vacuum-packaged meats	Bacon[a]	Poultry	Fish and seafood
Acinetobacter	−	XX	X	X	X	X	XX	X
Aeromonas	−	XX			X	X	X	X
Alcaligenes	−	X	X			X	X	X
Alteromonas	−	X			X		X	X
Arthrobacter[b]	−/+	X		X	X	X	X	
Bacillus	+	X		X		X	X	X
Bacteroides	−	X						
Brochothrix	+	X	X	X	XX			
Campylobacter	−						XX	
Chromobacterium	−						X	X
Citrobacter	−	X					X	
Clostridium	+	X					X	
Corynebacterium[b]	+	X	X	X	X	X	XX	X
Cytophaga	−							X
Enterobacter	−	X		X	X		X	X
Escherichia	−	X					X	X
Flavobacterium	−	X	X				XX	X
Hafnia	−	X			X			
Halobacterium	−							X
Kurthia[b]	+	X			X			
Lactobacillus	+	X		XX	XX			X
Leuconostoc	+	X	X	X	X			
Microbacterium[b]	+	X		X	X	X	X	X
Micrococcus	+	X	XX	X	X	X	X	
Moraxella	−	XX	X			X	X	X
Neisseria	−	X		X		X	X	
Pediococcus	+	X		X	X			
Photobacterium	−							X
Planococcus	+						X	
Plesiomonas	−						X	
Proteus	−	X					X	
Pseudomonas	−	XX			X		XX	XX
Salmonella	−	X					X	
Serratia	−	X		X	X		X	
Staphylococcus	+	X	X	X	X	X	X	
Streptococcus	+	X		XX	X		X	X
Streptomyces	+	X					X	
Vibrio	−					X		X
Yersinia[c]	−	X			X			

[a] Vacuum packaged not included.
[b] Belong to the coryneform group. The organisms once classified as *Brevibacterium* belong to this group.
[c] Especially *Y. enterocolitica.*
X = known to occur; XX = most frequently reported.

Table 9-4. Genera of molds most often found on meats, poultry, and seafood products (taken from the literature and a previous review (61).

Genus	Fresh and ref. meats	Poultry	Fish and shrimp	Processed and cured meats
Alternaria	X	X		X
Aspergillus	X	X	X	XX
Botrytis	X			X
Cladosporium	XX	X		X
Fusarium	X			X
Geotrichum	XX	X		X
Monascus	X			
Monilia	X			X
Mortierella	X			
Mucor	XX	X		X
Neurospora	X			
Oidium	X			X
Oospora	X		X	
Penicillium	X	X	X	XX
Rhizopus	XX	X		X
Scopulariopsis			X	X
Sporotrichum	XX			
Thamnidium	XX			X
Wallemia (Sporendonema)			X	
Zygorrhynchus				X

X = known to occur; XX = most frequently found.

Table 9-5. Yeast genera most often identified on meats, poultry, and seafood products (taken from the literature and a previous review (61).

Genus	Fresh and ref. meats	Poultry	Fish and shrimp	Processed and cured meats
Candida	XX	XX	XX	X
Cryptococcus			X	
Debaryomyces	X		X	XX
Hansenula			X	
Pichia			X	
Pullularia			X	
Rhodotorula	X	XX	XX	
Saccharomyces		X		
Sporobolomyces			X	
Torula		XX	XX	X
Torulopsis	X	X	X	X
Trichosporon		X	X	X

X = known to occur; XX = most frequently found.

listed in Table 9-3. Nutrient and moisture contents as stated above are adequate to support the growth of all organisms listed. While the O/R potential of whole meats is low, O/R conditions at the surfaces tend to be higher so that strict aerobes and facultative anaerobes, as well as strict anaerobes, generally find conditions suitable for growth. Antimicrobial constituents are not known to occur in products of the type in question. Upon examining the extrinsic parameters, temperature of incubation stands out as being of utmost importance in controlling the types of microorganisms that develop on meats, since these products are normally held at refrigerator temperatures. Essentially all studies on the spoilage of meats, poultry, and seafood carried out over the past 30 years or so have dealt with low-temperature-stored products.

SPOILAGE OF FRESH BEEF, PORK, AND RELATED MEATS

Most studies dealing with the spoilage of meats have been done with beef, and most of the discussion in this section is based upon beef studies. It should be pointed out, however, that pork, lamb, veal, and similar meats are presumed to spoil in a similar way.

Upon the slaughter of a well-rested beef animal, a series of events take place that lead to the production of meat. Lawrie (67) discussed these events in great detail and they are here presented only in outline form. The events following an animal's slaughter are: (1) its circulation ceases; the ability to resynthesize ATP (adenosine triphosphate) is lost; lack of ATP causes actin and myosin to combine to form actomyosin, which leads to a stiffening of muscles; (2) the oxygen supply fails, resulting in a reduction of the O/R potential; (3) the supply of vitamins and antioxidants ceases, resulting in a slow development of rancidity; (4) nervous and hormonal regulations cease, thereby causing the temperature of the animal to fall and fat to solidify; (5) respiration ceases, which stops ATP synthesis; (6) glycolysis begins, resulting in the conversion of most glycogen to lactic acid, which depresses pH from about 7.4 to its ultimate level of about 5.6; this pH depression also initiates protein denaturation, liberates and activates cathepsins, and completes rigor mortis; protein denaturation is accompanied by an exchange of divalent and monovalent cations on the muscle proteins; (7) the reticuloendothelial system ceases to scavenge, thus allowing microorganisms to grow unchecked; and (8) various metabolites accumulate that also aid protein denaturation.

The above events require between 24–36 h at the usual temperatures of holding freshly slaughtered beef (35–40°C). Meanwhile, part of the normal flora of this meat has come from the animal's own lymph nodes (72), the stick knife used for exsanguination, the hide of the animal, intestinal tract, dust, hands of handlers, cutting knives, storage bins, and the like. Upon prolonged storage at refrigerator temperatures, microbial spoilage begins.

In the event that the internal temperatures are not reduced to the refrigerator range, the spoilage that is likely to occur is caused by bacteria of internal sources. Chief among these are *Clostridium perfringens* and genera in the Enterobacteriaceae family (54). On the other hand, bacterial spoilage of refrigerator-stored meats is, by and large, a surface phenomenon reflective of external sources of the spoilage flora (54).

With respect to fungal spoilage of fresh meats, especially beef, the following genera of molds have been recovered from various spoilage conditions of whole beef: *Thamnidium, Mucor,* and *Rhizopus,* all of which produce "whiskers" on beef; *Cladosporium,* a common cause of "black spot"; *Penicillium,* which produces green patches; and *Sporotrichum* and *Chryso-sporium,* both of which produce "white spot." Molds apparently do not grow on meats if the storage temperature is below −5°C (75). Among genera of yeasts recovered from refrigerator-spoiled beef with any consistency are *Candida, Torulopsis,* and *Rhototorula,* with *C. lipolytica* and *C. zeylanoides* being the two most abundant species in spoiled ground beef (51).

Unlike the spoilage of fresh beef carcasses, ground beef or hamburger meat is spoiled exclusively by bacteria, with the following genera being the most important: *Pseudomonas, Alcaligenes, Acinetobacter, Moraxella,* and *Aeromonas.* Those generally agreed to be the primary cause of spoilage are *Pseudomonas* and *Acinetobacter-Moraxella* spp., with others playing relatively minor roles in the process (7, 8, 13, 38, 60, 63). Findings from a recent study suggest that *Acinetobacter* spp. may not be as abundant in spoiled beef as reported in the past (30).

A study of the aerobic gram-negative bacteria recovered from beef, lamb, pork, and fresh sausage revealed that all 231 polarly flagellated rods were pseudomonads and that, of 110 nonmotile organisms, 61 were *Moraxella* while 49 were *Acinetobacter* (24). The pseudomonads that cause meat spoilage at low temperatures generally do not match the named species in *Bergey's Manual.* Numerical taxonomic studies by Shaw and Latty (101, 102) led them to group most of their isolates into four clusters based upon carbon source utilization tests. Of 787 *Pseudomonas* strains isolated from meats, 89.7% were identified with 49.6% belonging to their cluster 2, 24.9% to cluster 1, and 11.1% to cluster 3 (102). The organisms in clusters 1 and 2 were nonfluorescent and egg-yolk negative, and resembled *P. fragi,* while those in cluster 3 were fluorescent and gelatinase positive. *P. fluorescens* biotype I strains were represented by 3.9%, biotype III by 0.9%, and *P. putida* by only one strain. The relative incidence of the clusters on beef, pork, and lamb and on fresh and spoiled meats was similar (102).

Beef rounds and quarters are known to undergo deep spoilage, usually near the bone, especially the "aitch" bone. This type of spoilage is often referred to as "bone taint" or "sours." Only bacteria have been implicated, with the genera *Clostridium* and *Streptococcus* being the primary causative agents (15).

Temperature of incubation is the primary reason why only a few genera of bacteria are found in spoiled meats as opposed to fresh. In one study, only four of the nine genera present in fresh ground beef could be found after the meat underwent frank spoilage at refrigerator temperatures (60). It was noted by Ayres (7) that after processing, more than 80% of the total population of freshly ground beef may be comprised of chromogenic bacteria, molds, yeasts, and sporeforming bacteria, but after spoilage only nonchromogenic, short gram-negative rods are found. While some of the bacteria found in fresh meats can be shown to grow at refrigerator temperatures on culture media, they apparently lack the capacity to compete successfully with the *Pseudomonas* and *Acinetobacter-Moraxella* types.

Beef cuts, such as steaks or roasts, tend to undergo surface spoilage, and whether or not the spoilage organisms are bacteria or molds depends upon available moisture. Freshly cut meats stored in a refrigerator with high humidity invariably undergo bacterial spoilage preferential to mold spoilage. The essential feature of this spoilage is surface sliminess in which the causative organisms can nearly always be found. The relatively high Eh, availability of moisture, and low temperature all favor the pseudomonads. It is sometimes possible to note discrete bacterial colonies on the surface of beef cuts, especially when the level of contamination is low. The slime layer results from the coalescence of surface colonies and is largely responsible for the tacky consistency of spoiled meats. Ayres (7) presented evidence that odors can be detected when the surface bacterial count is between log 7.0 and $7.5/cm^2$, followed by detectable slime with surface counts usually about log $7.5-8.0/cm^2$ (see Fig. 9-1). Molds tend to predominate in the spoilage of beef cuts when the surface is too dry for bacterial growth, or when beef has been treated with antibiotics such as the tetracyclines. Molds virtually never develop on meats when bacteria are allowed to grow freely. The reason for this appears to be that bacteria grow faster than molds, thus consuming available surface oxygen, which molds also require for their activities.

Unlike the case of beef cuts or beef quarters, mold growth is quite rare on ground beef except when antibacterial agents have been used as preservatives, or when the normal bacterial load has been reduced by long-term freezing. Among the early signs of spoilage of ground beef is the development of off-odors followed by tackiness, which indicate the presence of bacterial slime. The sliminess is due both to masses of bacterial growth and the softening or loosening of meat structural proteins.

In the spoilage of soy-extended ground meats, nothing indicates that the pattern differs from that of unextended ground meats even though their rate of spoilage is faster (see Chapter 4).

The precise roles played by spoilage microorganisms that result in the spoilage of meats are not fully understood at this time, but significant progress has been made over the past decade. Some of the earlier views

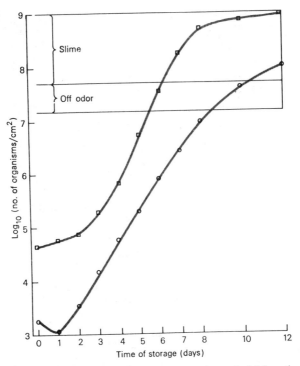

Fig. 9-1. The development of off-odor and slime on dressed chicken (squares) and packaged beef (circles) during storage at 5°C (7).

on the mechanism of meat spoilage are embodied in the many techniques proposed for its detection (see Table 9-6).

DETECTION AND MECHANISM OF MEAT SPOILAGE

It is reasonable to assume that reliable methods of determining meat spoilage should be based upon the cause and mechanism of spoilage. The chemical methods in Table 9-6 all embody the assumption that as meats undergo spoilage, some utilizable substrate is consumed, or some new product or products are created by the spoilage flora. It is well established that the spoilage of meats at low temperatures is accompanied by the production of off-odor compounds such as ammonia, H_2S, indole, amines, and others (see below). The drawbacks to the use of these methods are that not all spoilage organisms are equally capable of producing them. Inherent in some of these methods is the incorrect belief that low-temperature spoilage is accompanied by a breakdown of primary proteins (58). The physical and direct bacteriological methods all tend to show what is obvious—that meat that is clearly spoiled from the standpoint of organoleptic characteristics (odor, touch, appearance, and taste) is, indeed, spoiled! They apparently

Table 9-6.   Some methods proposed for detecting microbial spoilage in meats, poultry, and seafood.

Chemical methods

a. Measurement of H_2S production
b. Measurement of mercaptans produced
c. Determination of noncoagulable nitrogen
d. Determination of di- and trimethylamines
e. Determination of tyrosine complexes
f. Determination of indole and skatol
g. Determination of amino acids
h. Determination of volatile reducing substances
i. Determination of amino nitrogen
j. Determination of biochemical oxygen demand (BOD)
k. Determination of nitrate reduction
l. Measurement of total nitrogen
m. Measurement of catalase
n. Determination of creatinine content
o. Determination of dye-reducing capacity
p. Measurement of hypoxanthine
q. ATP measurement
r. Radiometric measurement of CO_2
s. Ethanol production (fish spoilage)
t. Measurement of lactic acid
u. Change in color

Physical methods

a. Measurement of pH changes
b. Measurement of refractive index of muscle juices
c. Determination of alteration in electrical conductivity
d. Measurement of surface tension
e. Measurement of UV illumination (fluorescence)
f. Determination of surface charges
g. Determination of cryoscopic properties
h. Impedance changes
i. Microcalorimetry

Direct bacteriological methods

a. Determination of total aerobes
b. Determination of total anaerobes
c. Determination of ratio of total aerobes to anaerobes
d. Determination of one or more of above at different temperatures
e. Determination of gram-negative endotoxins

Physio-chemical methods

a. Determination of extract-release volume (ERV)
b. Determination of water-holding capacity (WHC)
c. Determination of viscosity
d. Determination of meat swelling capacity

do not allow one to predict spoilage or shelf life, which a meat freshness test should ideally do.

The use of dye-reduction tests, especially resazurin, to assess freshness or spoilage in meat products continues to receive attention. While resazurin reduction was found unsuitable for raw meats because of inherent enzymic activity, the procedure was shown to correlate with bacterial numbers in about 600 samples of cooked meats (5). As noted in Chapter 5, homogenates from raw meat samples prepared by Stomacher rather than by mechanical blenders are relatively free of meat enzymes, and by use of this method, spoiled meats could be detected within 2 h by resazurin reduction (25). For deep-frozen, precooked shrimp, tissue exudates were employed for resazurin reduction and the method produced results in 4 h that did not differ significantly from plate counts (65). The problems associated with the use of dye-reduction tests are noted in Chapter 5.

The extract-release volume (ERV) technique, first described in 1964, has been shown to be of value in determining incipient spoilage in meats as well as in predicting refrigerator shelf life (56, 57, 59, 94). The ERV technique is based upon the volume of aqueous extract released by a homogenate of beef when allowed to pass through filter paper for a given period of time. By this method, beef of good organoleptic and microbial quality releases large volumes of extract while beef of poor microbial quality releases smaller volumes or none (Fig. 9-2). One of the more important aspects of this method is the information that it has provided concerning the mechanism of low-temperature beef spoilage.

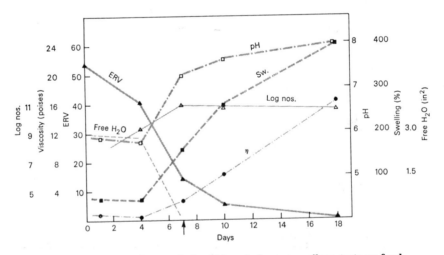

Fig. 9-2. The response of several physiochemical meat spoilage tests as fresh ground beef was held at 7°C until definite spoilage had occurred. The arrow indicates the first day off-odors were detected. ERV = extract-release volume; free H_2O = measurement of water-holding capacity (inversely related); Sw = meat swelling; η = viscosity; and log Nos. = total aerobic bacteria/g. (106, copyright © 1969 by Institute of Food Technologists)

The ERV method of detecting meat spoilage reveals two aspects of the spoilage mechanism not previously recognized. First, low-temperature meat spoilage occurs in the absence of any significant breakdown of primary proteins—at least not complete breakdown. Although this fact has been verified by total protein analyses on fresh and spoiled meats, it is also implicit in the operation of the method. That is, as meats undergo microbial spoilage, ERV is *decreased* rather than increased, which would be the case if complete hydrolysis of proteins occurred. The second aspect of meat spoilage revealed by ERV is the increase in hydration capacity of meat proteins by some as yet unknown mechanism, although amino sugar complexes produced by the spoilage flora have been shown to play a role (107). In the absence of complete protein breakdown, the question arises as to how the spoilage flora obtains its nutritional needs for growth.

When fresh meats are placed in storage at refrigerator temperatures, those organisms capable of growth at the particular temperature begin their growth. In the case of fresh meats that have an ultimate pH of around 5.6, enough glucose and other simple carbohydrates are present to support about 10^8 organisms/cm^2 (36). Among the heterogeneous fresh-meat flora, the organisms that grow the fastest and utilize glucose at refrigerator temperatures are the pseudomonads, and available surface O_2 has a definite effect on their ultimate growth (40). *Brochothrix thermosphacta* also utilizes glucose and glutamate but because of its slower growth rate, it is a poor competitor of the pseudomonads. Upon reaching a surface population of about 10^8/cm^2, the supply of simple carbohydrates is exhausted and off-odors may or may not be evident at this point depending upon the extent to which free amino acid utilization has occurred. Once simple carbohydrates have been exhausted, pseudomonads along with gram-negative psychrotrophs such as *Moraxella, Alcaligenes, Aeromonas, Serratia,* and *Enterobacter* utilize free amino acids and related simple nitrogenous compounds as sources of energy. *Acinetobacter* spp. utilize amino acids first, lactate next, and their growth is reduced at and below pH 5.7 (40).

The foul odors generally associated with spoiling meats owe their origin to free amino acids and related compounds (H_2S from sulfur-containing amino acids, NH_3 from many amino acids, and indole from tryptophane). Off-odors and off-flavors appear only when amino acids begin to be utilized (see below). In the case of dark, firm, and dry meats (DFD), which have ultimate pH > 6.0 and a considerably lower supply of simple carbohydrates, spoilage is more rapid and off-odors are detectable with cell numbers of around 10^6/cm^2 (89). With normal or DFD meats, the primary proteins are not attacked until the supply of the simpler constituents has been exhausted. It has been shown, for example, that the antigenicity of salt-soluble beef proteins is not destroyed under the usual conditions of low-temperature spoilage (76).

In the case of fish spoilage, it has been shown that raw fish press juice displays all the apparent aspects of fish spoilage as may be determined by use of the whole fish (73). This can be taken to indicate a general lack of attack upon insoluble proteins by the fish spoilage flora since these proteins were absent from the filtered press juice.

The same is apparently true for beef and related meats. Incipient spoilage is accompanied by a rise in pH, increase in bacterial numbers, increase in the hydration capacity of meat proteins, along with other changes. In ground beef, pH may rise as high as 8.5 in putrid meats, although at the time of incipient spoilage mean pH values of about 6.5 have been found (108). By plotting the growth curve of the spoilage flora, the usual phases of growth can be observed and the phase of decline may be ascribed to the exhaustion of utilizable nutrients by most of the flora and the accumulation of toxic by-products of bacterial metabolism. Precisely how the primary proteins of meat are destroyed at low temperatures is not well understood.

Dainty et al. (22) inoculated beef slime onto slices of raw beef and incubated them at 5°C. Off-odors and slime were noted after 7 days with counts at $2 \times 10^9/cm^2$. Proteolysis was not detected in either sarcoplasmic or myofibrillar fractions of the beef slices. No changes in the sarcoplasmic fractions could be detected even 2 days later when bacterial numbers reached $10^{10}/cm^2$. The first indication of breakdown of myofibrillar proteins occurred at this time with the appearance of a new band and the weakening of another. All myofibrillar bands disappeared after 11 days with weakening of several bands of the sarcoplasmic fraction. With naturally contaminated beef, odors and slime were first noted after 12 days when the numbers were $4 \times 10^8/cm^2$. Changes in myofibrillar proteins were not noted until 18 days of holding. By the use of pure culture studies, these workers showed that Shewan's Group I pseudomonads (see Appendix for the Shewan scheme) were active against myofibrillar proteins while Group II organisms were more active against sarcoplasmics. *Aeromonas* spp. were active on both myofibrillar and sarcoplasmic proteins. With pure cultures, protein changes were not detected until counts were above $3.2 \times 10^9/cm^2$. Borton et al. (12) showed earlier that *P. fragi* (a Group II pseudomonad) effected the loss of protein bands from inoculated pork muscle, but no indication was given as to the minimum numbers that were necessary.

For further information on the mechanism of meat spoilage, the following references should be consulted: 23, 37, 38, or 82.

SPOILAGE OF FRESH LIVERS

The events that occur in the spoilage of beef, pork, and lamb livers are not as well defined as for meats. The mean content of carbohydrate, NH_3, and pH of ten fresh lamb livers is presented in Table 9-7 (39). Based on the relatively high content of carbohydrates and mean pH of 6.41, these may be expected to undergo a fermentative spoilage with the pH decreasing below 6.0. This would undoubtedly occur if livers were comminuted or finely diced and stored at refrigerator temperatures, but most studies have been conducted on whole livers, where growth was assessed at the surface,

Table 9-7. pH and concentrations in 10 fresh livers of glycogen, glucose, lactic acid, and ammonia (39).

Component	Average concentration and range
Glucose	2.73 (0.68–6.33) mg/g
Glycogen	2.98 (0.70–5.43) mg/g
Lactic acid	4.14 (3.42–5.87) mg/g
Ammonia	7.52 (6.44–8.30) μmol/g
pH	6.41 (6.26–6.63)

from drip, or from deep tissue. In a study of the spoilage of diced beef livers, the initial pH of 6.3 decreased to about 5.9 after 7–10 days at 5°C and the predominant flora at spoilage consisted of lactic acid bacteria (104). In most other studies, the predominant flora at spoilage was found to consist essentially of the same types of organisms that are dominant in the spoilage of muscle meats. In pork livers held at 5°C for 7 days, the predominant organisms found in one study were *Pseudomonas, Alcaligenes, Escherichia,* lactic streptococci, and *B. thermosphacta* (34). In five beef livers stored at 2°C for 14 days, *Pseudomonas* constituted from 75 to 100% of the spoilage flora while the mean initial pH of 6.49 decreased to 5.93 over the 14-day period (44). In another study of beef, pork, and lamb livers, the predominant flora after 5 days at 2°C differed for the three products, with beef livers being dominated by streptococci, yeasts, coryneforms, and pseudomonads; lamb by coryneforms, micrococci, and streptococci; and pork livers by staphylococci, *Moraxella-Acinetobacter,* and streptococci (43). The mean initial pH of each of the three livers declined upon storage, although only slightly. In a study of spoilage of lamb livers by Gill and DeLacy (39), the spoiled surface flora was dominated by *Pseudomonas, Acinetobacter,* and *Enterobacter;* drip from the whole livers was dominated by *Pseudomonas* and *Enterobacter;* while *Enterobacter* and lactobacilli were dominant in the deep tissues. It was shown in this study that the initial pH of around 6.4 decreased to around 5.7 in antibiotic-treated samples, indicating that liver glycolytic events can lead to a decrease in pH in the absence of organisms even though these samples did contain $< 10^4$ organisms/cm^2. It was noted that the high glucose level was sufficient to allow visible surface colony growth before off-odors developed, and herein may lie the explanation for the dominance of the spoilage flora of livers by nonlactic types.

Since most psychrotrophic oxidative gram-negative bacteria grow at a faster rate and are more favored by the higher surface Eh than the lactic fermentative gram positives, their dominance in whole liver spoilage may not be unexpected. The higher concentration of carbohydrates would delay the onset of amino acid utilizers and consequently explain in part why pH does not increase with whole liver spoilage as it does for meats. In this regard, comminuted livers would be expected to support the growth of lactic acid bacteria because of the redistribution of the surface flora throughout the sample where the lactics would be favored by the high carbohydrate content and reduced Eh away from the surface. This would be somewhat analogous to the surface spoilage of meat carcasses, where the slower-growing yeasts and molds develop when conditions are not favorable for bacterial growth. Fungi never dominate the spoilage of fresh comminuted meats unless special steps are taken to inhibit bacteria. By this analogy, lactic acid bacteria are inconspicuous in the spoilage of whole livers because conditions favor the faster-growing, psychrotrophic gram-negative bacteria.

SPOILAGE OF VACUUM-PACKAGED MEATS

From the research of many groups over the past decade, it is clear that when vacuum-packaged meats undergo long-term refrigerator spoilage, very often the predominant organisms are lactobacilli or *B. thermosphacta* or both. Other organisms can be found and, indeed, others may predominate. Among the determining factors are the following: (1) whether the product is raw or cooked, (2) concentration of nitrites present, (3) relative load of psychrotrophic bacteria, (4) the degree to which the vacuum-package film excludes O_2, and (5) product pH.

Cooked or partially cooked meats along with DFD and dark-cutting meats have higher pH than raw and light-cutting meats, and the organisms that dominate these products during vacuum storage are generally different from those found in vacuum-packaged normal meats. In vacuum-packaged DFD meats held at 2°C for 6 weeks, the dominant flora consisted of *Y. enterocolitica, S. liquefaciens, A. putrefaciens,* and a *Lactobacillus* sp. (41). *A. putrefaciens* caused greening of product but a pH < 6.0 was inhibitory to its growth. When dark-cutting beef of pH 6.6 was vacuum packaged and stored at 0°–2°C, lactobacilli were dominant after 6 weeks but after 8 weeks psychrotrophic Enterobacteriaceae became dominant (93). Most of the Enterobacteriaceae resembled *S. liquefaciens* and the remainder resembled *Hafnia alvei.* In vacuum-packaged beef with pH 6.0, *Y. enterocolitica*-like organisms were found at levels of $10^7/g$ after 6 weeks at 0–2°C, but on meats with pH < 6.0, their numbers did not exceed $10^5/g$ even after 10 weeks (99). The high-pH meat also yielded *A. putrefaciens* with counts as high as log 6.58/g after 10 weeks.

When normal raw beef with an ultimate pH of about 5.6 is vacuum packaged, lactobacilli and other lactic acid bacteria predominate. When allowed to spoil aerobically, acidic/sour odors were noted when APC was about 10^7–$10^8/cm^2$ with approximately 15% of the flora being *Pseudomonas* spp.; but when vacuum-packaged samples spoiled, the product was accompanied by a slight increase in pH with general increase in ERV (112). After a 9-week storage at 0°–1°C, Hitchener et al. (49) found that 75% of the flora of vacuum-packaged raw beef consisted of catalase-negative organisms. Upon further characterization of 177 isolates, 18 were found to be *Leuconostoc mesenteroides,* 115 were hetero- and 44 were homofermentative lactobacilli. Using high-barrier oxygen film, the dominant flora of vacuum-packaged beef loin steaks after 12 and 24 days consisted of heterofermentative lactobacilli with *L. cellobiosus* being isolated from 92% of the steaks (116). In 59% of the samples, *L. cellobiosus* constituted 50% or more of the flora. The latter investigators found that when medium oxygen-barrier film was used, high percentages of organisms such as *Aeromonas, Enterobacter, Hafnia, Alteromonas, B. thermosphacta,* pseudomonads, and *Proteus morganii* were usually found.

When high concentrations of nitrites are present, they generally inhibit *B. thermosphacta* and psychrotrophic Enterobacteriaceae, and the lactic acid bacteria become dominant because they are relatively insensitive to nitrites (90). However, low concentrations of nitrites appear not to affect *B. thermosphacta* growth, especially in cooked, vacuum-packaged products. When Egan et al. (28) inoculated this organism and a homo- and a heterofermentative lactobacillus into corned beef and sliced ham containing 240 ppm nitrate and 20 ppm nitrite, *B. thermosphacta* grew with no detectable lag phase. It had a generation time of 12–16 h at 5°C, while the generation time for the heterofermentative lactobacillus was 13–16 h and for the homofermentative 18–22 h. Time to reach 10^8 cells/g was 9, 9–12, and 12–20 days, respectively. While off-flavors developed 2–3 days after the numbers attained 10^8/g for *B. thermosphacta,* the same did not occur for the homo- and heterofermentative lactobacilli until 11 and 21 days, respectively. The above, along with other work, suggests that the lactic acid bacteria are less significant than *B. thermosphacta* in the spoilage of vacuum-packaged luncheon meats (29). On the other hand, this organism has a longer lag phase and a slower growth rate than the lactobacilli (38). When these two groups are present in equal numbers, the lactobacilli generally dominate.

In earlier work employing steaks with pH 5.5–5.7 packaged in oxygen-impermeable bags and held at 38°F for 15 days, 90–95% of the APCs consisted of lactobacilli, with gram negatives and *B. thermosphacta* decreasing (96). In another study in which beef was vacuum packaged and stored at 0°–2°C for up to 9 weeks, from 10 to 30% of the flora developed on lactobacillus MRS agar, from 5 to 49% on a medium selective for *B. thermosphacta,* while for a medium selective for gram negatives from 4 to 61% developed. After 8 weeks, 81% of the total could develop on MRS, indicating that the flora consisted largely of lactic acid bacteria. Further proof of the lactic flora domination on vacuum-packaged meats was shown by a general lack of decrease in ERV compared to controls. These investigators noted that meats with an initially low level of microorganisms had a shelf life 1–2 days longer than meats with a higher initial count. In a study of vacuum-packaged beef and lamb by Hanna et al. (45, 46), the initial flora of both products was dominated by *Corynebacterium* spp. and *B. thermosphacta*, but after 28–35 days lactobacilli dominated the flora of beef while after 21 days pseudomonads and *Moraxella/Acintobacter* spp. were prominent among the flora of lamb along with lactobacilli.

The behavior of *B. thermosphacta* in vacuum-packaged meats and on meats stored under gas atmospheres is discussed in Chapters 3 and 4. While it grows on beef at pH 5.4 when incubated aerobically, it does not grow < pH 5.8 anaerobically (16). Under the latter conditions, the apparent minimum growth pH is 6.0. As noted above, *Alteromonas putrefaciens* is also pH sensitive and does not grow on beef of normal pH but grows on DFD meats. The spoilage of cured meats has been reviewed by Gardner (35).

Volatile components of vacuum-packaged meats, poultry, and seafood

The off-odors and off-flavors produced in vacuum-packaged meat products by the spoilage flora are summarized in Table 9-8. In general, short-chain fatty acids are produced both by lactobacilli and *B. thermosphacta,* and spoiled products may be expected to contain these compounds, which confer sharp off-odors. In vacuum-packaged luncheon meats, acetoin and diacetyl have been found to be the most significant relative to spoiled meat odors (111). Using a culture medium (APT) containing glucose and other simple carbohydrates, the formation by *B. thermosphacta* of isobutyric and isovaleric acids was favored by low glucose and near neutral pH, while acetoin, acetic acid, 2,3-butanediol, 3-methylbutanol, and 3-methylpropanol production were favored by high glucose and low pH (20, 21). According to these authors, acetoin is the major volatile compound produced on raw and cooked meats in O_2-containing atmospheres. This suggests that the volatile compounds produced by *B. thermosphacta* may be expected to vary between products with high and low glucose concentrations. The addition of 2% glucose to raw ground beef has been shown to decrease pH and delay off-odor and slime development without affecting the general spoilage flora (105), and while the studies noted were not conducted with vacuum-packaged meats, it would seem to be a way to shift the volatile components from short-chain fatty acids to acetoin and other compounds that derive from glucose. Since vacuum-packaged, high-pH meats have a much shorter shelf life, the addition of glucose could be of benefit in this regard.

In a study of spoiled vacuum-packaged steaks, a sulfide odor was evident with numbers of 10^7–10^8/cm^2 (42). The predominant organisms isolated were *Hafnia alvei*, lactobacilli, and *Pseudomonas*. *H. alvei* was the likely cause of the sulfide odor.

From the summary of volatiles in Table 9-8, it is evident that all organisms produced either dimethyl di- or trisulfide, or methyl mercaptan, except *B. thermosphacta*. Dimethyl disulfide was produced in chicken by eight of eleven cultures evaluated by Freeman et al. (32), ethanol by seven, and

Table 9-8. Summary of volatile compounds produced by the spoilage flora or specific spoilage organisms in meats, poultry, seafood, or culture media.

Organism/Inoculum	Substrate/Conditions	Principal Volatiles	Reference
Alteromonas putrefaciens	Sterile fish muscle, 1–2° C, 15 days	Dimethyl sulfide, dimethyl trisulfide, methyl mercaptan, trimethylamine, propionaldehyde, 1-penten-3-ol, H_2S, etc.	84
Achromobacter sp.	As above	Same as above except no dimethyl trisulfide or H_2S	84
P. fluorescens	As above	Methyl sulfide, dimethyl disulfide	84
P. perolens	As above	Dimethyl trisulfide, dimethyl disulfide, methyl mercaptan, 2-methoxy-3-isopropylpyrazine (potato-like odor)	85
Moraxella sp.	TSY agar, 2–4°C, 14 days	16 compounds including dimethyl disulfide, dimethyl trisulfide, methyl isobutyrate, and methyl-2-methyl butyrate	71
P. fluorescens	As above	15 compounds including all above except methyl isobutyrate	71
P. putrida	As above	14 compounds including same for Moraxella sp. above except methyl isobutyrate and methyl-2-methyl butyrate	71
B. thermosphacta	Inoculated vacuum-packaged corned beef, 5°C	7 compounds including diacetyl, acetoin, nonane, 3-methylbutanal, and 2-methylbutanol	111
	Aerobically stored, inoculated beef slices, 1°C, 14 days, pH 5.5–5.8	Acetoin, acetic acid, isobutyric/isovaleric acids. Acetic acid increased 4-fold after 28 days.	20
	As above; pH 6.2–6.6	Acetic acid, isolbutyric, isovaleric, and n-butyric acids	20
	APT broth, pH 6.5, 0.2% glucose	Acetoin, acetic acid, isobutyric and isovaleric acids	20
	APT broth, pH 6.5, no glucose	Same as above but no acetoin	20
B. thermosphacta (15 strains)	APT broth, pH 6.5, 0.2% glucose	Acetoin, acetic acid, isobutyric and isovaleric acids, traces of 3-methylbutanol	21
A. putrefaciens	Grown in radappertized chicken, 5 days, 10°C	H_2S, methyl mercaptan, dimethyl disulfide, methanol, ethanol	32
P. fragi	As above	Methanol, ethanol, methyl and ethyl acetate, dimethyl sulfide, methanol, ethanol	32
B. thermosphacta	As above	Methanol, ethanol	32
Flora	Spoiled chicken	11 compounds including H_2S, methanol, ethanol, methyl mercaptan, dimethyl sulfide, dimethyl disulfide	32

methanol and ethyl acetate by six each. *A. putrefaciens* consistently produces H_2S in vacuum-packaged meats on which it grows. From chicken breast muscle inoculated with *Pseudomonas* Group II strains and held at 2°C for 14 days, odors detected from chromatograph peaks were described by McMeekin (80) as being "sulfide-like," "evaporated milk," and "fruity."

SPOILAGE OF FRANKFURTERS, BOLOGNA, SAUSAGE, AND LUNCHEON MEATS

Unlike other meats covered in this chapter, these are prepared from various ingredients, any one or all of which may contribute microorganisms to the final product. Bacteria, yeasts, and molds may be found in and upon processed meats, but the former two groups are by far the most important in the microbial spoilage of these products.

Spoilage of these products is generally of three types: sliminess, souring, and greening. Slimy spoilage occurs on the outside of casings, especially of frankfurters, and may be seen in its early stages as discrete colonies which may later coalesce to form a uniform layer of gray slime. From the slimy material may be isolated yeasts, lactic acid bacteria of the genera *Lactobacillus, Streptococcus,* and *B. thermosphacta* (26, 27, 79). *L. viridescens* produces both sliminess and greening. Slime formation is favored by a moist surface and is usually confined to the outer casing. Removal of this material with hot water leaves the product essentially unchanged.

Souring generally takes place underneath the casing of these meats and results from the growth of lactobacilli, streptococci, and related organisms. The usual sources of these organisms to processed meats are milk solids. The souring results from the utilization of lactose and other sugars by these organisms with the production of acids. Sausage usually contains a more varied flora than most other processed meats due to the different seasoning agents employed, almost all of which contribute their own flora. *B. thermosphacta* has been found to be the most predominant spoilage organism for sausage by many investigators, including Dowdell and Board (26) and McLean and Sulzbacher (79).

Greening occurs more commonly on frankfurters than on other meats in this category, but may be seen on all from time to time. The heterofermentative species of lactobacilli and *Leuconostoc* have been found to be responsible for this condition. These organisms produce peroxides that act upon cured meat pigments and produce the green color. This reaction is made possible due to the inactivation of catalase by heat treatment or by the presence of nitrite. The accumulated H_2O_2 reacts with the meat pigments nitric oxide haemochromogen or nitric oxide myoglobin (see Table 9-9), producing a greenish oxidized porphyrin (100). This condition is caused by growth of the organisms in the interior core, where the low O/R potential allows H_2O_2 to accumulate. A small amount of oxygen favors greening and the green area is very often confined to small parts of the product. The one

Table 9-9. Pigments found in fresh, cured, or cooked meat (67).

Pigment	Mode of Formation	State of Iron	State of Haematin Nucleus	State of Globin	Color
1. Myoglobin	Reduction of metmyoglobin; de-oxygenation of oxymyoglobin	Fe^{++}	Intact	Native	Purplish red
2. Oxymyoglobin	Oxygenation of myoglobin	Fe^{++}	Intact	Native	Bright red
3. Metmyoglobin	Oxidation of myoglobin, oxymyoglobin	Fe^{+++}	Intact	Native	Brown
4. Nitric oxide myoglobin	Combination of myoglobin with nitric oxide	Fe^{++}	Intact	Native	Bright red
5. Metmyoglobin nitrite	Combination of metmyoglobin with excess nitrite	Fe^{+++}	Intact	Native	Red
6. Globin haemochromogen	Effect of heat, denaturing agents on myoglobin, oxymyoglobin; irradiation of globin haemichromogen	Fe^{++}	Intact	Denatured	Dull red
7. Globin haemichromogen	Effect of heat, denaturing agents on myoglobin, oxymyoglobin, metmyoglobin, haemochromogen	Fe^{+++}	Intact	Denatured	Brown
8. Nitric oxide haemochromogen	Effect of heat, salts on nitric oxide myoglobin	Fe^{++}	Intact	Denatured	Bright red
9. Sulphmyoglobin	Effect of H_2S and oxygen on myoglobin	Fe^{+++}	Intact but reduced	Denatured	Green
10. Choleglobin	Effect of hydrogen peroxide on myoglobin or oxymyoglobin; effect of ascorbic or other reducing agent on oxymyoglobin	Fe^{++} or Fe^{+++}	Intact but reduced	Denatured	Green
11. Verdohaem	Effect of reagents as in 9 in excess	Fe^{+++}	Porphyrin ring opened	Denatured	Green
12. Bile pigments	Effect of reagents as in 9 in large excess	Fe absent	Porphyin ring destroyed: chain of porphyrins	Absent	Yellow or colorless

Reprinted with permission from R. A. Lawrie, *Meat Science*, copyright 1966, Pergamon Press.

organism most frequently isolated from green meats is *L. viridescens*, first described by Niven et al. (91), and more recently isolated from anaerobically spoiled frankfurters (69) and both smoked pork loins and frankfurter sausage stored in atmospheres of CO_2 and N_2 (11). In spite of the discoloration, the green product is not known to be harmful if eaten.

Although mold spoilage of these meats is not common, it can and does occur under favorable conditions. When the products are moist and stored under conditions of high humidity, they tend to undergo bacterial and yeast spoilage. Mold spoilage is likely to occur only when the surfaces become dry or when the products are stored under other conditions that do not favor bacteria or yeasts.

SPOILAGE OF BACON AND CURED HAMS

The nature of these products and the procedures employed in preparing certain ones, such as smoking and brining, make them relatively insusceptible to spoilage by most bacteria. The most common form of bacon spoilage is moldiness, which may be due to *Aspergillus, Alternaria, Fusarium, Mucor, Rhizopus, Botrytis, Penicillium,* and other molds (Table 9-4). The high fat content and low a_w make it somewhat ideal for this type of spoilage. Bacteria of the genera *Streptococcus, Lactobacillus,* and *Micrococcus* are capable of growing well on certain types of bacon such as Wiltshire (53), and *S. faecalis* is often present on several types. Vacuum-packed bacon tends to undergo souring due primarily to micrococci and lactobacilli. Vacuum-packed, low-salt bacon stored above 20°C may be spoiled by staphylococci (115).

Cured hams undergo a type of spoilage different from that of fresh or smoked hams. This is due primarily to the fact that curing solutions pumped into the hams contain sugars that are fermented by the natural flora of the ham and also by those organisms pumped into the product in the curing solution, such as lactobacilli. The sugars are fermented to produce conditions referred to as "sours" of various types, depending upon their location within the ham. A large number of genera of bacteria have been implicated as the cause of ham "sours," among them *Acinetobacter, Bacillus, Pseudomonas, Lactobacillus, Proteus, Micrococcus, Clostridium,* and others. Gassiness is not unknown to occur in cured hams where members of the genus *Clostridium* have been found. The spoilage of canned hams is treated in Chapter 10.

In their study of vacuum-packed sliced bacon, Cavett (17) and Tonge et al. (115) found that when high-salt bacon was held at 20°C for 22 days, the catalase-positive cocci dominated the flora, while at 30°C the coagulase-negative staphylococci became dominant. In the case of low-salt bacon (5–7% NaCl versus 8–12% in high-salt bacon) held at 20°C, the micrococci as well as *S. faecalis* became dominant; while at 30°C the coagulase-negative staphylococci as well as *S. faecalis* and micrococci became dominant.

Spoilage of this type of product is characterized by a "cheesy," sour-scented, and putrid off-odor.

In a study of lean Wiltshire bacon stored aerobically at 5°C for 35 days or 10°C for 21 days, Gardner (33) found nitrates were reduced to nitrites when the microbial load reached about 10^9/g. The predominant organisms at this stage were micrococci, vibrios, and the yeast genera *Candida* and *Torulopsis*. Upon longer storage, microbial counts reached about 10^{10}/g with the disappearance of nitrites. At this stage, *Acinetobacter*, *Alcaligenes*, and *Arthrobacter-Corynebacterium* spp. became more important. Micrococci were always found while vibrios were found in all bacons with salt contents above 4%.

More information on the spoilage of some of these products can be obtained from the review by Gardner (35).

SPOILAGE OF POULTRY

Studies on the bacterial flora of fresh poultry by many investigators have revealed over twenty-five genera (Table 9-3). However, when these meats undergo low-temperature spoilage, almost all workers agree that the primary spoilage organisms belong to the genus *Pseudomonas* (8, 9, 88). In a study of 5,920 isolates from chicken carcasses (66), pseudomonads were found to constitute 30.5%, *Acinetobacter* 22.7%, *Flavobacterium* 13.9%, *Corynebacterium* 12.7%, with yeasts, Enterobacteriaceae, and others in lower numbers. Of the pseudomonads, these investigators found that 61.8% were fluorescent on King's medium and that 95.2% of all pseudomonads oxidized glucose. A previous characterization of pseudomonads on poultry undergoing spoilage was made by Barnes and Impey (9), who showed that the pigmented pseudomonads (Shewan's Group I) decreased from 34% to 16% from initial storage to the development of strong off-odors, while the nonpigmented actually increased from 11% to 58% (see section below on fish spoilage). *Acinetobacter* and other species of bacteria decreased along with the type I pseudomonads. A similar process occurs in spoiling fish.

Fungi are of considerably less importance in poultry spoilage except when antibiotics are employed to suppress bacterial growth. When antibiotics are employed, however, molds become the primary agents of spoilage (92). The genera *Candida*, *Rhodotorula*, and *Torula* are the most important yeasts found on poultry (Table 9-5). The essential feature of poultry spoilage is sliminess at the outer surfaces of the carcass or cuts. The visceral cavity very often displays sour odors or what is commonly called visceral taint. This is especially true of the spoilage of **New York dressed poultry,** where the viscera are left inside. The causative organisms here are also bacteria of the type noted above in addition to streptococci.

The primary reasons why poultry spoilage is mainly restricted to the surfaces are as follows. The inner portions of poultry tissue are generally sterile, or contain relatively few organisms, which generally do not grow at low temperatures. The spoilage flora, therefore, is restricted to the surfaces and hide where it is deposited from water, processing, and handling. The surfaces of fresh poultry stored in an environment of high humidity are very susceptible to the growth of aerobic bacteria such as pseudomonads. These organisms grow well on the surfaces where they form minute colonies that later coalesce to produce the sliminess characteristic of spoiled poultry. May et al. (78) showed that poultry skin supports the growth of the poultry spoilage flora better than even the muscle tissue. In the advanced stages of poultry spoilage, the surfaces will very often fluoresce when illuminated with ultraviolet light. The fluorescence is due to the presence of large numbers of fluorescent pseudomonads. Surface spoilage organisms can be recovered directly from the slime for plating, or one can prepare slides for viewing by smearing with portions of slime. Upon gram staining, one may note the uniform appearance of organisms indistinguishable from those listed above. Tetrazolium (2,3,5-triphenyltetrazolium chloride) can be used also to assess microbial activity on poultry surfaces. Upon spraying the eviscerated carcass with this compound, a red pigment develops in areas of high microbial activity. These areas generally consist of cut muscle surfaces and other damaged areas such as feather follicles (95).

As poultry undergoes spoilage, off-odors are generally noted before sliminess, with the former being first detected when log numbers/cm^2 are about 7.2–8.0. Sliminess generally occurs shortly after the appearance of off-odors with the log counts/cm^2 about 8 (8). Total aerobic plate counts/cm^2 of slimy surface rarely go higher than log 9.5. With the initial growth first confined to poultry surfaces, the tissue below the skin remains essentially free of bacteria for some time. Gradually, however, bacteria begin to enter the deep tissues, bringing about increased hydration of muscle proteins, much as occurs with beef. Whether autolysis plays an important role in the spoilage of inner poultry tissues is not clear at this time.

Vacuum packaging and CO_2-atmosphere storage are effective in delaying the spoilage of poultry. With raw cut-up poultry stored at 5°C in O_2-permeable film, vacuum packages, and CO_2-flushed high O_2-barrier film, it was unacceptable by day 9, between days 9–11, and after 17 days, respectively (114).

Alteromonas putrefaciens grows well at 5°C and produces potent off-odors in 7 days when growing on chicken muscle (83). Among odor producers in general, it has been noted that there is a selection of types that produce strong odors among the varied flora that exists on fresh poultry (80). The study noted was conducted with chicken breast muscle, which spoils differently from leg muscles since the latter have a higher pH. With chicken leg muscle stored at 2°C for 16 days, 47% of the flora consisted of Group I pseudomonads, 32% of Group II, 17% of *Acinetobacter/Moraxella*, and 4% of *A. putrefaciens* (81). All isolates of the latter produced sulfidelike odors and as may be noted from Table 9-8, this organism produces H_2S, methyl mercaptan, and dimethyl sulfide. It was not of significance in the spoilage of chicken breast muscle. Since Shewan's Group II pseudomonads grow faster than the pigment-producing Group I strains, it appears that the strong odor-producing capacity is a property of these strains. Group II pseudomonads have been shown to be consumers of free amino acids in chicken skin, while Group I types effected increases in the quantities of free amino acids and related nitrogenous compounds (2, 3).

When New York dressed poultry undergoes microbial spoilage, the organisms make their way through the gut walls and invade inner tissues of the intestinal cavity. The characteristic sharpness that is associated with the spoilage of this type of poultry is referred to as "visceral taint."

SPOILAGE OF FISH AND SHELLFISH

FISH. Both salt-water and fresh-water fish contain comparatively high levels of proteins and other nitrogenous constituents (Table 9-10). The carbohydrate content of these fish is nil, while fat content varies from very low to rather high values depending upon species. Of particular importance in fish flesh is the nature of the nitrogenous compounds. The relative percentage of total-N and protein-N are presented in Table 9-11, from which it can be seen that not all nitrogenous compounds in fish are in the form of proteins. Among the nonprotein nitrogen compounds are the free amino acids, volatile nitrogen bases such as ammonia and trimethylamine, creatine, taurine, the betaines, uric acid, anserine, carnosine, and histamine.

It is generally recognized that the internal flesh of healthy, live fish is

Table 9-10. Approximate percentage chemical composition of fish and shellfish (118).

Bony Fish	Water	Carbohydrates	Proteins	Fat	Ash
Bluefish	74.6	0	20.5	4.0	1.2
Cod	82.6	0	16.5	0.4	1.2
Haddock	80.7	0	18.2	0.1	1.4
Halibut	75.4	0	18.6	5.2	1.0
Herring (Atlantic)	67.2	0	18.3	12.5	2.7
Mackerel (Atlantic)	68.1	0	18.7	12.0	1.2
Salmon (Pacific)	63.4	0	17.4	16.5	1.0
Swordfish	75.8	0	19.2	4.0	1.3
Crustaceans					
Crab	80.0	0.6	16.1	1.6	1.7
Lobster	79.2	0.5	16.2	1.9	2.2
Mollusks					
Clams, meat	80.3	3.4	12.8	1.4	2.1
Oysters	80.5	5.6	9.8	2.1	2.0
Scallops	80.3	3.4	14.8	0.1	1.4

Table 9-11. Distribution of nitrogen in fish and shellfish flesh (55).

Species	Percentage total N	Percentage protein N	Ratio of protein N/total N
Cod (Atlantic)	2.83	2.47	0.87
Herring (Atlantic)	2.90	2.53	0.87
Sardine	3.46	2.97	0.86
Haddock	2.85	2.48	0.87
Lobster	2.72	2.04	0.75

sterile (48, 113), although a few reports to the contrary exist. Bacteria that exist on fresh fish are generally found in three places: the outer slime, gills, and the intestines of feeding fish.

The microorganisms known to cause fish spoilage are indicated in Tables 9-3, 9-4, and 9-5. Fresh iced fish are invariably spoiled by bacteria, while salted and dried fish are more likely to undergo fungal spoilage. The bacterial flora of spoiling fish is found to consist of asporogenous, gram-negative rods of the *Pseudomonas* and *Acinetobacter-Moraxella* types. Many fish spoilage bacteria are capable of good growth between 0°–1°C. Shaw and Shewan (103) found that a large number of *Pseudomonas* spp. are capable of causing fish spoilage at −3°C, although at a slow rate.

The spoilage of salt- and fresh-water fish appears to occur in essentially the same manner, with the chief differences being the requirement of the salt-water flora for a sea-water type of environment and the differences in chemical composition between various fish with respect to nonprotein nitrogenous constituents. The most susceptible part of fish is the gill region, including the gills. The earliest signs of organoleptic spoilage may be noted by examining the gills for the presence of off-odors. If feeding fish are not eviscerated immediately, intestinal bacteria soon make their way through the intestinal walls and into the flesh of the intestinal cavity. This process is believed to be aided by the action of proteolytic enzymes, which are from the intestines, and which may be natural enzymes inherent in the intestines of the fish, or enzymes of bacterial origin from the inside of the intestinal canal, or both. Fish spoilage bacteria apparently have little difficulty in growing in the slime and on the outer integument of fish. Slime is composed of mucopolysaccharide components, free amino acids, trimethylamine oxide, piperidine derivatives, and other related compounds. As is the case with poultry spoilage, plate counts are best done on the surface of fish with numbers of organisms expressed per cm^2 of examined surface.

It appears that the spoilage organisms first utilize the simpler compounds and in the process release various volatile off-odor components. According to Shewan (110), trimethylamine oxide, creatine, taurine, anserine, and related compounds along with certain amino acids decrease during fish spoilage with the production of trimethylamine, ammonia, histamine, hydrogen sulfide, indole, and other compounds (see Table 9-8). Fish flesh appears to differ from mammalian flesh in regard to autolysis. Flesh of the former type seems to undergo autolysis at more rapid rates. While the occurrence of this process along with microbial spoilage is presumed by some investigators to aid either the spoilage flora or the spoilage process (47), attempts to separate and isolate the events of the two have proved difficult. In a detailed study of fish isolates with respect to the capacity to cause typical fish spoilage by use of sterile fish muscle press juice, Lerke et al. (74) found that the spoilers belonged to the genera *Pseudomonas* and *Acinetobacter-Moraxella*, with none of the coryneforms, micrococci, or flavobacteria being spoilers. In characterizing the spoilers with respect to their ability to utilize certain compounds, these workers found that most spoilers were unable to degrade gelatin or digest egg albumin. This suggests that fish spoilage proceeds much as does that of beef—in the general absence of complete proteolysis by the spoilage flora. Pure culture inoculations of cod and haddock muscle blocks have failed to effect tissue softening (47). In those fish that contain high levels of lipids (herrings, mackerel, salmon, and others), these compounds undergo rancidity as microbial spoilage occurs. It should be noted that the skin of fish is rich in collagen. The scales of most fish are composed of a scleroprotein belonging to the keratin group, and it is quite probable that these are among the last parts of fish to be decomposed.

Studies on the interplay of the bacterial flora of fish undergoing spoilage indicate that *Pseudomonas* spp. of Shewan's Group II become the dominant types of all bacteria after 14 days at 5°C (Table 9-12; 68, 70). A similar result was found for poultry spoilage (9). H_2S producers also increase late in the spoilage process (47). While *Acinetobacter-Moraxella* spp. constituted the highest percentage of the initial flora, they could not be isolated after 14 days.

Table 9-12. Microbial population change in Pacific hake stored at 5°C (70).

	Microbial Population after Incubation (%)			
Microorganism	*0 Day*	*5 Days*	*8 Days*	*14 Days*
Pseudomonas				
Type I	14.0	7.3	2.7	15.1
Type II	14.0	52.4	53.4	77.4
Types III or IV	3.5	12.2	31.5	7.5
Acinetobacter-Moraxella				
Acinetobacter	31.6	17.0	8.2	0
Moraxella	19.3	9.8	2.7	0
Flavobacterium	17.6	0	0	0
Coliforms	0	1.2	1.4	0
Microbial count of sample	1.5×10^4	3.4×10^7	9.3×10^8	2.7×10^9
No. of microorganisms identified[a]	57	82	73	53

[a] All isolated colonies on initial isolation plates were picked and identified.

Studies on the skin flora of four different fish revealed the following as the most common organisms: *Pseudomonas/Alteromonas,* 32–60%; and *Moraxella/Acinetobacter,* 18–37% (50). The initial flora of herring fillets was dominated by *A. putrefaciens* and pseudomonads, and after spoilage in air, these organisms constituted 62–95% of the flora (86). When allowed to spoil in 100% CO_2 at 4°C, the herring fillets were dominated almost completely by lactobacilli (86). In the case of rock cod fillets stored in 80% CO_2 + 20% air at 4°C for 21 days, the flora consisted of 71–87% lactobacilli and some tan-colored pseudomonads (62).

Some of the volatile compounds produced in spoiling fish are noted in Table 9-8. Phenethyl alcohol has been shown to be produced consistently in fish by a specific organism designated "Achromobacter" by Chen et al. (19) and Chen and Levin (18). The compound along with phenol was recovered from a high-boiling fraction of haddock fillets held at 2°C. None of ten known *Acinetobacter* and only one of nine known *Moraxella* produced phenethyl alcohol under similar conditions. Ethanol, propanol, and isopropanol are produced by fish spoilers, and of 244 bacteria isolated from king salmon trout and tested in fish extracts, all produced ethanol; 241 or 98.8% produced isopropanol; and 227 or 93% produced propanol (4).

To detect spoilage of fish, reduction of trimethylamine-N-oxide (TMAO) to trimethylamine (TMA) has been employed with some success. However, it has been found that some fish spoilage organisms actually utilize TMAO as a terminal electron acceptor under anaerobic conditions, thus reducing the reliability of TMAO reduction as a spoilage indicator.

SHELLFISH. *Crustaceans:* The most widely consumed shellfish within this group are shrimp, lobsters, crabs, and crayfish, and the spoilage of all will be treated together. Unless otherwise specified, spoilage of each is presumed or known to be essentially the same. The chief differences in spoilage of these various foods are referable generally to the way in which they are handled and their specific chemical composition.

An inspection of Table 9-10 reveals that crustaceans differ from fish in having about 0.5% carbohydrate as opposed to none for the fish presented. Shrimp has been reported to have a higher content of free amino acids than fish (98) and to contain catheptic-like enzymes that rapidly break down proteins (31).

The bacterial flora of freshly caught crustaceans should be expected to reflect the waters from which these foods are caught, contaminants from the deck, handlers, and washing waters. Many of the organisms reported for fresh fish have been reported on these foods, with pseudomonads. *Acinetobacter-Moraxella,* and yeast spp. being predominant on microbially spoiled crustacean meats (15, 64). When shrimp was allowed to spoil at 0°C for 13 days, *Pseudomonas* spp. were the dominant spoilers, with only 2% of the spoilage flora being gram positives in contrast to 38% for the fresh product (77). *Moraxella* dominated spoilage at 5.6° and 11.1°C, while at 16.7° and 22.2°C *Proteus* was dominant (Table 9-13).

Table 9-13. The most predominant bacteria
in shrimp held to spoilage at five temperatures
(77).

Temp. (°C)	Days held	Organisms
0	13	Pseudomonas
5.6	9	Moraxella
11.1	7	Moraxella
16.7	5	Proteus
22.2	3	Proteus

The spoilage of crustacean meats appears to be quite similar to that of fish flesh. Spoilage would be expected to begin at the outer surfaces of these foods due to the anatomy of the organisms. It has been reported that the crustacean muscle contains over 300 mg of nitrogen/100 g of meat, which is considerably higher than that for fish (117). The presence of higher quantities of free amino acids in particular and of higher quantities of nitrogenous extractives in crustacean meats in general makes them quite susceptible to rapid attack by the spoilage flora. Initial spoilage of crustacean meats is accompanied by the production of large amounts of volatile base nitrogen much as is the case with fish. Some of the volatile base nitrogen arises from the reduction of trimethylamine oxide present in crustacean shellfish (lacking in most mollusks). Creatine is lacking among shellfish, both crustacean and molluscan, while arginine is prevalent. Shrimp microbial spoilage is accompanied by increased hydration capacity in a manner similar to that for meats or poultry (109).

Mollusks: The molluscan shellfish considered in this section are oysters, clams, squid, and scallops. These animals differ in their chemical composition from both teleost fish and crustacean shellfish in having a significant content of carbohydrate material and a lower total quantity of nitrogen in their flesh. The carbohydrate is largely in the form of glycogen, and with levels of the type that exist in molluscan meats, fermentative activities may be expected to occur as a part of the microbial spoilage. Molluscan meats contain high levels of nitrogen bases much as do other shellfish. Of particular interest in molluscan muscle tissue is a higher content of free arginine, aspartic, and glutamic acids than is found in fish. The most important difference in chemical composition between crustacean shellfish and molluscan shellfish is the higher content of carbohydrate in the latter. For example, clam meat and scallops have been reported to contain 3.4% and oysters 5.6% carbohydrate, mostly as glycogen. The higher content of carbohydrate materials in molluscan shellfish is responsible for the different spoilage pattern of these foods over other seafood.

The microbial flora of molluscan shellfish may be expected to vary considerably, depending upon the quality of the water from which these fish are taken and upon the quality of wash water and other factors. The following genera of bacteria have been recovered from spoiled oysters: *Serratia,*

Pseudomonas, Proteus, Clostridium, Bacillus, Escherichia, Enterobacter, Streptococcus, Lactobacillus, Flavobacterium, and *Micrococcus.* As spoilage sets in and progresses, *Pseudomonas* and *Acinetobacter-Moraxella* spp. predominate, with streptococci, lactobacilli, and yeasts dominating the late stages of spoilage.

Due to the relatively high level of glycogen, the spoilage of molluscan shellfish is basically fermentative. Several investigators, including Hunter and Linden (52) and Pottinger (97), have proposed the following pH scale as a basis for determining microbial quality in oysters:

pH 6.2–5.9 = good

pH 5.8 = "off"

pH 5.7–5.5 = musty

pH 5.2 and below = sour or putrid

A measure of pH decrease is apparently a better test of spoilage in oysters and other molluscan shellfish than volatile nitrogen bases. A measure of volatile acids was attempted by Beacham (10) and found to be unreliable as a test of oyster freshness. While pH is regarded by most investigators as being the best objective technique for examining the microbial quality of oysters, Abbey et al. (1) found that organoleptic evaluations and microbial counts were more desirable indices of microbial quality in this product.

Clams and scallops appear to display essentially the same patterns of spoilage as do oysters, but squid meat does not. In squid meat, volatile base nitrogen increases as spoilage occurs much in the same manner as for the crustacean shellfish (87).

REFERENCES

1. Abbey, A., R. A. Kohler, and S. D. Upham. 1957. Effect of aureomycin chlortetracycline in the processing and storage of freshly shucked oysters. *Food Technol.* 11:265–71.
2. Adamčič, M., and D. S. Clark. 1970. Bacteria-induced biochemical changes in chicken skin stored at 5°C. *J. Food Sci.* 35:103–6.
3. Adamčič, M., D. S. Clark, and M. Yaguchi. 1970. Effect of psychrotolerant bacteria on the amino acid content of chicken skin. *J. Food Sci.* 35:272–75.
4. Ahmed, A., and J. R. Matches. 1983. Alcohol production by fish spoilage bacteria. *J. Food Protect.* 46:1055–59.
5. Austin, B. L., and B. Thomas. 1972. Dye reduction tests on meat products. *J. Sci. Food Agric.* 23:542.
6. Ayres, J. C. 1955. Microbiological implications in handling, slaughtering and dressing of meat animals. *Adv. Food Res.* 6:109–61.
7. ———. 1960. The relationship of organisms of the genus *Pseudomonas* to the spoilage of meat, poultry and eggs. *J. Appl. Bacteriol.* 23:471–86.

8. Ayres, J. C., W. S. Ogilvy, and G. F. Stewart. 1950. Post mortem changes in stored meats. I. Microorganisms associated with development of slime on eviscerated cut-up poultry. *Food Technol.* 4:199–205.

9. Barnes, E. M., and C. S. Impey. 1968. Psychrophilic spoilage bacteria of poultry. *J. Appl. Bacteriol.* 31:97–107.

10. Beacham, L. M. 1946. A study of decomposition in canned oysters and clams. *J. Assoc. Offic. Agric. Chem.* 29:89–92.

11. Blickstad, E., and G. Molin. 1983. The microbial flora of smoked pork loin and frankfurter sausage stored in different gas atmospheres at 4°C. *J. Appl. Bacteriol.* 54:45–56.

12. Borton, R. J., L. J. Bratzler, and J. F. Price. 1970. Effects of four species of bacteria on porcine muscle. 2. Electrophoretic patterns of extracts of salt-soluble protein. *J. Food Sci.* 35:783–86.

13. Brown, A. D., and J. F. Weidemann. 1958. The taxonomy of the psychrophilic meat-spoilage bacteria: A reassessment. *J. Appl. Bacteriol.* 21:11–17.

14. Callow, E. H., and M. Ingram. 1955. Bone-taint. *Food.* Feb.

15. Campbell, L. L., Jr., and O. B. Williams. 1952. The bacteriology of Gulf Coast shrimp. IV. Bacteriological, chemical, and organoleptic changes with iced storage. *Food Technol.* 6:125–26.

16. Campbell, R. J., A. F. Egan, F. H. Grau, and B. J. Shay. 1979. The growth of *Microbacterium thermosphactum* on beef. *J. Appl. Bacteriol.* 47:505–9.

17. Cavett, J. J. 1962. The microbiology of vacuum packed sliced bacon. *J. Appl. Bacteriol.* 25:282–89.

18. Chen, T. C., and R. E. Levin. 1974. Taxonomic significance of phenethyl alcohol production by *Achromobacter* isolates from fishery sources. *Appl. Microbiol.* 28:681–87.

19. Chen, T. C., W. W. Nawar, and R. E. Levin. 1974. Identification of major high-boiling volatile compounds produced during refrigerated storage of haddock fillets. *Appl. Microbiol.* 28:679–80.

20. Dainty, R. H., and C. M. Hibbard. 1980. Aerobic metabolism of *Brochothrix thermosphacta* growing on meat surfaces and in laboratory media. *J. Appl. Bacteriol.* 48:387–96.

21. Dainty, R. H., and F. J. K. Hofman. 1983. The influence of glucose concentration and culture incubation time on end-product formation during aerobic growth of *Brochothrix thermosphacta*. *J. Appl. Bacteriol.* 55:233–39.

22. Dainty, R. H., B. G. Shaw, K. A. DeBoer, and E. S. J. Scheps. 1975. Protein changes caused by bacterial growth on beef. *J. Appl. Bacteriol.* 39:73–81.

23. Dainty, R. H., B. G. Shaw, and T. A. Roberts. 1983. Microbial and chemical changes in chill-stored red meats. In *Food microbiology: Advances and prospects,* ed. T. A. Roberts and F. A. Skinner, 151–78. New York and London: Academic Press.

24. Davidson, C. M., M. J. Dowdell, and R. G. Board. 1973. Properties of gram negative aerobes isolated from meats. *J. Food Sci.* 38:303–5.

25. Dodsworth, P. J., and A. G. Kempton. 1977. Rapid measurement of meat quality by resazurin reduction. II. Industrial application. *Can. Inst. Food Sci. Technol. J.* 10:158–60.

26. Dowdell, M. J., and R. G. Board. 1968. A microbiological survey of British fresh sausage. *J. Appl. Bacteriol.* 31:378–96.

27. Drake, S. D., J. B. Evans, and C. F. Niven, Jr. 1958. Microbial flora of packaged frankfurters and their radiation resistance. *Food Res.* 23:291–96.
28. Egan, A. F., A. L. Ford, and B. J. Shay. 1980. A comparison of *Microbacterium thermosphactum* and lactobacilli as spoilage organisms of vacuum-packaged sliced luncheon meats. *J. Food Sci.* 45:1745–48.
29. Egan, A. F., and B. J. Shay. 1982. Significance of lactobacilli and film permeability in the spoilage of vacuum-packaged beef. *J. Food Sci.* 47:1119–22, 1126.
30. Eribo, B. E., and J. M. Jay. 1985. Incidence of *Acinetobacter* spp. and other gram-negative, oxidase-negative bacteria in fresh and spoiled ground beef. *Appl. Environ. Microbiol.* 49:256–57.
31. Fieger, E. A., and A. F. Novak. 1961. Microbiology of shellfish deterioration. In *Fish as food,* vol. 1, ed. G. Borgstrom, 561–611. New York: Academic Press.
32. Freeman, L. R., G. J. Silverman, P. Angelini, C. Merritt, Jr., and W. B. Esselen. 1976. Volatiles produced by microorganisms isolated from refrigerated chicken at spoilage. *Appl. Environ. Microbiol.* 32:222–31.
33. Gardner, G. A. 1971. Microbiological and chemical changes in lean Wiltshire bacon during aerobic storage. *J. Appl. Bacteriol.* 34:645–54.
34. ———. 1971. A note on the aerobic microflora of fresh and frozen porcine liver stored at 5°C. *J. Food Technol.* 6:225–31.
35. ———. 1983. Microbial spoilage of cured meats. In *Food microbiology: Advances and prospects,* ed. T. A. Roberts and F. A. Skinner, 179–202. New York and London: Academic Press.
36. Gill, C. O. 1976. Substrate limitation of bacterial growth at meat surfaces. *J. Appl. Bacteriol.* 41:401–10.
37. Gill, C. O. 1982. Microbial interaction with meats. In *Meat Microbiology,* ed. M. H. Brown, 225–64. London: Applied Science.
38. ———. 1983. Meat spoilage and evaluation of the potential storage life of fresh meat. *J. Food Protect.* 46:444–52.
39. Gill, C. O., and K. M. DeLacy. 1982. Microbial spoilage of whole sheep livers. *Appl. Environ. Microbiol.* 43:1262–66.
40. Gill, C. O., and K. G. Newton. 1977. The development of aerobic spoilage flora on meat stored at chill temperatures. *J. Appl. Bacteriol.* 43:189–95.
41. ———. 1979. Spoilage of vacuum-packaged dark, firm, dry meat at chill temperatures. *Appl. Environ. Microbiol.* 37:362–64.
42. Hanna, M. O., G. C. Smith, L. C. Hall, and C. Vanderzant. 1979. Role of *Hafnia alvei* and a *Lactobacillus* species in the spoilage of vacuum-packaged strip loin steaks. *J. Food Protect.* 42:569–71.
43. Hanna, M. O., G. C. Smith, J. W. Savell, F. K. McKeith, and C. Vanderzant. 1982. Microbial flora of livers, kidneys and hearts from beef, pork, and lamb: Effects of refrigeration, freezing and thawing. *J. Food Protect.* 45:63–73.
44. ———. 1982. Effects of packaging methods on the microbial flora of livers and kidneys from beef or pork. *J. Food Protect.* 45:74–81.
45. Hanna, M. O., C. Vanderzant, Z. L. Carpenter, and G. C. Smith. 1977. Characteristics of psychrotrophic, gram-positive, catalase-positive, pleomorphic coccoid rods from vacuum-packaged wholesale cuts of beef. *J. Food Protect.* 40:94–97.
46. ———. 1977. Microbial flora of vacuum-packaged lamb with special reference to psychrotrophic, gram-positive, catalase-positive pleomorphic rods. *J. Food Protect.* 40:98–100.

47. Herbert, R. A., M. S. Hendrie, D. M. Gibson, and J. M. Shewan. 1971. Bacteria active in the spoilage of certain sea foods. *J. Appl. Bacteriol.* 34:41–50.
48. Hess, E. 1950. Bacterial fish spoilage and its control. *Food Technol.* 4:477–80.
49. Hitchener, B. J., A. F. Egan, and P. J. Rogers. 1982. Characteristics of lactic acid bacteria isolated from vacuum-packaged beef. *J. Appl. Bacteriol.* 52:31–37.
50. Hobbs, G. 1983. Microbial spoilage of fish. In *Food microbiology: Advances and prospects,* ed. T. A. Roberts and F. A. Skinner, 217–29. London: Academic Press.
51. Hsieh, D. Y., and J. M. Jay. 1984. Characterization and identification of yeasts from fresh and spoiled ground beef. *Int. J. Food Microbiol.* 1:141–47.
52. Hunter, A. C., and B. A. Linden. 1923. An investigation of oyster spoilage. *Amer. Food J.* 18:538–40.
53. Ingram, M. 1960. Bacterial multiplication in packed Wiltshire bacon. *J. Appl. Bacteriol.* 23:206–15.
54. Ingram, M., and R. H. Dainty. 1971. Changes caused by microbes in spoilage of meats. *J. Appl. Bacteriol.* 34:21–39.
55. Jacquot, R. 1961. Organic constituents of fish and other aquatic animal foods. In *Fish as food,* vol. 1, ed. G. Borgstrom, 145–209. New York: Academic Press.
56. Jay, J. M. 1964. Release of aqueous extracts by beef homogenates, and factors affecting release volume. *Food Technol.* 18:1633–36.
57. ———. 1964. Beef microbial quality determined by extract-release volume (ERV). *Food Technol.* 18:1637–41.
58. ———. 1966. Influence of postmortem conditions on muscle microbiology. In *The physiology and biochemistry of muscle as a food,* ch. 26, ed. E. J. Briskey et al. Madison: Univ. of Wisconsin Press.
59. ———. 1966. Relationship between the phenomena of extract-release volume and water-holding capacity of meats as simple and rapid methods for determining microbial quality of beef. *Hlth Lab. Sci.* 3:101–10.
60. ———. 1967. Nature, characteristics, and proteolytic properties of beef spoilage bacteria at low and high temperatures. *Appl. Microbiol.* 15:943–44.
61. ———. 1977. Meats, poultry, and seafoods. In *Food and Beverage Mycology,* ch. 5, ed. L. R. Beuchat. Westport, Conn.: AVI.
62. Johnson, A. R., and D. M. Ogrydziak. 1984. Genetic adaptation to elevated carbon dioxide atmospheres by *Pseudomonas*-like bacteria isolated from rock cod (*Sebastes* spp.). *Appl. Environ. Microbiol.* 48:486–90.
63. Kirsch, R. H., F. E. Berry, C. L. Baldwin, and E. M. Foster. 1952. The bacteriology of refrigerated ground meat. *Food Res.* 17:495–503.
64. Koburger, J. A., A. R. Norden, and G. M. Kampler. 1975. The microbial flora of rock shrimp—*Sicyonia brevirostris. J. Milk Food Technol.* 38:747–49.
65. Kümmerlin, R. 1982. Technical note: Resazurin test for microbiological control of deep-frozen shrimps. *J. Food Technol.* 17:513–15.
66. Lahellec, C., C. Meurier, and G. Bennejean. 1975. A study of 5,920 strains of psychrotrophic bacteria isolated from chickens. *J. Appl. Bacteriol.* 38:89–97.
67. Lawrie, R. A. 1966. *Meat science.* Chs. 4, 10. New York: Pergamon Press.
68. Laycock, R. A., and L. W. Regier. 1970. Pseudomonads and achromobacters in the spoilage of irradiated haddock of different pre-irradiation quality. *Appl. Microbiol.* 20:333–41.

69. Lee, B. H., and R. E. Simard. 1984. Three systems for biochemical characterization of lactobacilli associated with meat spoilage. *J. Food Protect.* 47:937–42.

70. Lee, J. S., and J. M. Harrison. 1968. Microbial flora of Pacific hake (*Merluccius productus*). *Appl. Microbiol.* 16:1937–38.

71. Lee, M. L., D. L. Smith, and L. R. Freeman. 1979. High-resolution gas chromatographic profiles of volatile organic compounds produced by microorganisms at refrigerated temperatures. *Appl. Environ. Microbiol.* 37:85–90.

72. Lepovetsky, B. C., H. H. Weiser, and F. E. Deatherage. 1953. A microbiological study of lymph nodes, bone marrow and muscle tissue obtained from slaughtered cattle. *Appl. Microbiol.* 1:57–59.

73. Lerke, P., R. Adams, and L. Farber. 1963. Bacteriology of spoilage of fish muscle. I. Sterile press juice as a suitable experimental medium. *Appl. Microbiol.* 11:458–62.

74. ———. 1965. Bacteriology of spoilage of fish muscle. III. Characteristics of spoilers. *Appl. Microbiol.* 13:625–30.

75. Lowry, P. D., and C. O. Gill. 1984. Temperature and water activity minima for growth of spoilage moulds from meat. *J. Appl. Bacteriol.* 56:193–99.

76. Margitic, S., and J. M. Jay. 1970. Antigenicity of salt-soluble beef muscle proteins held from freshness to spoilage at low temperatures. *J. Food Sci.* 35:252–55.

77. Matches, J. R. 1982. Effects of temperature on the decomposition of Pacific coast shrimp (*Pandalus jordani*). *J. Food Sci.* 47:1044–47, 1069.

78. May, K. N., J. D. Irby, and J. L. Carmon. 1961. Shelf life and bacterial counts of excised poultry tissue. *Food Technol.* 16:66–68.

79. McLean, R. A., and W. L. Sulzbacher. 1953. *Microbacterium thermosphactum* spec. nov., a non-heat resistant bacterium from fresh pork sausage. *J. Bacteriol.* 65:428–32.

80. McMeekin, T. A. 1975. Spoilage association of chicken breast muscle. *Appl. Microbiol.* 29:44–47.

81. ———. 1977. Spoilage association of chicken leg muscle. *Appl. Environ. Microbiol.* 33:1244–46.

82. ———. 1981. Microbial spoilage of meats. In *Developments in food microbiology,* ed. R. Davies, 1–40. London: Applied Science.

83. McMeekin, T. A., and J. T. Patterson. 1975. Characterization of hydrogen sulfide-producing bacteria isolated from meat and poultry plants. *Appl. Microbiol.* 29:165–69.

84. Miller, A., III, R. A. Scanlan, J. S. Lee, and L. M. Libbey. 1973. Volatile compounds produced in sterile fish muscle (*Sebastes melanops*) by *Pseudomonas putrefaciens, Pseudomonas fluorescens,* and an *Achromobacter* species. *Appl. Microbiol.* 26:18–21.

85. Miller, A., III, R. A. Scanlan, J. S. Lee, L. M. Libbey, and M. E. Morgan. 1973. Volatile compounds produced in sterile fish muscle (*Sebastes melanops*) by *Pseudomonas perolens. Appl. Microbiol.* 25:257–61.

86. Molin, G., and I.-M. Stenstrom. 1984. Effect of temperature on the microbial flora of herring fillets stored in air or carbon dioxide. *J. Appl. Bacteriol.* 56:275–82.

87. Motohiro, T., and E. Tanikawa. 1952. Studies on food poisoning of mollusk especially of squid and octopus meat. I. Chemical change and freshness tests

of squid and octopus meat during deterioration of freshness. *Bull. Fac. Fisheries Hokkaido Univ.* 3(2):142–53.

88. Nagel, C. W., K. L. Simpson, H. Ng, R. H. Vaughn, and G. F. Stewart. 1960. Microorganisms associated with spoilage of refrigerated poultry. *Food Technol.* 14:21–23.

89. Newton, K. G., and C. O. Gill. 1978. Storage quality of dark, firm, dry meat. *Appl. Environ. Microbiol.* 36:375–76.

90. Nielsen, H.-J. S. 1983. Influence of nitrite addition and gas permeability of packaging film on the microflora in a sliced vacuum-packed whole meat product under refrigerated storage. *J. Food Technol.* 18:573–85.

91. Niven, C. F., Jr., A. G. Castellani, and V. Allanson. 1949. A study of the lactic acid bacteria that cause surface discolorations of sausages. *J. Bacteriol.* 58:633–41.

92. Njoku-Obi, A. N., J. V. Spencer, E. A. Sauter, and M. W. Eklund. 1957. A study of the fungal flora of spoiled chlortetracycline treated chicken meat. *Appl. Microbiol.* 5:319–21.

93. Patterson, J. T., and P. A. Gibbs. 1977. Incidence and spoilage potential of isolates from vacuum-packaged meat of high pH value. *J. Appl. Bacteriol.* 43:25–38.

94. Pearson, D. 1968. Assessment of meat freshness in quality control employing chemical techniques: A review. *J. Sci. Food Agric.* 19:357–63.

95. Peel, J. L., and J. M. Gee. 1976. The role of micro-organisms in poultry taints. In *Microbiology in agriculture, fisheries and food,* ed. F. A. Skinner and J. G. Carr, 151–60. New York: Academic Press.

96. Pierson, M. D., D. L. Collins-Thompson, and Z. J. Ordal. 1970. Microbiological, sensory and pigment changes of aerobically and anaerobically packaged beef. *Food Technol.* 24:1171–75.

97. Pottinger, S. R. 1948. Some data on pH and the freshness of shucked eastern oysters. *Comm. Fisheries Rev.* 10(9):1–3.

98. Ranke, B. 1955. Über papier-chromatographische Untersuchungen des freien und eiweissgebundenen Aminosäuren-bestandes bei Krebsen und Fischen. *Arch. Fischereiwiss.* 6:109–13.

99. Seelye, R. J., and B. J. Yearbury. 1979. Isolation of *Yersinia enterocolitica*-resembling organisms and *Alteromonas putrefaciens* from vacuum-packed chilled beef cuts. *J. Appl. Bacteriol.* 46:493–99.

100. Sharpe, M. E. 1962. Lactobacilli in meat products. *Food Manuf.* 37:582–89.

101. Shaw, B. G., and J. B. Latty. 1982. A numerical taxonomic study of *Pseudomonas* strains from spoiled meat. *J. Appl. Bacteriol.* 52:219–28.

102. ———. 1984. A study of the relative incidence of different *Pseudomonas* groups on meat using a computer-assisted identification technique employing only carbon source tests. *J. Appl. Bacteriol.* 57:59–67.

103. Shaw, B. G., and J. M. Shewan. 1968. Psychrophilic spoilage bacteria of fish. *J. Appl. Bacteriol.* 31:89–96.

104. Shelef, L. A. 1975. Microbial spoilage of fresh refrigerated beef liver. *J. Appl. Bacteriol.* 39:273–80.

105. ———. 1977. Effect of glucose on the bacterial spoilage of beef. *J. Food Sci.* 42:1172–75.

106. Shelef, L. A., and J. M. Jay. 1969. Relationship between meat-swelling, viscosity,

extract-release volume, and water-holding capacity in evaluating beef microbial quality. *J. Food Sci.* 34:532–35.

107. ———. 1969. Relationship between amino sugars and meat microbial quality. *Appl. Microbiol.* 17:931–32.

108. ———. 1970. Use of a titrimetric method to assess the bacterial spoilage of fresh beef. *Appl. Microbiol.* 19:902–5.

109. ———. 1971. Hydration capacity as an index of shrimp microbial quality. *J. Food Sci.* 36:994–97.

110. Shewan, J. M. 1961. The microbiology of sea-water fish, pp. 487–560. In *Fish as food,* vol. 1, ed. G. Borgstrom. New York: Academic Press.

111. Stanley, G., K. J. Shaw, and A. F. Egan. 1981. Volatile compounds associated with spoilage of vacuum-packaged sliced luncheon meat by *Brochothrix thermosphacta. Appl. Environ. Microbiol.* 41:816–18.

112. Sutherland, J. P., J. T. Patterson, and J. G. Murray. 1975. Changes in the microbiology of vacuum-packaged beef. *J. Appl. Bacteriol.* 39:227–37.

113. Tarr, H. L. A. 1954. Microbiological deterioration of fish post mortem, its detection and control. *Bact. Revs.* 18:1–15.

114. Thomas, V. O., A. A. Kraft, R. E. Rust, and D. K. Hotchkiss. 1984. Effect of carbon dioxide flushing and packaging methods on the microbiology of packaged chicken. *J. Food Sci.* 49:1367–71.

115. Tonge, R. J., A. C. Baird-Parker, and J. J. Cavett. 1964. Chemical and microbiological changes during storage of vacuum packed sliced bacon. *J. Appl. Bacteriol.* 27:252–64.

116. Vanderzant, C., M. O. Hanna, J. G. Ehlers, J. W. Savell, G. C. Smith, D. B. Griffin, R. N. Terrell, K. D. Lind, and D. E. Galloway. 1982. Centralized packaging of beef loin steaks with different oxygen-barrier films: Microbiological characteristics. *J. Food Sci.* 47:1070–79.

117. Velankar, N. K., and T. K. Govindan. 1958. A preliminary study of the distribution of nonprotein nitrogen in some marine fishes and invertebrates. *Proc. Indian Acad. Sci.* B47:202–9.

118. Watt, B. K., and A. L. Merrill. 1950. Composition of foods—raw, processed, prepared. *Agricultural Handbook no. 8.* Washington, D.C.: U.S.D.A.

10.

SPOILAGE OF MISCELLANEOUS FOODS

This chapter covers the microbiological spoilage of the following groups of foods: eggs, cereals and flour, bakery products, dairy products, sugar and spices, nutmeats, beverages and fermented foods, salad dressings, and canned foods.

EGGS

The hen's egg is an excellent example of a product that normally is well protected by its intrinsic parameters. Externally, a fresh egg has three structures, each of which is effective to some degree in retarding the entry of microorganisms: the outer, waxy shell membrane; the shell; and the inner shell membrane (see Fig. 10-1). Internally, lysozyme is present in egg white. This enzyme has been shown to be quite effective against gram-positive bacteria. Egg white also contains avidin, which forms a complex with biotin, thereby making this vitamin unavailable to microorganisms. In addition, egg white has a high pH (about 9.3) and contains conalbumin, which forms a complex with iron, thus rendering it unavailable to micro-organisms. On the other hand, the nutrient content of the yolk material and its pH in fresh eggs (about 6.8) make it an excellent source of growth for most microorganisms.

Freshly laid eggs are generally sterile. However, in a relatively short period of time after laying, numerous microorganisms may be found on the outside and under the proper conditions may enter eggs, grow, and cause spoilage. Among the bacteria found are members of the following genera: *Pseudomonas, Acinetobacter, Proteus, Aeromonas, Alcaligenes, Escherichia,*

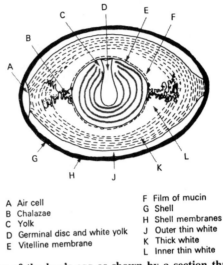

A Air cell	F Film of mucin
B Chalazae	G Shell
C Yolk	H Shell membranes
D Germinal disc and white yolk	J Outer thin white
E Vitelline membrane	K Thick white
	L Inner thin white

Fig. 10-1. Structure of the hen's egg as shown by a section through the long axis. (3, reproduced with permission of Elsevier Publishing Co.)

Micrococcus, Salmonella, Serratia, Enterobacter, Flavobacterium, and *Staphylococcus.* Among the molds generally found are members of the genera *Mucor, Penicillium, Hormodendron, Cladosporium,* and others, while *Torula* is the only yeast found with any degree of consistency.

The most common form of bacterial spoilage of eggs is a condition known as **rotting. Green rots** are caused by *Pseudomonas* spp., especially *Pseudomonas fluorescens;* **colorless rots** by *Pseudomonas, Acinetobacter,* and other species; **black rots** by *Proteus, Pseudomonas,* and *Aeromonas;* **pink rots** by *Pseudomonas;* **red rots** by *Serratia* spp., and **"custard" rots** by *Proteus vulgaris* and *P. intermedium* (7). Mold spoilage of eggs is generally referred to as **pinspots** from the appearance of mycelial growth on the inside upon candling. *Penicillium* and *Cladosporium* spp. are among the most common causes of pinspots and fungal rotting in eggs. Bacteria also cause a condition in eggs known as **mustiness.** *Pseudomonas graveolens* and *Proteus* spp. have been implicated in this condition, with *P. graveolens* producing the most characteristic spoilage pattern.

The entry of microorganisms into whole eggs is favored by high humidity. Under such conditions, growth of the microorganisms on the surface of eggs is favored, followed by penetration through the shell and inner membrane. The latter structure is the most important barrier to the penetration of bacteria into eggs, followed by the shell and the outer membrane (17). More bacteria are found in egg yolk than in egg white, and the reason for a general lack of microorganisms in egg white is quite possibly its content of antimicrobial substances. In addition, upon storage, the thick white loses water to the yolk, resulting in a thinning of yolk and a shrinking of the thick white. This phenomenon makes it possible for the yolk to come into direct contact with the inner membrane, where it may be infected directly by microorganisms. Once inside of the yolk, bacteria apparently grow in this nutritious medium, producing by-products of protein and amino acid metabolism such as H_2S and other foul-smelling compounds. The effect of significant growth is to cause the yolk to become "runny" and discolored. Molds generally multiply first in the region of the air sac, where oxygen favors the growth of these forms. Under conditions of high humidity, molds may be seen growing over the outer surface of eggs. Under conditions of low humidity and low temperatures, surface growth is not favored, but eggs lose water at a faster rate and thereby become undesirable as products of commerce.

The antimicrobial systems of eggs are noted in Chapter 3. In addition to those noted, hen egg albumen contains ovotransferrin, which chelates metal ions, particularly Fe^{3+}, and ovoflavoprotein, which binds riboflavin. At its normal pH of 9.0–10.0, egg albumen is cidal to gram-positive bacteria and yeasts at both 30° and 39.5°C (32). The addition of iron reduces the antimicrobial properties of egg albumen.

For a general review of the microbiology of the hen's egg, see Mayes and Takeballi (20).

CEREALS, FLOUR, AND DOUGH PRODUCTS

The microbial flora of wheat, rye, corn, and related products may be expected to be that of soil, storage environments, and those picked up during the processing of these commodities. While these products are high in proteins and carbohydrates, their low a_w is such as to restrict the growth of all microorganisms if stored properly. The microbial flora of flour is relatively low, since some of the bleaching agents reduce the load. When conditions of a_w favor growth, bacteria of the genus *Bacillus* and molds of several genera are usually the only ones that develop. Many aerobic sporeformers are capable of producing amylase, which enables them to utilize flour and related products as sources of energy, provided, of course, that sufficient moisture is present to allow growth to occur. With less moisture, mold growth occurs and may be seen as typical mycelial growth and spore formation. Members of the genus *Rhizopus* are common and may be recognized by their black spores.

The spoilage of fresh refrigerated dough products including buttermilk biscuits, dinner and sweet rolls, and pizza dough is caused mainly by lactic acid bacteria. In a study by Hesseltine et al. (11), 92% of isolates were Lactobacillaceae, with more than one-half belonging to the genus *Lactobacillus,* 36% to the genus *Leuconostoc,* and 3% to *Streptococcus.* Molds were found generally in low numbers in spoiled products. The fresh products showed lactic acid bacterial numbers as high as log 8.38/g.

BAKERY PRODUCTS

Commercially produced and properly handled bread generally lacks sufficient amounts of moisture to allow for the growth of any organisms except molds. One of the most common is *Rhizopus stolonifer,* often referred to as the "bread mold." The "red bread mold," *Neurospora sitophila,* may also be seen from time to time. Storage of bread under conditions of low humidity retards mold growth, and this type of spoilage is generally seen only where bread is stored at high humidities or where wrapped while still warm. Home-made breads may undergo a type of spoilage known as **ropiness,** which is caused by the growth of certain strains of *Bacillus subtilis (B. mesentericus).* The ropiness may be seen as stringiness by carefully breaking a batch of dough into two parts. The source of the organisms is flour, and their growth is favored by holding the dough for sufficient periods of time at suitable temperatures.

Cakes of all types rarely undergo bacterial spoilage due to their unusually high concentrations of sugars, which restrict the availability of water. The most common form of spoilage displayed by these products is moldiness. Common sources of spoilage molds are any and all cake ingredients, especially sugar, nuts, and spices. While the baking process is generally sufficient to destroy these organisms, many are added in icings, meringues, toppings, and so forth. Also, molds may enter baked cakes from handling and from the air. Growth of molds on the surface of cakes is favored by conditions of high humidity. On some fruit cakes, growth often originates underneath nuts and fruits if they are placed on the surface of such products after baking. Continued growth of molds on breads and cakes results in a hardening of the products.

DAIRY PRODUCTS

Dairy products such as milk, butter, cream, and cheese are all susceptible to microbial spoilage because of their chemical composition. Milk is an excellent growth medium for all of the common spoilage organisms including molds and yeasts. Fresh, nonpasteurized milk generally contains varying numbers of microorganisms, depending upon the care employed in milking, cleaning, and handling of milk utensils. Raw milk held at refrigerator temperatures for several days invariably shows the presence of several or all bacteria of the following genera: *Streptococcus, Leuconostoc, Lactobacillus, Microbacterium, Propionibacterium, Micrococcus,* coliforms, *Proteus, Pseudomonas, Bacillus,* and others. Those unable to grow at the usual low temperature of holding tend to be present in very low numbers. The pasteurization process eliminates all but thermoduric strains, primarily streptococci and lactobacilli, and sporeformers of the genus *Bacillus* (and clostridia if present in raw milk). The spoilage of pasteurized milk is caused by the growth of heat-resistant streptococci utilizing lactose to produce lactic acid, which depresses the pH to a point (about pH 4.5) where curdling takes place. If present, lactobacilli are able to grow at pH values below that required by *S. lactis*. These organisms continue the fermentative activities and may bring the pH to 4.0 or below. If mold spores are present, these organisms begin to grow at the surface of the sour milk and raise the pH toward neutrality, thus allowing the more proteolytic bacteria such as *Pseudomonas* spp. to grow and bring about the liquefaction of the milk curd.

The same general pattern outlined above may be expected to occur in raw milk, especially if held at refrigerator temperatures. Another condition sometimes seen in raw milk is referred to as **ropiness.** This condition is caused by the growth of *Alcaligenes viscolactis* and is favored by low-temperature holding of raw milk for several days. The rope consists of slime-layer material produced by the bacterial cells and it gives the product a stringy consistency. This condition is not as common today as it was in years past.

Butter contains around 15% water, 81% fat, and generally less than 0.5% carbohydrate and protein (Table 10-1). Although it is not a highly perishable product, it does undergo spoilage by bacteria and molds. The main source of microorganisms to butter is cream, whether sweet or sour, pasteurized or nonpasteurized. The flora of whole milk may be expected to be found in cream since as the fat droplets rise to the surface of milk, they carry up microorganisms. The processing of both raw and pasteurized creams to yield butter brings about a reduction in the numbers of all microorganisms, with values for finished cream ranging from several hundred to over 100,000/g having been reported for the finished salted butter (19).

Bacteria cause two principal types of spoilage in butter. The first is a condition known as **"surface taint"** or putridity. This condition is caused by *Pseudomonas putrefaciens* as a result of its growth on the surface of finished butter. It develops at temperatures within the range 4°–7°C and may become apparent within 7–10 days. The odor of this condition is apparently due to certain organic acids, especially isovaleric acid (5). The second most common bacterial spoilage condition of butter is **rancidity.**

Table 10-1. Percentage composition of nine miscellaneous foods (34).

Food	Water	Carbohydrates	Proteins	Fat	Ash
Beer (4% alcohol)	90.2	4.4	0.6	0.0	0.2
Bread, enriched white	34.5	52.3	8.2	3.3	1.7
Butter	15.5	0.4	0.6	81.0	2.5
Cake (pound)	19.3	49.3	7.1	23.5	0.8
Figbars	13.8	75.8	4.2	4.8	1.4
Jellies	34.5	65.0	0.2	0.0	0.3
Margarine	15.5	0.4	0.6	81.0	2.5
Mayonnaise	1.7	21.0	26.1	47.8	3.4
Peanut butter	16.0	3.0	1.5	78.0	1.5

This condition is caused by the hydrolysis of butterfat with the liberation of free fatty acids. It should be recognized, of course, that lipase from sources other than microorganisms can cause the effect. The causative organism is *Pseudomonas fragi* although *P. fluorescens* is sometimes found. Bacteria may cause three other less common spoilage conditions in butter. **Malty flavor** is reported to be due to the growth of *S. lactis* var. *maltigenes*. **Skunk-like** odor is reported to be caused by *Pseudomonas mephitica*, while black discolorations of butter have been reported to be caused by *P. nigrifaciens* (8).

Butter undergoes fungal spoilage rather commonly by species of the following genera: *Cladosporium, Alternaria, Aspergillus, Mucor, Rhizopus, Penicillium,* and *Geotrichum,* especially *G. candidum* (*Oospora lactis*). These organisms can be seen growing on the surface of butter, where they produce colorations referable to their particular spore colors. Black yeasts of the genus *Torula* also have been reported to cause discolorations on butter. The microscopic examination of moldy butter reveals the presence of mold mycelia some distances from the visible growth. The generally high lipid content and low water content make butter more susceptible to spoilage by molds than by bacteria.

Cottage cheese undergoes spoilage by bacteria, yeasts, and molds. The most common spoilage pattern displayed by bacteria is a condition known as **slimy curd**. *Alcaligenes* spp. have been reported to be among the most frequent causative organisms although *Pseudomonas, Proteus, Enterobacter,* and *Acinetobacter* spp. have been implicated. *Penicillium, Mucor, Alternaria,* and *Geotrichum* all grow well on cottage cheese, to which they impart stale, musty, moldy, and yeasty flavors (8). The shelf life of commercially produced cottage cheese in Alberta, Canada, was found to be limited by yeasts and molds (29). While 48% of fresh samples contained coliforms, these organisms did not increase upon storage in cottage cheese at 40°F for 16 days.

The low moisture content of ripened cheeses makes them insusceptible to spoilage by most organisms, although molds can and do grow on these products as would be expected. Some ripened cheeses have sufficiently low O/R potentials to support the growth of anaerobes. It is, therefore, not surprising to find that anaerobic bacteria sometimes cause the spoilage of these products when a_w permits growth to occur. *Clostridium* spp., especially *C. pasteurianum, C. butyricum,* and *C. sporogenes,* have been reported to cause **gassiness** of cheeses. One aerobic sporeformer, *B. polymyxa,* has been reported to cause gassiness. All of these organisms utilize lactic acid with the production of CO_2, which is responsible for the gassy condition of these products.

SUGARS, CANDIES, AND SPICES

These products rarely undergo microbial spoilage if properly prepared, processed, and stored, primarily because of the lack of sufficient moisture for growth. Both cane and beet sugars may be expected to contain micro-organisms. The important bacterial contaminants are members of the genera *Bacillus* and *Clostridium*, which sometimes cause trouble in the canning industry (see Chapter 14). If sugars are stored under conditions of extremely high humidity, growth of some of these organisms is possible, usually at the exposed surfaces. The successful growth of these organisms depends, of course, upon their getting an adequate supply of moisture and essential nutrients other than carbohydrates. *Torula* and osmophilic strains of *Saccharomyces* (formerly *Zygosaccharomyces* spp.) have been reported to cause trouble in high-moisture sugars. These organisms have been reported to cause inversion of sugar. One of the most troublesome organisms in sugar refineries is *Leuconostoc mesenteroides.* This organism hydrolyzes sucrose and synthesizes a glucose polymer referred to as **dextran.** This gummy and slimy polymer sometimes clogs the lines and pipes through which sucrose solutions pass.

Among candies that have been reported to undergo microbial spoilage are chocolate creams, which sometimes undergo explosions. The causative organisms have been reported to be *Clostridium* spp., especially *C. sporogenes,* which finds its way into these products through sugars, starch, and possibly other ingredients.

Although spices do not undergo microbial spoilage in the usual sense of the word, molds and a few bacteria do grow on those that do not contain antimicrobial principals, provided sufficient moisture is available. Prepared mustard has been reported to undergo spoilage by yeasts and by *Proteus* and *Bacillus* spp., usually with a gassy fermentation. The usual treatment of spices with propylene oxide reduces their content of microorganisms, and those that remain are essentially sporeformers and molds. No trouble should be encountered from microorganisms as long as the moisture level is kept low.

NUTMEATS

Due to the extremely high fat and low water content of products such as pecans and walnuts (Table 10-2), these products are quite refractory to spoilage by bacteria. Molds can and do grow upon them if they are stored

Table 10-2. Percentage composition of various nuts (34).

Nut	Water	Carbohydrates	Protein	Fat	Ash
Almonds (dried)	4.7	19.6	18.6	34.1	3.0
Brazil nuts	5.3	11.0	14.4	65.9	3.4
Cashews	3.6	27.0	18.5	48.2	2.7
Peanuts	2.6	23.6	26.9	44.2	2.7
Pecans	3.0	13.0	9.4	73.0	1.6
Mean	3.8	18.8	17.6	57.1	2.7

under conditions that permit sufficient moisture to be picked up. On examining nutmeats, molds of many genera may be found that are picked up by the products during collecting, cracking, sorting, and packaging. See Chapter 22 for a discussion of aflatoxins as related to nutmeats.

BEERS, WINES, AND FERMENTED FOODS

The products covered in this section are beers and ales, table wines, sauerkraut, pickles, and olives. All of these products are themselves the products of microbial actions.

The industrial spoilage of beers and ales is commonly referred to as beer infections. This condition is caused by yeasts and bacteria. The spoilage patterns of beers and ales may be classified into four groups: ropiness, sarcinae sickness, sourness, and turbidity. **Ropiness** is a condition in which the liquid becomes characteristically viscous and pours as an "oily" stream. It is caused by *Acetobacter, Lactobacillus, Pediococcus cerevisiae,* and *Gluconobacter oxydans* (formerly *Acetomonas*) (28, 35). **Sarcinae sickness** is caused by *P. cerevisiae,* which produces a honeylike odor. This characteristic odor is the result of diacetyl production by the spoilage organism in combination with the normal odor of beer. **Sourness** in beers is caused by *Acetobacter* spp. These organisms are capable of oxidizing ethanol to acetic acid, and the sourness that results is referable to increased levels of acetic acid. **Turbidity** and off-odors in beers are caused by *Zymomonas anaerobia* (formerly *Achromobacter anaerobium*) and several yeasts such as *Saccharomyces* spp. Growth of bacteria is possible in beers because of a normal pH range of 4–5 and a good content of utilizable nutrients.

An interesting gram-negative, anaerobic rod was isolated in 1978 by Lee et al. (15) at the Coors Brewery in Colorado from turbid beer and placed in the family Bacteroidaceae. Classified as *Pectinatus cerevisiiphilus,* it has been reported from breweries in Germany and Finland (9). One of the most interesting features of this organism is its possession of flagella that arise at varying sites rather than from only one side of the cell. From glucose it produces acetic, propionic, and succinic acids, with propionic being its main metabolic product (9). It is a mesophile with an optimum growth temperature of 32°C and a range of 15°–40°C. It has a G + C content of DNA of 39.8 moles %, and has been shown capable of fermenting many sugars.

With respect to spoiled packaged beer, one of the major contaminants found is *Saccharomyces diastaticus,* which is able to utilize dextrins that normal brewers' yeasts (*S. carlsbergensis* and *S. cerevisiae*) cannot (12). Pediococci, *Flavobacterium proteus* (formerly *Obesumbacterium*), and *Brettanomyces* are sometimes found in spoiled beer.

Table wines undergo spoilage by bacteria and yeasts, *Candida mycoderma* being the most important yeast. Growth of this organism occurs at the surface of wines, where a thin film is formed. The organisms attack alcohol and other constituents from this layer and create an appearance that is sometimes referred to as **wine flowers.** Among the bacteria that cause wine spoilage are members of the genus *Acetobacter,* which oxidize alcohol to acetic acid (produce vinegar). The most serious and the most common disease of table wines is referred to as **tourne disease** (26). Tourne disease is caused by a facultative anaerobe or anaerobe that utilizes sugars and seems to prefer conditions of low alcohol content. This type of spoilage is characterized by an increased volatile acidity, a silky type of cloudiness, and later in the course of spoilage, a "mousey" odor and taste.

Malo-lactic fermentation is a spoilage condition of great importance in wines. Malic and tartaric acids are two of the predominant organic acids in grape must and wine, and in the malo-lactic fermentation, contaminating bacteria degrade malic acid to lactic acid and CO_2:

$$\text{L}(-)\text{-Malic acid} \xrightarrow{\text{"malo-lactic enzyme"}} \text{L}(+)\text{-Lactic acid} + CO_2.$$

L-malic acid may be decarboxylated also to yield pyruvic acid (13). The effect of these conversions is to reduce the acid content and affect flavor. The malo-lactic fermentation (which may also occur in cider) can be carried out by many lactic acid bacteria including leuconostocs, pediococci, and lactobacilli (18, 27). While the function of the malo-lactic fermentation to the fermenting organism is not well understood, it has been shown that *L. oenos* is actually stimulated by the process (24). The decomposition in wines of tartaric acid is undesirable also, and this process can be achieved by some strains of *L. plantarum* in the following general manner:

$$\text{Tartaric acid} \rightarrow \text{Lactic acid} + \text{acetic acid} + CO_2.$$

The effect of the above is to reduce the acidity of wine. Unlike the malo-lactic fermentation, few lactic acid bacteria break down tartaric acid (27).

Root beer undergoes bacterial spoilage on occasion. Lehmann and Byrd (16) investigated spoiled root beer characterized by a musty odor and taste. The causative organism was found to be *Achromobacter* sp. By inoculating this organism into the normal product, these authors found that the characteristic spoilage appeared in 2 weeks.

Sauerkraut is the product of lactic acid fermentation of fresh cabbage, and while the finished product has a pH in the range of 3.1–3.7, it is still subject to spoilage by bacteria, yeasts, and molds. The microbial spoilage of sauerkraut generally falls into the following categories: soft kraut, slimy kraut, rotted kraut, and pink kraut. **Soft kraut** results when bacteria that normally do not initiate growth until the late stages of kraut production actually grow earlier. **Slimy kraut** is caused by the rapid growth of *Lactobacillus cucumeris* and *L. plantarum,* especially at elevated temperatures (26). **Rotted** sauerkraut may be caused by bacteria, molds, and/or yeasts, while **pink kraut** is caused by the surface growth of *Torula* spp., especially *T. glutinis*. Due to the high acidity, finished kraut is generally spoiled by molds growing on the surface. The growth of these organisms effects an increase in pH to levels where a large number of bacteria can grow that were previously inhibited by conditions of high acidity.

Pickles result from lactic acid fermentation of cucumbers. The finished product has a pH of around 4.0. These products undergo spoilage by bacteria and molds. **Pickle blackening** may be caused by *Bacillus nigrificans,* which produces a dark water-soluble pigment. *Enterobacter* spp., lactobacilli, and pediococci have all been implicated as causes of a condition known as "**bloaters,**" produced by gas formation within the individual pickles. **Pickle softening** is caused by pectolytic organisms of the genera *Bacillus, Fusarium, Penicillium, Phoma, Cladosporium, Alternaria, Mucor, Aspergillus,* and others. The actual softening of pickles may be caused by any one or several of these or related organisms. Pickle softening results from the production of pectinases, which break down the cementlike substance in the wall of the product.

Among the types of microbial spoilage that olives undergo, one of the most characteristic is **zapatera spoilage.** This condition, which sometimes occurs in brined olives, is characterized by a malodorous fermentation. The odor is due apparently to propionic acid, which is produced by certain species of *Propionibacterium* (25).

A **softening** condition of Spanish-type green olives has been found to be caused by the yeasts *Rhodotorula glutinis* var. *glutinis, R. minuta* var. *minuta,* and *R. rubra* (33). All of these organisms produce polygalacturonases, which effect olive tissue softening. Under appropriate cultural conditions, the organisms were shown to produce pectin methyl esterase as well as polygalacturonase. A **sloughing** type of spoilage of California ripe olives was shown by Patel and Vaughn (21) to be caused by *Cellulomonas flavigena.* This organism showed high cellulolytic activity, which was enhanced by the growth of other organisms such as *Xanthomonas, Enterobacter,* and *Escherichia* spp.

MAYONNAISE AND SALAD DRESSINGS

Mayonnaise can be defined as a semisolid emulsion of edible vegetable oil, egg yolk or whole egg, vinegar, and/or lemon juice, and other ingredients such as salt and other seasonings and glucose, in a finished product containing not less than 50% edible oil. The pH of this product ranges from 3.6 to 4.0, with acetic acid as the predominant acid representing 0.29–0.5% of total product with an a_w of 0.925. The aqueous phase contains 9–11% salt and 7–10% sugar (31). Salad dressings are quite similar in composition to mayonnaise, but the finished product contains at least 30% edible vegetable oil and has an a_w of 0.929, a pH of 3.2–3.9 with acetic acid usually the predominant acid accounting for 0.9–1.2% of total product. The aqueous phase contains 3.0–4.0% salt and 20–30% sugar (31). While the nutrient content of these products is suitable as food sources for many spoilage organisms, the pH, organic acids, and low a_w restrict spoilers to yeasts, a few bacteria, and molds. The yeast *Zygosaccharomyces bailii* is known to cause the spoilage of salad dressings, tomato catsup, carbonated beverages, and some wines. Yeasts of the genus *Saccharomyces* have been implicated in the spoilage of mayonnaise, salad dressing, and French dressing. One of the few bacteria reported to cause spoilage of products of this type is *Lactobacillus brevis,* which was reported to produce gas in salad dressing. Appleman et al. (1) investigated spoiled mayonnaise and recovered a strain of *B. subtilis* and a yeast that they believed to be the etiologic agents. *Bacillus vulgatus* has been recovered from spoiled thousand islands dressing, where it caused darkening and separation of the emulsion. In one particular study of the spoilage of thousand island dressing, pepper and paprika were shown to be the sources of *B. vulgatus* (22). Mold spoilage of products of this type occurs only at the surfaces when sufficient oxygen is available. Separation of the emulsion is generally one of the first signs of spoilage of these products although bubbles of gas and the rancid odor of butyric acid may precede emulsion separation. The spoilage organisms apparently attack the sugars fermentatively. It appears that the pH remains low, thereby preventing the activities of proteolytic and lipolytic organisms. It is not surprising to find yeasts and lactic acid bacteria under these conditions. In a study of seventeen samples of spoiled mayonnaise, mayonnaiselike, and blue cheese dressings, Kurtzman et al. (14) found high yeast counts in most samples and high lactobacilli counts in two. The pH of samples ranged from 3.6 to 4.1. Two-thirds of the spoiled samples yielded *Saccharomyces bailii*. Common in some samples was *L. fructivorans,* with aerobic spore-formers being found in only two samples. Of ten unspoiled samples tested, microorganisms were in low numbers or not detectable at all.

In regard to food-borne pathogenic bacteria, the interaction of low pH, acids, and low a_w is such that these products will not support growth of these types of organisms (31).

CANNED FOODS

Although the objective in the canning of foods is the destruction of micro-organisms, these products nevertheless undergo microbial spoilage under certain conditions. The main reasons for this are the following: under-processing, inadequate cooling, contamination of the can resulting from leakage through seams, and preprocess spoilage. Since some canned foods receive low-heat treatments, it is to be expected that a rather large number of different types of microorganisms may be found upon examining such foods.

As a guide to the type of spoilage that canned foods undergo, the following classification of canned foods based upon acidity is helpful.

LOW ACID. pH > 4.6. Meat and marine products, milk, some vegetables (corn, lima beans), meat and vegetable mixtures, and so on. Spoiled by thermophilic flat-sour group (*B. stearothermophilus, B. coagulans*), sulfide spoilers (*C. nigrificans, C. bifermentans*), and/or gaseous spoilers (*C. thermosaccharolyticum*). Mesophilic spoilers include putrefactive anaerobes (especially P.A. 3679 types). Spoilage and toxin production by *C. botulinum* types A and B may occur if these organisms are present. Medium-acid foods are those with pH range of 5.3–4.6, while low-acid foods are those with pH ⩾ 5.4.

ACID. pH 3.7–4.0 to 4.6. In this category are fruits such as tomatoes, pears, and figs. Thermophilic spoilers include *B. coagulans* types. Mesophiles include *B. polymyxa, B. macerans* (*B. betanigrificans*), *C. pasteurianium, C. butyricum,* lactobacilli, and others.

HIGH ACID. pH < 4.0–3.7. This category includes fruits and fruit and vegetable products—grapefruit, rhubarb, sauerkraut, pickles, and so forth. Generally spoiled by nonsporeforming mesophiles—yeasts, molds, and/or lactic acid bacteria.

Canned food spoilage organisms may be further characterized as follows:

1. Mesophilic organisms
   a. Putrefactive anaerobes
   b. Butyric anaerobes
   c. Aciduric flat sours
   d. Lactobacilli
   e. Yeasts
   f. Molds
2. Thermophilic organisms
   a. Flat-sour spores
   b. Thermophilic anaerobes producing sulfide
   c. Thermophilic anaerobes not producing sulfide

The canned food spoilage manifestations of these organisms are presented in Table 10-3.

With respect to the spoilage of high-acid and other canned foods by yeasts, molds, and bacteria, several of these organisms have been repeatedly associated with certain specific foods. The yeasts *Torula lactis-condensi* and *T. globosa* cause blowing or gaseous spoilage of sweetened condensed milk, which is not heat processed. The mold *Aspergillus repens* is associated with the formation of "buttons" on the surface of sweetened condensed milk. *Lactobacillus brevis* (*L. lycopersici*) causes a vigorous fermentation in tomato catsup, Worcestershire sauce, and similar products. *Leuconostoc mesenteroides* has been reported to cause gaseous spoilage of canned pineapples and ropiness in peaches. The mold *Byssochlamys fulva* causes spoilage

Table 10-3. Spoilage manifestations in acid and low-acid canned foods (30).

Type of organism	Appearance and manifestations of can	Condition of product
Acid products		
1. *B. thermoacidurans* (flat sour: tomato juice)	Can flat; little change in vacuum	Slight pH change; off-odor and flavor
2. Butyric anaerobes (tomatoes and tomato juice)	Can swells; may burst	Fermented; butyric odor
3. Nonsporeformers (mostly lactics)	Can swells, usually bursts, but swelling may be arrested	Acid odor
Low-acid products		
1. Flat sour	Can flat; possible loss of vacuum on storage	Appearance not usually altered; pH markedly lowered—sour; may have slightly abnormal odor; sometimes cloudy liquor
2. Thermophilic anaerobe	Can swells; may burst	Fermented, sour, cheesey, or butyric odor
3. Sulfide spoilage	Can flat; H_2S gas absorbed by product	Usually blackened; "rotten egg" odor
4. Putrefactive annaerobe	Can swells; may burst	May be partially digested; pH slightly above normal; typical putrid odor
5. Aerobic sporeformers (odd types)	Can flat; usually no swelling, except in cured meats when NO_3 and sugar are present	Coagulated evaporated milk, black beets

of bottled and canned fruits. Its actions cause disintegration of fruits as a result of pectin breakdown (2). *Torula stellata* has been reported to cause the spoilage of canned bitter lemon, and to grow at a pH of 2.5 (23).

Frozen concentrated orange juice sometimes undergoes spoilage by yeasts and bacteria. Hays and Riester (10) investigated samples of this product spoiled by bacteria. The orange juice was characterized as having a vinegary to buttermilk off-odor with an accompanying off-flavor. From the spoiled product were isolated *L. plantarum* var. *mobilis, L. brevis, Leuconostoc mesenteroides,* and *Leuconostoc dextranicum.* The spoilage characteristics could be reproduced by inoculating the above isolates into fresh orange juice.

Minimum growth temperatures of spoilage thermophiles are of some importance in diagnosing the cause of spoiled canned foods. *B. coagulans* (*B. thermoacidurans*) has been reported to grow only slowly at 25°C but grows well between 30° and 55°C. *B. stearothermophilus* does not grow at 37°, its optimum temperature being around 65°C with smooth variants showing a shorter generation time at this temperature than rough variants (6). *C. thermosaccharolyticum* does not grow at 30° but has been reported to grow at 37°C.

Also of importance in diagnosing the cause of canned food spoilage is the appearance of the unopened can or container. The ends of a can of food are normally flat or slightly concave. When microorganisms grow and produce gases, the can goes through a series of changes visible from the outside. The change is designated a **flipper** when one end can be made convex by striking or heating the can. A **springer** is a can with both ends bulged when one or both remain concave if pushed in, or when one end is pushed in the other pops out. A **soft swell** refers to a can with both ends bulged that may be dented by pressing with the fingers. A **hard swell** has both ends bulged so that neither end can be dented by hand. The above events tend to develop successively and become of value in predicting the type of spoilage that might be in effect. Flippers and springers may be incubated under wraps at a temperature appropriate to the pH and type of food in order to allow for further growth of any organisms that might be present. These effects on cans do not always represent microbial spoilage. Soft swells often represent microbial spoilage as do hard swells. In high-acid foods, however, hard swells are very often **hydrogen swells,** which result from the release of hydrogen gas by the action of food acids on the iron of the can. The other two most common gases in cans of spoiled foods are CO_2 and H_2S, both of which are the result of the metabolic activities of microorganisms. Hydrogen sulfide may be noted by its characteristic odor, while CO_2 and hydrogen may be determined by the following test. Construct an apparatus of glass or plastic tubing attached to a hollow punch fitted with a large rubber stopper. Into a test tube filled with dilute KOH, insert the free end of this apparatus and invert it in a beaker filled with dilute KOH. When an opening is made in one end of the can with the hollow punch, the gases will displace the dilute KOH inside the tube. Before removing the open end from the beaker, close the tube by placing the thumb over the end. To test for CO_2, shake the tube and note for a vacuum as evidenced by suction against the finger. To test for hydrogen, repeat above and apply a match near the top of the tube and then quickly remove thumb. A "pop" indicates the presence of hydrogen. Both gases may be found in some cans of spoiled foods.

"**Leakage-type**" spoilage of canned foods is characterized by a flora of nonsporeforming organisms that would not survive the heat treatment normally given heat-processed foods. These organisms enter cans at the start of cooling through faulty seams, which generally result from can abuse. The organisms that cause leakage-type spoilage can be found either on the cans

Table 10-4. Some features of canned food spoilage that results from understerilization and seam leakage (30).

	Understerilization	Leakage
Can	Flat or swelled; seams generally normal	Swelled; may show defects
Product Appearance	Sloppy or fermented	Frothy fermentation; viscous
Odor	Normal, sour or putrid, but generally consistent	Sour, fecal, generally varying from can to can
pH	Usually fairly constant	Wide variation
Microscopic and Cultural	Pure cultures, sporeformers; growth at 98°F and/or 131°F; may be characteristic on special media, e.g., acid agar for tomato juice	Mixed cultures, generally rods and cocci; growth only at usual temperatures
History	Spoilage usually confined to certain portions of pack. In acid products diagnosis may be less clearly defined. Similar organisms may be involved in understerilization and leakage	Spoilage scattered

or in the cooling water. This problem is minimized if the cannery cooling water contains < 100 bacteria/ml. This type of spoilage may be further differentiated from that caused by understerilization (see Table 10-4).

REFERENCES

1. Appleman, M. D., E. P. Hess, and S. C. Rittenberg. 1949. An investigation of a mayonnaise spoilage. *Food Technol.* 3:201–3.
2. Baumgartner, J. G., and A. C. Hersom. 1957. *Canned foods.* Princeton, N.J.: D. Van Nostrand.
3. Brooks, J., and H. P. Hale. 1959. The mechanical properties of the thick white of the hen's egg. *Biochem. Biophys. Acta* 32:237–50.
4. Cross, T. 1968. Thermophilic *Actinomycetes. J. Appl. Bacteriol.* 31:36–53.
5. Dunkley, W. L., G. Hunter, H. R. Thornton, and E. G. Hood. 1942. Studies on surface taint butter. II. An odorous compound in skim milk cultures of *Pseudomonas putrefaciens. Scientific Agr.* 22:347–55.
6. Fields, M. L. 1970. The flat sour bacteria. *Adv. Food Res.* 18:163–217.
7. Florian, M. L. E., and P. C. Trussell. 1957. Bacterial spoilage of shell eggs. IV. Identification of spoilage organisms. *Food Technol.* 11:56–60.
8. Foster, E. M., F. E. Nelson, M. L. Speck, R. N. Doetsch, and J. C. Olson, Jr. 1957. *Dairy microbiology.* Chs. 13, 14. Englewood Cliffs, N.J.: Prentice-Hall.
9. Haikara, A., L. Penttila, T.-M. Enari, and K. Lounatmaa. 1981. Microbiological,

biochemical, and electron microscopic characterization of a *Pectinatus* strain. *Appl. Environ. Microbiol.* 41:511–17.

10. Hays, G. L., and D. W. Riester. 1952. The control of "off-odor" spoilage in frozen concentrated orange juice. *Food Technol.* 6:386–89.

11. Hesseltine, C. W., R. R. Graves, R. Rogers, and H. R. Burmeister. 1969. Aerobic and facultative microflora of fresh and spoiled refrigerated dough products. *Appl. Microbiol.* 18:848–53.

12. Kleyn, J., and J. Hough. 1971. The microbiology of brewing. *Ann. Rev. Microbiol.* 25:583–608.

13. Kunkee, R. E. 1975. A second enzymatic activity for decomposition of malic acid by malo-lactic bacteria. In *Lactic acid bacteria in beverages and food,* ed. J. G. Carr et al., 29–42. New York: Academic Press.

14. Kurtzman, C. P., R. Rogers, and C. W. Hesseltine. 1971. Microbiological spoilage of mayonnaise and salad dressings. *Appl. Microbiol.* 21:870–74.

15. Lee, S. Y., M. S. Mabee, and N. O. Jangaard. 1978. *Pectinatus,* a new genus of the family *Bacteroidaceae. Int. J. Syst. Bacteriol.* 28:582–94.

16. Lehmann, D. L., and B. E. Byrd. 1953. A bacterium responsible for a musty odor and taste in root beer. *Food Res.* 18:76–78.

17. Lifshitz, A., R. G. Baker, and H. B. Naylor. 1964. The relative importance of chicken egg exterior structures in resisting bacterial penetration. *J. Food Sci.* 29:94–99.

18. London, J. 1976. The ecology and taxonomic status of the lactobacilli. *Ann. Rev. Microbiol.* 30:279–301.

19. Macy, H., S. T. Coulter, and W. B. Combs. 1932. Observations on the quantitative changes in the microflora during the manufacture and storage of butter. *Minn. Agric. Exp. Sta. Techn. Bull.* 82.

20. Mayes, F. J., and M. A. Takeballi. 1983. Microbial contamination of the hen's egg: A review. *J. Food Protect.* 46:1092–98.

21. Patel, I. B., and R. H. Vaughn. 1973. Cellulolytic bacteria associated with sloughing spoilage of California ripe olives. *Appl. Microbiol.* 25:62–69.

22. Pederson, C. S. 1930. Bacterial spoilage of a thousand island dressing. *J. Bacteriol.* 20:99–106.

23. Perigo, J. A., B. L. Gimbert, and T. E. Bashford. 1964. The effect of carbonation, benzoic acid, and pH on the growth rate of a soft drink spoilage yeast as determined by a turbidostatic continuous culture apparatus. *J. Appl. Bacteriol.* 27:315–32.

24. Pilone, G. J., and R. E. Kunkee. 1976. Stimulatory effect of malo-lactic fermentation on the growth rate of *Leuconostoc oenos. Appl. Environ. Microbiol.* 32:405–8.

25. Plastourgos, S., and R. H. Vaughn. 1957. Species of *Propionibacterium* associated with zapatera spoilage of olives. *Appl. Microbiol.* 5:267–71.

26. Prescott, S. C., and C. G. Dunn. 1959. *Industrial microbiology.* New York: McGraw-Hill.

27. Radler, F. 1975. The metabolism of organic acids by lactic acid bacteria. In *Lactic acid bacteria in beverages and food,* ed. J. G. Carr et al., 17–27. New York: Academic Press.

28. Rainbow, C. 1975. Beer spoilage lactic acid bacteria. In *Lactic acid bacteria in beverages and food,* ed. J. G. Carr et al., 149–58. New York: Academic Press.

29. Roth, L. A., L. F. L. Clegg, and M. E. Stiles. 1971. Coliforms and shelf life of commercially produced cottage cheese. *Can. Inst. Food Technol. J.* 4:107–11.
30. Schmitt, H. P. 1966. Commercial sterility in canned foods, its meaning and determination. *Assoc. Food Drug Off. of U.S., Quart. Bull.* 30:141–51.
31. Smittle, Richard B. 1977. Microbiology of mayonnaise and salad dressing: A review. *J. Food Protect.* 40:415–22.
32. Tranter, H. S., and R. G. Board. 1984. The influence of incubation temperature and pH on the antimicrobial properties of hen egg albumen. *J. Appl. Bacteriol.* 56:53–61.
33. Vaughn, R. H., T. Jakubczyk, J. D. MacMillan, T. E. Higgins, B. A. Dave, and V. M. Crampton. 1969. Some pink yeasts associated with softening of olives. *Appl. Microbiol.* 18:771–75.
34. Watt, B. K., and A. L. Merrill. 1950. Composition of foods—raw, processed, prepared. *Agricultural Handbook no. 8,* Washington, D.C.: U.S.D.A.
35. Williamson, D. H. 1959. Studies on lactobacilli causing ropiness in beer. *J. Appl. Bacteriol.* 22:392–402.

FOOD

PRESERVATION

11.

FOOD PRESERVATION WITH CHEMICALS

The use of chemicals to prevent or delay the spoilage of foods derives in part from the fact that such compounds have been used with great success in the treatment of diseases of man, animals, and plants. This is not to imply that any and all chemotherapeutic compounds can or should be used as food preservatives. On the other hand, there are some chemicals of value as food preservatives that would be ineffective or too toxic as chemotherapeutic compounds. With the exception of certain antibiotics, none of the presently used food preservatives find any real use as chemotherapeutic compounds in man and animals. While a large number of chemicals have been described that show potential as food preservatives, only a relatively small number are allowed in food products. This is due in large part to the strict rules of safety adhered to by the Food and Drug Administration (FDA), and to a lesser extent to the fact that not all compounds that show antimicrobial activity in vitro do so when added to certain foods. Below are described those compounds most widely used, their modes of action where known, and the types of foods in which they are used. Those chemical preservatives generally recognized as safe (GRAS) are summarized in Table 11-1.

BENZOIC ACID AND THE PARABENS

Benzoic acid (C_6H_5COOH) and its sodium salt ($C_7H_5NaO_2$) along with the esters of *p*-hydroxybenzoic acid (parabens) are considered together in this section. Sodium benzoate was the first chemical preservative permitted in foods by the U.S. Food and Drug Administration, and it continues in wide use today in a large number of foods. Its approved derivatives have structural formulas as noted:

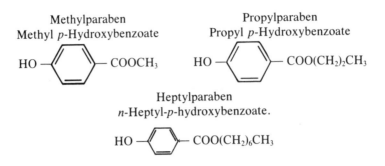

Methylparaben
Methyl *p*-Hydroxybenzoate

HO —⟨ ⟩— COOCH₃

Propylparaben
Propyl *p*-Hydroxybenzoate

HO —⟨ ⟩— COO(CH₂)₂CH₃

Heptylparaben
n-Heptyl-*p*-hydroxybenzoate.

HO —⟨ ⟩— COO(CH₂)₆CH₃

The antimicrobial activity of benzoate is related to pH, the greatest activity being at low pH values. The antimicrobial activity resides in the undissociated molecule (see below). These compounds are most active at the lowest pH values of foods and essentially ineffective at neutral values. The pK of benzoate is 4.20 and at a pH of 4.00, 60% of the compound is undissociated, while at a pH of 6.0 only 1.5% is undissociated. This results in the restriction of benzoic acid and its sodium salts to high-acid products such as apple cider, soft drinks, tomato catsup, and salad dressings. High acidity alone is generally sufficient to prevent growth of bacteria in these foods, but not that of certain molds and yeasts. As used in acidic foods, benzoate acts essentially as a mold and yeast inhibitor although it is effective against some bacteria in the 50–500 ppm range. Against yeasts and molds at around pH 5.0–6.0, from 100–500 ppm are effective in inhibiting the former, while for the latter from 30–300 ppm are inhibitory.

In foods such as fruit juices, benzoates may impart disagreeable tastes at the maximum level of 0.1%. The taste has been described as being "peppery" or burning.

As noted above, the three parabens that are permissible in foods in the United States are heptyl-, methyl-, and propyl-, while butyl- and ethylparabens are permitted in food in certain other countries. As esters of *p*-hydroxybenzoic acid, they differ from benzoate in their antimicrobial activity in being less sensitive to pH. Although not as many data have been presented on heptylparaben, it appears to be quite effective against microorganisms, with 10–100 ppm effecting complete inhibition of some gram-positive and gram-negative bacteria. Propylparaben is more effective than methylparaben on a ppm basis, with up to 1,000 ppm of the former and 1,000–4,000 ppm of the latter needed for bacterial inhibition, with gram-positive bacteria being more susceptible than gram negatives to the parabens in general (20). Heptylparaben has been reported to be effective against the malo-lactic bacteria. In a reduced-broth medium, 100 ppm propylparaben delayed germination and toxin production by *C. botulinum* type A, while 200 ppm effected inhibition up to 120 h at 37°C (100). In the case of methylparaben, 1,200 ppm were required for inhibition similar to that for the propyl- analog.

The parabens appear to be more effective against molds than against yeasts. As in the case of bacteria, the propyl- derivative appears to be the most effective where 100 ppm or less are capable of inhibiting some yeasts

Table 11-1. Summary of some GRAS chemical food preservatives.

Preservatives	Maximum tolerance	Organisms affected	Foods
Propionic acid/ propionates	0.32%	Molds	Bread, cakes, some cheeses, rope inhibitor in bread dough
Sorbic acid/ sorbates	0.2%	Molds	Hard cheeses, figs, syrups, salad dressings, jellies, cakes
Benzoic acid/ benzoates	0.1%	Yeasts and molds	Margarine, pickle relishes, apple cider, soft drinks, tomato catsup, salad dressings
Parabens[a]	0.1%[b]	Yeasts and molds	Bakery products, soft drinks, pickles, salad dressings
SO_2/sulfites	200–300 ppm	Insects, micro-organisms	Molasses, dried fruits, wine making, lemon juice (not to be used in meats or other foods recognized as sources of thiamine)
Ethylene/ propylene oxides[c]	700 ppm	Yeasts, molds, vermin	Fumigant for spices, nuts
Sodium diacetate	0.32%	Molds	Bread
Dehydroacetic acid	65 ppm	Insects	Pesticide on strawberries, squash
Sodium nitrite[c]	120 ppm	Clostridia	Meat-curing preparations
Caprylic acid	—	Molds	Cheese wraps
Ethyl formate	15–200 ppm[d]	Yeasts and molds	Dried fruits, nuts

GRAS (Generally Recognized As Safe) per Section 201 (32)(s) of the U.S. Federal Food, Drug, and Cosmetic Act as amended.
[a] Methyl-, propyl-, and heptyl-esters of *p*-hydroxybenzoic acid.
[b] Heptyl ester—12 ppm in beers; 20 ppm in noncarbonated and fruit-based beverages.
[c] May be involved in mutagenesis and/or carcinogenesis.
[d] As formic acid.

and molds, while for heptyl- and methylparabens, 50–200 and 500–1,000 ppm, respectively, are required.

Like benzoic acid and its sodium salt, the methyl- and propylparabens are permissible in foods up to 0.1% while heptylparaben is permitted in beers to a maximum of 12 ppm, and up to 20 ppm in fruit drinks and beverages. The pK for these compounds is around 8.47, and their antimicrobial activity is not increased to the same degree as for benzoate with the lowering of pH as noted above. They have been reported to be effective at pH values up to 8.0. For a more thorough review of these preservatives, see Davidson (20).

Similarities between the modes of action of benzoic and salicylic acids have been noted (8). Both compounds, when taken up by respiring microbial cells, were found to block the oxidation of glucose and pyruvate at the acetate level in *Proteus vulgaris*. With *P. vulgaris*, benzoic acid caused an increase in the rate of O_2 consumption during the first part of glucose oxidation (8). The benzoates, like propionate and sorbate, have been shown to act against microorganisms by inhibiting the cellular uptake of substrate molecules (36). The stage of endospore germination most sensitive to benzoate is noted in Fig. 11-1.

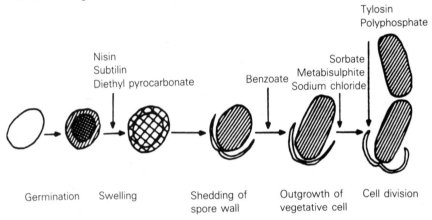

Fig. 11-1. Diagrammatic representation of growth of an endospore into vegetative cells showing stages arrested by minimum inhibitory concentrations of some food preservatives (40).

The undissociated form is essential to the antimicrobial activity of benzoate as well as for other lipophilic acids such as sorbate and propionate, as previously noted. In this state, these compounds are soluble in the cell membrane and act apparently as proton ionophores (41). As such, they facilitate proton leakage into cells and thereby increase energy output of cells to maintain their usual internal pH. With the disruption in membrane activity, amino acid transport is adversely affected (41).

SORBIC ACID

Sorbic acid ($CH_3CH{=}CHCH{=}CHCOOH$) is employed as a food preservative usually as the calcium, sodium, or potassium salt. These compounds are permissible in foods at levels not to exceed 0.2%. Like sodium benzoate, they are more effective in acid foods than in neutral foods and tend to be on par with the benzoates as fungal inhibitors. Sorbic acid works best below pH 6.0 and is generally ineffective > pH 6.5. These compounds are more effective than sodium benzoate between pH 4.0–6.0. At pH values of 3.0 and below, the sorbates are slightly more effective than the propionates but about the same as sodium benzoate. The pK of sorbate is 4.80 and at a pH of 4.0, 86% of the compound is undissociated while at a pH of 6.0 only 6% is undissociated. Sorbic acid can be employed in cakes at higher levels than propionates without imparting flavor to the product (79).

The sorbates are primarily effective against molds and yeasts but research during the past decade has shown them to be effective against a wide range of bacteria. In general, the catalase-positive cocci are more sensitive than the catalase negatives, and aerobes are more sensitive than anaerobes. The resistance of the lactic acid bacteria to sorbate, especially at pH 4.5 or above, permits its use as a fungistat in products that undergo lactic fermentations. Its effectiveness has been shown against *S. aureus,* salmonellae, coliforms, psychrotrophic spoilage bacteria (especially the pseudomonads), and *V. parahaemolyticus.* Against the latter organism, concentrations as low as 30 ppm have been shown to be effective. Shelf-life extensions have been obtained by use of sorbates on fresh poultry meat, vacuum-packaged poultry products, fresh fish, and perishable fruits. For further information, see nitrite-sorbate combinations below, and the review by Sofos and Busta (112).

The sorbates have been studied by a large number of groups for use in meat products in combination with nitrites. Bacon formulations that contain 120 ppm $NaNO_2$ without sorbate yield products that maintain their desirable organoleptic qualities in addition to being protected from *C. botulinum* growth. When 0.26% (2,600 ppm) potassium sorbate is added along with 40 ppm nitrite, no significant differences are found in the organoleptic qualities or in botulinal protection ([57] and Stevenson and Price, 1976, cited in [87]). The combination of 40 ppm $NaNO_2$ and 0.26% potassium sorbate (along with 550 ppm sodium ascorbate or sodium erythrobate) was proposed by the U.S. Department of Agriculture (USDA) in 1978 but postponed in 1979. The later action was prompted not by the failure of the reduced nitrite level in combination with sorbate but because of taste panel results that characterized finished bacon as having "chemical"-like flavors and producing prickly mouth sensations (3). The combination of sorbate plus reduced nitrite has been shown to be effective in a variety of cured meat products against not only *C. botulinum* but other bacteria such as *S. aureus.* (For further information, see nitrite section below, and reviews by Sofos and Busta [112], Tompkin [123], and Liewen and Marth [73].)

The widest use of sorbates is as fungistats in products such as cheeses, bakery products, fruit juices, beverages, salad dressings, and the like. With molds, inhibition has been reported to be due to inhibition of the dehydrogenase enzyme system (24), and to the inhibition of cellular uptake of substrate molecules such as amino acids (see benzoic acid section above), phosphate, organic acids, and so on (36). A number of other possible inhibitory mechanisms have been presented by various researchers. Against germinating endospores, sorbate prevents the outgrough of vegetative cells (Fig. 11-1).

With respect to toxicity, sorbic acid is metabolized in the body to CO_2 and H_2O in the same manner as fatty acids normally found in foods (26).

THE PROPIONATES

Propionic acid is a three-carbon organic acid with the following structure: CH_3CH_2COOH. This acid and its calcium and sodium salts are permitted in breads, cakes, certain cheeses, and other foods primarily as a mold inhibitor. Propionic acid is employed also as a "rope" inhibitor in bread dough. The tendency toward dissociation is low with this compound and its salts, and these compounds are consequently active in low-acid foods. They tend to be highly specific against molds, with the inhibitory action being primarily fungistatic rather than fungicidal.

With respect to the antimicrobial mode of action of propionates, they act in a manner similar to that of benzoate and sorbate. The pK of propionate is 4.87 and at a pH of 4.00, 88% of the compound is undissociated, while at a pH of 6.0, only 6.7% remains undissociated. The undissociated molecule of this lipophilic acid is necessary for its antimicrobial activity. The mode of action of propionic acid is noted above with benzoic acid. See also section below on medium-chain fatty acids and esters, and review by Doores (28) for further information.

SULFUR DIOXIDE AND SULFITES

Sulfur dioxide (SO_2) and the sodium and potassium salts of sulfite ($=SO_3$), bisulfite ($-HSO_3$), and metabisulfite ($=S_2O_5$) all appear to act similarly and are here treated together. Sulfur dioxide is used in its gaseous or liquid form, or in the form of one or more of its neutral or acid salts on dried fruits, in lemon juice, molasses, wines, fruit juices, and others. The parent compound has been used as a food preservative since ancient times. Its use as a meat preservative in the United States dates back to at least 1813; however, it is not permitted in meats or other foods recognizable as sources of thiamine. While SO_2 possesses antimicrobial activity, it is also used in certain foods as an antioxidant.

The predominant ionic species of sulfurous acid depends upon pH of milieu with SO_2 being favored by pH < 3.0, HSO_3^- by pH between 3.0 and 5.0, and $SO_3^=$ > pH 6.0 (86). SO_2 has pKs of 1.76 and 7.2. The sulfites react with various food constituents including nucleotides, sugars, disulfide bonds, and others.

With regard to its effect on microorganisms, SO_2 is bacteriostatic against *Acetobacter* spp. and the lactic acid bacteria at low pH, concentrations of 100 to 200 ppm being effective in fruit juices and beverages. It is bactericidal at higher concentrations. When added to temperature-abused comminuted pork, 100 ppm of SO_2 or higher were required to effect significant inhibition of spores of *C. botulinum* at target levels of 100 spores/g (127). The source of SO_2 was sodium metabisulfite. Employing the same salt to achieve an SO_2 concentration of 600 ppm, Banks and Board (2) found that growth of salmonellae and other Enterobacteriaceae were inhibited in British fresh sausage. The most sensitive bacteria were eight salmonellae serovars, which were inhibited by 15–109 ppm at pH 7.0, while *Serratia liquefaciens, S. marcescens,* and *Hafnia alvei* were the most resistant, requiring 185–270 ppm free SO_2 in broth.

Yeasts are intermediate to acetic and lactic acid bacteria and molds in their sensitivity to SO_2, and the more strongly aerobic species are generally more sensitive than the more fermentative species (64). Sulfurous acid at levels of 0.2–20 ppm was effective against some yeasts, including *Saccharomyces, Pichia,* and *Candida,* while *Zygosaccharomyces bailii* required up to 230 ppm for inhibition in certain fruit drinks at pH 3.1 (76). Yeasts can actually form SO_2 during juice fermentation—some *S. carlsbergensis* and *S. bayanus* strains produce up to 1,000 and 500 ppm, respectively (86). Molds such as *Botrytis* can be controlled on grapes by periodic gassing with SO_2, and bisulfite can be used to destroy aflatoxins (29). Both aflatoxins B_1 and B_2 can be reduced in corn (44, 82). Sodium bisulfite was found to be comparable to propionic acid in its antimicrobial activity in corn containing up to 40% moisture (44).

Although the actual mechanism of action of SO_2 is not known, several possibilities have been suggested, each supported by some experimental evidence. One suggestion is that the undissociated sulfurous acid or molecular SO_2 is responsible for the antimicrobial activity. Its greater effectiveness at low pH tends to support this. Vas and Ingram (129) suggested the lowering of pH of certain foods by addition of acid as a means of obtaining greater preservation with SO_2. It has been suggested that the antimicrobial action is due to the strong reducing power that allows these compounds to reduce oxygen tension to a point below that at which aerobic organisms can grow, or by direct action upon some enzyme system. SO_2 is also thought to be an enzyme poison, inhibiting growth of microorganisms by inhibiting essential enzymes. Its use in the drying of foods to inhibit enzymatic browning is based upon this assumption. Since the sulfites are known to act on disulfide bonds, it may be presumed that certain essential enzymes are affected and that inhibition ensues. The sulfites do not inhibit cellular transport. It may be noted from Fig. 11-1 that metabisulfite acts on germinating endospores during the outgrowth of vegetative cells.

NITRITES AND NITRATES

Sodium nitrate ($NaNO_3$) and sodium nitrite ($NaNO_2$) are used in curing formulae for meats since they stabilize red meat color, inhibit some spoilage and food poisoning organisms, and contribute to flavor development. The role of NO_2 in cured meat flavor has been reviewed by Gray and Pearson (43). NO_2 has been shown to disappear both on heating and storage. It should be recalled that many bacteria are capable of utilizing nitrate as an electron acceptor and in the process effect its reduction to nitrite. The nitrite ion is by far the more important of the two in preserved meats. This ion is highly reactive and is capable of serving both as a reducing and an oxidizing agent. In an acid environment, it ionizes to yield nitrous acid (3HONO). The latter further decomposes to yield nitric oxide (NO), which is the important product from the standpoint of color fixation in cured meats. Ascorbate or erythrobate acts also to reduce NO_2 to NO. Nitric oxide reacts with myoglobin under reducing conditions to produce the desirable red pigment **nitrosomyoglobin,** as shown in the following (see also Table 9-9):

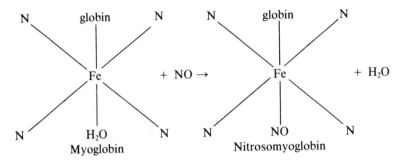

When the meat pigment exists in the form of **oxymyoglobin,** as would be the case for comminuted meats, this compound is first oxidized to **met-myoglobin** (brown color). Upon the reduction of the latter, nitric oxide reacts to yield nitrosomyoglobin. Since nitric oxide is known to be capable of reacting with other porphyrin-containing compounds such as catalase, peroxidases, cytochromes, and others, it is conceivable that some of the antibacterial effects of nitrites against aerobes may be due to this action (the mechanism is discussed below). It has been shown that the antibacterial effect of NO_2 increases as pH is lowered within the acid range, and this effect is accompanied by an overall increase in the undissociated HNO_2 (13).

Organisms affected

Although the single microorganism of greatest concern relative to nitrite inhibition is *C. botulinum,* the compound has been evaluated as an antimicrobial for other organisms. During the late 1940s it was evaluated as a fish preservative and found to be somewhat effective but generally only at low pH. It is effective against *S. aureus* at high concentrations and, again, the effectiveness increases as pH is lowered. The compound is generally ineffective against Enterobacteriaceae, including the salmonellae, and against the lactic acid bacteria, although some effects are noted in cured and in vacuum-packaged meats and are probably caused by the interaction of nitrite with other environmental parameters rather than to nitrite alone.

Nitrite is added to cheeses in some countries to control gassiness caused by *Clostridium butyricum* and *C. tyrobutyricum.* It is effective against other clostridia including *C. sporogenes* and *C. perfringens,* which are often employed in laboratory studies to assess potential antibotulinal effects not only of nitrites but of other inhibitors that might have value as nitrite adjuncts or sparing agents.

The Perigo factor

The almost total absence of botulism in cured, canned, and vacuum-packed meats and fish products led some investigators in the mid-1960s to seek reasons as to why meat products that contained viable endospores did not become toxic. Employing culture medium, it was shown in 1967 that about ten times more nitrite was needed to inhibit clostridia if it were added after instead of before the medium was autoclaved. It was concluded that the heating of the medium with nitrite produced a substance or agent about ten times more inhibitory than nitrite alone (89, 90). This agent is referred to as the **Perigo factor.** The existence of this factor or effect has been confirmed by some and questioned by others. While the Perigo factor may be questionable in cured and perishable cured meats, the evidence for an inhibitory factor in culture media involving nitrite, iron, and —SH groups is more conclusive (123).

The factor does not develop in all culture media, and heating to at least 100°C is necessary for its development, although some activity develops in meats when heated to as low as 70°C. The Perigo factor is dialyzable from some culture media and meat suspensions but not from other media (62). It is not found in filter-sterilized solutions of the same medium with nitrite added (104). It has been shown that if meat is added to a medium containing the Perigo factor, the inhibitory activity is lost (63). For this reason, some Canadian workers call the inhibitor that is formed in meat the "Perigo-type factor" (14).

It is this inhibitory or antibotulinal effect that results from the heat processing or smoking of certain meat and fish products containing nitrite that warrants the continued use of nitrite in such products. The antibotulinal activity of nitrite in cured meats is of greater public health importance than the facts of color and flavor development. For the latter, initial nitrite levels as low as 15 to 50 ppm have been reported to be adequate for various meat products including Thuringer sausage (25). Nitrite levels of 100 ppm or more have been found to make for maximum flavor and appearance in fermented sausages (70). The antibotulinal effect requires at least 120 ppm for bacon (9, 17), comminuted cured ham (16), and canned, shelf-stable luncheon meat (14). Many of these canned products are given a low heat process (F_o of 0.1–0.6).

Interaction with cure ingredients and other factors

The interplay of all ingredients and factors involved in heat-processed, cured meats on antibotulinal activity was noted over 20 years ago by Riemann

(98), and several other investigators have pointed out that curing salts in semipreserved meats are more effective in inhibiting heat-injured spores than noninjured (31, 103). With brine and pH alone, higher concentrations of the former are required for inhibition as pH increases, and Chang et al. (14) suggested that the inhibitory effect of salt in shelf-stable canned meats against heat-injured spores may be more important than the Perigo-type factor. With smoked salmon inoculated with 10^2 spores/g of *C. botulinum* types A and E and stored in O_2-impermeable film, 3.8 and 6.1% water-phase NaCl alone inhibited toxin production in 7 days by types E and A, respectively (87). With 100 ppm or more of NO_2, only 2.5% NaCl was required for inhibition of toxin production by type E, and for type A 3.5% NaCl + 150 ppm $NaNO_2$ was inhibitory. With longer incubations or larger spore inocula, more NaCl or $NaNO_2$ is needed.

The interplay of NaCl, $NaNO_2$, $NaNO_3$, isoascorbate, polyphosphate, thermal process temperatures, and temperature/time of storage on spore outgrowth and germination in pork slurries has been studied extensively by Roberts et al. (101), who found that significant reductions in toxin production could be achieved by increasing the individual factors noted. It is well known that low pH is antagonistic to growth and toxin production by *C. botulinum*, whether the acidity results from added acids or the growth of lactic acid bacteria. When 0.9% sucrose was added to bacon along with *Lactobacillus plantarum*, only one of forty-nine samples became toxic after 4 weeks, while with sucrose and no lactobacilli, fifty of fifty-two samples became toxic in 2 weeks (119). When 40 ppm nitrite was used alone, forty-seven of fifty samples became toxic after 2 weeks but when 40 ppm nitrite was accompanied by 0.9% sucrose and an inoculum of *L. plantarum*, none of thirty became toxic. While this was most likely a direct pH effect, other factors may have been involved (see section on lactic antagonism in Chapter 16). In more recent studies, bacon was prepared with 40 or 80 ppm $NaNO_2$ + 0.7% sucrose followed by inoculation with *Pediococcus acidilactici*. When inoculated with *C. botulinum* types A and B spores, vacuum-packaged, and incubated up to 56 days at 27°C, the bacon was found to have greater antibotulinal properties than control bacon prepared with 120 ppm $NaNO_2$ but not sucrose or lactic inoculum (118). Bacon prepared by the above formulation, called the Wisconsin process, was preferred by a sensory panel to that prepared by the conventional method (117). The Wisconsin process employs 550 ppm of sodium ascorbate or sodium erythrobate, as does the conventional process.

Nitrosamines

When nitrite reacts with secondary amines, **nitrosamines** are formed, and many are known to be carcinogenic. The generalized way in which nitrosamines may form is as follows:

$$R_2NH_2 + HONO \xrightarrow{\text{H}^+} R_2N\text{---}NO + H_2O$$

The amine dimethylamine reacts with nitrite to form N-nitrosodimethylamine:

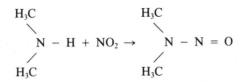

In addition to secondary amines, tertiary amines and quarternary ammonium compounds also yield nitrosamines with nitrite under acidic conditions. Nitrosamines have been found in cured meat and fish products at low levels (for reviews, see 19, 42).

It has been shown that lactobacilli, group D streptococci, clostridia, and other bacteria will nitrosate secondary amines with nitrite at neutral pH values (48). The fact that nitrosation occurred at near neutral pH values was taken to indicate that the process was enzymatic, although no cell-free enzyme was obtained (49). Several species of streptococci including *S. faecalis, S. faecium,* and *S. lactis* have been shown to be capable of forming nitrosamines, but the other lactic acid bacteria and pseudomonads tested did not (18). These investigators found no evidence for an enzymatic reaction. *S. aureus* and halobacteria obtained from Chinese salted marine fish (previously shown to contain nitrosamines) produced nitrosamines when inoculated into salted fish homogenates containing 40 ppm of nitrate and 5 ppm nitrite (34).

Nitrite-sorbate and other nitrite combinations

In an effort to reduce the potential hazard of N-nitrosamine formation in bacon, the USDA in 1978 reduced the input NO_2 level for bacon to 120 ppm and set a 10 ppb maximum level for nitrosamines. While 120 ppm nitrite along with 550 ppm sodium ascorbate or sodium erythrobate is adequate to reduce the botulism hazard, it is desirable to reduce nitrite levels even further if protection against botulinal toxin production can be achieved. To this end, a proposal to allow the use of 40 ppm nitrite in combination with 0.26% potassium sorbate for bacon was made in 1978 but rescinded a year later when taste panel studies revealed undesirable effects. Meanwhile, many groups of researchers have shown that 0.26% sorbate in combination with 40 or 80 ppm nitrite is effective in preventing botulinal toxin production. Extensive reviews of these studies have been provided to which the reader is referred for more detailed information (77, 114, 123).

In an early study of the efficacy of 40 ppm nitrite + sorbate to prevent or delay botulinal toxin production in commercial-type bacon, Ivey et al. (57) used an inoculum of 1,100 types A and B spores/g and incubated the product at 27°C for up to 110 days. Time for the appearance of toxic samples when neither nitrite nor sorbate was used was 19 days. With 40 ppm nitrite and no sorbate, toxic samples appeared in 27 days, and for samples containing 40 ppm nitrite + 0.26% sorbate or no nitrite and 0.26% sorbate, > 110 days were required for toxic samples. This reduced nitrite level resulted in lower levels of nitrosopyrrolidine in cooked bacon. Somewhat different findings were reported by Sofos et al. (Table 11-2), with 80 ppm nitrite being required for the absence of toxigenic samples after 60 days. In addition to its inhibitory effects on *C. botulinum*, sorbate slows the depletion of nitrite during storage (113).

The effect of isoascorbate is to enhance nitrite inhibition by sequestering iron, although under some conditions it may reduce nitrite efficiency by causing a more rapid depletion of residual nitrite (124, 126). EDTA at 500 ppm appears to be even more effective than erythrobate in potentiating the nitrite effect, but only limited studies have been reported. Another chelate, 8-hydroxyquinoline, has been evaluated as a nitrite-sparing agent. When 200 ppm were combined with 40 ppm nitrite, a *C. botulinum* spore mixture of types A and B strains was inhibited for 60 days at 27°C in comminuted pork (92).

In an evaluation of the interaction of nitrite and sorbate, the relative effectiveness of the combination has been shown to be dependent upon other cure ingredients and product parameters. Employing a liver-veal agar medium at pH 5.8–6.0, the germination rate of *C. botulinum* type E spores decreased to nearly zero with 1.0, 1.5, or 2.0% sorbate; but with the same concentrations at pH 7.0–7.2, germination and outgrowth of abnormally shaped cells occurred (108). When 500 ppm nitrite was added to the higher-pH medium along with sorbate, cell lysis was enhanced. These investigators also found that 500 ppm linoleic acid alone at the higher pH prevented emergence and elongation of spores. Potassium sorbate significantly decreased toxin production by types A and B spores in pork slurries when NaCl was increased or pH and storage temperature were reduced (102). For chicken frankfurters, a sorbate-betalains mixture was found to be as effective as a conventional nitrite system for inhibiting *C. perfringens* growth (128).

Mode of action

It appears that nitrite inhibits *C. botulinum* by interfering with iron-sulfur enzymes such as ferrodoxin and thus preventing the synthesis of ATP from

Table 11-2. Effect of nitrite and sorbate on toxin production in bacon inoculated with *C. botulinum* types A and B spores and held up to 60 days at 27°C (114).

Treatment	Percent Toxigenic
Control (no NO_2, no sorbate)	90.0
0.26% sorbate, no $NaNO_2$	58.8
0.26% sorbate + 40 ppm $NaNO_2$	22.0
0.26% sorbate + 80 ppm $NaNO_2$	0.0
No sorbate, 120 ppm $NaNO_2$	0.4

pyruvate. The first direct finding in this regard was that of Woods et al. (132), who showed that the phosphoroclastic system of *C. sporogenes* is inhibited by nitric oxide, and later that the same occurs in *C. botulinum,* resulting in the accumulation of pyruvic acid in the medium (131).

The phosphoroclastic reaction involves the breakdown of pyruvate with inorganic phosphate and coenzyme A to yield acetyl phosphate. In the presence of ADP, ATP is synthesized from acetyl phosphate with acetate as the other product. In the breakdown of pyruvate, electrons are transferred first to ferredoxin, and from ferredoxin to H^+ to form H_2 in a reaction catalyzed by hydrogenase. Ferrodoxin and hydrogenase are iron-sulfur (nonheme) proteins or enzymes (11).

Following the work of Woods and Wood (131), the next most significant finding was that of Reddy et al. (94), who subjected extracts of nitrite-ascorbate-treated *C. botulinum* to electron spin resonance and found that nitric oxide reacted with iron-sulfur complexes to form iron-nitrosyl complexes. The presence of the latter results in the destruction of iron-sulfur enzymes such as ferrodoxin.

The resistance of the lactic acid bacteria to nitrite inhibition is well known, but the basis is just now clear: these organisms lack ferrodoxin. The clostridia contain both ferrodoxin and hydrogenase, which function in electron transport in the anaerobic breakdown of pyruvate to yield ATP, H_2, and CO_2. The ferrodoxin in clostridia has a molecular weight of 6,000 and contains 8 Fe atoms/mole and 8-labile sulfide atoms/mole.

Although the first definitive experimental finding was reported in 1981 as noted above, earlier work pointed to iron-sulfur enzymes as the probable nitrite targets. Among the first were O'Leary and Solberg (85), who showed that a 91% decrease occurred in the concentration of free —SH groups of soluble cellular compounds of *C. perfringens* inhibited by nitrite. Two years later, Tompkin et al. (125) offered the hypothesis that nitric oxide reacted with iron in the vegetative cells of *C. botulinum,* perhaps the iron in ferrodoxin. The inhibition by nitrite of active transport and electron transport was noted by several investigators, and these effects are consistent with nitrite inhibition of nonheme enzymes such as ferrodoxin and hydrogenase (105, 133). The enhancement of inhibition in the presence of sequestering agents may be due to the reaction of sequestrants to substrate iron: more nitrite becomes available for nitric oxide production and reaction with microorganisms.

Summary of nitrite effects

The following summary of the overall role and effects of nitrite in cured meats emphasizes the antibotulinal activities.

When added to processed meats such as wieners, bacon, smoked fish, and canned cured meats followed by substerilizing heat treatments, nitrite has definite antibotulinal effects. It also forms desirable product color and enhances flavor in cured meat products. The antibotulinal effect consists of inhibition of vegetative cell growth and the prevention of germination and growth of spores that survive heat processing or smoking during post-processing storage. Clostridia other than *C. botulinum* are affected in a similar manner. While low initial levels of nitrite are adequate for color and flavor development, considerably higher levels are necessary for the antimicrobial effects.

When nitrite is heated in certain laboratory media, an antibotulinal factor or inhibitor is formed, the exact identity of which is not yet known. The inhibitory factor is the Perigo effect/factor or Perigo inhibitor. It does not form in filter-sterilized media. It develops in canned meats only when nitrite is present during heating. The initial level of nitrite is more important to antibotulinal activity than the residual level. Once formed, the Perigo factor is not affected greatly by pH changes.

Measurable preheating levels of nitrite decrease considerably during heating in meats and during postprocessing storage, more at higher storage temperatures than at lower.

The antibotulinal activity of nitrite is interdependent with pH, salt content, temperature of incubation, and numbers of botulinal spores. Heat-injured spores are more susceptible to inhibition than uninjured. Nitrite is more effective under Eh − than under Eh + conditions.

Nitrite does not decrease the heat resistance of spores. It is not affected by ascorbate in its antibotulinal actions but does act synergistically with ascorbate in pigment formation.

Lactic acid bacteria are relatively resistant to nitrite (see above).

Endospores remain viable in the presence of the antibotulinal effect and will germinate when transferred to nitrite-free media.

Nitrite has a pK of 3.29 and consequently exists as undissociated nitrous acid at low pH values. The maximum undissociated state and consequent greatest antibacterial activity of nitrous acid are between pH 4.5 and 5.5.

With respect to its depletion or disappearance in ham, Nordin (84) found the rate to be proportional to its concentration and to be exponentially related to both temperature and pH. The depletion rate doubled for every 12.2°C increase in temperature or 0.86 pH unit decrease, and was not affected by heat denaturation of the ham. These relationships did not apply at room temperature unless the product was first heat treated, suggesting that viable organisms aided in its depletion.

It appears that the antibotulinal activity of nitrite is due to its inhibition of nonheme, iron-sulfur enzymes.

NaCl AND SUGARS

These compounds are grouped together because of the similarity in their modes of action in preserving foods. NaCl has been employed as a food preservative since ancient times (see Chapter 1). The early food uses of salt were for the purpose of preserving meats. This use is based upon the fact that at high concentrations, salt exerts a drying effect upon both food and microorganisms. Nonmarine microorganisms may be thought of as normally possessing a degree of intracellular tonicity equivalent to that produced by about 0.85–0.9% NaCl. When microbial cells are suspended in salt (saline) of this concentration, the suspending menstrum can be said to be **isotonic** with respect to the cells. Since the amounts of NaCl and water are equal on both sides of the cell membrane, water moves across the cell membranes equally in both directions. When microbial cells are suspended in, say, a 5% saline solution, the concentration of water is greater inside the cells than outside (concentration of H_2O is highest where solute concentration is lowest). It should be recalled that in diffusion, water moves from its area of high concentration to its area of low concentration. In this case, water would pass out of the cells at a greater rate than it would enter. The result to the cell is **plasmolysis,** which results in growth inhibition and possibly death. This is essentially what is achieved when high concentrations of salt are added to fresh meats for the purpose of preservation. Both the microbial cells and those of the meat undergo plasmolysis (shrinkage), resulting in the drying of the meat as well as inhibition or death of microbial cells. To be effective, one must use enough salt to effect **hypertonic** conditions. The higher the concentration, the greater the preservative and drying effects. In the absence of refrigeration, fish and other meats may be effectively preserved by salting. The inhibitory effects of salt are not dependent upon pH as are some other chemical preservatives. Most nonmarine bacteria can be inhibited by 20% or less of NaCl, while some molds generally tolerate higher levels. Organisms that can grow in the presence of and require high concentrations of salt are referred to as **halophiles,** while those that can withstand but not grow in high concentrations are referred to as **halodurics.** The interaction of salt with nitrite and other agents in the inhibition of *C. botulinum* is discussed above under nitrites.

Sugars, such as sucrose, exert their preserving effect in essentially the same manner as salt. One of the main differences is in relative concentrations. It generally requires about six times more sucrose than NaCl to effect the same degree of inhibition. The most common uses of sugars as preserving agents are in the making of fruit preserves, candies, condensed milk, and the like. The shelf-stability of certain pies, cakes, and other such products is due in large part to the preserving effect of high concentrations of sugar, which, like salt, makes water unavailable to microorganisms.

Microorganisms differ in their response to hypertonic concentrations of sugars, with yeasts and molds being less susceptible than bacteria. Some yeasts and molds can grow in the presence of as much as 60% sucrose while most bacteria are inhibited by much lower levels. Organisms that are able to grow in high concentrations of sugars are designated **osmophiles,** while **osmoduric** microorganisms are those that are unable to grow but are able to withstand high levels of sugars. Some osmophilic yeasts such as *Saccharomyces rouxii* can grow in the presence of extremely high concentrations of sugars.

INDIRECT ANTIMICROBIALS

The compounds and products in this section are added to foods primarily for effects other than antimicrobial and are thus multifunctional food additives.

Antioxidants

Although used in foods primarily to prevent the auto-oxidation of lipids, the phenolic antioxidants listed in Table 11-3 have been shown to possess antimicrobial activity against a wide range of microorganisms including some viruses, mycoplasmas, and protozoa. These compounds have been evaluated extensively as nitrite-sparing agents in processed meats and in combination with other inhibitors, and several excellent reviews have been made (10, 37, 65).

Butylated hydroxyanisole (BHA), butylated hydroxytoluene (BHT), and *t*-butylhydroxyquinoline (TBHQ) are inhibitory to gram-positive and gram-negative bacteria as well as to yeasts and molds at concentrations ranging from about 10 to 1,000 ppm depending upon substrate. In general, higher concentrations are required to inhibit in foods than in culture media, especially in high-fat foods. BHA was about fifty times less effective against *Bacillus* spp. in strained chicken than in nutrient broth (110). BHA, BHT, TBHQ, and propyl gallate (PG) were all less effective in ground pork than in culture media (38). While strains of the same bacterial species may show wide variation in sensitivity to either of these antioxidants, it appears that BHA

Table 11-3. Some GRAS indirectly antimicrobial chemicals used in foods.

Compound	Primary use	Most susceptible organisms
Butylated hydroxyanisole (BHA)	Antioxidant	Bacteria, some fungi
Butylated hydroxytoluene (BHT)	Antioxidant	Bacteria, viruses, fungi
t-butylhydroxyquinoline (TBHQ)	Antioxidant	Bacteria, fungi
Propyl gallate (PG)	Antioxidant	Bacteria
Nordihydroguaiaretic acid	Antioxident	Bacteria
Ethylenediamine tetraacetic acid (EDTA)	Sequestrant/ stabilizer	Bacteria
Sodium citrate	Buffer/sequestrant	Bacteria
Lauric acid	Defoaming agent	Gram-positive bacteria
Monolaurin	Emulsifier	Gram-positive bacteria, yeasts
Diacetyl	Flavoring	Gram-negative bacteria, fungi
d- and *l*-carvone	Flavoring	Fungi, gram-positive bacteria
Phenylacetaldehyde	Flavoring	Fungi, gram-positive bacteria
Menthol	Flavoring	Bacteria, fungi
Vanillin, ethyl vanillin	Flavoring	Fungi
Spices/spice oils	Flavoring	Bacteria, fungi

and TBHQ are more inhibitory than BHT to bacteria and fungi, while the latter is more viristatic. To prevent growth of *C. botulinum* in a prereduced medium, 50 ppm BHA and 200 ppm BHT were required, while 200 ppm PG were ineffective (99). Employing 16 gram-negative and 8 gram-positive bacteria in culture media, Gailani and Fung (38) found the gram positives to be more susceptible than gram negatives to BHA, BHT, TBHQ, and PG with each being more effective in nutrient agar than in BHT broth. In nutrient agar the relative effectiveness was BHA > PG > TBHQ > BHT, while in BHI, TBHQ > PG > BHA > BHT.

Food-borne pathogens such as *B. cereus, V. parahaemolyticus,* salmonellae, and *S. aureus* are effectively inhibited at concentrations < 500 ppm, while some are sensitive to as little as 10 ppm. The pseudomonads, especially *P. aeruginosa,* are among the most resistant bacteria. Three toxin-producing penicillia were inhibited significantly in salami by BHA, TBHQ, and a combination of these two at 100 ppm, while BHT and PG were ineffective (74). Combinations of BHA/sorbate and BHT/monolaurin have been shown to be synergistic against *S. aureus* (10, 21), and BHA/sorbate against *S. typhimurium* (21). BHT/TBHQ has been shown to be synergistic against aflatoxin-producing penicillia (74).

Flavoring agents

Of the many agents used to impart aromas and flavors to foods, some possess definite antimicrobial effects. In general, flavor compounds tend to be more antifungal than antibacterial. The nonlactic, gram-positive bacteria are the most sensitive and the lactic acid bacteria are rather resistant. The essential oils and spices have received the most attention by food microbiologists, while the aroma compounds have been studied more for their use in cosmetics and soaps.

Of twenty-one flavoring compounds examined in one study, about one-half had minimal inhibitory concentrations (MIC) of 1,000 ppm or less against either bacteria or fungi (61). All were pH sensitive, with inhibition increasing as pH and temperature of incubation decreased. Some of these compounds are noted in Table 11-3.

One of the most effective flavoring agents is diacetyl, which imparts the aroma of butter (58). It is somewhat unique in being more effective against gram-negative bacteria and fungi than against gram-positive bacteria. In plate count agar at pH 6.0 and incubation at 30°C, all but one of twenty-five gram-negative bacteria and fifteen of sixteen yeasts and molds were inhibited by 300 ppm (59). At pH 6.0 and incubation at 5°C in nutrient broth, < 10 ppm inhibited *P. fluorescens, P. geniculata,* and *S. faecalis,* while under the same conditions except with incubation at 30°C, about 240 ppm were required to inhibit these and other organisms (61). It appears that diacetyl antagonizes arginine utilization by reacting with arginine-binding proteins of gram-negative bacteria. The greater resistance of gram-positive bacteria appears to be due to their lack of similar periplasmic binding proteins and their possession of larger amino acid pools. Another flavor compound that imparts the aroma of butter is 2,3-pentanedione, and it has been found to be inhibitory to a limited number of gram-positive bacteria and fungi at 500 ppm or less (61).

The agent *l*-carvone imparts spearmintlike and the agent *d*-carvone imparts carawaylike aromas, and both are antimicrobial with the *l*-isomer being more effective than the *d*-isomer, while both are more effective against fungi than bacteria at 1,000 ppm or less (61). Phenylacetaldehyde imparts a hyacinthlike aroma, and has been shown to be inhibitory to *S. aureus* at 100 ppm and *Candida albicans* at 500 ppm (61, 83). Menthol, which imparts a peppermintlike aroma, was found to inhibit *S. aureus* at 32 ppm, and *E. coli* and *C. albicans* at 500 ppm (61, 83). Vanillin and ethyl vanillin are inhibitory, especially to fungi at levels < 1,000 ppm.

Spices and essential oils

While used primarily as flavoring and seasoning agents in foods, many spices possess significant antimicrobial activity. In all instances, antimicrobial activity is due to specific chemicals or essential oils, some of which are noted in Chapter 3. The search for nitrite-sparing agents generated new interest in spices and spice extracts in the late 1970s (much of this work has been reviewed by Shelef, 109).

It would be difficult to predict what antimicrobial effects if any are derived from spices as they are used in foods, for the quantities employed differ widely depending upon taste and the relative effectiveness varies depending upon product composition. Because of the varying concentrations of the antimicrobial constituents in different spices, and because many studies have been conducted employing them on a dry weight basis, it is difficult to ascertain the MIC of given spices against specific organisms. Another reason for conflicting results by different investigators is the assay method employed. In general, higher MIC values are obtained when highly volatile compounds are evaluated on the surface of plating media than when they are tested in pour plates or broth. When eugenol was evaluated by surface plating onto PCA at pH 6, only nine of fourteen gram-negative and twelve of twenty gram-positive bacteria (including eight lactics) were inhibited by 493 ppm, while in nutrient broth at the same pH, MICs of 32 and 63 were obtained for *Torulopsis candida* and *Aspergillus niger,* and *S. aureus* and *E. coli,* respectively (61). It has been noted that spice extracts are less inhibitory in media than spices, and this is probably due to a slower release of volatiles by the latter (111). In spite of the difficulties of comparing results from study to study, the antimicrobial activity of spices is unquestioned and a large number of investigators have shown the effectiveness of at least twenty different spices or their extracts against most food-poisoning organisms including mycotoxigenic fungi (109).

In general, spices are less effective in foods than in culture media, and gram-positive bacteria are more sensitive than gram negatives, with the lactic acid bacteria being the most resistant among gram positives (134). While results concerning them are debatable, the fungi appear to be in general more sensitive than gram-negative bacteria. Some gram negatives, however, are highly sensitive. Antimicrobial substances vary in content from the allicin of garlic (with a range of 0.3–0.5%) to eugenol in cloves (16–18%) (109). When whole spices are employed, MIC values range from 1–5% for sensitive organisms. Sage and rosemary are among the most antimicrobial as reported by various researchers, and it has been reported that 0.3% in culture media inhibited twenty-one of twenty-four gram-positive bacteria and were more effective than allspice (111).

With respect to specific inhibitory levels of extracts and essential oils, Huhtanen (53) made ethanol extracts of thirty-three spices, tested them in broth against *C. botulinum*, and found that achiote and mace extracts produced an MIC of 31 ppm and were the most effective of the thirty-three. Next most effective were nutmeg, bay leaf, and white and black peppers with MICs of 125 ppm. Employing the essential oils of oregano, thyme, and sassafras, Beuchat (4) found that 100 ppm were cidal to *V. parahaemolyticus* in broth. Growth and aflatoxin production by *A. parasiticus* in broth were inhibited by 200–300 ppm of cinnamon and clove oils, by 150 ppm cinnamic aldehyde, and by 125 ppm eugenol (12).

The mechanisms by which spices inhibit microorganisms are unclear and may be presumed to be different for unrelated groups of spices. That the mechanism for oregano, rosemary, sage, and thyme may be similar is suggested by the finding that resistance development by some lactic acid bacteria to one was accompanied by resistance to the other three (134).

Medium-chain fatty acids and esters

Acetic, propionic, and sorbic acids are short-chain fatty acids used primarily as preservatives. Medium-chain fatty acids, however, are employed primarily as surface-active or emulsifying agents. The antimicrobial activity of the medium-chain fatty acids is best known from soaps, which are salts of fatty acids. Those most commonly employed are composed of twelve to sixteen carbons. For saturated fatty acids, the most antimicrobial chain length is C_{12}; for monounsaturated (containing 1 double bond) $C_{16:1}$; and for poly-unsaturated (containing more than one double bond) $C_{18:2}$ is the most antimicrobial (66). In general, fatty acids are effective primarily against gram-positive bacteria and yeasts. While the C_{12} to C_{16} chain lengths are the most active against bacteria, the C_{10} to C_{12} are most active against yeasts (66). Fatty acids and esters and the structure-function relationships among them have been reviewed and discussed by Kabara (65, 66). Saturated aliphatic acids effective against *C. botulinum* have been evaluated by Dymicky and Trenchard (32).

The monoesters of glycerol and the diesters of sucrose have been found to be more antimicrobial than the corresponding free fatty acids, and to compare favorably with sorbic acid and the parabens as antimicrobials (65).

Monolaurin is the most effective of the glycerol monoesters, while sucrose dicaprylate is the most effective of the sucrose diesters. Monolaurin (lauricidin) has been evaluated by a large number of investigators and found to be inhibitory to a variety of gram-positive bacteria and some yeasts at 5–100 ppm (10, 65). Unlike the short-chain fatty acids, which are most effective at low pH, monolaurin is effective over the range 5.0 to 8.0 (67).

Because the fatty acids and esters have a narrow range of effectiveness and GRAS substances such as EDTA, citrate, and phenolic antioxidants also have limitations as antimicrobial agents when used alone, Karara (65, 66) has stressed the "preservative system" approach to the control of microorganisms in foods by using combinations of chemicals to fit given food systems and preservation needs. By this approach, a preservative system might consist of three compounds—monolaurin/EDTA/BHA, for example. While EDTA possesses little antimicrobial activity by itself, it renders gram-negative bacteria more susceptible by rupturing the outer membrane and thus potentiating the effect of fatty acids or fatty acid esters. An antioxidant such as BHA would exert effects against bacteria and molds and serve as an antioxidant at the same time. By use of such a system, the development of resistant strains could be minimized and the pH of a food could become less important relative to the effectiveness of the inhibitory system.

ACETIC AND LACTIC ACIDS

These two organic acids are among the most widely employed as preservatives. In most instances, their origin to the subject foods is due to their production within the food by lactic acid bacteria. Products such as pickles, sauerkraut, and fermented milks, among others, are created by the fermentative activities by various lactic acid bacteria, which produce acetic, lactic, and other acids (see Chapter 16 for fermented foods, and the review by Doores, 28, for further information).

The antimicrobial effect of organic acids such as propionic and lactic is due both to the depression of pH below the growth range and metabolic inhibition by the undissociated acid molecules. In determining the quantity of organic acids in foods, **titratable acidity** is of more value than pH alone, since the latter is a measure of hydrogen-ion concentration and organic acids do not ionize completely. In measuring titratable acidity, the amount of acid that is capable of reacting with a known amount of base is determined. The titratable acidity of products such as sauerkraut is a better indicator of the amount of acidity present than pH.

ANTIBIOTICS

While no antibiotic is legally permissible as a food additive in the United States at the present time, two are approved for food use in many other countries (nisin and natamycin), and three others (tetracyclines, subtilin, and tylosin) have been studied and found effective for various food applications. The early history, efficacy, and applications of most were reviewed in 1966 (78), and all have been reviewed and discussed more recently (60). Detailed reviews on nisin have been provided by Hurst (54, 55) and Lipinska (75).

Three antibiotics have been investigated extensively as heat adjuncts for canned foods (subtilin, tylosin, and nisin). Nisin, however, is used most widely in cheeses. Chlortetracycline and oxytetracycline were widely studied for their application to fresh foods while natamycin is employed as a food fungistat.

While in general the use of chemical preservatives in foods is not popular among many consumers, the idea of employing antibiotics is even less popular. Some risks may be anticipated from the use of any food additive, but the risks should not outweigh the benefits overall. The general view in the United States at the present time is that the benefits to be gained by using antibiotics in foods do not outweigh the risks, some of which are known and some of which are presumed. Some fifteen considerations on the use of antibiotics as food preservatives were noted by Ingram et al. (56), and several of the key ones are summarized below:

1. The antibiotic agent should kill, not inhibit the flora, and should ideally decompose into innocuous products, or be destroyed on cooking for products that require cooking.
2. The antibiotic should not be inactivated by food components or products of microbial metabolism.
3. The antibiotic should not readily stimulate the appearance of resistant strains.
4. The antibiotic should not be used in foods if used therapeutically or as an animal feed additive.

It may be noted from the summary comparison in Table 11-4 that the tetracyclines are used both clinically and as feed additives while tylosin is used in animal feeds and only in the treatment of some poultry diseases. Neither nisin nor subtilin is used medically or in animal feeds, and while nisin is used in many countries, subtilin is not. The structural similarities of these two antibiotics may be noted from Fig. 11-2.

Nisin

This is a polypeptide antibiotic structurally related to subtilin, but unlike subtilin it does not contain tryptophane residues (Fig. 11-2). While the C-terminal amino acids are similar, the N-terminals are not. The first food use of nisin was by Hirsch et al. (51) to prevent the spoilage of Swiss cheese by *Clostridium butyricum*. It is clearly the most widely used antibiotic for food preservation, with around thirty-nine countries permitting its use in foods to varying degrees (for a list of countries, see Hurst, 54). It is not permitted in foods in the United States and Canada. Among some of its desirable properties as a food preservative are the following: (1) it is nontoxic, (2) it is produced naturally by *Streptococcus lactis* strains, (3) it is heat stable and has excellent storage stability, (4) it is destroyed by digestive enzymes, (5) it does not contribute to off-flavors or off-odors, and (6) it has a narrow spectrum of antimicrobial activity. The compound is effective against gram-positive bacteria, primarily sporeformers, and ineffective against fungi and gram-negative bacteria. *Enterococcus* (*Streptococcus*) *faecalis* is one of the most resistant gram positives.

A large amount of research has been carried out with nisin as a heat-adjunct in canned foods, or as an inhibitor of heat-shocked spores of *Bacillus*

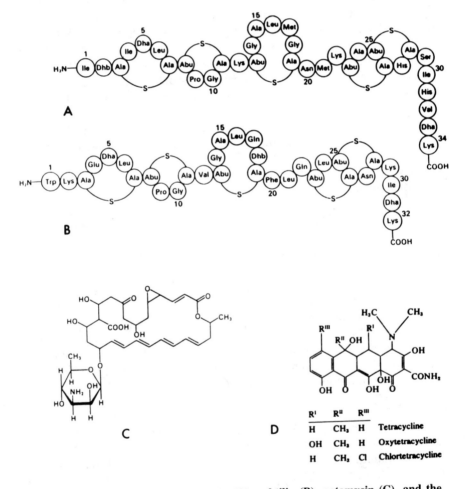

Fig. 11-2. Structural formulae of nisin (A), subtilin (B), natamycin (C), and the tetracyclines (D).

and *Clostridium* strains, and the MIC for preventing outgrowth of germinating spores ranges widely from 3 to > 5,000 IU/ml or < 1 to > 125 ppm (1 μg of pure nisin is about 40 IU or RU—Reading unit) (54). Depending upon the country and the particular food product, typical usable levels are in the range of about 2.5 to 100 ppm, although some countries do not impose concentration limits.

A conventional heat process for low-acid canned foods requires an F_o treatment of 6–8 (see Chapter 14) to inactivate the endospores of both *C. botulinum* and spoilage organisms. By adding nisin the heat process can be reduced to an F_o of 3 (to inactivate *C. botulinum* spores) resulting in increased product quality of low-acid canned foods. While the low-heat treatment will not destroy the endospores of spoilage organisms, nisin prevents their germination by acting early in the endospore germination cycle (Fig. 11-1). In addition to its use in certain canned foods, nisin is most often employed in dairy products—processed cheeses, condensed milk, pasteurized milk, and so on. Some countries permit its use in processed tomato products and canned fruits and vegetables (54). It is most stable in acidic foods.

Because of the effectiveness of nisin in preventing the outgrowth of germinating endospores of *C. botulinum* and the search to find safe substances that might replace nitrites in processed meats, this antibiotic has been studied as a possible replacement for nitrite. While some studies showed encouraging results employing *C. sporogenes* and other nonpathogenic organisms, a recent study employing *C. botulinum* types A and B spores in pork slurries indicated the inability of nisin at concentrations up to 550 ppm in combination with 60 ppm nitrite to inhibit spore outgrowth (93). Employed in culture media without added nitrite, the quantity of nisin required for 50% inhibition of *C. botulinum* type E spores was 1–2 ppm; 10–20 ppm for type B; and 20–40 ppm for type A (107). The latter authors found that higher levels were required for inhibition in cooked meat medium than in TPYG medium and suggested that nisin was approximately equivalent to nitrite in preventing the outgrowth of *C. botulinum* spores.

With respect to mode of action, nisin and subtilin may be presumed to act similarly since they are both polypeptide antibiotics with highly similar structures. They act at the same site on germinating endospores (Fig. 11-1). Some of the polypeptide antibiotics typically attack cell membranes and act possibly as surfactants or emulsifying agents on membrane lipids. These agents may be presumed to inhibit gram-positive bacteria by inhibiting cell wall murein synthesis since bacitracin (another polypeptide antibiotic that also inhibits gram-positive bacteria) is known to inhibit murein synthesis (47). That nisin affects murein synthesis has been shown by Reisinger et al. (95), and this finding is not inconsistent with its lack of toxicity for man. A similar lack of toxicity for subtilin may be presumed.

Natamycin

This antibiotic (also known as pimaricin, tennecetin, and myprozine) is a polyene that is quite effective against yeasts and molds but not bacteria.

Natamycin is the international nonproprietary name since it was isolated from *Streptomyces natalensis*. Its structural formula is presented in Fig. 11-2.

In granting the acceptance of natamycin as a food preservative, the joint FAO/WHO Expert Committee (35) took the following into consideration: (1) it does not affect bacteria, (2) it stimulates an unusually low level of resistance among fungi, (3) it is rarely involved in cross-resistance among other antifungal polyenes, and (4) DNA transfer between fungi does not occur to the extent that it does with some bacteria. Also, from Table 11-4 it may be noted that its use is limited as a clinical agent, and it is not used as a feed additive. Natamycin has been shown by numerous investigators to be very effective against both yeasts and molds, and many of these reports have been summarized (60).

The relative effectiveness of natamycin was compared to sorbic acid and four other antifungal antibiotics by Klis et al. (69) for the inhibition of sixteen different fungi (mostly molds), and while from 100 to 1,000 ppm sorbic acid were required for inhibition, from 1 to 25 ppm natamycin were effective against the same strains in the same media. To control fungi on strawberries and raspberries, natamycin was compared with rimocidin and nystatin, and it along with rimocidin was effective at levels of 10–20 ppm, while 50 ppm nystatin were required for effectiveness (1). In controlling fungi on salami, the spraying of fresh salami with a 0.25% solution was found to be effective by one group of investigators (50), but another researcher was unsuccessful in his attempts to prevent surface-mold growth on Italian dry sausages when they were dipped in a 2,000-ppm solution (53). Natamycin

Table 11-4. Summary comparison of some properties of the antibiotics discussed in this chapter (60).

Property	Tetracyclines	Subtilin	Tylosin	Nisin	Natamycin
Widely used in foods	No	No	No	Yes	Yes
First food use	1950	1950	1961	1951	1956
Chemical nature	Tetracycline	Poly-peptide	Macro-lide	Poly-peptide	Polyene
Used as heat adjunct	No	Yes	Yes	Yes	No
Heat stability	Sensitive	Stable	Stable	Stable	Stable
Microbial spectrum	G^+, G^-	G^+	G^+	G^+	Fungi
Used medically	Yes	No	Yes[a]	No	Yes[b]
Used in feeds	Yes	No	Yes	No	No

[a] In treating poultry diseases.
[b] Limited.

spray (2 × 1,000 ppm) was as good as or slightly better than 2.5% potassium sorbate.

Natamycin appears to act in the same manner as other polyene antibiotics— by binding to membrane sterols and inducing distortion of selective membrane permeability (45). Since bacteria do not possess membrane sterols, their lack of sensitivity to this agent is thus explained. Mycoplasmae, however, do have membrane sterols, but whether this antibiotic is effective against this group is unclear.

Tetracyclines

Chlortetracycline (CTC) and oxytetracycline (OTC) were approved by the FDA in 1955 and 1956, respectively, at a level of 7 ppm to control bacterial spoilage in uncooked refrigerated poultry, but these approvals were sub-sequently rescinded. The efficacy of this group of antibiotics in extending the shelf life of refrigerated foods was first established by Tarr and associates in Canada working with fish (120). Subsequent research by a large number of workers in many countries established the effectiveness of CTC and OTC in delaying bacterial spoilage of not only fish and seafoods but poultry, red meats, vegetables, raw milk, and other foods (for a review of food applications, see 60, 78). CTC is generally more effective than OTC. The surface treatment of refrigerated meats with 7–10 ppm typically results in shelf-life extensions of at least 3–5 days and a shift in ultimate spoilage flora from gram-negative bacteria to yeasts and molds. When CTC is combined with sorbate to delay spoilage of fish, the combination has been shown to be effective for up to 14 days. Rockfish fillets dipped in a solution of 5 ppm CTC and 1% sorbate had significantly lower APCs after vacuum-package storage at 2°C after 14 days than controls (81).

The tetracyclines are both heat sensitive and storage labile in foods, and these factors were important in their initial acceptance for food use. As may be noted from Table 11-4, they are used to treat diseases in man and animals and are used also in feed supplements. The risks associated with their use as food preservatives in developed countries seem clearly to outweigh the benefits.

Subtilin

This antibiotic was discovered and developed by scientists at the Western Regional Laboratory of the USDA, and its properties were described by Dimick et al. (27). As noted above, it is structurally similar to nisin (Fig. 11-2) even though it is produced by some strains of *B. subtilis*. Like nisin, it is effective against gram-positive bacteria, is stable to acid, and possesses enough heat resistance to withstand destruction at 121°C for 30 to 60 min. Subtilin is effective in canned foods at levels of 5 to 20 ppm in preventing the outgrowth of germinating endospores, and its site of action is the same as for nisin (Fig. 11-1). Like nisin, it is used neither in the treatment of human or animal infections nor as a feed additive. This antibiotic may be just as effective as nisin even though it has received little attention since the late 1950s. Its mode of action is discussed above along with that of nisin, and its development and evaluation have been reviewed (60).

Tylosin

This antibiotic is a nonpolyene macrolide as are the clinically useful antibiotics erythromycin, oleandomycin, and others. It is more inhibitory than nisin or subtilin. Denny et al. (23) were apparently the first to study its possible use in canned foods. When 1 ppm was added to cream-style corn containing flat-sour spores and given a "botulinal" cook, no spoilage of product occurred after 30 days with incubation at 54°C (22). Similar findings were made by others in the 1960s, and these have been summarized (60).

Unlike nisin, subtilin, and natamycin, tylosin is used in animal feeds and also to treat some diseases of poultry. As a macrolide, it is most effective against gram-positive bacteria. It inhibits protein synthesis by associating with the 50S ribosomal subunit, and shows at least partial cross-resistance with erythromycin.

ANTIFUNGAL AGENTS FOR FRUITS

Listed in Table 11-5 are some compounds applied to fruits after harvest to control fungi, primarily molds. **Benomyl** is applied uniformly over the entire surface of fruits, examples of which are noted in the table. It is applied at concentrations of 0.5–1.0 g/liter. It can penetrate the surface of some vegetables, and is used world-wide to control crown rot and anthracnose of bananas, and stem-end rots of citrus fruits. It is more effective than thi-

Table 11-5. Some chemical agents employed to control fungal spoilage of fresh fruits (33).

Compound	Fruits
Thiabendazole	Apples, pears, citrus fruits, pineapples
Benomyl	Apples, pears, bananas, citrus fruits, mangoes, papayas, peaches, cherries, pineapples
Biphenyl	Citrus fruits
SO₂ fumigation	Grapes
Sodium-α-phenylphenate	Apples, pears, citrus fruits, pineapples

bendazole and penetrates with greater ease. Both benomyl and thiabendazole are effective in controlling dry rot caused by *Fusarium* spp. To prevent the spread of *Botrytis* from grape to grape, SO_2 is employed for long-term storage. It is applied shortly after harvest and about once a week thereafter. A typical initial treatment consists of a 20-min application of a 1% preparation, and about 0.25% in subsequent treatments (the use of SO_2 in other foods is discussed above).

Biphenyl is used to control the decay of citrus fruits by penicillia for long-distance shipments and is generally impregnated into fruit wraps or sheets between fruit layers.

ETHYLENE AND PROPYLENE OXIDES

Ethylene and propylene oxides along with ethyl and methyl formate ($HCOOC_2H_5$ and $HCOOCH_3$, respectively) are treated together in this section because of their similar actions. The structures of the oxide compounds are as follows:

Ethylene oxide   Propylene oxide

The oxides exist as gases and are employed as fumigants in the food industry. The oxides are applied to dried fruits, nuts, spices, and so forth, primarily as antifungal compounds.

Ethylene oxide is an alkylating agent (91) and its antimicrobial activity is presumed to be related to this action in the following manner. In the presence of labile H atoms, the unstable three-membered ring of ethylene oxide splits. The H atom attaches itself to the oxygen, forming a hydroxyl ethyl radical, CH_2CH_2OH, which in turn attaches itself to the position in the organic molecule left vacant by the H atom. As a result, the hydroxyl ethyl group blocks reactive groups within microbial proteins, thus resulting in inhibition. Among the groups capable of supplying a labile H atom are the following: —COOH, —NH_2, —SH, and —OH. Ethylene oxide appears to affect endospores of *C. botulinum* by alkylation of guanine and adenine components of spore DNA (80, 130).

Ethylene oxide is used as a gaseous sterilant for flexible and semirigid containers for packaging aseptically processed foods. All of the gas dissipates from the containers following their removal from treatment chambers. With respect to its action on microorganisms, it is not much more effective against vegetative cells than it is against endospores, as can be seen from the D values given in Table 11-6.

Table 11-6. D values for four chemical sterilants of some food-borne microorganisms.

Organism	D^a	Conc.	Temp.[b]	Condition	Reference
Hydrogen peroxide					
C. botulinum 169B	0.03	35%	88		121
B. coagulans	1.8	26%	25		122
B. stearothermophilus	1.5	26%	25		122
B. subtilis ATCC 95244	1.5	20%	25		116
B. subtilis A	7.3	26%	25		121
Ethylene oxide					
C. botulinum 62A	11.5	700 mg/L	40	47% R.H.	106
C. botulinum 62A	7.4	700 mg/L	40	23% R.H.	130
C. sporogenes ATCC 7955	3.25	500 mg/L	54.4	40% R.H.	68
B. coagulans	7.0	700 mg/L	40	33% R.H.	7
B. coagulans	3.07	700 mg/L	60	33% R.H.	7
B. stearothermophilus ATCC 7953	2.63	500 mg/L	54.4	40% R.H.	68
L. brevis	5.88	700 mg/L	30	33% R.H.	7
M. radiodurans	3.00	500 mg/L	54.4	40% R.H.	68
Sodium hypochlorite					
A. niger conidiospores	0.61	20 ppm[c]	20	pH 3.0	15
A. niger conidiospores	1.04	20 ppm[c]	20	pH 5.0	15
A. niger conidiospores	1.31	20 ppm[c]	20	pH 7.0	15
Iodine ($\frac{1}{2}I_2$)					
A. niger conidiospores	0.86	20 ppm[c]	20	pH 3.0	15
A. niger conidiospores	1.15	20 ppm[c]	20	pH 5.0	15
A. niger conidiospores	2.04	20 ppm[c]	20	pH 7.0	15

[a] In minutes;
[b] °C;
[c] As Cl.

MISCELLANEOUS CHEMICAL PRESERVATIVES

Sodium diacetate ($CH_3COONa \cdot CH_3COOH \cdot xH_2O$), a derivative of acetic acid, is used in bread and cakes to prevent moldiness. Organic acids such as **citric,**

$$
\begin{array}{c}
CH_2COOH \\
| \\
HO - C - COOH, \\
| \\
CH_2COOH
\end{array}
$$

exert a preserving effect on foods such as soft drinks. **Hydrogen peroxide** (H_2O_2) has received limited use as a food preservative. In combination with heat, it has been used in milk pasteurization and sugar processing, but its widest use is as a sterilant for food-contact surfaces of olefin polymers and polyethylene in aseptic packaging systems (see Chapter 14). The D values of some food-borne microorganisms are presented in Table 11-6. **Ethanol** (C_2H_5OH) is present in flavoring extracts and effects preservation by virtue of its desiccant and denaturant properties. **Dehydroacetic acid,**

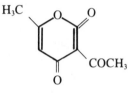

is used to preserve squash. **Diethylpyrocarbonate** has been used in bottled wines and soft drinks as a yeast inhibitor. It decomposes to form ethanol and CO_2 by either hydrolysis or alcoholysis (39):

Hydrolysis (reaction with water):

$$C_2H_5O - CO$$
$$\diagdown$$
$$O \xrightarrow{H_2O} 2C_2H_5OH + 2CO_2;$$
$$\diagup$$
$$C_2H_5O - CO$$

Alcoholysis (reaction with ethyl alcohol):

$$\begin{array}{cc} C_2H_5O - CO & C_2H_5O \\ \diagdown & \diagdown \\ O \xrightarrow{C_2H_5OH} & C = O + CO_2 + C_2H_5OH. \\ \diagup & \diagup \\ C_2H_5O - CO & C_2H_5O \end{array}$$

Saccharomyces cerevisiae and conidia of *Aspergillus niger* and *Byssochlamys fulva* have been shown to be destroyed by this compound during the first ½ h of exposure, while the ascospores of *B. fulva* required 4 to 6 h for maximal destruction (115). Cidal concentrations for yeasts range from about 20 to 1,000 ppm depending upon species or strain. *Lactobacillus plantarum* and *Leuconostoc mesenteroides* required 24 h or longer for destruction. Sporeforming bacteria are quite resistant to this compound. Sometimes urethane is formed when this compound is used and because it is a carcinogen, the use of diethylpyrocarbonate is no longer permissible in the United States.

Wood smoke imparts certain chemicals to smoked products that enable these products to resist microbial spoilage. One of the most important is formaldehyde (CH_2O), which has been known for many years to possess antimicrobial properties. This compound acts as a protein denaturant by virtue of its reaction with amino groups. Also in wood smoke are aliphatic acids, alcohols, ketones, phenols, higher aldehydes, tar, methanol, cresols, and other compounds (30), all of which may contribute to the antibacterial actions of meat smoking. Since a certain amount of heat is necessary to produce smoke, part of the shelf-stability of smoked products is due to heat destruction of surface organisms as well as to the drying that occurs. A study of the antibacterial activity of liquid smoke by Handford and Gibbs (46) revealed that little activity occurred at concentrations of smoke that produced acceptable smoked flavor. Employing an agar medium containing 1:1 dilution of smoked water, these investigators found that micrococci and staphylococci were slightly more inhibited than the lactic acid bacteria. The overall combined effect of smoking and vacuum packaging results in a reduction of numbers of catalase-positive bacteria on the smoked product, while the catalase-negative lactic acid bacteria are better able to withstand the low Eh conditions of vacuum-packaged products.

The **lactoperoxidase system** is an inhibitory system that occurs naturally in bovine milk. It consists of three components: lactoperoxidase, thiocyanate, and H_2O_2. All three components are required for antimicrobial effects, and the gram-negative psychrotrophs such as the pseudomonads are quite sensitive. The quantity of lactoperoxidase needed is 0.5–1.0 ppm, while bovine milk normally contains about 30 ppm (5). While both thiocyanate and H_2O_2 occur normally in milk, the quantities vary. For H_2O_2 about 100 U/ml are required in the inhibitory system, while only 1–2 U/ml normally occurs in milk. An effective level of thiocyanate is around 0.25 mM, while in milk the quantity varies between 0.02–0.25 mM (5).

When the lactoperoxidase system in raw milk was activated by adding thiocyanate to 0.25 mM along with an equimolar amount of H_2O_2, the shelf life was extended to 5 days compared to 48 h for controls (5). The system was more effective at 30 than at 4°C. The bactericidal effect increases with acidity, and the cytoplasmic membrane appears to be the cell target. In addition to the direct addition of H_2O_2, an exogenous source can be provided by the addition of glucose and glucose oxidase. To avoid the direct addition of glucose oxidase, this enzyme has been immobilized on glass beads so that glucose is generated only in the amounts needed by the use of immobilized β-galactosidase (6). The lactoperoxidase system can be used to preserve raw milk in countries where refrigeration is uncommon. The addition of about 12 ppm of SCN^- and 8 ppm of H_2O_2 should be harmless to the consumer (97). The system has been described in detail by Law and Reiter (72), Law and Mabbitt (71), and Reiter (96); and reviewed by Reiter and Harnulv (97).

REFERENCES

1. Ayres, J. C., A. A. Kraft, E. L. Denisen, and L. C. Peirce. 1964. The use of macrolide antifungal antibiotics in delaying spoilage of fresh small fruits and tomatoes. In *Microbial inhibitors in food,* ed. G. Molin, 185–98. Stockholm: Almquist & Wiksell.

2. Banks, J. G., and R. G. Board. 1982. Sulfite inhibition of *Enterobacteriaceae* including *Salmonella* in British fresh sausage and in culture systems. *J. Food Protect.* 45:1292–97, 1301.

3. Berry, B. W., and T. N. Blumer. 1981. Sensory, physical, and cooking characteristics of bacon processed with varying levels of sodium nitrite and potassium sorbate. *J. Food Sci.* 46:321–27.

4. Beuchat, L. R. 1976. Sensitivity of *Vibrio parahaemolyticus* to spices and organic acids. *J. Food Sci.* 41:899–902.

5. Björck, L. 1978. Antibacterial effect of the lactoperoxidase system on psychrotrophic bacteria in milk. *J. Dairy Res.* 45:109–18.

6. Björck, L., and C.-G. Rosen. 1976. An immobilized two-enzyme system for the activation of the lactoperoxidase antibacterial system in milk. *Biotechnol. Bioengin.* 18:1463–72.

7. Blake, D. F., and C. R. Stumbo. 1970. Ethylene oxide resistance of microorganisms important in spoilage of acid and high-acid foods. *J. Food Sci.* 35:26–29.

8. Bosund, I. 1962. The action of benzoic and salicylic acids on the metabolism of microorganisms. *Adv. Food Res.* 11:331–53.

9. Bowen, V. G., and R. H. Deibel. 1974. Effects of nitrite and ascorbate on botulinal toxin formation in wieners and bacon. In *Proceedings of the Meat Industry Research Conference,* 63–68. Chicago: American Meat Institute Foundation.

10. Branen, A. L., P. M. Davidson, and B. Katz. 1980. Antimicrobial properties of phenolic antioxidants and lipids. *Food Technol.* 34(5):42–53, 63.

11. Brock, T. D., D. W. Smith, and M. T. Madigan. 1984. *Biology of microorganisms.* 112–13. Englewood Cliffs, N.J.: Prentice-Hall.

12. Bullerman, L. B., F. Y. Lieu, and S. A. Seier. 1977. Inhibition of growth and aflatoxin production by cinnamon and clove oils, cinnamic aldehyde and eugenol. *J. Food Sci.* 42:1107–9, 1116.

13. Castellani, A. G., and C. F. Niven, Jr. 1955. Factors affecting the bacteriostatic action of sodium nitrate. *Appl. Microbiol.* 3:154–59.

14. Chang, P.-C., S. M. Akhtar, T. Burke, and H. Pivnick. 1974. Effect of sodium nitrite on *Clostridium botulinum* in canned luncheon meat: Evidence for a Perigo-type factor in the absence of nitrite. *Can. Inst. Food Sci. Technol. J.* 7:209–12.

15. Cheng, M. K. C., and R. E. Levin. 1970. Chemical destruction of *Aspergillus niger* conidiospores. *J. Food Sci.* 35:62–66.

16. Christiansen, L. N., R. W. Johnston, D. A. Kautter, J. W. Howard, and W. J. Aunan. 1973. Effect of nitrite and nitrate on toxin production by *Clostridium botulinum* and on nitrosamine formation in perishable canned comminuted cured meat. *Appl. Microbiol.* 25:357–62.

17. Christiansen, L. N., R. B. Tompkin, A. B. Shaparis, T. V. Kueper, R. W.

Johnston, D. A. Kautter, and O. J. Kolari. 1974. Effect of sodium nitrite on toxin production by *Clostridium botulinum* in bacon. *Appl. Microbiol.* 27:733–37.

18. Collins-Thompson, D. L., N. P. Sen, B. Aris, and L. Schwinghamer. 1972. Non-enzymic in vitro formation of nitrosamines by bacteria isolated from meat products. *Can. J. Microbiol.* 18:1968–71.

19. Crosby, N. T., and R. Sawyer. 1976. N-nitrosamines: A review of chemical and biological properties and their estimation in foodstuffs. *Adv. Food Res.* 22: 1–71.

20. Davidson, P. M. 1983. Phenolic compounds. In *Antimicrobials in foods*, ed. A. L. Branen and P. M. Davidson, 37–73. New York: Marcel Dekker.

21. Davidson, P. M., C. J. Brekke, and A. L. Branen. 1981. Antimicrobial activity of butylated hydroxyanisole, tertiary butylhydroquinone, and potassium sorbate in combination. *J. Food Sci.* 46:314–16.

22. Denny, C. B., J. M. Reed, and C. W. Bohrer. 1961. Effect of tylosin and heat on spoilage bacteria in canned corn and canned mushrooms. *Food Technol.* 15:338–40.

23. Denny, C. B., L. E. Sharpe, and C. W. Bohrer. 1961. Effects of tylosin and nisin on canned food spoilage bacteria. *Appl. Microbiol.* 9:108–10.

24. Desrosier, N. W. 1963. *The technology of food preservation.* Rev. ed., ch. 9. Westport, Conn.: AVI.

25. Dethmers, A. E., H. Rock, T. Fazio, and R. W. Johnston. 1975. Effect of added sodium nitrite and sodium nitrate on sensory quality and nitrosamine formation in thuringer sausage. *J. Food Sci.* 40:491–95.

26. Deuel, H. J., Jr., C. E. Calbert, L. Anisfeld, H. McKeehan, and H. D. Blunden. 1954. Sorbic acid as a fungistatic agent for foods. II. Metabolism of α,β-unsaturated fatty acids with emphasis on sorbic acid. *Food Res.* 19:13–19.

27. Dimick, K. P., G. Alderton, J. C. Lewis, H. D. Lightbody, and H. L. Fevold. 1947. Purification and properties of subtilin. *Arch. Biochem.* 15:1–11.

28. Doores, S. 1983. Organic acids. In *Antimicrobials in foods,* ed. A. L. Branen and P. M. Davidson, 75–107. New York: Marcel Dekker.

29. Doyle, M. P., and E. H. Marth. 1978. Bisulfite degrades aflatoxins. Effect of temperature and concentration of bisulfite. *J. Food Protect.* 41:774–80.

30. Draudt, H. N. 1963. The meat smoking process: A review. *Food Technol.* 17:1557–62.

31. Duncan, C. L., and E. M. Foster. 1968. Role of curing agents in the preservation of shelf-stable canned meat products. *Appl. Microbiol.* 16:401–5.

32. Dymicky, M., and H. Trenchard. 1982. Inhibition of *Clostridium botulinum* 62A by saturated n-aliphatic acids, n-alkyl formates, acetates, propionates and butyrates. *J. Food Protect.* 45:1117–19.

33. Eckert, J. W. 1979. Fungicidal and fungistatic agents: Control of pathogenic microorganisms on fresh fruits and vegetables after harvest. In *Food mycology,* ed. M. E. Rhodes, 164–99. Boston: Hall.

34. Fong, Y. Y., and W. C. Chan. 1973. Bacterial production of di-methyl nitrosamine in salted fish. *Nature* 243:421–22.

35. Food and Agriculture Organization/World Health Organization (FAO/WHO). 1976. *Evaluation of certain food additives.* WHO Technical Report Series 599.

36. Freese, E., C. W. Sheu, and E. Galliers. 1973. Function of lipophilic acids as antimicrobial food additives. *Nature* 241:321–25.

37. Fung, D. Y. C., C. C. S. Lin, and M. B. Gailani. 1985. Effect of phenolic antioxidants on microbial growth. *CRC Crit. Rev. Microbiol.* 12:153–83.
38. Gailani, M. B., and D. Y. C. Fung. 1984. Antimicrobial effects of selected antioxidants in laboratory media and in ground pork. *J. Food Protect.* 47:428–33.
39. Genth, H. 1964. On the action of diethylpyrocarbonate on microorganisms. In *Microbial inhibitors in food,* ed. G. Molin, 77–85. Stockholm: Almquist & Wiksell.
40. Gould, G. W. 1964. Effect of food preservatives on the growth of bacteria from spores. In *Microbial Inhibitors in Foods,* ed. G. Molin, 17–24. Stockholm: Almquist & Wiksell.
41. Gould, G. W., M. H. Brown, and B. C. Fletcher. 1983. Mechanisms of action of food preservation procedures. In *Food microbiology: Advances and prospects,* ed. T. A. Roberts and F. A. Skinner, 67–84. New York and London: Academic Press.
42. Gray, J. I. 1976. N-Nitrosamines and their precursors in bacon. A review. *J. Milk Food Technol.* 39:686–92.
43. Gray, J. I., and A. M. Pearson. 1984. Cured meat flavor. *Adv. Food Res.* 29:1–86.
44. Hagler, W. M., Jr., J. E. Hutchins, and P. B. Hamilton. 1982. Destruction of aflatoxin in corn with sodium bisulfite. *J. Food Protect.* 45:1287–91.
45. Hamilton-Miller, J. M. T. 1974. Fungal sterols and the mode of action of the polyene antibiotics. *Adv. Appl. Microbiol.* 17:109–34.
46. Handford, P. M., and B. M. Gibbs. 1964. Antibacterial effects of smoke constituents on bacteria isolated from bacon. In *Microbial inhibitors in food,* ed. G. Molin, 333–46. Stockholm: Almquist & Wiksell.
47. Hash, J. H. 1972. Antibiotic mechanisms. *Ann. Rev. Pharmacol.* 12:35–56.
48. Hawksworth, G., and M. J. Hill. 1971. The formation of nitrosamines by human intestinal bacteria. *Biochem. J.* 122:28–29P.
49. ———. 1971. Bacteria and the N-nitrosation of secondary amines. *Brit. J. Cancer.* 25:520–26.
50. Hechelman, H., and L. Leistner. 1969. Hemmung von unerwunschtem Schimmelpilzwachstum auf Rohwursten durch Delvocid (Pimaricin). *Fleischw.* 49:1639–41.
51. Hirsch, A., E. Grinsted, H. R. Chapman, and A. T. R. Mattick. 1951. Inhibition of an anaerobic sporeformer in Swiss-type cheese by a nisin-producing streptococcus. *J. Dairy Res.* 18:205–6.
52. Holley, R. A. 1981. Prevention of surface mold growth on Italian dry sausage by natamycin and potassium sorbate. *Appl. Environ. Microbiol.* 41:422–29.
53. Huhtanen, C. N. 1980. Inhibition of *Clostridium botulinum* by spice extracts and alphatic alcohols. *J. Food Protect.* 43:195–96, 200.
54. Hurst, A. 1981. Nisin. *Adv. Appl. Microbiol.* 27:85–123.
55. ———. 1983. Nisin and other inhibitory substances from lactic acid bacteria. In *Antimicrobials in foods,* ed. A. L. Branen and P. M. Davidson, 327–51. New York: Marcel Dekker.
56. Ingram, M., R. Buttiaux, and D. A. A. Mossel. 1964. General microbiological considerations in the choice of anti-microbial food preservatives. In *Microbial inhibitors in food,* ed. G. Molin, 381–92. Stockholm: Almquist & Wiksell.
57. Ivey, F. J., K. J. Shaver, L. N. Christiansen, and R. B. Tompkin. 1978. Effect

of potassium sorbate on toxinogenesis by *Clostridium botulinum* in bacon. *J. Food Protect.* 41:621–25.

58. Jay, J. M. 1982. Antimicrobial properties of diacetyl. *Appl. Environ. Microbiol.* 44:525–32.

59. ———. 1982. Effect of diacetyl on foodborne microorganisms. *J. Food Sci.* 47:1829–31.

60. ———. 1983. Antibiotics as food preservatives. In *Food microbiology,* ed. A. H. Rose, 117–43. New York and London: Academic Press.

61. Jay, J. M., and G. M. Rivers. 1984. Antimicrobial activity of some food flavoring compounds. *J. Food Safety* 6:129–39.

62. Johnston, M. A., and R. Loynes. 1971. Inhibition of *Clostridium botulinum* by sodium nitrite as affected by bacteriological media and meat suspensions. *Can. Inst. Food Technol. J.* 4:179–84.

63. Johnston, M. A., H. Pivnick, and J. M. Samson. 1969. Inhibition of *Clostridium botulinum* by sodium nitrite in a bacteriological medium and in meat. *Can. Inst. Food Technol. J.* 2:52–55.

64. Joslyn, M. A., and J. B. S. Braverman. 1954. The chemistry and technology of the pretreatment and preservation of fruit and vegetable products with sulfur dioxide and sulfites. *Adv. Food Res.* 5:97–160.

65. Kabara, J. J. 1981. Food-grade chemicals for use in designing food preservative systems. *J. Food Protect.* 44:633–47.

66. ———. 1983. Medium-chain fatty acids and esters. In *Antimicrobials in foods,* ed. A. L. Branen and P. M. Davidson, 109–39. New York: Marcel Dekker.

67. Kabara, J. J., R. Vrable, and M. S. F. Lie Ken Jie. 1977. Antimicrobial lipids: Natural and synthetic fatty acids and monoglycerides. *Lipids* 12:753–59.

68. Kereluk, K., R. A. Gammon, and R. S. Lloyd. 1970. Microbiological aspects of ethylene oxide sterilization. II. Microbial resistance to ethylene oxide. *Appl. Microbiol.* 19:152–56.

69. Klis, J. B., L. D. Witter, and Z. J. Ordal. 1964. The effect of several antifungal antibiotics on the growth of common food spoilage fungi. *Food Technol.* 13:124–28.

70. Kueper, T. V., and R. D. Trelease. 1974. Variables affecting botulinum toxin development and nitrosamine formation in fermented sausages. In *Procedings of the Meat Industry Research Conference,* 69–74. Chicago: American Meat Institute Foundation.

71. Law, B. A., and I. A. Mabbitt. 1983. New methods for controlling the spoilage of milk and milk products. In *Food microbiology: Advances and prospects,* ed. T. A. Roberts and F. A. Skinner, 131–50. New York and London: Academic Press.

72. Law, B. A., and B. Reiter. 1977. The isolation and bacteriostatic properties of lactoferrin from bovine milk whey. *J. Dairy Res.* 44:595–99.

73. Liewen, M. B., and E. H. Marth. 1985. Growth and inhibition of microorganisms in the presence of sorbic acid: A review. *J. Food Protect.* 48:364–75.

74. Lin, C. C. S., and D. Y. C. Fung. 1983. Effect of BHA, BHT, TBHQ, and PG on growth and toxigenesis of selected aspergilli. *J. Food Sci.* 48:576–80.

75. Lipinska, E. 1977. Nisin and its applications. In *Antibiotics and antibiosis in agriculture,* ed. M. Woodbine, 103–30. London: Butterworths.

76. Lloyd, A. C. 1975. Preservation of comminuted orange products. *J. Food Technol.* 10:565–67.

77. Marriott, N. G., R. V. Lechowich, and M. D. Pierson. 1981. Use of nitrite and nitrite-sparing agents in meats: A review. *J. Food Protect.* 44:881–85.
78. Marth, E. H. 1966. Antibiotics in foods—naturally occurring, developed, and added. *Residue Rev.* 12:65–161.
79. Melnick, D., H. W. Vahlteich, and A. Hackett. 1956. Sorbic acid as a fungistatic agent for foods. XI. Effectiveness of sorbic acid in protecting cakes. *Food Res.* 21:133–46.
80. Michael, G. T., and C. R. Stumbo. 1970. Ethylene oxide sterilization of *Salmonella senftenberg* and *Escherichia coli:* Death kinetics and mode of action. *J. Food Sci.* 35:631–34.
81. Miller, S. A., and W. D. Brown. 1984. Effectiveness of chlortetracycline in combination with potassium sorbate or tetrasodium ethylene-diaminetetraacetate for preservation of vacuum packed rockfish fillets. *J. Food Sci.* 49:188–91.
82. Moerck, K. E., P. McElfresh, A. Wohlman, and B. W. Hilton. 1980. Aflatoxin destruction in corn using sodium bisulfite, sodium hydroxide and aqueous ammonia. *J. Food Protect.* 43:571–74.
83. Morris, J. A., A. Khettry, and E. W. Seitz. 1979. Antimicrobial activity of aroma chemicals and essential oils. *J. Amer. Oil. Chem. Soc.* 56:595–603.
84. Nordin, H. R. 1969. The depletion of added sodium nitrite in ham. *Can. Inst. Food Sci. Technol. J.* 2:79–85.
85. O'Leary, V., and M. Solberg. 1976. Effect of sodium nitrite inhibition on intracellular thiol groups and on the activity of certain glycolytic enzymes in *Clostridium perfringens. Appl. Environ. Microbiol.* 31:208–12.
86. Ough, C. S. 1983. Sulfur dioxide and sulfites. In *Antimicrobials in foods,* ed. A. L. Branen and P. M. Davidson, 177–203. New York: Marcel Dekker.
87. Paquette, M. W., M. C. Robach, J. N. Sofos, and F. F. Busta. 1980. Effects of various concentrations of sodium nitrite and potassium sorbate on color and sensory qualities of commercially prepared bacon. *J. Food Sci.* 45:1293–96.
88. Pelroy, G. A., M. W. Eklund, R. N. Paranjpye, E. M. Suzuki, and M. E. Peterson. 1982. Inhibition of *Clostridium botulinum* types A and E toxin formation by sodium nitrite and sodium chloride in hot-process (smoked) salmon. *J. Food Protect.* 45:833–41.
89. Perigo, J. A., and T. A. Roberts. 1968. Inhibition of clostridia by nitrite. *J. Food Technol.* 3:91–94.
90. Perigo, J. A., E. Whiting, and T. E. Bashford. 1967. Observations on the inhibition of vegetative cells of *Clostridium sporogenes* by nitrite which has been autoclaved in a laboratory medium, discussed in the context of sublethally processed meats. *J. Food Technol.* 2:377–97.
91. Phillips, C. R. 1952. Relative resistance of bacterial spores and vegetative bacteria to disinfectants. *Bacteriol. Revs.* 16:135–38.
92. Pierson, M. D., and N. R. Reddy. 1982. Inhibition of *Clostridium botulinum* by antioxidants and related phenolic compounds in comminuted pork. *J. Food Sci.* 47:1926–29, 1935.
93. Rayman, K., N. Malik, and A. Hurst. 1983. Failure of nisin to inhibit outgrowth of *Clostridium botulinum* in a model cured meat system. *Appl. Environ. Microbiol.* 46:1450–52.
94. Reddy, D., J. R. Lancaster, Jr., and D. P. Cornforth. 1983. Nitrite inhibition

of *Clostridium botulinum*: Electron spin resonance detection of iron-nitric oxide complexes. *Science* 221:769–70.

95. Reisinger, P., H. Seidel, H. Tachesche, and W. P. Hammes. 1980. The effect of nisin on murein synthesis. *Arch. Microbiol.* 127:187–93.

96. Reiter, B. 1978. Review of the progress of dairy science: Antimicrobial systems in milk. *J. Dairy Res.* 45:131–47.

97. Reiter, B., and G. Harnulv. 1984. Lactoperoxidase antibacterial system: Natural occurrence, biological functions and practical applications. *J. Food Protect.* 47:724–32.

98. Riemann, H. 1963. Safe heat processing of canned cured meats with regard to bacterial spores. *Food Technol.* 17:39–49.

99. Robach, M. C., and M. D. Pierson. 1979. Inhibition of *Clostridium botulinum* types A and B by phenolic antioxidants. *J. Food Protect.* 42:858–61.

100. ———. 1978. Influence of para-hydroxybenzoic acid esters on the growth and toxin production of *Clostridium botulinum* 10755A. *J. Food Sci.* 43:787–89, 792.

101. Roberts, T. A., A. M. Gibson, and A. Robinson. 1981. Factors controlling the growth of *Clostridium botulinum* types A and B in pasteurized, cured meats. II. Growth in pork slurries prepared from "high" pH meat (range 6.3–6.8). *J. Food Technol.* 16:267–81.

102. ———. 1982. Factors controlling the growth of *Clostridium botulinum* types A and B in pasteurized, cured meats. III. The effect of potassium sorbate. *J. Food Technol.* 17:307–26.

103. Roberts, T. A., and M. Ingram. 1966. The effect of sodium chloride, potassium nitrate and sodium nitrite on the recovery of heated bacterial spores. *J. Food Technol.* 1:147–63.

104. Roberts, T. A., and J. L. Smart. 1974. Inhibition of spores of *Clostridium* spp. by sodium nitrite. *J. Appl. Bacteriol.* 37:261–64.

105. Rowe, J. J., J. M. Yarbrough, J. B. Rake, and R. G. Egon. 1979. Nitrite inhibition of aerobic bacteria. *Curr. Microbiol.* 2:51–54.

106. Savage, R. A., and C. R. Stumbo. 1971. Characteristics of progeny of ethylene oxide treated *Clostridium botulinum* type 62A spores. *J. Food Sci.* 36:182–84.

107. Scott, V. N., and S. L. Taylor. 1981. Effect of nisin on the outgrowth of *Clostridium botulinum* spores. *J. Food Sci.* 46:117–20, 126.

108. Seward, R. A., R. H. Deibel, and R. C. Lindsay. 1982. Effects of potassium sorbate and other antibotulinal agents on germination and outgrowth of *Clostridium botulinum* type E spores in microcultures. *Appl. Environ. Microbiol.* 44:1212–21.

109. Shelef, L. A. 1983. Antimicrobial effects of spices. *J. Food Safety* 6:29–44.

110. Shelef, L. A., and P. Liang. 1982. Antibacterial effects of butylated hydroxyanisole (BHA) against *Bacillus* species. *J. Food Sci.* 47:796–99.

111. Shelef, L. A., O. A. Naglik, and D. W. Bogen. 1980. Sensitivity of some common food-borne bacteria to the spices sage, rosemary, and allspice. *J. Food Sci.* 45:1042–44.

112. Sofos, J. N., and F. F. Busta. 1983. Sorbates. In *Antimicrobials in foods,* ed. A. L. Branen and P. M. Davidson, 141–75. New York: Marcel Dekker.

113. Sofos, J. N., F. F. Busta, and C. E. Allen. 1980. Influence of pH on *Clostridium*

botulinum control by sodium nitrite and sorbic acid in chicken emulsions. *J. Food Sci.* 45:7–12.

114. Sofos, J. N., F. F. Busta, K. Bhothipaksa, C. E. Allen, M. C. Robach, and M. W. Paquette. 1980. Effects of various concentrations of sodium nitrite and potassium sorbate on *Clostridium botulinum* toxin production in commercially prepared bacon. *J. Food Sci.* 45:1285–92.

115. Splittstoesser, D. F., and M. Wilkison. 1973. Some factors affecting the activity of diethylpyrocarbonate as a sterilant. *Appl. Microbiol.* 25:853–57.

116. Swartling, P., and B. Lindgren. 1968. The sterilizing effect against *Bacillus subtilis* spores of hydrogen peroxide at different temperatures and concentrations. *J. Dairy Res.* 35:423–28.

117. Tanaka, N., N. M. Gordon, R. C. Lindsay, L. M. Meske, M. P. Doyle, and E. Traisman. 1985. Sensory characteristics of reduced nitrite bacon manufactured by the Wisconsin process. *J. Food Protect.* 48:687–92.

118. Tanaka, N., L. Meske, M. P. Doyle, E. Traisman, D. W. Thayer, and R. W. Johnston. 1985. Plant trials of bacon made with lactic acid bacteria, sucrose and lowered sodium nitrite. *J. Food Protect.* 48:679–86.

119. Tanaka, N., E. Traisman, M. H. Lee, R. G. Cassens, and E. M. Foster. 1980. Inhibition of botulinum toxin formation in bacon by acid development. *J. Food Protect.* 43:450–57.

120. Tarr, H. L. A., B. A. Southcott, and H. M. Bissett. 1952. Experimental preservation of flesh foods with antibiotics. *Food Technol.* 6:363–68.

121. Toledo, R. T. 1975. Chemical sterilants for aseptic packaging. *Food Technol.* 29(5):102–7.

122. Toledo, R. T., F. E. Escher, and J. C. Ayres. 1973. Sporicidal properties of hydrogen peroxide against food spoilage organisms. *Appl. Microbiol.* 26:592–97.

123. Tompkin, R. B. 1983. Nitrite. In *Antimicrobials in foods,* ed. A. L. Branen and P. M. Davidson, 205–56. New York: Marcel Dekker.

124. Tompkin, R. B., L. N. Christiansen, and A. B. Shaparis. 1978. Enhancing nitrite inhibition of *Clostridium botulinum* with isoascorbate in perishable canned cured meat. *Appl. Environ. Microbiol.* 35:59–61.

125. ———. 1978. Causes of variation in botulinal inhibition in perishable canned cured meat. *Appl. Environ. Microbiol.* 35:886–89.

126. ———. 1979. Iron and the antibotulinal efficacy of nitrite. *Appl. Environ. Microbiol.* 37:351–53.

127. ———. 1980. Antibotulinal efficacy of sulfur dioxide in meat. *Appl. Environ. Microbiol.* 39:1096–99.

128. Vareltzis, K., E. M. Buck, and R. G. Labbe. 1984. Effectiveness of a betalains/potassium sorbate system versus sodium nitrite for color development and control of total aerobes, *Clostridium perfringens* and *Clostridium sporogenes* in chicken frankfurters. *J. Food Protect.* 47:532–36.

129. Vas, K., and M. Ingram. 1949. Preservation of fruit juices with less SO_2. *Food Manuf.* 24:414–16.

130. Winarno, F. G., and C. R. Stumbo. 1971. Mode of action of ethylene oxide on spores of *Clostridium botulinum* 62A. *J. Food Sci.* 36:892–95.

131. Woods, L. F. J., and J. M. Wood. 1982. A note on the effect of nitrite inhibition on the metabolism of *Clostridium botulinum*. *J. Appl. Bacteriol.* 52:109–10.

132. Woods, L. F. J., J. M. Wood, and P. A. Gibbs. 1981. The involvement of nitric oxide in the inhibition of the phosphoroclastic system in *Clostridium sporogenes* by sodium nitrite. *J. Gen. Microbiol.* 125:399–406.
133. Yarbrough, J. M., J. B. Rake, and R. G. Egon. 1980. Bacterial inhibitory effects of nitrite: Inhibition of active transport, but not of group translocation, and of intracellular enzymes. *Appl. Environ. Microbiol.* 39:831–34.
134. Zaika, L. L., J. C. Kissinger, and A. E. Wasserman. 1983. Inhibition of lactic acid bacteria by herbs. *J. Food Sci.* 48:1455–59.

12.

FOOD PRESERVATION
USING IRRADIATION

Although a patent was issued in 1929 for the use of radiation as a means of preserving foods, it was not until shortly after World War II that this method of food preservation received any serious consideration. While the application of radiation as a food preservation method has been somewhat slow in reaching its maximum potential use, the full application of this method presents some interesting challenges to food microbiologists and other food scientists.

Radiation may be defined as the emission and propagation of energy through space or through a material medium. The type of radiation of primary interest in food preservation is electromagnetic. The electromagnetic spectrum is presented in Fig. 12-1. The various radiations are separated on the basis of their wavelengths, with the shorter wavelengths being the most damaging to microorganisms. The electromagnetic spectrum may be further divided as follows with respect to these radiations of interest in food preservation: microwaves, ultraviolet rays, X rays, and gamma rays. The radiations of primary interest in food preservation are **ionizing radiations.** Ionizing radiations may be defined as those radiations that have wavelengths of 2,000Å or less—for example, alpha particles, beta rays, gamma rays, X rays, and cosmic rays. Their quanta contain enough energy actually to ionize molecules in their paths. Since they destroy microorganisms without appreciably raising temperature, the process is termed "cold sterilization."

In considering the application of radiation to foods, there are several useful concepts that should be clarified. A **Roentgen** is a unit of measure used for expressing exposure dose of X-ray or gamma radiation. A **milli-roentgen** is equal to 1/1,000 of a Roentgen. A **Curie** is a quantity of radioactive

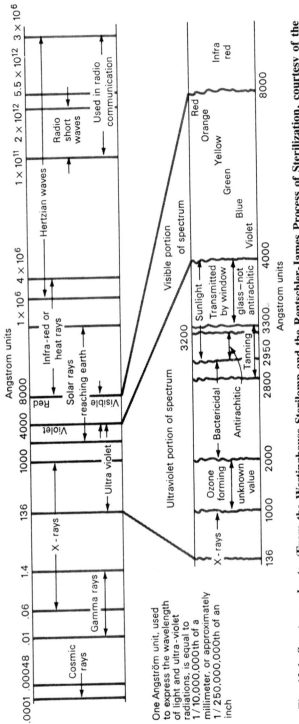

Fig. 12-1. Spectrum charts. (From the Westinghouse Sterilamp and the Rentschler-James Process of Sterilization, courtesy of the Westinghouse Electric & Manufacturing Co., Inc.)

substance in which 3.7×10^{10} radioactive disintegrations occur per second. For practical purposes, 1 g of pure radium possesses the radioactivity of 1 Curie of radium. The new unit for a Curie is the Becquerel (Bq). A **rad** is a unit equivalent to the absorption of 100 ergs/g of matter. A **kilorad** (krad) is equal to 1,000 rads, and a **megarad** (Mrad) is equal to 1 million rads. The new unit of absorbed dose is the Gray (1 Gy = 100 rads = 1 joule/kg; 1 kGy = 10^5 rads). The energy gained by an electron in moving through 1 volt is designated **ev** (electron volt). A **mev** is equal to 1 million electron volts. Both the rad and ev are measurements of the intensity of irradiation.

CHARACTERISTICS OF RADIATIONS OF INTEREST IN FOOD PRESERVATION

ULTRAVIOLET LIGHT (UV LIGHT). Ultraviolet light is a powerful bactericidal agent, with the most effective wavelength being about 2,600Å. It is nonionizing and is absorbed by proteins and nucleic acids, in which photochemical changes are produced that may lead to cell death. The mechanism of UV death in the bacterial cell is apparently due to the production of lethal mutations as a result of action upon cell nucleic acids. The poor penetrative capacities of UV light limit its food use to surface applications, where it may catalyze oxidative changes that lead to rancidity, discolorations, and other reactions. Small quantities of ozone may also be produced when UV light is used for the surface treatment of certain foods. UV light is sometimes used to treat the surfaces of baked fruit cakes and related products before wrapping.

BETA RAYS. Beta rays may be defined as a stream of electrons emitted from radioactive substances. Cathode rays are the same except that they are emitted from the cathode of an evacuated tube. These rays possess poor penetration power. Among the commercial sources of cathode rays are Van de Graaff generators and linear accelerators. The latter seem better suited for food preservation uses. There is some concern over the upper limit of energy level of cathode rays that can be employed without inducing radioactivity in certain constituents of foods.

GAMMA RAYS. These are electromagnetic radiations emitted from the excited nucleus of elements such as ^{60}Co and ^{137}Cs, which are of importance in food preservation. This is the cheapest form of radiation for food preservation, since the source elements are either by-products of atomic fission or atomic waste products. Gamma rays have excellent penetration power, as opposed to beta rays. ^{60}Co has a half-life of about 5 years, while the half-life for ^{137}Cs is about 30 years.

X RAYS. These rays are produced by the bombardment of heavy-metal targets with high-velocity electrons (cathode rays) within an evacuated tube. They are essentially the same as gamma rays in other respects.

MICROWAVES. Microwave energy may be illustrated in the following way (13). When electrically neutral foods are placed in an electromagnetic field, the charged asymmetric molecules are driven first one way and then another. During this process, each asymmetric molecule attempts to align itself with the rapidly changing alternating-current field. As the molecules oscillate about their axes while attempting to go to the proper positive and negative poles, intermolecular friction is created and manifested as a heating effect. This is microwave energy. Most food research has been carried out at two frequencies, 915 and 2450 megacycles. At the microwave frequency of 915 megacycles, the molecules oscillate back and forth 915 million times/sec (13). Microwaves lie between the infrared and radio frequency portion of the electromagnetic spectrum (see Fig. 12-1). The destruction of molds in bread, pasteurization of beer, and sterilization of wine have been achieved through the use of microwaves. The most successful application of microwave energy to date has been in the finish frying of potato chips. The general use of microwaves as a means of food preservation is limited by the heating effect that results from their use.

PRINCIPLES UNDERLYING THE DESTRUCTION OF MICROORGANISMS BY IRRADIATION

Several factors come into play when the effects of radiation on microorganisms are considered.

KINDS AND SPECIES OF ORGANISMS. With respect to kinds of microorganisms, gram-positive bacteria are more resistant to radiation than gram negatives. Sporeformers are in general more resistant than nonsporeformers, with the exception of *Micrococcus radiodurans,* which is one of the most radioresistant bacteria known (2). Among sporeformers, *B. larvae* has been reported to possess a higher degree of resistance than most aerobic sporeformers. Spores of *C. botulinum* type A appear to be the most resistant of all clostridial spores. Apart from *M. radiodurans,* one of the most resistant vegetative bacteria appears to be *Streptococcus faecium* R53. Among the more resistant vegetative forms are *Streptococcus faecalis,* micrococci in general, and the homofermentative lactobacilli. The bacteria most sensitive to radiation belong to the pseudomonad and flavobacteria groups, with other gram-negative bacteria being intermediate in radioresistance between these genera and the micrococci (see Fig. 12-2). Possible mechanisms of variation in radiosensitivity are discussed in Chapter 25.

With two exceptions, the radioresistance of bacterial endospores in general parallels that of heat resistance. The exceptions are *M. radiodurans* and

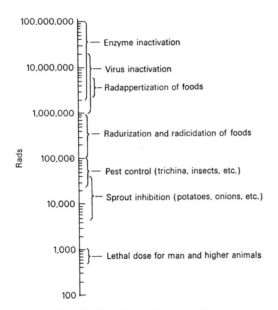

Fig. 12-2. Dose ranges of irradiation for various applications. (Adapted from Grü-newald, 18)

related species, which are more radioresistant than any endospores, and flat sour and thermophilic anaerobe (T.A.) spores, which are more sensitive to irradiation than to heat.

With respect to the radiosensitivity of molds and yeasts, the latter have been reported to be more resistant than the former, with both groups in general being less sensitive than gram-positive bacteria. Some *Candida* strains have been reported to possess resistance comparable to that of some bacterial endospores.

NUMBERS OF ORGANISMS. The numbers of organisms have the same effect upon the efficacy of radiations as in the case of heat, chemical disinfection, and certain other phenomena—the larger the number of cells, the less effective a given dose.

COMPOSITION OF SUSPENDING MENSTRUM (FOOD). Microorganisms are in general more sensitive to radiation when suspended in buffer solutions than in protein-containing media. For example, Midura et al. (31) found radiation D values for a strain of *C. perfringens* to be 0.23 in phosphate buffer, while in cooked-meat broth the D value was 0.30 Mrad. Proteins have been shown to exert a protective effect against radiations as well as against certain antimicrobial chemicals and heat. Several investigators have reported that the presence of nitrites tends to make bacterial endospores more sensitive to radiation.

PRESENCE OR ABSENCE OF OXYGEN. The radiation resistance of microorganisms is greater in the absence of oxygen than in its presence. Complete removal of oxygen from the cell suspension of *E. coli* has been reported to increase its radiation resistance up to threefold (34). The addition of reducing substances such as sulfhydryl compounds generally has the same effect in increasing radiation resistance as an anaerobic environment.

PHYSICAL STATE OF FOOD. The radiation resistance of dried cells is in general considerably higher than that for moist cells. This is most likely a direct consequence of the radiolysis of water by ionizing radiations, which is discussed later in this chapter. Radiation resistance of frozen cells has been reported to be greater than that of nonfrozen cells (25). Grecz et al. (16) found that the lethal effects of gamma radiation decreased by 47% when ground beef was irradiated at $-196°C$ as compared to 0°C.

AGE OF ORGANISMS. Bacteria tend to be most resistant to radiation in the lag phase just prior to active cell division. The cells become more radiation-sensitive as they enter and progress through the log phase and reach their minimum at the end of this phase.

PROCESSING OF FOODS FOR IRRADIATION

Prior to being exposed to ionizing radiations, there are several processing steps that must be carried out in much the same manner as for the freezing or canning of foods.

SELECTION OF FOODS. Foods to be irradiated should be carefully selected for freshness and overall desirable quality. Especially to be avoided are foods that are already in incipient spoilage.

CLEANING OF FOODS. All visible debris and dirt should be removed. This will reduce the numbers of microorganisms to be destroyed by the radiation treatment.

PACKING. It is obviously important that foods to be irradiated should be packed in containers that will afford protection against postirradiation contamination. Cans seem to be the best at the present time, although a large amount of research has been carried out on the feasibility of using plastic containers. Clear glass containers undergo color changes when exposed to doses of radiation of around 1 Mrad, and the subsequent color may be undesirable.

BLANCHING OR HEAT TREATMENT. Sterilizing doses of radiation are insufficient to destroy the natural enzymes of foods (see Fig. 12-2). In order to avoid undesirable postirradiation changes, it is necessary to destroy these enzymes.

The best method appears to be heat treatment—that is, the blanching of vegetables and mild heat treatment of meats prior to irradiation.

APPLICATION OF RADIATION

As previously stated, the two most widely used techniques of irradiating foods are gamma radiation from either ^{60}Co and ^{137}Cs, and the use of electron beams from linear accelerators.

GAMMA RADIATION. The advantage of gamma radiation is that ^{60}Co and ^{137}Cs are relatively inexpensive by-products of atomic fission. In a common experimental radiation chamber employing these elements, the radioactive material is placed on the top of an elevator that can be moved up for use and down under water when not in use. Materials to be irradiated are placed around the radioactive material (the source) at a suitable distance for the desired dosage. Once the chamber has been vacated by all personnel, the source is raised into position, and the gamma rays irradiate the food. Irradiation at desired temperatures may be achieved either by placing the samples in temperature-controlled containers, or by controlling the temperature of the entire concrete- and lead-walled chamber. Among the drawbacks to the use of radioactive material is that the isotope source emits rays in all directions and cannot be turned "on" or "off" as may be desirable (see Fig. 12-3). Also, the half-life of ^{60}Co (5.27 yr) requires that the source be changed periodically in order to maintain a given level of radioactive potential. This drawback is overcome by the use of ^{137}Cs, which has a half-life of around 30 yr.

ELECTRON BEAMS. The use of electron accelerators offers certain advantages over radioactive elements that make this form of radiation somewhat more attractive to potential commercial users. Koch and Eisenhower (24) have listed the following:

1. High efficiency for the direct deposition of energy of the primary electron beams means high plant-product capacity.
2. The efficient convertibility of electron power to X-ray power means the capability of handling very thick products that cannot be processed by electron or gamma-ray beams.
3. The easy variability of electron-beam current and energy means a flexibility in the choice of surface and depth treatments for a variety of food items, conditions, and seasons.
4. The monodirectional characteristic of the primary and secondary electrons and X rays at the higher energies permits a great flexibility in the food-package design.
5. The ability to program and to regulate automatically from one instant to the next with simple electronic detectors and circuits and various

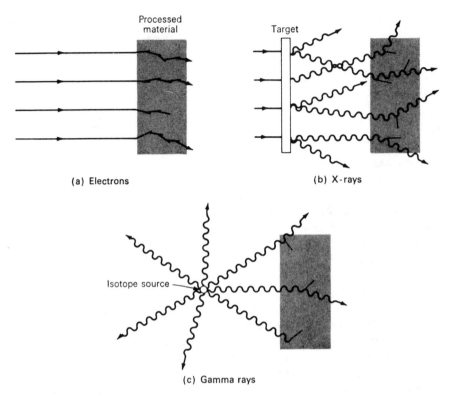

Fig. 12-3. The three basic techniques for radiation processing—interactions of electrons, X rays, and gamma rays in the medium. ([24], Koch and Eisenhower, 1965, *Radiation Preservation of Foods*, Publication 1273, Advisory Board on Military Personnel Supplies, National Academy of Sciences, National Research Council)

    beam parameters means the capability of efficiently processing small, intricate, or nonuniform shapes.

6. The ease with which an electron accelerator can be turned off or on means the ability to shut down during off-shifts or off-seasons without a maintenance problem and the ability to transport the radiation source without a massive radiation shield.

There appears to be a definite preference for electron accelerators in countries that lack atomic energy resources. The use of radioactive elements will probably remain popular in the United States for some time for reasons cited above.

RADAPPERTIZATION, RADICIDATION, AND RADURIZATION OF FOODS

Initially, the destruction of microorganisms in foods by ionizing radiation was referred to by terminology brought over from heat and chemical destruction of microorganisms. While microorganisms can indeed be destroyed by chemicals, heat, and radiation, there is, nevertheless, a lack of precision in the use of this terminology for radiation-treated foods. Consequently, in 1964 an international group of microbiologists suggested the following terminology for radiation treatment of foods (14):

Radappertization is equivalent to radiation sterilization or "commercial sterility," as it is understood in the canning industry, and typical levels of irradiation are 30–40 kGy.

Radicidation is equivalent to pasteurization—of milk, for example. Specifically, it refers to the reduction of the number of viable specific nonsporeforming *pathogens,* other than viruses, so that none is detectable by any standard method. Typical levels to achieve this process are 2.5–10 kGy.

Radurization may be considered equivalent to pasteurization. It refers to the enhancement of the keeping quality of a food by causing substantial reduction in the numbers of viable specific *spoilage* microbes by radiation. Common dose levels are 0.75–2.5 kGy for fresh meats, poultry, seafood, fruits, vegetables, and cereal grains.

RADAPPERTIZATION. Radappertization of any foods may be achieved by application of the proper dose of radiation under the proper conditions. The effect of this treatment on endospores and exotoxins of *C. botulinum* is of obvious interest. Type E spores have been reported to possess radiation D values on the order of 0.12–0.17 Mrad (15, 37). Types A and B spores were found by Kempe (23) to have D values of 0.279 and 0.238 Mrad, respectively. Type E spores are the most radiation sensitive of these three types.

The effect of temperature of irradiation on D values of *C. botulinum* spores is presented in Table 12-1. It can be seen that radiation resistance increases at the colder temperatures and decreases at higher temperatures. The different inoculum levels had no significant effect on D values whose calculations were based on a linear destruction rate. D values of four *C. botulinum* strains in three food products are presented in Table 12-2, from which it can be seen that each strain displayed different degrees of radiation resistance in each product. Also, irradiating in cured meat products produced the lowest D values. The possible significance of this fact is discussed in the previous chapter under nitrates and nitrites. The minimum radiation doses (MRD) in Mrad for the radappertization of nine meat and fish products are indicated below (4, 5, 21). With the exception of bacon (irradiated at ambient temperatures), each was treated at $-30°C \pm 10$:

Bacon	2.3	Shrimp	3.7
Beef	4.7	Codfish cakes	3.2
Chicken	4.5	Corned beef	2.5
Ham	3.7	Pork sausage	2.4–2.7
Pork	5.1		

Table 12-1. The effect of irradiation temperature on D values of two load levels of *C. botulinum* 33A in precooked ground beef. Data are based on linear spore destruction (17).

Temperature (°C)	D (Mrad)	
	ca. 5 × 10⁶ spores/can	ca. 2 × 10⁸/can
−196	0.577	0.595
−150	0.532	0.543
−100	0.483	0.486
−50	0.434	0.430
0	0.385	0.373
25	0.360	0.345
65	0.321	0.299

Table 12-2. Variations in radiation D values of 4 strains of *C. botulinum* at −30°C in three meat products as computed by the Schmidt equation (4).

Strain number	D (Mrad)		
	Codfish cake	Corned beef	Pork sausage
33A	0.203	0.129	0.109
77A	0.238	0.262	0.098
41B	0.245	0.192	0.184
53B	0.331	0.183	0.076

To achieve 12D treatments of meat products at about 30°C, the following kGy values are necessary (39): beef and chicken, 41.2–42.7; ham and codfish cake, 31.4–31.7; pork, 43.7; and corned beef and pork sausage, 25.5–26.9. Irradiation treatments of the types noted do not make the foods radioactive (39).

The radiation resistance of *C. botulinum* spores in aqueous media was studied by Roberts and Ingram (38), and these values are considerably lower than those obtained in meat products. On three type A strains, D ranged from 0.10 to 0.14; on two strains of type B, 0.10 to 0.11; on two strains of type E, 0.08 to 0.16; and the one type F strain examined by these authors showed a D value of 0.25. All strains were irradiated at 18–23°C and an exponential death rate was assumed in the D calculations.

With respect to the effect of radiation on *C. perfringens,* one each of five different strains (types A, B, C, E, and F) was found to have D values between 0.15 and 0.25 in an aqueous environment (38). 12D values for eight strains of this organism were found to range between 3.04 and 4.14 Mrad depending upon the strain and method of computing 12D doses (6).

As indicated in Fig. 12-2, viruses are considerably more resistant to radiation than bacteria. Radiation D values of thirty viruses were found by Sullivan et al. (40) to range between 0.39 and 0.53 Mrad in Eagle's minimal essential medium supplemented with 2% serum. The thirty viruses included coxsackie-, echo-, and poliovirus. Of five selected viruses subjected to ⁶⁰Co rays in distilled water, the D values ranged from 0.10 to 0.14 Mrad. D values of coxsackievirus B-2 in various menstra at −30 and −90°C are presented in Table 12-3. The use of a radiation 12D process for *C. botulinum* in meat products would result in the survival of virus particles unless previously destroyed by other methods such as heating.

Enzymes are also highly resistant to radiation, and a dose of from 2 to

Table 12-3. D values[a] of coxsackievirus B-2
at −30° and −90°C in various menstra (41).

Suspending menstrum	*D (Mrad)*	
	−30°C	*−90°C*
Eagle's minimal essential medium + 2% serum	0.69	0.64
Distilled water	—	0.53
Cooked ground beef	0.68	0.81
Raw ground beef	0.75	0.68

[a]A linear model was assumed in D calculations.
Copyright © 1973, American Society for Microbiology.

6 Mrad has been found to destroy only up to 75% of the proteolytic activity of ground beef (28). When blanching at 65 or 70°C was combined with radiation doses of 4.5 to 5.2, however, at least 95% of the beef proteolytic activity was destroyed.

The main drawbacks to the application of radiation to all foods are color changes and/or the production of off-odors. Consequently, those food products that undergo relatively minor changes in color and odor have received the greatest amount of attention for commercial radappertization. Bacon is one product that undergoes only slight changes in color and odor development following radappertization. Mean preference scores on radappertized versus control bacon were found to be rather close, with control bacon being scored just slightly higher (49). Acceptance scores on a larger variety of irradiated products were in the favorable range (21).

Radappertization of bacon is one way to reduce nitrosamines. When bacon containing 20 ppm $NaNO_2$ + 550 ppm sodium ascorbate was irradiated with 30 kGy, the resulting nitrosamine levels were similar to those in nitrite-free bacon (9).

RADICIDATION. Irradiation at levels of 2–5 kGy has been shown by many to be effective in destroying nonsporeforming and nonviral pathogens, and to present no health hazard (10, 22). The latter author notes that raw poultry meats should be given the highest priority since they are often contaminated with salmonellae and since radicidation is effective on prepackaged products, thus eliminating the possibilities of cross-contamination. The treatment of refrigerated and frozen chicken carcasses with 2.5 kGy was highly effective in destroying salmonellae (32). A radiation dosage up to 7 kGy (0.7 Mrad) has been approved by the WHO as being "unconditionally safe for human consumption" (10). When whole cacao beans were treated with 5 kGy, 99.9% of the bacterial flora was destroyed and *Penicillium citrinum* spores were reduced by about 5 logs/g, and at a level of 4 kGy, *Aspergillus flavus* spores were reduced by about 7 logs/g (36). It may be noted from Table 12-5 that fresh poultry, cod and red fish, and spices and condiments have been approved for radicidation in some countries.

RADURIZATION. Radurization as a means of extending the shelf life of seafood, vegetables, and fruits has been verified in many laboratories. Typical of what can be achieved by this treatment are data reported for various seafoods and meats (Table 12-4). The shelf life of shrimp, crab, haddock, and clams may be extended from two- to sixfold by radurization with doses of radiation from 100,000 to 400,000 rads. Similar results can be achieved for fish and shellfish under various conditions of packaging (35). As was pointed out earlier in this chapter, the gram-negative, nonsporeforming rods are among the most radiosensitive of all bacteria to radiation, and they are the main spoilage organisms for these foods. Among the organisms that survive radurization treatments are gram-positive bacteria of the following genera: *Micrococcus* (especially *M. roseus* and *M. radiodurans*), *Corynebacterium*, *Streptococcus, Microbacterium, Brochothrix, Lactobacillus,* and some yeasts. The gram-negative coccobacillary rods belonging to the genera *Moraxella*

Table 12-4.   Radiation D values determined in various substrates (see text or reference for more specifics).

Organism	D value (kGy)	Reference
Acinetobacter calcoaceticus	0.26	43
Clostridium botulinum, type E	1.2–1.7	10, 27
C. botulinum, type A	2.79	17
C. botulinum, type B	2.38	17
Escherichia coli	0.20	43
Bacillus pumulis spores, ATCC 27142	1.40	43
Enterobacter cloacae	0.18	43
Salmonella sp.	0.13	43
Staphylococcus aureus	0.16	43
Pseudomonas aeruginosa	0.13	43
Penicillium sp.	0.42	43
Aspergillus flavus spores	0.66[a]	36
Penicillium citrinum, NRRL 5452	0.88[a]	36
Clostridium bifermentans	1.4	28
C. botulinum 62A	1.0	28
C. botulinum type E Beluga	0.8	28
C. botulinum type F	2.5	28
C. butyricum	1.5	28
C. sordellii	1.5	28
C. sporogenes (PA 3679/S₂)	2.2	28
C. perfringens type A	1.2	28
Adenoviruses (4)[b]	4.1–4.9	30
Coxsackieviruses (7)[b]	4.1–5.0	30
Echoviruses (8)[b]	4.4–5.1	30
Polioviruses (6)[b]	4.1–5.4	30
Herpes simplex virus	4.3	30

[a] Average values.
[b] Number of strains tested.

and *Acinetobacter* have been found to possess degrees of radiation resistance higher than for all other gram negatives. In studies on ground beef subjected to doses of 272 krad, Tiwari and Maxcy (42) found that 73–75% of the surviving flora consisted of these related genera. In unirradiated meat, they constituted only around 8% of the flora. Of the two genera, the *Moraxella* spp. appear to be more resistant than *Acinetobacter* spp., with D_{10} values of 273 to 2,039 krad having been found (47). If this finding is correct, some of these organisms are among the most radiation-resistant of all bacteria. Among specific species, *M. nonliquefaciens* strains showed D_{10} values of 539 and 583 krad while the D_{10} for *M. osloensis* strains was 477 up to 1,000 krad.

In comparing the radio-sensitivity of some nonsporeforming bacteria in phosphate buffer at $-80°C$, Anellis et al. (3) found that *M. radiodurans* survived 1.8 Mrad, *Streptococcus faecium* strains survived 0.9–1.5, *S. faecalis* survived 0.6–0.9, and *S. lactis* did not survive 0.6 Mrad. *S. aureus, L. casei,* and *L. arabinosus* did not survive 0.3-Mrad exposures. It was shown that radiation sensitivity decreased as temperature of irradiation was lowered, as is the case for endospores.

The ultimate spoilage of radurized, low-temperature-stored foods is invariably caused by one or more of the *Acinetobacter-Moraxella* or lactic acid types noted above. The application of 2.5 kGy to ground beef destroyed all pseudomonads, Enterobacteriaceae, and *B. thermosphacta;* reduced APC from log 6.18/g to 1.78/g; but reduced lactic acid bacteria only by 3.4 log/g (33).

Radurization of fruits with doses of 2–3 kGy brings about an extension of shelf life of at least 14 days. It can be seen from Table 12-5 that radurization of fresh fruits is permitted by at least six countries, with some meats, poultry, and seafood permitted by several others. In general, shelf-life extension is not as great for radurized fruits as for meats and seafood because molds are generally more resistant to irradiation than the gram-negative bacteria that cause spoilage of the latter products.

Insect eggs and larvae can be destroyed by 1 kGy, and cysticerci of the pork tapeworm (*Taenia solium*) and the beef tapeworm (*T. saginata*) can be destroyed with even lower doses, with cysticercosis-infested carcasses being rendered free of parasites by exposure to 0.2–0.5 kGy (46).

LEGAL STATUS OF FOOD IRRADIATION

In 1963 the U.S. FDA approved the use of ^{60}Co and electron acceleration for the radappertization of bacon, and a year later, ^{137}Cs was approved. In 1968, however, these approvals were withdrawn pending additional information on product safety. Meanwhile, at least nineteen countries permit the irradiation of a variety of foods and food products, and some of these are summarized in Table 12-5. Of the products noted, fifteen are approved by the Netherlands, nine by the USSR, seven by Bulgaria, five each by

Table 12-5. Summary of foods and food products approved for irradiation by various countries and by WHO (16; 45).

Products	Objective	Dose range (kGy)	No. countries[a]
Potatoes	Sprout inhibition	0.1–0.15	17
Onions	Sprout inhibition	0.1–0.15	10
Garlic	Sprout inhibition	0.1–0.15	2
Mushrooms	Growth inhibition	2.5 max.	1
Wheat, wheat flour	Insect disinfestation	0.2–0.75	4
Dried fruits	Insect disinfestation	1.0	2
Cocoa beans	Insect disinfestation	0.7	1
Dry food concentrates	Insect disinfestation	0.7–1.0	1
Poultry, fresh	Radicidation[b]	7.0 max.	2
Cod and redfish	Radicidation	2.0–2.2	1
Spices/condiments	Radicidation	8.0–10.0	1
Semipreserved meats	Radurization	6.0–8.0	1
Fresh fruits[c]	Radurization	2.5	6
Asparagus	Radurization	2.0	1
Raw meats	Radurization	6.0–8.0	1
Cod and haddock fillets	Radurization	1.5 max.	1
Poultry (eviscerated)	Radurization	3.0–6.0	2
Shrimp	Radurization	0.5–1.0	1
Culinary prepared meat products	Radurization	8.0	1
Deep-frozen meals	Radappertization	25.0 min.	2
Fresh, tinned/liquid foodstuffs	Radappertization	25.0 min.	1

[a] Including WHO recommendations.
[b] For salmonellae.
[c] Includes tomatoes, peaches, apricots, strawberries, cherries, grapes, etc.

Canada and Italy, and only two by the United States. About twenty different food packaging materials are approved by the FDA for irradiation at levels of 10 or 60 kGy. In 1983, the FDA permitted spices and vegetable seasonings to be irradiated up to 10 kGy (U.S. *Federal Register,* 7/15/83). FDA permission was granted in 1985 for the irradiation of pork at up to 1 kGy to control *Trichinella spiralis* (U.S. *Federal Register,* 7/22/85). Sprout inhibition and insect disinfestation continue to be the most widely used direct applications of food irradiation.

As noted above, the WHO has given approval for radiation dosages up to 7 kGy (0.7 Mrad) as being unconditionally safe. In the early 1970s, Canada approved for test marketing a maximum dose of 1.5 kGy for fresh cod and haddock fillets. In 1983, the Codex Alimentarius Commission suggested 1.5 or 2.2 kGy for teleost fish and fish products (11). One of the obstacles to getting food irradiation approved on a wider scale in the United States is the way irradiation is defined. It is considered an additive rather than a process, which it is. This means that irradiated foods must be labeled as such. Another area of concern is the fate of *C. botulinum* spores (see below); and yet another is the concern that nonpathogens may become pathogens or that the virulence of pathogens may be increased after exposure to subradappertization doses. There is no evidence that the latter occurs (39).

When low-acid foods are irradiated at doses that do not effect the destruction of *C. botulinum* spores, legitimate questions about the safety of such foods are raised, especially when they are held under conditions that allow for growth and toxin production. Since these organisms would be destroyed by radappertization, only products subjected to radicidation and radurization are of concern here. In regard to the radurization of fish, Giddings (11) has pointed out that the lean whitefish species are the best candidates for irradiation, while high-fat fishes such as herring are not since they are more botulogenic. This investigator notes that when botulinal spores are found on edible lean whitefish, they occur at $< 1/g$. Since only spores of type E strains can grow on refrigerated foods, and since radurized fish would not be vacuum packaged, the chances of toxigenic products are reduced greatly after irradiation at levels of 1–2 kGy.

EFFECT OF IRRADIATION ON FOOD CONSTITUENTS

The undesirable changes that occur in certain irradiated foods may be caused directly by irradiation or indirectly as a result of postirradiation reactions. Water undergoes radiolysis when irradiated in the following manner:

$$3H_2O \xrightarrow{\text{radiolysis}} H + OH + H_2O_2 + H_2$$

In addition, free radicals are formed along the path of the primary electron and react with each other as diffusion occurs (8). Some of the products formed along the track escape and can then react with solute molecules. By irradiating under anaerobic conditions, off-flavors and odors are somewhat minimized due to the lack of oxygen to form peroxides. One of the best ways to minimize off-flavors is to irradiate at subfreezing temperatures (44). The effect of subfreezing temperatures is to reduce or halt radiolysis and its consequent reactants. Other ways to reduce side effects in foodstuffs are presented in Table 12-6.

Other than water, proteins and other nitrogenous compounds appear to be the most sensitive to irradiation effects in foods. The products of irradiation of amino acids, peptides, and proteins depend upon the radiation dose, temperature, amount of oxygen, amount of moisture present, etc. The following are among the products reported: NH_3, hydrogen, CO_2, H_2S, amides, and carbonyls. With respect to amino acids, the aromatics tend to be more sensitive than the others, and undergo changes in ring structure. Among the most sensitive to irradiation are methionine, cysteine, histidine,

Table 12-6. Methods for reducing side effects in foodstuffs exposed to ionizing radiations (12).

Method	Reasoning
Reducing temperature	Immobilization of free radicals
Reducing oxygen tension	Reduction of numbers of oxidative free radicals to activated molecules
Addition of free radical scavengers	Competition for free radicals by scavengers
Concurrent radiation distillation	Removal of volatile off-flavor, off-odor precursors
Reduction of dose	Obvious

arginine, and tyrosine. The amino acid most susceptible to electron beam irradiation is cystine; Johnson and Moser (20) reported that about 50% of this amino acid was lost when ground beef was irradiated. Tryptophan suffered a 10% loss while little or no destruction of the other amino acids occurred. Amino acids have been reported to be more stable to gamma irradiation than to electron beam irradiation.

Several investigators have reported that the irradiation of lipids and fats results in the production of carbonyls and other oxidation products such as peroxides, especially if irradiation and/or subsequent storage takes place in the presence of oxygen. The most noticeable organoleptic effect of lipid irradiation in air is the development of rancidity.

It has been observed that high levels of irradiation lead to the production of "irradiation odors" in certain foods, especially meats. Wick et al. (48) investigated the volatile components of raw ground beef irradiated with from 2 to 6 Mrads at room temperature and reported finding a large number of odorous compounds. Of the forty-five or more constituents identified by these investigators, there were seventeen sulfur-containing, fourteen hydrocarbons, and nine carbonyls, and five or more were basic and alcoholic in nature. The higher the level of irradiation, the greater the quantity of volatile constituents produced. It should be noted that many of these constituents have been identified in various extracts of nonirradiated, cooked ground beef.

With regard to B vitamins, Liuzzo et al. (27) found that levels of ^{60}Co irradiation between 0.2 and 0.6 Mrads effected partial destruction of the following B vitamins in oysters: thiamine, niacin, pyridoxine, biotin, and B_{12}. Riboflavin, pantothenic acid, and folic acid were reported to be increased by irradiation, probably owing to release of bound vitamins.

In addition to flavor and odor changes produced in certain foods by irradiation, certain detrimental effects have been reported for irradiated fruits and vegetables. One of the most serious is the softening of these products caused by the irradiation-degradation of pectin and cellulose, the structural polysaccharides of plants. This effect has been shown by Massey and Bourke (29) to be caused by radappertization doses of irradiation.

Ethylene synthesis in apples is affected by irradiation so that this product fails to mature as rapidly as nonirradiated controls (29). In green lemons, however, ethylene synthesis is stimulated upon irradiation, resulting in a faster ripening than in controls (30).

Among radiolytic products that develop upon irradiation are some that are antibacterial when exposed in culture media. When 15 kGy were applied to meats, no antimicrobial activity was found in the meats (7).

STORAGE STABILITY OF IRRADIATED FOODS

Foods subjected to radappertization doses of ionizing radiation may be expected to be as shelf-stable as commercially heat-sterilized foods. There are, however, two differences between foods processed by these two methods that affect storage stability: radappertization does not destroy inherent enzymes, which may continue to act, and some postirradiation changes may be expected to occur. Employing 4.5 Mrads and enzyme-inactivated chicken, bacon, and fresh and barbecued pork, Heiligman (19) found these products to be acceptable after storage for up to 24 months. Those stored at 70°C were more acceptable than those stored at 100°F. The effect of irradiation on beefsteak, ground beef, and pork sausage held at refrigerator temperatures for 12 years was reported by Licciardello et al. (26). These foods were packed with flavor preservatives and treated with 1.08 Mrads. The authors described the appearance of these meats as excellent after 12 years of storage. A slight irradiation odor was perceptible but was not considered objectionable. The meats were reported to have a sharp, bitter taste, which was presumed to be caused by the crystallization of the amino acid tyrosine. The free amino nitrogen content of the beefsteak was 75 and 175 mg %, respectively, before and after irradiation storage, and 67 and 160 mg % before and after storage for hamburger.

Foods subjected to radurization ultimately undergo spoilage from the surviving flora if stored at temperatures suitable for growth of the organisms in question. The normal spoilage flora of seafoods is so sensitive to ionizing radiations that 99% of the total flora of these products is generally destroyed by doses on the order of 0.25 Mrads. Ultimate spoilage of radurized products is the property of the few microorganisms that survive the radiation treatment.

For further information on all aspects of food irradiation, see reviews by Anderson (1), Ley (25), Tsuji (43), and Urbain (45).

REFERENCES

1. Anderson, A. W. 1983. Irradiation in the processing of food. In *Food microbiology,* ed. A. H. Rose, 145–71. New York: Academic Press.
2. Anderson, A. W., H. C. Nordan, R. F. Cain, G. Parrish, and D. Duggan. 1956. Studies on a radioresistant *Micrococcus*. I. Isolation, morphology, cultural characteristics, and resistance to gamma radiation. *Food Technol.* 10:575–78.

3. Anellis, A., D. Berkowitz, and D. Kemper. 1973. Comparative resistance of nonsporogenic bacteria to low-temperature gamma irradiation. *Appl. Microbiol.* 25:517–23.

4. Anellis, A., D. Berkowitz, W. Swantak, and C. Strojan. 1972. Radiation sterilization of prototype military foods: Low-temperature irradiation of codfish cake, corned beef, and pork sausage. *Appl. Microbiol.* 24:453–62.

5. Anellis, A., E. Shattuck, D. B. Rowley, E. W. Ross, Jr., D. N. Whaley, and V. R. Dowell, Jr. 1975. Low-temperature irradiation of beef and methods for evaluation of a radappertization process. *Appl. Microbiol.* 30:811–20.

6. Clifford, W. J., and A. Anellis. 1975. Radiation resistance of spores of some *Clostridium perfringens* strains. *Appl. Microbiol.* 29:861–63.

7. Dickson, J. S., and R. B. Maxcy. 1984. Effect of radiolytic products on bacteria in a food system. *J. Food Sci.* 49:577–80.

8. Doty, D. M. 1965. Chemical changes in irradiated meats. In *Radiation preservation of foods.* Pub. no. 1273, 121–25. Washington, D.C.: National Research Council, National Academy of Science.

9. Fiddler, W., R. A. Gates, J. W. Pensabene, J. G. Phillips, and E. Wierbicki. 1981. Investigations on nitrosamines in irradiation-sterilized bacon. *J. Agric. Food Chem.* 29:551–54.

10. Food and Agriculture Organization/IAEA/World Health Organization. 1977. *Wholesomeness of irradiated food.* Report of joint FAO/IAEA/WHO Expert Committee, WHO Technical Report Series 604.

11. Giddings, G. G. 1984. Radiation processing of fishery products. *Food Technol.* 38(4):61–65, 94–97.

12. Goldblith, S. A. 1963. Radiation preservation of foods—Two decades of research and development. In *Radiation research,* 155–67. Washington, D.C.: U.S. Dept. of Commerce, Office of Technical Services.

13. ———. 1966. Basic principles of microwaves and recent developments. *Adv. in Food Research* 15:277–301.

14. Goresline, H. E., M. Ingram, P. Macuch, G. Mocquot, D. A. A. Mossel, C. F. Niven, and F. S. Thatcher. 1964. Tentative classification of food irradiation processes with microbiological objectives. *Nature* 204:237–38.

15. Graikoski, J. T., and L. L. Kempe. 1962. Progress Report, Contract no. AT(11-1)-1095, U.S. Atomic Energy Commission.

16. Grecz, N., O. P. Snyder, A. A. Walker, and A. Anellis. 1965. Effect of temperature of liquid nitrogen on radiation resistance of spores of *Clostridium botulinum.* *Appl. Microbiol.* 13:527–36.

17. Grecz, N., A. A. Walker, A. Anellis, and D. Berkowitz. 1971. Effect of irradiation temperature in the range − 196 to 95°C on the resistance of spores of *Clostridium botulinum* 33A in cooked beef. *Can. J. Microbiol.* 17:135–42.

18. Grünewald, T. 1961. Behandlung von Lebensmitteln mit energiereichen Strahlen. *Ernährungs-Umschau* 8:239–44.

19. Heiligman, F. 1965. Storage stability of irradiated meats. *Food Technol.* 19:114–16.

20. Johnson, B., and K. Moser. 1967. Amino acid destruction in beef by high energy electron beam irradiation. In *Radiation preservation of foods,* Advances in Chemistry Series, 171–79. Washington, D.C.: American Chemical Society.

21. Josephson, E. S., A. Brynjolfsson, and E. Wierbicki. 1975. The use of ionizing radiation for preservation of food and feed products. In *Radiation research—*

Biomedical, chemical, and physical perspectives, ed. O. F. Nygaard. H. I. Adler, and W. K. Sinclair, 96–117. New York: Academic Press.

22. Kampelmacher, E. H. 1983. Irradiation for control of *Salmonella* and other pathogens in poultry and fresh meats. *Food Technol.* 37(4):117–19, 169.

23. Kempe, L. L. 1965. The potential problems of type E botulism in radiation-preserved seafoods. In *Radiation preservation of foods,* Pub. no. 1273, 211–15. Washington, D.C.: Research Council, National Academy of Science.

24. Koch, H. W., and E. H. Eisenhower. 1965. Electron accelerators for food processing. In *Radiation preservation of foods,* Pub. no. 1273, 149–80. Washington, D.C.: National Research Council, National Academy of Science.

25. Ley, F. J. 1983. New interest in the use of irradiation in the food industry. In *Food microbiology: Advances and prospects,* ed. T. A. Roberts and F. A. Skinner, 113–29. London: Adademic Press.

26. Licciardello, J. J., J. T. R. Nickerson, and S. A. Goldblith. 1966. Observations on radio-pasteurized meats after 12 years of storage at refrigerator temperatures above freezing. *Food Technol.* 20:1232.

27. Liuzzo, J. S., W. B. Barone, and A. F. Novak. 1966. Stability of B-vitamins in Gulf oysters preserved by gamma radiation. *Fed. Proc.* 25:722.

28. Losty, T., J. S. Roth, and G. Shults. 1973. Effect of irradiation and heating on proteolytic activity of meat samples. *J. Agr. Food Chem.* 21:275–77.

29. Massey, L. M., Jr., and J. B. Bourke. 1967. Some radiation-induced changes in fresh fruits and vegetables. In *Radiation preservation of foods,* Advances in Chemistry Series, 1–11. Washington, D.C.: American Chemical Society.

30. Maxie, E., and N. Sommer. 1965. Irradiation of fruits and vegetables. In *Radiation preservation of foods,* Pub. no. 1273, 39–52. Washington, D.C.: National Research Council, National Academy of Science.

31. Midura, T. F., L. L. Kempe, J. T. Graikoski, and N. A. Milone. 1965. Resistance of *Clostridium perfringens* type A spores to gamma-radiation. *Appl. Microbiol.* 13:244–47.

32. Mulder, R. W. A., S. Notermans, and E. H. Kampelmacher. 1977. Inactivation of salmonellae on chilled and deep frozen broiler carcasses by irradiation. *J. Appl. Bacteriol.* 42:179–85.

33. Niemand, J. G., H. J. van der Linde, and W. H. Holzapfel. 1983. Shelf-life extension of minced beef through combined treatments involving radurization. *J. Food Protect.* 46:791–96.

34. Niven, C. F., Jr. 1958. Microbiological aspects of radiation preservation of food. *Ann. Rev. Microbiol.* 12:507–24.

35. Novak, A. F., R. M. Grodner, and M. R. R. Rao. 1967. Radiation pasteurization of fish and shellfish. In *Radiation preservation of foods,* Advances in Chemistry Series, 142–51. Washington, D.C.: American Chemical Society.

36. Restaino, L., J. J. J. Myron, L. M. Lenovich, S. Bills, and K. Tscherneff. 1984. Antimicrobial effects of ionizing radiation on artificially and naturally contaminated cacao beans. *Appl. Environ. Microbiol.* 47:886–87.

37. Roberts, T. A., and M. Ingram. 1965. The resistance of spores of *Clostridium botulinum* Type E to heat and radiation. *J. Appl. Bacteriol.* 28:125–41.

38. ———. 1965. Radiation resistance of spores of *Clostridium* species in aqueous suspension. *J. Food Sci.* 30:879–85.

39. Rowley, D. B., and A. Brynjolfsson. 1980. Potential uses of irradiation in the processing of food. *Food Technol.* 34(10):75–77.

40. Sullivan, R., A. C. Fassolitis, E. P. Larkin, R. B. Read, Jr., and J. T. Peeler. 1971. Inactivation of thirty viruses by gamma radiation. *Appl. Microbiol.* 22:61–65.
41. Sullivan, R., P. V. Scarpino, A. C. Fassolitis, E. P. Larkin, and J. T. Peeler. 1973. Gamma radiation inactivation of coxsackievirus B-2. *Appl. Microbiol.* 26:14–17.
42. Tiwari, N. P., and R. B. Maxcy. 1972. *Moraxella-Acinetobacter* as contaminants of beef and occurrence in radurized product. *J. Food Sci.* 37:901–3.
43. Tsuji, K. 1983. Low-dose cobalt 60 irradiation for reduction of microbial contamination in raw materials for animal health products. *Food Technol.* 37(2):48–54.
44. Urbain, W. M. 1965. Radiation preservation of fresh meat and poultry. In *Radiation preservation of foods,* Pub. no. 1273, 87–98. Washington, D.C.: National Research Council, National Academy of Science.
45. ———. 1978. Food irradiation. *Adv. Food Res.* 24:155–227.
46. Verster, A., T. A. du Plessis, and L. W. van den Heever. 1977. The eradication of tapeworms in pork and beef carcasses by irradiation. *Radiat. Phys. Chem.* 9:769–71.
47. Welch, A. B., and R. B. Maxcy. 1975. Characterization of radiation-resistant vegetative bacteria in beef. *Appl. Microbiol.* 30:242–50.
48. Wick, E., E. Murray, J. Mizutani, and M. Koshika. 1967. Irradiation flavor and the volatile components of beef. In *Radiation preservation of foods,* Advances in Chemistry Series, 12–25. Washington, D.C.: American Chemical Society.
49. Wierbicki, E., M. Simon, and E. S. Josephson. 1965. Preservation of meats by sterilizing doses of ionizing radiation. In *Radiation preservation of foods,* Pub. no. 1273, 383–409. Washington, D.C.: National Research Council, National Academy of Science.

13.

FOOD PRESERVATION WITH
LOW TEMPERATURES

The use of low temperatures to preserve foods is based upon the fact that the activities of food-borne microorganisms can be slowed down and/or stopped at temperatures above freezing and generally stopped at subfreezing temperatures. The reason for this is that all metabolic reactions of microorganisms are enzyme catalyzed and that the rate of enzyme-catalyzed reactions is dependent on temperature. With a rise in temperature, there is an increase in reaction rate. The **temperature coefficient** (Q_{10}) may be generally defined as follows:

$$Q_{10} = \frac{(\text{Velocity at a given temp.} + 10°)}{\text{Velocity at T}}$$

The Q_{10} for most biological systems is 1.5 to 2.5, so that for each 10°C rise in temperature within the suitable range, there is a twofold increase in the rate of reaction. For every 10°C decrease in temperature, the reverse is true, of course. Since the basic feature of low-temperature food preservation consists of its effect upon spoilage organisms, most of the discussion that follows will be devoted to the effect of low temperatures on food-borne microorganisms. It should be remembered, however, that temperature is related to relative humidity (R.H.) and that subfreezing temperatures affect R.H. as well as pH and possibly other parameters of microbial growth as well.

The organisms that grow well at low temperatures are referred to in Chapter 2 as **psychrophiles.** This term was apparently coined by Schmidt-Nielsen in 1902 for microorganisms that can grow at 0°C (11). The existence of such organisms in foods, and especially in soils, was first noted and described by Forster in 1887. Eddy (4) suggested that the term psychrophile should be used only when a low optimum growth temperature is implied. In this regard, psychrophiles would be those organisms that display maximum growth temperatures below 35°C. For organisms able to grow at 5°C or less, Eddy suggested that they be called **psychrotrophs.** While more and more researchers have adopted the latter terminology, others continue to designate as psychrophile any organism capable of growth at or around 5°C.

Many of the bacteria listed in Table 9-3 for meat, poultry, and seafood products are psychrotrophic and are common to a large variety of other food products. Most psychrotrophic bacteria of importance in foods belong to the genus *Pseudomonas,* and to a lesser extent to the genera *Acinetobacter, Alcaligenes, Flavobacterium,* and others (5, 11, 21). Among the molds and yeasts, species and strains from a large number of genera are capable of growth at refrigerator temperatures. Among the low-temperature-growing molds are the genera *Penicillium, Mucor, Cladosporium, Botrytis,* and *Geotrichum.* Among yeasts, species and strains of *Debaryomyces, Torulopsis, Candida, Rhodotorula,* and others are known to be psychrotrophic.

There are at least two distinct low temperature ranges in which foods may be stored for preservation. **Chilling temperatures** are those between the usual refrigerator temperatures and room temperature, usually about 10°–15°C. These temperatures are suitable for the storage of certain vegetables and fruits, such as cucumbers, potatoes, limes, and so on. **Refrigerator temperatures** are those between 0°–2° and 5°–7°C and are suitable for the storage of a large number of perishable and semiperishable foods (see Table 13-1).

Table 13-1. Recommended storage temperatures, relative humidity, and approximate storage life of various fresh, dried, and processed foods.

Products	°F storage temp.	Percent relative humidity	Approx. storage life
Vegetables			
Artichokes, globe	31–32	90–95	1–2 weeks
Asparagus	32	90–95	3–4 weeks
Beans (green or snap)	45	85–90	8–10 days
Beans (lima)	32–40	85–90	10–15 days
Beets (bunch)	32	90–95	10–14 days
Brussels sprouts	32	90–95	3–4 weeks
Cabbage, late	32	90–95	3–4 months
Carrots (bunch)	32	90–95	10–14 days
Cauliflower	32	85–90	2–3 weeks
Celery	32	90–95	2–4 months
Corn, sweet	31–32	85–90	4–8 days
Cucumbers	45–50	90–95	10–14 days

Table 13-1, *continued.*

Products	°F storage temp.	Percent relative humidity	Approx. storage life
Endive	32	90–95	2–3 weeks
Lettuce	32	90–95	3–4 weeks
Onions	32	70–75	6–8 months
Peas, green	32	85–90	1–2 weeks
Potatoes, early crop	50–55	85–90	—
Potatoes, sweet	55–60	90–95	4–6 months
Radishes (spring, bunch)	32	90–95	10 days
Rhubarb	32	90–95	2–3 weeks
Rutabagas	32	90–95	2–4 months
Spinach	32	90–95	10–14 days
Squash, summer	32–40	85–95	10–14 days
Tomatoes, ripe	32	85–90	7 days
Dairy Products			
Butter	32–36	80–85	2 months
Cheese, process American, process Swiss	40–45	75	—
Eggs, shell	29–31	85–90	8–9 months
Eggs (dried, yolk)	35	minimum	6–12 months
Eggs (spray-dried albumin)	35	minimum	6 months
Meat, Poultry, and Fish			
Beef, fresh	32–34	88–92	1–6 weeks
Hams and shoulders, fresh	32–34	85–90	7–12 days
Hams and shoulders, cured	60–65	50–60	0–3 years
Poultry, fresh	32	—	1 week
Fish, fresh	33–40	90–95	5–20 days
Fruits			
Apples	30–32	85–90	2–7 months
Berries, blue	31–32	85–90	3–6 weeks
Berries, cranberries	36–40	85–90	1–3 months
Dried fruits	32	50–60	9–12 months
Figs (fresh)	28–32	85–90	5–7 days
Grapefruit	32–50	85–90	4–8 weeks
Grapes (American type)	31–32	85–90	3–8 weeks
Lemons	32, 55–58	85–90	1–4 months
Limes	48–50	85–90	6–8 weeks
Melons (honeydew, honeyball)	45–50	85–90	2–4 weeks
Melons, watermelons	36–40	85–90	2–3 weeks
Oranges	32–34	85–90	8–12 weeks
Peaches	31–32	85–90	2–4 weeks
Strawberries, fresh	31–32	85–90	7–10 days

Temperatures below 6°C will prevent the growth of most food-poisoning bacteria. Those demonstrated to grow at or below 6°C are *C. botulinum* type E and nonproteolytic B and F strains (see Chapter 19); strains of *V. parahaemolyticus* and *Y. enterocolitica* (see Chapter 21); and some *A. hydrophila* strains associated with gastroenteritis (see Chapter 22). While not a food-poisoning bacterium in the traditional sense, *Listeria monocytogenes* grows well at 4°C (see Chapter 22). The growth of all food spoilage organisms is effectively retarded below 6°C, and the mechanisms of this action are discussed in Chapter 23.

Freezer temperatures are those at or below −18°C. Under normal circumstances, these temperatures are sufficient to prevent the growth of all microorganisms, but as will be seen later, some can and do grow within the freezer range, but at an extremely slow rate.

TEMPERATURE GROWTH MINIMA OF FOOD-BORNE MICROORGANISMS

In their excellent review of the literature on the temperature growth minima of microorganisms, Michener and Elliott (15) summarized the findings of various authors who had reported the growth of microorganisms at and below −10°C. Of the thirteen organisms reported, there were six bacteria, four yeasts, and three molds. The yeasts grew at lower temperatures than the others, with one pink yeast reported to grow at −34°C and two others at −18°C. The lowest recorded temperature of growth for a bacterium is −20°C, and several have been reported to grow at about −12°C. The molds grew at about −12°C. Some foods that tend to support microbial growth at subfreezing temperatures are fruit juice concentrates, bacon, ice cream, and certain fruits.

The reported temperature minima for fecal indicator organisms vary between −2 to 10°C (15). In general, the enterococci grow better at refrigerator temperatures than do coliform bacteria. The lowest reported temperatures for growth of staphylococci and salmonellae in foods is 6.7°C (1). As noted above, *C. botulinum* type E has been reported to grow at a temperature lower than all other food-poisoning microorganisms; Schmidt et al. (17) reported 3.3°C for this organism in beef stew. Minimum temperature of growth and toxin production reported by many investigators for types A and B strains of *C. botulinum* is 10°C. The reader is referred to Chapter 17 for further discussion of the temperature growth minima of fecal indicator and food-poisoning microorganisms.

PREPARATION OF FOODS FOR FREEZING

The preparation of vegetables for freezing includes selecting, sorting, washing, blanching, and packaging prior to actual freezing. In selecting foods to be frozen, those in any state of detectable spoilage should be rejected. Meats, poultry, seafoods, eggs, and other foods should be as fresh as possible.

Blanching is achieved either by brief immersion of foods into hot water or the use of steam. Its primary functions are as follows: (1) inactivation of enzymes that might cause undesirable changes during freezing storage, (2) enhancement or fixing of the green color of certain vegetables, (3) reduction in the numbers of microorganisms on the foods, (4) facilitating the packing of leafy vegetables by inducing wilting, and (5) the displacement of entrapped air in the plant tissues. The method of blanching employed depends upon the products in question, size of packs, and other related information. When water is used, it is important that bacterial spores not be allowed to build up sufficiently to contaminate foods. With respect to the reduction in numbers of microorganisms by blanching, reductions of initial microbial loads as high as 99% have been claimed. It should be remembered that most vegetative bacterial cells can be destroyed at milk pasteurization temperatures (145°F for 30 min). This is especially true of most bacteria of importance in the spoilage of vegetables. While it is not the primary function of blanching to destroy microorganisms, the amount of heat necessary to effect destruction of most food enzymes is also sufficient to reduce vegetative cells significantly.

FREEZING OF FOODS AND FREEZING EFFECTS

The two basic ways to achieve the freezing of foods are quick and slow freezing. **Quick** or **fast freezing** is the process by which the temperature of foods is lowered to about −20°C within 30 min. This treatment may be achieved by direct immersion or indirect contact of foods with the refrigerant, and the use of air-blasts of frigid air blown across the foods being frozen.

Slow freezing refers to the process whereby the desired temperature is achieved within 3 to 72 h. This is essentially the type of freezing utilized in the home freezer.

Quick freezing possesses more advantages than slow freezing from the standpoint of overall product quality. The two methods are compared below (2):

Quick Freezing	*Slow Freezing*
1. Small ice crystals formed.	1. Large ice crystals formed.
2. Blocks or suppresses metabolism.	2. Breakdown of metabolic rapport.
3. Brief exposure to concentration of adverse constituents.	3. Longer exposure to adverse or injurious factors.
4. No adaptation to low temperatures.	4. Gradual adaptation.
5. Thermal shock (too brutal a transition).	5. No shock effect.
6. No protective effect.	6. Accumulation of concentrated solutes with beneficial effects.
7. Microorganisms frozen into crystals?	
8. Avoid internal metabolic imbalance.	

With respect to crystal formation upon freezing, slow freezing favors large extracellular crystals, while quick freezing favors the formation of small, intracellular ice crystals. Crystal growth is one of the factors that limits the freezer life of certain foods, since ice crystals grow in size and cause cell damage by disrupting membranes, cell walls, and internal structures to the point where the thawed product is quite unlike the original in texture and flavor. Upon thawing, foods frozen by the slow freezing method tend to lose more **drip** (drip for meats; **leakage** in the case of vegetables) than quick-frozen foods held for comparable periods of time. The overall advantages of small crystal formation to frozen food quality may be viewed also from the standpoint of what takes place when a food is frozen. During the freezing of foods, water is removed from solution and transformed into ice crystals of a variable but high degree of purity (7). In addition, the freezing of foods is accompanied by changes in properties such as pH, titratable acidity, ionic strength, viscosity, osmotic pressure, vapor pressure, freezing point, surface and interfacial tension, and O/R potential (see below). Some of the many complexities of this process are discussed by Fennema et al. (8).

STORAGE STABILITY OF FROZEN FOODS

A large number of microorganisms have been reported by many investigators to grow at and below 0°C as previously mentioned. In addition to factors inherent within these organisms, their growth at and below freezing temperatures is dependent upon several factors of foods, namely, nutrient content, pH, and the availability of liquid water. The a_w of foods may be expected to decrease as temperatures fall below the freezing point. The relationship between temperature and a_w of water and ice is presented in Table 13-2. For water at 0°C, a_w is 1.0 but falls to about 0.8 at -20°C and to 0.62 at about -50°C. Organisms that grow at subfreezing temperatures, then, must be able to grow at the reduced a_w levels, unless a_w is favorably

Table 13-2. Vapor pressures of water and ice at various temperatures (18).

°C	Liquid water mm. Hg	Ice mm. Hg	$a_w = \dfrac{p_{ice}}{p_{water}}$
0	4.579	4.579	1.00
-5	3.163	3.013	0.953
-10	2.149	1.950	0.907
-15	1.436	1.241	0.864
-20	0.943	0.776	0.823
-25	0.607	0.476	0.784
-30	0.383	0.286	0.75
-40	0.142	0.097	0.68
-50	0.048	0.030	0.62

affected by food constituents with respect to microbial growth. In fruit juice concentrates, which contain comparatively high levels of sugars, these compounds tend to maintain a_w at levels higher than would be expected in pure water, thereby making microbial growth possible even at subfreezing temperatures. The same type of effect can be achieved by the addition of glycerol to culture media. It should be borne in mind also that not all foods freeze at the same initial point (see Fig. 13-1). The initial freezing point of a given food is due in large part to the nature of its solute constituents and the relative concentration of those that have freezing-point depressing properties.

Although the metabolic activities of all microorganisms can be stopped at freezer temperatures, frozen foods may not be kept indefinitely if the thawed product is to retain the original flavor and texture. Most frozen foods are assigned a freezer life, and suggested maximum holding periods for various foods are presented in Table 13-3. The suggested maximum holding time for frozen foods is not based upon the microbiology of such foods but upon such factors as texture, flavor, tenderness, color, and overall nutritional quality upon thawing and subsequent cooking.

Some foods that are improperly wrapped during freezer storage undergo **freezer burn.** This condition is characterized by a browning of light-colored

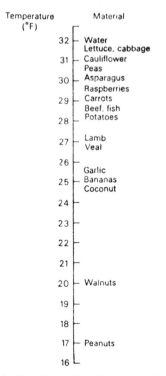

Fig. 13-1. Freezing point of selected foods (3).

Table 13-3. Storage life of frozen foods—average life in "good condition" for products processed and packaged in a normal manner (20).

Product	Months at			
	−10°F	0°F	+5°F	+10°F
Apricots[a]	24	18–24	—	6–8
Asparagus	16–18	8–12	—	4–6
Beans, green	16–18	8–12	—	4–6
Beans, lima	24	14–16	—	6–8
Broccoli	24	14–16	—	6–8
Brussels sprouts	16–18	8–12	—	4–6
Cauliflower	24	14–16	—	6–8
Corn, on cob	12–14	8–10	—	4–6
Corn, cut	36	24	—	12
Carrots	36	24	—	12
Fish, fatty	10–12	6–8	—	4
Fish, lean	14–16	10–12	—	6
Lobsters	10–12	8–10	—	3–4
Peaches[a]	24	18–24	—	6–8
Peas	24	14–16	—	6–8
Raspberries, sugared	24	18	—	8–10
Spinach	24	14–16	—	6–8
Strawberries, sliced	24	18	—	8–10
Beef				
roasts, steaks	—	12–14	8–10	—
cubed, small pieces	—	10–12	6–8	—
ground	—	8	4	—
Veal				
roasts, chops	—	10–12	6–8	—
thin cutlets, cubes	—	8–10	4–6	—
ground	—	6	3	—
Lamb				
roasts, chop	—	12–14	8–10	—
cubed	—	10–12	6	—
ground	—	8	4	—
Pork				
roasts, chops	—	6–12	3–5	—
ground, sausage	—	4	2	—
pork or ham, smoked	—	5–7	3	—
bacon	—	3	1	—
Lard				
unsalted	—	3–6	1–2	—
salted	—	1–3	1/2	—

Table 13-3, *continued.*

Product	Months at			
	−10°F	*0°F*	*+5°F*	*+10°F*
Variety meats				
beef or lamb liver, heart	—	4	2	—
veal liver, heart	—	3	1	—
pork liver, heart	—	2	1	—
tongue	—	4	2	—
kidneys	—	3	1	—
sweet breads	—	1	—	—
brains	—	1	—	—
oxtails	—	4	2	—
tripe	—	1	—	—
spiced sausage or delicatessen meats	—	2–3	1	—

[a] Contains ascorbic acid.
Courtesy of F. W. Williams Publications.

foods such as the skin of chicken meat. The browning results from the loss of moisture at the surface, leaving the product more porous than the original at the affected site. The condition is irreversible and is known to affect certain fruits, poultry, meats, and fish, both raw and cooked.

EFFECT OF FREEZING UPON MICROORGANISMS

In considering the effect of freezing upon those microorganisms that are unable to grow at freezing temperatures, it is well known that freezing is one means of preserving microbial cultures, with freeze-drying being perhaps the best method known. However, freezing temperatures have been shown to effect the killing of certain microorganisms of importance in foods. Ingram (12) has summarized the salient facts of what happens to certain microorganisms upon freezing:

1. There is a sudden mortality immediately on freezing, varying with species.
2. The proportion of cells surviving immediately after freezing is nearly independent of the rate of freezing.
3. The cells that are still viable immediately after freezing die gradually when stored in the frozen state.
4. This decline in numbers is relatively rapid at temperatures just below the freezing point, especially about −2°C, but less so at lower temperatures, and it is usually slow below −20°C.

Bacteria differ in their capacity to survive during freezing, with the cocci being generally more resistant than gram-negative rods. Of the food-poisoning bacteria, salmonellae are less resistant than *S. aureus* or vegetative cells of clostridia, while endospores and food-poisoning toxins are apparently unaffected by low temperatures (9). The effect of freezing several species of *Salmonella* to −25.5°C and holding up to 270 days is presented in Table 13-4. Although a significant reduction in viable numbers occurred over the 270-day storage period with most species, in no instance did all cells die off.

From the strict standpoint of food preservation, freezing should not be regarded as a means of destroying food-borne microorganisms. The type of organisms that lose their viability in this state differ from strain to strain and depend upon the type of freezing employed, the nature and composition of the food in question, the length of time of freezer storage, and other factors, such as temperature of freezing. Low freezing temperatures of about −20°C are less harmful to microorganisms than the median range of temperatures such as −10°C. For example, more microorganisms are destroyed at −4°C than at −15°C or below. Temperatures below −24°C seem to have no additional effect. Food constituents such as egg white, sucrose, corn syrup, fish, glycerol, and undenatured meat extracts have all been reported to increase freezing viability, especially of food-poisoning bacteria, while acid conditions have been reported to decrease cell viability (9).

To consider further the effects of freezing upon microorganisms, it would be of value to consider some of the events that are known to occur when cells freeze:

1. The water that freezes is the so-called free water. Upon freezing, the free water forms ice crystals. The growth of ice crystals occurs by accretion so that all of the free water of a cell might be represented by a relatively small number of ice crystals. In slow freezing, ice crystals are extracellular but in fast freezing they are intracellular. Bound water remains unfrozen. The freezing of cells depletes them of usable liquid water and thus dehydrates them.

Table 13-4. **The survival of pure cultures of enteric organisms in chicken chow mein at −25.5°C (10).**

Organism	Bacterial count (10^5/g) after storage for (days)								
	0	*2*	*5*	*9*	*14*	*28*	*50*	*92*	*270*
Salmonella newington	7.5	56.0	27.0	21.7	11.1	11.1	3.2	5.0	2.2
S typhi-murium	167.0	245.0	134.0	118.0	11.0	95.5	31.0	90.0	34.0
S. typhi	128.5	45.5	21.8	17.3	10.6	4.5	2.6	2.3	0.86
S. gallinarum	68.5	87.0	45.0	36.5	29.0	17.9	14.9	8.3	4.8
S. anatum	100.0	79.0	55.0	52.5	33.5	29.4	22.6	16.2	4.2
S. paratyphi B	23.0	205.0	118.0	93.0	92.0	42.8	24.3	38.8	19.0

1948 copyright © by Institute of Food Technologists.

2. Freezing results in an increase in the viscosity of cellular matter, a direct consequence of water being concentrated in the form of ice crystals.

3. Freezing results in a loss of cytoplasmic gases such as O_2 and CO_2. A loss of O_2 to aerobic cells suppresses respiratory reactions. Also, the more diffuse state of O_2 may make for greater oxidative activities within the cell.

4. Freezing causes changes in pH of cellular matter. Various authors have reported changes ranging from 0.3–2.0 pH units. Increases and decreases of pH upon freezing and thawing have been reported.

5. Freezing effects a concentration of cellular electrolytes. This effect is also a consequence of the concentration of water in the form of ice crystals.

6. Freezing causes a general alteration of the colloidal state of cellular protoplasm. It should be recalled that many of the constituents of cellular protoplasm such as proteins exist in a dynamic colloidal state in living cells. A proper amount of water is necessary to the well-being of this state.

7. Freezing causes some denaturation of cellular proteins. Precisely how this effect is achieved is not clear, but it is known that some —SH groups disappear upon freezing. It is also known that such groups as lipoproteins break apart from others upon freezing. The lowered water content along with the concentration of electrolytes no doubt affect this change in state of cellular proteins.

8. Freezing induces temperature shock in some microorganisms. This is true more for thermophiles and mesophiles than psychrophiles. It has been shown that more cells die when the temperature decline above freezing is sudden than when it is slow.

9. Freezing causes metabolic injury to some microbial cells such as certain *Pseudomonas* spp. It has been reported that some bacteria have increased nutritional requirements upon thawing from the frozen state and that as much as 40% of a culture may be affected in this way (see Chapter 5 for other effects of freeze injury).

The above should serve to illustrate the complex effects of the freezing process upon living cells such as bacteria and other microorganisms as well as upon foods. According to Mazur (14), the response of microorganisms to subzero temperatures appears to be largely determined by solute concentration and intracellular freezing, although there are only a few cases of clear demonstration of this conclusion.

Why are some bacteria killed by freezing but yet not all cells? Some small and microscopic organisms are unable to survive freezing as can most bacteria. Examples of these include the foot-and-mouth disease virus and the causative agent of trichinosis (*Trichinella spiralis*). Protozoa are generally killed when frozen below −5 or −10°C, if protective compounds are not present (14).

Effect of Thawing

Of great importance in the freezing survival of microorganisms is the process of thawing. It is well established that repeated freezing and thawing will destroy bacteria by disrupting cell membranes. It is also known that the faster the thaw, the greater the number of bacterial survivors. Just why this is so is not entirely clear. From the changes listed above that occur during freezing, it can be seen that the thawing process becomes complicated if it is to lead to the restoration of viable activity. Fennema (6) has pointed out that thawing is inherently slower than freezing and follows a pattern that is potentially more detrimental. Among the problems attendant on the thawing of specimens and products that transmit heat energy primarily by conduction are those summarized below (8):

1. Thawing is inherently slower than freezing when conducted under comparable temperature differentials.
2. In practice, the maximum temperature differential permissible during thawing is much less than that which is feasible during freezing.
3. The time-temperature pattern characteristic of thawing is potentially more detrimental than that of freezing. During thawing, the temperature rises rapidly to near the melting point and remains there throughout the long course of thawing, thus affording considerable opportunity for chemical reactions, recrystallization, and even microbial growth if thawing is extremely slow.

It has been stated that microorganisms do not die upon freezing per se, but rather, during the thawing process. Whether or not this is the case remains to be proven. As to why some organisms are able to survive freezing while others are not, Luyet (13) has suggested that it is a question of the ability of an organism to survive dehydration and to undergo dehydration when the medium freezes. Luyet has further stated that the small size of bacterial cells permits them to undergo dehydration upon freezing. With respect to survival after freeze-drying, this author has stated that it might be due to the fact that bacteria do not freeze at all but merely dry. (See Chapters 5 and 15 for further discussion of the effect of freeze-drying upon microorganisms.)

Most frozen foods processors advise against the refreezing of foods once they have thawed. While the reasons for this are more related to the texture, flavor, and other nutritional qualities of the frozen product, the microbiology of thawed frozen foods is pertinent. Some investigators have pointed out that foods thawed from the frozen state spoil faster than similar fresh products. There are textural changes associated with freezing that would seem to aid the invasion of surface organisms into deeper parts of the product and consequently facilitate the spoilage process. Upon thawing, surface condensation of water is known to occur. There is also, at the surface, a general concentration of water-soluble substances such as amino acids, minerals, B vitamins, and possibly other nutrients. Freezing has the effect of destroying many thermophilic and some mesophilic organisms, making for less competition among the survivors upon thawing. It is conceivable that a greater relative number of psychrotrophs on thawed foods might increase the spoilage rate. Some psychrotrophic bacteria have been reported to have Q_{10} values in excess of 4.0 at refrigerator temperatures. For example *P. fragi* has been reported to possess a Q_{10} of 4.3 at 0°C. Organisms of this type are capable of doubling their growth rate with only a 4–5 degree rise in temperature. Whether or not frozen thawed foods do in fact spoil faster than fresh foods would depend upon a large number of factors, such as the type of freezing, the relative numbers and types of organisms on the product prior to freezing, and the temperature at which the product is held to thaw. Although there are no known toxic effects associated with the refreezing of frozen and thawed foods, this act should be minimized in the interest of the overall nutritional quality of the products. One effect of freezing and thawing animal tissues is the release of lysosomal enzymes consisting of cathepsins, nucleases, phosphatases, glycosidases, and others (19). Once released, these enzymes may act to degrade macromolecules and thus make available simpler compounds that are more readily utilized by the spoilage flora.

More information on the microbiology of frozen foods can be obtained from Robinson (16).

REFERENCES

1. Angelotti, R., M. J. Foter, and K. H. Lewis. 1961. Time-temperature effects on salmonellae and staphylococci in foods. *Amer. J. Pub. Hlth* 51:76–88.
2. Borgstrom, G. 1961. Unsolved problems in frozen food microbiology. In *Proceedings, Low temperature microbiology symposium*, 197–251. Camden, N.J.: Campbell Soup Co.
3 Desrosier, N. W. 1963. *The technology of food preservation*. Ch. 4. Westport, Conn.: AVI.
4. Eddy, B. P. 1960. The use and meaning of the term "psychrophilic." *J. Appl. Bacteriol.* 23:189–90.
5. Farrell, J., and A. Rose. 1967. Temperature effects on microorganisms. *Ann. Rev. Microbiol.* 21:101–20.
6. Fennema, O. 1966. An over-all view of low temperature food preservation. *Cryobiol.* 3:197–213.
7. Fennema, O., and W. Powrie. 1964. Fundamentals of low-temperature food preservation. *Adv. Food Res.* 13:219–347.
8. Fennema, O. R., W. D. Powrie, and E. H. Marth. 1973. *Low-temperature preservation of foods and living matter*. New York: Marcel Dekker.
9. Georgala, D. L., and A. Hurst. 1963. The survival of food poisoning bacteria in frozen foods. *J. Appl. Bacteriol.* 26:346–58.
10. Gunderson, M. F., and K. D. Rose. 1948. Survival of bacteria in a precooked fresh-frozen food. *Food Res.* 13:254–63.

11. Ingraham, J. L., and J. L. Stokes. 1959. Psychrophilic bacteria. *Bacteriol. Revs.* 23:97–108.
12. Ingram, M. 1951. The effect of cold on microorganisms in relation to food. *Proc. Soc. Appl. Bacteriol.* 14:243.
13. Luyet, B. 1962. Recent developments in cryobiology and their significance in the study of freezing and freeze-drying of bacteria. In *Proceedings, Low temperature microbiology symposium,* 63–87. Camden, N.J.: Campbell Soup Co.
14. Mazur, P. 1966. Physical and chemical basis of injury in single-celled microorganisms subjected to freezing and thawing. In *Cryobiology,* ed. H. T. Meryman, Ch. 6. New York: Academic Press.
15. Michener, H., and R. Elliott. 1964. Minimum growth temperatures for food-poisoning, fecal-indicator, and psychrophilic microorganisms. *Adv. Food Res.* 13:349–96.
16. Robinson, R. K. 1985. *Microbiology of frozen foods.* New York: Elsevier.
17. Schmidt, C. F., R. V. Lechowich, and J. F. Folinazzo. 1961. Growth and toxin production by Type E *Clostridium botulinum* below 40°F. *J. Food Sci.* 26: 626–30.
18. Scott, W. J. 1962. Available water and microbial growth. In *Proceedings, Low temperature microbiology symposium,* 89–105. Camden, N.J.: Campbell Soup Co.
19. Tappel, A. L. 1966. Effects of low temperature and freezing on enzymes and enzyme systems. In *Cryobiology,* ed. H. T. Meryman, Ch. 4. New York: Academic Press.
20. Tressler, D. K. 1960. Storage life of selected frozen foods. Recommended maximum storage periods for meat. *Frozen Food Almanac, Quick Frozen Foods* 23:148.
21. Witter, L. D. 1961. Psychrophilic bacteria—A review. *J. Dairy Sci.* 44:983–1015.

14.

FOOD PRESERVATION WITH
HIGH TEMPERATURES

The use of high temperatures to preserve food is based on their destructive effects on microorganisms. By high temperatures are meant any and all temperatures above ambient. With respect to food preservation, there are two temperature catagories in common use: pasteurization and sterilization. **Pasteurization** by use of heat implies either the destruction of all disease-producing organisms (for example, pasteurization of milk) or the destruction or reduction in number of spoilage organisms in certain foods, as in the pasteurization of vinegar. The pasteurization of milk is achieved by heating at 145°F for 30 min, or at 161°F for 15 sec (high temperature short time—HTST method). These treatments are sufficient to destroy the most heat-resistant of the nonsporeforming pathogenic organisms—*Mycobacterium tuberculosis* and *Coxiella burnetti*. Milk pasteurization temperatures are sufficient to destroy, in addition, all yeasts, molds, gram-negative bacteria, and many gram positives. The two groups of organisms that survive milk pasteurization are placed into one of two groups: thermodurics and thermophiles. **Thermoduric** organisms are those that can survive heat treatment at relatively high temperatures but do not necessarily grow at these temperatures. The nonsporeforming organisms that survive milk pasteurization generally belong to the genera *Streptococcus* and *Lactobacillus,* and sometimes to other genera. **Thermophilic** organisms are those that not only survive relatively high temperatures but *require* high temperatures for their growth and metabolic activities. The genera *Bacillus* and *Clostridium* contain the thermophiles of greatest importance in foods.

    Sterilization means the destruction of all viable organisms as may be measured by an appropriate plating or enumerating technique. Canned foods are sometimes called "commercially sterile" to signify that no viable organisms can be detected by the usual cultural methods employed, or that the number of survivors is so low as to be of no significance under the conditions of canning and storage. Also, microorganisms may be present in canned foods that cannot grow in the product by reason of undesirable pH, Eh, or temperature of storage.

A more recent development in the processing of milk and milk products is the use of **ultra-high temperatures** (UHT). Milk so produced is a product in its own right and is to be distinguished from pasteurized milk. The primary features of the UHT treatment include: (1) its continuous nature, (2) its occurrence outside of the package necessitating aseptic storage and aseptic handling of the product downstream from the sterilizer, and (3) the very high temperatures (in the range of 140°–150°C) and the correspondingly short time (a few seconds) necessary to achieve commercial sterility (16). UHT-processed milks have higher consumer acceptability than the conventionally heated pasteurized products, and since they are commercially sterile they may be stored at room temperatures for up to 8 weeks without flavor changes.

FACTORS THAT AFFECT HEAT RESISTANCE IN MICROORGANISMS

It is well known that equal numbers of bacteria placed in physiologic saline and nutrient broth at the same pH are not destroyed with the same ease by heat. Some eleven factors or parameters of microorganisms and their environment have been studied for their effects on heat destruction and are presented below (13).

WATER. The heat resistance of microbial cells increases with decreasing humidity or moisture. Dried microbial cells placed into test tubes and then heated in a water bath are considerably more heat resistant than moist cells of the same type. Since it is well established that protein denaturation occurs at a faster rate when heated in water than in air, it is suggested that protein denaturation is either the mechanism of death by heat or is closely associated with it (see Chapter 24 for further discussion of the mechanism of heat death). The precise manner in which water facilitates heat denaturation of proteins is not entirely clear, but it has been pointed out that the heating of wet proteins causes the formation of free —SH groups with a consequent increase in the water-binding capacity of proteins. The presence of water allows for thermal breaking of peptide bonds, a process that requires more energy in the absence of water and consequently confers a greater refractivity to heat.

FAT. In the presence of fats, there is a general increase in the heat resistance of some microorganisms (see Table 14-1). This is sometimes referred to as

Table 14-1. The effect of the medium upon the thermal death point of *Escherichia coli*[a] (5).

Medium	Thermal death point (°C)
Cream	73
Whole milk	69
Skim milk	65
Whey	63
Bouillon (broth)	61

[a] Heating time: 10 min.
Courtesy of W. B. Saunders Co., Philadelphia.

fat protection and is presumed to increase heat resistance by directly affecting cell moisture. Sugiyama (27) demonstrated the heat-protective effect of long-chain fatty acids on *C. botulinum*. It appears that the long-chain fatty acids are better protectors than short-chain acids.

SALTS. The effect of salt on the heat resistance of microorganisms is variable and dependent upon the kind of salt, concentration employed, and other factors. It has been observed that some salts have a protective effect upon microorganisms while others tend to make cells more heat sensitive. It has been suggested that some salts may decrease water activity and thereby increase heat resistance by a mechanism similar to that of drying, while others may increase water activity (for example, Ca^{2+} and Mg^{2+}) and consequently increase sensitivity to heat. It has been shown that supplementation of the growth medium of *B. megaterium* spores with $CaCl_2$ yields spores with increased heat resistance, while the addition of L-glutamate, L-proline, or increased phosphate content decreases heat resistance (19).

CARBOHYDRATES. The presence of sugars in the suspending menstrum causes an increase in the heat resistance of microorganisms suspended therein. This effect is at least in part due to the decrease in water activity that is caused by high concentrations of sugars. There is great variation, however, among sugars and alcohols relative to their effect on heat resistance as may be seen in Table 14-2 for D values of *Salmonella senftenberg* 775W. At identical a_w values obtained by use of glycerol and sucrose, wide differences in heat sensitivity occur (3, 12). Corry (6) found that sucrose increased the heat resistance of *S. senftenberg* more than any of four other carbohydrates tested. The following decreasing order was found for the five tested substances: sucrose > glucose > sorbitol > fructose > glycerol.

pH. Microorganisms are most resistant to heat at their optimum pH of growth, which is generally about 7.0. As pH is lowered or raised from this optimum value, there is a consequent increase in heat sensitivity (Fig. 14-1). Advantage is taken of this fact in the heat processing of high-acid foods,

Table 14-2. Reported D values of *Salmonella senftenberg* 775W as functions of various parameters of growth and other conditions.

Temp. (C)	D Values	Conditions	Ref.
61	1.1 min	Liquid whole egg	(24)
61	1.19 min	Tryptose broth	(24)
60	9.5 min[a]	Liquid whole egg, pH ca. 5.5	(2)
60	9.0 min[a]	Liquid whole egg, pH ca. 6.6	(2)
60	4.6 min[a]	Liquid whole egg, pH ca. 7.4	(2)
60	0.36 min[a]	Liquid whole egg, pH ca. 8.5	(2)
65.6	34–35.3 sec	Milk	(23)
71.7	1.2 sec	Milk	(23)
70	360–480 min	Milk chocolate	(11)
55	4.8 min	TSB[b], log phase, grown 35°C	(22)
55	12.5 min	TSB[b], log phase, grown 44°C	(22)
55	14.6 min	TSB[b], stationary, grown 35°C	(22)
55	42.0 min	TSB[b], stationary, grown 44°C	(22)
57.2	13.5 min[a]	a_w 0.99 (4.9% glyc.), pH 6.9	(12)
57.2	31.5 min[a]	a_w 0.90 (33.9% glyc.), pH 6.9	(12)
57.2	14.5 min[a]	a_w 0.99 (15.4% sucro.), pH 6.9	(12)
57.2	62.0 min[a]	a_w 0.90 (58.6% sucro.), pH 6.9	(12)
60	0.2–6.5 min[c]	HIB[d], pH 7.4	(3)
60	2.5 min	a_w 0.90, HIB, glycerol	(3)
60	75.2 min	a_w 0.90, HIB, sucrose	(3)
65	0.29 min	0.1 M phosphate buf., pH 6.5	(6)
65	0.8 min	30% sucrose	(6)
65	43.0 min	70% sucrose	(6)
65	2.0 min	30% glucose	(6)
65	17.0 min	70% glucose	(6)
65	0.95 min	30% glycerol	(6)
65	0.70 min	70% glycerol	(6)
55	35 min	a_w 0.997, tryptone soya agar, pH 7.2	(14)

[a] Mean/average values.
[b] Trypticase soy broth.
[c] Total of 76 cultures.
[d] Heart infusion broth.

where considerably less heat is applied to achieve sterilization compared to foods at or near neutrality. The heat pasteurization of egg white provides an example of an alkaline food product that is neutralized prior to heat treatment, a practice which is not done with other foods. The pH of egg white is about 9.0. When this product is subjected to pasteurization conditions of 60–62°C for 3.5–4 min, coagulation of proteins occurs along with a marked increase in viscosity. These changes affect the volume and texture of cakes made from such pasteurized egg white. Cunningham and Lineweaver (7) have reported that egg white may be pasteurized the same as whole egg if the pH is reduced to about 7.0. This reduction of pH makes both

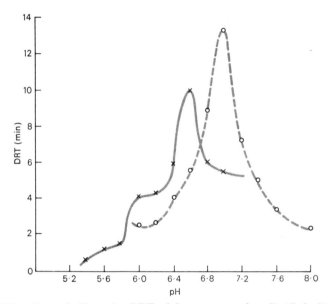

Fig. 14-1. The effect of pH on the DRT of *Streptococcus faecalis* (C & G) exposed to 60°C in citrate-phosphate buffer (crosses) and phosphate buffer (circles) solutions at various pH levels (29). DRT = decimal reduction time.

microorganisms and egg-white proteins more heat stable. The addition of salts of iron or aluminum increases the stability of the highly heat-labile egg protein conalbumin sufficiently to permit pasteurization at 60–62°C. Unlike their resistance to heat in other materials, bacteria are more resistant to heat in liquid whole egg at pH values of 5.4–5.6 than at values of 8.0–8.5 (Table 14-2). This is true when pH is lowered with an acid such as HCl. When organic acids such as acetic or lactic acid are used to lower pH, a decrease in heat resistance occurs.

PROTEINS AND OTHER SUBSTANCES. It is well known that proteins in the heating menstrum have a protective effect upon microorganisms. Consequently, high-protein-content foods must be heat processed to a greater degree than low-protein-content foods in order to achieve the same end results. For identical numbers of organisms, the presence of colloidal-size particles in the heating menstrum also offers protection against heat. For example, under identical conditions of pH, numbers of organisms, and so on, it would take longer to sterilize pea purée than nutrient broth.

NUMBERS OF ORGANISMS. The larger the number of organisms, the higher is the degree of heat resistance. This is well illustrated in Table 14-3. It has been assumed that the mechanism of heat protection by large microbial populations was due to the production of protective substances excreted by the cells, and some authors claim to have demonstrated the existence

Table 14-3.   Effect of the number of spores of *Clostridium botulinum* on the thermal death time at 100°C (5).

Number of Spores	Thermal Death Time (Minutes)
72,000,000,000	240
1,640,000,000	125
32,000,000	110
650,000	85
16,400	50
328	40

Courtesy of W. B. Saunders Co., Philadelphia.

of such substances. Since proteins are known to offer some protection against heat, many of the extracellular compounds in a culture would be expected to be protein in nature and consequently capable of affording some protection. Of perhaps equal importance in the higher heat resistance of large cell populations over smaller ones is the greater chance for the presence of organisms with differing degrees of natural heat resistance.

AGE OF ORGANISMS.   Bacterial cells tend to be most resistant to heat while in the stationary phase of growth (old cells) and less resistant during the logarithmic phase. This is true for *S. senftenberg* (see Table 14-2), whose stationary phase cells may be several times more resistant than log phase cells (22). Heat resistance has been reported to be high also at the beginning of the lag phase but decreases to a minimum as the cells enter the log phase. Old bacterial spores have been reported to be more heat resistant than young spores. The mechanism of increased heat resistance of less active microbial cells is undoubtedly complex and at this time is not well understood.

GROWTH TEMPERATURE.   The heat resistance of microorganisms tends to increase as the temperature of incubation increases. Lechowich and Ordal (16) showed that as sporulation temperature was increased for *B. subtilis* and *B. coagulans,* the thermal resistance of spores of both organisms also increased. Although the precise mechanism of this effect is unclear, it is conceivable that genetic selection favors the growth of the more heat-resistant strains at succeedingly high temperatures. *S. senftenberg* grown at 44°C has been found to be approximately three times more resistant than cultures grown at 35°C (Table 14-2).

INHIBITORY COMPOUNDS.   As might be expected, a decrease in heat resistance of most microorganisms occurs when heating takes place in the presence of heat-resistant antibiotics, SO_2, and other microbial inhibitors. The use of heat + antibiotics and heat + nitrite together has been found to be more effective in controlling the spoilage of certain foods than either alone. The practical effect of adding inhibitors to foods prior to heat treatment is to reduce the amount of heat that would be necessary if used alone (see Chapter 11).

TIME AND TEMPERATURE. One would expect that the longer the time of heating, the greater the killing effect of heat. All too often, though, there are exceptions to this basic rule. A more dependable rule is that the higher the temperature, the greater the killing effect of heat. This is illustrated in Table 14-4 for bacterial spores. As temperature increases, time necessary to achieve the same effect decreases.

Table 14-4. **Effect of temperature upon the thermal death times of spores (5).**

Temperature	Clostridium botulinum (60,000,000,000 spores suspended in buffer at pH 7)	A thermophile (150,000 spores per ml. of corn juice at pH 6.1)
	Minutes	
100°C	260	1140
105°C	120	
110°C	36	180
115°C	12	60
120°C	5	17

Courtesy of W. B. Saunders Co., Philadelphia.

These rules assume that heating effects are immediate and not mechanically obstructed or hindered. Also important is the size of the heating vessel or container and its composition (glass, metal, plastic). It should be obvious that it would take longer to effect pasteurization or sterilization in large containers than in smaller ones. The same would be true of containers with walls that do not conduct heat as readily as others.

RELATIVE HEAT RESISTANCE OF MICROORGANISMS

In general, the heat resistance of microorganisms is related to their optimum growth temperatures. Psychrophilic microorganisms are the most heat sensitive of the three temperature groups, followed by mesophiles and thermophiles. Sporeforming bacteria are more heat resistant than nonspore-formers, while thermophilic sporeformers are in general more heat resistant than mesophilic sporeformers. With respect to gram reaction, gram-positive bacteria tend to be more heat resistant than gram negative, with cocci in general being more resistant than nonsporeforming rods. Yeasts and molds tend to be fairly sensitive to heat, with yeast ascospores being only slightly more resistant than vegetative yeasts. The asexual spores of molds tend to be slightly more heat resistant than mold mycelia. Sclerotia are the most heat resistant of these types and sometimes survive and cause trouble in canned fruits.

The heat resistance of bacterial endospores is of special interest in the thermal processing of foods. These structures are produced by *Bacillus* and *Clostridium* spp. usually upon the exhaustion of nutrients essential for continued vegetative growth, although other factors appear to be involved. Only one spore is produced per cell, and it may occur in various parts of the vegetative cell and possess various shapes and sizes, all of which are of taxonomic value. The endospore is not only resistant to heat but to drying, cold, chemicals, and other adverse environmental factors. It is a highly refractive body that resists staining by ordinary methods. The refractivity is due in part to the spore coats, which consist of at least two layers—outer (exine) and inner (intine). Heat resistance is due also in part to the dehydrated nature of the cortex and spore core. Endospores are known to contain DNA, RNA, water, various enzymes, metal ions, and other compounds, especially dipicolinic acid (DPA). Although the precise mechanism of heat resistance of endospores is not yet fully understood, numerous authors have related this resistance to spore DNA and calcium content. DPA may constitute 5–15% of the dry weight of endospores, while these bodies may contain two to ten times more calcium than the corresponding vegetative cell. There is a general increase in heat resistance as the ratio of cations to DPA increases (17, 20). The addition of chelating agents with high affinities toward calcium and manganese has been shown to decrease the heat resistance of endospores (1), while the addition of calcium and manganese generally restores thermal resistance (9). The thermal death of endospores is accompanied by a release of DPA, divalent cations, and ninhydrin-positive material into the medium (8).

THERMAL DESTRUCTION OF MICROORGANISMS

In order to better understand the thermal destruction of microorganisms relative to food preservation and canning, it is necessary to understand certain basic concepts associated with this technology. Below are listed some of the more important concepts, but for a more extensive treatment of thermobacteriology, the excellent monograph by Stumbo (25) should be consulted.

THERMAL DEATH TIME (TDT). This is the time necessary to kill a given number of organisms at a specified temperature. By this method, the temperature is kept constant and the time necessary to kill all cells is determined. Of less importance is the **thermal death point**, which is the temperature necessary to kill a given number of microorganisms in a fixed time, usually 10 min. Various means have been proposed for determining TDT: the tube, can, "tank," flask, thermo-resistometer, unsealed tube, and capillary tube methods. The general procedure for determining TDT by these methods is to place a known number of cells or spores in a sufficient number of sealed containers in order to get the desired number of survivors for each test period. The organisms are then placed in an oil bath and heated for the required time period. At the end of the heating period, containers are removed and cooled quickly in cold water. The organisms are then placed on a suitable growth medium, or the entire heated containers are incubated if the organisms are suspended in a suitable growth substrate. The suspensions or containers are incubated at a temperature suitable for growth of the specific organisms. Death is defined as the inability of the organisms to form a visible colony.

D VALUE. This is the decimal reduction time, or the time required to destroy 90% of the organisms. This value is numerically equal to the number of min required for the survivor curve to traverse one log cycle (see Fig. 14-2). Mathematically, it is equal to the reciprocal of the slope of the survivor curve and is a measure of the death rate of an organism. When D is determined at 250°F, it is often expressed as D_r. The effect of pH on the D value of *C. botulinum* in various foods is presented in Table 14-5,

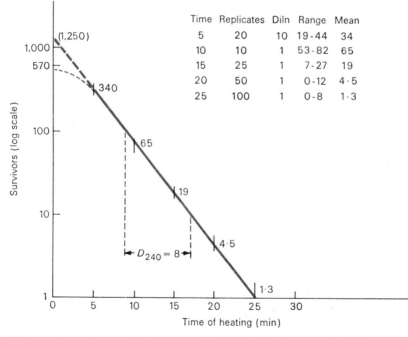

Fig. 14-2. Rate of destruction curve. Spores of strain F.S.7 heated at 240°F in canned pea brine pH 6.2. (10, courtesy of Butterworths Publishers, London)

Table 14-5. Effect of pH on D values for spores of *C. botulinum* 62A suspended in three food products at 240°F (30).

	D Value (in min.)		
pH	Spaghetti, tomato sauce, and cheese	Macaroni creole	Spanish rice
4.0	0.128	0.127	0.117
4.2	0.143	0.148	0.124
4.4	0.163	0.170	0.149
4.6	0.223	0.223	0.210
4.8	0.226	0.261	0.256
5.0	0.260	0.306	0.266
6.0	0.491	0.535	0.469
7.0	0.515	0.568	0.550

1965 copyright © by Institute of Food Technologists.

and D values for *S. senftenberg* 775W under various conditions are presented in Table 14-2. D values of 0.20 to 2.20 min at 150°F have been reported for *S. aureus* strains, D 150°F of 0.50–0.60 min for *Coxiella burnetti*, and D 150°F of 0.20–0.30 for *Mycobacterium hominis* (25). For pH-elevating strains of *Bacillus licheniformis* spores in tomatoes, a D 95°C of 5.1 min has been reported, while for *B. coagulans* a D 95°C of 13.7 min has been found (21).

z VALUE. The z value refers to the degrees Fahrenheit required for the thermal destruction curve to traverse one log cycle. Mathematically, this value is equal to the reciprocal of the slope of the TDT curve (see Fig. 14-3). While D reflects the resistance of an organism to a specific temperature, z provides information on the relative resistance of an organism to different destructive temperatures; it allows for the calculation of equivalent thermal processes at different temperatures. If, for example, 3.5 min at 140°F is considered an adequate process and $z = 8.0$, either 0.35 min at 148°F or 35 min at 132°F would be considered equivalent processes.

F VALUE. This value is the equivalent time, in minutes at 250°F, of all heat considered, with respect to its capacity to destroy spores or vegetative cells of a particular organism. The integrated lethal value of heat received by all points in a container during processing is designated F_s or F_o. This represents a measure of the capacity of a heat process to reduce the number of spores or vegetative cells of a given organism per container. When we assume instant heating and cooling throughout the container of spores, vegetative cells, or food, F_o may be derived as follows:

$$F_o = D_r(\log a - \log b)$$

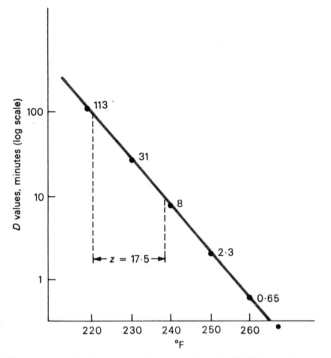

Fig. 14-3. Thermal death time curve. Spores of strain F.S.7 heated in canned pea brine pH 6.2. (10, courtesy of Butterworths Publishers, London)

where a = number of cells in the initial population, and b = number of cells in the final population.

THERMAL DEATH TIME CURVE. For the purpose of illustrating a thermal destruction curve and D value, data are employed from Gillespy (10) on the killing of flat sour spores at 240°F in canned pea brine at pH 6.2. Counts were determined at intervals of 5 min with the mean viable numbers indicated below:

Time (min.)	Mean viable count
5	340.0
10	65.0
15	19.0
20	4.5
25	1.3

The time of heating in minutes is plotted on semi-log paper along the linear axis, and the number of survivors is plotted along the log scale to produce the TDT curve presented in Fig. 14-2. The curve is essentially linear, indicating that the destruction of bacteria by heat is logarithmic and obeys a first order reaction. Although difficulty is encountered at times at either end of the TDT curve, process calculations in the canning industry are based upon a logarithmic order of death. From the data presented in Fig. 14-2, the D value is calculated to be 8 min, or $D_{240} = 8.0$.

D values may be used to reflect the relative resistance of spores or vegetative cells to heat. The most heat-resistant strains of *C. botulinum* types A and B spores have a D_r value of 0.21, while the most heat-resistant thermophilic spores have D_r values of around 4.0–5.0. Putrefactive anaerobe (P.A.) 3679 was found by Stumbo et al. (26) to have a D_r value of 2.47 in cream-style corn, while flat sour (F.S.) spores 617 were found to have a D_r of 0.84 in whole milk.

The approximate heat resistance of spores of thermophilic and mesophilic spoilage organisms may be compared by use of D_r values as below.

B. stearothermophilus:	4.0 –5.0
C. thermosaccharolyticum:	3.0 –4.0
C. nigrificans:	2.0 –3.0
C. botulinum (types A and B):	0.10–0.2
C. sporogenes (including P.A. 3679):	0.10–1.5
B. coagulans:	0.01–0.07

The effect of pH and suspending menstrum on D values of *C. botulinum* spores are presented in Table 14-5. As noted above, microorganisms are more resistant at and around neutrality and show different degrees of heat resistance in different foods.

In order to determine the z value, D values are plotted on the log scale and degrees F are plotted along the linear axis. From the data presented in Fig. 14-3, the z value is seen to be 17.5. Values of z for *C. botulinum* range from 14.7 to 16.3, while for P.A. 3679 the range of 16.6–20.5 has been reported. Some spores have been reported to have z values as high as 22. Peroxidase has been reported to have a z value of 47, while 50 has been reported for riboflavin and 56 for thiamine.

12-D CONCEPT. The 12-D concept refers to the process lethality requirement long in effect in the canning industry and implies that the minimum heat process should reduce the probability of survival of the most resistant *C. botulinum* spores to 10^{-12}. Since *C. botulinum* spores do not germinate and produce toxin below pH 4.6, this concept is observed only for foods above this pH value. An example from Stumbo (25) illustrates this concept from the standpoint of canning technology. If it is assumed that each container of food contains only one spore of *C. botulinum*, F_o may be calculated by use of the general survivor curve equation with the other assumptions noted above in mind:

$$F_o = D_r(\log a - \log b)$$
$$F_o = 0.21(\log 1 - \log 10^{-12})$$
$$F_o = 0.21 \times 12 = 2.52$$

Processing for 2.52 min at 250°F, then, should reduce the *C. botulinum* spores to one spore in one of one million million containers (10^{12}). When it is considered that some flat-sour spores have D_r values of about 4.0 and some canned foods receive F_o treatments of 6.0–8.0, the potential number of *C. botulinum* spores is reduced even more.

ASEPTIC PACKAGING

In traditional canning methods, nonsterile food is placed in nonsterile metal or glass containers followed by container closure and sterilization. In aseptic packaging, sterile food under aseptic conditions is placed in sterile containers, and the packages are sealed under aseptic conditions as well. While the methodology of aseptic packaging was patented in the early 1960s, the technology was little used until 1981 when the Food and Drug Administration approved the use of hydrogen peroxide for the sterilization of flexible multilayered packaging materials used in aseptic processing systems (28).

In general, any food that can be pumped through a heat exchanger can be aseptically packaged. The widest application has been to liquids such as fruit juices, and a wide variety of single-serve products of this type has resulted. The technology for foods that contain particulates has been more difficult to develop, with microbiological considerations only one of the many problems to overcome. In determining the sterilization process for foods pumped through heat exchangers, the fastest-moving components (those with the minimum holding time) are used, and where liquids and particulates are mixed, the latter will be the slower moving. Heat penetration rates are not similar for liquids and solids, making it more difficult to establish minimum process requirements that will effectively destroy both organisms and food enzymes.

Some of the advantages of aseptic packaging are these: (1) products such as fruit juices are more flavorful and lack the metallic taste of those processed in metal containers; (2) flexible multilayered cartons can be used instead of glass or metal containers; (3) the time a product is subjected to high temperatures is minimized when ultrahigh temperatures are used; (4) the technology allows the use of membrane filtration of certain liquids; and (5) various container headspace gases such as nitrogen may be used. Among the disadvantages are: the packages may not be equivalent to glass or metal containers in preventing the permeation of oxygen, and the output is lower than that for solid containers.

A wide variety of aseptic packaging techniques now exists, with more under development (4). Sterilization of packages is achieved in various ways, one of which involves the continuous feeding of rolls of packaging material into a machine where hot hydrogen peroxide is used to effect sterilization, followed by the forming, filling with food, and sealing of the containers (15). Sterility of the filling operation may be maintained by a positive pressure of air or gas such as nitrogen. Aseptically packaged fruit juices are shelf stable at ambient temperatures for 6 to 12 months or longer.

The spoilage of aseptically packaged foods may be expected to differ from foods in metal containers. While hydrogen swells occur in high-acid foods in the latter containers, aseptic packaging materials are nonmetallic. Seam leakage may be expected to be absent in aseptically packaged foods, but the permeation of oxygen by the nonmetal and nonglass containers may allow for other types of spoilage in low-acid foods.

REFERENCES

1. Amaha, M., and Z. J. Ordal. 1957. Effect of divalent cations in the sporulation medium on the thermal death rate of *Bacillus coagulans* var. *thermoacidurans*. *J. Bacteriol*. 74:596–604.
2. Anellis, A., J. Lubas, and M. M. Rayman. 1954. Heat resistance in liquid eggs of some strains of the genus *Salmonella*. *Food Res*. 19:377–95.
3. Baird-Parker, A. C., M. Boothroyd, and E. Jones. 1970. The effect of water activity on the heat resistance of heat sensitive and heat resistant strains of salmonellae. *J. Appl. Bacteriol*. 33:515–22.
4. Carlson, V. R. 1984. Current aseptic packaging techniques. *Food Technol*. 38(12):47–50.
5. Carpenter, P. L. 1967. *Microbiology*, 2d ed. Philadelphia: W. B. Saunders.
6. Corry, J. E. L. 1974. The effect of sugars and polyols on the heat resistance of salmonellae. *J. Appl. Bacteriol*. 37:31–43.
7. Cunningham, F. E., and H. Lineweaver. 1965. Stabilization of egg-white proteins to pasteurizing temperatures above 60°C. *Food Technol*. 19:1442–47.
8. El-Bisi, H. M., R. V. Lechowich, M. Amaha, and Z. J. Ordal. 1962. Chemical events during death of bacterial endospores by moist heat. *J. Food Sci*. 27:219–31.
9. El-Bisi, H. M., and Z. J. Ordal. 1956. The effect of certain sporulation conditions on the thermal death rate of *Bacillus coagulans* var. *thermoacidurans*. *J. Bacteriol*. 71:1–9.
10. Gillespy, T. G. 1962. The principles of heat sterilization. *Recent Adv. in Food Sci*. 2:93–105.
11. Goepfert, J. M., and R. A. Biggie. 1968. Heat resistance of *Salmonella typhimurium* and *Salmonella senftenberg* 775W in milk chocolate. *Appl. Microbiol*. 16:1939–40.
12. Goepfert, J. M., I. K. Iskander, and C. H. Amundson. 1970. Relation of the heat resistance of salmonellae to the water activity of the environment. *Appl. Microbiol*. 19:429–33.
13. Hansen, N. H., and H. Riemann. 1963. Factors affecting the heat resistance of nonsporing organisms. *J. Appl. Bacteriol*. 26:314–33.
14. Horner, K. J., and G. D. Anagnostopoulos. 1975. Effect of water activity on heat survival of *Staphylococcus aureus*, *Salmonella typhimurium* and *Salm. senftenberg*. *J. Appl. Bacteriol*. 38:9–17.

15. Ito, K. A., and K. E. Stevenson. 1984. Sterilization of packaging materials using aseptic systems. *Food Technol.* 38(3):60–62.
16. Jelen, P. 1982. Experience with direct and indirect UHT processing of milk— A Canadian viewpoint. *J. Food Protect.* 45:878–83.
17. Lechowich, R. V., and Z. J. Ordal. 1960. The influence of sporulation temperature on the thermal resistance and chemical composition of endospores. *Bacteriol. Proc.,* p. 44.
18. ———. 1962. The influence of the sporulation temperature on the heat resistance and chemical composition of bacterial spores. *Can. J. Microbiol.* 8:287–95.
19. Levinson, H. S., and M. T. Hyatt. 1964. Effect of sporulation medium on heat resistance, chemical composition, and germination of *Bacillus megaterium* spores. *J. Bacteriol.* 87:876–86.
20. Levinson, H. S., M. T. Hyatt, and F. E. Moore. 1961. Dependence of the heat resistance of bacterial spores on the calcium: dipicolinic acid ratio. *Biochem. Biophys. Res. Commun.* 5:417–19.
21. Montville, T. J., and G. M. Sapers. 1981. Thermal resistance of spores from pH elevating strains of *Bacillus licheniformis. J. Food Sci.* 46:1710–12.
22. Ng, H., H. G. Bayne, and J. A. Garibaldi. 1969. Heat resistance of *Salmonella:* The uniqueness of *Salmonella senftenberg* 775W. *Appl. Microbiol.* 17:78–82.
23. Read, R. B., Jr., J. G. Bradshaw, R. W. Dickerson, Jr., and J. T. Peeler. 1968. Thermal resistance of salmonellae isolated from dry milk. *Appl. Microbiol.* 16:998–1001.
24. Solowey, M., R. R. Sutton, and E. J. Calesnick. 1948. Heat resistance of *Salmonella* organisms isolated from spray-dried whole-egg powder. *Food Technol.* 2:9–14.
25. Stumbo, C. R. 1973. *Thermobacteriology in food processing,* 2d ed. New York: Academic Press.
26. Stumbo, C. R., J. R. Murphy, and J. Cochran. 1950. Nature of thermal death time curves for P.A. 3679 and *Clostridium botulinum. Food Technol.* 4:321–26.
27. Sugiyama, H. 1951. Studies on factors affecting the heat resistance of spores of *Clostridium botulinum. J. Bacteriol.* 62:81–96.
28. Tillotson, J. E. 1984. Aseptic packaging of fruit juices. *Food Technol.* 38(3):63–66.
29. White, H. R. 1963. The effect of variation in pH on the heat resistance of cultures of *Streptococcus faecalis. J. Appl. Bacteriol.* 26:91–99.
30. Xezones, H., and I. J. Hutchings. 1965. Thermal resistance of *Clostridium botulinum* (62A) spores as affected by fundamental food constituents. *Food Technol.* 19:1003–5.

15.

PRESERVATION OF FOODS BY DRYING

The preservation of foods by drying is based upon the fact that microorganisms and enzymes need water in order to be active. In preserving foods by this method, one seeks to lower the moisture content of foods to a point where the activities of food spoilage and food-poisoning microorganisms are inhibited. Dried, desiccated, or low moisture (LM) foods are those that generally do not contain more than 25% moisture and have an a_w between 0.00 and 0.60. These are the traditional dried foods. Freeze-dried foods are also in this category. Another category of shelf-stable foods are those that contain between 15 and 50% moisture and an a_w between 0.60 and 0.85. These are the intermediate moisture (IM) foods. Some of the microbiological aspects of IM and LM foods are dealt with in this chapter.

PREPARATION AND DRYING OF LOW-MOISTURE FOODS

The earliest uses of food desiccation consisted simply of exposing fresh foods to sunlight until drying had been achieved. Through this method of drying, which is referred to as sun drying, certain foods may be successfully preserved if the temperature and relative humidity (R.H.) allow. Fruits such as grapes, prunes, figs, and apricots may be dried by this method, which requires a large amount of space for large quantities of the product. The drying methods of greatest commercial importance consist of spray, drum, evaporation, and freeze drying.

Preparatory to drying, foods are handled in much the same manner as for freezing, with a few exceptions. In the drying of fruits such as prunes, alkali dipping is employed by immersing the fruits into hot lye solutions of between 0.1 and 1.5%. This is especially true when sun drying is employed. Light-colored fruits and certain vegetables are treated with SO_2 in order that levels of between 1,000 and 3,000 ppm may be absorbed. The latter treatment helps to maintain color, conserve certain vitamins, prevent storage changes, and reduce the microbial load. After drying, fruits are usually heat pasteurized at 150–185°F for 30–70 min.

Similar to the freezing preparation of vegetable foods, blanching or scalding is a vital step prior to dehydration. This may be achieved by immersion from 1 to 8 min, depending upon the particular type of product. The primary function of this step in drying is to destroy enzymes that may become active and bring about undesirable changes in the finished products. Leafy vegetables generally require less time than peas, beans, or carrots. For drying, temperatures of 140–145°F have been found to be safe for many vegetables. The moisture content of vegetables should be reduced below 4% in order to have satisfactory storage life and quality. Many vegetables may be made more stable if given a treatment with SO_2 or a sulfite. The drying of vegetables is usually achieved by use of tunnel, belt, or cabinet-type driers.

Meat is usually cooked before being dehydrated. The final moisture content after drying should be approximately 4% for beef and pork.

Milk is dried as either whole milk or nonfat skim milk. The dehydration may be accomplished by either the drum or spray method. The removal of about 60% water from whole milk results in the production of **evaporated** milk, which has about 11.5% lactose in solution. **Sweetened condensed** milk is produced by the addition of sucrose or glucose before evaporation so that the total average content of all sugar is about 54%, or over 64% in solution. The stability of sweetened condensed milk is due in part to the fact that the sugars tie up some of the water and make it unavailable for microbial growth.

Eggs may be dried as whole egg powder, yolks, or egg white. Dehydration stability is increased by reducing the glucose content prior to drying. Spray drying is the method most commonly employed.

In **freeze drying** (lyophilization, cryophilization), actual freezing is preceded by the blanching of vegetables and the precooking of meats. The rate at which a food material freezes or thaws is influenced by the following factors (11): (1) the temperature differential between the product and the cooling or heating medium, (2) the means of transferring heat energy to, from, and within the product (conduction, convection, radiation), (3) the type, size, and shape of the package, and (4) the size, shape, and thermal properties of the product. Rapid freezing has been shown to produce products that are more acceptable than slow freezing. As discussed in Chapter 12, rapid freezing allows for the formation of small ice crystals and consequently less mechanical damage to food structure. Upon thawing, fast-frozen foods take up more water and in general display characteristics more like the fresh product than slow-frozen foods. After freezing, the water in the form of ice is removed by sublimation. This process is achieved by various means of heating plus vacuum. The water content of protein foods can be placed into two groups: freezable and unfreezable. Unfreezable (bound) water has been defined as that which remains unfrozen below $-30°C$. The removal of freezable water takes place during the first phases of drying and this phase of drying may account for the removal of anywhere from 40 to 95% of the total moisture. The last water to be removed is generally bound water, some of which may be removed throughout the drying process. Unless heat treatment is given prior to freeze drying, freeze-dried foods retain their enzymes. In studies on freeze-dried meats, it has been shown that 40–80% of the enzyme activity is not destroyed and may be retained after 16-mon storage at $-20°C$ (24). The final product moisture level in freeze-dried foods may be about 2–8%, or have an a_w of 0.10–0.25 (38).

Freeze drying is generally preferred to high-temperature vacuum drying. Among the disadvantages of the latter compared to the former are the following (17):

1. pronounced shrinkage of solids
2. migration of dissolved constituents to the surface when drying solids
3. extensive denaturation of proteins
4. case-hardening: the formation of a relatively hard, impervious layer at the surface of a solid is caused by one or more of the first three changes; this impervious layer slows rates of both dehydration and reconstitution
5. formation of hard, impervious solids when drying liquid solution
6. undesirable chemical reactions in heat-sensitive materials
7. excessive loss of desirable volatile constituents
8. difficulty of rehydration as a result of one or more of the above changes

EFFECT OF DRYING UPON MICROORGANISMS

Although some microorganisms are destroyed in the process of drying, this process is not lethal per se to microorganisms, and indeed, many types may be recovered from dried foods, especially if poor-quality foods are used for drying and if proper practices are not followed in the drying steps.

As indicated in Chapter 3, bacteria require relatively high levels of moisture for their growth, with yeasts requiring less and molds still less. Since most bacteria require a_w values above 0.90 for growth, these organisms play no role in the spoilage of dried foods. With respect to the stability of dried foods, Scott (32) has related a_w levels to the probability of spoilage in the following manner. At a_w values of between 0.80 and 0.85, spoilage occurs readily by a variety of fungi in from 1 to 2 weeks. At a_w values of 0.75, spoilage is delayed, with fewer types of organisms in those products that spoil. At a_w 0.70, spoilage is greatly delayed and may not occur during prolonged holding. At a_w of 0.65, very few organisms are known to grow, and spoilage is most unlikely to occur for even up to 2 years. Some authors have suggested that dried foods to be held for several years should be processed so that the final a_w is between 0.65 and 0.75, with 0.70 suggested by most.

At a_w levels of about 0.90, the organisms most likely to grow are yeasts and molds. This value is near the minimum for most normal yeasts. Even though spoilage is all but prevented at $a_w < 0.65$, some molds are known to grow very slowly at a_w 0.60–0.62 (37). Osmophilic yeasts such as *Saccharomyces rouxii* strains have been reported to grow at an a_w of 0.65 under certain conditions. The most troublesome group of microorganisms in dried foods are the molds, with the *Aspergillus glaucus* group being the most notorious at low a_w values. The minimum a_w values reported for the germination and growth of molds and yeasts are presented in Table 15-1. Pitt and Christian (28) found the predominant spoilage molds of dried and high-moisture prunes to be members of the *A. glaucus* group and *Xeromyces bisporus*. Aleuriospores of *X. bisporus* were able to germinate in 120 days at an a_w of 0.605. Generally higher moisture levels were required for both asexual and sexual sporulation.

As a guide to the storage stability of dried foods, the **"alarm water"** content has been suggested. The "alarm water" content is the water content which should not be exceeded if mold growth is to be avoided. While these values may be used to advantage, they should be followed with caution, as a rise of only 1% may be disastrous in some instances (32). The "alarm water" content for some miscellaneous foods is presented in Table 15-2. In freeze-dried foods, the rule of thumb has been to reduce the moisture

Table 15-1. Minimum a_w reported for the germination and growth of various food spoilage yeasts and molds.[a]

Organism	Minimum a_w
Candida utilis	0.94
Botrytis cinerea	0.93
Rhizopus stolonifer (nigricans)	0.93
Mucor spinosus	0.93
Candida scottii	0.92
Trichosporon pullulans	0.91
Candida zeylanoides	0.90
Saccharomyces rouxii	0.90
Endomyces vernalis	0.89
Alternaria citri	0.84
Aspergillus glaucus	0.70
Aspergillus echinulatus	0.64
Saccharomyces rouxii (Z. barkeri)	0.62

[a]See Table 3-4, Chapter 3, for other organisms.

Table 15-2. The "alarm water" content for miscellaneous foods, assuming R.H. of 70% and a temperature of 20°C (27).

Foods	% water
Whole milk powder	ca. 8
Dehydrated whole eggs	10–11
Wheat flour	13–15
Rice	13–15
Milk powder (separated)	15
Fat-free dehydrated meat	15
Pulses	15
Dehydrated vegetables	14–20
Starch	18
Dehydrated fruit	18–25

level to 2%. Burke and Decareau (7) have pointed out that this low level is probably too severe for some foods that might keep well at higher levels of moisture without the extra expense of removing the last low levels of water.

Although drying destroys some microorganisms, bacterial endospores survive as do yeasts, molds, and many gram-negative and -positive bacteria. In their study of bacteria from chicken meat after freeze drying and rehydration at room temperature, May and Kelly (25) were able to recover about 32% of the original flora. These workers showed that *S. aureus* added prior to freeze drying could survive under certain conditions. Some or all food-borne parasites, such as *Trichinella spiralis*, have been reported to survive the drying process (10). The present goal is to produce dried foods with a total count of not more than 100,000/g. It is generally agreed that the coliform count of dried foods should be zero or nearly so, and no food-poisoning organisms should be allowed, with the possible exception of low numbers of *C. perfringens* and *S. faecalis*. With the exception of those that may be destroyed by blanching or precooking, relatively fewer organisms are destroyed during the freeze-drying process. More are destroyed during freezing than during dehydration. During freezing, between 5 and 10% of water remains "bound" to other constituents of the medium. This water is removed by drying. Death or injury from drying may result from: (1) denaturation in the still frozen, undried portions due to concentration resulting from freezing, (2) the act of removing the "bound" water, and/or (3) recrystallization of salts or hydrates formed from eutectic solutions (26). When death occurs during dehydration, the rate is highest during the early stages of drying. Young cultures have been reported to be more sensitive to drying than old cultures (12). The freeze-drying method is, of course, one of the best known ways of preserving microorganisms. Once the process has been completed, the cells may remain viable indefinitely. Upon examining the viability of 277 cultures of bacteria, yeasts, and molds that had been lyophilized for 21 years, Davis (9) found that only 3 failed to survive.

STORAGE STABILITY OF DRIED FOODS

In the absence of fungal growth, desiccated foods are subject to certain chemical changes that may result in the food becoming undesirable upon holding. In dried foods that contain fats and oxygen, oxidative rancidity is a common form of chemical spoilage. Foods that contain reducing sugars undergo a color change known as the **Maillard** reaction or nonenzymic browning. This process is brought about when the carbonyl groups of reducing sugars react with amino groups of proteins and amino acids, followed by a series of other more complicated reactions. Maillard-type browning is quite undesirable in fruits and vegetables not only because of the unnatural color but also because of the bitter taste imparted to susceptible foods.

Freeze-dried foods also undergo browning if the moisture content is above 2%. Thus, the moisture content should be held below 2% (13).

With regard to a_w, the maximal browning reaction rates in fruits and vegetable products occur in the 0.65–0.75 range, while for nonfat dry milk browning, it seems to occur most readily at about 0.70 (38).

Other chemical changes that take place in dried foods include a loss of vitamin C in vegetables, general discolorations, structural changes leading to the inability of the dried product to fully rehydrate, and toughness in the rehydrated, cooked product.

Conditions that favor one or more of the above changes in dried foods generally tend to favor all, so that preventative measures against one are also effective against others to varying degrees. At least four methods of minimizing chemical changes in dried foods have been offered: (1) Keep the moisture content as low as possible. Gooding (14) has pointed out that lowering the moisture content of cabbage from 5 to 3% doubles its storage life at 37°C. (2) Reduce the level of reducing sugars as low as possible. These compounds, of course, are directly involved in nonenzymic browning and their reduction has been shown to increase storage stability. (3) When blanching, use water in which the level of leached soluble solids is kept low. Gooding (14) has shown that the serial blanching of vegetables in the same water increases the chances of browning. The explanation given is that the various extracted solutes (presumably reducing sugars and amino acids) are impregnated on the surface of the treated products at relatively high levels. (4) Use of sulfur dioxide. The treatment of vegetables prior to dehydration with this gas protects vitamin C along with retarding the browning reaction. The precise mechanism of this gas in retarding the browning reaction is not well understood, but it apparently does not block reducing groups of hexoses. It has been suggested that it may act as a free radical acceptor.

One of the most important considerations in preventing fungal spoilage of dried foods is the R.H. of the storage environment. If improperly packed and stored under conditions of high R.H., dried foods will pick up moisture from the atmosphere until some degree of equilibrium has been established. Since the first part of the dried product to gain moisture is the surface, spoilage would be inevitable, as surface growth tends to be characteristic of molds due to their oxygen requirements.

INTERMEDIATE-MOISTURE FOODS (IMF)

As noted earlier in this chapter, intermediate-moisture foods (IMF) are characterized by a moisture content of around 15 to 50% and an a_w between 0.60 and 0.85. These foods are shelf stable at ambient temperatures for varying periods of time. While impetus was given to this class of foods during the early 1960s with the development and marketing of intermediate-moisture dog food, foods for human consumption that meet the basic criteria of this class have been produced for many years. These are referred to as "traditional IMFs" to distinguish them from the newer IMFs. In Table 15-3 are listed some traditional IMFs along with their a_w values. All of these foods have, of course, lowered a_w values, which are achieved by withdrawal of water by desorption, adsorption, and/or the addition of permissible additives such as salts and sugars. The newly developed IMFs are characterized not only by a_w values of 0.60–0.85 but by the use of additives such as glycerol, glycols, sorbitol, sucrose, and so forth, as humectants, and by their content of fungistats such as sorbate and benzoate. The remainder of this chapter is devoted to the newly developed IMFs.

The preparation of IMF

Since *S. aureus* is the only bacterium of public health importance that can grow at a_w values near 0.86, an IMF can be prepared by: (1) formulating the product so that its moisture content is between 15 and 50 percent, (2) adjusting the a_w to a value below 0.86 by use of humectants, and (3) adding an antifungal agent to inhibit the rather large number of yeasts and molds that are known to be capable of growth at a_w values above 0.70. Additional storage stability is achieved by reducing pH. While this is essentially all that one needs to produce an IMF, the actual process and the achievement of storage stability of the product are considerably more complicated.

The determination of the a_w of a food system is discussed in Chapter 3. One can use also Raoult's law of mole fractions where the number of moles

Table 15-3. Some traditional intermediate moisture foods (a_w values taken from the literature).

Food products	a_w range
Dried fruits	0.60–0.75
Cake and pastry	0.60–0.90
Frozen foods	0.60–0.90
Sugars, syrups	0.60–0.75
Some candies	0.60–0.65
Commercial pastry fillings	0.65–0.71
Cereals (some)	0.65–0.75
Fruit cake	0.73–0.83
Honey	0.75
Fruit juice concentrates	0.79–0.84
Jams	0.80–0.91
Sweetened condensed milk	0.83
Fermented sausages (some)	0.83–0.87
Maple syrup	0.90
Ripened cheeses (some)	0.96
Liverwurst	0.96

of water in a solution is divided by the total number of moles in the solution (3):

$$a_w = \frac{\text{Moles of } H_2O}{\text{Moles of } H_2O + \text{Moles of solute}}$$

For example, a liter of water contains 55.5 moles. Assuming that the water is pure,

$$a_w = \frac{55.5}{55.5 + 0} = 1.00$$

If, however, one mole of sucrose is added,

$$a_w = \frac{55.5}{55.5 + 1} = 0.98$$

This equation can be rearranged to solve for the number of moles of solute required to give a specified a_w value. While the foregoing is not incorrect, it is highly oversimplified, since food systems are complex by virtue of their content of ingredients that interact with water and with each other in ways that are difficult to predict. Sucrose, for example, decreases a_w more than expected by the above so that calculations based upon Raoult's law may be meaningless (4). The development of techniques and methods to predict a_w in IMF more accurately has been the concern of several investigators (3, 16, 31), while an extensive evaluation of available a_w-measuring instruments and techniques has been carried out by Labuza et al. (22).

In preparing IMF, water may be removed either by adsorption or desorption. By adsorption, food is first dried (often freeze dried), and then subjected to controlled rehumidification until the desired composition is achieved. By desorption, the food is placed in a solution of higher osmotic pressure so that at equilibrium, the desired a_w is reached (30). While identical a_w values may be achieved by these two methods, IMF produced by adsorption is more inhibitory to microorganisms than that produced by desorption (see below). When sorption isotherms of food materials are determined, adsorption isotherms sometimes reveal that less water is held than for desorption isotherms at the same a_w. The sorption isotherm of a food material is a plot of the amount of water adsorbed as a function of the relative humidity or activity of the vapor space surrounding the material. It is the amount of water that is held after equilibrium has been reached at a constant temperature (21). Sorption isotherms may be either adsorption or desorption, and when the former procedure results in the holding of more water than the latter, the difference is ascribed to an hysteresis effect. This, as well as other physical properties associated with the preparation of IMF, has been discussed by Labuza (21), Sloan et al. (34), and others, and will not be dealt with further here. The sorption properties of an IMF recipe, the interaction of each ingredient with water and with other ingredients, and the order of mixing of ingredients all add to the complications of the overall IMF preparation procedures, and both direct and indirect effects on the microbiology of these products may result.

The general techniques employed to change the water activity in producing an IMF are summarized below (20):

1. Moist infusion. Solid food pieces are soaked and/or cooked in an appropriate solution to give the final product the desired water level (desorption).
2. Dry infusion. Solid food pieces are first dehydrated, following which they are infused by soaking in a solution containing the desired osmotic agents (adsorption).
3. Component blending. All IMF components are weighed, blended, cooked, and extruded or otherwise combined to give the finished product the desired a_w
4. Osmotic drying. Foods are dehydrated by immersion in liquids with a water activity lower than that of the food. When salts and sugars are used, two simultaneous countercurrent flows develop: solute diffuses from solution into food while water diffuses out of food into solution.

The foods in Table 15-4 were prepared by moist infusion for military use. The 1-cm-thick slices equilibrated following cooking at 95°–100°C in water and holding overnight in a refrigerator. Equilibration is possible without cooking over prolonged periods under refrigeration (6). IMF deep-fried catfish, with raw samples of about 2 g each, has been prepared by the moist infusion method (8). Pet foods are more often prepared by component blending. The general composition of one such product is given in Table 15-5. The general way in which a product of this type is made is as follows. The meat and meat products are ground and mixed with liquid ingredients. The resulting slurry is cooked or heat treated and later mixed with the dry ingredient mix (salts, sugars, dry solids, and so on). Once the latter are mixed into the slurry, an additional cook or heat process may be applied prior to extrusion and packaging. The extruded material may be further shaped in the form of patties or packaged in loose form. The composition of a model IMF product called "Hennican" is given in Table 15-6. According to Acott and Labuza (1), this is an adaptation of pemmican, an Indian trail and winter storage food made of buffalo meat and berries. Hennican is the name given to the chicken-based IMF. Both moisture content and a_w of this system can be altered by adjustment of ingredient mix.

The humectants commonly used in pet food formulations are propylene glycol, polyhydric alcohols (sorbitol, for example), polyethylene glycols, glycerol, sugars (sucrose, fructose, lactose, glucose, and corn syrup), and

Table 15-4. Preparation of representative intermediate moisture foods by equilibration (6). The solutions in which the products were equilibrated are given in the six columns on the right side of the table.

Initial material	% H₂O	Processing	Equilibrated product % H₂O	a_w	Ratio: Initial wt. / Solution wt.	% Components of solution Glycerol	Water	NaCl	Sucrose	K-sorbate	Na-benzoate
Tuna, canned water pack pieces, 1 cm thick	60.0	cold soak	38.8	0.81	0.59	53.6	38.6	7.1	—	0.7	—
Carrots diced 0.9 cm, cooked	88.2	cook 95–98°C, refrig.	51.5	0.81	0.48	59.2	34.7	5.5	—	0.6	—
Macaroni, elbow cooked, drained	63.0	cook 95–98°C, refrig.	46.1	0.83	0.43	42.7	48.8	8.0	—	0.5	—
Pork loin, raw 1 cm thick	70.0	cook 95–98°C, refrig.	42.5	0.81	0.73	45.6	43.2	10.5	—	0.7	—
Pineapple canned, chunks	73.0	cold soak	43.0	0.85	0.46	55.0	21.5	—	23.0	0.5	—
Celery 0.6 cm cross cut, blanch	94.7	cold soak	39.6	0.83	0.52	68.4	25.2	5.9	—	0.5	—
Beef, ribeye 1 cm thick	70.8	cook 95–98°C, refrig.	—	0.86	2.35	87.9	—	10.1	—	—	2.0

Table 15-5. Typical composition of soft moist or intermediate moisture dog food (19).

Ingredient	%
Meat by-products	32.0
Soy flakes	33.0
Sugar	22.0
Skimmed milk, dry	2.5
Calcium and phosphorus	3.3
Propylene glycol	2.0
Sorbitol	2.0
Animal fat	1.0
Emulsifier	1.0
Salt	0.6
Potassium sorbate	0.3
Minerals, vitamins and color	0.3
	100.0%

Copyright © 1970, Institute of Food Technologists.

Table 15-6. Composition of IM food: Hennican[a] (1).

Components	Amount (wt basis, %)
Raisins	30
Water	23
Peanuts	15
Chicken (freeze dried)	15
Non-fat dry milk	11
Peanut butter	4
Honey	2

[a]Moisture content = 41 g water/100 g solids, a_w = 0.85.
Copyright © 1975, Institute of Food Technologists.

salts (NaCl, KCl, and so on). The commonly used mycostats are propylene glycol, K-sorbate, Na-benzoate, and others. The pH of these products may be as low as 5.4 and as high as 7.0.

Microbial aspects of IMF

The general a_w range of IMF products makes it unlikely that gram-negative bacteria will proliferate. This is true also for most gram-positive bacteria with the exception of cocci, some sporeformers, and lactobacilli. In addition to the inhibitory effect of lowered a_w, antimicrobial activity results from an interaction of pH, Eh, added preservatives (including some of the humectants), the competitive microflora, generally low storage temperatures, and the pasteurization or other heat processes applied during processing.

The fate of *S. aureus* S-6 in IM pork cubes with glycerol at 25°C is illustrated in Fig. 15-1. In this desorption IM pork at a_w 0.88, the numbers remained stationary for about 15 days and then increased slightly, while in the adsorption IM system at the same a_w the cells died off slowly during the first three weeks and thereafter more rapidly. At all a_w values below 0.88, the organisms died off, with the death rate considerably higher at 0.73 than at higher values (29). Findings similar to these have been reported by Haas et al. (15), who found that an inoculum of 10^5 staphylococci in a meat-sugar system at a_w 0.80 decreased to 3×10^3 after 6 days and to 3×10^2 after one month. Although growth of *S. aureus* has been reported to occur at an a_w of 0.83, enterotoxin is not produced below a_w 0.86 (35). It appears that enterotoxin A is produced at lower values of a_w than enterotoxin B (36).

Using the model IM Hennican at pH 5.6 and a_w 0.91, Boylan et al. (5) showed that the effectiveness of the IM system against *S. aureus* F265 was

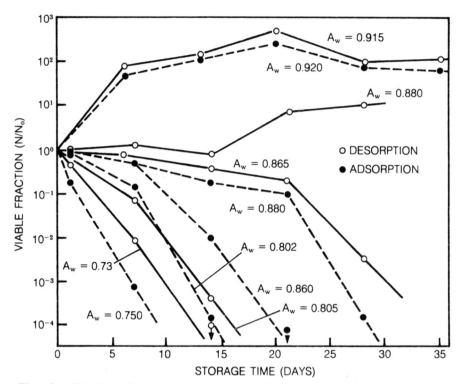

Fig. 15-1. Viability of *Staphylococcus aureus* in IMF systems: pork cubes and glyc-erol at 25°C. (29, copyright © 1973, Institute of Food Technologists)

a function of both pH and a_w. As noted above, adsorption systems are more destructive to microorganisms than desorption systems. Labuza et al. (23) found that the reported minimum a_ws apply in IMF systems when desorption systems are involved but that growth minima are much higher if the food is prepared by an adsorption method. *S. aureus* was inhibited at a_w 0.9 in adsorption while values between 0.75 and 0.84 were required for desorption systems. A similar effect was noted for molds, yeasts, and pseudomonads.

In regard to the effect of IMF systems on the heat destruction of bacteria, it is noted in Chapter 14 that heat resistance increases as a_w is lowered and that the degree of resistance is dependent upon the compounds employed to control a_w (see Table 14-2). In a study of the death rate of salmonellae and staphylococci in the IM range of about 0.8 at pasteurization temperatures (50–65°C), it has been found that cell death occurs under first order kinetics (18). These investigators confirmed the findings of many others that the heat destruction of vegetative cells is at a minimum in the IM range, especially when a solid menstrum is employed. Some D values for the thermal destruction of *Salmonella senftenberg* 775W at various a_w values are given in Table 14-2.

With respect to molds in IMF systems, these products would be made quite stable if a_w were reduced to around 0.70, but a dry-type product would then result. A large number of molds are capable of growth in the 0.80 range, and the shelf life of IM pet foods is generally limited by the growth of these organisms. The interaction of various IM parameters on the inhibition of molds was shown by Acott et al. (2). In their evaluation of seven chemical inhibitors used alone and in combination to inhibit *Aspergillus niger* and *A. glaucus* inocula, propylene glycol was the only approved agent that was effective alone. None of the agents tested could inhibit alone at a_w 0.88, but in combination the product was made shelf stable. All inhibitors were found to be more effective at pH 5.4 and a_w 0.85 than at pH 6.3. Growth of the two fungi occurred in 2 weeks in the a_w 0.85 formulation without inhibitors but did not occur until 25 weeks when K-sorbate and Ca-propionate were added (Table 15-7). Growth of *Staphylococcus epidermidis* was inhibited by both fungistats with the inhibition being greater at a_w 0.85 than at 0.88. This is probably an example of the combined effects of pH, a_w, and other growth parameters on the growth inhibition of microorganisms in IMF systems, as previously noted.

Storage stability of IMF

The undesirable chemical changes that occur in dried foods occur also in IM foods. Lipid oxidation and Maillard browning are at their optima in the general IMF ranges of a_w and percent moisture. A recent report, however, indicates that the maximum rate for Maillard browning occurs in the 0.4–0.5 a_w range, especially when glycerol is used as the humectant (38).

The storage of IMFs under the proper conditions of humidity is imperative in preventing moldiness and for overall shelf stability. The measurement of equilibrium relative humidity (ERH) is of importance in this regard. ERH

Table 15-7. Time for growth of microbes[a] in inoculated dog food with inhibitors, pH 5.4 (2).

	Storage conditions	
	$a_w = 0.85$	$a_w = 0.88$
Inhibitor	9-mon storage	6-mon storage
No inhibitor added	*A. niger*—2 wk *A. glaucus*—1 wk *S. epider.*—2 wk	*A. niger*—1 wk *A. glaucus*—1 wk *S. epider.*—½ wk
K-sorbate (0.3%)	No mold *S. epider.*—25 wk	*A. niger*—5 wk *S. epider.*—3½ wk
Ca-propionate (0.3%)	*A. niger*—25 wk *A. glaucus*—25 wk *S. epider.*—3½ wk	*A. glaucus*—2 wk *S. epider.*—1½ wk

[a]Mold—first visible sign; bacteria—2 log cycle increase.

is an expression of the desorbable water present in a food product, and is further defined by the following equation:

$$ERH = (P_{equ}/P_{sat})T, P = 1 \text{ atm}$$

where P_{equ} is partial pressure of water vapor in equilibrium with the sample in air at 1 atmosphere total pressure and temperature T; P_{sat} is the saturation partial vapor pressure of water in air at a total pressure of 1 atmosphere and temperature T (16). A food in moist air exchanges water until the equilibrium partial pressure at that temperature is equal to the partial pressure of water in the moist air, so that the ERH value is a direct measure of whether moisture will be sorbed or desorbed. In the case of foods packaged or wrapped in moisture-impermeable materials, the relative humidity of the food-enclosed atmosphere is determined by the ERH of the product, which in turn is controlled by the nature of the dissolved solids present, ratio of solids to moisture, and the like (33). Both traditional and newer IMF products have longer shelf stability under conditions of lower ERH.

In addition to the direct effect of packaging on ERH, gas impermeable packaging affects the Eh of packaged products with consequent inhibitory effects upon the growth of aerobic microorganisms.

REFERENCES

1. Acott, K. M., and T. P. Labuza. 1975. Inhibition of *Aspergillus niger* in an intermediate moisture food system. *J. Food Sci.* 40:137–39.
2. Acott, K. M., A. E. Sloan, and T. P. Labuza. 1976. Evaluation of antimicrobial agents in a microbial challenge study for an intermediate moisture dog food. *J. Food Sci.* 41:541–46.
3. Bone, D. 1973. Water activity in intermediate moisture foods. *Food Technol.* 27(4):71–76.
4. Bone, D. P. 1969. Water activity—Its chemistry and applications. *Food Prod. Dev.* 3(5):81–94.
5. Boylan, S. L., K. A. Acott, and T. P. Labuza. 1976. *Staphylococcus aureus* challenge study in an intermediate moisture food. *J. Food Sci.* 41:918–21.
6. Brockmann, M. C. 1970. Development of intermediate moisture foods for military use. *Food Technol.* 24:896–900.
7. Burke, R. F., and R. V. Decareau. 1964. Recent advances in the freeze-drying of food products. *Adv. Food Res.* 13:1–88.
8. Collins, J. L., and A. K. Yu. 1975. Stability and acceptance of intermediate moisture, deep-fried catfish. *J. Food Sci.* 40:858–63.
9. Davis, R. J. 1963. Viability and behavior of lyophilized cultures after storage for twenty-one years. *J. Bacteriol.* 85:486–87.
10. Desrosier, N. W. 1963. *The technology of food preservation.* Ch. 5. Westport, Conn.: AVI.
11. Fennema, O., and W. D. Powrie. 1964. Fundamentals of low-temperature food preservation. *Adv. Food Res.* 13:219–347.

12. Fry, R. M., and R. I. N. Greaves. 1951. The survival of bacteria during and after drying. *J. Hyg.* 49:220–46
13. Goldblith, S. A., and M. Karel. 1966. Stability of freeze-dried foods. In *Advances in freeze-drying,* ed. L. Rey, 191–210. Paris: Hermann.
14. Gooding, E. G. B. 1962. The storage behaviour of dehydrated foods. In *Recent advances in food science,* vol. 2, ed. J. Hawthorn and J. M. Leitch, 22–38. London: Butterworths.
15. Haas, G. J., D. Bennett, E. B. Herman, and D. Collette. 1975. Microbial stability of intermediate moisture foods. *Food Prod. Dev.* 9(4):86–94.
16. Hardman, T. M. 1976. Measurement of water activity. Critical appraisal of methods. In *Intermediate moisture foods,* ed. R. Davies, G. G. Birch, and K. J. Parker, 75–88. London: Applied Science.
17. Harper, J. C., and A. L. Tappel. 1957. Freeze-drying of food products. *Adv. Food Res.* 7:171–234.
18. Hsieh, F.-H., K. Acott, and T. P. Labuza. 1976. Death kinetics of pathogens in a pasta product. *J. Food Sci.* 41:516–19.
19. Kaplow, M. 1970. Commercial development of intermediate moisture foods. *Food Technol.* 24:889–93.
20. Karel, M. 1976. Technology and application of new intermediate moisture foods. In *Intermediate moisture foods,* ed. R. Davies, G. G. Birch, and K. J. Parker, 4–31. London: Applied Science.
21. Labuza, T. P. 1968. Sorption phenomena in foods. *Food Technol.* 22:263–72.
22. Labuza, T. P., K. Acott, S. R. Tatini, R. Y. Lee, I. Flink, and W. McCall. 1976. Water activity determination: A collaborative study of different methods. *J. Food Sci.* 41:910–17.
23. Labuza, T. P., S. Cassil, and A. J. Sinskey. 1972. Stability of intermediate moisture foods. 2. Microbiology. *J. Food Sci.* 37:160–62.
24. Matheson, N. A. 1962. Enzymes in dehydrated meat. In *Recent advances in food Science,* vol. 2, ed. J. Hawthorn and J. M. Leitch, 57–64. London: Butterworths.
25. May, K. N., and L. E. Kelly. 1965. Fate of bacteria in chicken meat during freeze-dehydration, rehydration, and storage. *Appl Microbiol.* 13:340–44.
26. Meryman, H. T. 1966. Freeze-drying. In *Cryobiology,* ed. H. T. Meryman, Ch. 13. New York: Academic Press.
27. Mossel, D. A. A., and M. Ingram. 1955. The physiology of the microbial spoilage of foods. *J. Appl. Bacteriol.* 18:232–68.
28. Pitt, J. I., and J. H. B. Christian. 1968. Water relations of xerophilic fungi isolated from prunes. *Appl. Microbiol.* 16:1853–58.
29. Plitman, M., Y. Park, R. Gomez, and A. J. Sinskey. 1973. Viability of *Staphylococcus aureus* in intermediate moisture meats. *J. Food Sci.* 38:1004–8.
30. Robson, J. N. 1976. Some introductory thoughts on intermediate moisture foods. In *Intermediate moisture foods,* ed. R. Davies, G. G. Birch, and K. J. Parker, 32–42. London: Applied Science.
31. Ross, K. D. 1975. Estimation of water activity in intermediate moisture foods. *Food Technol.* 29(3):26–34.
32. Scott, W. J. 1957. Water relations of food spoilage microorganisms. *Adv. Food Res.* 1:83–127.
33. Seiler, D. A. L. 1976. The stability of intermediate moisture foods with respect to mould growth. In *Intermediate moisture foods,* ed. R. Davies, G. G. Birch, and K. J. Parker, 166–81. London: Applied Science.

34. Sloan, A. E., P. T. Waletzko, and T. P. Labuza. 1976. Effect of order-of-mixing on a_w-lowering ability of food humectants. *J. Food Sci.* 41:536–40.
35. Tatini, S. R. 1973. Influence of food environments on growth of *Staphylococcus aureus* and production of various enterotoxins. *J. Milk Food Technol.* 36:559–63.
36. Troller, J. A. 1972. Effect of water activity on enterotoxin A production and growth of *Staphylococcus aureus*. *Appl. Microbiol.* 24:440–43.
37. ———. 1983. Effect of low moisture environments on the microbial stability of foods. In *Food microbiology*, ed. A. H. Rose, 173–98. New York: Academic Press.
38. Troller, J. A., and J. H. B. Christian. 1978. *Water activity and food*. New York: Academic Press.
39. Warmbier, H. C., R. A. Schnickels, and T. P. Labuza. 1976. Effect of glycerol on nonenzymatic browning in a solid intermediate moisture model food system. *J. Food Sci.* 41:528–31.

16.

FERMENTED FOODS
AND RELATED
PRODUCTS OF
FERMENTATION

There are numerous food products that owe their production and characteristics to the activities of microorganisms. Many of these, including such foods as ripened cheeses, pickles, sauerkraut, and fermented sausages, are preserved products in that their shelf life is extended considerably over that of the raw materials from which they are made. In addition to being made more shelf stable, all fermented foods have aroma and flavor characteristics that result directly or indirectly from the fermenting organisms. In some instances, the vitamin content of the fermented food is increased along with an increased digestibility of the raw materials. The fermentation process reduces the toxicity of some foods (for example, gari and peujeum), while others may become extremely toxic during fermentation (as in the case of bongkrek). From all indications, no other single group or category of foods or food products is as important as these are and have been relative to man's nutritional well-being throughout the world. Included in this chapter along with the classical fermented foods are such products as coffee beans, wines, and distilled spirits, for these and similar products either result from or are improved by microbial fermentation activities.

The microbial ecology of food and related fermentations has been studied for many years in the case of ripened cheeses, sauerkraut, wines, and so on, and the activities of the fermenting organisms are dependent upon the intrinsic and extrinsic parameters of growth discussed in Chapter 2. For example, when the natural raw materials are acidic and contain free sugars, yeasts develop readily and the alcohol that they produce restricts the activities of most other naturally contaminating organisms (for example, the fermentation of fruits to produce wines). If, on the other hand, the acidity of a plant product permits good bacterial growth and at the same time the product is high in simple sugars, lactic acid bacteria may be expected to develop and the addition of low levels of NaCl will ensure their growth preferential to yeasts (as in sauerkraut fermentation).

Products that contain polysaccharides but no significant levels of simple sugars are normally stable to the activities of yeasts and lactic acid bacteria due to the lack of amylase in these organisms. In order to effect their fermentation, an exogenous source of saccharifying enzymes must be supplied. The use of barley malt in the brewing and distilling industries is an example of this. The fermentation of sugars to ethanol that results from malting is then carried out by yeasts. The use of **koji** in the fermentation of soybean products is another example of the way in which alcoholic and lactic acid fermentations may be carried out on products that have low levels of sugars but high levels of starches and proteins. While the saccharifying enzymes of barley malt arise from germinating barley, the enzymes of koji are produced by *Aspergillus oryzae* growing on soaked or steamed rice or other cereals (the commercial product **takadiastase** is prepared by growing *A. oryzae* on wheat bran). The koji hydrolysates may be fermented by lactic acid bacteria and yeasts as is the case for soy sauce, or the koji enzymes may act directly upon soybeans in the production of products such as Japanese miso.

FERMENTATION—DEFINED AND CHARACTERIZED

Fermentation is the metabolic process in which carbohydrates and related compounds are oxidized with the release of energy in the absence of any external electron acceptors. The final electron acceptors are organic compounds produced directly from the breakdown of the carbohydrates. Consequently, only partial oxidation of the parent compound occurs and only a small amount of energy is released during the process. As fermenting organisms, the lactic acid bacteria lack functional heme-linked electron transport systems or cytochromes, and they obtain their energy by substrate-level phosphorylation while oxidizing carbohydrates: they do not have a functional Krebs cycle. The products of fermentation consist of some that are more reduced than others.

While the above is a molecular characterization, the word "fermentation" has had many shades of meaning in the past. According to one dictionary definition, it is ". . . a process of chemical change with effervescence . . . a state of agitation or unrest . . . any of various transformations of organic substances. . . ." The word came into use before Pasteur's studies on wines. Prescott and Dunn (88) and Doelle (25) have discussed the history of the concept of fermentation and the former authors note that in the broad sense in which the term is commonly used, it is "a process in which chemical changes are brought about in an organic substrate through the action of enzymes elaborated by microorganisms. . . ." It is in this broad context that the term is used in this chapter. In the brewing industry, a **top fermentation** refers to the use of a yeast strain that carries out its activity at the upper parts of a large vat such as in the production of ale, while a **bottom fermentation** requires the use of a yeast strain that will act in lower parts of the vat— such as in the production of lager beer.

LACTIC ACID BACTERIA

The lactic acid bacteria as presently constituted consist of the following four genera: *Lactobacillus, Leuconostoc, Pediococcus,* and *Streptococcus* (Groups D and N). These organisms are rather widespread in nature and the natural habitat of some is unclear. The common occurrence of leuconostocs on plants has been documented (73). The lactobacilli appear to be more widespread than the streptococci, and the pediococci tend to be restricted to plants, but exceptions exist for both of these genera (64). The common occurrence of some species of streptococci and lactobacilli in the oral cavity of man and animals is well known.

The history of man's knowledge of the lactic streptococci and their ecology have been reviewed by Sandine et al. (102). These authors believe that plant matter is the natural habitat of this group, but they note the lack of proof of a plant origin for *S. cremoris*. It has been suggested that plant streptococci may be the ancestral pool from which other species and strains developed (73).

Related to the above four genera of lactic acid bacteria in some respects but generally not considered to fit the group are the following: some *Aerococcus* spp; some *Erysipelothrix* spp; *Eubacterium, Microbacterium, Peptostreptococcus,* and *Propionibacterium* (55, 63).

While the lactic acid group is loosely defined with no precise boundaries, all members share the property of producing lactic acid from hexoses. Kluyver divided the lactic acid bacteria into two groups based upon end products of glucose metabolism. Those that produce lactic acid as the major or sole product of glucose fermentation are designated **homofermentative.** This pattern is summarized in Fig. 16-1*A*. The homofermentative pattern is observed when glucose is metabolized but not necessarily when pentoses are metabolized, for some homolactics produce acetic and lactic acids when utilizing pentoses. Also, the homofermentative character of homolactics may be shifted for some strains by altering cultural conditions such as glucose concentration, pH, and nutrient limitation (15, 63). The homolactics are able to extract about twice as much energy from a given quantity of glucose as are the heterolactics. Those lactics that produce equal molar amounts of lactate, carbon dioxide, and ethanol from hexoses are designated **heterofermentative** (Fig. 16-1*B*). All members of the genera *Pediococcus* and *Streptococcus* are homofermenters along with some of the lactobacilli, while all *Leuconostoc* spp., as well as some lactobacilli, are heterofermenters (Table 16-1). The heterolactics are more important than the homolactics in

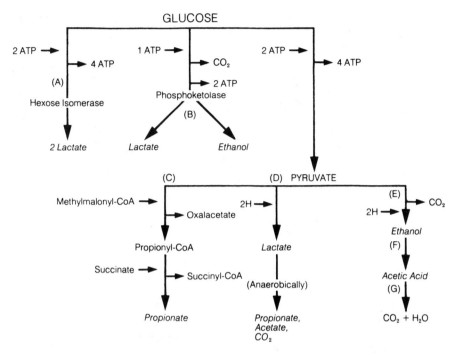

Fig. 16-1. Generalized pathways for the production of some fermentation products from glucose by various organisms. *A* = homofermentative lactics; *B* = heterofermentative lactics; *C & D* = *Propionibacterium* (see fig. 16-3); *E* = *Saccharomyces* spp.; *F* = *Acetobacter* spp.; *G* = *Acetobacter* "overoxidizers."

producing flavor and aroma components such as acetylaldehyde and diacetyl (Fig. 16-2).

The end product differences between homo- and heterofermenters when glucose is attacked are 'a result of basic genetic and physiological differences (Fig. 16-1). The homolactics possess the enzymes aldolase and hexose isomerase but lack phosphoketolase (Fig. 16-1*A*). They use the Embden-Meyerhof-Parnas (EMP) pathway toward their production of 2 lactates/glucose molecule. The heterolactics, on the other hand, have phosphoketolase but do not possess aldolase and hexose isomerase, and instead of the EMP pathway for glucose degradation, these organisms use the hexose monophosphate or pentose pathway (see Fig. 16-1*B*).

The genus *Lactobacillus* has been subdivided classically into three subgenera: *Betabacterium, Streptobacterium,* and *Thermobacterium.* All of the heterolactic lactobacilli in Table 16-1 are betabacteria. The streptobacteria (for example, *L. casei* and *L. plantarum*) produce up to 1.5% lactic acid with an optimal growth temperature of 30°C, while the thermobacteria (such as *L. acidophilus* and *L. bulgaricus*) can produce up to 3% lactic acid and have an optimal temperature of 40°C (68).

Table 16-1. Homo- and heterofermentative lactic acid bacteria with lactate configuration and %G + C (taken from the literature).

	Homofermentative			Heterofermentative	
Organisms	Lactate config.	%G + C	Organisms	Lactate config.	%G + C
Lactobacillus			*Lactobacillus*		
L. acidophilus	DL	36.7	L. brevis	DL	42.7–46.4
L. bulgaricus	D(−)	50.3	L. buchneri	DL	44.8
L. casei	L(+)	46.4	L. cellobiosus	DL	53
L. coryniformis	DL	45	L. confusus	DL	44.5–45.0
L. curvatus	DL	43.9	L. coprophilus	DL	41.0
L. delbrueckii	D(−)	50	L. fermentum	DL	53.4
L. helveticus	DL	39.3	L. hilgardii	DL	40.3
L. jugurti	DL	36.5–39.0	L. sanfrancisco	DL	38.1–39.7
L. jensenii	D(−)	36.1	L. trichodes	DL	42.7
L. lactis	D(−)	50.3	L. viridescens	DL	35.7–42.7
L. leichmannii	D(−)	50.8	*Leuconostoc*		
L. plantarum	DL	45	L. cremoris	D(−)	39–42
L. salivarius	L(+)	34.7	L. dextranicum	D(−)	38–39
L. xylosus	L(+)	39.4	L. lactis	D(−)	43–44
Pediococcus			L. mesenteroides	D(−)	39–42
P. acidilactici	DL	44.0	L. oenos	D(−)	39–40
P. cerevisiae[a]	DL		L. paramesenteroides	D(−)	38–39
P. pentosaceus	DL	38			
Streptococcus					
S. bovis	D(−)	38–42			
S. cremoris	D(−)	38–40			
S. diacetilactis	D(−)	35.1			
S. lactis	D(−)	38.4–38.6			
S. thermophilus	D(−)	40			

[a]*P. damnosus.*

In terms of their growth requirements, the lactic acid bacteria require preformed amino acids, B vitamins, and purine and pyrimidine bases— hence their use in microbiological assays for these compounds. Although they are mesophilic, some can grow at 5°C and some as high as 45°C. With respect to growth pH, some can grow as low as 3.2, some as high as 9.6, and most can grow in the pH range 4.0–4.5. The lactic acid bacteria are only weakly proteolytic and lipolytic (113).

In the past, the taxonomy of the lactic acid bacteria has been based largely upon the gram reaction, general lack of ability to produce catalase, the production of lactic acid of a given configuration, along with the ability to ferment various carbohydrates. More recently, the use of DNA base composition, DNA homology, cell wall peptidoglycan type, and immunologic specificity of enzymes have come into use. While these organisms may be thought of as having derived from a common ancestor, high degrees of

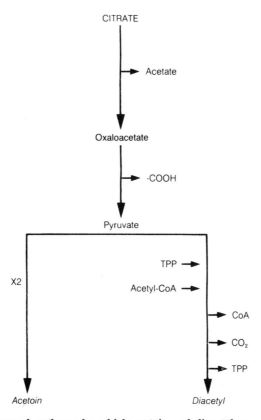

Fig. 16-2. The general pathway by which acetoin and diacetyl are produced from citrate by Group N streptococci and *Leuconostoc* spp. Pyruvate may be produced from lactate, and acetylCoA from acetate. For further details, see refs. 18, 111, 112.

diversity as well as high degrees of relatedness are found within the group. The DNA base composition (expressed as moles % G + C) of those lactics studied varies from a low of around 35 for *L. salivarius* to a high of 53 for *L. fermentum* (131). There is considerable overlap between the homo- and heterolactics (see Table 16-1). On the basis of DNA base composition, Miller et al. (69, 70) placed fifteen species of lactobacilli into three groups. Group I contains those within 32.4 to 38.3 moles % G + C and includes *L. jugurti, L. helveticus, L. salivarius, L. bulgaricus,* and *L. (jugurti) bulgaricus.* Group II contains 42.7 to 48.0 moles % G + C, and includes *L. buchneri, L. brevis, L. casei, L. viridescens,* and *L. plantarum;* while Group III contains 49.0 to 51.9 moles % G + C, and includes *L. lactis, L. leichmannii, L. delbrueckii, L. fermentum,* and *L. cellobiosus.* While the Group I species are all homofermenters, Groups II and III contain both homo- and heterolactic types. The diversity of relatedness of this group has been shown also by both DNA hybridization and immunologic studies. With respect to DNA hybridization, strains of *L. jugurti* and *L. helveticus* have been shown to share between 85 and 100% DNA homology; *L. leichmannii*, *L. delbrueckii*, *L. lactis*, and *L. bulgaricus* to share between 78 and 100%; and *L. fermentum* and *L. cellobiosus* to share from 77 to 100% DNA homology (63). Simonds et al. (108) have shown that *L. bulgaricus* shares 86% DNA homology with *L. lactis*, 4.8% with *L. helveticus*, and none with *L. jugurti*.

DNA-DNA hybridization studies in the genus *Leuconostoc* have been reported (52). Of forty-five strains representing six species, six homology groups were distinguished using four reference DNA preparations. Of nineteen strains of *L. mesenteroides*, three different hybridization groups were determined.

In regard to immunologic methods, Gasser and Gasser (36) showed that antisera prepared against NAD-dependent D-lactic dehydrogenases of three lactobacilli reacted against crude extracts of almost all other species of lactobacilli containing the enzyme. Extracts of *Leuconostoc* spp. cross reacted with anti-D-lactic dehydrogenases. In a somewhat similar manner, it has been shown that antisera prepared against purified *S. faecalis* fructose diphosphate aldolase reacted to varying degrees with the aldolases of all homofermentative lactics, including some other streptococci, pediococci, and lactobacilli (64).

The cell wall mucopeptides of lactics and other bacteria have been reviewed by Schleifer and Kandler (103) and Williams (131). While there appears to be wide variation within most of the lactic acid genera, the homofermentative lactobacilli of the subgenus *Thermobacterium* appear to be the most homogeneous in this regard in having L-lysine in the peptidoglycan peptide chain and D-aspartic acid as the interbridge peptide. The Group N streptococci have similar wall mucopeptides.

The measurement of molar growth yields provides information on fermenting organisms relative to their fermentation substrates and pathways. By this concept, the μg dry weight of cells produced per μmole substrate fermented is determined as the **molar yield constant,** indicated by Y. It is tacitly assumed that essentially none of the substrate carbon is used for cell biosynthesis, that oxygen does not serve as an electron or hydrogen acceptor, and that all of the energy derived from the metabolism of the substrate is coupled to cell biosynthesis (45). When the substrate is glucose, for example, the molar yield constant for glucose, Y_G, is determined by:

$$Y_G = \frac{\text{g dry weight of cells}}{\text{moles glucose fermented}}$$

If the ATP yield or moles ATP produced per mole substrate used is known for a given substrate, the amount of dry weight of cells produced/mole of ATP formed can be determined by:

$$Y_{ATP} = \frac{\text{g dry weight of cells/moles ATP formed}}{\text{moles substrate fermented}}$$

A large number of fermenting organisms have been examined during growth and found to have $Y_{ATP} = 10.5$ or close thereto. This value is assumed to be a constant, so that an organism that ferments glucose by the EMP pathway to produce 2 ATP/mole of glucose fermented should have $Y_G = 21$ (that is, it should produce 21 g of cells dry weight/mole of glucose). This has been verified for *Streptococcus faecalis, Saccharomyces cerevisiae, Saccharomyces rosei,* and *L. plantarum* on glucose (all $Y_G = 21$, $Y_{ATP} = 10.5$, within experimental error). A study by Brown and Collins (15) indicates that Y_G and Y_{ATP} values for *S. diacetilactis* and *S. cremoris* differ when cells are grown aerobically on a partially defined medium with low and higher levels of glucose, and further when grown on a complex medium. On a partially defined medium with low glucose levels (1 to 7 μmol/ml), values for *S. diacetilactis* were $Y_G = 35.3$, $Y_{ATP} = 15.6$, while for *S. cremoris* $Y_G = 31.4$ and $Y_{ATP} = 13.9$. On the same medium with higher glucose levels (1 to 15 μmol/ml), Y_G for *S. diacetilactis* was 21. Y_{ATP} values for these two organisms on the complex medium with 2 μmol glucose/ml were 21.5 and 18.9 for *S. diacetilactis* and *S. cremoris,* respectively. Anaerobic molar growth yields for streptococcal species on low levels of glucose have been studied by Johnson and Collins (56). *Zymomonas mobilis* utilizes the Entner-Doudoroff pathway to produce only 1 ATP/mole of glucose fermented ($Y_G = 8.3$, $Y_{ATP} = 8.3$). If and when the produced lactate is metabolized further, the molar growth yield would be higher. *Bifidobacterium bifidum* produces 2.5–3 ATP/mole of glucose fermented with $Y_G = 37$, $Y_{ATP} = 13$ (116).

In addition to the use of molar growth yields to compare organisms on the same energy substrate, this concept can be applied to assess the metabolic routes used by various organisms in attacking a variety of carbohydrates (for further information, see 30, 82, 104, 116).

Molecular genetics have been employed by McKay and co-workers to stabilize lactose fermentation by *S. lactis.* The genes responsible for lactose fermentation by some lactic streptococci are plasmid borne, and loss of the plasmid results in the loss of lactose fermentation. In an effort to make lactose fermentation more stable, lac^+ genes from *S. lactis* were cloned into a cloning vector, which was incorporated into a *Streptococcus sanguis* strain (48). Thus, the *lac* genes from *S. lactis* were transformed into *S. sanguis* via a vector plasmid, or transformation could be effected by use of appropriate fragments of DNA through which the genes were integrated into the chromosome of the host cells (49). In the latter state, lactose fermentation would be a more stable property than when the *lac* genes are plasmid borne.

PRODUCTS OF FERMENTATION

A selected list of fermented food products is described briefly below. For more detailed information on these and other fermented products, see Beuchat (14), Pederson (84), Rose (99), and Steinkraus (114).

Dairy products

Some of the many fermented foods and products produced and utilized world-wide are listed in Table 16-2. The commercial and sometimes the home production of many of these is begun by use of appropriate **starter** cultures. A lactic starter is a basic starter culture with widespread use in the dairy industry. For cheese making of all kinds, lactic acid production is essential, and the lactic starter is employed for this purpose. Lactic starters are also used for butter, cultured buttermilk, cottage cheese, and cultured sour cream and are often referred to by product such as butter starter, buttermilk starter, and so on. Lactic starters always include bacteria that convert lactose to lactic acid, usually *S. lactis, S. cremoris,* or *S. diacetilactis.* Where flavor and aroma compounds such as diacetyl are desired, the lactic starter will include a heterolactic such as *Leuconostoc citrovorum, S. diacetilactis,* or *L. dextranicum* (for biosynthetic pathways, see Fig. 16-2 and reference 18). Starter cultures may consist of single or mixed strains. They may be produced in quantity and preserved by freezing in liquid nitrogen (38), or by freeze drying. The streptococci generally make up around 90% of a mixed dairy starter population, and a good starter culture can convert most of the lactose to lactic acid. The titratable acidity may increase to 0.8–1.0%, calculated as lactic acid, and the pH usually drops to 4.3–4.5 (31). Some of the many products that result from the use of lactic and other starters are discussed below.

Butter, buttermilk, and **sour cream** are produced generally by inoculating pasteurized cream or milk with a lactic starter culture and holding until the desired amount of acidity is attained. In the case of butter where cream is inoculated, the acidified cream is then churned to yield butter, which is washed, salted, and packaged (84). Buttermilk, as the name suggests, is the milk that remains after cream is churned for the production of butter. The commercial product, however, is usually prepared by inoculating skim milk with a lactic or buttermilk starter culture and holding until souring occurs. The resulting curd is broken up into fine particles by agitation and this product is termed **cultured buttermilk.** Cultured sour cream is produced generally by fermenting pasteurized and homogenized light cream with a lactic starter. These products owe their tart flavor to lactic acid, and their buttery aroma and taste to diacetyl.

Yogurt (yoghurt) is produced with a yogurt starter, which is a mixed culture of *Streptococcus thermophilus* and *L. bulgaricus* in a 1:1 ratio. The coccus grows faster than the rod and is primarily responsible for acid production, while the rod adds flavor and aroma. The associative growth of the two organisms results in lactic acid production at a rate greater than that produced by either when growing alone, and more acetaldehyde (the chief volatile flavor component of yogurt) is produced by *L. bulgaricus* when growing in association with *S. thermophilus* (see 91).

The product is prepared by first reducing the water content of either whole or skim milk by at least one-fourth. This may be done in a vacuum

Table 16-2. Some of the many known fermented foods and related products (compiled from various sources).

Foods and products	Raw ingredients	Fermenting organisms	Commonly produced
Dairy products			
Acidophilus milk	Milk	*Lactobacillus acidophilus*	Many countries
Bulgarian buttermilk	Milk	*Lactobacillus bulgaricus*	Balkans, other areas
Cheeses (ripened)	Milk curd	Lactic starters; others	Worldwide
Kefir	Milk	*Streptococcus lactis, L. bulgaricus, Torula* spp.	Southwestern Asia
Kumiss	Raw mare's milk	*L. bulgaricus, Lactobacillus leichmannii, Torulla* spp.	Russia
Taette	Milk	*S. lactis* var. *taette*	Scandinavian peninsula
Tarhana[a]	Wheat meal and yogurt	Lactics	Turkey
Yogurt[b]	Milk, milk solids	*Streptococcus thermophilus, L. bulgaricus*	Worldwide
Meat and fishery products			
Country-cured hams	Pork hams	*Aspergillus, Penicillium* spp.	Southern U.S.A.
Dry sausages[c]	Pork, beef	*Pediococcus cerevisiae*	Europe, U.S.A.
Lebanon bologna	Beef	*P. cerevisiae*	U.S.A.
Burong dalag	Dalag fish and rice	*Leuconostoc mesenteroides, P. cerevisiae, L. plantarum*	Philippines
Izushi	Fresh fish, rice, vegetables	*Lactobacillus* spp.	Japan
Fish sauces[d]	Small fish	halophilic *Bacillus* spp.	Southeast Asia
Katsuobushi	Skipjack tuna	*Aspergillus glaucus*	Japan
Nonbeverage plant products			
Bongkrek	Coconut presscake	*Rhizopus oligosporus*	Indonesia
Cocoa beans	Cacao fruits (pods)	*Candida krusei, Geotrichum* spp.	Africa, South America

Table 16-2, continued.

Foods and products	Raw ingredients	Fermenting organisms	Commonly produced
Coffee beans	Coffee cherries	Erwinia dissolvens, Saccharomyces spp.	Brazil, Congo, Hawaii, India
Gari	Cassava	Corynebacterium manihot, Geotrichum spp.	West Africa
Kenkey	Corn	Aspergillus spp., Penicillium spp., Lactobacilli, yeasts	Ghana, Nigeria
Kimchi	Cabbage and other veg.	Lactic acid bacteria	Korea
Miso	Soybeans	Aspergillus oryzae, Saccharomyces rouxii	Japan
Ogi	Corn	L. plantarum, S. lactis, Saccharomyces rouxii	Nigeria
Olives	Green olives	L. mesenteroides, L. plantarum	Worldwide
Ontjom[e]	Peanut presscake	Neurospora sitophila	Indonesia
Peujeum	Cassava	Molds	Indonesia
Pickles	Cucumbers	P. cerevisiae, L. plantarum	Worldwide
Poi	Taro roots	Lactics	Hawaii
Sauerkraut	Cabbage	L. mesenteroides, L. plantarum	Worldwide
Soy sauce (shoyu)	Soybeans	A. oryzae; or A. soyae; S. rouxii, L. delbrueckii	Japan
Sufu	Soybeans	Mucor spp.	China and Taiwan
Tao-si	Soybeans	A. oryzae	Philippines
Tempeh	Soybeans	Rhizopus oligosporus; R. oryzae	Indonesia, New Guinea, Surinam

Beverages and related products

Arrack	Rice	Yeasts, bacteria	Far East
Beer and ale	Cereal wort	Saccharomyces cerevisiae; S. carlsbergensis	Worldwide
Binuburan	Rice	Yeasts	Philippines
Bourbon whiskey	Corn, rye	S. cerevisiae	U.S.A.
Bouza beer	Wheat grains	Yeasts	Egypt
Cider	Apples; others	Saccharomyces spp.	Worldwide
Kaffir beer	Kaffircorn	Yeasts, molds, lactics	Nyasaland

Magon	Corn	Lactobacillus spp.	Bantus of South Africa
Mezcal	Century plant	Yeasts	Mexico
Oo	Rice	Yeasts	Thailand
Pulque[f]	Agave juice	Yeasts and lactics	Mexico, U.S. Southwest
Sake	Rice	Saccharomyces sake	Japan
Scotch whiskey	Barley	S. cerevisiae	Scotland
Teekwass	Tea leaves	Acetobacter xylinum, Schizosaccharomyces pombe	
Thumba	Millet	Endomycopsis fibuliges	West Bengal
Tibi	Dried figs; raisins	Betabacterium vermiforme, Saccharomyces intermedium	
Vinegar	Cider, wine	Acetobacter spp.	Worldwide
Wines	Grapes, other fruits	Saccharomyces ellipsoideus strains	Worldwide
Palm wine	Palm sap	Acetobacter spp., lactics, yeasts	Nigeria
Breads			
Idli	Rice and bean flour	Leuconostoc mesenteroides	Southern India
Rolls, cakes, etc.	Wheat flours	S. cerevisiae	Worldwide
San Francisco sourdough bread	Wheat flour	S. exiguus, L. sanfrancisco	Northern California
Sour pumpernickel	Wheat flour	L. mesenteroides	Switzerland, other areas

[a]Similar to Kishk in Syria and Kushuk in Iran.
[b]Also yogurt (matzoon in Armenia; Leben in Egypt; Naja in Bulgaria; Gioddu in Italy; Dadhi in India).
[c]Such as Genoa, Milano, Siciliano.
[d]See text for specific names.
[e]N. sitophila is used to make red ontjom; R. oligosporus for white ontjom.
[f]Distilled to produce tequila.

pan following sterilization of the milk. Approximately 5% by weight of milk solids or condensed milk is usually added. The concentrated milk is then heated to 82°–93°C for 30–60 min and cooled to around 45°C (84). The yogurt starter is now added at a level of around 2% by volume and incubated at 45°C for 3–5 h followed by cooling to 5°C. The titratable acidity of a good finished product is around 0.85–0.90%, and to get this amount of acidity the fermenting product should be removed from 45°C when the titratable acidity is around 0.65–0.70% (20). Good yogurt keeps well at 5°C for 1 to 2 weeks. As noted above, the coccus grows first during the fermentation followed by the rod so that after around 3 h the numbers of the two organisms should be approximately equal. Higher amounts of acidity such as 4% can be achieved by allowing the product to ferment longer, with the effect that the rods will exceed the cocci in number. The streptococci tend to be inhibited at pH values of 4.2–4.4, while the lactobacilli can tolerate pHs in the 3.5–3.8 range. The lactic acid of yogurt is produced more from the glucose moiety of lactose than the galactose moiety. Goodenough and Kleyn (43) found only a trace of glucose throughout yogurt fermentation, while galactose increased from an initial trace to 1.2%. Samples of commercial yogurts showed only traces of glucose, while galactose varied from around 1.5 to 2.5%.

Freshly produced yogurt typically contains around 10^9 organisms/g but during storage, numbers may decrease to 10^6/g, especially when stored at 5°C for up to 60 days (47). The rod generally decreases more rapidly than the coccus. The addition of fruits to yogurt appears not to affect the numbers of fermenting organisms (47).

The antimicrobial qualities of yogurt, buttermilk, sour cream, and cottage cheese were examined by Goel et al. (40) who inoculated *Enterobacter aerogenes* and *Escherichia coli* separately into commercial products and studied the fate of these organisms when the products were stored at 7.2°C. A sharp decline of both coliforms was noted in yogurt and buttermilk after 24 h. Neither could be found in yogurt generally beyond 3 days. While the numbers of coliforms were reduced also in sour cream, they were not reduced as rapidly as in yogurt. Some cottage cheese samples actually supported an increase in coliform numbers, probably because these products had higher pH values. The initial pH ranges for the products studied by these workers were as follows: 3.65–4.40 for yogurts; 4.1–4.9 for buttermilks; 4.18–4.70 for sour creams; and 4.80–5.10 for cottage cheese samples. In another study, commercially produced yogurts in Ontario were found to contain the desired 1:1 ratio of coccus to rod in only 15% of 152 products examined (7). Staphylococci were found in 27.6% and coliforms in around 14% of these yogurts. Twenty-six percent of the samples had yeast counts > 1,000/g and almost 12% had psychrotroph counts > 1,000/g. In his study of commercial unflavored yogurt in Great Britain, Davis (20) found *S. thermophilus* and *L. bulgaricus* counts to range from a low of around 82 million to a high of over 1 billion/g, and final pH to range from 3.75 to 4.20.

Kefir is prepared by the use of kefir grains, which contain *S. lactis, L. bulgaricus,* and a lactose-fermenting yeast held together by layers of coagulated protein. Acid production is controlled by the bacteria, while the yeast produces alcohol. The final concentration of lactic acid and alcohol may be as high as 1%. **Kumiss** is similar to kefir except that mare's milk is used, the culture organisms do not form grains, and the alcohol content may reach 2%.

Acidophilus milk is produced by the inoculation into sterile skim milk of an intestinal implantable strain of *L. acidophilus.* The inoculum of 1–2% is added, followed by holding of the product at 37°C until a smooth curd develops. **Bulgarian buttermilk** is produced in a similar manner by the use of *L. bulgaricus* as the inoculum or starter, but unlike *L. acidophilus, L. bulgaricus* is not implantable in the human intestines.

All **cheeses** result from a lactic fermentation of milk. In general, the process of manufacture consists of the following important steps. (1) Milk is prepared and inoculated with an appropriate lactic starter. The starter produces lactic acid, which along with added rennin, gives rise to curd formation. The starter for cheese production may differ depending upon the amount of heat applied to the curds. *Streptococcus thermophilus* is employed for acid production in cooked curds, since it is more heat tolerant than either of the other more commonly used lactic starters; or a combination of *S. thermophilus* and *S. lactis* is employed for curds that receive an intermediate cook. (2) The curd is shrunk and pressed, followed by salting, and, in the case of ripened cheeses, allowed to ripen under conditions appropriate to the cheese in question. While most ripened cheeses are the product of metabolic activities of the lactic acid bacteria, several well-known cheeses owe their particular character to other related organisms. In the case of **Swiss cheese,** a mixed culture of *L. bulgaricus* and *S. thermophilus* is usually employed along with a culture of *Propionibacterium shermanii,* which is added to function during the ripening process in flavor development and eye formation. (See Fig. 16-1*C, D* for a summary of propionibacteria pathways and Fig. 16-3 for pathway in detail.) These organisms have been reviewed extensively by Hettinga and Reinbold (51). For blue cheeses such as **Roquefort,** the curd is inoculated with spores of *Penicillium roqueforti,* which effect ripening and impart the blue-veined appearance characteristic of this type of cheese. In a similar fashion, either the milk or the surface of **Camembert** cheese is inoculated with spores of *Penicillium camemberti.*

There are over 400 varieties of cheeses representing fewer than twenty distinct types, and these are grouped or classified according to texture or moisture content, whether ripened or unripened, and if ripened, whether by bacteria or molds. The three textural classes of cheeses are hard, semihard,

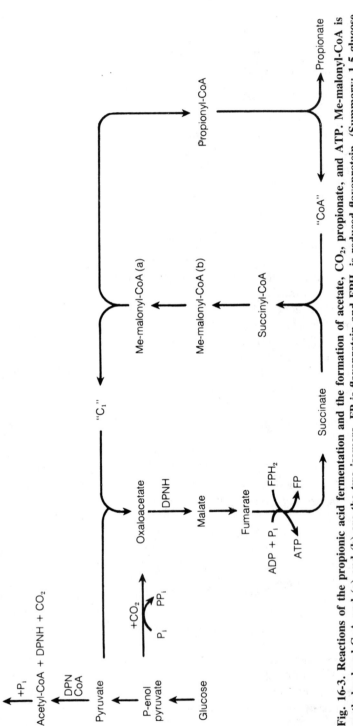

Fig. 16-3. Reactions of the propionic acid fermentation and the formation of acetate, CO_2, propionate, and ATP. Me-malonyl-CoA is methylmalonyl-CoA and (a) and (b) are the two isomers. FP is flavoprotein and FPH_2 is reduced flavoprotein. (Summary: 1.5 glucose + 6 Pi + 6 ADF → 6 ATP + $2H_2O$ + CO_2 + acetate + 2 propionate). (Allen et al., 5; copyright © 1964, American Society for Microbiology)

and soft. Examples of hard cheeses include Cheddar, Provolone, Romano, Edam, and so forth. All hard cheeses are ripened by bacteria over periods ranging from 2 to 16 mon. Semihard cheeses include Muenster, Gouda, and so forth, and are ripened by bacteria over periods of 1 to 8 mon. Blue and Roquefort are two examples of semihard cheeses that are mold ripened for 2 to 12 mon. Limburger is an example of a soft bacteria-ripened cheese, while Brie and Camembert are examples of soft mold-ripened cheeses. Among unripened cheeses are cottage, cream, and Neufchatel.

Several other less widely produced fermented dairy products are listed in Table 16-2.

Meat and fishery products

Fermented sausages are produced generally as dry or semidry products although some are intermediate. Dry or Italian-type sausages contain 30–40% moisture, are generally not smoked or heat processed, and are eaten usually without cooking (84). In their preparation, curing and seasoning agents are added to ground meat, followed by its stuffing into casings and incubation for varying periods of time at 80°–95°F. Incubation times are shorter when starter cultures are employed. The curing mixtures include glucose as substrate for the fermenters, and nitrates and/or nitrites as color stabilizers. When only nitrates are used, it is necessary that the sausage contain bacteria that reduce nitrates to nitrites, usually micrococci present in the sausage flora or added to the mix. Following the incubation, during which fermentation occurs, the products are placed in drying rooms with a relative humidity of 55–65% for periods ranging from 10 to 100 days, or, in the case of Hungarian salami, up to 6 mon (75). Genoa and Milano salamis are other examples of dry sausages.

In a recent study of dry sausages, the pH was found to decrease from an initial of 5.8 to 4.8 during the first 15 days of ripening and remained constant thereafter (24). Nine different brands of commercially produced dry sausages were found by these investigators to have pH values ranging from 4.5 to 5.2, with a mean of 4.87. With respect to the changes that occur in the flora of fermenting dry sausage when starters are not used, Urbaniak and Pezacki (122) found the homofermenters to predominate overall, with *L. plantarum* being the most commonly isolated species. Heterofermenters such as *L. brevis* and *L. buchneri* increased during the 6-day incubation period as a result of changes in pH and Eh brought about by the homofermenters.

Semidry sausages are prepared essentially as above but are subjected to less drying. They contain about 50% moisture and are finished by heating to an internal temperature of 140–154°F during smoking. Thuringer, cervelat, and summer sausage and Lebanon bologna are some examples of semidry sausages. "Summer sausage" refers to those traditionally of Northern European origin, made during colder months, stored, aged, and then eaten during summer months. They may be dry or semidry.

Lebanon bologna is typical of a semidry sausage. This product, originally produced in the Lebanon, Pennsylvania, area, is an all-beef, heavily smoked and spiced product that may be prepared by use of a *Pediococcus cerevisiae* starter. The product is made by the addition of approximately 3% NaCl along with sugar, seasoning, and either nitrate, nitrite, or both to raw cubed beef. The salted beef is allowed to age at refrigerator temperatures for about 10 days during which time the growth of naturally occurring lactic acid bacteria or the starter organisms is encouraged and gram negatives are inhibited. A higher level of microbial activity along with some drying occurs during the smoking step at higher temperatures. A controlled production process for this product has been studied (79), and it consists of aging salted beef at 5°C for 10 days and smoking at 35°C with high R.H. for 4 days. Fermentation may be carried out either by the natural flora of the meat or by use of a commercial starter of *P. cerevisiae* or *P. acidilactici*. The amount of acidity produced in Lebanon bologna may reach 0.8–1.2% (84).

The hazard of eating improperly prepared homemade, fermented sausage is pointed up by an outbreak of trichinosis. Of the fifty persons who actually consumed the raw summer sausage, twenty-three became ill with trichinosis (87). The sausage was made on two different days in three batches according to a family recipe that called for smoking at cooler smoking temperatures, believed to produce a better flavored product. All three batches of sausages contained home-raised beef. In addition, two batches eaten by victims contained USDA-inspected pork in one case and home-raised pork in the other, but *Trichinella spiralis* larvae were found only in the USDA-inspected pork. This organism can be destroyed by a heat treatment that results in internal temperatures of at least 137°F.

Fermented sausages produced without use of starters have been found to contain large numbers of lactobacilli such as *L. plantarum* (23). The use of a *P. cerevisiae* starter leads to the production of a more desirable product (22, 50). In their study of commercially produced fermented sausages, Smith and Palumbo (109) found total aerobic plate counts to be in the 10^7–10^8/g range with a predominance of lactic acid types. When starter cultures were used, the final pH of the products ranged from 4.0 to 4.5, while those produced without starters ranged between 4.6 and 5.0. For summer-type sausages, pH values of 4.5 to 4.7 have been reported for a 72-h fermentation (2). These investigators found that fermentation at 30° and 37°C led to a lower final pH than at 22°C and that the final pH was directly related to the amount of lactic acid produced. The pH of fermented sausage may actually increase by 0.1–0.2 units during long periods of drying due to uneven buffering produced by increases in amounts of basic compounds (129). The ultimate pH attained following fermentation depends upon the type of sugar added. While glucose is most widely used, sucrose has been found to be an equally effective fermentable sugar for low pH production (1). The effect of a commercial frozen concentrate starter (*P. acidilactici*) in fermenting various sugars added to a sausage preparation is illustrated in Fig. 16-4.

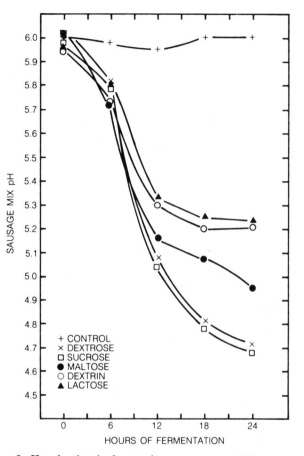

Fig. 16-4. Rate of pH reduction in fermenting sausage containing 0% or 1% of various carbohydrates. (Acton et al., 1; copyright © 1977, Institute of Food Technologists)

Prior to the late 1950s, the production of fermented sausages was facilitated by either back inoculations, or a producer took the chance of the desired organisms being present in the raw materials. The manufacture of these as well as of many other fermented foods has been, until recently, more of an art than a science. With the advent of pure culture starters, not only has production time been shortened but more uniform and safer products can be produced (27). While the use of starter cultures has been in effect for many years in the dairy industry, their use in many nondairy products world-wide is a recent development with great promise for the future. *Micrococcus aurantiacus* has been employed along with starters in the production of some European sausages (75).

Molds are known to contribute to the quality of dry European-type sausages such as Italian salami. In an extensive study of the fungi of ripened meat products, Ayres et al. (8) found nine species of penicillia and seven of aspergilli on fermented sausages and concluded that these organisms play a role in the preservation of products of this type. Fewer species of other mold genera were also found.

Country-cured hams are dry-cured hams produced in the southern United States, and during the curing and ripening period of 6 mon to 2 yr, heavy mold growth occurs on the surfaces. Although Ayres et al. (8) noted that the presence of molds is incidental and that a satisfactory cure does not depend upon their presence, it seems quite likely that some aspects of flavor development of these products derive from the heavy growth of such organisms, and to a lesser extent from yeasts. Heavy mold growth obviates the activities of food-poisoning and food-spoilage bacteria, and in this sense the mold flora aids in preservation. The above investigators found aspergilli and penicillia to be the predominant types of molds on country-cured hams.

The processing of country-cured hams of the above type takes place during the early winter and consists of rubbing sugar cure into the flesh side and onto the hock end. This is followed some time later by rubbing NaCl into all parts of the ham not covered by skin. The hams are then wrapped in paper and individually placed in cotton fabric bags and left lying flat for several days between 32° and 40°C. The hams are hung shank end down in ham houses for 6 weeks or longer and may be given a hickory smoke during this time, although smoking is said not to be essential to a desirable product.

Italian-type country-cured hams are produced with NaCl as the only cure. Curing is carried out for about a month, followed by washing, drying, and ripening for 6–12 mon or longer (39). While halophilic and halotolerant bacteria increase as Italian hams ripen, the microflora in general is thought to play only a minor role (39). For more detailed information on meat starter cultures and formulations for fermented sausages along with cure ingredients for country-style hams, see Pearson and Tauber (83) and Bacus (10).

Fish sauces are popular products in Southeast Asian countries, where they are known by various names such as *ngapi* (Burma), *nuoc-mam* (Cambodia and Vietnam), *nam-pla* (Laos and Thailand), *ketjap-ikan* (Indonesia), and so on. The production of some of these sauces begins with the addition of salt to uneviscerated fish at a ratio of approximately 1:3, salt to fish. The salted fish are then transferred either to fermentation tanks generally constructed of concrete and built into the ground, or placed in earthenware pots and buried in the ground. The tanks or pots are filled and sealed off for at least 6 mon to allow the fish to liquefy. The liquid is collected, filtered, and transferred to earthenware containers and ripened in the sun for 1 to 3 mon. The finished product is described as being clear dark-brown in color with a distinct aroma and flavor (100). In a study of fermenting Thai fish sauce by the latter authors, pH from start to finish ranged from 6.2 to 6.6 with NaCl content around 30% over the 12-mon fermentation period (100). These parameters, along with the relatively high fermentation temperature, result in the growth of halophilic aerobic sporeformers as the

predominant microorganisms of these products. Lower numbers of streptococci, micrococci, and staphylococci were found; and they along with
the *Bacillus* spp. were apparently involved in the development of flavor
and aroma. Some part of the liquefaction that occurs is undoubtedly due
to the activities of fish proteases. While the temperature and pH of the
fermentation are well within the growth range of a large number of undesirable
organisms, the safety of products of this type is due to the 30–33% NaCl.

Fish pastes are also common in Southern Asia, but the role of fermenting
microorganisms in these products appears to be minimal. Among the many
other fermented fish, fish-paste, and fish-sauce products are the following:
mam-tom of China; *mam-ruoc* of Cambodia; *bladchan* of Indonesia; *shiokara*
of Japan; *belachan* of Malaya; *bagoong* of the Philippines; *kapi, hoi-dong,*
and *pla-mam* of Thailand; *fessik* of Africa; and *nam-pla* of Thailand. Some
of these as well as other fish products of Southeast Asia have been reviewed
and discussed by van Veen and Steinkraus (127) and Sundhagul et al. (119).
See Table 16-2 for other related products.

Nonbeverage food products of plant origin

Sauerkraut is a fermentation product of fresh cabbage. The starter for
sauerkraut production is usually the normal mixed flora of cabbage. The
addition of 2.25–2.5% salt restricts the activities of gram-negative bacteria
while the lactic acid rods and cocci are flavored. *Leuconostoc mesenteroides*
and *L. plantarum* are the two most desirable lactic acid bacteria in sauerkraut
production, with the former having the shortest generation time and the
shortest life span. The activities of the coccus usually cease when the acid
content increases to 0.7–1.0%. The final stages of kraut production are
effected by *L. plantarum* and *L. brevis. P. cerevisiae* and *S. faecalis* may
also contribute to product development. Final total acidity is generally 1.6–
1.8%, with lactic acid at 1.0–1.3%.

Pickles are fermentation products of fresh cucumbers, and as is the case
for sauerkraut production, the starter culture normally consists of the normal
mixed flora of cucumbers. In the natural production of pickles the following
lactic acid bacteria are involved in the process in order of increasing prevalence: *L. mesenteroides, S. faecalis, P. cerevisiae, L. brevis,* and *L. plantarum* (26). Of these the pediococci and *L. plantarum* are the most involved,
with *L. brevis* being undesirable because of its capacity to produce gas. *L.
plantarum* is the most essential species in pickle production as it is for
sauerkraut.

In the production of pickles, selected cucumbers are placed in wooden
brine tanks with initial brine strengths as low as 5% NaCl (20° salinometer).
Brine strength is increased gradually during the course of the 6–9-week
fermentation until it reaches around 60° salinometer (15.9% NaCl). In addition
to exerting an inhibitory effect on the undesirable gram-negative bacteria,
the salt extracts from the cucumbers water and water-soluble constituents
such as sugars, which are converted by the lactic acid bacteria to lactic

acid. The product that results is a salt-stock pickle from which pickles such as sour, mixed sour, chowchow, and so forth, may be made.

The general technique of producing brine-cured pickles briefly outlined above has been in use for many years, but it often leads to serious economic loss because of pickle spoilage from such conditions as bloaters, softness, off-colors, and so on. More recently, the controlled fermentation of cucumbers brined in bulk has been achieved, and this process not only reduces economic losses of the type noted above but leads to a more uniform product over a shorter period of time (26). In brief, the controlled fermentation method employs a chlorinated brine of 25° salinometer, acidification with acetic acid, the addition of sodium acetate, and inoculation with *P. cerevisiae* and *L. plantarum,* or the latter alone. The course of the 10- to 12-day fermentation is represented in Fig. 16-5 (for more detailed information, see Etchells et al., 26).

Olives to be fermented (Spanish, Greek, or Sicilian) are done so by the natural flora of green olives, which consists of a variety of bacteria, yeasts, and molds. The olive fermentation is quite similar to that of sauerkraut except that it is slower, involves a lye treatment, and may require the addition of starters. The lactic acid bacteria become prominent during the intermediate stage of fermentation. *L. mesenteroides* and *P. cerevisiae* are

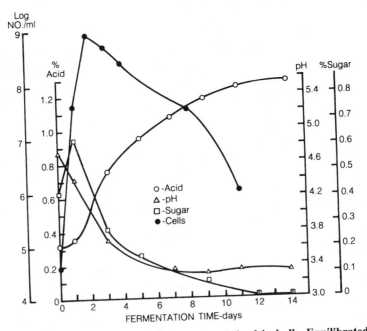

Fig. 16-5. Controlled fermentation of cucumbers brined in bulk. Equilibrated brine strength during fermentation, 6.4% NaCl; incubation temperature, 27°C. (26, copyright © 1975 by Academic Press)

the first lactics to become prominent, and these are followed by lactobacilli with *L. plantarum* and *L. brevis* being the most important (128).

The olive fermentation is preceded by a treatment of green olives with from 1.6 to 2.0% lye, depending upon type of olive, at 21–24°C for 4–7 h for the purpose of removing some of the bitter principal. Following the complete removal of lye by soaking and washing, the green olives are placed in oak barrels and brined so as to maintain a constant 28°–30° salinometer level. Inoculation with *L. plantarum* may be necessary because of destruction of organisms during the lye treatment. The fermentation may take as long as 6 to 10 mon, and the final product has a pH of 3.8–4.0 following up to 1% lactic acid production.

Soy sauce or shoyu is produced in a two-stage manner. The first stage, the koji (analogous to malting in the brewing industry), consists of inoculating either soybeans or a mixture of beans and wheat flour with *Aspergillus oryzae* or *A. soyae* and allowing them to stand for 3 days. This results in the production of large amounts of fermentable sugars, peptides, and amino acids. The second stage, the moromi, consists of adding the fungal-covered product to around 18% NaCl and incubating at room temperatures for at least a year. The liquid that is obtained at this time is soy sauce. During the incubation of the moromi, lactic acid bacteria, *L. delbrueckii* in particular, and yeasts such as *Saccharomyces rouxii* carry out an anaerobic fermentation of the koji hydrolysate. Pure cultures of *A. oryzae* for the koji and *L. delbrueckii* and *S. rouxii* for the moromi stages have been shown to produce good quality soy sauce (132).

Tempeh is a fermented soybean product. While there are many variations in its production, the general principle of the Indonesian method for tempeh consists of soaking soybeans overnight in order to remove the seedcoats or hulls. Once seedcoats are removed, the beans are cooked in boiling water for about 30 min and spread on a bamboo tray to cool and surface dry. Small pieces of tempeh from a previous fermentation are incorporated as starter followed by wrapping with banana leaves. The wrapped packages are kept at room temperature for 1–2 days during which mold growth occurs and binds the beans together as a cake—the tempeh. An excellent product can be made by storing in perforated plastic bags and tubes with fermentations completed in 24 h at 31°C (29). The desirable organism in the fermentation is *Rhizopus oligosporus*, especially for wheat tempeh. Good soybean tempeh can be made with *R. oryzae* or *R. arrhizus*. During the fermentation, the pH of soybeans rises from around 5.0 to values as high as 7.5.

Miso is a fermented soybean product common in Japan. It is prepared by mixing or grinding steamed or cooked soybeans with koji and salt, and allowing fermentation to take place usually over a 4–12-mon period. White or sweet miso may be fermented for only a week, while the higher quality dark brown product (*mame*) may ferment for 2 yr (106). In Israel, Ilany-Feigenbaum et al. (54) have prepared miso-type products by using defatted soybean flakes instead of whole soybeans and fermenting for around 3 mon.

The koji for these products was made by growing *A. oryzae* on corn, wheat, barley, millet or oats, potatoes, sugar beets, or bananas, and the authors found that the miso-type products compared favorably to Japanese-prepared miso. Because of the possibility that *A. oryzae* may produce toxic substances, koji was prepared by fermenting rice with *Rhizopus oligosporus* at 25°C for 90 days and the product was found to be an acceptable alternative to *A. oryzae* as a koji fungus (107).

Ogi is a staple cereal of the Yorubas of Nigeria and is the first native food given to babies at weaning. It is produced generally by soaking corn grains in warm water for 2 to 3 days followed by wet-milling and sieving through a screen mesh. The sieved material is allowed to sediment and ferment, and is marketed as wet cakes wrapped in leaves. Various food dishes are made from the fermented cakes or the ogi (12). During the steeping of corn, *Corynebacterium* spp. become prominent and appear to be responsible for the diastatic action necessary for the growth of yeasts and lactic acid bacteria (4). Along with the corynebacteria, *S. cerevisiae* and *L. plantarum* have been found to be prominent in the traditional ogi fermentation, as are *Cephalosporium*, *Fusarium*, *Aspergillus*, and *Penicillium*. Most of the acid produced is lactic, which depresses the pH of desirable products to around 3.8. The corynebacteria develop early and their activities cease after the first day, while those of the lactobacilli and yeasts continue beyond the first day of fermentation. More recently, a new process for making ogi has been developed, tested, and found to produce a product of better quality than the traditional (11). By the new method, corn is dry-milled into whole corn and dehulled corn flour. Upon the addition of water, the mixture is cooked, cooled, and then inoculated with a mixed culture (starter) of the following three organisms: *L. plantarum*, *S. lactis*, and *Saccharomyces rouxii*. The inoculated preparation is incubated at 32°C for 28 h, during which time the pH of the corn drops from 6.1 to 3.8. This process eliminates the need for starch-hydrolyzing bacteria. In addition to the shorter fermentation time, there is also less chance for faulty fermentations.

Gari is a staple food of West Africa prepared from the root of the cassava plant. Cassava roots contain a cyanogenic glucoside that makes them poisonous if eaten fresh or raw. The roots are rendered safe by a fermentation during which the toxic glucoside decomposes with the liberation of gaseous hydrocyanic acid. In the home preparation of gari, the outer peel and the thick cortex of the cassava roots are removed, followed by grinding or grating the remainder. The pulp is pressed to remove the remaining juice, and placed in bags for 3–4 days to allow fermentation to occur. The fermented product is cooked by frying. The fermentation has been found to occur in a two-stage manner. In the first stage, *Corynebacterium manihot* ferments starch, with the production of acids and the consequent lowering of pH. Under acidic conditions, the cyanogenic glucoside then undergoes spontaneous hydrolysis with the liberation of gaseous hydrocyanic acid (17). The acidic conditions favor the growth of *Geotricum* spp., and these organisms appear to be responsible for the characteristic taste and aroma of gari.

Bongkrek is an example of a fermented food product that has in the past led to a large number of deaths. Bongkrek or semaji is a coconut presscake product of central Indonesia, and it is the homemade product that may become toxic. The safe products fermented by *Rhizopus oligosporus* are finished cakes covered with and penetrated by the white fungus. In order to obtain the desirable fungal growth, it appears to be essential that conditions permit good growth within the first 1–2 days of incubation. If, however, bacterial growth is favored during this time, and if the bacterium *Pseudomonas cocovenenans* is present, it grows and produces two toxic substances— toxoflavin and bongkrekic acid (125, 126). Both of these compounds show antifungal and antibacterial activity, are toxic for man and animals, and are heat stable. Production of both is favored by growth of the organisms on coconut (toxoflavin can be produced in complex culture media). The structural formulae of the two antibiotics are indicated below. Toxoflavin acts as an electron carrier and bongkrekic acid inhibits oxidative phosphorylation in mitochondria.

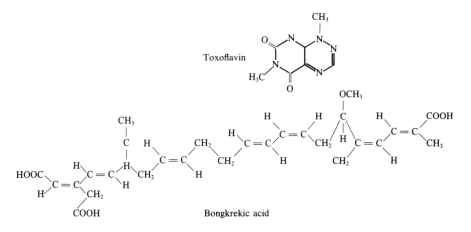

Toxoflavin

Bongkrekic acid

The latter antibiotic has been shown to be cidal to all seventeen molds studied by Subik and Behun (117) by preventing spore germination and mycelial outgrowth. The growth of *P. cocovenenans* in the preparation of bongkrek is not favored if the acidity of starting materials is kept at or below pH 5.5 (124).

Ontjom is a somewhat similar but more popular fermented product of Indonesia made from peanut presscake. Peanut presscake is the material that remains after oil has been extracted from peanuts. The presscake is soaked in water for about 24 h, steamed, and pressed into molds. The molds are covered with banana leaves and inoculated with either *Neurospora sitophila* or *Rhizopus oligosporus*. The product is ready for consumption

1 or 2 days later. A more detailed description of ontjom fermentation and the nutritive value of this product has been provided by Beuchat (13).

Beverage and distilled products

Beer and **ale** are examples of malt beverages that are produced by brewing. An essential step in the brewing process is the fermentation of carbohydrates to ethanol. Since most of the carbohydrates in grains used for brewing exist as starches, and since the fermenting yeasts do not produce amylases to degrade the starch, a necessary part of beer brewing includes a step whereby malt or other exogenous sources of amylase are provided for the hydrolysis of starches to sugars. The malt is first prepared by allowing barley grains to germinate. This serves as a source of amylases (fungal amylases may be used also). Both β- and α-amylases are involved, with the latter acting to liquefy starch and the former to increase sugar formation (46). In brief, the brewing process begins with the mixing of malt, malt adjuncts, hops, and water. Malt adjuncts include certain grains, grain products, sugars, and other carbohydrate products to serve as fermentable substances. Hops are added as sources of pyrogallol and catechol tannins, resins, essential oils, and other constituents for the purpose of precipitating unstable proteins during the boiling of wort and to provide for biological stability, bitterness, and aroma. The process by which the malt and malt adjuncts are dissolved and heated and the starches digested is called mashing. The soluble part of the mashed materials is called wort (compare with koji). In some breweries, lactobacilli are introduced into the mash to lower the pH of wort through lactic acid production. The species generally used for this purpose is *L. delbrueckii* (58).

Wort and hops are mixed and boiled for 1.5–2.5 h for the purpose of enzyme inactivation, extraction of soluble hop substances, precipitation of coagulable proteins, concentration, and sterilization. Following the boiling of wort and hops, the wort is separated, cooled, and fermented. The fermentation of the sugar-laden wort is carried out by the inoculation of *S. cerevisiae* (Fig. 16-1*E*). Ale results from the activities of top-fermenting yeasts, which depress the pH to around 3.8, while bottom-fermenting yeasts (*S. carlsbergensis* strains) give rise to lager and other beers with pH values of 4.1–4.2. A top fermentation is complete in 5 to 7 days while a bottom fermentation requires 7 to 12 days (88). The freshly fermented product is aged and finished by the addition of CO_2 to a final content of 0.45–0.52% before it is ready for commerce. The pasteurization of beer, at 140°F or higher, may be carried out for the purpose of destroying spoilage organisms. When lactic acid bacteria are present in beers, the lactobacilli are found more commonly in top fermentations, while pediococci are found in bottom fermentations (58).

Distilled spirits are alcoholic products that result from the distillation of yeast fermentations of grain, grain products, molasses, or fruit or fruit products. Whiskeys, gin, vodka, rum, cordials, and liqueurs are examples of distilled spirits. While the process for producing most products of these types is quite similar to that for beers, the content of alcohol in the final products is considerably higher than for beers. **Rye** and **bourbon** are examples of whiskeys. In the former, rye and rye malt or rye and barley malt are used in different ratios, but at least 51% rye is required by law. Bourbon is made from corn, barley malt or wheat malt, and usually another grain in different proportions, but at least 51% corn is required by law. A sour wort is maintained in order to keep down undesirable organisms, the souring occurring naturally or by the addition of acid. The mash is generally soured by inoculating with a homolactic such as *L. delbrueckii,* which is capable of lowering the pH to around 3.8 in 6–10 h (84). The malt enzymes (diastases) convert the starches of the cooked grains to dextrins and sugars, and upon completion of diastatic action and lactic acid production, the mash is heated to destroy all microorganisms. It is then cooled to 75–80°F, and pitched (inoculated) with a suitable strain of *S. cerevisiae* for the production of ethanol. Upon completion of fermentation, the liquid is distilled to recover the alcohol and other volatiles, and these are handled and stored under special conditions relative to the type of product being made. **Scotch whiskey** is made primarily from barley and is produced from barley malt dried in kilns over peat fires. **Rum** is produced from the distillate of fermented sugar cane or molasses. **Brandy** is a product prepared by distilling grape or other fruit wines.

Wines are normal alcoholic fermentations of sound grapes followed by aging. A large number of other fruits such as peaches, pears, and so forth, may be fermented for wines, but in these instances the wine is named by the fruit, such as "pear wine," "peach wine," and the like. Since fruits already contain fermentable sugars, the use of exogenous sources of amylases is not necessary as it is when grains are used for beers or whiskeys. Wine making begins with the selection of suitable grapes, which are crushed and then treated with a sulfite such as potassium metabisulfite to retard the growth of acetic acid bacteria, wild yeasts, and molds. The pressed juice, called must, is inoculated with a suitable wine strain of *S. ellipsoideus.* The fermentation is allowed to continue for 3 to 5 days at temperatures between 70 and 90°F, and good yeast strains may produce up to 14–18% ethanol (84). Following fermentation, the wine is racked—that is, drawn off from the lees or sediment, which among other things contains potassium bitartrate (cream of tartar). The clearing and development of flavor occur during the storage and aging process. Red wines are made by initially fermenting the crushed grape must "on the skins" during which pigment is extracted into the juice, while white wines are prepared generally from the juice of white grapes. **Champagne,** a sparkling wine made by a secondary fermentation of wine, is produced by adding sugar, citric acid, and a champagne yeast starter to bottles of previously prepared, selected table wine. The bottles are corked, clamped, and stored horizontally at suitable temperatures for about 6 mon. They are then removed, agitated, and aged for an additional period of up to 4 yr. The final sedimentation of yeast cells and tartrates is accelerated by reducing the temperature of the wine to around 25°C and holding for 1 to 2 weeks. Clarification of the champagne is brought about by working the sediment down the bottle onto the cork over a period of 2 to 6 weeks by frequent rotation of the bottle. Finally, the sediment is frozen and disgorged upon removal of the cork. (See references such as Prescott and Dunn (88) for more details of the production and classification of the various types of wine.)

Palm wine or Nigerian palm wine is an alcoholic beverage consumed throughout the tropics and is produced by a natural fermentation of palm sap. The sap is sweet, dirty brown in color, and contains 10–12% sugar, mainly sucrose. The fermentation process results in the sap becoming milky-white in appearance due to the presence of large numbers of fermenting bacteria and yeasts. This product is unique in that the microorganisms are alive when the wine is consumed. The fermentation has been reviewed and studied by Faparusi and Bassir (28) and Okafor (77), who found the following genera of bacteria to be the most predominant in finished products: *Micrococcus, Leuconostoc, Streptococcus, Lactobacillus,* and *Acetobacter.* The predominant yeasts found are *Saccharomyces* and *Candida* spp., with the former being the more common (76). The fermentation occurs over a 36–48 h period during which the pH of sap falls from 7.0–7.2 to < 4.5. Fermentation products consist of organic acids in addition to ethanol. During the early phases of fermentation, *Serratia* and *Enterobacter* spp. increase in numbers, followed by lactobacilli and leuconostocs. After a 48-h fermentation, *Acetobacter* spp. begin to appear (28, 78).

Sake is an alcoholic beverage commonly produced in Japan. The substrate is the starch from steamed rice and its hydrolysis to sugars is carried out by *A. oryzae* to yield the koji. Fermentation is carried out by *Saccharomyces sake* over periods of 30–40 days resulting in a product containing 12–15% alcohol and around 0.3% lactic acid (84). The latter is produced by hetero- and homolactic lactobacilli.

Cider, in the United States, is a product that represents a mild fermentation of apple juice by naturally occurring yeasts. In making apple cider, the fruits are selected, washed, and ground into a pulp. The pulp "cheeses" are pressed to release the juice. The juice is strained and placed in a storage tank, where sedimentation of particulate matter occurs, usually for 12–36 h or several days if the temperature is kept at 40°F or below. The clarified juice is cider. If pasteurization is desired, this is accomplished by heating at 170°F for 10 min. The chemical preservative most often used is sodium sorbate at a level of 0.10%. Preservation may be effected also by chilling or freezing. The finished product contains small amounts of ethanol in addition to acetaldehyde. The holding of nonpasteurized or unpreserved cider at suitable temperatures invariably leads to the development of cider vinegar, which indicates the presence of acetic acid bacteria in these products.

The pathway employed by acetic acid bacteria is summarized in Fig. 16-1F, G.

In their study of the ecology of the acetic acid bacteria in cider manufacture, Passmore and Carr (80) found six species of *Acetobacter* and noted that those that display a preference for sugars tend to be found early in the cider process, while those that are more acid-tolerant and capable of oxidizing alcohols appear after the yeasts have converted most of the sugars to ethanol. *Zymomonas* spp., gram-negative bacteria that ferment glucose to ethanol, have been isolated from ciders, but they are presumed to be present in low numbers.

Coffee beans, which develop as berries or cherries in their natural state, have an outer pulpy and mucilaginous envelope that must be removed before the beans can be dried and roasted. The wet method of removal of this layer seems to produce the most desirable product, and it consists of depulping and demucilaging followed by drying. While depulping is done mechanically, demucilaging is accomplished by natural fermentation. The mucilage layer is composed largely of pectic substances (32), and pectinolytic microorganisms are important in their removal. *Erwinia dissolvens* has been found to be the single most important bacterium during the demucilaging fermentation in Hawaiian (33) and Congo coffee cherries (123), although Pederson and Breed (85) indicated that the fermentation of coffee berries from Mexico and Colombia was carried out by typical lactic acid bacteria (leuconostocs and lactobacilli). Agate and Bhat (3) in their study of coffee cherries from the Mysore State of India found that the following pectinolytic yeasts predominated and played important roles in the loosening and removal of the mucilaginous layers: *Saccharomyces marxianus, S. bayanus, S. ellipsoideus,* and *Schizosaccharomyces* spp. Molds are common on green coffee beans, and in one study 99.1% of products from thirty-one countries contained these organisms, generally on the surface (71). Seven species of aspergilli dominated the flora, with *A. ochraceus* being the most frequently recovered from beans before surface disinfection, followed by *A. niger* and species of the *A. glaucus* group. The toxigenic molds, *A. flavus* and *A. versicolor,* were found, as were *P. cyclopium, P. citrinum,* and *P. expansum,* but the penicillia were less frequently found than the aspergilli (71). Microorganisms do not contribute to the development of flavor and aroma in coffee beans as they do in cocoa beans.

Cocoa beans (actually cacao beans—cocoa is the powder and chocolate is the manufactured product), from which chocolate is derived, are obtained from the fruits or pods of the cacao plant in parts of Africa, Asia, and South America. The beans are extracted from the fruits and fermented in piles, boxes, or tanks for 2–12 days, depending upon the type and size of beans. During the fermentation, high temperatures (45–50°C) and large quantities of liquid develop. Following sun or air drying, during which the water content is reduced to < 7.5%, the beans are roasted to develop the characteristic flavor and aroma of chocolate. The fermentation occurs in two phases. In the first, sugars from the acidic pulp (about pH 3.6) are converted to alcohol. The second phase consists of the alcohol being oxidized to acetic acid. In a study of Brazilian cocoa beans by Camargo et al. (16), the flora on the first day of fermentation at 21°C consisted of yeasts. On the third day, the temperature had risen to 49°C and the yeast count had decreased to no more than 10% of the total flora. Over the 7-day fermentation, the pH increased from 3.9 to 7.1. The cessation of yeast and bacterial activity around the third day is due in part to the unfavorable temperature, lack of fermentable sugars, and the increase in alcohol. While some decrease in acetic acid bacteria occurs because of high temperature, not all of these organisms are destroyed. Yeasts and acetic acid bacteria are the most important fermenters of cocoa beans. Of the 142 yeasts isolated by Camargo et al., 105 were asporogenous and 37 were ascosporogenous. *Candida krusei* (a thermotolerant yeast) was the most frequently encountered species, and it became predominant after the second day. High numbers of *Geotrichum candidum* and *Candida mycoderma* were found, and both were shown to be pectinolytic. On the other hand, lactic acid bacteria were more abundant than yeasts on fermented cocoa beans in the state of Bahia (Brazil) after 48 h (81). The homolactics were the most abundant of three genera and eight species represented. The only heterolactics were *Leuconostoc mesenteroides* and *L. brevis*. The lactic acid bacteria are thought to be responsible for the acidity of cocoa from this area (81). The gram-negative bacterium, *Zymomonas mobilis,* has been tried without success as a starter in cocoa bean fermentation. This organism can convert sugars to ethanol and the latter to acetic acid. Heterofermentative lactics and *Acetobacter* spp. are involved in the fermentation (96).

While yeasts play important roles in producing alcohol in cocoa bean fermentation, their presence appears even more essential to the development of the final, desirable chocolate flavor of roasted beans. Levanon and Rossetini (61) found that the endoenzymes released by autolyzing yeasts are responsible for the development of chocolate precursor compounds. The acetic acid apparently makes the bean tegument permeable to the yeast enzymes. It has been shown that chocolate aroma occurs only after cocoa beans are roasted and that the roasting of unfermented beans does not produce the characteristic aroma (97). Reducing sugars and free amino acids are in some way involved in the final chocolate aroma development (98).

Breads

San Francisco sourdough bread is similar to sourdough breads produced in various countries. Historically, the starter for sourdough breads consists of the natural flora of baker's barm (sour ferment or mother sponge, with a portion of each inoculated dough saved as starter for the next batch). The barm generally contains a mixture of yeasts and lactic acid bacteria.

In the case of San Francisco sourdough bread, the yeast has been identified as *Saccharomyces exiguus* (118), and the responsible bacterium as *Lactobacillus sanfrancisco* (59). The souring is caused by the acids produced by the bacterium, while the yeast is responsible for the leavening action, even though some CO_2 is produced by the bacterium. The pH of these sourdoughs ranges from 3.8 to 4.5. Both acetic and lactic acids are produced, with the former accounting for 20–30% of the total acidity (59).

Idli is a fermented bread-type product common in southern India. It is made from rice and black gram mungo (urd beans). These two ingredients are soaked in water separately for 3–10 h and then ground in varying proportions, mixed, and allowed to ferment overnight. The fermented and raised product is cooked by steaming and served hot. It is said to resemble a steamed, sourdough bread (115). During the fermentation, the initial pH of around 6.0 falls to values of 4.3–5.3. In a particular study, a batter pH of 4.70 after a 20-h fermentation was associated with 2.5% lactic acid, based on dry grain weight (72). In their studies of idli, Steinkraus et al. (115) found total bacterial counts of 10^8–10^9/g after 20–22 h of fermentation. Most of the organisms consisted of gram-positive cocci or short rods with *L. mesenteroides* being the single most abundant species followed by *S. faecalis*. The leavening action of idli is produced by *L. mesenteroides*. This is the only known instance of a lactic acid bacterium having this role in a naturally fermented bread (72). The latter authors confirmed the work of others before them in finding the urd beans to be a more important source of lactic acid bacteria than rice. *L. mesenteroides* reaches its peak at around 24 h, with *S. faecalis* becoming active only after about 20 h. Other probable fermenters include *L. delbruecki, L. fermenti,* and *Bacillus* spp. (94). Only after idli has fermented for more than 30 h does *P. cerevisiae* become active. The product is not fermented generally beyond 24 h because maximum leavening action occurs at this time and decreases with longer incubations. When idli is allowed to ferment longer, more acidity is produced. It has been found that total acidity (expressed as g lactic acid/g of dry grains) increased from 2.71% after 24 h to 3.70% after 71 h, while pH decreased from 4.55 to 4.10 over the same period (72). (A review of idli fermentation has been made by Reddy et al., 94.)

SINGLE-CELL PROTEIN

The cultivation of unicellular microorganisms as a direct source of food for man was suggested in the early 1900s. The expression "single-cell protein" (SCP) was coined at the Massachusetts Institute of Technology around 1966 to depict the idea of microorganisms as food sources (110). Although SCP is a misnomer in that proteins are not the only food constituent represented by microbial cells, it obviates the need to refer to each product generically as in "algal protein," "yeast cell protein," and so on. Although SCP as a potential and real source of food for man differs from the other products covered in this chapter, with the exception of that from algal cells it is produced in a similar manner—by fermentation.

Rationale for SCP production

It is imperative that new food sources be found in order that future generations of mankind be adequately fed. A food source that is nutritionally complete and that requires a minimum of land, time, and cost to produce is highly desirable. In addition to meeting these criteria, SCP can be produced on a variety of waste materials. Among the overall advantages of SCP over plant and animal sources of proteins are the following (57): (1) microorganisms have a very short generation time, and can thus provide a rapid mass increase; (2) microorganisms can be easily modified genetically—to produce cells that bring about desirable results; (3) the protein content is high; (4) the production of SCP can be based upon raw materials readily available in large quantities; and (5) SCP production can be carried out in continuous culture and thus be independent of climatic changes.

The greater speed and efficiency of microbial protein production compared to plant and animal sources may be illustrated as follows: a 1,000-lb steer produces about 1 lb of new protein/day; soybeans (prorated over a growing season) produce about 80 lb; while yeasts produce about 50 tons!

Organisms and fermentation substrates

A large number of algae, yeasts, molds, and bacteria have been studied as SCP sources. Among the most promising genera and species are the following: **Algae**—*Spirulina maxima, Chlorella* spp., and *Scenedesmus* spp. **Yeasts**—*Candida guilliermondii, C. utilis, C. lipolytica,* and *C. tropicalis; Debaryomyces kloeckeri; Torulopsis candida, T. methanosorbosa; Pichia* spp.; *Kluyveromyces fragilis; Hansenula polymorpha; Rhodotorula* spp.; and *Saccharomyces* spp. **Filamentous fungi**—*Agaricus* spp.; *Aspergillus* spp.; *Fusarium* spp.; *Penicillium* spp.; *Endomycopsis fibuligera;* and *Trichosporon cutaneum.* **Bacteria**—*Bacillus* spp.; *Cellulomonas* spp.; *Acinetobacter calcoaceticus; Nocardia* spp.; *Methylomonas* spp.; *Aeromonas hydrophilia; Alcaligenes eutrophus (Hydrogenomonas eutropha), Mycobacterium* sp.; and *Rhodopseudomonas* sp. Of these groups, yeasts have received by far the most attention.

The choice of a given organism is dictated in large part by the type of substrate or waste material in question. The cyanobacterium *Spirulina maxima* grows in shallow waters high in bicarbonate at a temperature of 30°C and pH 8.5–11.0. It can be harvested from pond waters and dried for food use. This alga has been eaten by the people of the Chad Republic for many years (110). Other algal cells require only sunlight, CO_2, minerals, water, and proper growth temperatures. However, the large-scale use of algal cells as SCP sources is said to be practical only in areas below 35° latitude, where sunlight is available most of the year (62).

Bacteria, yeasts, and molds can be grown on a wide variety of fermentable materials including food processing wastes (such as cheese whey and brewery, potato processing, cannery, and coffee wastes); industrial wastes (such as sulfite liquor in the paper industry and combustion gases); and cellulosic wastes (including bagasse, newsprint mill, and barley straw). In the case of cellulosic wastes, it is necessary to use organisms that can utilize cellulose, such as a *Cellulomonas* sp. or *Trichoderma viride*. A mixed culture of *Cellulomonas* and *Alcaligenes* has been employed. For starchy materials, a combination of *Endomycopsis fibuligera* and a *Candida* sp. such as *C. utilis* has been employed, in which the former effects hydrolysis of starches and the latter subsists upon the hydrolyzed products to produce biomass. Some other representative substrates and fermenting organisms are listed in Table 16-3.

Table 16-3. Some examples of substrate materials that support the growth of microorganisms in the production of SCP.

Substrates	*Microorganisms*
CO_2 and sunlight	*Chlorella pyrenoidosa* *Scenedesmus quadricauda* *Spirulina maxima*
n-Alkanes, kerosene	*Candida intermedia, C. lipolytica,* *C. tropicalis* *Nocardia* sp.
Methane	*Methylomonas* sp. *(Methanomonas)* *Methylococcus capsulatus* *Trichoderma* sp.
H_2 and CO_2	*Alcaligenes eutrophus (Hydrogenomonas* *eutropha)*
Gas oil	*Acinetobacter calcoaceticus (Micrococcus* *cerificans)* *Candida lipolytica*
Methanol	*Methylomonas methanica (Methanomonas* *methanica)*
Ethanol	*Candida utilis* *Acinetobacter calcoaceticus*
Sulfite liquor wastes	*Candida utilis*
Cellulose	*Cellulomonas* sp. *Trichoderma viride*
Starches	*Endomycopsis fibuligera*
Sugars	*Saccharomyces cerevisiae* *Candida utilis* *Kluyveromyces fragilis*

SCP products

While the products of fermentation of other foods in this chapter consist of an altered or modified substrate, the product of SCP fermentation consists solely of the microbial cells that grow upon the substrate. The cells may be used directly as a protein source in animal feed formulations, thereby freeing animal feed, such as corn, for human consumption; or they may be used as a protein source or food ingredient for human food. In the case of animal feed or feed supplements, the dried cells may be used without further processing. As noted above, whole cells of *Spirulina maxima* are consumed by man in at least one part of Africa.

For human use, the most likely products are SCP concentrates or isolates that may be further processed into textured or functional SCP products. To produce functional protein fibers, cells are mechanically disrupted, cell walls removed by centrifugation, proteins precipitated from disrupted cells, and the resulting protein extruded from syringelike orifices into suitable menstra such as acetate buffer, $HClO_4$, acetic acid, and the like. The SCP fibers may now be used to form textured protein products. Baker's yeast protein is the only product of this type presently approved for human food ingredient use in the United States, but as Litchfield (62) has noted, it is reasonable to assume that similar products from other food-grade yeasts such as *Candida utilis* and *Kluyveromyces fragilis* will ultimately be approved.

The nutrition and safety of SCP

Chemical analyses of the microorganisms evaluated for SCP reveal that they are comparable in amino acid content and type to plant and animal sources with the possible exception of methionine, which is lower in some SCP sources. All are relatively high in nitrogen. For example, the approximate percentage composition of N on a dry weight basis is as follows: bacteria 12-13, yeast 8-9, algae 8-10, and filamentous fungi 5-8 (57). In addition to proteins, microorganisms contain adequate levels of carbohydrates, lipids, and minerals, and are excellent sources of B vitamins. The fat content varies among these sources, with algal cells containing the highest levels and bacteria the lowest. On a dry weight basis, nucleic acids average 3-8% for algae, 6-12% for yeasts, and 8-16% for bacteria (57). B vitamins are high in all SCP sources. The digestibility of SCP in experimental animals has been found to be lower than for animal proteins such as casein. A thorough review of the chemical composition of SCP from a large variety of microorganisms has been made (19, 130).

Success has been achieved in rat-feeding studies with a variety of SCP products, but human-feeding studies have been less successful, except in the case of certain yeast cell products. Gastrointestinal disturbances are common complaints in man following the consumption of algal and bacterial SCP, and these and other problems associated with the consumption of SCP have been reviewed elsewhere (130). When gram-negative bacteria are used as SCP sources for human use, the endotoxins must be removed or detoxified.

The high nucleic acid content of SCP leads to kidney stone formation and/or gout in man. As noted above, the nucleic acid content of bacterial SCP may be as high as 16%, while the recommended daily intake for man is about 2 g. The problems are caused by an accumulation of uric acid, which is sparingly soluble in plasma. Upon the breakdown of nucleic acids, purine and pyrimidine bases are released. Adenine and guanine (purines) are metabolized to uric acid. Lower animals can degrade uric acid to the soluble compound allantoin (they possess the enzyme uricase) and consequently the consumption of high levels of nucleic acids does not present metabolic problems to these animals as it does to man. While high nucleic acid contents presented problems in the early development and use of SCP, these compounds can be reduced to levels below 2% by techniques such as acid precipitation, acid or alkaline hydrolysis, or by use of endogenous and bovine pancreatic RNAses (62). (For further information on SCP, see References 19, 57, 62, 110, 120, 130.)

LACTIC ANTAGONISM

Numerous investigators have observed the inhibition by lactic acid bacteria of food-spoilage and food-poisoning organisms when they grow together in the same food product or culture medium. Lactic antagonism has been observed for nearly 60 years, and its early history has been reviewed by Marth (67), Hurst (53), Babel (9), and Sandine et al. (101). Renewed interest in this phenomenon during the 1970s was due in part to the possibility that it could be exploited as one means of abating the use of chemical preservatives in foods.

The lactic acid bacteria often associated with culture antagonism are streptococci, lactobacilli, and pediococci, with *S. diacetylactis, S. ther-mophilus, L. plantarum, L. brevis, L. casei,* and *P. cerevisiae* among the many species studied. They have been shown to be inhibitory in a variety of foods to *S. aureus,* salmonellae, and most food-spoilage gram-negative bacteria. Production of the inhibitory factor(s) by *S. lactis* subsp. *diacetylactis* was better in yeast extract dextrose broth than eight other media studied (93). The effect of *P. cerevisiae* and *L. plantarum* on the growth of *S. aureus* in cooked, mechanically deboned poultry meat is presented in Fig. 16-6.

In spite of a large number of claims to the contrary, the precise mechanism by which lactic cultures effect microbial inhibition is not yet clear. Among the factors that have been identified are antibiotics, colicins, H_2O_2, organic acids, depressed pH, and nutrient depletion. It is generally agreed that none of these alone explains the antagonism. Nisin is an antibiotic produced by some strains of *S. lactis* (see Chapter 11), but it is not the antagonistic

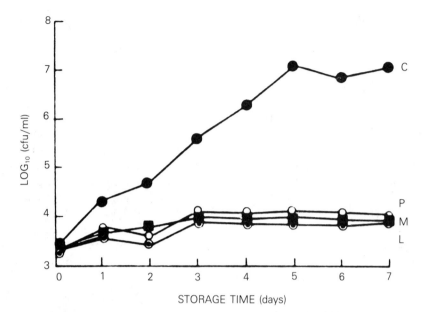

Fig. 16-6. Growth of *S. aureus* in pure culture (c) and in association with *L. plantarum* (L), *P. cerevisiae* (P), and the mixture (M) in cooked MDPM at 15°C. Lactic-acid bacteria were added at a concentration of 10^9 cells/g. (90, copyright © 1978, International Association of Milk, Food and Environmental Sanitarians)

factor. Several investigators have noted that antagonism is increased at low pH and low temperatures. The antagonistic factor is reported often to be a low-molecular-weight substance. While some have found heat-labile factors, others have reported heat-stable factors (see Pinheiro et al., 86). Although definitive data are lacking, diacetyl may be at least one component of the antagonistic system. It is produced by many lactic acid bacteria, has a low molecular weight, is heat labile, and its antimicrobial properties are enhanced at low pH and low temperatures (see Chapter 11).

APPARENT HEALTH BENEFITS OF FERMENTED FOODS

The topic of health-promoting effects of certain fermented foods and/or the organisms of fermentation is beset by findings both for and against such effects. While some studies that appear to be well designed support health benefits, other equally well designed studies do not. The three areas of concern are the possible benefits to lactose-intolerant individuals, the lowering of serum cholesterol, and anticancer activity. More information on these as well as others can be found in the reviews by Deeth and Tamine (21), Friend and Shahani (34), Richardson (95), and Shahani and Ayebo (105).

Lactose intolerance

Lactose intolerance (lactose malabsorption, intestinal hypolactemia) is the normal state for adult mammals and most adult humans, and many more groups are intolerant to lactose than are tolerant (60). Among the relatively few groups that have a majority of adults who tolerate lactose are northern Europeans, white Americans, and members of two nomadic pastoral tribes in Africa (60). When lactose malabsorbers consume certain quantities of milk or ice cream, they immediately experience flatulence and diarrhea. The condition is due to the absence or reduced amounts of intestinal lactase, and this allows the bacteria in the colon to utilize lactose with the production of gases. The breath hydrogen test for lactose intolerance is based upon the increased levels of H_2 produced by anaerobic and facultatively anaerobic bacteria utilizing the nonabsorbed lactose.

A large number of investigators have found that lactose malabsorbers can consume certain fermented dairy products without harmful effects while other studies found no beneficial effects. When beneficial effects are found, they are attributed to the reduced level of lactose in the fermented product and to the production of β-galactosidase by the fermenting organisms following ingestion of the products. In one study, the lactose content of yogurt after storage for 11 days decreased about one-half (to about 2.3 g/100 g from 4.8 g/100 g in nonfermented milk). During the same period, galactose increased from traces in milk to 1.3 g/100 g in yogurt, and similar results were found for acidophilus and bifidus milks (6). In a study employing rats, the animals were fed experimental diets containing yogurt, pasteurized yogurt, and simulated yogurt for 7 days. Those that received natural yogurt were able to absorb galactose more efficiently and also had higher levels of intestinal lactase (42). The yogurt bacteria remained viable in the gut for up to 3 h. When eight lactose malabsorbers ingested yogurt or acidophilus milk, they did not experience any of the symptoms that resulted when low-fat milk was ingested (6).

"Sweet" acidophilus milk has been reported by some to prevent symptoms of lactose intolerance, while others have found this product to be ineffective. It was developed by Prof. M. L. Speck and co-workers, and it consists of normal pasteurized milk to which is added large numbers of viable *L. acidophilus* cells as frozen concentrates. As long as the milk remains under refrigeration, the organisms do not grow, but when drunk the consumer gets the benefit of viable *L. acidophilus* cells. It is "sweet" because it lacks the tartness of traditional acidophilus milk. When eighteen lactase-deficient patients ingested unaltered milk for one week followed by "sweet" acidophilus milk for an additional week, they were as intolerant to the latter product as to the unaltered milk (74). In a study with rats, the yogurt bacteria had little effect in preventing the malabsorption of lactose (35). The indigenous lactics in the gut tended to be suppressed by yogurt, and the rat lactobacillus flora changed from one that was predominantly heterofermentative to one that was predominantly homofermentative.

It appears that several factors may be important in the contradictory findings noted above: the strains of lactic acid bacteria employed, the basic differences between the digestive tracts of animals and man; and the degree of lactose intolerance in test subjects.

CHOLESTEROL. Impetus for studies on the effect of fermented milks on cholesterol came from a study of Maasai tribesmen in Africa who, in spite of consuming substantial amounts of meat, have low serum cholesterol and a very low incidence of coronary diseases. This was associated with their common consumption of 4 to 5 liters/day of fermented whole milk (66). Subsequent studies by a large number of groups leave unanswered the true effect of organisms of fermentation on serum cholesterol levels in man, although the weight of evidence tends to support a positive effect. The published findings through 1977 have been reviewed (95).

In a study by Mann (65) using twenty-six human subjects, large dietary intakes of yogurt were found to lower cholesteremia, and the findings suggested that yogurt contains a factor that inhibits the synthesis of cholesterol from acetate. This factor may be either 3-hydroxy-3-methylglutaric acid and/or orotic acid (95). Rats were employed in a study to evaluate the effect of rat chow plus thermophilus milk and methanol solubles of thermophilus milk on liver cholesterol, and the investigators found that both products significantly reduced liver cholesterol levels compared to controls (92). In another study with rats fed for 4 weeks with a stock diet plus 10% milk fermented by *L. acidophilus,* significantly lower serum cholesterol was found than when those rats were fed two other diets not containing fermented milk (44). While in some studies the lowered cholesterol levels are believed to result from decreased synthesis, in others the bacteria were found to remove cholesterol or its precursors from the gastrointestinal tract. In a study by Gilliland et al. (37), two strains of *L. acidophilus* (recovered from swine) had the ability to grow in the presence of bile. One strain assimilated cholesterol from laboratory culture media in the presence of bile under anaerobic conditions and significantly inhibited increases in serum cholesterol levels in pigs that were fed a high-cholesterol diet. The other strain did not remove cholesterol from laboratory media and did not reduce serum cholesterol when fed to pigs. These investigators thus presented evidence that some strains of *L. acidophilus* reduce serum cholesterol by acting directly on cholesterol in the gastrointestinal tract.

A total of sixty-eight volunteers (ages 18 to 26) in groups of ten or thirteen were put on a regimen consisting of the following supplements: raw milk; whole milk; skim milk; yogurt; buttermilk; and "sweet" acidophilus milk. The regimen was maintained for 3 weeks, and the findings suggested that cultured buttermilk, yogurt, and acidophilus milk had no noticeable effect on serum cholesterol (121). From a study using rats fed for 4 weeks with chow plus skim milk fermented by *S. thermophilus, L. bulgaricus,* and *L. acidophilus* along with appropriate controls, no significant changes in plasma or whole body cholesterol were found (89).

Anticancer effects

Apparently the first observation of anticancer activity of lactic acid bacteria was that of I. G. Bogdanov and co-workers in Russia in 1962 (105), who demonstrated an effect against a sarcoma and a carcinoma. Anticancer activities have been demonstrated in animal models by a large number of investigators who variously employed yogurt and yogurt extracts *L. acidophilus, L. bulgaricus,* and *L. casei* in addition to extracts of these organisms. The specifics of these findings have been reviewed by Shahani and Ayebo (105) and Friend and Shahani (34).

To study the effect of oral supplements of *L. acidophilus* on fecal bacterial enzyme activity, Goldin and Gorbach (41) used twenty-one human subjects. The enzymes assayed were β-glucuronidase, nitroreductase, and azoreductase because they can convert indirectly acting carcinogens to proximal carcinogens. The feeding regimen consisted of a 4-week control period followed by 4 weeks of plain milk; 4 weeks of control; 4 weeks of milk containing 2×10^6/ml of viable *L. acidophilus;* and 4 weeks of control. Reductions of two- to fourfold in activities of the three fecal enzymes were observed in all subjects only during the period of lactobacillus feeding, and fecal enzyme levels returned to normal during the final 4-week control period. Similar but more limited studies have been reported by others. Findings of the type noted may prove to be significant in colon cancer where the body of evidence supports a role for diet.

REFERENCES

1. Acton, J. C., R. L. Dick, and E. L. Norris. 1977. Utilization of various carbohydrates in fermented sausage. *J. Food Sci.* 42:174–78.
2. Acton, J. C., J. G. Williams, and M. G. Johnson. 1972. Effect of fermentation temperature on changes in meat properties and flavor of summer sausage. *J. Milk Food Technol.* 35:264–68.
3. Agate, A. D., and J. V. Bhat. 1966. Role of pectinolytic yeasts in the degradation of mucilage layer of *Coffea robusta* cherries. *Appl. Microbiol.* 14:256–60.
4. Akinrele, I. A. 1970. Fermentation studies on maize during the preparation of a traditional African starch-cake food. *J. Sci. Food Agric.* 21:619–25.
5. Allen, S. H. G., R. W. Killermeyer, R. L. Stjernholm, and H. G. Wood. 1964. Purification and properties of enzymes involved in the propionic acid fermentation. *J. Bacteriol.* 87:171–87.
6. Alm, L. 1982. Effect of fermentation on lactose, glucose, and galactose content in milk and suitability of fermented milk products for lactose intolerant individuals. *J. Dairy Sci.* 65:346–52.

7. Arnott, D. R., C. L. Duitschaever, and D. H. Bullock. 1974. Microbiological evaluation of yogurt produced commercially in Ontario. *J. Milk Food Technol.* 37:11–13.

8. Ayres, J. C., D. A. Lillard, and L. Leistner. 1967. Mold ripened meat products. In *Proceedings 20th annual reciprocal meat conference,* 156–68. Chicago: National Live Stock and Meat Board.

9. Babel, F. J. 1977. Antibiosis by lactic culture bacteria. *J. Dairy Sci.* 60:815–21.

10. Bacus, J. 1984. *Utilization of microorganisms in meat processing: A handbook for meat plant operators.* New York: Wiley.

11. Banigo, E. O. I., J. M. deMan, and C. L. Duitschaever. 1974. Utilization of high-lysine corn for the manufacture of ogi using a new, improved processing system. *Cereal Chem.* 51:559–72.

12. Banigo, E. O. I., and H. G. Muller. 1972. Manufacture of ogi (a Nigerian fermented cereal porridge): Comparative evaluation of corn, sorghum and millet. *Can. Inst. Food Sci. Technol. J.* 5:217–21.

13. Beuchat, L. R. 1976. Fungal fermentation of peanut press cake. *Econ. Bot.* 30:227–34.

14. ———. 1978. Traditional fermented food products. In *Food and beverage mycology,* ed. L. R. Beuchat, 224–53. Westport, Conn.: AVI.

15. Brown, W. V., and E. B. Collins. 1977. End products and fermentation balances for lactic streptococci grown aerobically on low concentrations of glucose. *Appl. Environ. Microbiol.* 33:38–42.

16. Camargo, R. de, J. Leme, Jr., and A. M. Filho. 1963. General observations on the microflora of fermenting cocoa beans (*Theobroma cacao*) in Bahia (Brazil). *Food Technol.* 17:1328–30.

17. Collard, P., and S. Levi. 1959. A two-stage fermentation of cassava. *Nature* 183:620–21.

18. Collins, E. B. 1972. Biosynthesis of flavor compounds by microorganisms. *J. Dairy Sci.* 55:1022–28.

19. Cooney, C. L., C. Rha, and S. R. Tannenbaum. 1980. Single-cell protein: Engineering, economics, and utilization in foods. *Adv. Food Res.* 26:1–52.

20. Davis, J. G. 1975. The microbiology of yoghourt. In *Lactic acid bacteria in beverages and food,* ed. J. G. Carr et al., 245–63. New York: Academic Press.

21. Deeth, H. C., and A. Y. Tamime. 1981. Yogurt: Nutritive and therapeutic aspects. *J. Food Protect.* 44:78–86.

22. Deibel, R. H., and C. F. Niven, Jr. 1957. *Pediococcus cerevisiae:* A starter culture for summer sausage. *Bacteriol. Proc.,* 14–15.

23. Deibel, R. H., C. F. Niven, Jr., and G. D. Wilson. 1961. Microbiology of meat curing. III. Some microbiological and related technological aspects in the manufacture of fermented sausages. *Appl. Microbiol.* 9:156–61.

24. DeKetelaere, A., D. Demeyer, P. Vandekerckhove, and I. Vervaeke. 1974. Stoichiometry of carbohydrate fermentation during dry sausage ripening. *J. Food Sci.* 39:297–300.

25. Doelle, H. W. 1975. *Bacterial metabolism.* Ch. 9. New York: Academic Press.

26. Etchells, J. L., H. P. Fleming, and T. A. Bell. 1975. Factors influencing the growth of lactic acid bacteria during the fermentation of brined cucumbers. In *Lactic acid bacteria in beverages and food,* ed. J. G. Carr et al., 281–305. New York: Academic Press.

27. Everson, C. W., W. E. Danner, and P. A. Hammes. 1970. Improved starter culture for semi-dry sausage. *Food Technol.* 24:42–44.
28. Faparusi, S. I., and O. Bassir. 1971. Microflora of fermenting palm-wine. *J. Food Sci. Technol.* 8:206–10.
29. Filho, A. M., and C. W. Hesseltine. 1964. Tempeh fermentation: Package and tray fermentations. *Food Technol.* 18:761–65.
30. Forrest, W. W., and D. J. Walker. 1971. The generation and utilization of energy during growth. *Adv. Microbiol. Physiol.* 5:213–74.
31. Foster, E. M., F. E. Nelson, M. L. Speck, R. N. Doetsch, and J. C. Olson. 1957. *Dairy microbiology.* Englewood Cliffs, N.J.: Prentice-Hall.
32. Frank, H. A., and A. S. Dela Cruz. 1964. Role of incidental microflora in natural decomposition of mucilage layer in Kona coffee cherries. *J. Food Sci.* 29:850–53.
33. Frank, H. A., N. A. Lum, and A. S. Dela Cruz. 1965. Bacteria responsible for mucilage-layer decomposition in Kona coffee cherries. *Appl. Microbiol.* 13:201–7.
34. Friend, B. A., and K. M. Shahani. 1984. Antitumor properties of lactobacilli and dairy products fermented by lactobacilli. *J. Food Protect.* 47:717–23.
35. Garvie, E. I., C. B. Cole, R. Fuller, and D. Hewitt. 1984. The effect of yoghurt on some components of the gut microflora and on the metabolism of lactose in the rat. *J. Appl. Bacteriol.* 56:237–45.
36. Gasser, F., and C. Gasser. 1971. Immunological relationships among lactic dehydrogenases in the genera *Lactobacillus* and *Leuconostoc*. *J. Bacteriol.* 106:113–25.
37. Gilliland, S. E., C. R. Nelson, and C. Maxwell. 1985. Assimilation of cholesterol by *Lactobacillus acidophilus*. *Appl. Environ. Microbiol.* 49:377–81.
38. Gilliland, S. E., and M. L. Speck. 1974. Frozen concentrated cultures of lactic starter bacteria: A review. *J. Milk Food Technol.* 37:107–11.
39. Giolitti, G., C. A. Cantoni, M. A. Bianchi, and P. Renon. 1971. Microbiology and chemical changes in raw hams of Italian type. *J. Appl. Bacteriol.* 34:51–61.
40. Goel, M. C., D. C. Kulshrestha, E. H. Marth, D. W. Francis, J. G. Bradshaw, and R. B. Read, Jr. 1971. Fate of coliforms in yogurt, buttermilk, sour cream, and cottage cheese during refrigerated storage. *J. Milk Food Technol.* 34:54–58.
41. Goldin, B. R., and S. L. Gorbach. 1984. The effect of milk and lactobacillus feeding on human intestinal bacterial enzyme activity. *Amer. J. Clin. Nutr.* 39:756–61.
42. Goodenough, E. R., and D. H. Kleyn. 1976. Influence of viable yogurt microflora on digestion of lactose by the rat. *J. Dairy Sci.* 59:601–6.
43. ———. 1976. Qualitative and quantitative changes in carbohydrates during the manufacture of yoghurt. *J. Dairy Sci.* 59:45–47.
44. Grunewald, K. K. 1982. Serum cholesterol levels in rats fed skim milk fermented by *Lactobacillus acidophilus*. *J. Food Sci.* 47:2078–79.
45. Gunsalus, I. C., and C. W. Shuster. 1961. Energy yielding metabolism in bacteria. In *The bacteria*, ed. I. C. Gunsalus and R. Y. Stanier, vol. 2, 1–58. New York: Academic Press.
46. Haas, G. J. 1976. Alcoholic beverages and fermented food. In *Industrial Microbiology*, ed. B. M. Miller and W. Litsky, 165–91. New York: McGraw-Hill.

47. Hamann, W. T., and E. H. Marth. 1984. Survival of *Streptococcus thermophilus* and *Lactobacillus bulgaricus* in commercial and experimental yogurts. *J. Food Protect.* 47:781–86.

48. Harlander, S. K., and L. L. McKay. 1984. Transformation of *Streptococcus sanguis* Challis with *Streptococcus lactis* plasmid DNA. *Appl. Environ. Microbiol.* 48:342–46.

49. Harlander, S. K., L. L. McKay, and C. F. Schachtels. 1984. Molecular cloning of the lactose-metabolizing genes from *Streptococcus lactis. Appl. Environ. Microbiol.* 48:347–51.

50. Harris, D. A., L. Chaiet, R. P. Dudley, and P. Ebert. 1957. The development of a commercial starter culture for summer sausages. *Bacteriol. Proc.,* 15.

51. Hettinga, D. H., and G. W. Reinbold. 1972. The propionic-acid bacteria—A review. *J. Milk Food Technol.* 35:295–301, 358–72, 436–47.

52. Hontebeyrie, M., and F. Gasser. 1977. Deoxyribonucleic acid homologies in the genus *Leuconostoc. Intern. J. Syst. Bacteriol.* 27:9–14.

53. Hurst, A. 1973. Microbial antagonism in foods. *Can. Inst. Food Sci. Technol. J.* 6:80–90.

54. Ilany-Feigenbaum, J. Diamant, S. Laxer, and A. Pinsky. 1969. Japanese miso-type products prepared by using defatted soybean flakes and various carbohydrate-containing foods. *Food Technol.* 23:554–56.

55. Ingram, M. 1975. The lactic acid bacteria—A broad view. In *Lactic acid bacteria in beverages and food,* ed. J. G. Carr et al., 1–13. New York: Academic Press.

56. Johnson, M. G., and E. B. Collins. 1973. Synthesis of lipoic acid by *Streptococcus faecalis* 10C1 and end-products produced anaerobically from low concentrations of glucose. *J. Gen. Microbiol.* 78:47–55.

57. Kihlberg, R. 1972. The microbe as a source of food. *Ann. Rev. Microbiol.* 26:427–66.

58. Kleyn, J., and J. Hough. 1971. The microbiology of brewing. *Ann. Rev. Microbiol.* 25:583–608.

59. Kline, L., and T. F. Sugihara. 1971. Microorganisms of the San Francisco sour dough bread process. II. Isolation and characterization of undescribed bacterial species responsible for the souring activity. *Appl. Microbiol.* 21:459–65.

60. Kretchmer, N. 1972. Lactose and lactase. *Sci. Am.* 227(10):71–78.

61. Levanon, Y., and S. M. O. Rossetini. 1965. A laboratory study of farm processing of cocoa beans for industrial use. *J. Food Sci.* 30:719–22.

62. Litchfield, J. H. 1977. Single-cell proteins. *Food Technol.* 31(5):175–79.

63. London, J. 1976. The ecology and taxonomic status of the lactobacilli. *Ann. Rev. Microbiol.* 30:279–301.

64. London, J., and K. Kline. 1973. Aldolase of lactic acid bacteria: A case history in the use of an enzyme as an evolutionary marker. *Bacteriol. Rev.* 37:453–78.

65. Mann, G. V. 1977. A factor in yogurt which lowers cholesteremia in man. *Atherosclerosis* 26:335–40.

66. Mann, G. V., and A. Spoerry. 1974. Studies of a surfactant and cholesteremia in the Masai. *Amer. J. Clin. Nutr.* 27:464–69.

67. Marth, E. H. 1966. Antibiotics in foods—Naturally occurring, developed, and added. *Residue Rev.* 12:65–161.

68. Marth, E. H. 1974. Fermentations. In *Fundamentals of dairy chemistry,* ed. B. H. Webb et al., Ch. 13. Westport, Conn.: AVI.

69. Miller, A., III, W. E. Sandine, and P. R. Elliker. 1970. Deoxyribonucleic acid

base composition of lactobacilli determined by thermal denaturation. *J. Bacteriol.* 102:278–80.

70. ———. 1971. Deoxyribonucleic acid homology in the genus *Lactobacillus. Can. J. Microbiol.* 17:625–34.

71. Mislivec, P. B., V. R. Bruce, and R. Gibson. 1983. Incidence of toxigenic and other molds in green coffee beans. *J. Food Protect.* 46:969–73.

72. Mukherjee, S. K., M. N. Albury, C. S. Pederson, A. G. van Veen, and K. H. Steinkraus. 1965. Role of *Leuconostoc mesenteroides* in leavening the batter of idli, a fermented food of India. *Appl. Microbiol.* 13:227–31.

73. Mundt, J. O. 1975. Unidentified streptococci from plants. *Int. J. Syst. Bacteriol.* 25:281–85.

74. Newcomer, A. D., H. S. Park, P. C. O'Brien, and D. B. McGill. 1983. Response of patients with irritable bowel syndrome and lactase deficiency using unfermented acidophilus milk. *Amer. J. Clin. Nutr.* 38:257–63.

75. Niinivaara, F. P., M. S. Pohja, and S. E. Komulainen. 1964. Some aspects about using bacterial pure cultures in the manufacture of fermented sausages. *Food Technol.* 18:147–53.

76. Okafor, N. 1972. Palm-wine yeasts from parts of Nigeria. *J. Sci. Food Agric.* 23:1399–1407.

77. ———. 1975. Microbiology of Nigerian palm wine with particular reference to bacteria. *J. Appl. Bacteriol.* 38:81–88.

78. ———. 1975. Preliminary microbiological studies on the preservation of palm wine. *J. Appl. Bacteriol.* 38:1–7.

79. Palumbo, S. A., J. L. Smith, and S. A. Kerman. 1973. Lebanon bologna. I. Manufacture and processing. *J. Milk Food Technol.* 36:497–503.

80. Passmore, S. M., and J. G. Carr. 1975. The ecology of the acetic acid bacteria with particular reference to cider manufacture. *J. Appl. Bacteriol.* 38:151–58.

81. Passos, F. M. L., D. O. Silva, A. Lopez, C. L. L. F. Ferreira, and W. V. Guimaraes. 1984. Characterization and distribution of lactic acid bacteria from traditional cocoa bean fermentations in Bahia. *J. Food Sci.* 49:205–8.

82. Payne, W. J. 1970. Energy yields and growth of heterotrophs. *Ann. Rev. Microbiol.* 24:17–52.

83. Pearson, A. M., and F. W. Tauber. 1984. *Processed meats.* 2d ed., Ch. 10. Westport, Conn.: AVI.

84. Pederson, C. S. 1979. *Microbiology of food fermentations.* 2d ed. Westport, Conn.: AVI.

85. Pederson, C. S., and R. S. Breed. 1946. Fermentation of coffee. *Food Res.* 11:99–106.

86. Pinheiro, A. J. R., B. J. Liska, and C. E. Parmalee. 1968. Properties of substances inhibitory to *Pseudomonas fragi* produced by *Streptococcus citrovorous* and *Streptococcus diacetilactis. J. Dairy Sci.* 51:183–87.

87. Potter, M. E., M. B. Kruse, M. A. Matthews, R. O. Hill, and R. J. Martin. 1976. A sausage-associated outbreak of trichinosis in Illinois. *Amer. J. Pub. Hlth* 66:1194–96.

88. Prescott, S. C., and C. G. Dunn. 1957. *Industrial microbiology.* New York: McGraw-Hill.

89. Pulusani, S. R., and D. R. Rao. 1983. Whole body, liver and plasma cholesterol levels in rats fed thermophilus, bulgaricus and acidophilus milks. *J. Food Sci.* 48:280–81.

90. Raccach, M., and R. C. Baker. 1978. Lactic acid bacteria as an antispoilage and safety factor in cooked, mechanically deboned poultry meat. *J. Food Protect.* 41:703–5.

91. Radke-Mitchell, L., and W. E. Sandine. 1984. Associative growth and differential enumeration of *Streptococcus thermophilus* and *Lactobacillus bulgaricus:* A review. *J. Food Protect.* 47:245–48.

92. Rao, D. R., C. B. Chawan, and S. R. Pulusani. 1981. Influence of milk and thermophilus milk on plasma cholesterol levels and hepatic cholesterogenesis in rats. *J. Food Sci.* 46:1339–41.

93. Reddy, N. S., and S. Ranganathan. 1983. Nutritional factors affecting growth and production of antimicrobial substances by *Streptococcus lactis* subsp. *diacetylactis* S_1-67/C. *J. Food Protect.* 46:514–17.

94. Reddy, N. R., S. K. Sathe, M. D. Pierson, and D. K. Salunkha. 1981. Idli, an Indian fermented food: A review. *J. Food Qual.* 5:89–101.

95. Richardson, T. 1978. The hypocholesteremic effect of milk—A review. *J. Food Protect.* 41:226–35.

96. Roelofsen, P. A. 1958. Fermentation, drying, and storage of cacao beans. *Adv. Food Res.* 8:225–96.

97. Rohan, T. A. 1964. The precursors of chocolate aroma: A comparative study of fermented and unfermented cocoa beans. *J. Food Sci.* 29:456–59.

98. Rohan, T. A., and T. Stewart. 1966. The precursors of chocolate aroma: Changes in the sugars during the roasting of cocoa beans. *J. Food Sci.* 31:206–9.

99. Rose, A. H. 1982. *Fermented foods.* Economic Microbiology Series, 7. New York: Academic Press.

100. Saisithi, P., B.-O. Kasemsarn, J. Liston, and A. M. Dollar. 1966. Microbiology and chemistry of fermented fish. *J. Food Sci.* 31:105–10.

101. Sandine, W. E., K. S. Muralidhara, P. R. Eiliker, and D. C. England. 1972. Lactic acid bacteria in food and health: A review with special reference to enteropathogenic *Escherichia coli* as well as certain enteric diseases and their treatment with antibiotics and lactobacilli. *J. Milk Food Technol.* 35:691–702.

102. Sandine, W. E., P. C. Radich, and P. R. Elliker. 1972. Ecology of the lactic streptococci. A review. *J. Milk Food Technol.* 35:176–85.

103. Schleifer, K. H., and O. Kandler. 1972. Peptidoglycan types of bacterial cell walls and their taxonomic implications. *Bacteriol. Rev.* 36:407–77.

104. Senez, J. C. 1962. Some considerations on the energetics of bacterial growth. *Bateriol. Rev.* 26:95–107.

105. Shahani, K. M., and A. D. Ayebo. 1980. Role of dietary lactobacilli in gastrointestinal microecology. *Amer. J. Clin. Nutr.* 33:2448–57.

106. Shibasaki, K., and C. W. Hesseltine. 1962. Miso fermentation. *Econ. Bot.* 16:180–95.

107. Shieh, Y.-S. G., and L. R. Beuchat. 1982. Microbial changes in fermented peanut and soybean pastes containing kojis prepared using *Aspergillus oryzae* and *Rhizopus oligosporus*. *J. Food Sci.* 47:518–22.

108. Simonds, J., P. A. Hansen, and S. Lakshmanan. 1971. Deoxyribonucleic acid hybridization among strains of lactobacilli. *J. Bacteriol.* 107:382–84.

109. Smith, J. L., and S. A. Palumbo. 1973. Microbiology of Lebanon bologna. *Appl. Microbiol.* 26:489–96.

110. Snyder, H. E. 1970. Microbial sources of protein. *Adv. Food Res.* 18:85–140.

111. Speckman, R. A., and E. B. Collins. 1968. Diacetyl biosynthesis in *Streptococcus diacetilactis* and *Leuconostoc citrovorum*. *J. Bacteriol*. 95:174–80.
112. ———. 1973. Incorporation of radioactive acetate into diacetyl by *Streptococcus diacetilactis*. *Appl. Microbiol*. 26:744–46.
113. Stamer, J. R. 1976. Lactic acid bacteria. In *Food microbiology: Public health and spoilage aspects*. ed. M. P. deFigueiredo and, D. F. Splittstoesser, 404–26. Westport, Conn.: AVI.
114. Steinkraus, K. H., ed. 1983. *Handbook of indigenous fermented foods*. New York: Marcel Dekker.
115. Steinkraus, K. H., A. G. van Veen, and D. B. Thiebeau. 1967. Studies on idle—As Indian fermented black gram-rice food. *Food Technol*. 21:916–19.
116. Stouthamer, A. H. 1969. Determination and significance of molar growth yields. *Meth. Microbiol*. 1:629–63.
117. Subik, J., and M. Behun. 1974. Effect of bongkrekic acid on growth and metabolism of filamentous fungi. *Arch. Microbiol*. 97:81–88.
118. Sugihara, T. F., L. Kline, and M. W. Miller. 1971. Microorganisms of the San Francisco sour dough bread process. I. Yeasts responsible for the leavening action. *Appl. Microbiol*. 21:456–58.
119. Sundhagul, M., W. Daengsubha, and P. Suyanandana. 1975. Thailand's traditional fermented food products: A brief description. *Thai J. Agr. Sci*. 8:205–19.
120. Tannenbaum, S. R., and D. I. C. Wang, eds. 1975. *Single-cell protein*. Vol. 2. Cambridge, MA.: M.I.T. Press.
121. Thompson, L. U., D. J. A. Jenkins, M. A. Vic Amer, R. Reichert, A. Jenkins, and J. Kamulsky. 1982. The effect of fermented and unfermented milks on serum cholesterol. *Amer. J. Clin. Nutr*. 36:1106–11.
122. Urbaniak, L., and W. Pezacki. 1975. Die Milchsäure bildende Rohwurst-Mikroflora und ihre technologisch bedingte Veränderung. *Fleischwirtschaft* 55:229–37.
123. Van Pee, W., and J. M. Castelein. 1972. Study of the pectinolytic microflora, particularly the *Enterobacteriaceae*, from fermenting coffee in the Congo. *J. Food Sci*. 37:171–74.
124. van Veen, A. G. 1967. The bongkrek toxins. In *Biochemistry of some foodborne microbial toxins*, ed. R. I. Mateles and G. N. Wogan, 43–50. Cambridge, MA.: M.I.T. Press.
125. van Veen, A. G., and W. K. Mertens. 1934. Die Gifstoffe der sogenannten Bongkrek-vergiftungen auf Java. *Rec. Trav. Chim*. 53:257–68.
126. ———. 1934. Das Toxoflavin, der gelbe Gifstoff der Bongkrek. *Rec. Trav. Chim*. 53:398–404.
127. van Veen, A. G., and K. H. Steinkraus. 1970. Nutritive value and wholesomeness of fermented foods. *J. Agr. Food Chem*. 18:576–78.
128. Vaughn, R. H. 1975. Lactic acid fermentation of olives with special reference to California conditions. In *Lactic acid bacteria in beverages and food*, ed. J. G. Carr et al., 307–23. New York: Academic Press.
129. Wardlaw, F. B., G. C. Skelley, M. G. Johnson, and J. C. Acton. 1973. Changes in meat components during fermentation, heat processing and drying of a summer sausage. *J. Food Sci*. 38:1228–31.
130. Waslien, C. I. 1976. Unusual sources of proteins for man. *CRC Crit. Rev. Food Sci. Nutri*. 6:77–151.

131. Williams, R. A. D. 1975. A review of biochemical techniques in the classification of lactobacilli. In *Lactic acid bacteria in beverages and food,* ed. J. G. Carr et al., 351–67. New York: Academic Press.
132. Yong, F. M., and B. J. B. Wood. 1974. Microbiology and biochemistry of the soy sauce fermentation. *Adv. Appl. Microbiol.* 17:157–94.

VI

INDICATOR AND

FOOD-BORNE PATHOGENS

17.

INDICES OF FOOD
SANITARY QUALITY;
MICROBIOLOGICAL
STANDARDS AND CRITERIA

Contamination of foods by disease-producing microorganisms has been known and studied since around 1880. Since that time, numerous instances of food-borne diseases have been recorded in addition to those commonly referred to as food poisoning. Prior to the development of the pasteurization process, pathogens causing brucellosis, scarlet fever, typhoid fever, diphtheria, and other diseases were commonplace in milk. Diseases of animals transmissible to man, such as tuberculosis, brucellosis, and so on, were also commonplace in meats prior to the mandatory federal inspection of slaughter animals. A third source of food-borne diseases is contamination of foods by food handlers. This source is one of the most serious in food-poisoning outbreaks, especially with convenience foods. Also of great current importance are pathogenic and food-poisoning organisms that tend to be associated with certain animals, organisms such as salmonellae and *C. perfringens*.

Among the requirements for foods to be of good sanitary quality, they must be shown to be free of hazardous microorganisms, or those present should be at a safe low level. In general, it would not be feasible to examine each food or food product for the presence of hazardous organisms. The practice that has been in effect for many years and continues to be followed is to determine the sanitary quality of foods by their content of certain indicator organisms. The indicators of sanitary quality now employed for foods consist of two groups of bacteria: coliforms and enterococci. In addition, total numbers are also useful in this regard, and are here treated in this manner.

COLIFORM BACTERIA AS INDICATORS OF FOOD SANITARY QUALITY

The use of *Escherichia coli* as an indicator of water-borne pathogens was apparently first suggested in 1892 by F. Schardinger. This organism was proposed because it is generally found in the intestinal contents of man and animals. A year later, Theobald Smith noted that since this organism is so uniformly present in the intestinal tract, its presence outside the intestines may be regarded as due to contamination with fecal discharges of man or animals. This marked the beginning of the use of the coliforms as indicators of pathogens in water, a practice which has continued.

The primary coliform bacteria consist of *E. coli* and *Enterobacter aerogenes* (see below). Both of these organisms are short, gram-negative rods that ferment lactose with the production of gas. *E. coli* has as its primary habitat the intestinal tract of man and other animals—hence, *coli* for colon. Of these organisms, *E. coli* is normally found in the gastrointestinal tract of animals, where they normally do not cause disease. From his studies on the flora of the intestinal tract of man, Haenel (24) placed the percentage of coliforms at less than 1. He found 10^8–10^9/g of these organisms to be common in adult feces. Although *E. aerogenes* may sometimes be found in the intestinal tract of man, this organism is normally associated with vegetation. In a study of 6,577 strains of coliforms from various sources, Griffin and Stuart (23) concluded that the occurrence of *E. coli* outside the intestinal tract and *E. aerogenes* and intermediates in places other than nonfecal materials were adventitious. The coliform organisms are well established as fecal indicators for water. Their use as indicators of food sanitary quality derives from their successful use for water. The finding of large numbers of these organisms in foods and water is taken to indicate fecal pollution or contamination. Since the water-borne diseases are generally intestinal diseases, the existence of pollution is taken to indicate the *possibility* that the etiologic agents of these diseases may be present. Whether or not intestinal pathogens are present, the presence of fecal matter in foods or water is undesirable. McCoy (41) has stated that in the examination of foods, the presence of intestinal inhabitants should be taken to indicate a lack of cleanliness, not of safety. He asserts that the safety of foods can be assessed only by examining for the presence of pathogens.

With respect to the overall choice of indicator organisms, Buttiaux and Mossel (6) stated that they should possess the following properties: (1) ideally the bacteria selected should demonstrate specificity—occur only in intestinal environments; (2) they should occur in very high numbers in feces, so as still to be encountered in high dilutions; (3) they should possess a high resistance to the extraenteral environment, the pollution of which is to be indicated; and (4) they should permit relatively easy and fully reliable detection even when present in very low numbers.

Since *E. coli* is more indicative of fecal pollution than *E. aerogenes,* it is sometimes desirable to determine the incidence of this organism in a coliform population. For this purpose, the IMViC formula is employed where I = indole production; M = methyl red reaction; V = Voges-Proskauer reaction (production of acetoin); and C = citrate utilization. The two organisms may then be differentiated as follows:

	I	M	V	C
E. coli	+	+	−	−
E. aerogenes	−	−	+	+

While the coliforms are generally regarded as being *E. coli* and *E. aerogenes*, it should be noted that the genera *Citrobacter* and *Klebsiella* come under this functional classification. *Citrobacter* spp. have been referred to as intermediate coliforms, and delayed lactose fermentation by some strains is known. All strains are MR + and VP −. Most are citrate + while indole production varies. *Klebsiella* isolates are highly variable with respect to IMViC reactions. Although *K. pneumoniae,* the most common klebsiellae in the U.S., is generally MR − VP + and C +, variations are known to occur especially in the MR and indole reactions. *E. coli* isolates tend to be rather consistent in their reactions, with the MR reaction being the most stable. The IMViC reactions + + − − designate *E. coli* variety I, while variety II strains are − + − −. Some recent studies with coliform organisms give indications that this indicator group is not as easily definable as was once believed. When 765 coliforms recovered from frozen-blanched vegetables were subjected to elevated temperature tests, only 7 and 2% were *E. coli* types I and II, respectively (55). The majority of the isolates gave the IMViC reaction − − + + and suggested that coliforms from products of the type examined have no special sanitary significance.

Fecal coliforms are distinguished most readily from nonfecal coliforms by use of elevated temperature incubations. This is accomplished by inoculating tubes of EC broth from gas-positive lauryl sulfate tryptose (LST) broth cultures and incubating in a water bath at a temperature between 44 and 46°C, usually 44.5° or 45.5°C. While a higher percentage of *E. coli* isolates are positive at elevated temperatures than other coliforms, some *Enterobacter* and *Klebsiella* isolates are fecal coliforms by this criterion. In general, about 90% of fecal coliforms are *E. coli* strains. They produce gas in EC broth at elevated temperatures while nonfecals generally do not. In a study of clinical/pathogenic and environmental strains, 93% of the former and 85% of the latter were fecal-coliform positive. With *K. pneumoniae* 85% of 58 known pathogenic strains but only 16% of 120 environmental isolates were positive (1). These authors indicated that the occurrence of fecal coliform–positive klebsiellae in environmental water samples strongly supports their sanitary significance. Whether or not this position can be transferred to foods is unclear at this time. The superiority of EC broth for fecal coliforms was shown by Fishbein (16). In another study, a preference for

45.5°C over 44.5°C was indicated (17). Two to threefold fewer false positives at 45.5° were found compared to 44.5°C by the latter investigators. The Canadian requirement for fecal coliform determinations is 45.0°C. In a study comparing 44.5°, 45.0°, and 45.5°C, 21% false negatives were found using 45.5°C for ground meats, and the continued use of elevated temperatures as a valid criterion for fecal coliforms was questioned (60). Fecal coliforms can be determined by use of other media at elevated temperatures, and some of these techniques have been reviewed (13). Methods for determining coliforms or fecal coliforms should be obtained from a standard reference in Table 5-1.

As stated above, the coliform index was originally proposed for and used as a test of water quality, followed by its use in the dairy industry. Although it may be a bit risky to transfer a procedure from one type of product to another and expect the same performance, the coliform index has found wide use in assessing sanitary quality in foods, and indeed, coliform standards have been established for some foods and recommended for a large number of others. The main difference between water and foods is that the former generally does not support growth of these organisms while many foods do (see below). For a review of *E. coli* and other indicators for foods and water, see Chordash and Insalata (9).

With respect to enteric viruses, there exists a body of data indicating that waters that meet coliform or *E. coli* standards for shellfish may also contain enteroviruses at significant levels. Gulf Coast waters off the Texas and Louisiana coasts contained echovirus 4 and polio viruses 1 and 3 even though coliform and *E. coli* indices were at acceptable levels (19). Similar findings have been reported for viruses in oysters harvested from two reefs off the Mississippi Gulf Coast (14). Enteric viruses have been recovered from waters that met coliform, residual chlorine, and turbidity standards (32). Food-borne viruses are discussed further in Chapter 22.

Giardiasis is contracted from waters containing the etiologic agent *Giardia lamblia,* and may be transmitted by foods (2). This protozoan has been found in waters that met coliform standards (11).

THE GROWTH OF COLIFORMS. Like most nonpathogenic gram-negative bacteria, coliforms grow well on a large number of media and in many foods. These organisms have been reported to grow at as low as -2°C and as high as about 50°C. In foods, growth is poor or very slow at about 5°C although several authors have reported the growth of coliforms at 3–6°C. With respect to pH, these organisms have been reported to grow over a rather wide range, with values from 4.4 to 9.0. *E. coli* can be grown on a medium containing only an organic carbon source such as glucose and a source of nitrogen such as $(NH_4)_2SO_4$, along with other minerals. Consequently these organisms grow well on nutrient agar and produce visible colonies within 24 h at 37°C. They may therefore be expected to grow in a large number of foods under the proper conditions. They are capable of growth in the presence of bile salts, which inhibit the growth of gram-positive bacteria. Advantage is taken of the latter fact in their isolation from various sources. Unlike most bacteria, they have the capacity to ferment lactose with the production of gas, and this characteristic alone is sufficient to make presumptive determinations of coliforms. By incorporating lactose and bile salts into culture media (such as MacConkey agar), it becomes possible to differentiate between these organisms and most others that may be present in foods as well as water. The general ease with which the coliforms can be cultivated and differentiated makes them somewhat ideal as indicators of sanitary quality. Their identification is sometimes complicated by the presence of atypical strains. The aberrant lactose fermenters, however, appear to be of questionable sanitary significance (23).

DISTRIBUTION OF COLIFORMS. As already noted, the primary habitat of *E. coli* is the intestinal tract of many warm-blooded animals, although it may at times be absent from the gut of hogs, while that of *E. aerogenes* is vegetation (see above) and occasionally the intestinal tract. It is not difficult, however, to sometimes demonstrate the presence of these organisms in air, in dust, on the hands, and in and on many foods. The problem is not simply the presence of coliforms, but their relative numbers. Most market vegetables can be shown to harbor small numbers of lactose-fermenting, gram-negative rods of the coliform type; but if these products have been harvested and handled properly, the numbers tend to be quite low and consequently of no real significance from the standpoint of public health. The bacteria formerly classified as paracolons and characterized by the slow fermentation of lactose are often associated with fresh vegetables (see *Citrobacter*, Chapter 2). The author has found large numbers of these types on raw, preroasted squash seeds. While generally regarded as being nonpathogens, they have been incriminated in mild food-poisoning outbreaks where large numbers of cells were required.

One of the attractive properties of *E. coli* as a fecal indicator for water is its period of survival in water. It generally dies off at about the same time as the more common intestinal bacterial pathogens, although some reports indicate that some bacterial pathogens are more resistant than this organism in water. It is not, however, as resistant as intestinal viruses that some investigators feel its presence should also indicate. From studies by a number of workers, Buttiaux and Mossel (6) concluded that various pathogens may persist after *E. coli* is destroyed in frozen foods, under refrigerated conditions, in radiated foods, and in treated waters. These authors have reported that only in acid food does *E. coli* have particular value as an indicator organism due to its relative resistance to low pH values.

COLIFORM STANDARDS FOR FOODS. While the presence of large numbers of coliforms in foods is highly undesirable, it would be virtually impossible to eliminate all of these organisms from fresh and frozen foods. The basic questions at this point are these: (1) Under proper conditions of harvesting, handling, storage, and transport of foods, what is the lowest possible and feasible number of coliforms to maintain? (2) At what point does a good product become unsafe with respect to numbers of coliforms? In the case of water and dairy products, there is a long history behind the safety of allowable coliform numbers. Coliform standards for water, dairy products, and other foods covered by some federal, state, and city laws now in effect are as follows: not over 10/ml for Grade A pasteurized milk and milk products, including cultured products; not over 10 for certified raw, and not over 1 for certified pasteurized milk; not over 10 for precooked and partially cooked frozen foods; not over 100 for crabmeat; and not over 100 for custard-filled items (see later section in this chapter). It can be seen that low numbers of coliforms are permitted in many instances ranging from 1 to not over 100/g or ml. Implicit in these standards are answers to the two questions raised above, that is, feasibility and safety.

On the basis of present knowledge, it appears that the coliform index as an index of sanitary quality is applicable to at least some foods. The use of this index for foods brings objections from some investigators who feel that the enterococci rather than coliforms better reflect the sanitary quality of some foods.

ENTEROCOCCI AS INDICATORS OF FOOD SANITARY QUALITY

The enterococci are members of the genus *Streptococcus* (see Chapter 3), which consists of gram-positive cocci that produce long or short chains and differ from most other gram-positive cocci in being catalase negative. Although there is some confusion as to which species of streptococci should be included under the term enterococci, it is generally agreed that the enterococci are all members of Lancefield's serologic Group D streptococci. The Group D streptococci consist of four species and three subspecies:

1. *S. faecalis*
   subsp. *faecalis*
   subsp. *liquefaciens*
   subsp. *zymogenes*
2. *S. faecium*
3. *S. bovis*
4. *S. equinus*

Numbers 1 and 2 above constitute the enterococcus group. Numbers 3 and 4 actually belong to the viridans group but possess antigen D in common with the enterococci. **Fecal streptococci** have been characterized as being "all *Streptococcus* spp. consistently present in significant numbers in fresh fecal excreta . . ." (27). Culturally, the enterococci grow at 10°C, in the presence of methylene blue, in 6.5% NaCl, and at pH 9.6, while the other two Group D streptococci do not.

The above description of enterococci has been valid for several decades. While it may remain valid, proposals have been made to place *S. faecalis* and *S. faecium* in the newly proposed genus *Enterococcus*. All members of this genus are gram-positive, facultatively anaerobic, catalase-negative cocci that occur mostly in pairs or short chains. They grow at 10°C, and generally at 45°C; survive heating at 60°C for 30 min; grow in 6.5% (wt/vol) NaCl, at pH 9.6, and in 40% (vol/vol) bile; and produce L-lactic acid as the major end product of glucose fermentation (10). As proposed, seven species have been assigned to the genus *Enterococcus: E. avium, E. casseliflavus, E. durans, E. faecalis, E. faecium, E. gallinarum,* and *E. malodorus.* The differential biochemical characteristics of these species have been presented by Collins et al. (10). If this proposal stands, a redefinition of fecal streptococci/enterococci will be in order. Meanwhile, the remainder of this section deals with enterococci as streptococci and does not take into account the proposed taxonomic changes.

Ostrolenk et al. (48) were apparently the first to show the feasibility of employing the enterococci as indicators of pollution. The first comparative study of coliforms and enterococci as indicators for foods was apparently that of Burton (5).

CHARACTERISTICS AND GROWTH OF ENTEROCOCCI. The enterococci are typical of gram-positive bacteria in that they are more fastidious in their nutritional requirements than gram-negative bacteria. These organisms are unable to grow on simple, sugar-salts media and require varying numbers of B vitamins and other organic constituents. With respect to growth temperature, they have been reported to grow at from 0 to over 50°C, with at least five authors having reported growth between 0° and 6°C. While these organisms are basically mesophilic in nature, many strains may be said to be facultative psychrophiles. The enterococci grow over a wider pH range than coliforms. As may be seen from Table 17-1, *S. faecalis* and *S. faecium* types grow in the presence of 6.5% NaCl and in the presence of 40% bile. Table 17-1 also presents other physiologic characteristics that are useful in differentiating one species or variety from another. With respect to oxygen tension, the enterococci are like most other streptococci in that they are aerobic but tend to grow best under somewhat reduced conditions. Organisms that behave in this manner are generally referred to as *microaerophiles.*

DISTRIBUTION OF ENTEROCOCCI IN NATURE. Much like the coliforms, especially *E. coli,* the enterococci are primarily of fecal origin, with *S. faecalis* and its varieties being associated more with the human intestinal canal than that of other animals. *S. faecium* and *S. durans* tend to be associated more with the intestinal tract of swine than *S. faecalis,* while *S. bovis* and *S.*

Table 17-1. Group D streptococci: Selected differential physiological characters.

	Division 1	Division 2		Division 3	
	S. faecalis and varieties	S. faecium	S. durans	S. bovis	S. equinus
β Hemolysis	−/+	−	+/−	−	−
Growth					
10°	+	+	+	−	−
45°	+	+	+	+	+
50°	+	+[a]	−	−	−
pH 9.6	+	+	+/−	−	−
6.5% NaCl	+/−	+/−	+/−	−	−
40% bile	+	+	+	+	+
Resists 60°C for 30 min.	+	+	+/−	−	−[a]
NH_3 from arginine	+	+	+	−	−
Gelatin liquefied	−/+	−	−	−	−
Tolerates 0.04% Pot. tellurite	+	−	−	−	−
Acid from:					
Glycerol (anaerobic)	+[a]	−	−	−	−
Mannitol	+	+	−[a]	−/+	−
Sorbitol	+[a]	−[a]	−	−/+	−
L-arabinose	−	+[a]	−	+/−	−
Lactose	+	+	+	+	−
Sucrose	+[a]	+/−	−	+	+[a]
Raffinose	−[a]	−[a]	−	+	−[a]
Melibiose	−	+[a]	−[a]	+	−[a]
Melezitose	+[a]	−	−	−	−
Starch hydrolyzed	−	−	−	+[a]	−[a]
Tetrazolium reduced at pH 6.0	+	−	−	+/−	−

+ = positive result
− = negative result
+/− = variation between strains, majority positive
−/+ = variation between strains, majority negative
[a]Occasional strains atypical.
Division 1 and Division 2 fulfill the criteria for the "enterococcus group" of Sherman (1938).
Reproduced by permission from P. M. F. Shattock, "Enterococci," in *Chemical and Biological Hazards in Food.* J. C. Ayres, A. A. Kraft, A. T. Snyder, and H. W. Walker, editors. © 1963 by the Iowa State University Press, Ames.

equinus tend to be associated more with the intestinal tract of cattle and horses, respectively. All samples of human and pig feces examined by Buttiaux (6) contained these organisms while 66 and 88% of the cow and sheep feces, respectively, were found to contain these organisms. This is a somewhat higher carriage rate than for coliforms. *S. faecalis* appears to be more specific for the intestinal tract of man than other enterococcal species, yet this species is not generally accorded the same relationship to the enterococci as is *E. coli*. In addition to their fecal sources, these organisms exist on plants, insects, and in soils (45, 46). Their sources to insects and plants may be through animal fecal matter. They have been regarded as temporary residents of plants and are disseminated among plants by the action of insects and wind and reach the soil by these agencies, and by rain, and gravity (45). Although *S. faecalis* is often of fecal origin and its presence in foods may be taken to indicate fecal contamination, some *S. faecalis* strains appear to be common microflora of plants with no sanitary significance when found in foods. Mundt (46) studied *S. faecalis* from humans, plants, and other sources and found that the nonfecal indicators could be distinguished from the more fecal types by their reaction in litmus milk and their fermentation reactions in melezitose and melibiose broths. In another study of 2,334 strains from dried and frozen foods, this investigator found that a high percentage of strains bore a close similarity to the plant-resident streptococci and would therefore not be of any sanitary significance (47). When used as indicators of sanitary quality of foods, it is necessary to ascertain whether isolates are of the plant type or whether they represent those of human origin. The enterococci may also be found in dust. In general, they are rather widely distributed in nature, especially in slaughterhouses and curing rooms where pork products such as cured hams are handled.

With respect to the use of enterococci as indicators of water pollution, some investigators who have studied their persistence in water have found that this group dies off at a faster rate than coliforms, while others have found the reverse to be true. Leininger and McCleskey (37) noted that this group does not multiply in water as coliforms sometimes do. Their more exacting growth requirements may be taken to indicate a less competitive role for these organisms in water environments. In sewage, Litsky et al. (39) found coliforms and enterococci to exist in high numbers but found approximately thirteen times more coliforms than enterococci.

RELATIONSHIP OF ENTEROCOCCI TO SANITARY QUALITY OF FOODS. A large number of investigators have examined the use of an enterococcus index for food safety and feel that it is a better index of food sanitary quality than the coliform index, especially for frozen foods. In his study of frozen chicken pies obtained from one processor, Hartman (26) found that enterococcus counts were more closely related to total counts than to coliform counts, while coliform counts were more closely related to enterococcus counts than to total counts (see Table 17-2). The same author's findings were in agreement with those of Hahn and Appleman (25), Larkin et al. (34, 35, 36), and Kereluk and Gunderson (31), and were supported by Raj et al. (51) that in frozen foods, enterococci occur in greater numbers than coliforms. This fact is further illustrated in Tables 17-3 and 17-4 for frozen precooked fish sticks.

In a study of 376 samples of commercial frozen vegetables, Burton (5) found that coliforms were more efficient indicators than enterococci prior

Table 17-2. Relationships between coliform, enterococcus, and total counts in 456 chicken pies obtained from one processor (26).

A. Numbers of samples with various coliform: total count ratios

Coliforms per g	Total count per g			
	Below 10^3	*10^3 to 10^4*	*10^4 to 10^5*	*Over 10^5*
Below 3	0	*94*	82	3
3 to 9	0	41	*93*	10
10^1 to 10^2	0	25	60	*25*
Over 10^2	0	0	8	*12*

B. Numbers of samples with various enterococcus: total count ratios

Enterococci per g	Total count per g			
	Below 10^3	*10^3 to 10^4*	*10^4 to 10^5*	*Over 10^5*
Below 10^2	0	15	1	0
10^2 to 10^3	0	*104*	41	2
Over^3 to 10^4	0	36	*136*	15
Over 10^4	0	0	23	*32*

C. Numbers of samples with various coliform: enterococcus ratios

Coliforms per g	Enterococci per g			
	Below 10^2	*10^2 to 10^3*	*10^3 to 10^4*	*Over 10^4*
Below 3	*11*	*84*	76	4
3 to 9	4	44	*86*	10
10^1 to 10^2	1	19	63	*26*
Over 10^2	0	1	7	*12*

Italicized figures denote similar trends in all three comparisons where increases in one count were, in general, associated with concomitant increases in the other two counts.

to freezing and storage, while enterococci were superior indicators in frozen foods. In samples stored at $-20°C$ for 1–3 mon, 81% of enterococci and 75% of coliforms survived. After 1 year, 89% of enterococci survived while only 60% of the coliforms survived. Larkin et al. (35) showed that enterococci remained relatively constant for 400 days when stored at freezing temperatures (see Table 17-3). The same authors inoculated *E. coli* and *S. faecalis* into orange juice concentrate, followed by storage at $-10°F$ for 51 days. During this period of time, there was no appreciable decrease in the numbers of *S. faecalis* whereas *E. coli* fluctuated. The relative occurrence of coliforms, enterococci, and staphylococci in 0.1-g portions of 391 precooked frozen foods is presented in Table 17-5. It should be noted that both enterococci and staphylococci could be recovered from more samples than coliforms. In this particular study, enterococci were found in 86 and staphylococci in 40 of the samples in which coliforms were not detected. Cuthbert et al. (12) studied the viability of *E. coli* and *S. faecalis* in soils and found that

Table 17-3. Comparison of enterococci and coliform most probable number (MPN) counts in frozen precooked fish sticks (51).

Sample no.	Enterococci MPN count/ 100 g	Coliforms MPN count/ 100 g
1	86,000	6
2	18,600	19
3	86,000	0
4	46,000	300
5	48,000	150
6	46,000	28
7	46,000	150
8	18,600	7
9	8,600	0
10	4,600	186
11	4,600	186
12	48,000	1,280
13	8,600	46
14	4,600	480
15	48,000	240
16	10,750	1,075
17	10,750	17,000
18	60,000	23,250
19	10,750	2,275
Average	32,339	2,457

Table 17-4. The effect of $-6°F$ storage on the longevity of coliforms and enterococci in precooked frozen fish sticks (31).

Days in storage at $-6°F$	Most probable number[a]	
	Coliform	Enterococci
0	5,600,000	15,000,000
7	6,000,000	20,000,000
14	1,400,000	13,000,000
20	760,000	11,300,000
35	440,000	11,200,000
49	600,000	20,000,000
63	88,000	11,000,000
77	395,000	15,000,000
91	125,000	41,000,000
119	50,000	5,400,000
133	136,000	7,400,000
179	130,000	5,600,000
207	55,000	3,500,000
242	14,000	4,000,000
273	21,000	4,000,000
289	42,000	3,200,000
347	20,000	2,300,000
410	8,000	1,600,000
446	260	2,300,000
481	66	5,000,000

[a]Average of four determinations.

in limestone soils, these organisms persisted for several weeks, though in acid peat soils they persisted for only a few days. In general, the enterococci have a greater potential for survival with remoteness from the sources of pollution (4).

From fourteen groups of dried foods, Mundt (47) recovered streptococci

Table 17-5. Presence of coliforms, enterococci, and staphylococci in 0.1-g portions (precooked frozen foods) (38).

Processing plant		Samples examined	Samples positive for		
			Coliforms	Enterococci	Staphylococci
Military	No.	234	10	96	50
	(%)	(100)	(4.3)	(41.0)	(21.3)
Commercial	No.	157	65	131	65
	(%)	(100)	(41.4)	(83.4)	(41.4)

Table 17-6. **Comparison of coliforms and enterococci as indicators of food sanitary quality.**

Characteristic	Coliforms	Enterococci
Morphology	Rods	Cocci
Gram reaction	Negative	Positive
Incidence in intestinal tract	10^7–10^9/g feces	10^5–10^8/g feces
Incidence in fecal matter of various animal species	Absent from some	Present in most
Specificity to intestinal tract	Generally specific	Generally less specific
Occurrence outside of intestinal tract	Common in low nos.	Common in higher nos.
Ease of isolation and identification	Relatively easy	More difficult
Response to adverse environmental conditions	Less resistant	More resistant
Response to freezing	Less resistant	More resistant
Relative survival in frozen foods	Generally low	High
Relative survival in dried foods	Low	High
Incidence in fresh vegetables	Low	Generally high
Incidence in fresh meats	Generally low	Generally low
Incidence in cured meats	Low or absent	Generally high
Relationship to food-borne intestinal pathogens	Generally high	Lower
Relationship to nonintestinal food-borne pathogens	Low	Low

from 57% while 87% of thirteen different frozen vegetables yielded these organisms. As noted above, many of these isolates were of the plant-resident type (see Chapter 4 for the incidence of streptococci in other foods).

Enterococci and coliforms as sanitary indicators are compared in Table 17-6.

OTHER INDICATORS

The determination of any or all members of the family Enterobacteriaceae as indicators of food sanitary quality has received the attention of more and more food scientists. Some advantages of using this group include: (1) since the definition of the Enterobacteriaceae is well established, problems due to ill-defined taxonomy of coliforms are resolved; (2) the possibility of obtaining false negative strains or of slow lactose fermenters as the predominant enteric contaminants is eliminated; and (3) the reliability of monitoring can be increased by the addition of direct testing for pathogens (44). In addition to *Escherichia, Citrobacter, Enterobacter,* and *Klebsiella* spp., the Enterobacteriaceae also contain the genera *Serratia, Proteus, Edward-*

siella, and other genera that commonly exist in the intestinal tract. While most of the Enterobacteriaceae exist in the intestinal canal, some (for example, *Proteus* and *Serratia*) are widespread in other environments.

The bifidobacteria were first suggested by Mossel in 1958 as being good indicators of sanitary quality, especially of waters. The problem that prevented much attention to this group was the lack of a relatively simple and reliable method for their enumeration, but the development of the YN-6 medium has made it possible to enumerate this group with much more consistency and ease. About seventeen species of the genus *Bifidobacterium* are recognized and these organisms are common to the gastrointestinal tract of man and pigs. They are found occasionally in cattle and sheep but apparently not in other mammals. The indicator types are sorbitol-fermenting, and *B. breve* and *B. adolescentis* appear to be exclusively associated with human feces (40). Compared to fecal coliforms, their survival in water is shorter and ratios of bifidobacteria/*E. coli* decrease with increasing distance from the pollution source (52). Studies on a tropical rainforest watershed in Puerto Rico indicated that *E. coli* could survive, remain physiologically active, and regrow at rates that were dependent on nutrient levels of ambient waters, thus becoming normal flora in tropical freshwater environments. On the other hand, *B. adolescentis* did not survive as well and thus proved to be a better indicator of recent fecal contamination (7). The latter suggests that bifidobacteria can be used to determine the recency of human pollution in waters that are not chlorinated, and perhaps to distinguish between pollution from man and other animals. Their value as indicators of sanitary quality for foods remains to be determined.

Among other organisms suggested as fecal indicators are *Ristella* spp., lactobacilli, clostridia, and Group D streptococci (all members). *Ristella* or *Bacteroides* spp. have been reported to be the second largest group of organisms in the human adult's intestinal tract (24). Anaerobic lactobacilli have been found to represent over 50% of the human intestinal flora—generally 10^9–10^{10}/g of fecal matter (24). While the clostridia are basically soil forms, they may at times be found in the intestinal tract in varying numbers. Their existence in places outside the intestinal tract makes them unattractive as fecal indicators along with the lactobacilli. Although the Group D streptococci are also found outside the intestinal tract, some have suggested that they are better indicators than coliforms for certain foods.

TOTAL COUNTS AS INDICATORS OF FOOD SANITARY QUALITY

As discussed in Chapter 4, total counts (more often aerobic plate counts, APC) on food products not only reflect handling history, state of decomposition, or degree of freshness; they may in some instances reflect on the sanitary quality of foods. As Silliker pointed out (54), total counts most effectively evaluate the sanitary quality of foods that do not support microbial growth. In foods of this type (dried or frozen, for example), total counts may be taken to indicate the type of sanitary control exercised in their production, transport, and storage. The same may be said of fresh foods when it is desired to set a standard to be used as a guide to storage life (28). Total counts as indicators of sanitary quality in fresh foods are made somewhat untenable by the fact that part of the total count flora may represent pathogens. The latter, however, is minimized by food storage at temperatures that do not permit the growth of pathogens, or by other conditions that are generally unfavorable to the growth of pathogens (see the section on the microbial ecology of staphylococcal growth in Chapter 18). It should be noted that low total counts do not always represent safe products. In their study of commercial frozen-egg preparations, Montford and Thatcher (43) were able to isolate salmonellae from one preparation with a total count of only 380/g and from several others with total counts below 5,000/g. Even though the total counts of these preparations were quite low, all were shown to contain coliforms. It is also possible to have low-count foods in which toxin-producing organisms have grown and produced toxins that remain stable to conditions that may not favor the continued survival of the cells. The sanitary quality of foods such as sauerkraut, fermented milks, and related products cannot be ascertained by total plate counts, since these products are produced by the activities of microorganisms.

In a critical discussion of coliforms, enterococci, and total plate counts as indices of food sanitary quality, Levine (38) pointed out that all too often a sanitary index becomes one of attainment rather than safety. This author further noted that total plate counts would probably give as much information as any other microbiological index suggested to date as an index of sanitary processing or proper storage of food products. A more recent study of a large number of ready-to-eat foods suggests that the APC is the most suitable method for evaluating the microbial quality of foods and that where food safety is of concern, a search for specific pathogens should be made (42).

In adopting microbiological standards for foods, the primary concerns are those of product safety and shelf life. It might well be that total plate counts, rather than other indicators, applied primarily to plant sanitation and critical process steps rather than merely to the finished products would be the most suitable approach to this problem. Also, not all dangerous contamination can be controlled by the use of total counts or indicator organisms. In these instances, it may be necessary to process foods in such a way that certain pathogens are specifically destroyed as, for example, in the pasteurization of whole egg. Of obvious value in deciding which method is best for assessing food sanitary quality would be more knowledge about the microbial ecology of microorganisms in foods and the use of standardized isolation and plating procedures. This is discussed further in the next section.

MICROBIOLOGICAL STANDARDS AND CRITERIA

The microbiological criterion for foods as defined by the Codex Alimentarius Commission in 1980 and as summarized by Silliker (55) is composed of four basic components:

1. a statement of the contaminant of concern (species/groups of organisms, or their toxins)
2. the analytical methodology to be used for detecting/enumerating the above
3. the sampling plan—number of samples from a given lot or site
4. the decision criteria—microbiological limit(s) appropriate to the product

In general, a microbiological standard is a microbiological criterion that has the connotation of law attached to it. Specifications, limits, and guidelines are criteria established sometimes for use in commerce and by regulatory agencies without the connotation of law (see ref. 29 for further details). In the remaining parts of this section, microbiological standards and criteria are used somewhat interchangeably.

Since the safety and keeping quality of fresh foods are related to microbial content, microbiological standards have been proposed for a variety of foods and some of these have been adopted. Microbiological standards for pasteurized and certified milk have been in effect since around the turn of the century while standards for other foods have appeared at a much slower rate. Elliott and Michener (15) presented the principal precautionary arguments of various individuals relative to the adoption of microbiological standards as follows:

1. A single set of microbiological standards should not be applied to foods as a miscellaneous group, such as "frozen foods" or "precooked foods."
2. Microbiological standards should be applied first to the more hazardous types of foods on an individual basis, after sufficient data are accumulated on expected bacterial levels, with consideration of variations in composition, processing procedures, and time of frozen storage.
3. When standards are chosen, there should be a definite relation between the standard and the hazard against which it is meant to protect the public.
4. Methods of sampling and analysis should be carefully studied for reliability and reproducibility among laboratories, and chosen methods should be specified in detail as part of the standard.
5. Tolerances should be included in the standard to account for inaccuracies of sampling and analysis.
6. At first, the standard should be applied on a tentative basis to allow for voluntary compliance before becoming a strictly enforced regulation.

7. Microbiological standards will be expensive to enforce.
8. If standards are unwisely chosen they will not stand in courts of law.

The above authors have also presented views of some of those in favor of adopting microbiological standards for foods:

1. Bacteriological standards are a convenience and a necessity.
2. Bacteriological standards enhance plant sanitation.
3. Low bacterial counts are attainable.
4. Low bacterial counts are associated with safe foods.
5. Bacterial counts reflect sanitation level.
6. Bacterial counts reflect degree of decomposition.
7. Low bacterial counts will enhance shelf life.

Among the arguments advanced by those who oppose the adoption of bacterial standards are the following (15):

1. There is no need for bacteriological standards; present control is adequate.
2. Bacteriological standards will not free foods of danger from pathogens.
3. A fecal indicator standard has limitations.
4. Total count is unrelated to danger or to spoilage.
5. Methods of sampling and analysis are inadequate.
6. Existing laboratory facilities and personnel are inadequate.
7. Processing and storage influence viable counts.
8. Excessive sanitation will introduce a food-poisoning hazard.
9. Foods will be overcooked or preservatives will be introduced to meet a standard.
10. More background information is needed.
11. Bacterial standards will be hard to defend in court.

In their examination of the advantages and objectives of microbiological standards, Shiffman and Kronick (53) found that the arguments fell into the following four categories:

1. Scientific or theoretical validity—are the standards based on verified data?
2. Technical—are standards technically feasible?
3. Administrative—can standards be employed in a regulatory food program?
4. Legal—will standards have legal acceptance?

Greenberg et al. (22) called attention to the increased costs that will result if standards are adopted widely. According to these authors, the increased costs will result from: (1) increased sampling, (2) lower action levels, (3) lengthened sampling time, (4) disposing of noncompliance products, (5) increased processing costs, and (6) reduced competition. They suggested the institution of an adequate audit system for testing products direct from the manufacturer and the field in lieu of lot standards. This practice could be aided by more widespread uses of the hazard analysis, critical control point (HACCP) concept at the manufacturing level. As described by Bauman (3), HACCP is a preventive system of control, especially with respect to microbiological hazards. It includes a careful analysis of ingredients, products, and processes in an effort to determine those components or areas that must be maintained under very strict control to assure that the end product meets the microbiological specifications that have been developed. **Hazard analysis** is the identification of sensitive ingredients, critical process points, and relevant human factors as they affect product safety; and **critical control points** are those processing determiners whose loss of control would result in an unacceptable food safety risk (3).

For meats and poultry products in particular, Tompkin (58) has indicated that the development of good manufacturing practices (GMP) along with an HACCP program are better alternatives to microbiological testing of finished products except for an occasional analysis for a specific purpose. The application of the HACCP concept in the production of frozen (49) and canned foods (30) has been discussed elsewhere.

It should be noted that the basic idea behind the adoption of microbiological standards for foods is protection of the consumer—protection from unnecessary food-poisoning hazards as well as from deteriorating products. Foster (18) called attention to some of the more basic thinking behind the adoption of microbiological standards. This author noted that high numbers of bacteria are associated with potential food hazards, especially in raw foods. He also noted that high counts in processed foods must generally be viewed with suspicion. When high counts are found in precooked foods, a process which should effectively reduce the number of organisms to a low level, there is every reason to be suspicious of such products.

Microbiological standards have been suggested by many investigators and public health officials, beginning in 1903, when A. Marxer first suggested a microbial standard for hamburger meat of 10^6/g. While most authors have suggested total viable count standards, some have stressed the presence, absence, or numbers of indicator or pathogenic organisms as being perhaps more important than total numbers. While it is true that aliquots from the same batch of food tested under similar conditions sometimes yield variable results, the variations in previously suggested standards from one author to another no doubt reflect different plate count techniques as well as varying attitudes toward the significance of microorganisms in foods. It is important that standard procedures be employed in examining foods, and for this reason the references in Table 5-1 are recommended. This not only makes more meaningful the findings of all investigators, but also clarifies the feasibility and desirability of standards for some of the many foods not now covered.

Microbial standards and food safety

As noted above, microbial standards should result in foods that have a longer shelf life and foods that are free of microbial hazards. That the latter is not necessarily true was shown by Miskimin et al. (42) and Solberg et al. (56), who studied over 1,000 foods consisting of 853 ready-to-eat and 180 raw products. These investigators applied arbitrary standards for APC, coliforms, and *E. coli,* and tested the efficacy of the standards to assess safety of the foods with respect to *S. aureus, C. perfringens,* and salmonellae. An APC standard of $< 10^6/g$ for raw foods resulted in 47% of the samples being accepted even though one or more of the three pathogens were present, while 5% were rejected from which pathogens were not isolated, for a total of 52% wrong decisions. An APC of $< 10^5/g$ for ready-to-eat foods resulted in only 5% being accepted that contained pathogens while 10% that did not yield pathogens were rejected. In a somewhat similar fashion, coliform standards of $< 10^2/g$ resulted in a total of 34% wrong decisions for raw and 15% for ready-to-eat foods. The lowest percentage of wrong decisions for ready-to-eat foods (13%) occurred with an *E. coli* standard of $< 3/g$, while 30% of the decisions were wrong when the same standard was applied to raw foods. It was noted that even though the three pathogens were found in both types of foods, no food-borne outbreaks were reported over the 4-yr period of the study, during which time more than 16 million meals were consumed (56).

While fecal coliforms are not precise indicators of food safety, their validity as indicators for the presence of salmonellae in irrigation water has been found to be high (21). These investigators found that when the fecal coliform density/100 ml was above 1,000, salmonellae occurrence exceeded a 96% frequency, while a frequency of only about 54% was associated with fecal coliform levels of $< 1,000/100$ ml.

A study of 2,211 food-borne outbreaks by the U.S. Center for Disease Control over the 9-yr period 1967–1975 revealed that ground beef was involved in 78 outbreaks, frankfurters in 22, and sliced luncheon meat in 11 for a total of about 5% of the 2,211 (20). It was concluded from this survey that with rare exceptions, the proximate causes of the food-borne outbreaks involved mishandling of food by the food service industry and by homemakers. A high percentage of cases was due to homemaker negligence. It could be argued that microbiological standards would do little if anything to prevent outbreaks at the home level if foods are improperly cooked or stored at improper temperatures. Even though microbial standards were not in effect to cover the foods surveyed above, when the reported outbreaks are considered in light of the high volume of sales, it is apparent that these three meat products are not high-risk foods (20). Along similar lines, the USDA surveyed beef patties, frankfurters, and sliced luncheon meat for the presence of salmonellae over the period 1972–75 and found only 3 positive, raw finished beef patties out of 735 examined (8). No salmonellae were isolated from 690 cooked finished frankfurters or from 456 cooked sliced luncheon meats. In view of findings such as the above, opponents of microbiological standards would question the need for such standards in regard to food safety, especially for ground beef, frankfurters, and sliced luncheon meats.

Microbiological standards/criteria now in effect for various foods

Milk was the first food product for which microbiological standards were adopted in the United States. Organizations such as the National Canners Association and the American Bottlers of Carbonated Beverages have for many years set standards for certain relevant groups of organisms on product ingredients. The Association of Food and Drug Officials of the United States (AFDOUS), the USDA, and other agencies (especially the ICMSF of the International Association of Microbiological Societies) have given much consideration to the adoption of standards for foods and continue to be concerned with this problem. Since foods that are produced in one country may be shipped to and sold in other countries, this aspect of food microbiology has also received the attention of the Food and Agricultural Organization (FAO) and the World Health Organization (WHO). The category of foods most likely to come first under microbiological standards nationwide and worldwide are the precooked foods, especially precooked frozen foods.

Presented below are those foods and food ingredients that are covered under microbiological standards of various organizations along with federal, state, and city standards in effect (after W. C. Frazier, 1968, courtesy of McGraw-Hill Publishing Company).

1. Standards for Starch and Sugar (National Canners Association).
    A. *Total thermophilic spore count:* Of the five samples from a lot of sugar or starch none shall contain more than 150 spores per 10 g, and the average for all samples shall not exceed 125 spores per 10 g.
    B. *Flat/sour spores:* Of the five samples none shall contain more than 75 spores per 10 g, and the average for all samples shall not exceed 50 spores per 10 g.
    C. *Thermophilic anaerobe spores:* Not more than three (60%) of the five samples shall contain these spores, and in any one sample not more than four (65 + %) of the six tubes shall be positive.
    D. *Sulfide spoilage spores:* Not more than two (40%) of the five samples shall contain these spores, and in any one sample there shall be no more than five colonies per 10 g (equivalent to two colonies in the six tubes).

2. Standard for "Bottlers" Granulated Sugar, Effective July 1, 1953 (American Bottlers of Carbonated Beverages).
   A. *Mesophilic bacteria:* Not more than 200 per 10 g.
   B. *Yeasts:* Not more than 10 per 10 g.
   C. *Molds:* Not more than 10 per 10 g.
3. Standard for "Bottlers" Liquid Sugar, Effective in 1959 (American Bottlers of Carbonated Beverages). All figures based on dry-sugar equivalent (D.S.E.).
   A. *Mesophilic bacteria: (a)* Last 20 samples average 100 organisms or less per 10 g D.S.E.; *(b)* 95% of last 20 counts show 200 or less per 10 g; *(c)* 1 of 20 samples may run over 200; other counts as in *(a)* or *(b)*.
   B. *Yeasts: (a)* Last 20 samples average 10 organisms or less per 10 g D.S.E.; *(b)* 95% of last 20 counts show 18 or less per 10 g; *(c)* 1 of 20 samples may run over 18; other counts as in *(a)* and *(b)*.
   C. *Molds:* Standards like those for yeasts.
4. Standards for Dairy Products.
   A. From 1965 recommendations of the U.S. Public Health Service.
      a. *Grade A raw milk for pasteurization:* Not to exceed 100,000 bacteria per milliliter prior to commingling with other producer milk; and not exceeding 300,000 per milliliter as commingled milk prior to pasteurization.
      b. *Grade A pasteurized milk and milk products* (except cultured products): Not over 20,000 bacteria per milliliter, and not over 10 coliforms per milliliter.
      c. *Grade A pasteurized cultured products:* Not over 10 coliforms per milliliter.
      NOTE: Enforcement procedures for *(a)*, *(b)*, and *(c)* require three-out-of-five compliance by samples. Whenever two of four successive samples do not meet the standard, a fifth sample is tested; and if this exceeds any standard, the permit from the health authority may be suspended. It may be reinstated after compliance by four successive samples has been demonstrated.
   B. *Certified milk* (American Association of Medical Milk Commissions, Inc.)
      a. Certified milk (raw): Bacterial plate count not exceeding 10,000 colonies per milliliter; coliform colony count not exceeding 10 per milliliter.
      b. Certified milk (pasteurized): Bacterial plate count not exceeding 10,000 colonies per milliliter before pasteurization and 500 per milliliter in route samples. Milk not exceeding 10 coliforms per milliliter before pasteurization and 1 coliform per milliliter in route samples.
   C. *Milk for manufacturing and processing* (U.S. Dep. Agr. 1955)
      a. Class 1: Direct microscopic clump count (DMC) not over 200,000 per milliliter.

b. Class 2: DMC not over 3 million per milliliter.

c. Milk for Grade A dry milk products: must comply with requirements for Grade A raw milk for pasteurization (see above).

D. *Dry milk*

a. Grade A dry milk products: at no time a standard plate count over 30,000 per gram, or coliform count over 90 per gram (U.S. Public Health Service).

b. Standards of Agricultural Marketing Service (U.S. Dep. Agr.):

(1) Instant nonfat: U.S. Extra Grade, a standard plate count not over 35,000 per gram, and coliform count not over 90 per gram.

(2) Nonfat (roller or spray): U.S. Extra Grade, a standard plate count not over 50,000 per gram; U.S. Standard Grade, not over 100,000 per gram.

(3) Nonfat (roller or spray): Direct microscopic clump count not over 200 million per gram; and must meet the requirements of U.S. Standard Guide. U.S. Extra Grade, such as used for school lunches, has an upper limit of 75 million per gram.

c. Standards of American Dry Milk Institute, Inc.

Kind of dry milk	Grade	Process	Maximal SPC,[a] nos/g
Nonfat	Extra	Spray or roller[b]	50,000
Nonfat	Standard	Spray or roller	100,000
Nonfat, instant	Extra		35,000
Whole milk	Premium	Gas-packed, spray	30,000
Whole milk	Extra	Bulk spray or roller	50,000
Whole milk	Standard	Bulk spray or roller	100,000
Buttermilk	Extra	Spray or roller	50,000
Buttermilk	Standard	Spray or roller	200,000

[a] Standard plate count.
[b] Atmospheric roller throughout table.

E. *Frozen desserts*

States and cities that have bacterial standards usually specify a maximal count of 50,000 to 100,000 per milliliter or gram. The U.S. Public Health Ordinance and Code sets the limit at 50,000 and recommends bacteriological standards for cream and milk used as ingredients. Few localities have coliform standards.

5. Standard for Tomato Juice and Tomato Products—Mold-count Tolerances (Food and Drug Administration).

The percentage of positive fields tolerated is 2% for tomato juice and 40% for other comminuted tomato products, such as catsup, purée, paste, etc. A microscopic field is considered positive when aggregate length of not more than three mold filaments present exceeds one-sixth of the diameter of the field (Howard mold count method). This method also has been applied to raw and frozen fruits of various kinds, especially to berries.

MICROBIOLOGICAL CRITERIA/STANDARDS FOR FRESH OR FROZEN SHELLFISH. The ICMSF (29) has recommended criteria not only for these products but for the condition of the overlying waters. With respect to the latter, the limit is for not more than 70 coliforms/100 ml, and no more than 10% of samples may exceed a coliform MPN of 230/ml by the five-tube method. The market standard for oysters from approved waters is that they contain no more than MPN 230/100 g of homogenized shell liquor and meats, and by SPC (at 35°C) no more than 500,000 bacteria/g. If two or more successive samples from the same shipper exceed either standard, corrective actions should be taken.

At least six states as of November 1981 (59) had adopted essentially the above as a standard (Hawaii, Kentucky, Maryland, Massachusetts, Virginia, and Washington), while six others employ this criterion as a guideline (Arizona, Colorado, Delaware, Nebraska, Ohio, and Rhode Island). Maryland and Virginia have microbiological standards for fresh crabmeat with SPC for both states being 100,000/g but for *E. coli,* Maryland's limit is 36/100 g while that for Virginia is 50/100 g.

STANDARDS FOR MISCELLANEOUS FOOD PRODUCTS. New Jersey has in place microbiological standards specifying/g not more than 100,000 APC, $\leq$ 100 coliforms, $\leq$ *E. coli,* and negative for *S. aureus,* covering chopped chicken liver and the following salads: coleslaw, chicken, egg, macaroni, potato, shrimp, tuna, and turkey. New York City has similar standards for protein salads and for baked goods containing custard, toppings, or fillings made with milk, eggs, and the like. For these products, the limits/g are: APC $\leq$ 100,000, coliforms $\leq$ 100, hemolytic staphylococci $\leq$ 100, and negative for *E. coli, S. aureus, C. perfringens,* salmonellae, and shigellae.

Canada has limits and sampling plans for specific indicators or pathogens for various products including salmonellae in skim milk powder ($n = 20$, $c = 0$, and $m = 0$); coliforms in cottage cheese ($n = 5$, $c = 1$, $m = 10$, and $M = 50$); and salmonellae in frog legs ($n = 5$, $c = 0$, and $m = 0$) (50).

SAMPLING PLANS

A sampling plan is basically a statement of the criteria of acceptance applied to a lot based upon appropriate examinations of a required number of sample units by specified methods. It consists of a sampling procedure and decision criteria and may be a two-class or a three-class plan.

A two-class plan consists of the following specifications: n, c, m; while a three-class plan requires n, c, m, and M, where

n = the number of sample units (packages, beef patties, and so forth) from a lot that must be examined to satisfy a given sampling plan.

c = the maximum acceptable number, or the maximum allowable number of sample units that may exceed the microbiological criterion m. When this number is exceeded, the lot is rejected.

m = the maximum number or level of relevant bacteria/g; values above this level are either marginally acceptable or unacceptable. It is used to separate acceptable from unacceptable foods in a two-class plan, or, in a three-class plan, to separate good quality from marginally acceptable quality foods. The level of the organism in question that is acceptable and attainable in the food product is m. In the presence/absence situations for two-class plans, it is common to assign $m = 0$. For three-class plans, m is usually some nonzero value.

M = a quantity that is used to separate marginally acceptable quality from unacceptable quality foods. It is used only in three-class plans. Values at or above M in any sample are unacceptable relative to either health hazard, sanitary indicators, or spoilage potential.

A two-class plan is the simpler of the two and in its simplest form may be used to accept or reject a large batch (lot) of food in a presence/absence decision by a plan such as $n = 5$, $c = 0$, where $n = 5$ means that five individual units of the lot will be examined microbiologically for, say, the presence of salmonellae, and $c = 0$ means that all five units must be free of the organisms by the method of examination in order for the lot to be acceptable. If any unit is positive for salmonellae, the entire lot is rejected. If it is desired that two of the five samples may contain coliforms, in a presence/absence test, for example, the sampling plan would be $n = 5$, $c = 2$. By this plan, if three or more of the five-unit samples contained coliforms, the entire lot would be rejected. While presence/absence situations generally obtain for salmonellae, an allowable upper limit for indicator organisms such as coliforms is more often the case. If it is desired to allow up to 100 coliforms/g in two of the five units, the sampling plan would be $n = 5$, $c = 2$, $m = 10^2$. After the five units have been examined for coliforms, the lot is acceptable if no more than two of the five contain as many as 10^2 coliforms/g, but is rejected if three or more of the five contain 10^2 coliforms/g. This particular sampling plan may be made more stringent by increasing n (for example, $n = 10$, $c = 2$, $m = 10^2$), or by reducing c (for example, $n = 5$, $c = 1$, $m = 10^2$). On the other hand, it can be made more lenient for a given size n by increasing c.

While a two-class plan may be used to designate acceptable-unacceptable foods, a three-class plan is required to designate acceptable/marginally acceptable/unacceptable foods. To illustrate a typical three-class plan, assume that for a given food product SPC shall not exceed 10^6/g *(M)* nor be higher than 10^5/g from three or more of five units examined. The specifications are thus $n = 5, c = 2, m = 10^5, M = 10^6$. If any of the five units exceeds 10^6/g, the entire lot is rejected (unacceptable). If not more than c sample units give results above m, the lot is acceptable. Unlike two-class plans, the three-class plan distinguishes values between m and M (marginally acceptable).

With either two- or three-class attributes plans, the numbers n and c may be employed to find the probability of acceptance (P_a) of lots of foods by reference to appropriate tables (see ref. 29). The decision to employ a two-class or three-class plan may be determined by whether presence/absence tests are desirable, in which case a two-class plan is required, or whether count or concentration tests are desired, in which case a three-class plan is preferred. The latter offers the advantages of being less affected by nonrandom variations between sample units, and of being able to measure the frequency of values in the m to M range. The ICMSF report and recommendations (29) should be consulted for further details on the background, uses, and interpretations of sampling plans. Further information may also be obtained from Kilsby (33).

MICROBIOLOGICAL GUIDELINES

Some states and cities in the United States have established guidelines for various food products, and those for raw meats are summarized in Table 17-7. In a 1981 survey, Wehr (59) found that the percentage of states with criteria was essentially unchanged from 1977 to 1981 (31 and 32%, respec-

Table 17-7. Guidelines of some states and cities in the United States for red meats and poultry as of November 1981 (59).

Product	$APC \times 10^6$	E. coli	Coliforms	City/State
Ground meat	1	—	—	Rhode Island
Ground meat	1	—	2,000	Eau Claire, Wisconsin
Ground meat	<10	—	—	La Crosse, Wisconsin
Raw meat/poultry	1	—	1,000	Iowa
Ground beef	5	—	1,000	Georgia[a]
Ground beef	10	—	1,000	Iowa[b]
Ground beef	5	50	—	Montana
Ground beef	3.5	—	100	Utah
Raw meat	0.1	—	100	Massachusetts
Meats	5	10	100	New Hampshire[c]
Ground beef, lamb, veal; pork and beef sausage	15	—	10,000	Ohio

[a]100 S. aureus.
[b]100 yeasts and molds.
[c]1,000 C. perfringens; 5,000 B. cereus.

tively). Five agencies that had criteria for ground beef in 1977 had none 5 years later. Oregon was the first state to adopt a standard for ground beef in 1973, but the law was repealed in 1977. That particular criterion specified an APC not to exceed 5×10^6/g and not more than 50 *E. coli*/g for raw frozen meats. For SPC, ICMSF recommends a three-class sampling plan for frozen comminuted meat: $n = 5$, $c = 3$, $m = 10^6$, and $M = 10^7$ (see Table 17-8). The Canadian government in 1977 adopted provisional guidelines employing three-class sampling plans for fresh and frozen ground beef as follows: APC (at 35°C) for nonfrozen, $n = 5$, $c = 3$, $m = 10^7$, $M = 5 \times 10^7$, and for *E. coli* for nonfrozen, $n = 5$, $c = 3$, $m = 10^2$, and $M = 10^3$; APC at 35°C for frozen, $n = 5$, $c = 2$, $m = 10^6$, and $M = 10^7$, and for *E. coli* for frozen, $n = 5$, $c = 3$, $m = 10^2$, and $M = 10^3$ (50). In spite of ICMSF recommendations, U.S. guidelines and standards continue to be based on single sample analyses, as may be noted from Table 17-7. The use of an appropriate sampling plan to assess microbiological limits is a better alternative. It may be anticipated that at some time in the future, all agencies that establish criteria will express them in a manner similar to those in Table 17-8.

Table 17-8. Sampling plans and ICMSF-recommended microbiological limits for five food products (29). For sampling methods and case number, see reference.

Product	Test	Class plan	n	c	m	M
Breaded precooked fish sticks, fish portions, fish cakes	SPC	3	5	2	10^6	10^7
	Fecal coliforms (MPN)	3	5	2	4	400
	S. aureus	3	5	2	10^3	2×10^3
Blanched frozen vegetables	SPC	3	5	3	10^5	10^6
	E. coli	3	5	2	10	10^2
Frozen entrees, precooked, and vegetables in sauce	SPC	3	5	2	10^5	10^6
	Coliforms	3	5	2	10^2	10^4
	E. coli	3	5	2	<3	10^2
Ice cream (without added ingredients)	SPC	3	5	2	10^4	2.5×10^5
	Coliforms	3	5	2	10	10^3
	S. aureus	3	5	1	10	10^2
	Salmonellae[a]	2	10	0	0	—
Comminuted meat (frozen)	SPC	3	5	3	10^6	10^7
	Salmonellae	2	5	1(0)	0	—

[a]Investigational sampling only.

REFERENCES

1. Bagley, S. T., and R. J. Seidler. 1977. Significance of fecal coliform-positive *Klebsiella. Appl. Environ. Microbiol.* 33:1141–48.

2. Barnard, R. J., and G. J. Jackson. 1984. *Giardia lamblia*—The transfer of human infections by foods. In *Giardia and giardiasis: Biology, pathogenesis, and epidemiology,* ed. S. L. Erlandsen and E. A. Meyer, 365–78. New York: Plenum.

3. Bauman, H. E. 1974. The HAACP concept and microbiological hazard categories. *Food Technol.* 28(9):30–33.

4. Burman, N. P. 1961. Some observations on coli-aerogenes bacteria and streptococci in water. *J. Appl. Bacteriol.* 24:368–76.

5. Burton, M. C. 1949. Comparison of coliform and enterococcus organisms as indices of pollution in frozen foods. *Food Res.* 14:434–38.

6. Buttiaux, R., and D. A. A. Mossel. 1961. The significance of various organisms of faecal origin in foods and drinking water. *J. Appl. Bacteriol.* 24:353–64.

7. Carrillo, M., E. Estrada, and T. C. Hazen. 1985. Survival and enumeration of the fecal indicators *Bifidobacterium adolescentis* and *Escherichia coli* in a tropical rain forest watershed. *Appl. Environ. Microbiol.* 50:468–76.

8. Center for Disease Control. 1975. Microbiologic standards for raw ground beef, cold cuts, and frankfurters. *Morb. Mort. Wkly. Rept.* 24(27):229–30.

9. Chordash, R. A., and N. F. Insalata. 1978. Incidence and pathological significance of *Escherichia coli* and other sanitary indicator organisms in food and water. *Food Technol.* 32(10):54, 56–58, 62–63.

10. Collins, M. D., D. Jones, J. A. E. Farrow, R. Kilpper-Balz, and K. H. Schleifer. 1984. *Enterococcus avium* nom. rev., comb. nov.; *E. casseliflavus* nom. rev., comb. nov.; *E. durans* nom. rev., comb. nov.; *E. gallinarum* comb. nov.; and *E. malodoratus* sp. nov. *Int. J. Syst. Bacteriol.* 34:220–23.

11. Craun, G. F. 1979. Waterborne giardiasis in the United States: A review. *Amer. J. Pub. Hlth* 69:817–19.

12. Cuthbert, W. A., J. J. Panes, and E. C. Hill. 1955. Survival of *Bacterium coli* type I and *Streptococcus faecalis* in soil. *J. Appl. Bacteriol.* 18:408–14.

13. deFigueiredo, M. P., and J. M. Jay. 1976. Coliforms, enterococci, and other microbial inhibitors. In *Food microbiology: Public health and spoilage aspects,* ed. M. P. deFigueiredo and D. F. Splittstoesser, 271–97. Westport, Conn: AVI.

14. Ellender, R. D., J. B. Mapp, B. L. Middlebrooks, D. W. Cook, and E. W. Cake. 1980. Natural enterovirus and fecal coliform contamination of Gulf Coast oysters. *J. Food Protect.* 43:105–10.

15. Elliott, H. P., and H. D. Michener. 1961. Microbiological standards and handling codes for chilled and frozen foods: A review. *Appl. Microbiol.* 9:452–68.

16. Fishbein, M. 1962. The aerogenic response of *Escherichia coli* and strains of *Aerobacter* in EC broth and selected sugar broths at elevated temperatures. *Appl. Microbiol.* 10:79–85.

17. Fishbein, M., and B. F. Surkiewicz. 1964. Comparison of the recovery of *Escherichia coli* from frozen foods and nutmeats by confirmatory incubation in EC medium at 44.5 and 45.5°C. *Appl. Microbiol.* 12:127–31.

18. Foster, E. M. 1966. Significance of bacterial counts in food evaluation. *Assoc. Food Drug Off. Quart. Bull.* 30:20–27.

19. Fugate, K. J., D. O. Cliver, and M. T. Hatch. 1975. Enteroviruses and potential bacterial indicators in Gulf Coast oysters. *J. Milk Food Technol.* 38:100–104.

20. Gangarosa, E. J., and J. M. Hughes. 1977. A public health perspective of microbial standards for meats. *Assoc. Food Drug Off. Quart. Bull.* 41(1):23–28.

21. Geldreich, E. E., and R. H. Bordner. 1971. Fecal contamination of fruits and vegetables during cultivation and processing for market. A review. *J. Milk Food Technol.* 34:184–95

22. Greenberg, R. A., R. B. Tompkin, and R. S. Geister. 1974. Who will pay for microbiological quality standards? *Food Technol.* 28(10):48–49.

23. Griffin, A. M., and C. A. Stuart. 1940. An ecological study of the coliform bacteria. *J. Bacteriol.* 40:83–100.

24. Haenel, H. 1961. Some rules in the ecology of the intestinal microflora of man. *J. Appl. Bacteriol.* 24:242–51.

25. Hahn, S. S., and M. D. Appleman. 1952. Microbiology of frozen orange concentrate. I. Survival of enteric organisms in frozen orange concentrate. *Food Technol.* 6:156–57.

26. Hartman, P. A. 1960. Enterococcus: coliform ratios in frozen chicken pies. *Appl. Microbiol.* 8:114–16.

27. Hartman, P. A., G. W. Reinbold, and D. S. Saraswat. 1966. Indicator organisms— a review. I. Taxonomy of the fecal streptococci. *Intern. J. System. Bacteriol.* 16:197–221.

28. Ingram, M. 1961. Microbiological standards for foods. *Food Technol.* 15(2):4– 16.

29. International Commission on the Microbiological Specification of Foods. 1982. *Microorganisms in food.* Vol. 2, *Sampling for microbiological analysis: Principles and specific applications.* Toronto: Univ. Toronto Press.

30. Ito, K. 1974. Microbiological critical control points in canned foods. *Food Technol.* 28(9):46–47.

31. Kereluk, K., and M. F. Gunderson. 1959. Studies on the bacteriological quality of frozen meats. IV. Longevity studies on the coliform bacteria and enterococci at low temperatures. *Appl. Microbiol.* 7:327–28.

32. Keswick, B. H., C. P. Gerba, R. L. DuPont, and J. B. Rose. 1984. Detection of enteric viruses in treated drinking water. *Appl. Environ. Microbiol.* 47:1290– 94.

33. Kilsby, D. C. 1982. Sampling schemes and limits. In *Meat microbiology,* ed. M. H. Brown, 387–421. London: Applied Science Publishers.

34. Larkin, E. P., W. Litsky, and J. E. Fuller. 1955. Fecal streptococci in frozen foods. I. A bacteriological survey of some commercially frozen foods. *Appl. Microbiol.* 3:98–101.

35. ———. 1955. Fecal streptococci in frozen foods. II. Effect of freezing storage on *Escherichia coli* and some fecal streptococci inoculated onto green beans. *Appl. Microbiol.* 3:102–4.

36. ———. 1955. Fecal streptococci in frozen foods. III. Effect of freezing storage on *Escherichia coli, Streptococcus faecalis. Appl. Microbiol.* 3:104–6.

37. Leininger, H. V., and C. S. McCleskey. 1953. Bacterial indicators of pollution in surface waters. *Appl. Microbiol.* 1:119–24.

38. Levine, M. 1961. Facts and fancies of bacterial indices in standards for water and foods. *Food Technol.* 15(11):29–34.

39. Litsky, W., M. J. Rosenbaum, and R. L. France. 1953. A comparison of the most probable numbers of coliform bacteria and enterococci in raw sewage. *Appl. Microbiol.* 1:247–50.

40. Mara, D. D., and J. I. Oragui. 1983. Sorbitol-fermenting bifidobacteria as specific indicators of human faecal pollution. *J. Appl. Bacteriol.* 55:349–57.

41. McCoy, J. H. 1961. The safety and cleanliness of waters and foods. *J. Appl. Bacteriol.* 24:365–67.
42. Miskimin, D. K., K. A. Berkowitz, M. Solberg, W. E. Riha, Jr., W. C. Franke, R. L. Buchanan, and V. O'Leary. 1976. Relationships between indicator organisms and specific pathogens in potentially hazardous foods. *J. Food Sci.* 41:1001–6.
43. Montford, J., and F. S. Thatcher. 1961. Comparison of four methods of isolating salmonellae from foods, and elaboration of a preferred procedure. *J. Food Sci.* 26:510–17.
44. Mossel, D. A. A., G. A. Harrewijn, and C. F. M. Nesselrooy-van Zadelhoff. 1974. Standardization of the selective inhibitory effect of surface active compounds used in media for the detection of *Enterobacteriaceae* in foods and water. *Hlth Lab. Sci.* 11:260–67.
45. Mundt, J. O. 1961. Occurrence of enterococci: Bud, blossom, and soil studies. *Appl. Microbiol.* 9:541–44.
46. ———. 1973. Litmus milk reaction as a distinguishing feature between *Streptococcus faecalis* of human and nonhuman origins. *J Milk Food Technol.* 36:364–67.
47. ———. 1976. Streptococci in dried and frozen foods. *J. Milk Food Technol.* 39:413–16.
48. Ostrolenk, M., N. Kramer, and R. C. Cleverdon. 1947. Comparative studies of enterococci and *Escherichia coli* as indices of pollution. *J. Bacteriol.* 53:197–203.
49. Peterson, A. C., and R. E. Gunnerson. 1974. Microbiological critical control points in frozen foods. *Food Technol.* 28(9):37–44.
50. Pivnick, H. 1978. Canadian microbiological standards for foods. *Food Technol.* 32(1):58–62.
51. Raj, H., W. J. Wiebe, and J. Liston. 1961. Detection and enumeration of fecal indicator organisms in frozen sea foods. *Appl. Microbiol.* 9:295–308.
52. Resnick, I. G., and M. A. Levin. 1981. Assessment of bifidobacteria as indicators of human fecal pollution. *Appl. Environ. Microbiol.* 42:433–38.
53. Shiffman, M. A., and D. Kronick, 1963. The development of microbiological standards for foods. *J. Milk Food Technol.* 26:110–14.
54. Silliker, J. H. 1963. Total counts as indexes of food quality. In *Microbiological quality of foods,* ed. L. W. Slanetz et al., 102–12. New York: Academic Press.
55. ———. 1982. Selecting methodology to meet industry's microbiological goals for the 1980s. *Food Technol.* 36(12):65–70.
56. Solberg, M., D. K. Miskimin, B. A. Martin, G. Page, S. Goldner, and M. Libfeld. 1977. Indicator organisms, foodborne pathogens and food safety. *Assoc. Food Drug Off. Quart. Bull.* 41(1):9–21.
57. Splittstoesser, D. F. 1983. Indicator organisms on frozen blanched vegetables. *Food Technol.* 37(6):105–6.
58. Tompkin, R. B. 1983. Indicator organisms in meat and poultry products. *Food Technol.* 37(6):107–10.
59. Wehr, H. M. 1982. Attitudes and policies of governmental agencies on microbial criteria for foods—An update. *Food Technol.* 36(9):45–54, 92.
60. Weiss, K. F., N. Chopra, P. Stotland, G. W. Riedel, and S. Malcolm. 1983. Recovery of fecal coliforms and of *Escherichia coli* at 44.5, 45.0, and 45.5° C. *J. Food Protect.* 46:172–77.

18.

STAPHYLOCOCCAL GASTROENTERITIS

The staphylococcal food-poisoning/food-intoxication syndrome was first studied in 1894 by J. Denys and later in 1914 by Barber, who produced in himself the signs and symptoms of the disease by consuming milk that had been contaminated with a culture of *Staphylococcus aureus*. The capacity of some strains of *S. aureus* to produce food poisoning was proved conclusively in 1930 by G. M. Dack et al. (24), who showed that the symptoms could be produced by feeding culture filtrates of *S. aureus*. While some authors refer to food-associated illness of this type as food intoxication rather than food poisoning, the designation **gastroenteritis** obviates the need to indicate whether the illness is an intoxication or an infection.

Staphylococcal gastroenteritis is usually produced by some strains of *S. aureus* that produce coagulase. This enzyme is elaborated by growing cells and is identified by its capacity to clot blood plasma. Some coagulase-negative strains are associated with the gastroenteritis syndrome, but coagulase-positive strains are by far the most frequently involved. All symptoms of staphylococcal gastroenteritis are caused by an extracellular substance designated as **enterotoxin,** and these specific enterotoxins are not known to be produced by any other microorganisms.

An extensive literature exists on staphylococci and the food-poisoning syndrome, much of which goes beyond the scope of this chapter. For more extensive information, other references should be consulted including Bryan (14), Bergdoll (9), Minor and Marth (60), and Smith et al. (83).

POTENTIALLY PATHOGENIC SPECIES/STRAINS

The taxonomy of the genus *Staphylococcus* has been the subject of intense study during the past fifteen years, resulting in the assignment of more species to the genus. In the seventh edition of *Bergey's Manual* (1948), only two species were recognized—*S. aureus* and *S. epidermidis*. The eighth edition (1974) recognized these two and one additional species—*S. saprophiticus* (see ref. 49 for complete history). It appears that at least fifteen species will be recognized when the next edition of the *Manual* is published. The enlargement of the genus has come about in part because of the application of more sensitive and extensive methods of analysis such as DNA-DNA hybridization. The expanded genus results in part from the classification of staphylococci by Baird-Parker (5), who established six subgroups with the classical *S. aureus* strains placed in subgroup I, and from the work of Kloos and others in biotyping host-adapted strains of staphylococci (49). Most human strains are placed in biotype A, most poultry and porcine strains in biotype B, and most bovine strains in biotype C.

The species and strains of current potential interest in food microbiology are summarized in Table 18-1. It may be noted that *S. hyicus* subsp. *hyicus* and *S. intermedius* may produce both coagulase and thermostable nuclease, both of which are important in determining the potential enterotoxigenicity of a staphylococcal isolate. In one study, *S. hyicus* subsp. *hyicus* filtrates from four strains produced emetic responses in monkeys but were negative for either of the known enterotoxins, suggesting that one or more new enterotoxins may have been involved (41). These authors found one strain that produced enterotoxin B (SEB) but was more closely related to *S. epidermidis* than *S. aureus*. Evidence to date suggests that *S. intermedius* and *S. hyicus* subsp. *hyicus* cannot be overlooked when examining foods for staphylococci of health significance. While in the past organisms resembling *S. aureus* that produce coagulase, thermostable nuclease, protein A, or the clumping factor could be reasonably presumed to be *S. aureus,* additional

Table 18-1. Species and strains of staphylococci important in foods. (From Devriese et al., 29; and Devriese, 28)

Characteristic	*S. aureus*	*S. hyicus* subsp. *hyicus*	*S. hyicus* subsp. *chromogens*	*S. intermedius*
Coagulase (rabbit plasma)	+	+[a]	−	+
Thermostable nuclease	+	+	−[b]	+
Clumping factor	+	−	−	+[a]
Pigment	+	−	+	−
α, β, or δ hemolysins	+	−	−	+[a]
Phosphatase	+	+	+	+
Sensitivity to acriflavine[c]	>10	<5	<5	<5

[a]11 to 89% of strains.
[b]Slight or none.
[c]MIC, μg/ml.

characteristics need to be determined before the identity of this species is certain. Susceptibility to acriflavine appears to be diagnostic of *S. aureus,* with > 10 μg/ml required for its inhibition, while with the other three species noted in Table 18-1, inhibition was achieved by from 0.4 to 3.2 μg/ml (28).

HABITAT AND DISTRIBUTION

In man, the main reservoir of *S. aureus* is the nasal cavity. From this source, the organisms find their way to the skin and into wounds either directly or indirectly. While the nasal carriage rate varies, it is generally about 50% for adults and somewhat higher among children. The most common skin sources are the arms, hands, and face, where the carriage rate runs between 5 and 30% (30). In addition to skin and nasal cavities, *S. aureus* may be found in the eyes, throat, and in the intestinal tract. From these sources, the organism finds its way into air and dust, onto clothing, and in other places from which it may contaminate foods. The two most important sources to foods are nasal carriers and individuals whose hands and arms are inflicted with boils and carbuncles, who are permitted to handle foods.

Most domesticated animals harbor *S. aureus.* Staphylococcal mastitis is not unknown among dairy herds, and if milk from infected cows is consumed or used for cheese making, the chances of contracting food intoxication are excellent. There is little doubt that many strains of this organism that cause bovine mastitis are of human origin. However, some are designated as "animal strains." In one study, staphylococcal strains isolated from parts of raw pork products were essentially all of the animal strain type. However, during the manufacture of pickled pork products, these animal strains were gradually replaced by human strains during the production process, to a point where none of the original animal strains could be detected in finished products (80).

INCIDENCE IN FOODS

In general, staphylococci may be expected to exist, at least in low numbers, in any or all food products that are of animal origin; or in those that are handled directly by man, unless heat processing steps are applied to effect their destruction. They have been found in a large number of commercial foods by many investigators (see Chapter 4 for foods and relative numbers of *S. aureus* found).

NUTRITIONAL REQUIREMENTS FOR GROWTH

Staphylococci are typical of other gram-positive bacteria in having a requirement for certain organic compounds in their nutrition. Amino acids are required as nitrogen sources, and thiamine and nicotinic acid are required among the B vitamins. When grown anaerobically, they appear to require uracil. In one minimal medium for aerobic growth and enterotoxin production, monosodium glutamate serves as the C, N, and energy sources. This medium contains only three amino acids (arginine, cystine, and phenylalanine) and four vitamins (pantothenate, biotin, niacin, and thiamine) in addition to inorganic salts (59). Arginine appears to be essential for enterotoxin B production (92).

TEMPERATURE GROWTH RANGE

Although it is a mesophile, some strains of *S. aureus* can grow at a temperature as low as 6.7°C (3). The latter authors found three food-poisoning strains that grew in custard at 114°F but decreased at 116°–120°F with time of incubation. They grew in chicken à la king at 112°F but failed to grow in ham salad at the same temperature. In general, growth occurs over the range of 7°–47.8°C, and enterotoxins are produced between 10° and 46°C, with the optimum between 40° and 45°C (see 83). These minimum and maximum temperatures of growth and toxin production assume optimal conditions relative to the other parameters, and the ways in which they interact to raise minimum growth or lower maximum growth temperatures are noted below.

EFFECT OF SALTS AND OTHER CHEMICALS

While *S. aureus* grows well in culture media without NaCl, it can grow well in 7–10% concentrations, and some strains can grow in 20%. The maximum concentrations that permit growth actually depend on other parameters such as temperature, pH, a_w, and Eh (see below).

S. aureus has a high degree of tolerance to compounds such as tellurite, mercuric chloride, neomycin, polymyxin, and sodium azide, all of which have been used as selective agents in culture media. *S. aureus* can be differentiated from other staphylococcal species by its greater resistance to acriflavine (see above). In the case of borate, *S. aureus* is sensitive while *S. epidermidis* is resistant (46). With novobiocin, *S. saprophyticum* is resistant whereas *S. aureus* and *S. epidermidis* are not. The capacity to tolerate high levels of NaCl and certain other compounds is shared by members of the genus *Micrococcus,* which are widely distributed in nature and occur in foods generally in greater numbers than staphylococci, thus making the recovery of the latter more difficult. The effect of other chemicals on *S. aureus* is presented in Chapter 11.

EFFECT OF pH, a_w, AND OTHER PARAMETERS

Regarding pH, *S. aureus* can grow over the range of 4.0–9.8 but its optimum is in the range of 6–7. As is the case with the other growth parameters, the precise minimum growth pH is dependent upon the degree to which all other parameters are at optimal levels.

With respect to a_w, the staphylococci are unique in being able to grow at values lower than for any other nonhalophilic bacteria. Growth has been demonstrated as low as 0.83 under otherwise ideal conditions, although 0.86 is the generally recognized minimum a_w. The interrelationship of a_w, pH, NaCl level, and temperature of incubation is discussed below.

NaCl AND pH. Using a protein hydrolysate medium incubated at 37°C for 8 days, growth and enterotoxin C production occurred over the pH range 4.00–9.83 with no NaCl. With 4% NaCl, the pH range was restricted to 4.4–9.43 (Table 18-2). Toxin was produced at 10% NaCl with a pH of 5.45 or higher, but none was produced at 12% NaCl (34).

Table 18-2. The effect of pH and NaCl on the production of enterotoxin C by an inoculum of 10^8 cells/ml of *S. aureus* 137 in a protein hydrolysate medium incubated at 37°C for 8 days (34).

pH range	4.00–9.83	4.4–9.43	4.50–8.55	5.45–7.30	4.50–8.55
NaCl content (%)	0	4	8	10[a]	12
Enterotoxin production	+	+	+ ·	+	−

[a]Enterotoxin was detected also with an inoculum of 3.6×10^6 at pH 6.38–7.30.
Copyright © 1971, American Society for Microbiology.

It has been shown that *S. aureus* growth is inhibited in broth at pH 4.8 and 5% NaCl (50). In another study, growth and enterotoxin B production by strain S-6 occurred in 10% NaCl at pH 6.9, but not with 4% at pH 5.1 (36). The general effect of increasing NaCl concentration is to raise the minimum pH of growth. At pH 7.0 and 37°C, enterotoxin B was inhibited by 6% or more NaCl (Fig. 18-1).

pH, a_w, AND TEMPERATURE. No growth of a mixture of *S. aureus* strains occurred in brain heart infusion (BHI) broth containing NaCl and sucrose as humectants either at pH 4.3, a_w of 0.85, or at 8°C. No growth occurred with a combination of pH < 5.5, 12°C, and a_w of 0.90 or 0.93; and no growth occurred at pH < 4.9, 12°C, and a_w of 0.96 (65).

NaNO$_2$, Eh, pH, AND TEMPERATURE OF GROWTH. *S. aureus* strain S-6 grew and produced enterotoxin B in cured ham under anaerobic conditions with a brine content up to 9.2%, but not below pH 5.30 and 30°C, or below pH 5.58 at 10°C. Under aerobic conditions, enterotoxin production occurred sooner than under anaerobic conditions. As the concentration of HNO$_2$ increased, enterotoxin production decreased (35).

STAPHYLOCOCCAL ENTEROTOXINS—TYPES AND INCIDENCE

By use of serologic methods, seven different enterotoxins are recognized and designated A, B, C_1, C_2, C_3, D, and E. Enterotoxin C_3 (SEC_3) is chemically and serologically related to but not identical to SEC_1 and SEC_2 (74). The latter two show some cross reactivity. Preliminary reports during the early 1980s that the toxic shock syndrome toxin was SEF were subsequently not confirmed.

The relative incidence of the enterotoxins is presented in Table 18-3. In general, SEA is recovered from food-poisoning outbreaks more often than any of the others, with SED being second most frequent. The fewest number of outbreaks are associated with SEE. The incidence of SEA among 3,109 and SED among 1,055 strains from different sources, and by a large number of investigators, was 23 and 14%, respectively (84). For SEB, SEC, and SEE, 11, 10, and 3%, respectively, were found among 3,367, 1,581, and 1,072 strains.

The relative incidence of specific enterotoxins among strains recovered from various sources varies widely. While from human specimens in the United States over 50% of isolates secrete SEA alone or in combination (16), from human isolates in Sri Lanka SEA producers constituted only 7.8% (67). Unlike other reports, the latter study found more SEB producers than any other types. Wide variations are found among *S. aureus* strains isolated from foods. While in one study Harvey et al. (40) found SED to be associated more with poultry isolates than human strains, in another study these investigators found no SED producers among fifty-five poultry isolates (38). In yet another study, 2 of 3 atypical *S. aureus* isolates that produced a slow, weak positive or negative coagulase reaction, and were negative for the anaerobic fermentation of mannitol, produced SED (32). The isolates were from poultry. From Nigerian ready-to-eat foods, about

Table 18-3. Percent incidence of staphylococcal enterotoxins alone and in combination from various sources.

Source	No. cultures	Percent enterotoxic	Enterotoxins A	B	C	D	E	Reference
Human specimens	582	—	54.5	28.1	8.4	41.0	—	15
Raw milk	236	10	1.8	0.8	1.2	6.8	—	15
Frozen foods	260	30	3.4	3.0	7.4	10.4	—	15
Food-poisoning outbreaks	80	96.2	77.8	10.0	7.4	37.5	—	15
Foods	200	62.5	47.5	3.5	12.0	18.5	6.5	70
Poultry	139	25.2	1.4	0	0.7	23.7	0	40
Humans	293	39	7.8	17.7	7.2	6.8	0.7	67
Poultry	55	62	60.0	1.8	3.6	0	0	38

39% of 248 isolates were enterotoxigenic, with 44% of these producing SED (1). Of 48 isolates from dairy and 134 from meat products, 45.8 and 48.5%, respectively, were enterotoxigenic (70), and of 80 strains from food-poisoning outbreaks, 96.2% produced SEA (16).

Regarding the percentage of strains that are enterotoxigenic, widely different percentages have been found depending on the source of isolates. Only 10% of 236 raw milk isolates were enterotoxigenic (16) while 62.5% of 200 food isolates were positive (70). In a study of *S. aureus* from chicken livers, 40% were enterotoxigenic (37). In another study, 33% of 36 food isolates were enterotoxigenic (82).

Attempts to associate enterotoxigenicity with other biochemical properties of staphylococci such as gelatinase, phosphatase, lysozyme, lecithinase, lipase, and DNAse production, or the fermentation of various carbohydrates have been unsuccessful. Enterotoxigenic strains appear to be about the same as other coagulase-positive strains in these respects. Attempts to relate enterotoxigenesis with specific bacteriophage types have been unsuccessful also. Most enterotoxigenic strains belong to phage Group III, but all phage groups are known to contain toxigenic strains. Of fifty-four strains from clinical specimens that produced SEA, 5.5, 1.9, and 27.8% belonged respectively to phage Groups I, II, and III, with 20.4% being untypable (15). Among poultry isolates, 49% were found to be phage untypable (38). In a study of bruised poultry tissue, essentially the same phage types were found on the hands and in lesions of handlers as in the bruised tissues, suggesting that the bruised poultry tissue served as the source of staphylococci to the handlers (76).

Chemical and physical properties of enterotoxins

These properties are summarized in Table 18-4. All are simple proteins that, upon hydrolysis, yield eighteen amino acids, with aspartic, glutamic, lysine, and valine being the most abundant. The amino acid sequence of SEB was determined first (42). Its N-terminal is glutamic acid, and lysine is the C-terminal amino acid. SEA, SEB, and SEE are composed of 239 to 296 amino acid residues. SEC_3 contains 236 amino acid residues and the N-terminal is serine, while the N-terminal of SEC_1 is glutamic acid (74). The disulfide bond in enterotoxin B is not essential for biological activity and conformation (25). Biological activity of SEA is destroyed when the abnormal tyrosyl residues are modified (21). The effect of acetylation, succinylation, guanidination, and carbamylation on the biological activity of SEB from strain S-6 has been studied (20). It was found that guanidination of 90% of the lysine residues had no effect on emetic activity or on the combining power of antigen-antibody reactions of SEB. The toxin was reduced, however, when acetylated, succinylated, and carbamylated. The investigators concluded that acetylation and succinylation decreased the net positive charge of the enterotoxin that is contributed by amino groups. The normal positive charge of the enterotoxin is thought to play an important

Table 18-4. Some properties of the enterotoxins of *S. aureus*. (Compiled from refs. 7, 42, 74, 78, and other sources)

	Enterotoxin						
	A	*B*	*C₁*	*C₂*	*C₃*	*D*	*E*
Emetic dose (ED_{50}) (Monkey) (μg/animal)	5	5	5	5–10	$<10^a$	20	10–20
Nitrogen content (%)	16.5	16.10	16.2	16.0	—	—	—
Sedimentation coefficient (s_{20}^o,w), S	3.04	2.89 2.78	3.00	2.90	—	—	2.60
Diffusion coefficient (D_{20}^o,w), $\times$ 10^{-7} cm^2 sec^{-1}	7.94	7.72 8.22	8.10	8.10	—	—	—
Reduced viscosity (ml/g)	4.07	3.92 3.81	3.4	3.7	—	—	—
Molecular weight	27,800	28,366	34,100	34,000	26,900	27,300	29,600
Partial specific volume	0.726	0.743 0.726	0.732	0.742	—	—	—
Isoelectric point	6.8	8.60	8.6	7.0	8.15	7.4	7.0
Maximum absorption (mμ)	277	277	277	277	—	278	277
Extinction ($E_{1cm}^{1\%}$)	14.3	14.00 14.40	12.1	12.1	—	10.8	12.5

[a]per os. 0.05 μg/kg by IV route (74).

role in both its emetic activity and in its combining with specific antibody. In their activate states, the enterotoxins are resistant to proteolytic enzymes such as trypsin, chymotrypsin, rennin, and papain, but sensitive to pepsin at a pH of about 2 (7). While the various enterotoxins differ in certain physiochemical properties, each has about the same potency. Although biological activity and serologic reactivity are generally associated, it has been shown that aerologically negative enterotoxin may be biologically active.

The enterotoxins are quite heat resistant, as noted above. The biological activity of SEB was retained after heating for 16 h at 60°C and pH 7.3 (78). Heating of one preparation of SEC for 30 min at 60°C resulted in no change in serologic reactions (11). The heating of SEA at 80°C for 3 min or at 100°C for 1 min caused it to lose its capacity to react serologically (7).

The thermal inactivation of SEA based on cat emetic response was shown by Denny et al. (27) to be 11 min at 250°F (F_{250}^{48} = 11 min). When monkeys were employed, thermal inactivation was F_{250}^{46} = 8 min. These enterotoxin preparations consisted of a 13.5-fold concentration of casamino acid culture filtrate employing strains 196-E. Using double-gel-diffusion assay, Read and Bradshaw (73) found the heat inactivation of 99+% pure SEB in veronal buffer to be F_{250}^{58} = 16.4 min. The end point for enterotoxin inactivation by gel diffusion was identical to that by intravenous injection of cats. The slope of the thermal inactivation curve for SEA in beef bouillon at pH 6.2

was found to be around 27.8°C (50°F) using three different toxin concentrations (5, 20, and 60 μg/ml, 26). Some D values for the thermal destruction of SEB are presented in Table 18-5. Crude toxin preparations have been found to be more resistant than purified toxins (73). It may be noted from Table 18-5 that staphylococcal thermonuclease displays heat resistance similar to that of SEB (see Chapter 6 for more information on this enzyme). In one study, SEB was found to be more heat sensitive at 80°C than at 100° or 110°C (77). The thermal destruction was more pronounced at 80°C than at either 60° or 100°C when heating was carried out in the presence of meat proteins. In spite of the generally high degree of heat resistance, present thermal process treatments for low-acid foods are adequate to destroy these toxins (27).

S. *aureus* cells are considerably more sensitive to heat than the enterotoxins, as may be noted from D values presented in Table 18-6 from various heating menstra. The cells are quite sensitive in Ringers solution at pH 7.2 ($D_{140°F}$ = 0.11), and much more resistant in milk at pH 6.9 ($D_{140°F}$ = 10.0). In frankfurters, heating to 71.1°C was found to be destructive to several strains of *S. aureus* (68), and microwave heating for 2 min was destructive to over 2 million cells/g (91).

The maximum growth temperature and heat resistance of *S. aureus* strain MF 31 were shown to be affected when the cells were grown in heart infusion broth containing soy sauce and monosodium glutamate (MSG). Without these ingredients in the broth, maximum growth temperature was 44°C but with them the maximum was above 46 (43). The most interesting effect of MSG was on $D_{60°C}$ values determined in Tris buffer at pH 7.2. With cells grown at 37°C, the mean D 60°C value in buffer was 2.0 min, but when 5% MSG and 5% NaCl were added to the buffer, D 60°C was 15.5 min. Employing cells grown at 46°C, the respective D 60°C values were 7.75 and 53.0 min in buffer and buffer-MSG-NaCl. It is well known that heat resistance increases along with increasing growth temperature, but changes of this magnitude in vegetative cells is unusual.

Table 18-5. D values for the heat destruction of staphylococcal enterotoxin B and staphylococcal heat-stable nuclease.

Conditions	D(C)	Reference
Veronal buffer	D_{110} = 29.7[a]	73
Veronal buffer	D_{110} = 23.5[b]	73
Veronal buffer	D_{121} = 11.4[a]	73
Veronal buffer	D_{121} = 9.9[b]	73
Veronal buffer, pH 7.4	D_{110} = 18	51
Beef broth, pH 7.4	D_{110} = 60	51
Staph. nuclease	D_{130} = 16.5	31

[a]Crude toxin.
[b]99+ percent purified.

Table 18-6. D and z values for the thermal destruction of *S. aureus* 196E in various heating menstra at 140°F.

Products	D(F)	z	Reference
Chicken à la king	5.37	10.5	4
Custard	7.82	10.5	4
Green pea soup	6.7–6.9	8.1	86
Skim milk	3.1–3.4	9.2	86
0.5 percent NaCl	2.2–2.5	10.3	86
Beef bouillon	2.2–2.6	10.5	86
Skim milk alone	5.34	—	47
Raw skim milk + 10% sugar	4.11	—	47
Raw skim milk + 25% sugar	6.71	—	47
Raw skim milk + 45% sugar	15.08	—	47
Raw skim milk + 6% fat	4.27	—	47
Raw skim milk + 10% fat	4.20	—	47
Tris buffer, pH 7.2	2.0	—	43
Tris buffer, pH 7.2, 5.8% NaCl or 5% MSG	7.0	—	43
Tris buffer, pH 7.2 + 5.8% NaCl + 5% MSG	15.5	—	43

Production

In general, enterotoxin production tends to be favored by optimum growth conditions of pH, temperature, Eh, and so on, and these conditions are discussed above. It is well established that staphylococci can grow under conditions that do not favor enterotoxin production.

With respect to a_w, enterotoxin production (except for SEA) occurs over a slightly narrower range than growth. In precooked bacon incubated aerobically at 37°C, *S. aureus* A100 grew rapidly at a_w as low as 0.84 and produced SEA (52). The production of the individual enterotoxins is more inherent to the toxin than to the strain that produces them (71). SEA but not SEB has been shown to be produced by L-phase cells (22). In pork, SEA production occurred at a_w 0.86 but not at 0.83; and in beef at 0.88 but not at 0.86 (85). SEA can be produced under conditions of a_w that do not favor SEB (88). In general, SEB production is sensitive to a_w while SEC is sensitive to both a_w and temperature. Regarding NaCl and pH, enterotoxin production has been recorded at pH 4.0 in the absence of NaCl (see Table 18-2). The effect of NaCl on SEB synthesis by strain S-6 at pH 7.0 and 37°C is presented in Fig. 18-1. Although this strain produces both SEA and SEB, neither was produced above 10% NaCl (71). In general SEA production is less sensitive to pH than SEB. The buffering of a culture medium at pH 7.0 leads to more SEB than when the medium is unbuffered or buffered in the acid range (58). A similar result was noted at a controlled pH of 6.5 rather than 7.0 (45).

With respect to growth temperature, SEB production in ham at 10°C has been recorded (35), as well as small amounts of SEA, SEB, SEC, and SED

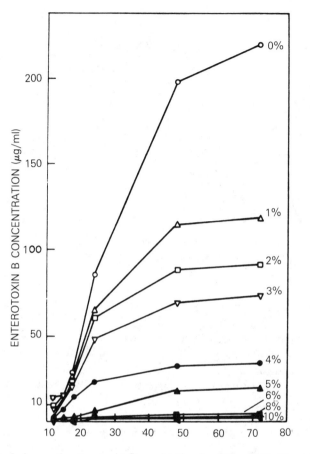

Fig. 18-1. Staphylococcal enterotoxin B production in different NaCl concentrations in 4% NZ-Amine NAK medium at pH 7.0 and 37°C. (71, copyright © 1982, International Association of Milk, Food and Environmental Sanitarians)

in cooked ground beef, ham, and bologna at 10°C (85). Production has been observed at 46°C, but the optimum temperature for SEB and SEC is 40°C in a protein hydrolysate medium (90), and for SEE, 40°C at pH 6.0 (87). The growth of *S. aureus* on cooked beef at 45.5°C for 24 h has been demonstrated, but at 46.6°C the initial inoculum decreased by 2 log cycles over the same period (12). The optimum for SEB in a culture medium at pH 7.0 was 39.4°C (71). Thus, the optimum temperature for enterotoxin production is in the 40–45°C range.

Staphylococcal enterotoxins have been reported to appear in cultures as early as 4–6 h (see Fig. 18-2) and to increase proportionately through the stationary phase (53) and into the transitional phase (see Fig. 18-3). Enterotoxin production has been shown to occur during all phases of growth (23) although earlier studies revealed that with strain S-6, 95% of SEB was released

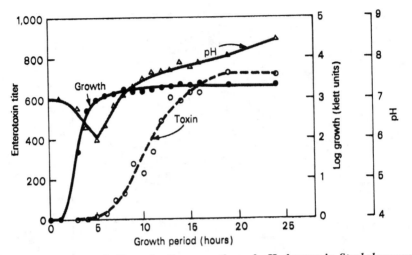

Fig. 18-2. Enterotoxin B production, growth, and pH changes in *Staphylococcus aureus* at 37°C (57).

during the latter part of the log phase of growth (55, 56). The latter authors have also demonstrated the production of SEB in nongrowing cells. On the other hand, strain S-6 was found not to produce detectable quantities of SEB during exponential growth (61), but did during the postexponential phase in association with total protein synthesis. Chloramphenicol inhibited the appearance of enterotoxin, suggesting that the presence of toxin was dependent upon *de novo* protein synthesis.

The maximum amount of enterotoxin A that can be produced in culture media under ideal conditions is about 5–6 μg/ml, while levels of 350 and 60 μg/ml or more of SEB and SEC, respectively, can be produced (75). In protein hydrolysate media, up to 500 μg/ml of SEB may be produced (8). A study by Chesbro et al. (17) suggests that SEB is heterogeneous. They found that two electrophoretically distinct toxins can be identified in cultures, one produced in early to mid-log and the other in mid–late log phase growth. Analyses of old cultures should show the presence of both.

SEB production in unbuffered media has been found to be repressed by excess glucose in the medium (61). Streptomycin, actinomycin D, acriflavine, Tween 80, and other compounds have been found to inhibit SEB synthesis in broth (33). SEB production is inhibited by 2-deoxyglucose and the inhibition is not restored by glucose, indicating that this toxin, at least, is not under catabolite control (44). While actinomycin D has been shown to inhibit SEB synthesis in strain S-6, the inhibition occurred about 1 h after cellular synthesis ceased. The latter was immediately and completely inhibited. A possible conclusion from this finding is that the mRNA responsible for enterotoxin synthesis is more stable than that for cellular synthesis (48).

The minimum number of cells of *S. aureus* required to produce the minimum level of enterotoxin considered necessary to cause the gastroenteritis

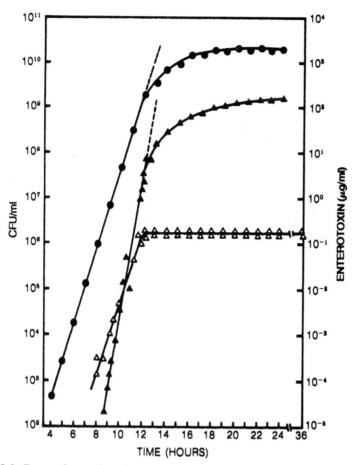

Fig. 18-3. Rates of growth and enterotoxins A and B synthesis by *Staphylococcus aureus* S-6. Symbols: ● CFU/ml; △, enterotoxin A; ▲, enterotoxin B. (23, copyright © 1974, American Society for Microbiology)

syndrome in man (1 ng/g) appears to differ for substrates and for the particular enterotoxin. In milk, SEA and SED were detected with counts of 10^7 but not below this level (63). Employing a strain of *S. aureus* that produces SEA, SEB, and SED, SEB and SED were detected when the count reached 6×10^6/ml and the enterotoxin level was 1 ng/ml, while SEA at a level of 4 ng/ml was detected with a count of 3×10^7 cfu/ml (64). In imitation cheese with pH of 5.56–5.90 and a_w of 0.94–0.97, enterotoxins were first detected at the following counts: SEA at 4×10^6/g; SEC at 1×10^8; SED at 3×10^6; SEE at 5×10^6; and SEC and SEE at 3×10^6/g (6). In precooked bacon, SEA was produced by strain A100 with cells $> 10^6$/g (81). In meat products and vanilla custard, SEA was produced with $\geq \log_{10} 7.2$ cells/g, but in certain vegetable products no toxin was detected with counts up to $\log_{10} 10.00$/g (66).

SEA has been determined to be under the control of a chromosomal gene (69, 79) but the gene is not a stable chromosomal entity. In some wild-type strains of *S. aureus,* the gene for SEA is carried by temperate polymorphic phages that integrate with the bacterial chromosome (10). This explains why SEA is produced generally under all conditions that support growth of SEA-producing strains, and perhaps why this enterotoxin causes more food-poisoning outbreaks than any other, at least in the United States. Chromosomal genes are more stable than plasmid-borne genes. In strain S-6, the SEA gene has been shown to occur on the chromosome very close to the alpha-hemolysin gene, but for strain FRI-196E it appears to be in another position (69). In the case of SEB, it appears to be plasmid-borne although a chromosomal location has been reported at least for some strains. SEB and SEC$_1$ genes are carried on a penicillinase-bearing plasmid, where they may serve either as structural genes for the synthesis of these two enterotoxins, or as essential regulatory genes that switch on cryptic structural genes in recipient cells (2).

With respect to enterotoxin synthesis, evidence has been presented for SEB that the kinetic precursor of the extracellular product is a larger membrane-bound form designated pSEB (89). The latter investigators believe that the temporary sequestering on the membrane may be critical to the mechanism that facilitates the transfer of SEB through the cell wall. After its release from the membrane, the enterotoxin appears to be transiently sequestered by the cell wall before its ultimate release into the extracellular environment. A similar pattern of synthesis, sequestering, and release has been shown for SEA (19). Regarding intracellular synthesis, evidence has been presented that SEB and SEC are synthesized in one way while SEA, SED, and SEE are assembled in another (9).

Detection in foods

Serologic and in vitro methods for detecting enterotoxins are presented in Chapter 6, and in vivo and related methods in Chapter 7. For the extraction of foods for enterotoxins, an appropriate reference in Table 5-1 should be consulted.

THE GASTROENTERITIS SYNDROME

Upon the ingestion of contaminated food, the symptoms of staphylococcal food poisoning usually develop within 4 h, although a range of from 1 to 6 h has been reported. The symptoms consist of nausea, vomiting, abdominal cramps (which are usually quite severe), diarrhea, sweating, headache, prostration, and sometimes a fall in body temperature. The symptoms generally last from 24 to 48 h, and the mortality rate is very low or nil. The usual treatment for healthy persons consists of bed rest and maintenance of fluid balance. Upon cessation of symptoms, the victim possesses no demonstrable immunity to recurring attacks, although animals become resistant to enterotoxin after repeated oral doses (8). Since the symptoms are referable to the ingestion of preformed enterotoxin, it is conceivable that stool cultures might be negative for the organisms, although this is rare. Proof of staphylococcal food poisoning is established by recovering coagulase-positive, enterotoxigenic staphylococci from leftover food and from the stool cultures of victims. Attempts should be made to extract enterotoxin from suspect foods, especially when the number of recoverable viable cells is low.

The minimum quantity of enterotoxin needed to cause illness in man is about 1 ng/g. Data obtained from the use of three human volunteers showed that a dosage of 20–35 μg of pure SEB can produce a syndrome (72). From sixteen incidents of staphylococcal food poisoning, enterotoxin levels of < 0.01–0.25 μg/g of food were found (39).

The pathogenesis of enterotoxins in man is not yet clear. They act upon the intestines to induce vomiting and diarrhea, and the same effects can be achieved by IV injections. When SEA was administered IV to monkeys, an initial state of lymphopenia was induced, which lasted for 1 to 2 days and was followed by the release of new immature cells that had greater DNA synthesis activity (93).

INCIDENCE AND VEHICLE FOODS

The incidence of staphylococci in a variety of foods is presented in Chapter 4 (see Tables 4-1, 4-2, 4-3, 4-4, and 4-5). They may be expected to occur in a wide variety of foods not given heat treatments for their destruction.

With regard to vehicle foods for staphylococcal enteritis, a large number has been incriminated in outbreaks, usually products made by hand and improperly refrigerated after being prepared. Of selected cases of staphylococcal food poisoning over the period 1961–73, meat products were involved in over 40% of the cases (Table 18-7). Of the 251 meat products, 76 were baked ham (14). The precise number of cases in the United States is not known, but various investigators have placed the number in the tens of thousands/year. The problem here is one of reporting, and all too often the small outbreaks that occur in homes are not reported to public health officials. A large percentage of the reported cases of all types are those that result from banquets, generally involving large numbers of persons.

An unusual outbreak was caused by SEA and SED and traced to wild mushrooms in vinegar (54). The food contained 10 ng SEA and 1 ng of SED/g.

ECOLOGY OF *S. AUREUS* GROWTH

In general, the staphylococci do not compete well with the normal flora of most foods, and this is especially true for those that contain large numbers

Table 18-7. Foods incriminated in staphylococcal food-poisoning outbreaks, 1961–73.
(Summarized from Bryan, 14)

Food Products		No. of Outbreaks
Meat		251
Ham products	137	
Beef products	60	
Uncured pork products	27	
Others/combinations	27	
Poultry		102
Turkey products	52	
Chicken	50	
Custards and cream-filled pastries		55
Fish and shellfish		34
Salads (nonmeat)		31
Eggs and egg products		17
Milk and milk products		14
Vegetables		9
Cereal products		6
Miscellaneous products		59

Condensed from a table in *Food Microbiology: Public
Health and Spoilage Aspects,* by deFigueiredo and
Splittstoesser, published by AVI Publishing Co., Westport,
Conn. 06880, and used by permission of publisher.

of lactic acid bacteria where conditions permit the growth of the latter
organisms (see Chapter 16). A large number of investigators have shown
the inability of *S. aureus* to compete in both fresh and frozen foods. At
temperatures that favor staphylococcal growth, the normal food saprophytic
flora offers protection against staphylococcal growth through antagonism,
competition for nutrients, and modification of the environment to conditions
less favorable to *S. aureus*. Bacteria known to be antagonistic to *S. aureus*
growth include *Acinetobacter, Aeromonas, Bacillus, Pseudomonas, S. epidermidis,* the Enterobacteriaceae, the Lactobacillaceae, streptococci, and
others (62). SEA has been shown to be resistant to a variety of environmental
stresses, but growth of several lactic acid bacteria did lead to its reduction
and to a suggestion that toxin reduction might have resulted from specific
enzymes or other metabolites of the lactic acid bacteria (18). The generally
higher incidence of food poisonings of all types over the past decade may
simply reflect the general improvement in food plant sanitation and the fact
that improving technology enables food producers to produce more low-
count foods. More studies on the ecology of staphylococcal growth are
necessary before this picture can be made clearer.

PREVENTION OF STAPHYLOCOCCAL AND OTHER FOOD-POISONING SYNDROMES

When susceptible foods are produced with low numbers of staphylococci, they will remain free of enterotoxins and other food-poisoning hazards if kept either *below* 40° or *above* 140°F until consumed. For the years 1961–72, over 700 food-borne disease outbreaks were investigated by Bryan (13) relative to the factors that contributed to the outbreaks, and of the sixteen factors identified, the five most frequently involved were:

1. inadequate refrigeration
2. preparing foods far in advance of planned service
3. infected persons practicing poor personal hygiene
4. inadequate cooking or heat processing
5. holding food in warming devices at bacterial growth temperatures

Inadequate refrigeration alone comprised 25.5% of the contributing factors. The five listed above contributed to 68 percent of outbreaks. Susceptible foods should not be held within the staphylococcal growth range for more than 3–4 h.

REFERENCES

1. Adesiyun, A. A. 1984. Enterotoxigenicity of *Staphylococcus aureus* strains isolated from Nigerian ready-to-eat foods. *J. Food Protect.* 47:438–40.
2. Altboum, Z., I. Hertman, and S. Sarid. 1985. Penicillinase plasmid-linked genetic determinants for enterotoxins B and C_1 production in *Staphylococcus aureus*. *Infect. Immun.* 47:514–21.
3. Angelotti, R., M. J. Foter, and K. H. Lewis. 1961. Time-temperature effects on salmonellae and staphylococci in foods. *Amer. J. Pub. Hlth* 51:76–88.
4. ———. 1960. Time-temperature effects on salmonellae and staphylococci in foods. II. Behavior at warm holding temperatures. Thermal-Death-Time Studies. Cincinnati: Public Health Service, U.S. Dept. Health, Education and Welfare.
5. Baird-Parker, A. C. 1965. The classification of staphylococci and micrococci from world-wide sources. *J. Gen. Microbiol.* 38:363–87.
6. Bennett, R. W., and W. T. Amos. 1983. *Staphylococcus aureus* growth and toxin production in imitation cheeses. *J. Food Sci.* 48:1670–73.
7. Bergdoll, M. S. 1967. The staphylococcal enterotoxins. In *Biochemistry of some foodborne microbial toxins*, ed. R. I. Mateles and G. N. Wogan, 1–25. Cambridge, Mass.: MIT Press.
8. ———. 1972. The enterotoxins. In *The staphylococci*, ed. J. O. Cohen, 301–31. New York: Wiley-Interscience.
9. ———. 1979. Staphylococcal intoxications. In *Food-borne infections and intoxications,* ed. H. Riemann and F. L. Bryan, 443–94. New York: Academic Press.

10. Betley, M. J., and J. J. Mekalanos. 1985. Staphylococcal enterotoxin A is encoded by phage. *Science* 229:185–87.

11. Borja, C. R., and M. S. Bergdoll. 1967. Purification and partial characterization of enterotoxin C produced by *Staphylococcus aureus* strain 137. *J. Biochem.* 6:1457–73.

12. Brown, D. F., and R. M. Twedt. 1972. Assessment of the sanitary effectiveness of holding temperatures on beef cooked at low temperature. *Appl. Microbiol.* 24:599–603.

13. Bryan, F. L. 1974. Microbiological food hazards today—based on epidemiological information. *Food Technol.* 28(9):52–59.

14. ———. 1976. *Staphylococcus aureus*. In *Food microbiology: Public health and spoilage aspects*, ed. M. P. deFigueiredo and D. F. Splittstoesser, 12–128. Westport, Conn.: AVI.

15. Casman, E. P. 1965. Staphylococcal enterotoxin. In *The staphylococci: Ecologic perspectives*, Annals of the New York Academy of Science, vol. 28, 128:124–33.

16. Casman, E. P., R. W. Bennett, A. E. Dorsey, and J. A. Issa. 1967. Identification of a fourth staphylococcal enterotoxin, enterotoxin D. *J. Bacteriol.* 94:1875–82.

17. Chesbro, W., D. Carpenter, and G. J. Silverman. 1976. Heterogeneity of *Staphylococcus aureus* enterotoxin B as a function of growth stage: Implications for surveillance of foods. *Appl. Environ. Microbiol.* 31:581–89.

18. Chordash, R. A., and N. N. Potter. 1976. Stability of staphylococcal enterotoxin A to selected conditions encountered in foods. *J. Food Sci.* 41:906–9.

19. Christianson, K. K., R. K. Tweten, and J. J. Iandolo. 1985. Transport and processing of staphylococcal enterotoxin A. *Appl. Environ. Microbiol.* 50:696–97.

20. Chu, F. S., E. Crary, and M. S. Bergdoll. 1969. Chemical modification of amino groups in staphylococcal enterotoxin B. *Biochem.* 8:2890–96.

21. Chu, F. S., K. Thadhani, E. J. Schantz, and M. S. Bergdoll. 1966. Purification and characterization of staphylococcal enterotoxin A. *Biochem.* 5:3281–89.

22. Czop, J. K., and M. S. Bergdoll. 1970. Synthesis of enterotoxins by L-forms of *Staphylococcus aureus*. *Infect. Immun.* 1:169–73.

23. ———. 1974. Staphylococcal enterotoxin synthesis during the exponential, transitional, and stationary growth phases. *Infect. Immun.* 9:229–35.

24. Dack, G. M., W. E. Cary, O. Woolpert, and H. Wiggers. 1930. An outbreak of food poisoning proved to be due to a yellow hemolytic staphylococcus. *J. Prev. Med.* 4:167–75.

25. Dalidowicz, J. E., S. J. Silverman, E. J. Schantz, D. Stefanye, and L. Spero. 1966. Chemical and biological properties of reduced and alkylated staphylococcal enterotoxin B. *Biochem.* 5:2375–81.

26. Denny, C. B., J. Y. Humber, and C. W. Bohrer. 1971. Effect of toxin concentration on the heat inactivation of staphylococcal enterotoxin A in beef bouillon and in phosphate buffer. *Appl. Microbiol.* 21:1064–66.

27. Denny, C. B., P. L. Tan, and C. W. Bohrer. 1966. Heat inactivation of staphylococcal enterotoxin. *J. Food Sci.* 31:762–67.

28. Devriese, L. A. 1981. Baird-Parker medium supplemented with acriflavine, polymyxins and sulphonamide for the selective isolation of *Staphylococcus aureus* from heavily contaminated materials. *J. Appl. Bacteriol.* 50:351–57.

29. Devriese, L. A., V. Hájek, P. Oeding, A. A. Meyer, and K. H. Schleifer. 1978. *Staphylococcus hyicus* (Sompolinsky 1953) comb. nov. and *Staphylococcus hyicus* subsp. *chromogens* subsp. nov. *Int. J. Syst. Bacteriol.* 38:482–90.

30. Elek, S. D. 1959. *Staphylococcus pyogenes and its relation to disease.* Edinburgh and London: Livingstone.

31. Erickson, A., and R. H. Deibel. 1973. Production and heat stability of staphylococcal nuclease. *Appl. Microbiol.* 25:332–36.

32. Evans, J. B., G. A. Ananaba, C. A. Pate, and M. S. Bergdoll. 1983. Enterotoxin production by atypical *Staphylococcus aureus* from poultry. *J. Appl. Bacteriol.* 54:257–61.

33. Friedman, M. E. 1966. Inhibition of staphylococcal enterotoxin B formation in broth cultures. *J. Bacteriol.* 92:277–78.

34. Genigeorgis, C., M. S. Foda, A. Mantis, and W. W. Sadler. 1971. Effect of sodium chloride and pH on enterotoxin C production. *Appl. Microbiol.* 21:862–66.

35. Genigeorgis, C., H. Riemann, and W. W. Sadler. 1969. Production of enterotoxin B in cured meats. *J. Food Sci.* 34:62–68.

36. Genigeorgis, C., and W. W. Sadler. 1966. Effect of sodium chloride and pH on enterotoxin B production. *J. Bacteriol.* 92:1383–87.

37. ———. 1966. Characterization of strains of *Staphylococcus aureus* isolated from livers of commercially slaughtered poultry. *Poultry Sci.* 45:973–80.

38. Gibbs, P. A., J. T. Patterson, and J. Harvey. 1978. Biochemical characteristics and enterotoxigenicity of *Staphylococcus aureus* strains isolated from poultry. *J. Appl. Bacteriol.* 44:57–74.

39. Gilbert, R. J., and A. A. Wieneke. 1973. Staphylococcal food poisoning with special reference to the detection of enterotoxin in food. In *The microbiological safety of food*, ed. B. C. Hobbs and J. H. B. Christian, 273–85. New York: Academic Press.

40. Harvey, J., J. T. Patterson, and P. A. Gibbs. 1982. Enterotoxigenicity of *Staphylococcus aureus* strains isolated from poultry: Raw poultry carcasses as a potential food-poisoning hazard. *J. Appl. Bacteriol.* 52:251–58.

41. Hoover, D. G., S. R. Tatini, and J. B. Maltais. 1983. Characterization of staphylococci. *Appl. Environ. Microbiol.* 46:649–60.

42. Huang, I.-Y., and M. S. Bergdoll. 1970. The primary structure of staphylococcal enterotoxin B. III. The cyanogen bromide peptides of reduced and aminoethylated enterotoxin B and the complete amino acid sequence. *J. Biol. Chem.* 245:3518–25.

43. Hurst, A., and A. Hughes. 1983. The protective effect of some food ingredients on *Staphylococcus aureus* MF 31. *J. Appl. Bacteriol.* 55:81–88.

44. Iandolo, J. J., and W. M. Shafer. 1977. Regulation of staphylococcal enterotoxin B. *Infect. Immun.* 16:610–16.

45. Jarvis, A. W., R. C. Lawrence, and G. G. Pritchard. 1973. Production of staphylococcal enterotoxins A, B, and C under conditions of controlled pH and aeration. *Infect. Immun.* 7:847–54.

46. Jay, J. M. 1970. Effect of borate on the growth of coagulase-positive and coagulase-negative staphylococci. *Infect. Immun.* 1:78–79.

47. Kadan, R. S., W. H. Martin, and R. Mickelsen. 1963. Effects of ingredients used in condensed and frozen dairy products on thermal resistance of potentially pathogenic staphylococci. *Appl. Microbiol.* 11:45–49.

48. Katsuno, S., and M. Kondo. 1973. Regulation of staphylococcal enterotoxin B synthesis and its relation to other extracellular proteins. *Jap. J. Med. Sci. Biol.* 26:26–29.
49. Kloos, W. E. 1980. Natural populations of the genus *Staphylococcus. Ann. Rev. Microbiol.* 34:559–92.
50. Lechowich, R. V., J. B. Evans, and C. F. Niven, Jr. 1956. Effect of curing ingredients and procedures on the survival and growth of staphylococci in and on cured meats. *Appl. Microbiol.* 4:360–63.
51. Lee, I. C., K. E. Stevenson, and L. G. Harmon. 1977. Effect of beef broth protein on the thermal inactivation of staphylococcal enterotoxin B. *Appl. Environ. Microbiol.* 33:341–44.
52. Lee, R. Y., G. J. Silverman, and D. T. Munsey. 1981. Growth and enterotoxin A production by *Staphylococcus aureus* in precooked bacon in the intermediate moisture range. *J. Food Sci.* 46:1687–92.
53. Lilly, H. D., R. A. McLean, and J. A. Alford. 1967. Effects of curing salts and temperature on production of staphylococcal enterotoxin. *Bacteriol. Proc.,* 12.
54. Lindroth, S., E. Strandberg, A. Pessa, and M. J. Pellinen. 1983. A study of the growth potential of *Staphylococcus aureus* in *Boletus edulis,* a wild edible mushroom, prompted by a food poisoning outbreak. *J. Food Sci.* 48:282–83.
55. Markus, Z., and G. J. Silverman. 1968. Enterotoxin B production by nongrowing cells of *Staphylococcus aureus. J. Bacteriol.* 96:1446–47.
56. ———. 1969. Enterotoxin B synthesis by replicating and nonreplicating cells of *Staphylococcus aureus. J. Bacteriol.* 97:506–12.
57. McLean, R. A., H. D. Lilly, and J. A. Alford. 1968. Effects of meat-curing salts and temperature on production of staphylococcal enterotoxin B. *J. Bacteriol.* 95:1207–11.
58. Metzger, J. F., A. D. Johnson, W. S. Collins, II, and V. McGann. 1973. *Staphylococcus aureus* enterotoxin B release (excretion) under controlled conditions of fermentation. *Appl. Microbiol.* 25:770–73.
59. Miller, R. D., and D. Y. C. Fung. 1973. Amino acid requirements for the production of enterotoxin B by *Staphylococcus aureus* S-6 in a chemically defined medium. *Appl. Microbiol.* 25:800–806.
60. Minor, T. E., and E. H. Marth. 1976. *Staphylococci and their significance in foods.* New York: Elsevier.
61. Morse, S. A., R. A. Mah, and W. J. Dobrogosz. 1969. Regulation of staphylococcal enterotoxin B. *J. Bacteriol.* 98:4–9.
62. Mossel, D. A. A. 1975. Occurrence, prevention, and monitoring of microbial quality loss of foods and dairy products. *CRC Crit. Rev. Environ. Control* 5:1–140.
63. Noleto, A. L., and M. S. Bergdoll. 1980. Staphylococcal enterotoxin production in the presence of nonenterotoxigenic staphylococci. *Appl. Environ. Microbiol.* 39:1167–71.
64. ———. 1982. Production of enterotoxin by a *Staphylococcus aureus* strain that produces three identifiable enterotoxins. *J. Food Protect.* 45:1096–97.
65. Notermans, S., and C. J. Heuvelman. 1983. Combined effect of water activity, pH and sub-optimal temperature on growth and enterotoxin production of *Staphylococcus aureus. J. Food Sci.* 48:1832–35, 1840.

66. Notermans, S., and R. L. M. van Otterdijk. 1985. Production of enterotoxin A by *Staphylococcus aureus* in food. *Int. J. Food Microbiol.* 2:145–49.
67. Palasuntheram, C., and M. S. Beauchamp. 1982. Enterotoxigenic staphylococci in Sri Lanka. *J. Appl. Bacteriol.* 52:39–41.
68. Palumbo, S. A., J. L. Smith, and J. C. Kissinger. 1977. Destruction of *Staphylococcus aureus* during frankfurter processing. *Appl. Environ. Microbiol.* 34:740–44.
69. Pattee, P. A., and B. A. Glatz. 1980. Identification of a chromosomal determinant of enterotoxin A production in *Staphylococcus aureus*. *Appl. Environ. Microbiol.* 39:186–93.
70. Payne, D. N., and J. M. Wood. 1974. The incidence of enterotoxin production in strains of *Staphylococcus aureus* isolated from foods. *J. Appl. Bacteriol.* 37:319–25.
71. Pereira, J. L., S. P. Salzberg, and M. S. Bergdoll. 1982. Effect of temperature, pH and sodium chloride concentrations on production of staphylococcal enterotoxins A and B. *J. Food Protect.* 45:1306–9.
72. Raj, H. D., and M. S. Bergdoll. 1969. Effect of enterotoxin B on human volunteers. *J. Bacteriol.* 98:833–34.
73. Read, R. B., and J. G. Bradshaw. 1966. Thermal inactivation of staphylococcal enterotoxin B in veronal buffer. *Appl. Microbiol.* 14:130–32.
74. Reiser, R. F., R. N. Robbins, A. L. Noleto, G. P. Khoe, and M. S. Bergdoll. 1984. Identification, purification, and some physiochemical properties of staphylococcal enterotoxin C$_3$. *Infect. Immun.* 45:625–30.
75. Reiser, R. F., and K. F. Weiss. 1969. Production of staphylococcal enterotoxins A, B, and C in various media. *Appl. Microbiol.* 18:1041–43.
76. Roskey, C. T., and M. K. Hamdy. 1972. Bruised poultry tissue as a possible source of staphylococcal infection. *Appl. Microbiol.* 23:683–87.
77. Satterlee, L. D., and A. A. Kraft. 1969. Effect of meat and isolated meat proteins on the thermal inactivation of staphylococcal enterotoxin B. *Appl Microbiol.* 17:906–9.
78. Schantz, E. J., W. G. Roessler, J. Wagman, L. Spero, D. A. Dunnery, and M. S. Bergdoll. 1965. Purification of staphylococcal enterotoxin B. *J. Biochem.* 4:1011-16.
79. Shafer, W. M., and J. J. Iandolo. 1978. Staphylococcal enterotoxin A: A chromosomal gene product. *Appl. Environ. Microbiol.* 36:389–91.
80. Siems, H., D. Kusch, H.-J. Sinell, and F. Untermann. 1971. Vorkommen and Eigenschaften von Staphylokokken in verschiedenen Produktionsstufen bei der Fleischverarbeitung. *Fleischwirts.* 51:1529–33.
81. Silverman, G. J., D. T. Munsey, C. Lee, and E. Ebert. 1983. Interrelationship between water activity, temperature and 5.5 percent oxygen on growth and enterotoxin A secretion by *Staphylococcus aureus* in precooked bacon. *J. Food Sci.* 48:1783–86, 1795.
82. Šimkovičová, M., and R. J. Gilbert. 1971. Serological detection of enterotoxin from food-poisoning strains of *Staphylococcus aureus*. *J. Med. Microbiol.* 4:19–30.
83. Smith, J. L., R. L. Buchanan, and S. A. Palumbo. 1983. Effect of food environment on staphylococcal enterotoxin synthesis: A review. *J. Food Protect.* 46:545–55.

84. Sperber, W. H. 1977. The identification of staphylococci in clinical and food microbiology laboratories. *CRC Crit. Rev. Clin. Lab. Sci.* 7:121–84.

85. Tatini, S. R. 1973. Influence of food environments on growth of *Staphylococcus aureus* and production of various enterotoxins. *J. Milk Food Technol.* 36:559–63.

86. Thomas, C. T., J. C. White, and K. Longree. 1966. Thermal resistance of salmonellae and staphylococci in foods. *Appl. Microbiol.* 14:815–20.

87. Thota, F. H., S. R. Tatini, and R. W. Bennett. 1973. Effects of temperature, pH and NaCl on production of staphylococcal enterotoxins E and F. *Bacteriol. Proc.*, 1.

88. Troller, J. A. 1972. Effect of water activity on enterotoxin A production and growth of *Staphylococcus aureus*. *Appl. Microbiol.* 24:440–43.

89. Tweten, R. K., and J. J. Iandolo. 1983. Transport and processing of staphylococcal enterotoxin B. *J. Bacteriol.* 153:297–303.

90. Vandenbosch, L. L., D. Y. C. Fung, and M. Widomski. 1973. Optimum temperature for enterotoxin production by *Staphylococcus aureus* S-6 and 137 in liquid medium. *Appl. Microbiol.* 25:498–500.

91. Woodburn, M., M. Bennion, and G. E. Vail. 1962. Destruction of salmonellae and staphylococci in precooked poultry products by heat treatment before freezing. *Food Technol.* 16:98–100.

92. Wu, C.-H., and M. S. Bergdoll. 1971. Stimulation of enterotoxin B production. *Infect. Immun.* 3:784–92.

93. Zehavi-Willner, T., E. Shenberg, and A. Barnea. 1984. In vivo effect of staphylococcal enterotoxin A on peripheral blood lymphocytes. *Infect. Immun.* 44:401–5.

19.

FOOD POISONING CAUSED
BY GRAM-POSITIVE
SPOREFORMING BACTERIA

At least three gram-positive sporeforming rods are known to cause bacterial food poisoning: *Clostridium perfringens (welchii), C. botulinum,* and *Bacillus cereus.* The incidence of food poisoning caused by each of these organisms is related to certain specific foods as is food poisoning in general.

CLOSTRIDIUM PERFRINGENS FOOD POISONING

The causative organism of this syndrome is a gram-positive, anaerobic sporeforming rod widely distributed in nature. Based upon their ability to produce certain exotoxins, five types are recognized: types A, B, C, D, and E. The food-poisoning strains belong to type A as do the classical gas gangrene strains, but unlike the latter, the food-poisoning strains are generally heat-resistant and produce only traces of alpha toxin. Some type C strains produce enterotoxin and may cause a food-poisoning syndrome. The classical food-poisoning strains differ from type C strains in not producing beta toxin. The latter, which have been recovered from enteritis necroticans, are compared to type A heat-sensitive and heat-resistant strains in Table 19-1.

While *C. perfringens* has been associated with gastroenteritis since 1895, the first clear-cut demonstration of its etiological status in food poisoning was made by McClung (58), who investigated four outbreaks in which chicken was incriminated. The first detailed report of the characteristics of this food-poisoning syndrome was that of Hobbs et al. (38) in Great Britain. While the British workers were more aware of this organism as a cause of food poisoning during the 1940s and 1950s, few incidents were recorded in the United States prior to the publication by Angelotti et al. (6) of methods

Table 19-1. Toxins of *Clostridium welchii* types A and C.

Cl. welchii	α	β	γ	δ	ε	θ	ι	κ	λ	μ	ν
						Toxins					
Heat-sensitive type A	+ + +	−	−	−	−	+ +	−	+ +	−	+ or −	+
Heat-resistant type A	± or tr	−	−	−	−	−	−	+ or −	−	+ + + or −	−
Heat-resistant type C	+	+	+	−	−	−	−	−	−	−	+

B. C. Hobbs, 1962. *Bacterial Food Poisoning*. London: Royal Society of Health.

of recovery and quantitation of the organism. It is now known that *C. perfringens* food poisoning is widespread in the United States and many other countries.

Distribution of C. perfringens

The food-poisoning strains of *C. perfringens* exist in soils, water, foods, dust, spices, and the intestinal tract of man and other animals. Various authors have reported the incidence of the heat-resistant, nonhemolytic strains to range from 2–6% in the general population. Between 20 and 30% of healthy hospital personnel and their families have been found to carry these organisms in their feces, while the carrier rate of victims after 2 weeks may be 50% or as high as 88% (19). The heat-sensitive types are common to the intestinal tract of all humans. *C. perfringens* gets into meats either directly from slaughter animals or by subsequent contamination of slaughtered meat from containers, handlers, or dust. Since it is a sporeformer, it can withstand the adverse environmental conditions of drying, heating, and certain toxic compounds.

Characteristics of the organism

Food poisoning as well as most other strains of *C. perfringens* grow well on a variety of media if incubated under anaerobic conditions, or if provided with sufficient reducing capacity. Strains of *C. perfringens* isolated from horse muscle grew without increased lag phase at an Eh of −45 or lower, while more positive Eh values had the effect of increasing the lag phase (7). Although it is not difficult to obtain growth of these organisms on various media, sporulation occurs with difficulty and requires the use of special media such as those described by Duncan and Strong (20), or the employment of special techniques such as dialysis sacs (87).

With respect to growth temperatures, *C. perfringens* is mesophilic with an optimum between 37° and 45°C. The lowest temperature for growth is around 20°C, and the highest is around 50°C. Optimum growth in thioglycollate medium for six strains was found to occur between 30 and 40°C, while the optimum for sporulation in Ellner's medium was 37°–40°C (82). Growth at 45°C under otherwise optimal conditions leads to generation times as short as 7 min. Regarding pH, many strains grow over the range 5.5–8.0 but generally not below 5.0 or above 8.5. The lowest reported a_w values for growth and germination of spores lie between 0.97 and 0.95 with sucrose or NaCl, or about 0.93 with glycerol employing a fluid thioglycollate base (43). Spore production appears to require higher a_w values than the above minima. While growth of type A was demonstrated at pH 5.5 by Labbe and Duncan (47), no sporulation or toxin production occurred. A pH of 8.5 appears to be the highest for growth. *C. perfringens* is not as strict an anaerobe as are some other clostridia. Its growth at an initial Eh of +320 mv has been observed (78). At least thirteen amino acids are required for growth along with biotin, pantothenate, pyridoxal, adenine, and other related compounds. It is heterofermentative, and a large number of carbohydrates are attacked. Growth is inhibited by around 5% NaCl.

The endospores of food-poisoning strains differ in their resistance to heat, with some being typical of other mesophilic sporeformers and some being highly resistant. A D 100°C value of 0.31 for *C. perfringens* (ATCC 3624) and a value of 17.6 for strain NCTC 8238 have been reported (113). For eight strains that produced reactions in rabbits, D 100°C values ranged from 0.70 to 38.37; strains that did not produce rabbit reactions were more heat sensitive (103).

In view of the practice of cooking roasts in water baths for long times at low temperatures (LTLT), the heat destruction of vegetative cells of *C. perfringens* has been studied by several groups. For strain ATCC 13124 in autoclaved ground beef, D 56.8°C was 48.3 min, essentially similar to the D 56.8°C or D 47.9°C for phospholipase C (26). Employing strain NCTC 8798, D values for cells were found to increase with increasing growth temperatures in autoclaved ground beef. For cells grown at 37°C, D 59°C was 3.1 min; cells grown at 45°C had D 59°C of 7.2; and cells grown at 49°C had D 59°C of 10.6 min (83). While the wide differences in heat resistance between the two strains noted may in part be due to strain differences, the effect of fat in the heating menstrum may also have played a role. With beef roasts cooked in plastic bags in a water bath at 60°–61°C, holding the product to an internal temperature of 60°C for at least 12 min eliminated salmonellae and reduced the *C. perfringens* population by about 3 log cycles. To effect a 12-log reduction of numbers for roasts weighing 1.5 kg, holding at 60°C for 2.3 h or longer was necessary (90). The thermal destruction of *C. perfringens* enterotoxin in buffer and gravy at 61°C required 25.4 and 23.8 min, respectively (10).

While the wide variations in heat resistance recorded for *C. perfringens* spores may be due to many factors, similar variations have not been recorded for *C. botulinum,* especially types A and B. The latter organisms have no history in the intestinal tract of man, while *C. perfringens* strains are common inhabitants of this environment as well as soils. An organism inhabiting environments as diverse as these may be expected to show wide variations among its strains. Another factor that is important in heat resistance of bacterial spores is that of the chemical environment. Alderton and Snell (2) have pointed out that spore heat resistance is largely an inducible property, chemically reversible between a sensitive and resistant state. Using this hypothesis, it has been shown that spores can be made more heat resistant by treating them in Ca-acetate solutions—for example, 0.1 or 0.5 M at pH 8.5 for 140 h at 50°C. The heat resistance of endospores may be increased 5- to 10-fold by this method (1). On the other hand, heat resistance may be decreased by holding spores in 0.1 N HCl at 25°C for 16 h, or as a result of the exposure of endospores to the natural acid conditions of some foods. It is not inconceivable that the high variability of heat resistance of *C. perfringens* spores may be a more or less direct result of immediate environmental history.

The freezing survival of *C. perfringens* in chicken gravy was studied by Strong and Canada (101), who found that only around 4% of cells survived when frozen to − 17.7°C for 180 days. Dried spores, on the other hand, displayed a survival rate of about 40% after 90 days but only about 11% after 180 days.

For epidemiologic studies, serotyping has been employed, but because of the many serovars there appears to be no consistent relationship between outbreaks and given serovars (see Hatheway et al., 35). The bacteriocin typing of type A has been achieved, and of ninety strains involved in food outbreaks, all were typable by a set of eight bacteriocins and 85.6% consisted of bacteriocin types 1–6 (85).

The enterotoxin

The causative factor of *C. perfringens* food poisoning is an enterotoxin. It is unusual in that it is a spore-specific protein; its production occurs together with that of sporulation. All known food-poisoning cases by this organism are caused by type A strains. An unrelated disease, necrotic enteritis, is caused by beta toxin produced by type C strains and is only rarely reported outside of New Guinea. While necrotic enteritis due to type C has been associated with a mortality rate of 35–40%, food poisoning due to type A strains has been fatal only in elderly or otherwise debilitated persons. Some type C strains have been shown to produce enterotoxin but its role in disease is unclear at this time (88).

The enterotoxin of type A strains was demonstrated by Duncan and Strong (21). The purified enterotoxin has a molecular weight of 36,000 daltons and an isoelectric point of 4.3 (37). It is heat sensitive (biological activity destroyed at 60°C for 10 min), pronase sensitive but resistant to trypsin, chymotrypsin, and papain (100). L-forms of *C. perfringens* produce the toxin, and in one study they were shown to produce as much enterotoxin as classical forms (57).

The enterotoxin is synthesized by sporulating cells in association with late stages of sporulation. The peak for toxin production is just before lysis of the cell's sporangium, and the enterotoxin is released along with spores. Conditions that favor sporulation also favor enterotoxin production, and this was demonstrated with raffinose (50), caffeine, and theobromine (49). The latter two compounds increased enterotoxin from undetectable levels to 450 μg/ml of cell extract protein. The enterotoxin has been shown to be similar to spore structural proteins covalently associated with the spore coat. Cells sporulate freely in the intestinal tract and in a wide variety of foods (15). In culture media, the enterotoxin is normally produced only when endospore formation is permitted (see Fig. 19-1), but vegetative cells are known to produce enterotoxin but at extremely low levels (34). A single gene has been shown to be responsible for the enterotoxin trait (19, 22), and enterotoxin and a spore-coat protein have been shown to be controlled by a stable mRNA (48).

The enterotoxin may appear in a growth and sporulation medium about 3 h after inoculation with vegetative cells (18), and from 1 to 100 μg/ml of enterotoxin production has been shown for three strains of *C. perfringens* in Duncan-Strong (DS) medium after 24–36 h (28). It has been suggested that preformed enterotoxin may exist in some foods and in infrequent cases contribute to the early onset of symptoms (72). Purified enterotoxin has been shown to contain up to 3,500 mouse LD/mg N.

The enterotoxin may be detected in the feces of victims. From one case, 13–16 μg/g feces were found and from another victim with a milder case, 3–4 μg/g were detected (89).

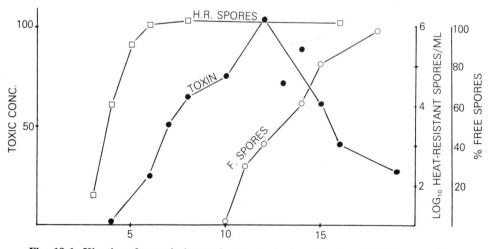

Fig. 19-1. Kinetics of sporulation and enterotoxin formation by *Clostridium perfringens* type A. (Redrawn from Labbe, 46; copyright © 1980 by Institute of Food Technologists)

With respect to mode of action, it has been demonstrated that binding of the enterotoxin is necessary for the expression of biological activity (63). The toxin has been shown to bind to isolated rabbit intestinal epithelial cells and tissue homogenates from liver and kidney but not to brain tissue (60). Its binding is specific, and time and temperature dependent, and it appears to be irreversible (Fig. 19-2). Once bound, the enterotoxin acts directly on the cell membrane via receptors and subsequently becomes "trapped" (61). Its entrapment in the membrane causes loss of structural integrity and function, with resulting fluid accumulation. In rabbits the activity of enterotoxin was shown to affect the intestinal tract in the following order: ileum > jejunum > duodenum (62). According to McDonel (59), in the normal ileum, there is a net absorption of water, Na^+, Cl^-, and glucose, while at the same time there is a net efflux of K^+ and bicarbonate ions.

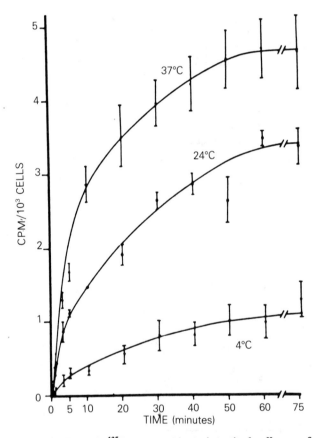

Fig. 19-2. Specific binding of (^{125}I) enterotoxin to intestinal cells as a function of temperature and time. Error bars show standard error of the mean, which if not shown was smaller than the symbol. (60, copyright © 1980, American Chemical Society)

Under the influence of enterotoxin, the net movement of water, Na$^+$, and Cl$^-$ is reversed while net movement of K$^+$ and bicarbonate is unaffected (Fig. 19-3). By use of cell cultures and isolated tissues, it has been shown that enterotoxin decreases O$_2$ consumption, causes rounding of cells, induces holes in outer cell membranes, inhibits macromolecular synthesis, and causes

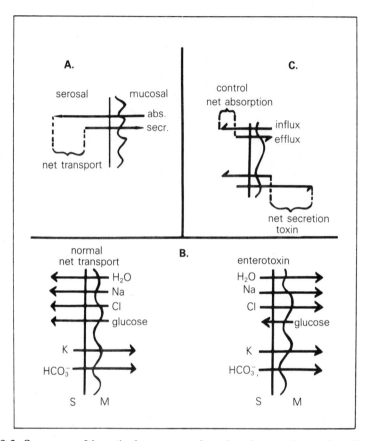

Fig. 19-3. Summary of intestinal transport alterations in rate due to the effects of *Clostridium perfringens* **enterotoxin. A. Most substances are simultaneously absorbed from the luminal (mucosal) side of the intestine to the blood (serosal) side, and in the opposite direction. Net absorption occurs when the absorptive flux (M-S) exceeds the secretory (S-M) flux. B. In the normal ileum, net absorption of water, sodium, chloride, and glucose occurs, while there is a net efflux of potassium and bicarbonate ions. Under the influence of enterotoxin, the movement of water, sodium, and chloride is reversed while net movement of potassium and bicarbonate is unaffected. Glucose continues to be absorbed but at a significantly reduced rate. C. When net sodium transport is resolved into its component fluxes, it is found that uptake of sodium is the same under control and enterotoxin-treated conditions. Net secretion due to enterotoxin is the result of a significant increase in the secretory flux. (From McDonel, 59; used with permission of** *Amer. J. Clin. Nutri.***)**

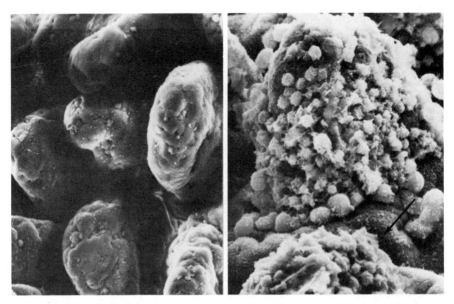

Fig. 19-4. Scanning electron micrographs (× 370) of control rabbit ileum (left) showing typical morphology of villi; and micrograph (× 900) of rabbit ileum treated with *Clostridium perfringens* enterotoxin (right) for 90 min showing toxin-damaged villus tips. (Courtesy of J. L. McDonel)

cell death (61). The toxin also causes erythema in test animals, increased capillary permeability, and exhibits parasympathomimetric properties. Its effect on rabbit ileum is presented in Fig. 19-4. These and other effects on tissue culture systems and bioassays are treated further in Chapter 7.

Vehicle foods and symptoms

Upon the ingestion of contaminated foods, symptoms appear between 6 and 24 h, especially between 8 and 12 h. The symptoms are characterized by acute abdominal pain and diarrhea with nausea, fever, and vomiting being rare. Except in the elderly or in debilitated persons, the illness is of short duration, one day or less. The fatality rate is quite low and no immunity seems to occur, although circulating antibodies to the enterotoxin may be found in some persons with a history of the syndrome.

The true incidence of *C. perfringens* food poisoning is unknown. Because of the relative mildness of the disease, it is quite likely that only those outbreaks and cases that affect groups of people are ever reported and recorded. The confirmed outbreaks reported to the U.S. Centers for Disease Control for the years 1973–75 totaled forty, with 2,706 cases.

The foods involved in *C. perfringens* outbreaks are often meat dishes prepared one day and eaten the next. The heat preparation of such foods is presumably inadequate to destroy the heat-resistant endospores and upon cooling and rewarming, the endospores germinate and grow. Meat dishes are most often the cause of this syndrome, although nonmeat dishes may be contaminated by meat gravy. The greater involvement of meat dishes may in part be due to the slower cooling rate of such cooked foods and also to the higher incidence of food-poisoning strains in meats. Strong et al. (102) found the overall incidence of the organism to be about 6% in 510 American foods. The incidence for various foods was 2.7% for commercially prepared frozen foods, 3.8% for fruits and vegetables, 5% for spices, 1.8% for home-prepared foods, and 16.4% for raw meat, poultry, and fish. Hobbs et al. (38) found that 14–24% of veal, pork, and beef samples examined contained heat-resistant endospores, but all 17 samples of lamb were negative. An outbreak of food poisoning involving 375 persons where 140 became ill was shown to be caused by both *C. perfringens* and *Salmonella typhimurium* (80). *C. perfringens* has been demonstrated to grow in a large number of foods. A study of retail, frozen precooked foods revealed that one-half were positive for vegetative cells, while 15% contained endospores (109). The latter investigators inoculated meat products with the organism and stored them at −29°C for up to 42 days. While spore survival was high, vegetative cells were virtually eliminated during the holding period. The survival of inoculated cells in raw ground beef was studied by Goepfert and Kim (32), who found decreased numbers upon storage at temperatures between 1° and 12.5°C. The raw beef contained a natural flora, and the above finding suggests that *C. perfringens* is unable to compete under these conditions. For the recovery of *C. perfringens* from foods, an appropriate reference in Table 4-1 or the review by Walker (112), should be consulted.

Prevention

The *C. perfringens* gastroenteritis syndrome may be prevented by proper attention to the leading causes of food poisoning of all types noted in the previous chapter. Since this syndrome often occurs in institutional cafeterias, some special precautions should be taken. Upon investigating a *C. perfringens* food-poisoning outbreak in a school lunchroom in which 80% of students and teachers became ill, Bryan et al. (11) constructed a time-temperature chart in an effort to determine when, where, and how the turkey became the vehicle (Fig. 19-5). It was concluded that meat and gravy, but not dressing, were responsible for the illness. As a means of preventing recurrences of such episodes, these authors suggested nine points for the preparation of turkey and dressing, and they are summarized below.

1. Cook turkeys until internal breast temperature reaches at least 165°F, preferably higher.
2. Thoroughly wash and sanitize all containers and equipment that previously had contact with raw turkeys.
3. Wash hands and use disposable plastic gloves when deboning, deicing, or otherwise handling cooked turkey.
4. Separate turkey meat and stock before chilling.

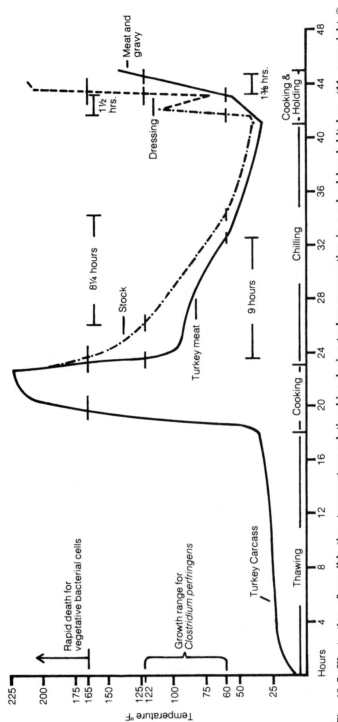

Fig. 19-5. Illustration of possible time-temperature relationships during turkey preparation in a school lunch kitchen. (11, copyright © 1971, Intern. Assoc. Milk, Food & Environ. Sanitarians, Inc.)

5. Chill the turkey and stock as rapidly as possible after cooking.
6. Use shallow pans for storing stock and deboned turkey in refrigerators.
7. Bring stock to a rolling boil before making gravy or dressing.
8. Bake dressing until all portions reach 165°F or higher.
9. Just prior to serving, heat turkey pieces submerged in gravy until largest portions of meat reach 165°F.

BOTULISM

Unlike *C. perfringens* food poisoning, in which large numbers of viable cells must be ingested, the symptoms of botulism are caused by the ingestion of a highly toxic, soluble exotoxin produced by the organism while growing in foods.

The first recorded case of botulism was in 1793, and the etiologic agent of the disease was first isolated in 1895 by E. Van Ermengen. The outbreak studied by Van Ermengen, which occurred in Belgium, involved thirty-four cases and three deaths. The causative organism was named *Bacillus botulinus* from the Latin *botulus* meaning sausage. Botulism is caused by certain strains of *Clostridium botulinum,* a gram-positive, anaerobic sporeforming rod with oval to cylindrical, terminal to subterminal spores. On the basis of the serological specificity of their toxins, seven types are recognized: A, B, C, D, E, F, and G. Types A, B, E, F, and G cause disease in man; type C causes botulism in fowls, cattle, mink, and other animals; and type D is associated with forage poisoning of cattle, especially in South Africa. The types are also differentiated on the basis of their proteolytic activity. Types A and G are proteolytic, as are some types B and F strains. Type E is nonproteolytic, as are some B and F strains (see Table 19-2). The proteolytic activity of type G is slower than that for type A, and its toxin requires trypsin potentiation (94).

A vast literature exists on botulism, and the information that follows is in no way meant to be complete. For more information, see Smith (92), Sperber (97), and Sakaguchi (84).

Distribution of *C. botulinum*

This organism is indigenous to soils and waters. In the United States, type A occurs more frequently in soils in the western states, and type B is found more frequently in the eastern states and in Europe. Soils and manure from various countries have been reported to contain 18% type A and 7% type B spores. Cultivated soil samples examined showed 7% to contain type A and 6% type B endospores. Type E spores tend to be confined more to waters, especially marine waters. In a study of mud samples from the harbor of Copenhagen, Pederson (79) found 84% to contain type E spores, while 26% of soil samples taken from a city park contained the organism. From a study of 684 environmental samples from Denmark, the Faroe Islands, Iceland, Greenland, and Bangladesh, 90% of aquatic samples from Denmark

Table 19-2. Summary comparison of *C. botulinum* strains and their toxins.

Property	Serologic types						
	A	B	B	E	F	F	G
Year discovered	1904	1896	1960	1936	1960	1965	1969
Proteolytic (+), nonproteolytic (−)	+	+	−	−	+	−	+ (weak)
Primary habitat	Terrestrial	Terrestrial	Aquatic	Aquatic	Aquatic	Aquatic	Terrestrial
Minimum growth temp. (°C)	~10	~10	3.3	3.3	~10	3.3	~12
Maximum growth temp. (°C)	~50	~50	~45	~45	~50	~45	n.d.
Minimum pH for growth (see text)	4.7	4.7	4.7	4.8	4.8	4.8	4.8
Minimum a_w for growth	0.94	0.94	~0.97	~0.97	0.94?	~0.97	n.d.
Thermal D values for endospores (°C)	$D_{110} = 2.72-2.89$[a]	$D_{110} = 1.34-1.37$[a]	n.d.	$D_{80} = 0.80$[b]	$D_{110} = 1.45-1.82$[c]	$D_{82.2} = 0.25-0.84$[d]	$D_{110} = 0.45-0.54$[e]
Radiation D values of spores (kGy)	1.2-1.5	1.1-1.3	n.d.	1.2	1.1; 2.5	1.5	n.d.
Maximum NaCl for growth (%)	~10	~10	5-6	5-6	8-10	5-6	n.d
Relative frequency of food outbreaks	High	High	n.d.	Highest for seafoods	1 outbreak	1 outbreak	None
H₂S production	+	+	−	−	+	−	+ +
Casein hydrolysis	+	+	−	−	+	−	+
Lipase production	+	+	+	+	+	+	−
Glucose fermentation	+	+	+	+	+	+	−
Mannose fermentation	−	−	+	+	−	+	−
Propionic acid produced	+	+	n.d.	n.d.	+	n.d.	n.d.

+ = positive
+ + = strongly positive
− = negative
n.d. = no data

[a]In phosphate buffer (74).
[b]Strain 8-E (54).
[c](54).
[d](53).
[e](55).

and 86% of marine samples from Greenland contained type E (40). This strain was not found in Danish soils and woodlands, while type B was. Based on these results, Huss (40) suggested that type E is a truly aquatic organism that proliferates in dead aquatic animals and sediments and is disseminated by water currents and migrating fish. Type E spores have been known for some time to exist in waters off the shores of northern Japan. Prior to 1960, the existence of these organisms in Great Lakes and Gulf Coast waters was not known, but their presence in these waters as well as in the Gulf of Maine and the gulfs of Venezuela and Darien has been established. Ten percent of soil samples tested in Russia were found positive for *C. botulinum*, with type E strains being predominant. A study of sediment samples from the upper Chesapeake Bay area of the U.S. Atlantic Coast revealed the presence of types A and E spores in 4 of 33 samples (86). The investigators believed these organisms to be randomly distributed in sediment and to be autochthonous. From Lake Michigan, 9% of fish caught contained type E spores, while from Green Bay 57% of fish contained these organisms (9). In another study (25), 6.2% of 500 commercially dressed fish taken from Lake Michigan near the Two Rivers, Wisconsin, area yielded type E, while only 0.4% of 427 laboratory-dressed fish were positive. The authors found the type E organisms to exist in relatively low numbers on freshly caught fish but to increase after evisceration. In a study of the incidence of *C. botulinum* on whitefish chubs in smoking plants, Pace et al. (77) found the highest incidence (21%) at the brine tank processing stage. The buildup of microorganisms on foods through successive processing stages is discussed more fully in Chapter 4.

As to the overall incidence of *C. botulinum* in soils, it has been suggested that the numbers/g are probably less than one. It appears that the non-proteolytic types are associated more with waters than soils and it may be noted from Table 19-2 that the discovery of these types occurred between 1960 and 1969. The late recognition is probably a consequence of the low heat resistance of the nonproteolytics, which would be destroyed if specimens were given their usual heat treatment for spore recovery from vegetative cells.

The first type F strains were isolated by Moller and Scheibel (70) from a homemade liver paste incriminated in an outbreak of botulism, involving one death, on the Danish island Langeland. Since that time, Craig and Pilcher (14) isolated type F spores from salmon caught in the Columbia River; Eklund and Poysky (24) found type F spores in marine sediments taken off the coasts of Oregon and California; Williams-Walls (115) isolated two proteolytic strains from crabs collected from the York River in Virginia; and Midura et al. (67) isolated the organism from venison jerky in California.

The type G strain was isolated first in 1969 from soil samples in Argentina by Gimenez and Ciccarelli (31), and it was isolated more recently from five human corpses in Switzerland (96). These deaths were not food associated. It has not been incriminated in food-poisoning outbreaks to date, and the reason for this might be due to the fact that this strain produces considerably less neurotoxin than type A. It has been shown that type G produced 40 LD_{50}/ml of toxin in media in which type A normally produces 10,000 to 1,000,000 LD_{50}/ml, but that under certain conditions the organism could be induced to produce up to 90,000 LD_{50}/ml of medium (13).

Growth of *C. botulinum* strains

Some of the growth and other characteristics of the strains that cause botulism in man are summarized in Table 19-2. The discussion that follows emphasizes the differences between the proteolytic and nonproteolytic strains irrespective of serologic type. The proteolytic strains, unlike the nonproteolytics, digest casein and produce H_2S. The latter, on the other hand, ferment mannose while the proteolytics do not. The proteolytics and nonproteolytics have been shown to form single groups relative to somatic antigens as evaluated by agglutination (95). The absorption of antiserum by any one of a group removes antibodies from all three of that group. Proteolytic strains are placed in Group I, nonproteolytic E, B, and F are placed in Group III, and type G is placed in Group IV (92).

The nutritional requirements of these organisms are complex with amino acids, B vitamins, and minerals being required. Synthetic media have been devised that support growth and toxin production of most types. The proteolytic strains tend not to be favored in their growth by carbohydrates, while the nonproteolytics are. At the same time, the nonproteolytics tend to be more fermentative than the proteolytic types.

The proteolytics generally do not grow below 12.5°C although a few reports exist in which growth was detected at 10°C. The upper range for types A and proteolytic B, and presumably for the other proteolytic types, is about 50°C. On the other hand, the nonproteolytic strains can grow as low as 3.3°C with the maximum about 5 degrees below that for proteolytics. As is noted in Chapter 3, minimum and maximum temperatures of growth of these organisms are dependent upon the state of other growth parameters, and the minima and maxima noted may be presumed to be at totally optimal conditions relative to pH, a_w, and the like. In a study of the minimum temperature for growth and toxin production by nonproteolytic types B and F in broth and crabmeat, both grew and produced toxin at 4°C in broth, but in crabmeat growth and toxin production occurred only at 26°C and not at 12°C or lower (93). A type G strain grew and produced toxin both in broth and crabmeat at 12°C but not at 8°C (93).

The minimum pH that permits growth and toxin production of *C. botulinum* strains has been the subject of many studies. It is generally recognized that growth does not occur at or below pH 4.5 and it is this fact that determines the degree of heat treatment given to foods with pH values below this level (see Chapter 14). Because of the existence of botulinal toxins in some high-acid, home-canned foods, this area has been the subject of recent studies. In one study, no growth of types A and B occurred in tomato juice at pH around 4.8, but when the product was inoculated with *Aspergillus gracilis,* toxin was produced at pH 4.2 in association with the mycelial mat (75). In another study with the starting pH of tomato juice at 5.8, the pH on the underside of the mold mat increased to 7.0 after 9 days, and to 7.8 after 19 days (39). The tomato juice was inoculated with type A botulinal spores, a *Cladosporium* sp., and a *Penicillium* sp. The topmost 0.5 ml of product showed pH increases from 5.3 to 6.4 or 7.5 after 9 and 19 days, respectively. One type B strain was shown to produce gas in tomato juice at pH 5.24 after 30 days, and at pH 5.37 after 6 days. In food systems consisting of whole shrimp, shrimp purée, tomato purée, and tomato and shrimp purée acidified to pH 4.2 and 4.6 with acetic or citric acid, none of three type E strains grew or produced toxin at 26°C after 8 weeks (81). Growth and toxin production of a type E strain at pH 4.20 and 26°C in 8 weeks was demonstrated when citric acid but not acetic acid was used to control the pH of a culture medium (110). In general, the pH minima are similar for proteolytic and nonproteolytic strains.

The minimum a_w that permits growth and toxin production of types A and proteolytic B strains is 0.94, and this value seems to be well established. The minimum for type E is a bit higher, around 0.97. While all strains have not been studied equally, it is possible that the other nonproteolytic strains have a minimum similar to that of type E. The way in which a_w is achieved in culture media affects the minimum values obtained. When glycerol is used as humectant, a_w values tend to be a bit lower than when NaCl or glucose is used (see Sperber, 98). Salt at a level of about 10%, or 50% sucrose, will inhibit growth of types A and B, and 3–5% salt has been found to inhibit toxin production in smoked fish chubs (12). Lower levels of salt are required when nitrites are present (see Chapter 11).

With respect to heat resistance, the proteolytic strains are much more resistant than nonproteolytics (Table 19-2). Although the values noted in the table suggest that type A is the most heat resistant, followed by proteolytic F and then proteolytic B, these data should only be taken as representative since heating menstra, previous history of strains, and other factors are known to affect heat resistance, as noted in Chapter 14. Of those noted, all were determined in phosphate buffer. Among type E, the Alaska and Beluga strains appear to be more heat resistant than others, and in ground whitefish chubs, D 80°C of 2.1 and 4.3 have been reported (16), while in crabmeat, D 82.2°C of 0.51 and 0.74 have been reported, respectively, for Alaska and Beluga (56). With regard to smoked whitefish chubs, it was determined in one study that heating to an internal temperature of 180°C for 30 min produced a nontoxigenic product (25), while in another study, 10 or 1.2% of 858 freshly smoked chubs given the same heat treatment were contaminated, mostly with type E strains (77). (The heat destruction of bacterial endospores is dealt with further in Chapter 14.)

With regard to type G, the Argentine and Swiss strains both produce two kinds of spores—heat labile and heat resistant. The former, which are destroyed at around 80°C after 10 min, represent about 99% of the spores in a culture of the Swiss strains; while in the Argentine strain only about 1 in 10,000 endospores are heat resistant (55). The D 230°F of two heat-resistant strains in phosphate buffer was 0.45 to 0.54 min, while for two heat-labile strains, D 180°F was 1.8 to 5.9 min (55). The more heat-resistant spores of type G have not yet been propagated.

Unlike heat, radiation seems to affect the endospores of proteolytic and nonproteolytic strains similarly, with D values of 1.1 to 2.5 kGy having been reported (see Chapter 12). However, the D value of one nonproteolytic type F strain was found to be 1.5 kGy, which was similar to the D value for a type A strain; but a proteolytic type F strain produced a D of 1.16 kGy (5).

Ecology of *C. botulinum* growth

It appears that this organism cannot grow and produce its toxins in competition with large numbers of other microorganisms. Toxin-containing foods are generally devoid of other types of organisms because of heat treatments. In the presence of yeasts, however, *C. botulinum* has been reported to grow and produce toxin at a pH as low as 4.0. While a synergistic effect between clostridia and lactic acid bacteria has been reported on the one hand, lactobacilli will antagonize growth and toxin production—indirect evidence for this is the absence of botulinal toxins in milk. Yeasts are presumed to produce growth factors needed by the clostridia to grow at low pH, while the lactic acid bacteria may aid growth by reducing the O/R potential or inhibit growth by "lactic antagonism" (see Chapter 16). In one study, type A was inhibited by soil isolates of *C. sporogenes, C. perfringens*, and *B. cereus* (91). Some *C. perfringens* strains produced an inhibitor that was effective on eleven type A strains, on seven type B proteolytic and one nonproteolytic strains, and on five type E and seven type F strains (44). It is possible for *C. botulinum* spores to germinate and grow in certain canned foods whose pH is < 4.5 when *Bacillus coagulans* is present. In a study with tomato juice of pH 4.5 inoculated with *B. coagulans,* the pH increased after 6 days at 35°C to 5.07, and to 5.40 after 21 days, thus making it possible for *C. botulinum* to grow (4). Kautter et al. (44) found that type E strains are inhibited by other nontoxic organisms whose biochemical properties and morphological characteristics were similar to type E. These organisms were shown to effect inhibition of type E strains by producing a bacteriocin-like substance designated "boticin E." In a more detailed study, proteolytic A, B, and F strains were found to be resistant to boticin E elaborated by a nontoxic type E, but toxic E cells were susceptible (3). The boticin was found to be sporostatic for nonproteolytic types B, E, F, and nontoxigenic type E.

Vacuum-packaged foods such as bacon are capable of supporting growth and toxin production by *C. botulinum* strains without causing noticeable off-odors, and types A and E spores have been shown to germinate and produce toxin in smoked fish (108). Fish inoculated with type A was offensively spoiled, while type E strains caused a less drastic type of spoilage. The proteolytic strains generally produce more offensive by-products than the nonproteolytics. In a study of type E toxin formation and cell growth in turkey rolls incubated at 30°C, type E spores germinated and produced toxin within 24 h (65). The appearance of toxin coincided with cell growth for 2 weeks, after which time toxins outlasted viable cells. These findings suggest that it is possible to find type E toxin in foods in the absence of type E cells. Toxin could not be demonstrated after 56 days of incubation.

A report on the ecology of type F showed that the absence of this strain in mud samples during certain times of the year was associated with the presence of *Bacillus licheniformis* in the samples during these periods, when the bacillus was apparently inhibiting type F strains (114).

Nature of the botulinal toxins

The neurotoxins are formed within the organism and released upon autolysis. They are produced by cells growing under optimal conditions, though resting cells have been reported to form toxin as well. The botulinal toxins are the most toxic substances known, with purified type A reported to contain about 30,000,000 mouse LD_{50}/mg. When grown in cellophane sacs suspended in culture media, higher yields may be attained. The first of these toxins to be purified was type A, which was achieved by C. Lamanna et al. and by A. Abrams et al., both in 1946. The purification of B, E, and F has been achieved.

It appears that all botulinal strains synthesize the neurotoxin as a single polypeptide chain (unnicked) with a molecular weight of about 150,000 daltons. The proteolytic strains A, B, and F produce endogenous proteases that cleave the 150,000-dalton unit to form a heavy chain of about 100,000 daltons and a light chain of about 50,000, the two being held together by at least one disulfide bond (23). The nonproteolytic types B, E, and F release into the environment the 150,000-dalton chain, which is referred to as a **progenitor** toxin. It is made toxic by treatment with exogenous proteases (nicking) such as trypsin (17). If the double-chained molecule is reduced to individual chains, each lacks toxicity and neither is known to possess enzymatic activity. In culture fluids, botulinal toxins actually exist as complexes with nontoxic culture proteins, which may be considered the natural state of botulinal toxins (105). The molecular weight of these complexes may be as high as 900,000 daltons.

As a prerequisite to its neurotoxic activity, the botulinal toxin must attach to neural tissue, and its specific nerve receptors have been the subject of many studies. It has been demonstrated that treatment of botulinal toxin with formalin (toxoiding) results in the destruction of receptor-binding activity. The gangliosides have been shown to react with botulinal toxin and reduce its in vivo toxicity. It appears that for the binding of toxin, two sialic acid residues on the inner galactose of the gangliosides are essential, and an additional sialic acid at the nonreducing end of the ganglioside also aids toxin binding (23). Once attached, the toxin exerts its effects by presynaptically blocking the release of acetylcholine from cholinergic nerve endings.

Type A toxin has been reported to be more lethal than B or E. Type B has been reported to have associated with it a much lower case mortality than type A, and case recoveries from type B have occurred even when appreciable amounts of toxin could be demonstrated in the blood.

Symptoms of botulism can be produced by either parenteral or oral administration of the toxins. They may be absorbed into the blood stream through the respiratory mucous membranes as well as through the walls of the stomach and intestines. The toxins are not completely inactivated by the proteolytic enzymes of the stomach, and, indeed, those produced by nonproteolytics may be activated. It has been shown that the high molecular weight complexes or the progenitor possess higher resistance to acid and pepsin (104). While the derivative toxin was rapidly inactivated, the progenitor was shown to be resistant to rat intestinal juice in vitro. The progenitor was more stable in the stomach of rats. Similar findings were made by Ohishi et al. (76), who showed that progenitor toxins of nonproteolytics were more toxic orally in mice than the dissociated toxic components of the derived toxins. It appears that the nontoxic component of the progenitor provides protection to the toxin activity. After botulinal toxins are absorbed into the bloodstream, they enter the peripheral nervous system where they affect nerves as noted above.

Unlike the staphylococcal enterotoxins and heat-stable toxins of other food-borne pathogens, the botulinal toxins are heat-sensitive and may be destroyed by heating at 80°C (176°F) for 10 min, or boiling temperatures for a few min.

The adult botulism syndrome—incidence and vehicle foods

Upon the ingestion of toxin-containing foods, symptoms of botulism may develop anywhere between 12 and 72 h later. Even longer incubation periods are not unknown. Symptoms consist of nausea, vomiting, fatigue, dizziness, and headache; dryness of skin, mouth, and throat; constipation, lack of fever, paralysis of muscles, double vision, and finally, respiratory failure and death. The duration of the illness is from 1 to 10 or more days, depending upon host resistance and other factors. The mortality rate varies between 30 and 65%, with the rate being generally lower in European countries than in the United States. All symptoms are caused by the exotoxin, and treatment consists of administering specific antisera as early as possible. Although it is assumed that the tasting of toxin-containing foods allows for absorption from the oral cavity, Lamanna et al. (51) found that mice and monkeys are more susceptible to the toxins when administered by stomach tube than by exposure to the mouth. The botulinal toxins are neurotoxins and attach irreversibly to nerves. Early treatment by use of antisera brightens the prognosis.

Prior to 1963, most cases of botulism in the United States in which the vehicle foods were identified were traced to home-canned vegetables and were caused by types A and B toxins. In almost 70% of the 640 cases reported for the period 1899–1967, the vehicle food was not identified. Among the 640 cases, 17.8% were associated with vegetables, 4.1% fruits, 3.6% fish, 2.2% condiments, 1.4% meats and poultry, and 1.1% for all others. Reported food-borne cases in the United States for the years 1960–84 are shown in Fig. 19-6. The three largest U.S. outbreaks involved fifty-

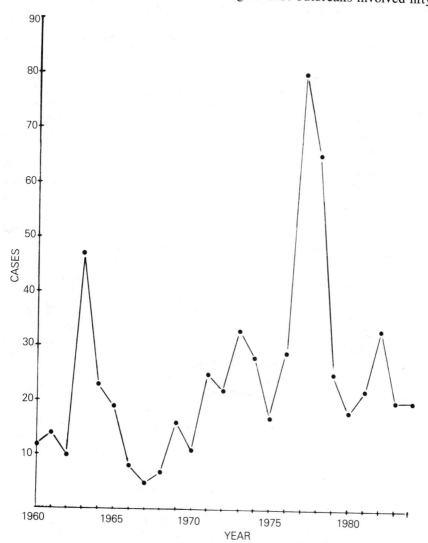

Fig. 19-6. Reported cases of food-borne botulism in the United States for the period 1960–84. (Redrawn from *Morbidity and Mortality Weekly Report*, U.S. Centers for Disease Control)

eight, thirty-four, and twenty-eight cases each. The fifty-eight cases occurred in a restaurant in Pontiac, Michigan, in 1977, following consumption of a hot sauce prepared from home-canned jalapeño peppers. No deaths occurred, and type B toxin was identified. The outbreak of thirty-four cases occurred in 1978 in New Mexico, with potato salad or bean salad incriminated. The outbreak of twenty-eight cases occurred in 1983 in the state of Illinois, with sautéed onions as the apparent food source. The total number of cases from all sources in the United States rarely exceeds 50 per year, with the highest 10-year period being 1930–39, when 384 cases were reported from noncommercial foods. Between 1899 and 1963, 1,561 cases were reported from noncommercial foods, while 219 were reported from commercial foods between 1906 and 1963, with 24 in 1963 alone.

Of 404 verified cases of type E botulism through 1963, 304 or 75% occurred in Japan (27). No outbreak of botulism was recorded in Japan prior to 1951. For the period May 1951 through January 1960, 166 cases were recorded, with 58 deaths for a mortality rate of 35%. Most of these outbreaks were traced to a home-prepared food called *izushi,* a preserved food consisting of raw fish, vegetables, cooked rice, malted rice (*koji*), and a small amount of salt and vinegar. This preparation is packed tightly in a wooden tub equipped with a lid and held for 3 weeks or longer to permit lactic acid fermentation. During this time, the O/R potential is lowered, thus allowing for the growth of anaerobes.

Sixty-two outbreaks of botulism resulting from commercially canned foods were recorded for the period 1899–1973 (52), with forty-one prior to 1930. Between 1941 and 1982, seven outbreaks occurred in the United States involving commercially canned foods in metal containers, with seventeen cases and eight deaths (73). Three of these outbreaks were caused by type A and the remainder by type E. In five of the outbreaks, can leakage or underprocessing occurred (73). Canned mushrooms have been incriminated in several botulism outbreaks. A study in 1973 and 1974 turned up thirty cans of mushrooms containing botulinal toxin (twenty-nine were type B). An additional eleven cans contained viable spores of *C. botulinum* without preformed toxin (52). The capacity of the commercial mushroom (*Agaricus bisporus*) to support the growth of inoculated spores of *C. botulinum* was studied by Sugiyama and Yang (107). Following inoculation of various parts of mushrooms, they were sealed with plastic film and incubated. Toxin was detected as early as 3–4 days later, when products were incubated at 20°C. The type A strains appeared to be more active than type B, even though type B strains appear more often in canned products. While the plastic film used to wrap the inoculated mushrooms allowed for gas exchange, the respiration of the fresh mushrooms apparently consumed oxygen at a faster rate than it entered the film. No toxin was detected in products stored at refrigerator temperatures.

One of the recorded outbreaks of botulism (five cases with one death) due to type F involved homemade liver paste. The only U.S. outbreak occurred in 1966 from home-prepared venison jerky, with three clinical cases (67).

It is quite clear that the greatest hazards of botulism come from home-prepared and home-canned foods that are improperly handled or given insufficient heat treatments to destroy botulinal spores. Such foods are often consumed without heating. The best preventative measure is the heating of suspect foods to boiling temperatures for a few minutes, which is sufficient to destroy the neurotoxins.

Infant botulism

First recognized as such in California in 1976 (66), infant botulism has since been confirmed in most states in the United States and in some other countries. In the adult form of botulism, preformed toxins are ingested, while in infant botulism viable botulinal spores are ingested and upon germination in the intestinal tract, toxin is synthesized. While it is possible that in some adults under special conditions botulinal endospores may germinate and produce small quantities of toxin, the colonized intestinal tract does not favor spore germination. Infants over one year of age tend not to be affected by this syndrome because of the establishment of a more normal intestinal flora. The disease is mild in some infants while in others it can be rather severe. High numbers of spores are found in the feces of infants during the acute phase of the disease and as recovery progresses, numbers of organisms abate.

This syndrome is diagnosed by demonstrating botulinal toxins in infant stools and by use of the mouse lethality test. Since *Clostridium difficle* produces mouse-lethal toxins in the intestinal tract of infants, it is necessary to differentiate between these toxins and that of *C. botulinum* (30).

Infants get viable spores from infant foods and possibly from their environment. Vehicle foods are those that do not undergo heat processing to destroy endospores, and the two most common products are syrup and honey. Of ninety samples of honey examined, nine contained viable spores. Six of these had been fed to babies who developed infant botulism (68). Of the nine, seven were type B and two were type A. Of 910 infant foods from ten product classes, only two classes were positive for spores—honey and corn syrup (45). Of 100 honey samples, two contained type A, while eight of forty corn syrup samples yielded type B. Reported cases in the United States through 1984 are shown in Fig. 19-7. The sixty-one cases for 1982, which occurred among infants aged 2–48 weeks, involved type A and B toxins equally.

Animal models for the study of this syndrome involve 8–11-day-old mice (106), and 7–13-day-old rats (69). In the mouse model, botulinal toxin was found in the lumen of the large intestine, and it was not associated with the ileum. The sensitivity of these animal models is noted in Chapter 7.

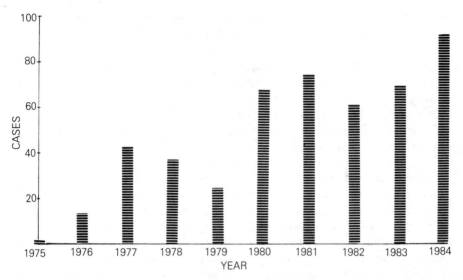

Fig. 19-7. Reported cases of infant botulism in the United States for the period 1975–84. (Redrawn from *Morbidity and Mortality Weekly Report*, U.S. Centers for Disease Control)

BACILLUS CEREUS GASTROENTERITIS

Bacillus cereus is an aerobic, sporeforming rod normally present in soil, dust, and water. It has been associated with food poisoning in Europe since at least 1906. Among the first to report this syndrome with precision was U. Plazikowski. His findings were confirmed by several other European workers in the early 1950s. The first documented outbreak in the United States occurred in 1969, while the first in Great Britain occurred in 1971.

This bacterium has a minimum growth temperature around 10°–12°C and a maximum of about 48°–50°C. Growth has been demonstrated over the pH range 4.9–9.3 (33). Its spores possess a resistance to heat typical of other mesophiles. It grows rapidly in foods held in the 30°–40°C range.

B. cereus toxins

Food-poisoning strains produce the following toxins and extracellular products: lecithinase, proteases, hemolysin, β-lactamase, mouse lethal toxin, cereolysin, emetic enterotoxin, and diarrheagenic enterotoxin. The enterotoxins are responsible for the two food-poisoning syndromes caused by this organism.

The **diarrheagenic** toxin induces vascular permeability in the skin of rabbits, elicits fluid accumulation in the rabbit ileal loop, and causes diarrhea in Rhesus monkeys (see Chapter 7). It has been shown to be a protein of molecular weight of about 50,000 with an isoelectric point of 4.9 (111). Its production is favored by low dissolved O_2, and most is produced during the logarithmic phase of growth (99). It is heat labile and sensitive to trypsin and pronase. Its production is favored over the pH range 6.0–8.5, with the optimum between 7.0–7.5, and is produced over the temperature range 18–43°C (41). Unlike the enterotoxin of *C. perfringens,* it is a vegetative growth metabolite and can be separated from phospholipase and the heat-labile cereolysin (111). The purified toxin shows mice lethality. It induces diarrhea apparently by stimulating the adenylate cyclase-cAMP system.

The **emetic** (vomiting-type) toxin is distinctly different from the diarrheagenic in having a molecular weight of 5,000 daltons (111) and in being heat- and pH-stable. It is insensitive to trypsin and pepsin (41). The emetic toxin strains grow over the range 15°–50°C, with an optimum between 35°–40°C (42). While the emetic syndrome is most often associated with rice dishes, growth of the emetic toxin strains in rice is not favored in general over other *B. cereus* strains, although higher populations and more extensive germination have been noted in this product (42).

Diarrheal syndrome

This syndrome is rather mild, with symptoms developing within 8–16 h, more commonly within 12–13 h, and lasting for 6–12 (36). Symptoms consist of nausea (with vomiting being rare), cramplike abdominal pains, tenesmus, and watery stools. Fever is generally absent. The similarity between this syndrome and that of *C. perfringens* food poisoning has been noted (29).

Vehicle foods consist of cereal dishes that contain corn and corn starch, mashed potatoes, vegetables, minced meat, liver sausage, meat loaf, milk, cooked meat, Indonesian rice dishes, puddings, soups, and others (29). Reported outbreaks between 1950 and 1978 have been summarized by Gilbert (29), and when plate counts on leftover foods were recorded, they ranged from 10^5 to 9.5×10^8/g, with many in the 10^7–10^8/g range. The first well-studied outbreaks were those investigated by Hauge (36), which were traced to vanilla sauce, and the counts ranged from 2.5×10^7 to 1×10^8/g. From meat loaf involved in a U.S. outbreak in 1969, 7×10^7/g were found (64). Serovars found in diarrheal outbreaks include types 1, 6, 8, 9, 10, and 12. Serovars 1, 8, and 12 have been associated with this as well as with the emetic syndrome (29).

Emetic syndrome

This form of *B. cereus* food poisoning is more severe and acute than that described above. The incubation period ranges from 1 to 6 h, with 2 to 5 being most common (71). Its similarity to the staphylococcal food-poisoning syndrome has been noted (29). It is often associated with fried or boiled rice dishes. In addition to these, pasteurized cream, spaghetti, mashed potatoes, and vegetable sprouts have been incriminated (29). Outbreaks have been reported from Great Britain, Canada, Australia, the Netherlands, Finland, Japan, and the United States. The first U.S. outbreak was reported in 1975, with mashed potatoes as the vehicle food.

The numbers of organisms necessary to cause this syndrome seem to be higher than for the diarrheal syndrome, with numbers as high as $2 \times 10^9/g$ having been found (see 29). *B. cereus* serovars associated with the emetic syndrome include 1, 3, 4, 5, 8, 12, and 19 (29).

For the recovery and quantitation of *B. cereus* enterotoxins, see Chapters 6 and 7. The incidence of this organism in certain foods is noted in Chapter 4; for more information, see Gilbert (29) and Johnson (41).

REFERENCES

1. Alderton, G., K. A. Ito, and J. K. Chen. 1976. Chemical manipulation of the heat resistance of *Clostridium botulinum* spores. *Appl. Environ. Microbiol.* 31:492–98.
2. Alderton, G., and N. Snell. 1969. Bacterial spores: Chemical sensitization to heat. *Science* 163:1212–13.
3. Anastasio, K. L., J. A. Soucheck, and H. Sugiyama. 1971. Boticinogeny and actions of the bacteriocin. *J. Bacteriol.* 107:143–49.
4. Anderson, R. E. 1984. Growth and corresponding elevation of tomato juice pH by *Bacillus coagulans*. *J. Food Sci.* 49:647, 649.
5. Anellis, A., and D. Berkowitz. 1977. Comparative dose-survival curves of representative *Clostridium botulinum* type F spores with type A and B spores. *Appl. Environ. Microbiol.* 34:600–601.
6. Angelotti, R., H. H. Hall, M. Foter, and K. H. Lewis. 1962. Quantitation of *Clostridium perfringens* in foods. *Appl. Microbiol.* 10:193–99.
7. Barnes, E., and M. Ingram. 1956. The effect of redox potential on the growth of *Clostridium welchii* strains isolated from horse muscle. *J. Appl. Bacteriol.* 19:117–28.
8. Benjamin, M. J. W., D. M. Wheather, and P. A. Shepherd. 1956. Inhibition and stimulation of growth and gas production by clostridia. *J. Appl. Bacteriol.* 19:159–63.
9. Bott, T. L., J. S. Deffner, E. McCoy, and E. M. Foster. 1966. *Clostridium botulinum* type E in fish from the Great Lakes. *J. Bacteriol.* 91:919–24.
10. Bradshaw, J. G., G. N. Stelma, V. I. Jones, J. T. Peeler, J. G. Wimsatt, J. J. Corwin, and R. M. Twedt. 1982. Thermal inactivation of *Clostridium perfringens* enterotoxin in buffer and in chicken gravy. *J. Food Sci.* 47:914–16.
11. Bryan, F. L., T. W. McKinely, and B. Mixon. 1971. Use of time-temperature evaluations in detecting the responsible vehicle and contributing factors of foodborne disease outbreaks. *J. Milk Food Technol.* 34:576–82.
12. Christiansen, L. N., J. Deffner, E. M. Foster, and H. Sugiyama. 1968. Survival and outgrowth of *Clostridium botulinum* type E spores in smoked fish. *Appl. Microbiol.* 16:133–37.
13. Ciccarelli, A. S., D. N. Whaley, L. M. McCroskey, D. F. Gimenez, V. R. Dowell, Jr., and C. L. Hatheway. 1977. Cultural and physiological characteristics of *Clostridium botulinum* type G and the susceptibility of certain animals to its toxin. *Appl. Environ. Microbiol.* 34:843–48.
14. Craig, J., and K. Pilcher. 1966. *Clostridium botulinum* type F: Isolation from salmon from the Columbia River. *Science* 153:311–12.

15. Craven, S. E. 1980. Growth and sporulation of *Clostridium perfringens* in foods. *Food Technol.* 34(4):80–87, 95.
16. Crisley, F. D., J. T. Peeler, R. Angelotti, and H. E. Hall. 1968. Thermal resistance of spores of five strains of *Clostridium botulinum* type E in ground whitefish chubs. *J. Food Sci.* 33:411–16.
17. DasGupta, B. R., and H. Sugiyama. 1976. Molecular forms of neurotoxins in proteolytic *Clostridium botulinum* type B cultures. *Infect. Immun.* 14:680–86.
18. Duncan, C. L. 1973. Time of enterotoxin formation and release during sporulation of *Clostridium perfringens* type A. *J. Bacteriol.* 113:932–36.
19. ———. 1976. *Clostridium perfringens.* In *Food microbiology: Public health and spoilage aspects,* ed. M. P. deFigueiredo and D. F. Splittstoesser, 170–97. Westport, Conn.: AVI.
20. Duncan, C. L., and D. H. Strong. 1968. Improved medium for sporulation of *Clostridium perfringens. Appl. Microbiol.* 16:82–89.
21. ———. 1969. Ileal loop fluid accumulation and production of diarrhea in rabbits by cell-free products of *Clostridium perfringens. J. Bacteriol.* 100:86–94.
22. Duncan, C. L., D. H. Strong, and M. Sebald. 1972. Sporulation and enterotoxin production by mutants of *Clostridium perfringens. J. Bacteriol.* 110:378–91.
23. Eidels, L., R. L. Proia, and D. A. Hart. 1983. Membrane receptors for bacterial toxins. *Microbiol. Rev.* 47:596–620.
24. Eklund, M., and F. Poysky. 1965. *Clostridium botulinum* type E from marine sediments. *Science* 149:306.
25. Fantasia, L. D., and A. P. Duran. 1969. Incidence of *Clostridium botulinum* Type E in commercially and laboratory dressed white fish chubs. *Food Technol.* 23:793–94.
26. Foegeding, P. M., and F. F. Busta. 1980. *Clostridium perfringens* cells and phospholipase C activity at constant and linearly rising temperatures. *J. Food Sci.* 45:918–24.
27. Foster, E. M., J. Deffner, T. L. Bott, and E. McCoy. 1965. *Clostridium botulinum* food poisoning. *J. Milk Food Technol.* 28:86–91.
28. Genigeorgis, C., G. Sakaguchi, and H. Riemann. 1973. Assay methods for *Clostridium perfringens* type A enterotoxin. *Appl. Microbiol.* 26:111–15.
29. Gilbert, R. J. 1979. *Bacillus cereus* gastroenteritis. In *Food-borne infections and intoxications,* ed. H. Riemann and F. L. Bryan, 495–518. New York: Academic Press.
30. Gilligan, P. H., L. Brown, and R. E. Berman. 1983. Differentiation of *Clostridium difficile* toxin from *Clostridium botulinum* toxin by the mouse lethality test. *Appl. Environ. Microbiol.* 45:347–49.
31. Gimenez, D. F., and A. S. Ciccarelli. 1970. Another type of *Clostridium botulinum. Zentral. Bakteriol. Orig. A.* 215:221–24.
32. Goepfert, J. M., and H. U. Kim. 1975. Behavior of selected foodborne pathogens in raw ground beef. *J. Milk Food Technol.* 38:449–52.
33. Goepfert, J. M., W. M. Spira, and H. U. Kim. 1972. *Bacillus cereus:* Food poisoning organism. A review. *J. Milk Food Technol.* 35:213–27.
34. Granum, P. E., W. Telle, Ø. Olsvik, and A. Stavn. 1984. Enterotoxin formation by *Clostridium perfringens* during sporulation and vegetative growth. *Intern. J. Food Microbiol.* 1:43–49.
35. Hatheway, C. L., D. N. Whaley, and V. R. Dowell, Jr. 1980. Epidemiological

aspects of *Clostridium perfringens* foodborne illness. *Food Technol.* 34(4):77–79, 90.

36. Hauge, S. 1955. Food poisoning caused by aerobic spore-forming bacilli. *J. Appl. Bacteriol.* 18:591–95.

37. Hauschild, A. H. W., and R. Hilsheimer. 1971. Purification and characteristics of the enterotoxin of *Clostridium perfringens* type A. *Can. J. Microbiol.* 17:1425–33.

38. Hobbs, B., M. Smith, C. Oakley, G. Warrack, and J. Cruickshank. 1953. *Clostridium welchii* food poisoning. *J. Hyg.* 51:75–101.

39. Huhtanen, C. N., J. Naghski, C. S. Custer, and R. W. Russell. 1976. Growth and toxin production by *Clostridium botulinum* in moldy tomato juice. *Appl. Environ. Microbiol.* 32:711–15.

40. Huss, H. H. 1980. Distribution of *Clostridium botulinum. Appl. Environ. Microbiol.* 39:764–69.

41. Johnson, K. M. 1984. *Bacillus cereus* foodborne illness—An update. *J. Food Protect.* 47:145–53.

42. Johnson, K. M., C. L. Nelson, and F. F. Busta. 1983. Influence of temperature on germination and growth of spores of emetic and diarrheal strains of *Bacillus cereus* in a broth medium and in rice. *J. Food Sci.* 48:286–87.

43. Kang, C. K., M. Woodburn, A. Pagenkopf, and R. Cheney. 1969. Growth, sporulation, and germination of *Clostridium perfringens* in media of controlled water activity. *Appl. Microbiol.* 18:798–805.

44. Kautter, D. A., S. M. Harmon, R. K. Lynt, Jr., and T. Lilly, Jr. 1966. Antagonistic effect on *Clostridium botulinum* type E by organisms resembling it. *Appl. Microbiol.* 14:616–22.

45. Kautter, D. A., T. Lilly, Jr., H. M. Solomon, and R. K. Lynt. 1982. *Clostridium botulinum* spores in infant foods: A survey. *J. Food Protect.* 45:1028–29.

46. Labbe, R. G. 1980. Relationship between sporulation and enterotoxin production in *Clostridium perfringens* type A. *Food Technol.* 34(4):88–90.

47. Labbe, R. G., and C. L. Duncan. 1974. Sporulation and enterotoxin production by *Clostridium perfringens* type A under conditions of controlled pH and temperature. *Can. J. Microbiol.* 20:1493–1501.

48. ———. 1977. Evidence for stable messenger ribonucleic acid during sporulation and enterotoxin synthesis by *Clostridium perfringens* type A. *J. Bacteriol.* 129:843–49.

49. Labbe, R. G., and L. L. Nolan. 1981. Stimulation of *Clostridium perfringens* enterotoxin formation by caffeine and theobromine. *Infect. Immun.* 34:50–54.

50. Labbe, R. G., and D. K. Rey. 1979. Raffinose increases sporulation and enterotoxin production by *Clostridium perfringens* type A. *Appl. Environ. Microbiol.* 37:1196–1200.

51. Lamanna, C., R. A. Hillowalla, and C. C. Alling. 1967. Buccal exposure to botulinal toxin. *J. Infect. Dis.* 117:327–31.

52. Lynt, R. K., D. A. Kautter, and R. B. Read, Jr. 1975. Botulism in commercially canned foods. *J. Milk Food Technol.* 38:546–50.

53. Lynt, R. K., D. A. Kautter, and H. M. Solomon. 1979. Heat resistance of nonproteolytic *Clostridium botulinum* type F in phosphate buffer and crabmeat. *J. Food Sci.* 44:108–11.

54. ———. 1982. Heat resistance of proteolytic *Clostridium botulinum* type F in phosphate buffer and crabmeat. *J. Food Sci.* 47:204–6, 230.
55. Lynt, R. K., H. M. Solomon, and D. A. Kautter. 1984. Heat resistance of *Clostridium botulinum* type G in phosphate buffer. *J. Food Protect.* 47:463–66.
56. Lynt, R. K., H. M. Solomon, T. Lilly, Jr., and D. A. Kautter. 1977. Thermal death time of *Clostridium botulinum* type E in meat of the blue crab. *J. Food Sci.* 42:1022–25, 1037.
57. Mahony, D. E. 1977. Stable L-forms of *Clostridium perfringens:* Growth, toxin production, and pathogenicity. *Infect. Immun.* 15:19–25.
58. McClung, L. 1945. Human food poisoning due to growth of *Clostridium perfringens (C. welchii)* in freshly cooked chicken: Preliminary note. *J. Bacteriol.* 50:229–31.
59. McDonel, J. L. 1979. The molecular mode of action of *Clostridium perfringens* enterotoxin. *Amer. J. Clin. Nutri.* 32:210–18.
60. ———. 1980. Binding of *Clostridium perfringens* (^{125}I)enterotoxin to rabbit intestinal cells. *Biochem.* 19:4801–7.
61. ———. 1980. Mechanism of action of *Clostridium perfringens* enterotoxin. *Food Technol.* 34(4):91–95.
62. McDonel, J. L., and C. L. Duncan. 1977. Regional localization of activity of *Clostridium perfringens* type A enterotoxin in the rabbit ileum, jejunum, and duodenum. *J. Infect. Dis.* 136:661–66.
63. McDonel, J. L., and B. A. McClane. 1979. Binding versus biological activity of *Clostridium perfringens* enterotoxin in Vero cells. *Biochem. Biophys. Res. Comm.* 87:497–504.
64. Midura, T., M. Gerber, R. Wood, and A. R. Leonard. 1970. Outbreak of food poisoning caused by *Bacillus cereus. Publ. Hlth Rept.* 85:45–47.
65. Midura, T., C. Taclindo, Jr., G. S. Mygaard, H. L. Bodily, and R. M. Wood. 1968. Use of immunofluorescence and animal tests to detect growth and toxin production by *Clostridium botulinum* type E in food. *Appl. Microbiol.* 16:102–5.
66. Midura, T. F., and S. S. Arnon. 1976. Infant botulism: Identification of *Clostridium botulinum* and its toxins in faeces. *Lancet* 2:934–36.
67. Midura, T. F., G. S. Nygaard, R. M. Wood, and H. L. Bodily. 1972. *Clostridium botulinum* type F: Isolation from venison jerky. *Appl. Microbiol.* 24:165–67.
68. Midura, T. F., S. Snowden, R. M. Wood, and S. S. Arnon. 1979. Isolation of *Clostridium botulinum* from honey. *J. Clin. Microbiol.* 9:282–83.
69. Moberg, L. J., and H. Sugiyama. 1980. The rat as an animal model for infant botulism. *Infect. Immun.* 29:819–21.
70. Moller, V., and I. Scheibel. 1960. Preliminary report on the isolation of an apparently new type of *Cl. botulinum. Acta Path. Microbiol. Scan.* 48:80.
71. Mortimer, P. R., and G. McCann. 1974. Food-poisoning episodes associated with *Bacillus cereus* in fried rice. *Lancet* 1:1043–45.
72. Naik, H. S., and C. L. Duncan. 1977. Enterotoxin formation in foods by *Clostridium perfringens* type A. *J. Food Safety* 1:7–18.
73. NFPA/CMI Task Force. 1984. Botulism risk from post-processing contamination of commercially canned foods in metal containers. *J. Food Protect.* 47:801–16.

74. Odlaug, T. E., and I. J. Pflug. 1977. Thermal destruction of *Clostridium botulinum* spores suspended in tomato juice in aluminum thermal death time tubes. *Appl. Environ. Microbiol.* 34:23–29.

75. ———. 1979. *Clostridium botulinum* growth and toxin production in tomato juice containing *Aspergillus gracilis*. *Appl. Environ. Microbiol.* 37:496–504.

76. Ohishi, I., S. Sugii, and G. Sakaguchi. 1977. Oral toxicities of *Clostridium botulinum* toxins in response to molecular size. *Infect. Immun.* 16:107–9.

77. Pace, P. J., E. R. Krumbiegel, R. Angelotti, and H. J. Wisniewski. 1967. Demonstration and isolation of *Clostridium botulinum* types from whitefish chubs collected at fish smoking plants of the Milwaukee area. *Appl. Microbiol.* 15:877–84.

78. Pearson, C. B., and H. W. Walker. 1976. Effect of oxidation-reduction potential upon growth and sporulation of *Clostridium perfringens*. *J. Milk Food Technol.* 39:421–25.

79. Pederson, H. O. 1955. On type E botulism. *J. Appl. Bacteriol.* 18:619–29.

80. Peterson, D., H. Anderson, and R. Detels. 1966. Three outbreaks of foodborne disease with dual etiology. *Pub. Hlth Repts.* 81:899–904.

81. Post, L. S., T. L. Amoroso, and M. Solberg. 1985. Inhibition of *Clostridium botulinum* type E in model acidified food systems. *J. Food Sci.* 50:966–68.

82. Rey, C. R., H. W. Walker, and P. L. Rohrbaugh. 1975. The influence of temperature on growth, sporulation, and heat resistance of spores of six strains of *Clostridium perfringens*. *J. Milk Food Technol.* 38:461–65.

83. Roy, R. J., F. F. Busta, and D. R. Thompson. 1981. Thermal inactivation of *Clostridium perfringens* after growth at several constant and linearly rising temperatures. *J. Food Sci.* 46:1586–91.

84. Sakaguchi, G. 1983. *Clostridium botulinum* toxins. *Pharmac. Ther.* 19:165–94.

85. Satija, K. C., and K. G. Narayan. 1980. Passive bacteriocin typing of strains of *Clostridium perfringens* type A causing food poisoning for epidemiologic studies. *J. Infect. Dis.* 142:899–902.

86. Saylor, G. S., J. D. Nelson, Jr., A. Justice, and R. R. Colwell. 1976. Incidence of *Salmonella* spp., *Clostridium botulinum*, and *Vibrio parahaemolyticus* in an estuary. *Appl. Environ. Microbiol.* 31:723–30.

87. Schneider, M., N. Grecz, and A. Anellis. 1963. Sporulation of *Clostridium botulinum* types A, B, and E, *Clostridium perfringens*, and putrefactive anaerobe 3679 in dialysis sacs. *J. Bacteriol.* 85:126–33.

88. Skjelkvåle, R., and C. L. Duncan. 1975. Enterotoxin formation by different toxigenic types of *Clostridium perfringens*. *Infect. Immun.* 11:563–75.

89. Skjelkvåle, R., and T. Uemura. 1977. Detection of enterotoxin in faeces and anti-enterotoxin in serum after *Clostridium perfringens* food-poisoning. *J. Appl. Bacteriol.* 42:355–63.

90. Smith, A. M., D. A. Evans, and E. M. Buck. 1981. Growth and survival of *Clostridium perfringens* in rare beef prepared in a water bath. *J. Food Protect.* 44:9–14.

91. Smith, L. DS. 1975. Inhibition of *Clostridium botulinum* by strains of *Clostridium perfringens* isolated from soil. *Appl. Microbiol.* 30:319–23.

92. ———. 1977. *Botulism—The organism, its toxins, the disease.* Springfield, Ill.: Thomas.

93. Solomon, H. M., D. A. Kautter, and R. K. Lynt. 1982. Effect of low temperatures on growth of nonproteolytic *Clostridium botulinum* types B and F and proteolytic type G in crabmeat and broth. *J. Food Protect.* 45:516–18.
94. ———. 1985. Common characteristics of the Swiss and Argentine strains of *Clostridium botulinum* type G. *J. Food Protect.* 48:7–10.
95. Solomon, H. M., R. K. Lynt, Jr., D. A. Kautter, and T. Lilly, Jr. 1971. Antigenic relationships among the proteolytic and nonproteolytic strains of *Clostridium botulinum. Appl. Microbiol.* 21:295–99.
96. Sonnabend, O., W. Sonnabend, R. Heinzle, T. Sigrist, R. Dirnhofer, and U. Krech. 1981. Isolation of *Clostridium botulinum* type G and identification of type G botulinal toxin in humans: Report of five sudden unexpected deaths. *J. Infect. Dis.* 143:22–27.
97. Sperber, W. H. 1982. Requirements of *Clostridium botulinum* for growth and toxin production. *Food Technol.* 36(12):89–94.
98. ———. 1983. Influence of water activity on foodborne bacteria—A review. *J. Food Protect.* 46:142–50.
99. Spira, W. M., and G. J. Silverman. 1979. Effects of glucose, pH, and dissolved-oxygen tension on *Bacillus cereus* growth and permeability factor production in batch culture. *Appl. Environ. Microbiol.* 37:109–16.
100. Stark, R. L., and C. L. Duncan. 1971. Biological characteristics of *Clostridium perfringens* type A enterotoxin. *Infect. Immun.* 4:89–96.
101. Strong, D. H., and J. C. Canada. 1964. Survival of *Clostridium perfringens* in frozen chicken gravy. *J. Food Sci.* 29:479–82.
102. Strong, D. H., J. C. Canada, and B. Griffiths. 1963. Incidence of *Clostridium perfringens* in American foods. *Appl. Microbiol.* 11:42–44.
103. Strong, D. H., C. L. Duncan, and G. Perna. 1971. *Clostridium perfringens* type A food poisoning. II. Response of the rabbit ileum as an indication of enteropathogenicity of strains of *Clostridium perfringens* in human beings. *Infect. Immun.* 3:171–78.
104. Sugii, S., I. Ohishi, and G. Sakaguchi. 1977. Correlation between oral toxicity and in vitro stability of *Clostridium botulinum* types A and B toxins of different molecular sizes. *Infect. Immun.* 16:910–14.
105. Sugiyama, H. 1980. *Clostridium botulinum* neurotoxin. *Microbiol. Rev.* 44:419–48.
106. Sugiyama, H., and D. C. Mills. 1978. Intraintestinal toxin in infant mice challenged intragastrically with *Clostridium botulinum* spores. *Infect. Immun.* 21:59–63.
107. Sugiyama, H., and K. H. Yang. 1975. Growth potential of *Clostridium botulinum* in fresh mushrooms packaged in semipermeable plastic film. *Appl. Microbiol.* 30:964–69.
108. Thatcher, F. S., J. Robinson, and I. Erdman. 1962. The "vacuum pack" method of packaging foods in relation to the formation of the botulinum and staphylococcal toxins. *J. Appl. Bacteriol.* 25:120–24.
109. Trakulchang, S. P., and A. A. Kraft. 1977. Survival of *Clostridium perfringens* in refrigerated and frozen meat and poultry items. *J. Food Sci.* 42:518–21.
110. Tsang, N., L. S. Post, and M. Solberg. 1985. Growth and toxin production by *Clostridium botulinum* in model acidified systems. *J. Food Sci.* 50:961–65.
111. Turnbull, P. C. B., J. M. Kramer, K. Jorgensen, R. J. Gilbert, and J. Melling.

1979. Properties and production characteristics of vomiting, diarrheal, and necrotizing toxins of *Bacillus cereus*. *Amer. J. Clin. Nutri.* 32:219–28.

112. Walker, H. W. 1975. Food-borne illness from *Clostridium perfringens*. *CRC Crit. Rev. Food Sci. Nutri.* 7:71–104.

113. Weiss, K. F., and D. H. Strong. 1967. Some properties of heat-resistant and heat-sensitive strains of *Clostridium perfringens*. I. Heat resistance and toxigenicity. *J. Bacteriol.* 93:21–26.

114. Wentz, M., R. Scott, and J. Vennes. 1967. *Clostridium botulinum* type F: Seasonal inhibition by *Bacillus licheniformis*. *Science* 155:89–90.

115. Williams-Walls, N. J. 1968. *Clostridium botulinum* type F: Isolation from crabs. *Science* 162:375–76.

20.

FOOD-BORNE GASTROENTERITIS
CAUSED BY *SALMONELLA*
AND *ESCHERICHIA*

Among the gram-negative rods known to cause food-borne gastroenteritis, the most important are members of the genus *Salmonella*. This syndrome and those caused by *Escherichia coli* are presented in this chapter. The general incidence of these organisms in foods is discussed in Chapter 4.

SALMONELLOSIS

The salmonellae are small, gram-negative, nonsporing rods that are indistinguishable from *E. coli* under the microscope or on ordinary nutrient media. They are widely distributed in nature, with man and animals being their primary reservoirs. *Salmonella* food poisoning results from the ingestion of foods containing appropriate strains of this genus in significant numbers.

All species and strains of *Salmonella* may be presumed to be pathogenic for man, and the disease syndromes divide themselves into several distinct clinical types. Typhoid fever, caused by *S. typhi,* is the most severe of all diseases caused by this genus and a classic example of an enteric fever. In the same general category are the paratyphoid fevers caused by *S. paratyphi* A, *S. paratyphi* B, and others. The paratyphoid syndrome tends to be milder than that of typhoid. In the latter, the period of incubation is longer, a higher body temperature is produced, the organisms may be isolated from the blood and sometimes urine, and the mortality rate is higher. Blood cultures are often positive in the paratyphoid syndrome. The etiologic agents of the typhoid and paratyphoid syndromes are specifically pathogenic for man.

The third disease entity caused by salmonellae is gastroenteritis. This syndrome differs from the enteric fevers in having an incubation period as short as 8 h, generally negative blood cultures, and a lack of host specificity among the numerous serovars (serotypes) capable of causing this disease. While some of the gastroenteritis or food-poisoning strains can be identified on the basis of biochemical and cultural characteristics, by far the largest numbers are identified on the basis of antigenic analysis.

The salmonellae may be divided into three groups based on host predilections (13): **primarily adapted to man**—the typhoid and paratyphoid agents are the prime examples of this group; **primarily adapted to particular animal hosts**—included in this group are *S. choleraesuis* and serovars of *S. enteritidis* such as *S. pullorum, S. gallinarum, S. dublin,* and so on; and **unadapted**—this group includes over 2,000 serovars that may cause illness in man and other animals and generally do not show any host preference. Food-borne gastroenteritis is caused primarily by members of this group.

A more recent scheme divides salmonellae into five subgenera (5). By this scheme, all but about 200 of the 2,000 or so serovars are placed in subgenus I, which is characterized, largely, as follows: ONPG, lactose, and gelatinase negative; dulcitol and mucate positive; no growth in the presence of KCN; and natural habitat or warm-blooded animals.

Classification of *Salmonella*

The classification of these organisms by antigenic analysis is based upon the original work of Kauffmann and White and is often referred to as the Kauffmann-White Scheme. Classification by this scheme makes use of both somatic and flagellar antigens. Somatic antigens are designated O antigens, while flagellar antigens are designated H antigens. The K antigens are capsular antigens that lie at the periphery of the cell and prevent access of anti-O agglutinins (antibodies) to their homologous somatic antigens. The K antigen differs from ordinary O antigens in being destroyed by heating for 1 h at 60°C and by dilute acids and phenol. The use of H, O, and K antigens as the basis of classification of *Salmonella* spp. is based upon the fact that each antigen possesses its own genetically determined specificity.

When classification is made by use of antigenic patterns, species and serovars are placed in groups designated A, B, C, and so on, according to similarities in content of one or more O antigens. Thus, *S. hirschfeldii, S. choleraesuis, S. oranienberg,* and *S. montevideo* are placed in Group C_1 because they all possess O antigens 6 and 7 in common. *S. newport* is placed in Group C_2 due to its possession of O antigens K and 8 (see Table 20-1). For further classification, the flagellar or H antigens are employed. These antigens are of two types: specific phase or phase 1, and group phase or phase 2. Phase-1 antigens are shared with only a few other species or varieties of *Salmonella,* while phase 2 may be more widely distributed among several species. Any given culture of *Salmonella* may consist of organisms in only one phase, or of organisms in both flagellar phases. The H antigens of phase 1 are designated with small letters, and those of phase

2 are designated by arabic numerals. Thus, the complete antigenic analysis of *S. choleraesuis* is as follows: 6, 7, c, 1, 5, where 6 and 7 refer to O antigens, c to phase-1 flagellar antigens, and 1 and 5 to phase-2 flagellar antigens (see Table 20-1). *Salmonella* subgroups of this type are referred to as serovars. With a relatively small number of O, phase-1, and phase-2 antigens, a large number of permutations are possible, allowing for the possibility of a large number of serovars. Over 2,000 *Salmonella* serovars are presently known, and the number increases yearly.

The naming of *Salmonella* is now done by international agreement. Under this system, a serovar is named after the place where it was first isolated— *S. london, S. miami, S. richmond,* and so on. Prior to the adoption of this convention, species and subtypes were named in various ways—for example, *S. typhimurium* as the cause of typhoid fever in mice. Most food-borne salmonellae are serovars of *S. enteritidis.*

Distribution of *Salmonella*

The primary habitat of *Salmonella* spp. is the intestinal tract of animals such as birds, reptiles, farm animals, man, and occasionally insects. Although their primary habitat is the intestinal tract, they may be found in other parts of the body from time to time. As intestinal forms, these organisms are excreted in feces from which they may be transmitted by insects and other living creatures to a large number of places. As intestinal forms, they may also be found in water, especially polluted water. When polluted water and foods that have been contaminated by insects or by other means are consumed by man and other animals, these organisms are once again shed

Table 20-1. **Antigenic structure of some of the more common salmonellae.**

| | | | H antigens | |
| | | | phase 1 | phase 2 |
Group	Species/serovars	O antigens[a]		
A	S. paratyphi A	1, 2, 12	a	(1,5)
B	S. schottmuelleri	1, 4, (5), 12	b	1,2
	S. typhimurium	1, 4, (5), 12	i	1,2
C₁	S. hirschfeldii	6, 7, (Vi)	c	1,5
	S. choleraesuis	6, 7	(c)	1,5
	S. oranienburg	6, 7	m,t	—
	S. montevideo	6, 7	g,m,s (p)	(1,2,7)
C₂	S. newport	6, 8	e,h	1,2
D	S. typhi	9, 12, (Vi)	d	—
	S. enteritidis	1, 9, 12	g,m	(1,7)
	S. gallinarum	1, 9, 12	—	—
E₁	S. anatum	3, 10	e,h	1,6

[a]The underlined antigens are associated with phage conversion. ( ) = May be absent.

through fecal matter with a continuation of the cycle. The augmentation of this cycle through the international shipment of animal products and feeds is in large part responsible for the present world-wide distribution of salmonellosis and its consequent problems.

While *Salmonella* spp. have been recovered repeatedly from a large number of different animals, their incidence in various parts of animals has been shown to vary. In a study of slaughterhouse pigs, Kampelmacher (47) found these organisms in spleen, liver, bile, mesenteric and portal lymph nodes, diaphragm, and pillar, as well as in feces. A higher incidence was found in lymph nodes than in feces. The frequent occurrence of *Salmonella* spp. among susceptible animal populations is due in part to the contamination of *Salmonella*-free animals by animals within the population that are carriers of these organisms or are infected by them. A **carrier** is defined as a person or an animal that repeatedly sheds *Salmonella* spp., usually through feces, without showing any signs or symptoms of the disease. Upon examining poultry at slaughter, Sadler and Corstvet (84) found an intestinal carrier rate of 3–5%. During and immediately after slaughter, carcass contamination from fecal matter may be expected to occur. In an examination of the rumen contents of healthy cattle after slaughter, Grau and Brownlie (30) found 45% to contain salmonellae. Some 57% of samples taken from the environment of cattle in transit to slaughter were positive for these organisms. From 53.1 to 61.9% of inspected broiler carcasses have been found to be contaminated with *Salmonella* spp. These organisms appear not to be normal flora of poultry but are acquired from the environment via insects, rodents, feeds, other animals, and man (63).

Equally serious is the contamination of eggs and egg products. Reports from various countries indicate that 2.6–7.0% of eggs are contaminated, mostly with *S. typhimurium* (78). Duck eggs have been reported to have an even higher contamination rate, reaching 20%. Another common source of these organisms to animal populations is animal feed (Fig. 20-1). Some investigators feel that this source is perhaps the most important in terms of the overall control of salmonellosis. In a study of animal feeds in England for the years 1958–1960, Taylor (95) isolated *Salmonella* serovars from meat and bone products 855 times; from mixtures, mashes, and the like 158 times; from fish products 100 times; and from vegetable products 17 times. *S. senftenberg* was isolated most frequently, followed by *S. anatum* and *S. cubana*. *S. senftenberg* has been shown to be the most heat-resistant of all *Salmonella* serovars, and its higher incidence in animal feeds may be due to the fact that most others are destroyed by heat in the processing of these products. A very large number of different serovars was found in bile, liver, spleen, and lymph glands of slaughter animals by Taylor. Among the many serovars reported by this author, *S. dublin* was found to be associated with bovines more than with any other animals. The presence of salmonellae in animal feeds has been shown by many investigators.

The most common food vehicles of salmonellosis in man are eggs, poultry,

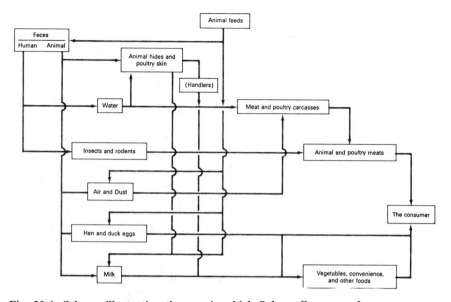

Fig. 20-1. Scheme illustrating the way in which *Salmonella* spp. and serovars are disseminated.

meat, and meat products. In a study of sixty-one outbreaks of *Salmonella* food poisoning for the period 1963–65, Steele and Galton (94) found that eggs and egg products accounted for twenty-three, chicken and turkey for sixteen, beef and pork for eight, ice cream for three, potato salad for two, and other miscellaneous foods for nine. The most common food vehicles involved in 12,836 cases of salmonellosis from thirty-seven states in 1967 were beef, turkey, eggs and egg products, and milk. Of 7,907 salmonellae isolations made by CDC in 1966, 70% were from raw and processed food sources. Turkey and chicken sources accounted for 42%. Of the food-borne disease outbreaks with known etiology traced to poultry for the period 1972–74, 44% were salmonellae (40). In a study of the incidence of salmonellae in sixty-nine packs of raw chicken pieces, 34.8% were positive and eleven serovars were represented with *S. muenchen* being the most common (21). From a study in Venezuela, forty-one of forty-five chicken carcasses studied yielded salmonellae consisting of eleven serovars with *S. anatum* being most frequently isolated (79). Poultry products are important sources of salmonellosis outbreaks, and the wide distribution of these organisms among slaughter animals makes all meats potential sources. Even when such meats are cooked sufficiently to destroy salmonellae, in the raw state they may still serve as sources to other foods such as vegetables, salads, and the like by cross contamination.

The fresh food item in a typical supermarket most likely to contain viable salmonellae is pork sausage. A study of forty producing plants in 1969 revealed an overall incidence of 28.6% positive samples of 566 examined

(43). Ten years later, the overall incidence of positive samples had decreased to 12.4% of 603 examined. Of the forty matched plants, the incidence decreased in twenty, increased in thirteen, and remained the same in seven. In commercially prepared and packaged foods, salmonellae were found in 17 of 247 examined (1). Among the contaminated foods were cake mixes, cookie doughs, dinner rolls, and cornbread mixes. Foods of this type usually become contaminated from infected eggs, bulk egg products, or by contact with rodents, flies, or even man. Salmonellae have been found in coconut meal, salad dressing, mayonnaise, milk, and many other foods. In a study of health foods, none of plant origin yielded salmonellae, but from two of three lots of beef liver powder from the same manufacturer were isolated *S. minnesota, S. anatum,* and *S. derby* (97). Salmonellae may be found in pet foods and on pets, especially pet turtles.

Of the various serovars in food-borne outbreaks, *S. typhimurium* is invariably the most commonly encountered throughout the world. Of 1,713 serovars isolated from foods during 1963–65, Steele and Galton (94) found the following 5 to be the most common: *S. infantis* (215), *S. oranienberg* (192), *S. typhimurium* (171), *S. montevideo* (151), and *S. heidelberg* (147). In general, the incidence of *Salmonella* serovars in foods closely parallels their incidence in man and animals. The 7 most frequently isolated from man in England and Wales for 1956–60 were as follows (95): *S. typhimurium* (15,808), *S. heidelberg* (1,013), *S. enteritidis* (734), *S. newport* (667), *S. thompson* (623), *S. saint-paul* (282), and *S. anatum* (279). For the period April 1962 to April 1963, the 9 most prevalent serovars identified from animal sources in the United States were: *S. typhimurium, S. heidelberg, S. anatum, S. choleraesuis, S. infantis, S. montevideo, S. derby, S. saint-paul,* and *S. oranienberg*. Of 17 serovars isolated from 603 samples of fresh pork sausage in 1979, the 3 most frequently found were *S. derby* (23%), *S. anatum* (14%), and *S. agona* (11%, 43). The 5 most frequently isolated from human sources in 1972 and 1980 are listed in Table 20-2. The pattern of human isolations of salmonellae in the United States for the period 1955

Table 20-2. The most frequently isolated salmonellae from humans in the United States in 1972 and 1980.

	% of isolates	
Serovars	*1972*[a]	*1980*[b]
S. typhimurium	25.8	34.8
S. heidelberg	5.6	6.6
S. enteritidis	6.5	6.3
S. newport	8.4	5.5
S. infantis	6.3	4.8

[a](98).
[b]A total of 30,004 isolations (*Morbidity Mortality Weekly Reports* 30:377–378, 1981).

through 1982 is depicted in Fig. 20-2. Forty percent of the 34,766 reported cases for 1982 affected children < 5 years of age. The increase in isolations from adults is accounted for in large part by the consumption of improperly pasteurized and raw milk. The specific serovar associated with raw milk consumption (including certified raw milk) is *S. dublin*. Persons from whom this serovar was isolated over the past 5 years or so were older and had underlying illnesses requiring longer periods of hospitalization.

Among trends noted in the United States and Canada over the period 1969–77 was a decline in isolations of *S. thompson* and increases in *S. agona* isolates (9). The latter was first isolated from fish meal, and while no isolations were made in 1969, 1,461 were made in 1976. The incidence

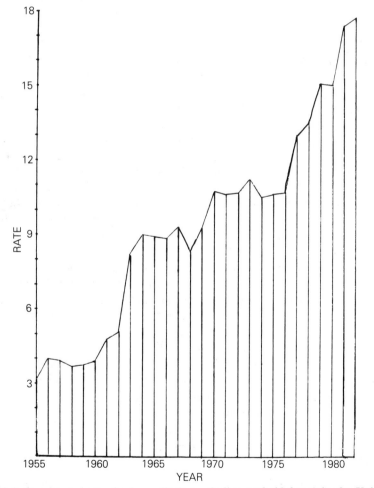

Fig. 20-2. Reported cases of salmonellosis (excluding typhoid fever) in the United States for the period 1955–82. Rate = cases/100,000 population. (Redrawn from *Morbidity and Mortality Weekly Report,* **Vol. 31, no. 54, 1983)**

of some serovars tends to be associated with single products, and an example of this is *S. eastbourne,* which was frequently isolated from contaminated chocolate. Bryan (9) has noted the association of some serovars with animal sources. Frequently isolated from chickens and turkeys are *S. heidelberg* and *S. saint-paul;* from cattle *S. dublin; S. johannesburg* from chickens; and *S. choleraesuis* from swine.

For the years 1971–83, antimicrobial-resistant salmonellae were examined from fifty-two outbreaks in the United States. The case fatality rate was higher for persons infected with antimicrobial-resistant strains than those with antimicrobial-sensitive strains, and food animals were the source of eleven of sixteen resistant and only six of thirteen sensitive strains (39). Raw milk was the source of resistant strains in four outbreaks and beef in another. In seven outbreaks, the resistant serovars were *S. typhimurium;* and *S. newport* in two outbreaks.

Growth and destruction of salmonellae

These organisms are typical of other gram-negative bacteria in being able to grow on a large number of culture media and produce visible colonies well within 24 h at about 37°C. They are generally unable to ferment lactose, sucrose, or salicin, although glucose and certain other monosaccharides are fermented, with the production of gas. Although they normally utilize amino acids as N-sources, in the case of *S. typhimurium,* nitrate, nitrite, and NH_3 will serve as sole sources of nitrogen (74). While lactose fermentation is not usual for these organisms, some serovars can utilize this sugar.

The pH for optimum growth is around neutrality, with values above 9.0 and below 4.0 being bactericidal (78). A minimum growth pH of 4.05 has been recorded for some (with HCl and citric acids), but depending upon the acid used to lower pH, the minimum may be as high as 5.5 (11). The effect of acid used to lower pH on minimum growth is presented in Table 20-3. Aeration was found to favor growth at the lower pH values. The parameters of pH, a_w, nutrient content, and temperature are all interrelated for salmonellae as they are for most bacteria (98). For best growth, the salmonellae require pH between 6.6 and 8.2. The lowest temperatures at which growth has been reported are 5.3°C for *S. heidelberg* and 6.2°C for *S. typhimurium* (61). Temperatures of around 45°C have been reported by several authors to be the upper limit for growth. Regarding available moisture, growth inhibition has been reported for a_w values below 0.94 in media with neutral pH, with higher a_w values being required as pH is decreased toward growth minima.

Unlike the staphylococci, the salmonellae are unable to tolerate high salt concentrations. Brine above 9% is reported to be bactericidal. Nitrite is effective, with the effect being greatest at the lower pH values. This suggests that the inhibitory effect of this compound is referable to the undissociated HNO_2 molecule. The survival of *Salmonella* spp. in mayonnaise was studied by Lerche (57), who found that they were destroyed in this product if the

Table 20-3. Minimum pH at which salmonellae would initiate growth under optimum laboratory conditions.[a] (11)

Acid	pH
Hydrochloric	4.05
Citric	4.05
Tartaric	4.10
Gluconic	4.20
Fumaric	4.30
Malic	4.30
Lactic	4.40
Succinic	4.60
Glutaric	4.70
Adipic	5.10
Pimelic	5.10
Acetic	5.40
Propionic	5.50

[a]Tryptone–yeast extract–glucose broth was inoculated with 10^4 cells per milliliter of *Salmonella anatum, S. tennessee,* or *S. senftenberg.*

pH was below 4.0. It was found that several days may be required for destruction if the level of contamination is high, but within 24 h for low numbers of cells. *S. thompson* and *S. typhimurium* were found to be more resistant to acid destruction than *S. senftenberg.*

With respect to heat destruction, all salmonellae are readily destroyed at milk pasteurization temperatures. Thermal D values for the destruction of *S. senftenberg* 775W under various conditions are given in Chapter 14. Shrimpton et al. (91) reported that *S. senftenberg* 775W required 2.5 min for a 10^4–10^5 reduction in numbers at 54.4°C in liquid whole egg. This strain is the most heat-resistant of all salmonellae serovars. The above treatment of liquid whole egg has been shown to produce a *Salmonella*-free product and destroy egg alpha-amylase (see Chapter 14 for the heat pasteurization of egg white). It has been suggested (6) that the alpha-amylase test may be used as a means of determining the adequacy of heat pasteurization of liquid egg (compare with the pasteurization of milk and the enzyme phosphatase). In a study on the heat resistance of *S. senftenberg* 775W, Ng et al. (70) found this strain to be more heat sensitive in the log phase than in the stationary phase of growth. These authors also found that cells grown at 44°C were more heat resistant than those grown at either 15° or 35°C.

With respect to the destruction of *Salmonella* in baked foods, Beloian and Schlosser (4) found that baked foods reaching a temperature of 160°F or higher in the slowest heating region can be considered *Salmonella* free. These authors employed *S. senftenberg* 775W at a concentration of 7,000–10,000 cells/ml placed in reconstituted dried egg. With respect to the heat destruction of this strain in poultry, it is recommended that internal temperatures of at least 160°F be attained (65). Although *S. senftenberg* 775W has been reported to be thirty times more heat resistant than *S. typhimurium* (70), the latter organism has been found to be more resistant to dry heat than the former (29). These authors tested dry heat resistance in milk chocolate.

The destruction of *S. pullorum* in turkeys was investigated by Rogers and Gunderson (81), who found that it required 4 h and 55 min to destroy an initial inoculum of 115,000,000 in 10–11-lb. turkeys with an internal temperature of 160°F, and for 18-lb. turkeys with an initial inoculum of 320,000,000 organisms, 6 h and 20 min were required for destruction. The salmonellae are quite sensitive to ionizing radiation, with doses of 0.5 to 0.75 Mrads being sufficient to eliminate them from most foods and feed (78). The decimal reduction dose has been reported to range from 0.04–0.07 Mrad for *Salmonella* spp. in frozen eggs. The effect of various foods on the radiosensitivity of salmonellae is shown in a study by Ley et al. (58). These investigators found that for frozen whole egg, 0.5 Mrad gave a 10^7 reduction in the numbers of *S. typhimurium*, while 0.65 Mrad was required to give a 10^5 reduction in frozen horsemeat, between 0.5–0.75 Mrad for a 10^5–10^8 reduction in bone meal, and only 0.45 Mrad to give a 10^3 reduction of *S. typhimurium* in desiccated coconut.

In dry foods, *S. montevideo* was found to be more resistant than *S. heidelberg* when inoculated ino dry milk, cocoa powder, poultry feed, meat, and bone meal (46). Survival was greater at a_w 0.43 and 0.52 than at a_w 0.75.

The *Salmonella* food-poisoning syndrome

This syndrome is caused by the ingestion of foods that contain significant numbers of non–host specific species or serotypes of the genus *Salmonella*. From the time of ingestion of food, symptoms usually develop in 12–14 h, although shorter and longer times have been reported. The symptoms consist of nausea, vomiting, abdominal pain (not as severe as with staphylococcal food poisoning), headache, chills, and diarrhea. These symptoms are usually accompanied by prostration, muscular weakness, faintness, moderate fever, restlessness, and drowsiness. Symptoms usually persist for 2–3 days. The average mortality rate is 4.1%, varying from 5.8% during the first year of life, to 2% between the first and fiftieth year, and 15% in persons over fifty (78). Among the different species of *Salmonella, S. choleraesuis* has been reported to produce the highest mortality rate—21%.

While these organisms generally disappear rapidly from the intestinal tract, up to 5% of patients may become carriers of the organisms upon recovery from this disease.

Numbers of cells on the order of 10^7–10^9/g are generally necessary for salmonellosis. That outbreaks may occur in which relatively low numbers of cells are found has been noted (17). From three outbreaks, numbers of cells found were as low as 100/100 g (*S. eastbourne* in chocolate) to 15,000/g (*S. cubana* in a carmine dye solution). In general, minimum numbers for gastroenteritis range between 10^5–10^6/g for *S. bareilly* and *S. newport*, to 10^9–10^{10} for *S. pullorum* (7).

Salmonella toxins

It appears that the pathogenesis of salmonellosis may involve two toxins—an enterotoxin and a cytotoxin. The enterotoxin was first demonstrated in 1975 by Koupal and Deibel (54). Using *S. enteritidis* and the suckling mouse assay, a difficult-to-separate cell envelope–associated toxin was prepared, producing results in the suckling mouse assay similar to those elicited by the heat-stable and heat-labile enterotoxins of *E. coli* (see Chapter 7 and the section on *E. coli* in this chapter). Loop assays including the infant rabbit gave inconsistent results. Employing *S. typhimurium,* an enterotoxin was produced in BHI broth and a 2% Casamino acids medium and measured by the rabbit ileal loop assay (90). These two media were found to be the best of several that were tried. Meanwhile, another group of researchers recovered from salmonellae a toxin that induced rabbit skin permeability and acted in a manner similar to the enterotoxins of *Vibrio cholerae* and *E. coli* (87). The toxin in question produced positive responses in the rabbit ileal loop assay. It was later shown that the vascular permeability factor could be neutralized with monospecific cholera antitoxin, and that the toxin induced elongation in Chinese hamster ovary (CHO) cells similar to the toxin of *V. cholerae* (88). In their attempts to obtain larger quantities of salmonellae enterotoxin, Koo and Peterson (49) studied the influence of nutritional factors and found that glycerol, biotin, and Mn^{2+} enhanced production, while glucose was found to be a poor carbon source. It was later determined that more enterotoxin is produced during the stationary phase of growth, at pH 6–7 or higher, at 37°C, and with increased aeration (50). The enterotoxin was found to be heat-labile at 100°C, to have a molecular weight > 110,000 daltons, and an isoelectric point of approximately 4.3–4.8 (42). Mitomycin C added 3 h after inoculation increased the quantities in culture filtrates due to bacteriophage induction and subsequent cell lysis. The weight of the evidence suggests that the salmonellae toxin described above is an enterotoxin that acts in a manner similar to those of *E. coli* by elevating intestinal cAMP. Unlike the *E. coli* enterotoxins, it is produced in much lower quantities and is more difficult to separate from producing cells. Further, the salmonellae enterotoxin is quite similar to cholera toxin (CT) in biologic and antigenic characteristics. Heated CT (procholeragenoid) administered parenterally protects against loop fluid responses by viable salmonellae cells (75).

While the enterotoxin described above was shown to affect the adenylate cyclase system and induce fluid accumulation in animal models, the intestinal toxicity associated with salmonellosis was not explained. Although the enterotoxins of *E. coli* induce fluid accumulation, tissue toxicity that is normally associated with shigellosis and salmonellosis does not occur. In other words, the pathogenesis of salmonellosis resembles that of shigellosis more than the enterotoxin-mediated syndromes. This led Koo et al. (52) to examine salmonellae extracts for a cytotoxin. Some European workers were actually the first to show cytotoxic activity of salmonellae extracts as early as 1962. When extracts from salmonellae were added to isolated rabbit intestinal epithelial cells and to Vero cells, protein synthesis was inhibited (51, 52). These investigators presented evidence to support the view that the cellular damage that occurs to the intestinal mucosa during salmonellosis is caused by a cytotoxin. Once damaged, the intestinal mucosa is more easily invaded by the infecting organisms, which can then bring about additional tissue damage. While the pathogenesis of shigellosis and salmonellosis appear to be closely related, more tissue destruction occurs in the former than in the latter. This may be due to an additional toxin— one that is more destructive—or to other factors. That prostaglandins may play a role is suggested by the finding that both salmonellae enterotoxin and CT induced their synthesis in intestinal epithelial cells when an anti-inflammatory agent (indomethacin) was employed (20).

Virulence for mice by three serovars (*S. dublin, S. typhimurium,* and *S. enteritidis*) has been associated with plasmids of 36 to 60 Mdal (69), but how virulence factors are controlled by plasmid genes is unclear. The pathogenic mechanisms of *Salmonella* require much more study.

Animal models and tissue culture systems for the assay of salmonellae toxins are further described in Chapter 7.

Incidence and vehicle foods

The precise incidence of salmonellae food poisoning is not known since small outbreaks are often not reported to public health authorities. Some authorities believe as many as 2 million human cases of salmonellosis may occur each year in the United States alone. As is the case for staphylococcal food poisoning, the largest outbreaks of salmonellosis usually occur at banquets or similar functions. However, the outbreak that occurred during the spring of 1985 was exceptional. Over 15,000 cases of salmonellosis in at least four states were traced to pasteurized milk produced at a plant in Illinois. Most of the milk contained 2% fat, and *S. typhimurium* was recovered from some unopened containers as well as from victims. Although the cause is still under investigation, it appears that the organisms entered the pasteurized product by cross contamination of raw milk. The next largest on record, which occurred in 1974 on the Navajo Nation Indian Reservation, affected an estimated 3,400 persons (41). The vehicle food was potato salad served to about 11,000 individuals at a free barbecue. It was prepared and stored for up to 16 h at improper holding temperatures prior to serving, and the serovar isolated was *S. newport.* Among the unusual outbreaks was one traced to marijuana in four midwestern states. The serovar isolated was *S. muenchen* and counts as high as $10^7/g$ were obtained from marijuana samples.

In general, vehicle foods for staphylococcal food poisoning may also serve as sources of salmonellosis. Common to both are foods that are prepared by hand and consumed sometime later without subsequent heating. Vehicle foods for salmonellae often contain uncooked eggs. The five most common vehicle foods in the United States and Canada are beef, turkey, homemade ice cream (containing eggs), pork, and chicken (9). Turkey is the most common source in Canada.

Food-borne salmonellosis continues to increase. The reasons for this have been addressed by Kampelmacher (47): (1) the increase in mass food preparation, which favors spread of *Salmonella;* (2) inappropriate methods of storing food, which, because of modern living conditions, is sometimes accumulated in excessive amounts; (3) the increasing habit of eating raw or insufficiently heated foods, partly because of overreliance on food inspection; (4) increasing international food trade; and (5) decreased resistance to infection resulting from improved standards of general hygiene.

Recovery of *Salmonella* from foods

In view of the federal regulation prohibiting the presence of *Salmonella* spp. in foods, the recovery of these organisms from food presents a difficult problem to food microbiologists and food scientists. How can one be certain that a 1,000-lb. lot of powdered eggs is free of *Salmonella* if there might be only 1 organism/100 g? The problem is made all the more difficult by the fact that foods normally contain larger numbers of microorganisms other than *Salmonella*, such as *Proteus, Pseudomonas, Acinetobacter,* and *Alcaligenes* spp., all of which may develop on some of the media employed for the recovery of salmonellae. The problems encountered in the recovery of these organisms from foods are similar to those that one encounters in the recovery of staphylococci from foods; that is, a generally high ratio of total numbers of other organisms to pathogens. Where the ratio of total flora to salmonellae is rather low, any one of a large number of media may be suitable for recovering these organisms. Special methods are necessary, however, when few salmonellae exist in the presence of a high total count.

In view of the generally low ratio of salmonellae to total flora, it is necessary to inhibit or kill nonsalmonellae types, while the salmonellae are allowed to increase to numbers high enough to improve the chances of finding them on plates or in broth. The fluorescent antibody technique and other salmonellae methods are discussed in Chapter 6. For a review of salmonellae methodology, see D'Aoust et al. (16); and for a comparison of rapid and conventional methods, see Flowers (27).

Prevention and control of salmonellosis

An overall view of the primary sources and transmission of *Salmonella* spp. to man is presented in Fig. 20-1. As previously noted, the primary source of these organisms is the intestinal tract of man and other animals. From human feces, salmonellae may enter water from which meats, poultry, and other foods may become contaminated when such water is used. Insects and rodents may also become contaminated with these organisms from polluted waters and disseminate them directly to either prepared or raw foods. Animal fecal matter is of greater importance than human, and it may be noted from Fig. 20-1 that animal hides and poultry products may become contaminated from this source. *Salmonella* spp. are maintained within an animal population by means of nonsymptomatic animal infections, and by means of animal feeds, which have been shown by many investigators to harbor comparatively large numbers of these organisms. Both of these sources serve to keep slaughter animals reinfected in a cyclical manner.

Secondary contamination is another of the more important means of transmission of salmonellae to man. The presence of these organisms on eggs, meats, and in the air makes their presence in certain foods inevitable through the agency of handlers and direct contact of noncontaminated foods with contaminated foods (37). Prost and Riemann (78) have pointed out that this secondary contamination may occur at various stages of food preparation, and is most frequently of animal origin, although human sources can be of some importance.

Among the leading causes of outbreaks are: improper cooling of cooked foods, lapse of a day or more between preparation and serving of foods, inadequate cooking or heating, ingestion of contaminated or raw ingredients, and cross contamination (9).

In view of the world-wide distribution of salmonellosis involving numerous serovars from many different animals, the ultimate control of this problem consists of freeing animals and man of the organisms. This is obviously a difficult task. It is not impossible, however, when it is considered that only 33 of the more than 2,000 species and serovars account for almost 90% of human isolates and approximately 80% of nonhuman isolates (60). The reinfection of animals through animal feeds can be controlled by treatment of feeds so that they are rendered salmonellae-free. One of the more promising ways of achieving the latter consists of heat treatments.

At the consumer level, the *Salmonella* carrier is presumed to play an important role, although just how important is not known. The proper cooking of vehicle foods, their proper handling, and subsequent storage at temperatures below the growth range of these organisms will obviously do a great deal to lessen the incidence of salmonellosis at this level. As long as these organisms remain among the animal population, however, the potential hazards of salmonellosis in man will likewise remain.

E. COLI GASTROENTERITIS SYNDROMES

Although sporadic reports of *Escherichia coli*–related gastroenteritis of food origin appeared prior to the 1970s, it was the 1971 outbreak in the United States traced to imported cheese that focused attention on this organism as a food-borne pathogen. This interest coincided with that of medical microbiologists who were interested in *E. coli* as a cause of infant diarrhea. The latter work led to the development of specific and accurate in vitro and bioassay methods for assessing toxic components (see Chapters 6 and 7), and to a better understanding of the virulence mechanisms of this organism. Although only a relatively few food-borne outbreaks have been documented, more is known about the pathogenesis of *E. coli* gastroenteritis than that of salmonellosis. *E. coli* as an indicator organism is discussed in Chapter 17, culture and isolation methods are covered in Chapter 5, and chemical and bioassay methods for its enterotoxins are covered in Chapters 6 and 7. The early history of this syndrome can be found in Bryan (8), Sack (82), and Mehlman et al. (64). A more recent review is that of Kornacki and Marth (53).

Strains and distribution
The first studies on *E. coli*–associated diarrheal disease involved nursery epidemics in the mid-1940s that brought mortality rates as high as 50%. During the mid 1950s, some *E. coli* isolates were shown to produce responses in the rabbit ileal loop test similar to those of *V. cholerae*. These findings led to studies of *E. coli* as a possible etiologic agent of cholera-like diseases in India. The first reports of enterotoxigenic strains from young animals with diarrhea appeared in 1967, and in 1970 the production of two enterotoxins by virulent strains was reported (see review by Sack, 82).

The *E. coli* strains of importance as potential food-borne pathogens are among the fecals (with the possible exception of hemorrhagic colitis agents), and their incidence in some foods is noted in Chapter 4. In general, these strains have a wide distribution in food environments in *low* numbers. The microbiological criteria for raw/uncooked foods acknowledges this fact (see Chapter 17). It should be recalled that as an indicator, *E. coli* in foods in sufficient numbers is taken to indicate the possibility of fecal contamination and the possible presence of other enteropathogens such as salmonellae. As a potential food-borne pathogen, the generally acceptable low numbers take on new meaning, especially when conditions permit their proliferation. While the latter can be said for coagulase-positive staphylococci, unlike fecal *E. coli,* they are not employed as indicators of sanitary quality.

E. coli strains involved in gastroenteritis may be placed into four groups: enteropathogenic (EPEC), enterotoxigenic (ETEC), enteroinvasive (EIEC), and facultatively enteropathogenic (FEEC). EPEC strains usually do not produce either of the two enterotoxins but some do (see below). Just how they cause disease is not entirely clear. Some EPEC strains produce a Shigalike (similar to that produced by *Shigella dysenteriae* 1) cytotoxin that can be demonstrated by the use of Vero cells, and the cytotoxin is produced more frequently by these strains than by other *E. coli* isolates. In one study, 79% of twenty-nine EPEC were positive for the Shigalike cytotoxin while only 24% of eighty-three from healthy individuals were positive (12). ETEC strains produce one or both enterotoxins as well as colonizing factor antigens

(CFA) represented by fimbriae (or pili), which mediate binding of cells to epithelial cells. Of 240 *E. coli* isolates from cheeses, ground beef, seafood, and sausage, 19 (8%) produced the heat-labile toxin, and of these, 8 also produced the heat-stable toxin while 11 produced only the heat-labile toxin (83). FEEC strains are associated wih sporadic diarrheal outbreaks.

Three antigens are employed to serotype these strains: O (heat-stable somatic antigens); K (heat-labile somatic antigens); and H (heat-labile flagellar antigens). About 164, 100, and 56 of O, K, and H, respectively, are known (73). Only about thirty serovars have been associated wih diarrheal disease, and the first was 0111, which was isolated from children with diarrhea and constituted as high as 80–100% of fecal flora (82). About 1% of the cultivable fecal flora of man consists of *E. coli* carried normally in the lower bowel, while the pathogens (except hemorrhagic colitis strains) colonize the upper bowel.

With respect to serotypes/serovars, there appears to be little consistency from individual to individual. A study of this organism in the fecal flora of thirteen healthy adults in England revealed wide variations in serotypes, with several individuals carrying one predominant type while several yielded as many as twelve different types (36). The O antigenic types carried by four to six persons were: 01, 02, 018, 068, 088, 0107, 0116, 0126, and 0132. The serovars responsible for diarrheal disease in thirty-three children and adults varied rather widely, with only six being recovered from two to four individuals: 06, 015, 025, 078, 0126, and 0128 (82). Some of the latter types are known to be nontoxic. In a study of the incidence of pathogenic strains among 219 food handlers, 14 (6.4%) were carriers of a total of eight different serovars (34). From other reports, the incidence in the feces of healthy adults and children range from 1.8 to 15.1%. Among EPEC strains, only one of thirty-four of serotype 014; nine of forty-five of serotype 0114; and eighteen of eighty-two of serotype 0128 produced either enterotoxin (89). In one study, the most common serotype isolated from foods was 0149 (83).

The K88-positive strain of porcine origin has been shown to effect surface colonization in piglets, and to be lethal to 50% of piglets while the K88-negative variant was lethal to only 3% (72). The K88 antigen is a short, piluslike structure that is highly antigenic. K99 is similar but pathogenic in cows and lambs but not pigs. Although K88 produces the heat-labile toxin, it has been suggested that an immunofluorescence test for in vivo-produced pilus antigens be used for the diagnosis of ETEC infections (66). The K88 antigen is plasmid borne (92). It was observed by earlier investigators that a poor correlation exists between serotypes and ileal loop responses of *E. coli* strains. While serotyping is of value epidemiologically, it appears to be of little value as a predictor of *E. coli* virulence.

Plasmids from *E. coli* strains have been studied more than from any other bacteria. Strains from patients have been shown to contain more conjugative plasmids than those from healthy individuals. ETEC strains usually carry five or more plasmids with antibiotic resistance, enterotoxins, and adherence antigens on separate plasmids (24). Among specific plasmids in *E. coli* is ColV, which controls an iron-sequestering mechanism—possibly enterochelin or enterobactin—and serum resistance (24).

The enterotoxins

As noted above, the ETEC strains produce two enterotoxins—heat labile (LT) and heat stable (ST). The maximum amount of ST was produced after 7 h of growth in one study in a Casamino acids yeast extract medium containing 0.2% glucose (56). In a synthetic medium, ST appeared as early as 8 h but maximal production required 24 h with aeration (93). While they are generally produced under all conditions that allow cell growth, the release of LT in particular from cells in enriched media was favored at pH of 7.5–8.5 (68).

While most strains produce both enterotoxins, some are known that produce only one. The enterotoxins along with the colonization factor antigen and hemolysin are plasmid mediated (100). The enterotoxins are encoded on Ent plasmids that also may carry genes for resistance to several anti-microbials (62). However, an LT-like toxin produced by *E. coli* strain SA53 is apparently controlled by chromosomal genes (31). EIEC strains apparently carry a 140-Mdal (megadalton) plasmid (35). In the case of ST_a, it is encoded within a transposon (92).

The ST toxin can withstand 100°C for 15 min while LT is destroyed at 60°C in about 30 min. LT is labile in acid while ST is resistant. LT is a protein with a molecular weight of about 91,000 daltons, and cistrons encoding for its production on plasmid DNA have been mapped (15). It is associated more with man and porcines than other animals. LT is composed of two protomers. The A protomer has a molecular weight of about 25,500 daltons and is synthesized as a single polypeptide chain that, when nicked (with trypsin), becomes an enzymatically active A_1 polypeptide chain of 21,000 daltons linked by a disulfide bond to an A_2-like chain (compare with botulinal toxins in Chapter 19). The B protomer has a molecular weight of about 59,000 and consists of five noncovalently linked individual polypeptide B chains of 11,800 daltons (see Eidels et al., 23).

There are two types of *E. coli* ST, both of which retain heat stability. ST_a (STI) has a molecular weight of 1,972, is methanol soluble, is an acidic polypeptide (lacks basic amino acids), is active in neonatal piglets 1–3 days old, and is active in infant mice but generally not in weaned pigs (10, 23). Biological activity of ST_a is lost upon treatment with 2-mercaptoethanol or dithiothreitol, indicating the presence and need of disulfide bridges (93). ST_a's from human and bovine sources all contain ten different amino acids in a sequence of eighteen with the same C-terminal (tyrosine) and N-terminal (asparagine) groups, and antisera from one cross reacts with all (19). ST_a has been chemically synthesized and, when compared to that produced by *E. coli,* exhibited the same general properties (48). The ST from human, bovine, and porcine sources is essentially similar and may have originated from a single, widely disseminated transposon (85). There is some evidence that ST_a may be a family of at least two peptide toxins.

ST_b (STII) is methanol insoluble, active in weaned pigs and rabbit ileal loop tests, but not in infant mice (10). ST_a and ST_b are genetically and immunologically distinct (23).

E. coli LT has a molecular weight of 84,000–91,000 daltons. It possesses enzymatic activity similar to that of CT but ST does not. LT shares some features in common with the enterotoxin of *V. cholerae* (CT). Antisera to CT neutralized LT from human and porcine strains in the rabbit ileal loop test (33), and immunization with CT induced equal protection against subsequent challenges against both CT and LT (77), thus establishing that the two toxins share common antigenic determinants. LT differs from CT in amino acid composition, but they do share many other common features (55).

Mode of action of enterotoxins

The *E. coli* gastroenteritis syndrome is caused by the ingestion of 10^6–10^{10} viable cells/g that must colonize the small intestines and produce enterotoxin(s). The colonizing factors are generally fimbriae or pili. The syndrome is characterized primarily by nonbloody diarrhea without inflammatory exudates in stools. The diarrhea is watery and similar to that caused by *V. cholerae*. Diarrhea results from enterotoxin activation of intestinal adenylate cyclase, which increases cyclic 3′,5′-adenosine monophosphate (cAMP), first shown for *E. coli* enterotoxins by Evans et al. (25).

With respect to LT, the B protomer mediates binding of the molecule to intestinal cells. LT binds to gangliosides, especially monosialogangliosides (GM_1) (23). CT also binds to GM_1 ganglioside, and CT and LT are known to share antigenic determinants among corresponding protomers although they do not cross react. Upon binding, the A polypeptide chain (of the A protomer) catalyzes ADP ribosylation of adenylate cyclase, which induces increases in intracellular cAMP.

Regarding ST, ST_a binds irreversibly to a specific high-affinity nonganglioside receptor and initiates a transmembrane signal to active particulate guanylate cyclase (28, 32). The increased levels of mucosal cGMP lead to loss of fluids and electrolytes. ST differs from CT in that only the particulate form of intestinal guanylate cyclase is stimulated by ST (28). ST_a differs from LT in that the former stimulates guanylate cyclase while the latter and CT both activate adenylate cyclase. Stimulation of guanylate cyclase by ST_a is tissue specific, and only the intestinal form of the particulate enzyme responds to ST_a. The precise role of ST_b is not understood but it does not activate guanylate cyclase. Genes controlling its production have been genetically mapped (62) and subcloned from its plasmid and sequenced (76).

Food-borne and related outbreaks

Reports incriminating *E. coli* as the etiologic agent of food poisoning involving a variety of foods such as cream pie, mashed potatoes, cream puffs, and creamed fish date from around 1900. The feeding of volunteers with serotypes 055:B5, 0111:B4, and 0127:B8 at levels of 10^6–10^8 organisms was shown to produce gastroenteritis symptoms (26, 45, 102).

Most of the documented outbreaks are summarized in Table 20-4. Not included are four to six outbreaks reported from the USSR, Eastern Europe, and Japan (see Bryan, 8). As noted above, the U.S. outbreaks in 1971 were well documented. An EIEC strain was the etiologic agent. The cheese involved was imported from France and sold under several names (Brie, Camembert, Coulommiers), but all had been made in the same way. The serovar was 0124:B17 (59). Median time for onset of symptoms was 18 h with a mean duration of 2 days. Common symptoms were diarrhea, fever, and nausea, in order of decreasing frequency. Less common symptoms included cramps, chills, vomiting, aches, and headaches. Among those who ate the soft-ripened cheese, the attack rate was 94%. A total of 107 outbreaks occurred in thirteen states and the District of Columbia. While some of the cheese contained 10^3–10^5 coliforms/g, Enterobacteriaceae at levels of 10^6–10^7/g were found (67). Imported Brie cheese was the common source of the outbreak in 1983 (see Table 20-4).

Table 20-4. Summary of some outbreaks of *E. coli* gastroenteritis from foods and other sources.

Year	Location	Vehicle food/source	No. victims/no. at risk	Toxin/ strain type	Serotype	Reference
1947	England	Salmon (apparently)	47/300		0124	38
1961	Romania	Substitute coffee drink	10/50		086:B7;H34	14
1963	Japan	Ohagi	17/31			2
1966	Japan	Vegetables	244/435			2
1967	Japan	Sushi	835/1,736		027	86
1971	14 American states	Imported cheeses	387/?	EIEC	0124:B17	59
1980	Wisconsin	Food handler	500/>3,000	LT,ST	06:H16	96
1981	Texas	Not identified	282/3,000	LT	025:H+	103
1983	4 American states	Imported Brie cheese	>15/?	ST	027:H20	99

Hemorrhagic colitis

This syndrome was first characterized in 1982 in outbreaks in the states of Michigan and Oregon affecting at least forty-seven individuals. It was associated with eating any of three sandwiches at fast-food restaurants of the same chain in both states with all sandwiches containing ground beef (80).

Of 43 patients studied, all had bloody diarrhea and severe abdominal cramps with 63% experiencing nausea, 49% vomiting, but only 7% fever. The mean incubation period was 3.8–3.9 days, and symptoms lasted for 3 to > 7 days (80). A noninvasive, nontoxigenic *E. coli* serotype 0157:H7 was recovered from nine of twelve stools collected within 4 days, and from a frozen raw beef patty (80). This rare serotype was not recovered from stools collected 7 or more days after onset of illness (101). The only previous isolation of this serotype in the United States was from a sporadic case of hemorrhagic colitis in 1975. An outbreak of hemorrhagic colitis occurred in Ottawa, Canada, in November 1982, involving 31 of 353 residents of a home for the aged. The serotype was 0157:H7, and the apparent source was kitchen-prepared food.

Among the unusual features of *E. coli* 0157:H7 are its lack of production of LT or ST; its lack of invasiveness or toxigenicity by cell culture assays, although it produces a Vero cell cytotoxin; and its site of pathology in the colon rather than the small intestines. It is more heat sensitive than salmonellae (D 60°C = 45 sec. in ground beef) but it survived in ground beef for 9 months held at −20°C with little change in numbers (18). The organism grows poorly in the temperature range 44°–45°C with no growth at 45.5°C (18), suggesting that its presence may not be detected by fecal coliform assays. It is nonfluorogenic in the MUG assay (18; see Chapter 6). Its capacity to colonize chicken cecae suggests that chickens may serve as hosts or reservoirs (3). The cytotoxin has been shown to result from lysogenization by a bacteriophage (71).

Travelers' diarrhea

E. coli is well established as one of the leading causes of acute watery diarrhea that often occurs among new arrivals in certain foreign countries. In a study of thirty-five Peace Corps volunteers during their first 5 weeks in rural Thailand, 57% developed the syndrome and 50% had evidence of infection by ETEC strains (22). The shipboard outbreak of gastroenteritis of 1976 was caused by serotype 025:K98:NM that produced only LT. Strains that produced only LT have been isolated from victims traveling in Mexico (100). EPEC and ST-producing strains have been isolated from victims.

Among other organisms associated with this syndrome are rotaviruses, Norwalk agent (virus), *Entamoeba histolytica*, *Yersinia enterocolitica*, *Giardia lamblia*, *Campylobacter jejuni/coli*, *Shigella* spp., and possibly *Aeromonas hydrophila*, *Klebsiella pneumoniae*, and *Enterobacter cloacae*. Some of these are discussed further in the next chapter.

REFERENCES

1. Adinarayanan, N., V. D. Foltz, and F. McKinley. 1965. Incidence of Salmonellae in prepared and packaged foods. *J. Infect. Dis.* 115:19–26.
2. Akahane, S. 1973. Epidemiological studies on outbreaks of food poisonings due to *Escherichia coli* 0124:K72:H-. *J. Jap. Assoc. Infect. Dis.* 47:63–76.

3. Beery, J. T., M. P. Doyle, and J. L. Schoeni. 1985. Colonization of chicken cecae by *Escherichia coli* associated with hemorrhagic colitis. *Appl. Environ. Microbiol.* 49:310–15.
4. Beloian, A., and G. C. Schlosser. 1963. Adequacy of cooking procedures for the destruction of salmonellae. *Amer. J. Pub. Hlth* 53:782–91.
5. *Bergey's manual of systematic bacteriology.* 1984. Vol. 1. Ed. N. R. Krieg. Baltimore: Williams & Wilkins.
6. Brooks, J. 1962. Alpha amylase in whole eggs and its sensitivity to pasteurization temperatures. *J. Hyg.* 60:145–51.
7. Bryan, F. L. 1977. Diseases transmitted by foods contaminated by wastewater. *J. Food Protect.* 40:45–56.
8. ———. 1979. Infections and intoxications caused by other bacteria. In *Food-borne infections and intoxications,* ed. H. Riemann and F. L. Bryan, 211–97. New York: Academic Press.
9. ———. 1981. Current trends in foodborne salmonellosis in the United States and Canada. *J. Food Protect.* 44:394–402.
10. Burgess, M. N., R. J. Bywater, C. M. Cowley, N. A. Mullan, and P. M. Newsome. 1978. Biological evaluation of a methanol-soluble, heat-stable *Escherichia coli* enterotoxin in infant mice, pigs, rabbits, and calves. *Infect. Immun.* 21:526–31.
11. Chung, K. C., and J. M. Goepfert. 1970. Growth of *Salmonella* at low pH. *J. Food Sci.* 35:326–28.
12. Cleary, T. G., J. J. Mathewson, E. Faris, and L. K. Pickering. 1985. Shiga-like cytotoxin production by enteropathogenic *Escherichia coli* serogroups. *Infect. Immun.* 47:335–37.
13. Committee on Salmonella. 1969. *An evaluation of the salmonella problem.* Publ. #1683, Nat'l Acad. Sci. Washington, D.C.
14. Costin, I. D., D. Volculescu, and V. Gorcea. 1964. An outbreak of food poisoning in adults associated with serotype 086:B7:H34. *J. Path. Microbiol.* 27:68–78.
15. Dallas, W. S., D. M. Gill, and S. Falkow. 1979. Cistrons encoding *Escherichia coli* heat-labile toxin. *J. Bacteriol.* 139:850–58.
16. D'Aoust, J.-Y., H. J. Beckers, M. Boothroyd, A. Mates, C. R. McKee, A. B. Moran, P. Sado, G. E. Spain, W. H. Sperber, P. Vassiliadis, D. E. Wagner, and C. Wiberg. 1983. ICMSF methods studies. XIV. Comparative study on recovery of *Salmonella* from refrigerated pre-enrichment and enrichment broth cultures. *J. Food Protect.* 46:391–99.
17. D'Aoust, J. Y., and H. Pivnick. 1976. Small infectious doses of *Salmonella*. *Lancet* 1:866.
18. Doyle, M. P., and J. L. Schoeni. 1984. Survival and growth characteristics of *Escherichia coli* associated with hemorrhagic colitis. *Appl. Environ. Microbiol.* 48:855–56.
19. Dreyfus, L. A., J. C. Frantz, and D. C. Robertson. 1983. Chemical properties of heat-stable enterotoxins produced by enterotoxigenic *Escherichia coli* of different host origins. *Infect. Immun.* 42:539–48.
20. Duebbert, I. E., and J. W. Peterson. 1984. Enterotoxin-induced fluid accumulation during experimental salmonellosis and cholera: Requirement for prostaglandin synthesis by intestinal cells. *Bacteriol. Proc.,* 17.
21. Duitschaever, C. L. 1977. Incidence of *Salmonella* in retailed raw cut-up chicken. *J. Food Protect.* 40:191–92.
22. Echeverria, P., N. R. Blacklow, L. B. Sanford, and G. G. Cukor. 1981. Travelers'

diarrhea among American Peace Corps volunteers in rural Thailand. *J. Infect. Dis.* 143:767–71.

23. Eidels, L., R. L. Proia, and D. A. Hart. 1983. Membrane receptors for bacterial toxins. *Microbiol. Rev.* 47:596–620.

24. Elwell, L. P., and P. L. Shipley. 1980. Plasmid-mediated factors associated with virulence of bacteria to animals. *Ann. Rev. Microbiol.* 34:465–96.

25. Evans, D. J., Jr., L. C. Chen, G. T. Curlin, and D. G. Evans. 1972. Stimulation of adenyl cyclase by *Escherichia coli* enterotoxin. *Nature* 236:137–38.

26. Ferguson, W. W., and R. C. June. 1952. Experiments on feeding adult volunteers with *Escherichia coli* 111:B4, a coliform organism associated with infant diarrhea. *Amer. J. Hyg.* 55:155–69.

27. Flowers, R. S. 1985. Comparison of rapid *Salmonella* screening methods and the conventional culture method. *Food Technol.* 39(3):103–8.

28. Frantz, J. C., L. Jaso-Friedman, and D. C. Robertson. 1984. Binding of *Escherichia coli* heat-stable enterotoxin to rat intestinal cells and brush border membranes. *Infect. Immun.* 43:622–30.

29. Goepfert, J. M., and R. A. Biggie. 1968. Heat resistance of *Salmonella typhimurium* and *Salmonella senftenberg* 775W in milk chocolate. *Appl. Microbiol.* 16:1939–40.

30. Grau, F. H., and L. E. Brownlie. 1968. Effect of some pre-slaughter treatments on the *Salmonella* population in the bovine rumen and feces. *J. Appl. Bacteriol.* 31:157–63.

31. Green, B. A., R. J. Neill, W. T. Ruyechan, and R. K. Holmes. 1983. Evidence that a new enterotoxin of *Escherichia coli* which activates adenylate cyclase in eucaryotic target cells is not plasmid mediated. *Infect. Immun.* 41:383–90.

32. Guerrant, R. L., J. M. Hughes, B. Chang, D. C. Robertson, and F. Murad. 1980. Activation of intestinal guanylate cyclase by heat-stable enterotoxin of *Escherichia coli:* Studies of tissue specificity, potential receptors, and intermediates. *J. Infect. Dis.* 142:220–28.

33. Gyles, C. L. 1974. Relationships among heat-labile enterotoxins of *Escherichia coli* and *Vibrio cholerae. J. Infect. Dis.* 129:277–83.

34. Hall, H. E., and G. H. Hauser. 1966. Examination of feces from food handlers for salmonellae, shigellae, enteropathogenic *Escherichia coli,* and *Clostridium perfringens. Appl. Microbiol.* 14:928–33.

35. Harris, J. R., I. K. Wachsmuth, B. R. Davis, and M. L. Cohen. 1982. High-molecular-weight plasmid correlates with *Escherichia coli* enteroinvasiveness. *Infect. Immun.* 37:1295–98.

36. Hartley, C. L., H. M. Clements, and K. B. Linton. 1977. *Escherichia coli* in the faecal flora of man. *J. Appl. Bacteriol.* 43:261–69.

37. Hobbs, B. C. 1961. Public health significance of *Salmonella* carriers in livestock and birds. *J. Appl. Bacteriol.* 24:340–52.

38. Hobbs, B. C., M. E. M. Thomas, and J. Taylor. 1949. School outbreak of gastroenteritis associated with a pathogenic paracolon bacillus. *Lancet* 2:530–32.

39. Holmberg, S. D., J. G. Wells, and M. L. Cohen. 1984. Animal-to-man transmission of antimicrobial-resistant *Salmonella:* Investigations of U.S. outbreaks. *Science* 225:833–35.

40. Horwitz, M. A., and E. J. Gangarosa. 1976. Foodborne disease outbreaks traced to poultry, United States, 1966–1974. *J. Milk Food Technol.* 39:859–63.

41. Horwitz, M. A., R. A. Pollard, M. H. Merson, and S. M. Martin. 1977. A large outbreak of foodborne salmonellosis on the Navajo Nation Indian Reservation, epidemiology and secondary transmission. *Amer. J. Pub. Hlth* 67:1071–76.
42. Houston, C. W., C. W. Koo, and J. W. Peterson. 1981. Characterization of *Salmonella* toxin released by mitomycin C-treated cells. *Infect. Immun.* 32:916–26.
43. Johnston, R. W., S. S. Green, J. Chiu, M. Pratt, and J. Rivera. 1982. Incidence of *Salmonella* in fresh pork sausage in 1979 compared with 1969. *J. Food Sci.* 47:1369–71.
44. Jones, G. W., and J. M. Rutter. 1972. Role of the K88 antigen in the pathogenesis of neonatal diarrhea caused by *Escherichia coli* in piglets. *Infect. Immun.* 6:918–27.
45. June, R. C., W. W. Ferguson, and M. T. Waifel. 1953. Experiments in feeding adult volunteers with *Escherichia coli* 055:B5, a coliform organism associated with infant diarrhea. *Amer. J. Hyg.* 57:222–36.
46. Juven, B. J., N. A. Cox, J. S. Bailey, J. E. Thomson, O. W. Charles, and J. V. Shutze. 1984. Survival of *Salmonella* in dry food and feed. *J. Food Protect.* 47:445–48.
47. Kampelmacher, E. H. 1963. The role of salmonellae in foodborne diseases. In *Microbiological quality of foods,* ed. L. W. Slanetz et al., 84–101. New York: Academic Press.
48. Klipstein, F. A., R. F. Engert, and R. A. Houghten. 1983. Properties of synthetically produced *Escherichia coli* heat-stable enterotoxin. *Infect. Immun.* 39:117–21.
49. Koo, F. C. W., and J. W. Peterson. 1981. The influence of nutritional factors on synthesis of *Salmonella* toxin. *J. Food Safety* 3:215–32.
50. ———. 1983. Effects of cultural conditions on the synthesis of *Salmonella* toxin. *J. Food Safety* 5:61–71.
51. ———. 1983. Cell-free extracts of *Salmonella* inhibit protein synthesis and cause cytotoxicity in eukaryotic cells. *Toxicon* 21:309–20.
52. Koo, F. C. W., J. W. Peterson, C. W. Houston, and N. C. Molina. 1984. Pathogenesis of experimental salmonellosis: Inhibition of protein synthesis by cytotoxin. *Infect. Immun.* 43:93–100.
53. Kornacki, J. L., and E. H. Marth. 1982. Foodborne illness caused by *Escherichia coli:* A review. *J. Food Protect.* 45:1051–67.
54. Koupal, L. R., and R. H. Deibel. 1975. Assay, characterization, and localization of an enterotoxin produced by *Salmonella. Infect. Immun.* 11:14–22.
55. Kunkel, S. L., and D. C. Robertson. 1979. Purification and chemical characterization of the heat-labile enterotoxin produced by enterotoxigenic *Escherichia coli. Infect. Immun.* 25:586–96.
56. Lallier, R., S. Lariviere, and S. St-Pierre. 1980. *Escherichia coli* heat-stable enterotoxin: Rapid method of purification and some characteristics of the toxin. *Infect. Immun.* 28:469–74.
57. Lerche, M. 1961. Zur Lebenfahigkeit von Salmonellabakterien in Mayonnaise und Fleischsalat. *Wien. tierarztl. Mschr.* 6:348–61.
58. Ley, F. J., B. M. Freeman, and B. C. Hobbs. 1963. The use of gamma radiation for the elimination of salmonellae from various foods. *J. Hyg.* 61:515–29.
59. Marier, R., J. G. Wells, R. C. Swanson, W. Callahan, and I. J. Mehlman. 1973. An outbreak of enteropathogenic *Escherichia coli* foodborne disease traced to imported French cheese. *Lancet* 2:1376–78.

60. Martin, W. J., and W. H. Ewing. 1969. Prevalence of serotypes of *Salmonella*. *Appl. Microbiol.* 17:111–17.

61. Matches, J. R., and J. Liston. 1968. Low temperature growth of *Salmonella*. *J. Food Sci.* 33:641–45.

62. Mazaitis, A. J., R. Maas, and W. K. Maas. 1981. Structure of a naturally occurring plasmid with genes for enterotoxin production and drug resistance. *J. Bacteriol.* 145:97–105.

63. Mead, G. C. 1982. Microbiology of poultry and game birds. In *Meat microbiology*, ed. M. H. Brown, 67–101. London: Applied Science.

64. Mehlman, I. J., M. Fishbein, S. L. Gorbach, A. C. Sanders, E. L. Eide, and J. C. Olson, Jr. 1976. Pathogenicity of *Escherichia coli* recovered from food. *J. Assoc. Off. Anal. Chem.* 59:67–80.

65. Milone, N. A., and J. A. Watson. 1970. Thermal inactivation of *Salmonella senftenberg* 775W in poultry meat. *Hlth Lab. Sci.* 7:199–225.

66. Moon, H. W., E. M. Kohler, H. A. Schneider, and S. C. Whipp. 1980. Prevalence of pilus antigens, enterotoxin types, and enteropathogenicity among K88-negative enterotoxigenic *Escherichia coli* from neonatal pigs. *Infect. Immun.* 27:222–30.

67. Mossell, D. A. A. 1974. Bacteriological safety of foods. *Lancet* 1:173.

68. Mundell, D. H., C. R. Anselmo, and R. M. Wishnow. 1976. Factors influencing heat-labile *Escherichia coli* enterotoxin activity. *Infect. Immun.* 14:383–88.

69. Nakamura, M., S. Sato, T. Ohya, S. Suzuki, and S. Ikeda. 1985. Possible relationship of a 36-megadalton *Salmonella enteritidis* plasmid to virulence in mice. *Infect. Immun.* 47:831–33.

70. Ng, H., H. G. Bayne, and J. A. Garibaldi. 1969. Heat resistance of *Salmonella:* The uniqueness of *Salmonella senftenberg* 775W. *Appl. Microbiol.* 17:78–82.

71. O'Brien, A. D., J. W. Newland, S. F. Miller, R. K. Holmes, H. W. Smith, and S. B. Formal. 1984. Shiga-like toxin-converting phages from *Escherichia coli* strains that cause hemorrhagic colitis or infantile diarrhea. *Science* 226:694–96.

72. Ørskov, I., and F. Ørskov. 1966. Episome-carried surface antigen K88 of *Escherichia coli*. I. Transmission of the determinant of the K88 antigen and influence on the transfer of chromosomal markers. *J. Bacteriol.* 91:69–75.

73. Ørskov, I., F. Ørskov, B. Jann, and K. Jann. 1977. Serology, chemistry, and genetics of O and K antigens of *Escherichia coli*. *Bacteriol. Rev.* 41:667–710.

74. Page, G. V., and M. Solberg. 1980. Nitrogen assimilation by *Salmonella typhimurium* in a chemically defined minimal medium containing nitrate, nitrite, or ammonia. *J. Food Sci.* 45:75–76 and 83.

75. Peterson, J. W. 1980. *Salmonella* toxin. *Pharm. Ther.* 11:719–24.

76. Picken, R. N., A. J. Mazaitis, W. K. Maas, M. Rey, and H. Heyneker. 1983. Nucleotide sequence of the gene for heat-stable enterotoxin II of *Escherichia coli*. *Infect. Immun.* 42:269–75.

77. Pierce, N. F. 1977. Protection against challenge with *Escherichia coli* heat-labile enterotoxin by immunization of rats with cholera toxin/toxoid. *Infect. Immun.* 18:338–41.

78. Prost, E., and H. Riemann. 1967. Food-borne salmonellosis. *Ann. Rev. Microbiol.* 21:495–528.

79. Rengel, A., and S. Mendoza. 1984. Isolation of *Salmonella* from raw chicken in Venezuela. *J. Food Protect.* 47:213–16.

80. Riley, L. W., R. S. Remis, S. D. Helgerson, H. B. McGee, J. G. Wells, B. R. Davis, R. J. Hebert, E. S. Olcott, L. M. Johnson, N. T. Hargrett, P. A.

Blake, and M. L. Cohen. 1983. Hemorrhagic colitis associated with a rare *Escherichia coli* serotype. *N. Eng. J. Med.* 308:681–85.

81. Rogers, R. E., and M. F. Gunderson. 1958. Roasting of frozen stuffed turkeys. I. Survival of *Salmonella pullorum* in inoculated stuffing. *Food Res.* 23:87–95.

82. Sack, R. B. 1975. Human diarrheal disease caused by enterotoxigenic *Escherichia coli. Ann. Rev. Microbiol.* 29:333–53.

83. Sack, R. B., D. A. Sack, I. J. Mehlman, F. Ørskov, and I. Ørskov. 1977. Enterotoxigenic *Escherichia coli* isolated from food. *J. Infect. Dis.* 135:313–17.

84. Sadler, W. W., and R. E. Corstvet. 1965. Second survey of market poultry for *Salmonella* infections. *Appl. Microbiol.* 13:348–51.

85. Saeed, A. M. K., N. S. Magnuson, N. Sriranganathan, D. Burger, and W. Cosand. 1984. Molecular homogeneity of heat-stable enterotoxins produced by bovine enterotoxigenic *Escherichia coli. Infect. Immun.* 45:242–47.

86. Sakai, S., T. Maruyama, T. Itoh, K. Saitch, and H. Zen-Yoji. 1971. An outbreak of enterocolitis ascribed to the infection of *Escherichia coli* 011:K(B):H27. *Ann. Rep. Tokyo Metropol. Res. Lab. Publ. Hlth.*

87. Sandefur, P. D., and J. W. Peterson. 1978. Isolation of skin permeability factors from culture filtrates of *Salmonella typhimurium. Infect. Immun.* 14:671–79.

88. ———. 1977. Neutralization of Salmonella toxin-induced elongation of Chinese hamster ovary cells by cholera antitoxin. *Infect. Immun.* 15:988–92.

89. Scotland, S. M., N. P. Day, A. Cravioto, L. V. Thomas, and B. Rowe. 1981. Production of heat-labile or heat-stable enterotoxins by strains of *Escherichia coli* belonging to serogroups 044, 0114, and 0128. *Infect. Immun.* 31:500–503.

90. Sedlock, D. M., L. R. Koupal, and R. H. Deibel. 1978. Production and partial purification of *Salmonella* enterotoxin. *Infect. Immun.* 20:375–80.

91. Shrimpton, D. H., J. B. Monsey, B. C. Hobbs, and M. E. Smith. 1962. A laboratory determination of the destruction of alpha amylase and salmonellae in whole egg by heat pasteurization. *J. Hyg.* 60:153–62.

92. So, M., and B. J. McCarthy. 1980. Nucleotide sequence of the bacterial transposon Tn1681 encoding a heat-stable (ST) toxin and its identification in enterotoxigenic *Escherichia coli* strains. *Proc. Natl. Acad. Sci., USA* 77:4011–15.

93. Staples, S. J., S. E. Asher, and R. A. Giannella. 1980. Purification and characterization of heat-stable enterotoxin produced by a strain of *E. coli* pathogenic for man. *J. Biol. Chem.* 255:4716–21.

94. Steele, J. H., and M. M. Galton. 1967. Epidemiology of foodborne salmonellosis. *Hlth Lab. Sci.* 4:207–12.

95. Taylor, J. 1967. *Salmonella* and salmonellosis. In *Food poisoning,* 15–32. London: Royal Soc. Hlth.

96. Taylor, W. R., W. L. Schell, J. G. Wells, K. Choi, D. E. Kinnunen, P. T. Heiser, and A. G. Helstad. 1982. A foodborne outbreak of enterotoxigenic *Escherichia coli* diarrhea. *N. Eng. J. Med.* 306:1093–95.

97. Thomason, B. M., W. B. Cherry, and D. J. Dodd. 1977. Salmonellae in health foods. *Appl. Environ. Microbiol.* 34:602–3.

98. Troller, J. A. 1976. *Salmonella* and *Shigella.* In *Food microbiology: Public health and spoilage aspects,* ed. M. P. deFigueiredo and D. F. Splittstoesser, 129–55. Westport, Conn.: AVI.

99. United States Centers for Disease Control. 1984. Update: Gastrointestinal

illness associated with imported semi-soft cheese. *Morb. Mort. W. Rep.* 33:16 and 22.

100. Wachsmuth, K., J. Wells, P. Shipley, and R. Ryder. 1979. Heat-labile enterotoxin production in isolates from a shipboard outbreak of human diarrheal illness. *Infect. Immun.* 24:793–97.

101. Wells, J. G., B. R. Davis, K. Wachsmuth, L. W. Riley, R. S. Remis, R. Sokolow, and G. K. Morris. 1983. Laboratory investigation of hemorrhagic colitis outbreaks associated with a rare *Escherichia coli* serotype. *J. Clin. Microbiol.* 18:512–20.

102. Wentworth, F. H., D. W. Broek, C. S. Stulberg, and R. H. Page. 1956. Clinical bacteriological and serological observations of two human volunteers following ingestion of *E. coli* 0127:B8. *Proc. Soc. Exptl. Biol. Med.* 91:586–88.

103. Wood, L. V., W. H. Wolfe, G. Ruiz-Palacios, W. S. Foshee, L. I. Gorman, F. McCleskey, J. A. Wright, and H. L. DuPont. 1983. An outbreak of gastroenteritis due to a heat-labile enterotoxin-producing strain of *Escherichia coli*. *Infect. Immun.* 41:931–34.

21.

FOOD-BORNE GASTROENTERITIS
CAUSED BY *VIBRIO, YERSINIA,*
AND *CAMPYLOBACTER* SPECIES

VIBRIO PARAHAEMOLYTICUS

While most other known food-poisoning syndromes may be contracted from a variety of foods, *V. parahaemolyticus* gastroenteritis is contracted almost solely from seafood. When other foods are involved, they represent cross contamination from seafood products. Another unique feature of this syndrome is the natural habit of the etiologic agent—the sea. In addition to its role in gastroenteritis, *V. parahaemolyticus* is known to cause extraintestinal infections in man.

The genus *Vibrio* consists of at least twenty-eight species, and three that are often associated with *V. parahaemolyticus* in aquatic environments and seafood are *V. vulnificus, V. alginolyticus,* and *V. cholerae.* Some of the distinguishing features of these species are noted in Table 21-1, and the syndromes caused by each are described below. Thorough reviews of these and related organisms have been provided by Colwell (28) and Joseph et al. (52).

V. parahaemolyticus is common in oceanic and coastal waters. Its detection is related to water temperatures, with numbers of organisms being undetectable until water temperature rises to around 19°–20°C. A study of the Rhode River area of the Chesapeake Bay showed that the organisms survive in sediment during the winter and later are released into the water column, where they associate with the zooplankton from April to early June (53). In ocean waters, they tend to be associated more with shellfish than with other forms (66). They have been demonstrated to adsorb onto chitin particles and copepods, whereas organisms such as *Escherichia coli* and *Pseudomonas fluorescens* do not (53). This species is generally not found in the open

Table 21-1. **Some colonial and biochemical differences between** *V. parahaemolyticus* **and three other** *Vibrio* **spp. (8).**

Species	*V. parahaemolyticus*	*V. alginolyticus*	*V. vulnificus*	*V. cholerae*
Lateral flagella on solid media	+	+	−	−
Rod shape	S	S	C	d
VP	−	+[a]	−	v
Growth in 10% NaCl	−	+	−	−
Growth in 6% NaCl	+	+	+	−
Swarming	−	+	−	−
Production of acetoin/diacetyl	−	+	−	+
Sucrose	−	+	−	+
Cellobiose	−	−	+	−
Utilization of putrescine	+	d	−	−
Color on TCBS agar	G	Y	G	Y

[a]24 h.
S = straight
C = curved
G = green
Y = yellow
d = 11 to 90% of strains positive.
v = variable; strain instability.

oceans (52), and it cannot tolerate the hydrostatic pressures of ocean depths (96).

Growth conditions

V. parahaemolyticus can grow in the presence of 1–8% NaCl, with best growth occurring in the 2–4% range (95). It dies off in distilled water (11). It does not grow at 4°C, but growth between 5° and 9°C has been demonstrated at pH 7.2–7.3 and 3% NaCl, or at pH 7.6 and 7% NaCl (Table 21-2). Its growth at 9.5–10°C in food products has been demonstrated (11), although the minimum for growth in open waters has been found to be 10°C (53). The upper growth temperature is 44°C, with an optimum between 30° and 35°C (96). Growth has been observed over the pH range 4.8–11.0, with 7.6–8.6 being optimum (9, 11, 96). It may be noted from Table 21-2 that the minimum growth pH is related to temperature and NaCl content, with moderate growth of one strain observed at pH 4.8 when the temperature was 30°C and NaCl content was 3%; but minimum pH was 5.2 when NaCl content was 7% (9). Similar results were found for five other strains. Under optimal conditions, this organism has a generation time of 9–13 min (compared to about 20 min for *E. coli*). Optimum a_w for growth corresponding to shortest generation time was found to be 0.992 (2.9% NaCl in tryptic soy

Table 21-2. **Minimum pH of growth of** *V.*
parahaemolyticus **ATCC 107914 in TSB with**
3% and 7% NaCl at different temperatures
(9).

	Minimum pH at NaCl conc.	
Temp. (°C)	3%	7%
5	7.3	7.6
9	7.2	7.1
13	5.2	6.0
21	4.9	5.3
30	4.8	5.2

broth). Employing the latter medium at 29°C and various solutes to control a_w, minimum values were 0.937 (glycerol), 0.945 (KCl), 0.948 (NaCl), 0.957 (sucrose), 0.983 (glucose), and 0.986 with propylene glycol (10). The organism is heat sensitive, with D47°C values ranging from 0.8 to 65.1 min having been reported (12). With one strain, destruction of 500 cells/ml in shrimp homogenates was achieved at 60°C in 1 min, but with 2×10^5 cells/ml some survived 80°C for 15 min (127). Cells are most heat resistant when grown at high temperatures in the presence of about 7% NaCl.

Virulence properties

The most widely used in vitro test of potential virulence for *V. parahae-molyticus* is the Kanagawa reaction, with most all virulent strains being positive (K^+) and most avirulent strains being negative (K^-). About 1% of sea isolates and about 100% of those from gastroenteritis patients are K^+ (92). K^+ strains produce a thermostable direct hemolysin (toxin), K^- strains produce a heat-labile hemolysin, and some strains produce both. The Kanagawa reaction is determined generally by use in human red blood cells in Wagatsuma's agar medium. In addition to human red blood cells, those of the dog and rat are lysed; those of the rabbit and sheep give weak reactions; and those of the horse are not lysed (96). To determine the K reaction, the culture is surface plated, incubated at 37°C for 18–24 h, and read for the presence of β-hemolysis. Of 2,720 *V. parahaemolyticus* isolates from diarrheal patients, 96% were K^+ whereas only 1% of 650 fish isolates were K^+ (23). In general, isolates from waters are K^-.

The thermostable direct hemolysin has a molecular weight of 42,000 daltons and is a cardiotrophic, cytotoxic protein that is lethal to mice (48) and induces a positive response in the rabbit ileal loop assay (see Chapter 7). Its mean mouse LD_{50} by intraperitoneal (IP) injection is 1.5 μg, and the rabbit ileal loop dose is 200 μg (132). The hemolysin is under pH control and was found to be produced only when pH is 5.5–5.6 (27). That K^+ hemolysin may aid cells in obtaining iron stems from the observation that lysed erythrocyte extracts enhanced the virulence of the organism for mice (58). The membrane receptors of the thermostable direct hemolysin are gangliosides G_{T1} and G_{D1a}, with the former binding hemolysin more firmly than the latter (113). The resistance of horse erythrocytes to the hemolysin apparently is due to their absence of these gangliosides (113).

A synthetic medium has been developed for the production of both the thermostable direct and the heat-labile hemolysins, and serine and glutamic acid were found to be indispensable (57). The heat stability of the thermostable direct hemolysin is such that it can remain in foods after its production. In Tris buffer at pH 7.0, D120°C and D130°C values of 34 and 13 min, respectively, were found for semipurified toxin; whereas in shrimp D_{120} and D_{130} values were 21.9 and 10.4 min, respectively (21). Hemolysin was detected when cell counts reached 10^6/g, and its heat resistance was greater at pH 5.5–6.5 than at 7.0–8.0 (21).

The thermostable direct hemolysin gene (tdh) is chromosomal (114), and it has been cloned in E. coli (56, 114). When the tdh gene was introduced into a K^- strain, it produced extracellular hemolysin (77). The nucleotide sequence of the tdh gene has been determined (77), and a specific tdh gene probe constructed, which consists of a 406-base pair (76). Employing this probe, 141 V. parahaemolyticus strains were tested. All K^+ strains were gene positive—86% of them were weak positives—and 16% of K^- strains reacted with the probe. All gene-positive strains produced the thermostable direct hemolysin as assessed by an ELISA method. Of 129 other vibrios tested with the gene probe including 19 named Vibrio spp., only V. hollisae was positive (76). The transfer of R-plasmids from E. coli to V. parahaemolyticus has been demonstrated (40).

At least twelve O-antigens and fifty-nine K-antigens have been identified, but no correlations have been made between these and K^+ and K^- strains, and the value of serotyping as an epidemiological aid has been minimal.

Since not all K^+ strains produce positive responses in the rabbit ileal loop assay (22), and because some K^- strains are associated with gastroenteritis and sometimes are the only strains isolated (116), the precise virulence mechanisms of this organism are not known. The thermostable direct hemolysin undoubtedly is responsible for the deaths that occur, but it plays no role in the diarrhea that characterizes the syndrome (49). While it has been shown to penetrate the intestinal epithelium of suckling rabbits, it is unclear whether this occurs in adult animals (24). Twedt et al. (116) have noted that pathogenicity depends upon something other than the thermostable direct toxin, but the identity of that factor(s) is not known at this time.

Gastroenteritis syndrome and vehicle foods

The identity of V. parahaemolyticus as a food-borne gastroenteritis agent was made first by Fujino in 1951 (35). While the incidence of this illness is rather low in the United States and some European countries, it is the leading cause of food poisoning in Japan, accounting for 24% of bacterial food poisoning for 1965–74 (14, 95). In 1973–75, three outbreaks were recorded in the United States, with 224 cases. A synopsis of some outbreaks is presented in Table 21-3. Among the more than 81,000 cases reported during this period, thirty-one deaths were recorded (14).

With regard to symptomatology, findings from the Louisiana outbreak (noted in Table 21-3) illustrate the typical features. The mean incubation period in that outbreak was 16.7 h, with a range of 3–76 h. The symptoms lasted from 1 to 8 days, with a mean of about 4.6 days. Symptoms (along with percent incidence of each) were diarrhea (95), cramps (92), weakness (90), nausea (72), chills (55), headache (48), and vomiting (12). Both sexes were equally affected, the age of victims ranging from 13 to 78 years.

No illness occurred among fourteen volunteers who ingested $> 10^9$ cells, but illness did occur in one person from the accidental ingestion of $\sim 10^7$ K^+ cells (95). In another study, 2×10^5 to 3×10^7 K^+ cells produced symptoms in volunteers, whereas 10^{10} cells of K^- strains did not (96, 116). As noted above, some K^- strains have been associated with outbreaks (7, 95).

Table 21-3. Synopsis of some gastroenteritis outbreaks caused by *V. parahaemolyticus.*

Year	Location/incidence/comments	Reference
1951	Japan. The outbreak occurred following consumption of *shirasu* (boiled and semidried young sardines). There were 272 victims, including 20 deaths.	(95)
1956	Japan. Salted cucumber was involved in a hospital outbreak with 120 victims.	(95)
1960	Japan. An explosive outbreak affecting thousands of people occurred along the Pacific coast. Horse mackerel was thought to be the vehicle.	(95)
1971	Maryland. Steamed crabs and crab salad were incriminated in 3 outbreaks in which 425 of about 745 people became ill. Isolations from food and victims revealed K^+ strain 04:K11 as the causative agent. This was first documented outbreak in America.	(72)
1974	Mexico. K^+ strains were recovered from two victims who ate raw fish.	(37)
1976	Guam. Some 122 cases developed from eating octopus aboard a ship.	(118)
1976	Louisiana. Some 100 cases resulted from eating boiled shrimp at a picnic.	(118)
1978	Louisiana. Boiled shrimp was the vehicle food in this outbreak involving about 67% of 1,700 persons at risk. K^+ strains were recovered from leftover shrimp and other foods as well as from the stools of 7 of 15 victims.	(121)

Vehicle foods for outbreaks are seafood, such as oysters, shrimps, crabs, lobsters, clams, and related shellfish. Cross contamination may lead to other foods as vehicles.

For the recovery and enumeration of *V. parahaemolyticus* from foods, an appropriate reference from Table 5-1 should be consulted.

OTHER VIBRIOS

Vibrio cholerae

The *V. cholerae* that causes epidemic cholera belongs to serovar 0 Group 1. The strains of *V. cholerae* that are biochemically similar to the epidemic strains but that do not agglutinate in *V. cholerae* 0 Group 1 antiserum and are not associated with the epidemic disease are referred to as non-01 or nonagglutinating vibrios (NAGs). The non-01 strains have been shown to cause gastroenteritis, soft-tissue infections, and septicemia in man. As in the case of *V. parahaemolyticus*, they are found in brackish surface waters during the warm-weather months. Kaper et al. (55) consider them to be autochthonous (indigenous) estuarine bacterial species in Chesapeake Bay waters.

Among the earliest information linking non-01 *V. cholerae* to gastroenteritis in the United States are findings from twenty-six of twenty-eight patients with acute diarrheal illness between 1972 and 1975. While some had systemic infections, 50% of the twenty-eight yielded non-cholera vibrios from stools and no other pathogens (51). In another retrospective study of non-01 *V. cholerae* cultures submitted to the U.S. Centers for Disease Control (CDC) in 1979, nine were from domestically acquired cases of gastroenteritis and each patient had eaten raw oysters within 72 h of symptoms (73). One of these isolates produced a heat-labile toxin, whereas none produced heat-stable toxins.

At least five documented gastroenteritis outbreaks of non-01 *V. cholerae* have been reported. Those in Czechoslovakia and Australia (1965 and 1973, respectively) were traced to potatoes and to egg and asparagus salads, and practically all victims experienced diarrhea. The third outbreak occurred in the Sudan, and well water was the source. Incubation periods from these three outbreaks ranged from 5 h to 4 days (see Blake et al., 13). The fourth outbreak occurred in Florida in 1979 and involved eleven persons who ate raw oysters. Eight experienced diarrheal illness within 48 h after eating oysters, and the other three developed symptoms 12, 15, and 30 h after eating. The fifth outbreak, which occurred in 1980 mainly among U.S. soldiers in Venice, Italy, was traced to raw oysters. Of about fifty persons at risk, twenty-four developed gastroenteritis. The mean incubation period was 21.5 h with a range of 0.5 h to 5 days; and the symptoms (and percent complaining) were diarrhea (91.7), abdominal pain (50), cramps (45.8), nausea (41.7), vomiting (29.2), and dizziness (20.8). All victims recovered in 1–5 days and non-01 strains were recovered from the stools of four.

With regard to distribution, non-01 strains of *V. cholerae* have been found in the Orient and Mexico in stools of diarrheal patients along with enteropathogenic *E. coli*. *V. cholerae* non-01 was isolated from 385 persons with diarrhea in Mexico City in 1966–67 (14). The organism causes diarrhea by an enterotoxin similar to cholera toxin (CT) and perhaps by other mechanisms (see below). A strain isolated from a patient produced a CT-like toxin, positive responses in the Chinese hamster ovary (CHO) and skin permeability assays, was invasive in rabbit ileal mucosa, but was Sereny-test negative (91).

From Chesapeake Bay, sixty-five non-01s were isolated in one study (55). Throughout the year their numbers in waters were generally low, from 1 to 10 cells/l. They were found only in areas where salinity ranged between 4 and 17%. Their presence was not correlated with fecal *E. coli,* whereas the presence of the latter did correlate with *Salmonella* (55). Of those examined, 87% produced positive responses in Y-1 adrenal, rabbit ileal loop, and mouse lethality assays. Investigations conducted on waters along the Texas, Louisiana, and Florida coasts reveal that both 01 and non-01 *V. cholerae* are fairly common. Of 150 water samples collected along a Florida estuary, 57% were positive for *V. cholerae* (29). Of 753 isolates examined, 20 were 01 and 733 were non-01 types. Of the 20 01 strains, 8 were Ogawa and 12 were Inaba serovars, and they were found primarily at a sewage treatment plant. The highest numbers of both 01 and non-01 strains occurred in August and November (29). Neither the fecal coliform nor the total coliform index was an adequate indicator of the presence of *V. cholerae,* but the former was more useful than the latter. Along the Santa Cruz, California, coast, the highest numbers of non-01s occurred during the summer months and were associated with high coliform counts (59).

Non-01 strains produce a cytotoxin and a hemolysin with a molecular weight of 60,000, which is immunologically related to the hemolysin of the El Tor strain (131). The role of this hemolysin in gastroenteritis is unclear at this time.

The prevalence of non-01 *V. cholerae* in seafood is not known, nor is their true significance as food-borne gastroenteritis agents. Unlike the other vibrios, they do not require NaCl for growth and their isolation on ordinary culture media is possible.

Vibrio vulnificus

This organism is found in seawater and seafood. It is isolated more often from oysters and clams than from crustacean shellfish products. It has been isolated from seawater from the coast of Miami, Florida, to Cape Cod, Massachusetts, with most (84%) isolated from clams (80). Upon injection into mice, 82% of tested strains were lethal. Like *V. alginolyticus* (see below), it causes soft-tissue infections and primary septicaemia in man. The strains are highly invasive and produce hemolysin and a cytotoxin (see

Colwell, 28). Infections have been seen in the United States, Japan, and Belgium. Of those seen in the United States, all but three occurred between May and October, and most patients were males over 40 years of age. This organism is believed to be a significant pathogen in individuals with higher than normal levels of iron (as, for example, in hepatitis and chronic cirrhosis), although its virulence is not explained entirely by its capacity to sequester iron.

Vibrio alginolyticus

This species is a normal inhabitant of seawater and has been found to cause soft-tissue and ear infections in man. Human pathogenicity was first confirmed in 1973, although its possible role in wound infections was noted by Twedt et al. in 1969 (117). Extraintestinal infections have been reported from several countries including the United States. Wound infections occur usually on body extremities, with most patients being males with a history of exposure to seawater.

The prevalence of V. alginolyticus along with that of V. parahaemolyticus was studied in coastal waters of the state of Washington by Baross and Liston (6), who found higher numbers of organisms in invertebrates and sediment samples than in water where the numbers were quite variable. Numbers found in oysters correlated with the temperature of overlying waters, with highest numbers associated with warmer waters. Whether or not this organism is of any significance as a food-borne pathogen is not clear at this time.

For more information on vibrio infections, see the reviews by Blake et al. (14), Colwell (28), and Joseph et al. (52).

YERSINIA ENTEROCOLITICA

In the genus Yersinia, which belongs to the family Enterobacteriaceae, seven species and five biovars are recognized, including Y. pestis, the cause of plague (8). The species of primary interest in foods is Y. enterocolitica. First isolated in New York state in 1933 by M. B. Coleman (47), this gram-negative rod is somewhat unique in that it is motile below 30°C but not at 37°C. It produces colonies of 1.0 mm or less on nutrient agar, is oxidase negative, ferments glucose with little or no gas, lacks phenylalanine deaminase, is urease positive, and is unique as a pathogen in being psychrotrophic. It is often present in the environment with at least three other yersiniae noted in Table 21-4 (see below). The early history of Y. enterocolitica was reviewed by Bottone (18). General reviews relative to foods have been published by Stern and Pierson (110), Swaminathan et al. (112), and Zink et al. (135).

Growth requirements

Growth of Y. enterocolitica has been observed over the temperature range of $-2°-45°C$, with an optimum between 22° and 29°C. For biochemical

Table 21-4.   Species of *Yersinia* associated with *Y. enterocolitica* in the environment and in foods, and minimum biochemical differences between them.

Species	VP	Sucrose	Rhamnose	Raffinose	Melibiose
Y. enterocolitica	+	+	−	−	−
Y. kristensenii	−	−	−	−	−
Y. frederikensii	+	+	+	−	−
Y. intermedia	+	+	+	+	+

VP = Voges-Proskauer reaction
+ = positive reaction
− = negative reaction

reactions, 29°C appears to be the optimum. The upper limit for growth of some strains is 40°C, and not all grow below 4°–5°C. Growth at 0°–2°C in milk after 20 days has been observed. Growth at 0°–1°C on pork and chicken has been observed (64), and three strains were found to grow on raw beef held for 10 days at 0°–1°C (43). The addition of NaCl to growth media raises the minimum growth temperature. In brain heart infusion (BHI) broth containing 7% NaCl, growth did not occur at 3° or 25°C after 10 days. At pH 7.2, growth of one strain was observed at 3°C and very slight growth at pH 9.0 at the same temperature, while no growth occurred at pH 4.6 and 9.6 (111). Although 7% NaCl was inhibitory at 3°C, growth occurred at 5% NaCl. With no salt, growth was observed at 3°C over the pH range 4.6–9.0 (111, 112). Clinical strains were less affected by these parameters than were environment isolates.

Y. enterocolitica is destroyed in 1–3 min at 60°C (42). It is rather resistant to freezing, with numbers decreasing only slightly in chicken after 90 days at − 18°C (64). The calculated $D62.8°C$ for twenty-one strains in milk ranged from 0.7 to 17.8 seconds, and none survived pasteurization (34).

Distribution

Y. enterocolitica and the related species noted in Table 21-4 are widely distributed in the terrestrial environment and in lake, well, and stream waters, which are sources of the organisms to warm-blooded animals. It is more animal adapted and is found more often among human isolates than the other species in Table 21-4. Of 149 strains of human origin, 81, 12, 5.4, and 2% were, respectively, *Y. enterocolitica, Yersinia intermedia, Yersinia frederiksenii,* and *Yersinia kristensenii* (100). *Y. intermedia* and *Y. frederiksenii* are found mainly in fresh waters, fish, foods, and only occasionally are isolated from man. *Y. kristensenii* is found mainly in soils and other environmental samples as well as in foods but rarely isolated from man (8). Like *Y. enterocolitica,* this species produces a heat-stable enterotoxin. Many of the *Y. enterocolitica*-like isolates of Hanna et al. (44) were rhamnose-positive and consequently are classified as *Y. intermedia* and/or *Y. frederiksenii;* and all grow at 4°C. Rhamnose-positive yersiniae are not known to cause infections in man (63).

Animals from which *Y. enterocolitica* has been isolated include cats, birds, dogs, beavers, guinea pigs, rats, camels, horses, chickens, raccoons, chinchillas, deer, cattle, swine, lambs, fish, and oysters. It is widely believed that swine constitutes the single most common source of *Y. enterocolitica* in humans. Of forty-three samples of pork obtained from a slaughterhouse and examined for *Y. enterocolitica, Y. intermedia, Y. kristensenii,* and *Y. frederiksenii,* eight were positive and all four species were found (45). Along with *Klebsiella pneumoniae, Y. enterocolitica* was recovered from crabs collected near Kodiak Island, Alaska, and was shown to be pathogenic (33).

Serovars and biovars

The most commonly occurring *Y. enterocolitica* serovars (serotypes) in human infections are 0:3, 0:5,27, 0:8, and 0:9. Each of forty-nine isolates belonging to these serovars produced a positive HeLa cell response while only five of thirty-nine other serovars were positive (74). Most pathogenic strains in the United States are 0:8 (biovars 2 and 3); and except for occasional isolations in Canada, it is rarely reported from other continents. In Canada, Africa, Europe, and Japan, serovar 0:3 (biovar 4) is the most common (115). The second most common in Europe and Africa is 0:9, which has been reported also from Japan. Serovar 0:3 (biovar 4, phage type 9b) was practically the only type found in the province of Quebec, Canada, and it was predominant in Ontario (115). The next most common were 0:5,27 and 0:6,30. From human infections in Canada, 0:3 represented 85% of 256 isolates, whereas for nonhuman sources, 0:5,27 represented 27% of 22 isolates (115). Six isolates of 0:8 recovered from porcine tongues were lethal to adult mice (31), and only 0:8 was found by Mora and Pai (74) to be Sereny positive. Employing HeLa cells, the following serovars were found to be infective: 0:1, 0:2, 0:3, 0:4, 0:5, 0:8, 0:9, and 0:21 (98). Serovar 0:8 strains are not only virulent in man but they possess mice lethality and invasiveness by the Sereny test. The four most common biovars of *Y. enterocolitica* are indicated in Table 21-5. It appears that only biovars 2, 3, and 4 carry the virulence plasmid (62).

Table 21-5. The four most common biovars of *Y. enterocolitica*.

Substrate/product	Biovars			
	1	*2*	*3*	*4*
Lipase (Tween 80)	+	−	−	−
DNAse	−	−	−	+
Indole	+	+	−	−
D-xylose	+	+	+	−

Virulence factors

Y. enterocolitica produces a heat-stable (ST) enterotoxin that survives 100°C for 20 min. It is not affected by proteases and lipases and has a molecular weight of 9,000–9,700 daltons, and biological activity is lost upon treatment with 2-mercaptoethanol (78, 79). When subjected to isoelectric focusing, two active fractions with pI's of 3.29 (ST-1) and 3.00 (ST-2) have been found (78). Antiserum from guinea pigs immunized with the purified ST neutralized the activity of *Y. enterocolitica* ST and *E. coli* ST (78). Like *E. coli* ST, it elicits positive responses in suckling mice and rabbit ileal loop assays, and negative responses in the CHO and Y-1 adrenal cell assays (see Chapter 7). It is methanol soluble (19), and stimulates guanylate cyclase and the cAMP response in intestines but not adenylate cyclase (79, 92). It is produced only at or below 30°C (84), and its production is favored in the pH range 7–8. Of forty-six milk isolates, only three produced ST in milk at 25°C and none at 4°C (34). At 25°C, > 24 h were required for ST production. It appears to be chromosomal rather than plasmid mediated (63).

In a study of 232 human isolates, 94% produced enterotoxin, while only 32% of 44 from raw milk and 18% of 55 from other foods were enterotoxigenic (85). Of the serovars 0:3, 0:8, 0:5,27, 0:6,30, and 0:9, 97% of 196 were enterotoxigenic. Ninety percent of the rhamnose-positive strains studied by Pai et al. (85) produced enterotoxin, indicating that not all isolates were *Y. enterocolitica* and that some of the other species produce enterotoxin, as previously noted. It has been found that most natural waters in the United States contain rhamnose-positive strains that are either serologically untypable or react with multiple serovars (47). In another study, forty-three strains of *Y. enterocolitica* from children with gastroenteritis and eighteen laboratory strains were examined for ST production and all clinical and seven laboratory strains produced ST as assessed by the infant mouse assay, and all were negative in the Y-1 adrenal cell assay (84).

Although pathogenic strains of *Y. enterocolitica* produce ST, it appears that this agent is not critical to virulence. Some evidence for the lack of importance of ST was provided by Schiemann (97), who demonstrated positive HeLa-cell and Sereny-test responses, with a 0:3 strain that did not produce enterotoxin. On the other hand, each of forty-nine isolates belonging to serovar 0:3 and the other four virulent serovars produced ST (74). Virulence appears to be a result of tissue invasiveness for this organism. The latter has been shown to be mediated by a 40–48-Mdal plasmid (134). The 44-Mdal plasmid of *Y. enterocolitica* and a 47-Mdal plasmid of a *Y. pestis* strain have been shown to share 55% DNA sequence homology over about 80% of the plasmid genomes (86). In addition to tissue invasiveness, the 40–50-Mdal plasmids are responsible for calcium-dependent growth at 37°C, autoagglutination in tissue culture medium, adult mouse lethality for serovar 0:8 strains, suckling mouse lethality, HEp-2 cell adherence, and adherence in at least three outer membrane proteins (46, 92). These plasmids appear not to be responsible for enterotoxigenicity, HEp-2 cell invasiveness, or expression of fimbrial proteins. Other plasmids exist in yersiniae ranging from 3 to 36 Mdal, but they are not virulence associated (63). Of ten strains representing six serovars that contained 42–44-Mdal plasmids, all were lethal to suckling mice, whereas those without these plasmids were not (4). The feeding of virulence plasmid–bearing strains to thirst-stressed mice was found to be lethal, while plasmidless strains had no effect on mice (63). The same plasmid is responsible for other virulence-associated properties of this organism, including autoagglutination (62), calcium-dependent growth (36), production of V and W antigens (26), and serum resistance (83). The V and W antigens of 0:8 (biovar 2) were immunologically identical to those of *Y. pestis* and *Yersinia pseudotuberculosis* (26). The serum resistance factors encoded for by the virulence plasmid consist of outer membrane proteins synthesized when cells are grown at 37°C but not at 25°C (70, 83). The latter cells adhered more to Henle monolayers than the 37°C-grown, thus making the latter more resistant to serum killing.

Although its role in virulence is unclear, some strains of *Y. enterocolitica* have been shown to produce a broad-spectrum, mannose-resistant adhesion at 20°C that agglutinated erythrocytes of at least ten animal species (68). The hemagglutination is associated with fimbriae, which were not produced when cells were cultured at 37°C. Of 21 0:3 and 0:8 serovars, 7 produced the agglutinating fimbriae, while only 46 of 115 from a variety of sources did.

To determine which of three in vitro tests best correlated with virulence, thirty-four strains were tested for calcium dependency, autoagglutination, and the presence of 40–48-Mdal plasmids. With Ca^{2+} dependency, twenty-nine of thirty-one strains were positive, whereas all of thirty-four were positive by the other two tests (106). These authors favored autoagglutination in tissue culture medium at 35°C as being perhaps the single best test. Prpic et al. (88) also found autoagglutination to be the best in vitro method, followed by Ca^{2+} dependency. In addition to *Y. enterocolitica* strains, virulent *Y. pestis* and *Y. pseudotuberculosis* autoagglutinate in tissue culture medium, whereas avirulent strains do not (62). Schiemann and Devenish (98) have suggested that the two most important factors involved in virulence of this organism are the presence of V and W antigens and the presence of an invasive factor demonstrable by HeLa cell infectivity, but general support of this position is wanting.

When iron-dextran was administered IP to mice, the median lethal dose of *Y. enterocolitica* serovars 0:3 and 0:9 was reduced by about tenfold, while Desferal (desferrioxamine B mesylate) reduced the lethal dose > 100,000-fold (90). The 0:8 strains were less affected by these compounds, suggesting their lower requirement for iron.

Incidence of *Y. enterocolitica* in foods

This organism has been isolated from cakes, vacuum-packaged meats, seafood, vegetables, milk, and other food products. It has been isolated also from beef, lamb, and pork (64). Of all sources, swine appears to be the major source of strains pathogenic for man.

From thirty-one porcine tongues from freshly slaughtered animals, twenty-one strains were isolated and represented six serovars with 0:8 the most common and 0:6,30 the second most commonly isolated (31). The other serovars recovered were 0:3, 0:13,7, 0:18, and 0:46. Of one hundred milk samples examined in the United States, twelve raw and one pasteurized yielded *Y. enterocolitica* (75). In eastern France, 81% of seventy-five samples of raw milk contained *Y. enterocolitica* following enrichment, with serovar 0:5 being the most predominant (128). In Australia, thirty-five isolates were recovered from raw goat's milk, with 71% being rhamnose positive (50).

Gastroenteritis syndrome and incidence

In addition to gastroenteritis, this organism has been associated with human pseudoappendicitis, mesenteric lymphadenitis, terminal ileitis, reactive arthritis, peritonitis, colon and neck abscesses, cholecystis, and erythema nodosum. It has been recovered from urine, blood, cerebrospinal fluid, and the eyes of infected individuals. It is, of course, recovered from the stools of gastroenteritis victims. Only the gastroenteritis syndrome is addressed below.

There is a seasonal incidence associated with this syndrome, with the fewest outbreaks occurring during the spring and the greatest number in October and November. The incidence is highest in the very young and the old. In an outbreak studied by Gutman et al. (41), the symptoms (and percent complaining of them) were fever (87), diarrhea (69), severe abdominal pain (62), vomiting (56), pharyngitis (31), and headache (18). The outbreak led to two appendectomies and two deaths.

A synopsis of some outbreaks in which *Y. enterocolitica* was shown or suspected as being the etiologic agent is presented in Table 21-6. Milk (raw, improperly pasteurized, or recontaminated) was the vehicle food in most. The first documented outbreak in the United States occurred in 1976 in New York state, with serovar 0:8 as the responsible strain, and chocolate milk prepared by adding chocolate syrup to previously pasteurized milk was the vehicle food (13).

Symptoms of the gastroenteritis syndrome develop several days following ingestion of contaminated foods, and are characterized by abdominal pain and diarrhea as noted above. Children appear to be more susceptible than adults, and the organisms may be present in stools for up to 40 days following illness (3). As noted above, a variety of systemic involvements may occur as a consequence of the gastroenteritis syndrome.

CAMPYLOBACTER JEJUNI

The genus *Campylobacter* is composed of organisms once classified as *Vibrio* spp. and known primarily to veterinary microbiologists until recent years. The genus contains at least eight species, and the interest of veter-

Table 21-6. Synopsis of some outbreaks of yersiniosis.

Year	Location	Vehicle	Characteristics of outbreaks	Serovar	Reference
1972	Japan	Unknown	47% of 182 school children and 1 teacher infected	0:3	(3)
1972	Japan	Unknown	53% of 993 children and adults at a primary school affected	0:3	(3)
1972	Japan	Unknown	198 of 1,086 junior high school pupils infected		(133)
1972	North Carolina	Dog/puppies	16 of 21 persons in 4 families were infected		(41)
1975	North Carolina	Food handler	Two common-source outbreaks occurred in nursery schools. The children also ate snow covered with maple syrup		(81)
1975	Montreal	Raw milk	57 elementary school children and 1 adult infected. Serovar 0:5,27 was recovered from milk; 0:6,30 from victims		(25)
1976	New York state	Chocolate milk	Some 218 school children were infected. Chocolate syrup added to pasteurized milk in open vat was the apparent vehicle	0:8	(13)
1981	New York state	Powdered milk and chow mein	About 35% of 455 teenage campers and staff were infected from dissolved powdered milk and/or chow mein. Five had appendectomies.	0:8	(101)
1982	Washington state	Commercial tofu	Of the 87 victims, 56 had positive stools. Water used in processing was the apparent source of the organisms	0:8, 0:Tacoma	(4)
1982	Connecticut	Pasteurized milk	53 of 300 became victims with the attack rate being greatest among the 6- to 13-year-olds. Twenty of 52 stools were positive for the serovar noted.	0:8	(2)
1982	Tennessee, Arkansas, Mississippi	Pasteurized mik	More than 172 victims resulted from milk pasteurized in Tennessee. Seventeen patients underwent appendectomies, and 41% of victims were under age 5.	0:13,0:18	(125)

inarians in this group stems from the role of these organisms in spontaneous abortions in cattle and sheep and other animal pathologies. This bacterium is microaerophilic, requiring small amounts of oxygen (3–6%) for growth. Growth is actually inhibited in 21% oxygen. Carbon dioxide (about 10%) is required for good growth. Its metabolism is respiratory. In addition to *C. jejuni, C. coli* and *C. intestinalis* are infectious for man. More detailed information on the campylobacters can be found in Smibert (103).

The species of interest in foods is *C. jejuni,* and to a lesser extent *C. coli,* which, like all members of the genus, are slender, spirally curved rods that possess a single polar flagellum at one or both ends of the cell. They are oxidase and catalase positive, will not grow in the presence of 3.5% NaCl, do not grow at 25° but do at 42°C, and are inhibited by nalidixic acid. *C. jejuni* will hydrolyze hippurate (8). They grow best at 42°C but grow slower than the Enterobacteriaceae. Because of their small cell size, they can be separated from most other gram-negative bacteria by use of a 0.65-μm filter (16). *C. jejuni* is heat sensitive, with D55°C for a composite of equal numbers of five strains being 1.09 min in peptone and 2.25 min in ground, autoclaved chicken (15). With internal heating of ground beef to 70°C, 10^7 cells/g could not be detected after about 10 min (109). It appears to be sensitive to freezing, with about 10^5 cells/chicken carcass being greatly reduced or eliminated at − 18°C, and for artificially contaminated hamburger meat, the numbers were reduced by 1 log cycle over a 7-day period (39).

Distribution

Unlike *Y. enterocolitica* and *V. parahaemolyticus, C. jejuni* is not an environmental organism but rather is one that is associated with warm-blooded animals. A large percentage of all major meat animals have been shown to contain these organisms in their feces, with poultry being prominent. A synopsis of some reports on their prevalence in some animal specimens as well as in poultry edible parts is presented in Table 21-7. Its prevalence in fecal samples often ranges from around 30 to 100%. Reports on isolations by various investigators have been summarized by Blaser (16), and the specimens and percent positive for *C. jejuni* are as follows: chicken intestinal contents (39–83), swine feces (66–87), sheep feces (up to 73), swine intestinal contents (61), sheep carcasses (24), swine carcasses (22), eviscerated chicken (72–80), and eviscerated turkey (94). The prevalence of *C. jejuni* and *C. coli* in 396 frozen and 405 fresh meats was examined. About 12% of fresh meats were positive but only 2.3% of the frozen, suggesting the lethal effects of freezing on the organisms (107). A higher percentage of chicken livers was positive (30% of fresh and 15% of frozen) than any of the other meats, which included beef, pork, and lamb livers as well as muscle meats from these animals. Over 2,000 samples of a variety of retail-store meats were examined for *C. jejuni/coli* and *C. coli* by nine different laboratories (108). The organisms were found on 29.7% of chicken samples, 4.2% of pork sausage, 3.6% of ground beef, and about 5.1% of 1,800 red meats. Only *C. coli* was recovered from pork products. A higher incidence was noted

Table 21-7.   Incidence of *C. jejuni* in some foods and specimens.

Year	Foods/specimens	% pos./ no. tested	Numbers/comments	Location	Reference
1982	Feces from un-weaned calves	30/30	Mean no. was $\log_{10}$ 3.26	New Zealand	(38)
1982	Fecal swabs from turkeys	Up to 76% pos.	Turkeys were 15–19 days old	Texas	(1)
1982	Cloacal swabs	41/327	82% of chicken isolates were biovar 1	Australia	(99)
1982	Broiler carcasses	45/40	Mean/carcass log 4.93	Australia	(99)
1983	Cecal cultures	All of 600	From turkeys	Colorado	(67)
1983	Chicken stomachs	50/20	From ready-for-market chickens	Netherlands	(82)
1983	Chicken ceca	92/25	Mean no. was log 4.11	Netherlands	(82)
1983	Chicken ceca	72/60	From 2 plants, range was log 3.72–7.25/g	California	(130)
1983	Chicken hearts	65/20	From ready-for-market chickens	Netherlands	(82)
1983	Chicken carcasses	49/120	From ready-for-market chickens	Netherlands	(82)
1983	Chicken livers	73/40	From ready-for-market chickens	Netherlands	(82)
1983	Chicken livers	69/36	Count range was log 2.00–4.15/g	California	(130)
1983	Chicken wings	67/36	Count range was log 2.00–4.26/g	California	(130)
1983	Turkey wings (fresh)	64/184	Mean count was 740/wing	California	(89)
1983	Turkey wings (frozen)	56/81	Mean count was 890/wing	California	(89)
1984	Fresh eggs	0.9/226	Only 2 eggs from fecal pos. hens contained organisms on shell	Wisconsin	(30)

in June and September (8.6%) than in December and March (4.5 and 3.9%, respectively).

Fecal specimens from humans with diarrhea yield *C. jejuni,* and this bacterium may be the single most common cause of acute diarrhea in man. Of 8,097 specimens submitted to eight hospital laboratories over a 15-month period in different parts of the United States, this organism was recovered from 4.6%, salmonellae from 2.3%, and shigellae from 1% (17). The peak isolations for *C. jejuni* were in the age group 10–29 years. Peak isolations occur during the summer months, and it has been noted that 3–14% of diarrheal patients in developed countries yield stool specimens that contain *C. jejuni* (16). Peak isolations from individually caged hens occurred in October and late April–early May (30). In the latter study, 8.1% of the hens were chronic excreters of the organism, whereas 33% were negative even though they were likely exposed.

The prevalence of *C. jejuni* on some poultry products is noted in Table 21-7. The numbers reported range from log 2.00 to 4.26/g. Once this organism is established in a chicken house, most of the flock becomes infected over time. A recent study revealed that the organism appeared in all chicken inhabitants within a week once it was found among any of the inhabitants (104). In addition to poultry, the other primary source of this organism is raw milk. Some recorded outbreaks of *C. jejuni* enteritis from raw milk are noted in Table 21-8. Since the organism exists in cow feces, it is not surprising that it may be found in raw milk, and the degree of contamination would be expected to vary depending upon milking procedures. In a survey of 108 samples from bulk tanks of raw milk in Wisconsin, only 1 was positive for *C. jejuni,* whereas the feces of 64% of the cows in a grade A herd were positive (32).

Virulence properties

At least some strains of *C. jejuni* produce a heat-labile enterotoxin (CJT) that shares some common properties with the enterotoxins of *V. cholerae* (CT) and *E. coli* (LT). CJT increases cAMP levels, induces changes in CHO cells, and induces fluid accumulation in rat ileal loops (94). Maximal production of CJT in a special medium was achieved at 42°C for 24 h, and the amount produced was enhanced by polymyxin (60). The quantities produced by strains varied widely from none to about 50 ng/ml CJT protein. The amount of toxin was doubled as measured by Y-1 adrenal cell assay when cells were first exposed to lincomycin and then polymyxin (71). CJT is neutralized by CT and *E. coli* LT antisera, indicating immunological homology with these two enterotoxins (60). The *C. jejuni* LT appears to share the same cell receptors as CT and *E. coli* LT, and it contains a B subunit immunologically related to the B subunits of CT and LT of *E. coli* (61).

C. jejuni enteritis appears to be caused in part by the invasive abilities of the organism. Evidence for this comes from the nature of the clinical symptoms, the rapid development of high agglutinin titers after infection, recovery of the organism from peripheral blood during the acute phase of the disease, and the finding that *C. jejuni* can penetrate HeLa cells (69).

Plasmids have been demonstrated in *C. jejuni* cells. Of seventeen strains studied, eleven were found to carry plasmids ranging from 1.6 to 70 Mdal, but their role and function in disease is unclear (20).

A serotyping scheme has been developed for *C. jejuni*. From chickens

Table 21-8. **Synopsis of some outbreaks of** *Campylobacter* **enteritis.**

Year	Location	Vehicle	Synopsis	Reference
1938	Illinois	Contaminated pasteurized milk	This presumptive outbreak involved 357 cases at 2 institutions. The outbreaks stopped after milk was boiled.	(65)
1978	Vermont	Water	The town's water supply was contaminated. About 2,000 of 10,000 persons were infected. Swabs from 5 of 9 victims revealed the agent.	(119)
1978	Colorado	Raw milk	Three of 5 family members were infected. Organism was recovered from stools of all victims as well as from cow feces.	(120)
1979	Iowa	Barbecued chicken	There were 8 victims of 11 who ate undercooked barbecued chicken.	(122)
1979	Scotland	Raw milk	There were 648 cases following an electrical failure at the dairy plant. The incubation period ranged from 1 to 13 days.	(129)
1980	England	Raw milk	About 75 of 300 college students were affected. The organism was found in milk samples, and 46 students had antibodies to *C. jejuni*.	(93)
1981	Kansas	Raw milk	There were over 264 cases. Fifty-two percent of 116 persons in households that had one or more ill family members yielded the agent.	(123)
1981	Oregon	Raw milk	Of 167 who drank infected milk, 77 became ill. Agent was found in stools.	(124)
1981	Georgia	Raw milk	There were 50 victims in 30 households but the organism was not found in milk.	(87)
1982	Connecticut	Cake icing	*C. jejuni* was isolated from 16 of 41 victims.	(17)
1984	California	Raw milk	Twelve of 35 children and adults became infected after drinking certified raw milk.	(126)

and humans, 82 and 98%, respectively, of isolates belonged to biovar 1 (99).

Enteritis syndrome

From the first U.S. outbreak of *C. jejuni*, traced to a water supply, in which about 2,000 individuals contracted infections, the symptoms (and percent of individuals affected) were as follows: abdominal pain or cramps (88), diarrhea (83), malaise (76), headache (54), and fever (52). Symptoms lasted from 1 to 4 days. In the more severe cases, bloody stools may occur, and the diarrhea may resemble ulcerative colitis, while the abdominal pain may mimic acute appendicitis (16). The incubation period for enteritis is highly variable. It is usually 48 to 82 h but may be as long as 7 to 10 days or more.

PREVENTION

V. parahaemolyticus, Y. enterocolitica, and *C. jejuni* are all heat-sensitive bacteria that are destroyed by milk pasteurization temperatures. The avoidance of raw seafood products and care in preventing cross contamination with contaminated raw materials will eliminate or drastically reduce the incidence of food-borne gastroenteritis caused by *V. parahaemolyticus* and *Y. enterocolitica*. To prevent wound infections by vibrios, individuals with body nicks or abrasions should avoid entering seawaters. Yersinosis can be avoided or certainly minimized by not drinking water that has not been purified, and by avoiding raw or underprocessed milk. Campylobacteriosis can be avoided by not eating undercooked or unpasteurized foods of animal origin, especially milk.

REFERENCES

1. Acuff, G. R., C. Vanderzant, F. A. Gardner, and F. A. Colan. 1982. Examination of turkey eggs, poults and brooder house facilities for *Campylobacter jejuni. J. Food Protect.* 45:1279–81.
2. Amsterdam, L., P. Bourbeau, and P. Checko. 1984. An outbreak of *Yersinia enterocolitica* associated with pasteurized milk in a self-contained communal group in Connecticut. *Abstracts of the American Society of Microbiology,* C-78.
3. Asakawa, Y., S. Akahane, N. Kagata, and M. Noguchi. 1973. Two community outbreaks of human infection with *Yersinia enterocolitica. J. Hyg.* 71:715–23.
4. Aulisio, C. C. G., W. E. Hill, J. T. Stanfield, and R. L. Sellers, Jr. 1983. Evaluation of virulence factor testing and characteristics of pathogenicity of *Yersinia enterocolitica. Infect. Immun.* 40:330–35.
5. Aulisio, C. C. G., J. T. Stanfield, S. D. Weagant, and W. E. Hill. 1983. Yersinosis associated with tofu consumption: Serological, biochemical and pathogenicity studies of *Yersinia enterocolitica* isolates. *J. Food Protect.* 46:226–30.
6. Baross, J., and J. Liston. 1970. Occurrence of *Vibrio parahaemolyticus* and

related hemolytic vibrios in marine environments of Washington state. *Appl. Microbiol.* 20:179–86.

7. Barrow, G. I., and D. C. Miller. 1976. *Vibrio parahaemolyticus* and seafoods. In *Microbiology in agriculture, fisheries and food,* ed. F. A. Skinner and J. G. Carr, 181–95. New York: Academic Press.

8. *Bergey's manual of systematic bacteriology,* vol. 1. 1984. Ed. N. R. Krieg. Baltimore: Williams & Wilkins.

9. Beuchat, L. R. 1973. Interacting effects of pH, temperature, and salt concentration on growth and survival of *Vibrio parahaemolyticus. Appl. Microbiol.* 25:844–46.

10. ———. 1974. Combined effects of water activity, solute, and temperature on the growth of *Vibrio parahaemolyticus. Appl. Microbiol.* 27:1075–80.

11. ———. 1975. Environmental factors affecting survival and growth of *Vibrio parahaemolyticus.* A review. *J. Milk Food Technol.* 38:476–80.

12. Beuchat, L. R., and R. E. Worthington. 1976. Relationships between heat resistance and phospholipid fatty acid composition of *Vibrio parahaemolyticus. Appl. Environ. Microbiol.* 31:389–94.

13. Black, R. E., R. J. Jackson, T. Tsai, M. Medvesky, M. Shayegani, J. C. Feeley, K. I. E. MacLeod, and A. M. Wakelee. 1978. Epidemic *Yersinia enterocolitica* infection due to contaminated chocolate milk. *N. Eng. J. Med.* 298:76–79.

14. Blake, P. A., R. E. Weaver, and D. G. Hollis. 1980. Diseases of humans (other than cholera) caused by vibrios. *Ann. Rev. Microbiol.* 34:341–67.

15. Blankenship, L. C., and S. E. Craven. 1982. *Campylobacter jejuni* survival in chicken meat as a function of temperature. *Appl. Environ. Microbiol.* 44:88–92.

16. Blaser, M. J. 1982. *Campylobacter jejuni* and food. *Food Technol.* 36(3):89–92.

17. Blaser, M. J., P. Checko, C. Bopp, A. Bruce, and J. M. Hughes. 1982. *Campylobacter* enteritis associated with foodborne transmission. *Amer. J. Epidemiol.* 116:886–94.

18. Bottone, E. J. 1977. *Yersinia enterocolitica:* A panoramic view of a charismatic microorganism. *CRC Crit. Rev. Microbiol.* 5:211–41.

19. Boyce, J. M., D. J. Evans, Jr., D. G. Evans, and H. L. DuPont. 1979. Production of heat-stable, methanol-soluble enterotoxin by *Yersinia enterocolitica. Infect. Immun.* 25:532–37.

20. Bradbury, W. C., M. A. Marko, J. N. Hennessy, and J. L. Penner. 1983. Occurrence of plasmid DNA in serologically defined strains of *Campylobacter jejuni* and *Campylobacter coli. Infect. Immun.* 40:460–63.

21. Bradshaw, J. G., D. B. Shah, A. J. Wehby, J. T. Peeler, and R. M. Twedt. 1984. Thermal inactivation of the Kanagawa hemolysin of *Vibrio parahaemolyticus* in buffer and shrimp. *J. Food Sci.* 49:183–87.

22. Brown, D. F., P. L. Spaulding, and R. M. Twedt. 1977. Enteropathogenicity of *Vibrio parahaemolyticus* in the ligated rabbit ileum. *Appl. Environ. Microbiol.* 33:10–14.

23. Bryan, F. L. 1979. Infections and intoxications caused by other bacteria. In *Food-borne infections and intoxications,* ed. H. Riemann and F. L. Bryan, 211–97. New York: Academic Press.

24. Calia, F. M., and D. E. Johnson. 1975. Bacteremia in suckling rabbits after oral challenge with *Vibrio parahaemolyticus. Infect. Immun.* 11:1222–25.

25. Canada Diseases Weekly Report. 1976. *Yersinia enterocolitica* gastroenteritis outbreak—Montreal. Report 2(11):41–44; report 2(19):73–74.
26. Carter, P. B., R. J. Zahorchak, and R. R. Brubaker. 1980. Plague virulence antigens from *Yersinia enterocolitica*. *Infect. Immun.* 28:638–40.
27. Cherwonogrodzky, J. W., and A. G. Clark. 1981. Effect of pH on the production of the Kanagawa hemolysin by *Vibrio parahaemolyticus*. *Infect. Immun.* 34:115–19.
28. Colwell, R. R., ed. 1984. *Vibrios in the environment*. New York: Wiley.
29. DePaola, A., M. W. Presnell, R. E. Becker, M. L. Motes, Jr., S. R. Zywno, J. F. Musselman, J. Taylor, and L. Williams. 1984. Distribution of *Vibrio cholerae* in the Apalachicola (Florida) Bay estuary. *J. Food Protect.* 47:549–53.
30. Doyle, M. P. 1984. Association of *Campylobacter jejuni* with laying hens and eggs. *Appl. Environ. Microbiol.* 47:533–36.
31. Doyle, M. P., M. B. Hugdahl, and S. L. Taylor. 1981. Isolation of virulent *Yersinia enterocolitica* from porcine tongues. *Appl. Environ. Microbiol.* 42:661–66.
32. Doyle, M. P., and D. J. Roman. 1982. Prevalence and survival of *Campylobacter jejuni* in unpasteurized milk. *Appl. Environ. Microbiol.* 44:1154–58.
33. Faghri, M. A., C. L. Pennington, L. B. Cronholm, and R. M. Atlas. 1984. Bacteria associated with crabs from cold waters, with emphasis on the occurrence of potential human pathogens. *Appl. Environ. Microbiol.* 47:1054–61.
34. Francis, D. W., P. L. Spaulding, and J. Lovett. 1980. Enterotoxin production and thermal resistance of *Yersinia enterocolitica* in milk. *Appl. Environ. Micobiol.* 40:174–76.
35. Fujino, T., G. Sakaguchi, R. Sakazaki, and Y. Takeda. 1974. *International symposium on Vibrio parahaemolyticus*. Tokyo: Saikon.
36. Gemski, P., J. R. Lazere, and T. Casey. 1980. Plasmid associated with pathogenicity and calcium dependency of *Yersinia enterocolitica*. *Infect. Immun.* 27:682–85.
37. Gil-Recasens, M. E., A. M. Peral-Lopez, and G. Ruiz-Reyes. 1974. Aislamiento de *Vibrio parahaemolyticus* en casos de gastroenteritis y en mariscos crudos en la ciudad de Puebla. *Rev. Lat. Amer. Microbiol.* 16:85–88.
38. Gill, C. O., and L. M. Harris. 1982. Contamination of red-meat carcasses by *Campylobacter fetus* subsp. *jejuni*. *Appl. Environ. Microbiol.* 43:977–80.
39. ———. 1984. Hamburgers and broiler chickens as potential sources of human *Campylobacter* enteritis. *J. Food Protect.* 47:96–99.
40. Guerry, P., and R. R. Colwell. 1977. Isolation of cryptic plasmid deoxyribonucleic acid from Kanagawa-positive strains of *Vibrio parahaemolyticus*. *Infect. Immun.* 16:328–34.
41. Gutman, L. T., E. A. Ottesen, T. J. Quan, P. S. Noce, and S. L. Katz. 1973. An inter-familial outbreak of *Yersinia enterocolitica* enteritis. *N. Eng. J. Med.* 288:1372–77.
42. Hanna, M. O., J. C. Stewart, Z. L. Carpenter, and C. Vanderzant. 1977. Heat resistance of *Yersinia enterocolitica* in skim milk. *J. Food Sci.* 42:1134, 1136.
43. Hanna, M. O., J. C. Stewart, D. L. Zink, Z. L. Carpenter, and C. Vanderzant. 1977. Development of *Yersinia enterocolitica* on raw and cooked beef and pork at different temperatures. *J. Food Sci.* 42:1180–84.
44. Hanna, M. O., D. L. Zink, Z. L. Carpenter, and C. Vanderzant. 1976. *Yersinia*

enterocolitica-like organisms from vacuum-packaged beef and lamb. *J. Food Sci.* 41:1254–56.

45. Harmon, M. C., B. Swaminathan, and J. C. Forrest. 1984. Isolation of *Yersinia enterocolitica* and related species from porcine samples obtained from an abattoir. *J. Appl. Bacteriol.* 56:421–27.

46. Heesemann, J., B. Algermissen, and R. Laufs. 1984. Genetically manipulated virulence of *Yersinia enterocolitica*. *Infect. Immun.* 46:105–10.

47. Highsmith, A. K., J. C. Feeley, and G. K. Morris. 1977. *Yersinia enterocolitica:* A review of the bacterium and recommended laboratory methodology. *Hlth Lab. Sci.* 14:253–60.

48. Honda, T., K. Goshima, Y. Takeda, Y. Sugino, and T. Miwatani. 1976. Demonstration of the cardiotoxicity of the thermostable direct hemolysin (lethal toxin) produced by *Vibrio parahaemolyticus*. *Infect. Immun.* 13:163–71.

49. Honda, T., M. Shimizu, Y. Takeda, and T. Miwatani. 1976. Isolation of a factor causing morphological changes of Chinese hamster ovary cells from the culture filtrate of *Vibrio parahaemolyticus*. *Infect. Immun.* 14:1028–33.

50. Hughes, D., and N. Jensen. 1981. *Yersinia enterocolitica* in raw goat's milk. *Appl. Environ. Microbiol.* 41:309–10.

51. Hughes, J. M., D. G. Hollis, E. J. Gangarosa, and R. E. Weaver. 1978. Noncholera vibrio infections in the United States: Clinical, epidemiologic and laboratory features. *Ann. Intern. Med.* 88:602–6.

52. Joseph, S. W., R. R. Colwell, and J. B. Kaper. 1982. *Vibrio parahaemolyticus* and related halophilic vibrios. *CRC Crit. Rev. Microbiol.* 10:77–124.

53. Kaneko, T., and R. R. Colwell. 1973. Ecology of *Vibrio parahaemolyticus* in Chesapeake Bay. *J. Bacteriol.* 113:24–32.

54. ———. 1975. Adsorption of *Vibrio parahaemolyticus* onto chitin and copepods. *Appl. Microbiol.* 29:269–74.

55. Kaper, J., H. Lockman, R. R. Colwell, and S. W. Joseph. 1979. Ecology, serology, and enterotoxin production of *Vibrio cholerae* in Chesapeake Bay. *Appl. Environ. Microbiol.* 37:91–103.

56. Kaper, J. B., R. K. Campen, R. J. Seidler, M. M. Baldini, and S. Falkow. 1984. Cloning of the thermostable direct or Kanagawa phenomenon-associated hemolysin of *Vibrio parahaemolyticus*. *Infect. Immun.* 45:290–92.

57. Karunsagar, I. 1981. Production of hemolysin by *Vibrio parahaemolyticus* in a chemically defined medium. *Appl. Environ. Microbiol.* 41:1274–75.

58. Karunsagar, I., S. W. Joseph, R. M. Twedt, H. Hada, and R. R. Colwell. 1984. Enhancement of *Vibrio parahaemolyticus* virulence by lysed erythrocyte factor and iron. *Infect. Immun.* 46:141–44.

59. Kenyon, J. E., D. R. Piexoto, B. Austin, and D. C. Gillies. 1984. Seasonal variation in numbers of *Vibrio cholerae* (non-01) isolated from California coastal waters. *Appl. Environ. Microbiol.* 47:1243–45.

60. Klipstein, F. A., and R. F. Engert. 1984. Properties of crude *Campylobacter jejuni* heat-labile enterotoxin. *Infect. Immun.* 45:314–19.

61. ———. 1985. Immunological relationship of the B subunits of *Campylobacter jejuni* and *Escherichia coli* heat-labile enterotoxins. *Infect. Immun.* 48:629–33.

62. Laird, W. J., and D. C. Cavanaugh. 1980. Correlation of auto-agglutination and virulence of yersiniae. *J. Clin. Microbiol.* 11:430–32.

63. Lee, W. H., R. E. Smith, J. M. Damare, M. E. Harris, and R. W. Johnston.

1981. Evaluation of virulence test procedures for *Yersinia enterocolitica* recovered from foods. *J. Appl. Bacteriol.* 50:529–39.

64. Leistner, L., H. Hechelmann, and R. Albert. 1975. Nachweis von *Yersinia enterocolitica* in Faeces und Fleisch von Schweinen, Hindern und Geflugel. *Fleischwirtschaft* 55:1599–1602.

65. Levy, A. J. 1946. A gastro-enteritis outbreak probably due to a bovine strain of *Vibrio. Yale J. Biol. Med.* 18:243–58.

66. Liston, J. 1973. *Vibrio parahaemolyticus.* In *Microbial safety of fishery products,* ed. C. O. Chichester and H. D. Graham, 203–13. New York: Academic Press.

67. Luechtefeld, N. W., and W. L. L. Wang. 1981. *Campylobacter fetus* subsp. *jejuni* in a turkey processing plant. *J. Clin. Microbiol.* 13:266–68.

68. MacLagan, R. M., and D. C. Old. 1980. Hemagglutinins and fimbriae in different serotypes and biotypes of *Yersinia enterocolitica. J. Appl. Bacteriol.* 49:353–60.

69. Manninen, K. I., J. F. Prescott, and I. R. Dohoo. 1982. Pathogenicity of *Campylobacter jejuni* isolates from animals and humans. *Infect. Immun.* 38:46–52.

70. Martinez, R. J. 1983. Plasmid-mediated and temperature-regulated surface properties of *Yersinia enterocolitica. Infect. Immun.* 41:921–30.

71. McCardell, B. A., J. M. Madden, and E. C. Lee. 1984. *Campylobacter jejuni* and *Campylobacter coli* production of a cytotonic toxin immunologically similar to cholera toxin. *J. Food Protect.* 47:943–49.

72. Molenda, J. R., W. G. Johnson, M. Fishbein, B. Wentz, I. J. Mehlman, and T. A. Dadisman, Jr. 1972. *Vibrio parahaemolyticus* gastroenteritis in Maryland: Laboratory aspects. *Appl. Microbiol.* 24:444–48.

73. Morris, J. G., R. Wilson, B. R. Davis, I. K. Wachsmuth, C. F. Riddle, H. G. Wathen, R. A. Pollard, and P. A. Blake. 1981. Non-0 group 1 *Vibrio cholerae* gastroenteritis in the United States. *Ann. Intern. Med.* 94:656–58.

74. Mors, V., and C. H. Pai. 1980. Pathogenic propeprties of *Yersinia enterocolitica. Infect. Immun.* 28:292–94.

75. Moustafa, M. K., A. A.-H. Ahmed, and E. H. Marth. 1983. Occurrence of *Yersinia enterocolitica* in raw and pasteurized milk. *J. Food Protect.* 46:276–78.

76. Nishibuchi, M., M. Ishibashi, Y. Takeda, and J. B. Kaper. 1985. Detection of the thermostable direct hemolysin gene and related DNA sequences in *Vibrio parahaemolyticus* and other *Vibrio* species by the DNA colony hybridization test. *Infect. Immun.* 49:481–86.

77. Nishibuchi, M., and J. B. Kaper. 1985. Nucleotide sequence of the thermostable direct hemolysin gene of *Vibrio parahaemolyticus. J. Bacteriol.* 162:558–64.

78. Okamoto, K., T. Inoue, H. Ichikawa, Y. Kawamoto, and A. Miyama. 1981. Partial purification and characterization of heat-stable enterotoxin produced by *Yersinia enterocolitica. Infect. Immun.* 31:554–59.

79. Okamoto, K., T. Inoue, K. Shimizu, S. Hara, and A. Miyama. 1982. Further purification and characterization of heat-stable enterotoxin produced by *Yersinia enterocolitica. Infect. Immun.* 35:958–64.

80. Oliver, J. D., R. A. Warner, and D. R. Cleland. 1983. Distribution of *Vibrio vulnificus* and other lactose-fermenting vibrios in the marine environment. *Appl. Environ. Microbiol.* 45:985–98.

81. Olsovsky, Z., V. Olsakova, S. Chobot, and V. Sviridov. 1975. Mass occurrence

of *Yersinia enterocolitica* in two establishments of collective care of children. *J. Hyg. Epid. Microbiol. Immunol.* 19:22–29.

82. Oosterom, J., S. Notermans, H. Karman, and G. B. Engels. 1983. Origin and prevalence of *Campylobacter jejuni* in poultry processing. *J. Food Protect.* 46:339–44.

83. Pai, C. H., and L. DeStephano. 1982. Serum resistance associated with virulence in *Yersinia enterocolitica. Infect. Immun.* 35:605–11.

84. Pai, C. H., and V. Mors. 1978. Production of enterotoxin by *Yersinia enterocolitica. Infect. Immun.* 19:908–11.

85. Pai, C. H., V. Mors, and S. Toma. 1978. Prevalence of enterotoxigenicity in human and nonhuman isolates of *Yersinia enterocolitica. Infect. Immun.* 22:334–38.

86. Portnoy, D. A., and S. Falkow. 1981. Virulence-associated plasmids from *Yersinia enterocolitica* and *Yersinia pestis. J. Bacteriol.* 148:877–83.

87. Potter, M. E., M. J. Blaser, R. K. Sikes, A. F. Kaufmann, and J. G. Wells. 1983. Human *Campylobacter* infection associated with certified raw milk. *Amer. J. Epidemiol.* 117:475–83.

88. Prpic, J. K., R. M. Robins-Browne, and R. B. Davey. 1985. In vitro assessment of virulence in *Yersinia enterocolitica* and related species. *J. Clin. Microbiol.* 22:105–10.

89. Rayes, H. M., C. A. Genigeorgis, and T. B. Farver. 1983. Prevalence of *Campylobacter jejuni* on turkey wings at the supermarket level. *J. Food Protect.* 46:292–94.

90. Robins-Browne, R. M., and J. K. Prpic. 1985. Effects of iron and desferrioxamine on infections with *Yersinia enterocolitica. Infect. Immun.* 47:774–79.

91. Robins-Browne, R. M., C. S. Still, M. Isaacson, H. J. Koornhof, P. C. Appelbaum, and J. N. Scragg. 1977. Pathogenic mechanisms of a nonagglutinable *Vibrio cholerae* strain: Demonstration of invasive and enterotoxigenic properties. *Infect. Immun.* 18:542–45.

92. Robins-Browne, R. M., C. S. Still, M. D. Miliotis, and H. J. Koornhof. 1979. Mechanism of action of *Yersinia enterocolitica* enterotoxin. *Infect. Immun.* 25:680–84.

93. Robinson, D. A., and D. M. Jones. 1981. Milk-borne campylobacter infection. *Brit. Med. J.* 282:1374–76.

94. Ruiz-Palacios, G. M., J. Torres, E. Escamilla, B. R. Ruiz-Palacios, and J. Tamayo. 1983. Cholera-like enterotoxin produced by *Campylobacter jejuni. Lancet* 2:250–53.

95. Sakazaki, R. 1979. *Vibrio* infections. In *Food-borne infections and intoxications,* ed. H. Riemann and F. L. Bryan, 173–209. New York: Academic Press.

96. ———. 1983. *Vibrio parahaemolyticus* as a food-spoilage organism. In *Food microbiology,* ed. A. H. Rose, 225–41. New York: Academic Press.

97. Schiemann, D. A. 1981. An enterotoxin-negative strain of *Yersinia enterocolitica* serotype 0:3 is capable of producing diarrhea in mice. *Infect. Immun.* 32:571–74.

98. Schiemann, D. A., and J. A. Devenish. 1982. Relationship of HeLa cell infectivity to biochemical, serological, and virulence characteristics of *Yersinia enterocolitica. Infect. Immun.* 35:497–506.

99. Shanker, S., J. A. Rosenfield, G. R. Davey, and T. C. Sorrell. 1982. *Campylobacter*

jejuni: Incidence in processed broilers and biotype distribution in human and broiler isolates. *Appl. Environ. Microbiol.* 43:1219–20.

100. Shayegani, M., I. Deforge, D. M. McGlynn, and T. Root. 1981. Characteristics of *Yersinia enterocolitica* and related species isolated from human, animal and environmental sources. *J. Clin. Microbiol.* 14:304–12.

101. Shayegani, M., D. Morse, I. DeForge, T. Root, L. M. Parsons, and P. Maupin. 1982. Foodborne outbreak of *Yersinia enterocolitica* in Sullivan County, New York with pathogenicity studies in the isolates. *Bacteriol. Proc.,* C-175.

102. Skurnik, M. 1984. Lack of correlation between the presence of plasmids and fimbriae in *Yersinia enterocolitica* and *Yersinia pseudotuberculosis. J. Appl. Bacteriol.* 56:355–63.

103. Smibert, R. M. 1978. The genus *Campylobacter. Ann. Rev. Microbiol.* 32:673–709.

104. Smitherman, R. E., C. A. Genigeorgis, and T. B. Farver. 1984. Preliminary observations on the occurrence of *Campylobacter jejuni* at four California chicken ranches. *J. Food Protect.* 47:293–98.

105. Stern, N. J. 1981. Recovery rate of *Campylobacter fetus* ssp. *jejuni* on eviscerated pork, lamb, and beef carcasses. *J. Food Sci.* 46:1291, 1293.

106. Stern, N. J., and J. M. Damare. 1982. Comparison of selected *Yersinia enterocolitica* indicator tests for potential virulence. *J. Food Sci.* 47:582–88.

107. Stern, N. J., S. S. Green, N. Thaker, D. J. Krout, and J. Chiu. 1984. Recovery of *Campylobacter jejuni* from fresh and frozen meat and poultry collected at slaughter. *J. Food Protect.* 47:372–74.

108. Stern, N. J., M. P. Hernandez, L. Blankenship, K. E. Deibel, S. Doores, M. P. Doyle, H. Ng, M. D. Pierson, J. N. Sofos, W. H. Sveum, and D. C. Westhoff. 1985. Prevalence and distribution of *Campylobacter jejuni* and *Campylobacter coli* in retail meats. *J. Food Protect.* 48:595–99.

109. Stern, N. J., and A. W. Kotula. 1982. Survival of *Campylobacter jejuni* inoculated into ground beef. *Appl. Environ. Microbiol.* 44:1150–53.

110. Stern, N. J., and M. D. Pierson. 1979. *Yersinia enterocolitica:* A review of the psychrotrophic water and foodborne pathogen. *J. Food Sci.* 44:1736–42.

111. Stern, N. J., M. D. Pierson, and A. W. Kotula. 1980. Effects of pH and sodium chloride on *Yersinia enterocolitica* growth at room and refrigeration temperatures. *J. Food Sci.* 45:64–67.

112. Swaminathan, B., M. C. Harmon, and I. J. Mehlman. 1982. *Yersinia enterocolitica. J. Appl. Bacteriol.* 52:151–83.

113. Takeda, Y. 1983. Thermostable direct hemolysin of *Vibrio parahaemolyticus. Pharmac. Ther.* 19:123–46.

114. Taniguchi, H., H. Ohta, M. Ogawa, and Y. Mizuguchi. 1985. Cloning and expression in *Escherichia coli* of *Vibrio parahaemolyticus* thermostable direct hemolysin and thermolabile hemolysin genes. *J. Bacteriol.* 162:510–15.

115. Toma, S., and L. Lafleur. 1974. Survey on the incidence of *Yersinia enterocolitica* infection in Canada. *Appl. Microbiol.* 28:469–73.

116. Twedt, R. M., J. T. Peeler, and P. L. Spaulding. 1980. Effective ileal loop dose of Kanagawa-positive *Vibrio parahaemolyticus. Appl. Environ. Microbiol.* 40:1012–16.

117. Twedt, R. M., P. L. Spaulding, and H. E. Hall. 1969. Morphological, cultural, biochemical, and serological comparison of Japanese strains of *Vibrio para-*

haemolyticus with related cultures isolated in the United States. *J. Bacteriol.* 98:511–18.

118. United States Centers for Disease Control. 1976. Foodborne and waterborne disease outbreaks. *Annual Summary*, HEW Pub. No. (CDC) 76-8185.

119. ———. 1978. Waterborne *Campylobacter* gastroenteritis—Vermont. *Morb. Mort. W. Rept.* 27:207.

120. ———. 1978. *Campylobacter* enteritis—Colorado. *Morb. Mort. W. Rept.* 27:226, 231.

121. ———. 1978. *Vibrio parahaemolyticus* foodborne outbreak—Louisiana. *Morb. Mort. W. Rept.* 27:345–46.

122. ———. 1979. *Campylobacter* enteritis—Iowa. *Morb. Mort. W. Rept.* 28:565–66.

123. ———. 1981. Outbreak of *Campylobacter* enteritis associated with raw milk—Kansas. *Morb. Mort. W. Rept.* 30:218–20.

124. ———. 1981. Raw-milk-associated illness—Oregon, California. *Morb. Mort. W. Rept.* 30:80–81.

125. ———. 1982. Multi-state outbreak of yersiniosis. *Morb. Mort. W. Rept.* 31:505–6.

126. ———. 1984. *Campylobacter* outbreak associated with certified raw milk products. *Morb. Mort. W. Rept.* 33:562.

127. Vanderzant, C., and R. Nickelson. 1972. Survival of *Vibrio parahaemolyticus* in shrimp tissue under various environmental conditions. *Appl. Microbiol.* 23:34–37.

128. Vidon, D. J. M., and C. L. Delmas. 1981. Incidence of *Yersinia enterocolitica* in raw milk in Eastern France. *Appl. Environ. Microbiol.* 41:353–59.

129. Wallace, J. M. 1980. Milk-associated *Campylobacter* infection. *Hlth Bull.* 38:57–61.

130. Wempe, J. M., C. A. Genigeorgis, T. B. Farver, and H. I. Yusufu. 1983. Prevalence of *Campylobacter jejuni* to two California chicken processing plants. *Appl. Environ. Microbiol.* 45:355–59.

131. Yamamoto, K., M. Al-Omani, T. Honda, Y. Takeda, and T. Miwatani. 1984. Non-01 *Vibrio cholerae* hemolysin: Purification, partial characterization, and immunological relatedness to El Tor hemolysin. *Infect. Immun.* 45:192–96.

132. Zen-Yoji, H., Y. Kudoh, H. Igarashi, K. Ohta, and K. Fukai. 1975. Further studies on characterization and biological activities of an enteropathogenic toxin of *Vibrio parahaemolyticus. Toxicon* 13:134–35.

133. Zen-Yoji, H., T. Maruyama, S. Sakai, S. Kimura, T. Mizuno, and T. Momose. 1973. An outbreak of enteritis due to *Yersinia enterocolitica* occurring at a junior high school. *Japan. J. Microbiol.* 17:220–22.

134. Zink, D. L., J. C. Feeley, J. G. Wells, C. Vanderzant, J. C. Vickery, W. C. Roof, and G. A. O'Donovan. 1980. Plasmid-mediated tissue invasiveness in *Yersinia enterocolitica. Nature* 283:224–26.

135. Zink, D. L., R. V. Lachica, and J. R. Dubel. 1982. *Yersinia enterocolitica* and *Yersinia enterocolitica*-like species: Their pathogenicity and significance in foods. *J. Food Safety* 4:223–41.

22.

OTHER PROVEN AND
SUSPECTED FOOD-BORNE PATHOGENS

MYCOTOXINS

A very large number of molds have been demonstrated to produce toxic substances designated mycotoxins. Some are mutagenic and carcinogenic, some display specific organ toxicity, and some are toxic by other mechanisms. While the clear-cut toxicity of many mycotoxins for man has not been demonstrated, the effect of these compounds on experimental animals and their effect in in vitro assay systems leaves little doubt about their real and potential toxicity for man. As of 1980, fourteen mycotoxins were recognized as being carcinogens, with the aflatoxins being the most potent (116). It is generally accepted that about 93% of mutagenic compounds are carcinogens. With mycotoxins, microbial assay systems reveal an 85% level of correlation between carcinogenicity and mutagenesis (116).

Mycotoxins are produced as secondary metabolites. The primary metabolites of fungi as well as for other organisms are those compounds that are essential for growth. Secondary metabolites are formed during the end of the exponential growth phase and have no apparent significance to the producing organism relative to growth or metabolism. In general, it appears that they are formed when large pools of primary metabolic precursors such as amino acids, acetate, pyruvate, and so on, accumulate; and the synthesis of mycotoxins represents one way the fungus has of reducing the pool of metabolic precursors that its metabolic needs no longer require. In the case of aflatoxin synthesis, it begins at the onset of the stationary phase of growth and occurs along with that of lipid synthesis (111). The biosynthetic pathways of the aflatoxins have been reviewed by Magoon et al. (83). Of the three primary routes of secondary metabolism in fungi (polyketide, terpenoid, and that utilizing essential amino acids), mycotoxins are synthesized via the polyketide route.

Presented below are some of the mycotoxins demonstrated to occur in foods. More extensive treatments of this subject have been presented by Davis and Diener (31), Ayres et al. (3), Busby and Wogan (17), Stark (116), and Bullerman (12). Methods for detecting mycotoxins along with detection limits are covered in Chapter 6. See Chu (23) for review of immunoassay methods.

Aflatoxins

These are clearly the most widely studied of all mycotoxins. Knowledge of their existence dates from 1960, when more than 100,000 turkey poults died in England after eating peanut meal imported from Africa and South America. From the poisonous feed were isolated *Aspergillus flavus* and a toxin produced by this organism that was designated aflatoxin (*Aspergillus flavus* toxin—A-fla-toxin). Studies on the nature of the toxic substances revealed the four components presented below.

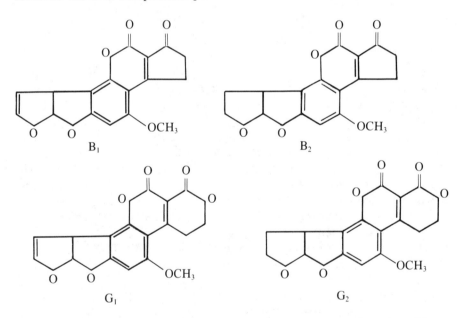

It was later determined that *A. parasiticus* produces aflatoxins. These compounds are highly substituted coumarins, and at least eighteen closely related toxins are known. Aflatoxin B_1 (AFB$_1$) is produced by all aflatoxin-positive strains, and it is the most potent of all. AFM$_1$ is a hydroxylated product of AFB$_1$ and appears in milk, urine, and feces as metabolic products (37). AFL, AFLH$_1$, AFQ$_1$, and AFP$_1$ are all derived from AFB$_1$. AFB$_2$ is the 2,3–dehydro form of AFB$_1$; and AFG$_2$ is the 2,3-dihydro form of AFG$_1$. The toxicity of the six most potent aflatoxins decreases in the following order: $B_1 > M_1 > G_1 > B_2 > M_2 \neq G_2$ (3). When viewed under UV light, six of the toxins fluoresce as noted.

B_1 and B_2—blue
G_1—green
G_2—green-blue
M_1—blue-violet
M_2—violet

REQUIREMENTS FOR GROWTH AND TOXIN PRODUCTION. No aflatoxins were produced by twenty-five isolates of *A. flavus/parasiticus* on wort agar at 2°, 7°, 41°, or 46°C within 8 days. and none was produced < 7.5 or > 40°C even under otherwise favorable conditions (105). In another study employing Sabouraud's agar, maximal growth of *A. flavus* and *A. parasiticus* occurred at 33°C when pH was 5.0 and a_w was 0.99 (60). At 15°C, growth occurred at a_w 0.95 but not at 0.90, while at 27° and 33°C, slight growth was observed at an a_w of 0.85. The optimum temperature for toxin production has been found by many to be between 24° and 28°C. In one study, maximal growth of *A. parasiticus* was 35°C, but the highest level of toxin was produced at 25°C (110).

The limiting moisture content for AFB_1 and AFB_2 on corn was 17.5% at a temperature of 24°C or higher with up to 50 ng/g being produced (123). No toxin was produced at 13°C. Overall, toxin production has been observed over the a_w range of 0.93–0.98, with limiting values variously reported as being 0.71–0.94 (84). In another study, no detectable quantities of AFB_1 were formed by *A. parasiticus* at a_w of 0.83 and 10°C (93). The optimum temperature at a_w 0.94 was 24°C (see Fig. 22-1). Growth without demonstrable toxin appeared possible at a_w 0.83 on malt agar-containing sucrose. It has been observed by several investigators that rice supports the production of high levels of aflatoxins at favorable temperatures but none is produced at 5°C on either rice or Cheddar cheese (95).

Overall, the minimal and maximal parameters that control growth and toxin production by these eucaryotic organisms are not easy to define, in part because of their diverse habitats in nature and in part because of their eucaryotic status. It seems clear that growth can occur without toxin production.

It has been reported that AFG_1 is produced at lower growth temperatures than AFB_1, and while some investigators have found more AFB_1 than AFG_1 at around 30°C, others have found equal production. With regard to *A. flavus* and *A. parasiticus,* the former generally produces only AFB and AFG (31). Aeration favors aflatoxin production, and amounts of 2 mg/g can be produced on natural substrates such as rice, corn, soybeans, and the like (31). Up to 200–300 mg/liter can be produced in broth containing appropriate levels of Zn^{2+}. Aflatoxin synthesis can be inhibited by caffeine,

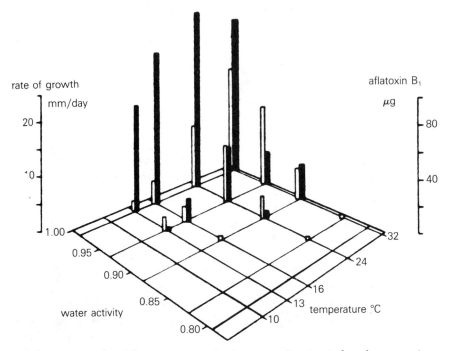

rate of growth

mm/day

aflatoxin B₁

μg

Fig. 22-1. Growth and aflatoxin B₁ production on malt extract-glycerine agar at various water activity values and temperatures. White columns = rate of growth; black columns = average AFB₁ production. (93, copyright © 1976, International Association of Milk, Food and Environmental Sanitarians)

which apparently inhibits the uptake of carbohydrates (8). The release of AFB₁ by *A. flavus* appears to involve an energy-dependent transport system (1). The degradation of AFB₁ and AFG₁ by bisulfite is discussed in Chapter 11.

PRODUCTION AND OCCURRENCE IN FOODS. With respect to production in foods, aflatoxin has been demonstrated on fresh beef, ham, and bacon inoculated with toxigenic cultures and stored at 15°, 20°, and 30°C (13), and on country-cured hams during aging when temperatures approached 30°C, but not at temperatures < 15°C or R.H. < 75% (14). They have been found in a wide variety of foods, including milk, beer, cocoa, raisins, soybean meal, and so on (see below). In fermented sausage at 25°C, 160 and 426 ppm of AFG₁ were produced in 10 and 18 days, respectively, and ten times more AFG₁ was found than B₁ (80). Aflatoxins have been produced in whole-rye and whole-wheat breads, in tilsit cheese, and in apple juice at 22°C (43). They have been demonstrated in the upper layer of 3-month-old Cheddar cheese held at room temperature (81) and on brick cheese at 12.8°C by *A. parasiticus* after 1 week, but not by *A. flavus* (110). From a 5-year survey of around 500 samples of Virginia corn and wheat, aflatoxins were detected in about

25% of corn samples for every crop year, with 18–61% of samples containing 20 ng/g or more, and 5–29% containing > 100 ng/g (112). The average quantity detected over the 5-year period was 21–137 ng/g (Table 22-1). Neither aflatoxins nor zearalenone and ochratoxin A were detected in any of the wheat samples.

The effect of temperature cycling between 5° and 25°C on production in rice and cheese has been investigated. *A. parasiticus* produced more toxin under cycling temperatures than at either 15°, 18°, or 25°C, while *A. flavus* produced less under these conditions (95). On Cheddar cheese, however, less aflatoxin was produced than at 25°C, and these investigators noted that cheese is not a good substrate for aflatoxin production if it is held much below the optimum temperature.

Aflatoxin production has been demonstrated to occur on an endless number of food products in addition to those noted above. Under optimal conditions of growth, some toxin can be detected within 24 h—otherwise within 4–10 days (22). On peanuts, Hesseltine (59) has made the following observations: (1) growth and formation of aflatoxin occur mostly during the curing of peanuts after removal from soil; (2) in a toxic lot of peanuts, only comparatively few kernels contain toxin, and success in detecting the toxin depends on collecting a relatively large sample, such as 1 kg, for assay; (3) the toxin will vary greatly in amount even within a single kernel; and (4) the two most important factors affecting aflatoxin formation are moisture and temperature.

The U.S. FDA has established allowable action levels of aflatoxins in foods as follows: 20 ppb for food, feeds, Brazil nuts, peanuts, peanut products, and pistachio nuts; and 0.5 ppb for milk (see Labuza, 77, for further information on regulations).

RELATIVE TOXICITY AND MODE OF ACTION. For the expression of mutagenicity, mammalian metabolizing systems are essential for aflatoxins, especially AFB_1. Also essential is their binding with nucleic acids, especially DNA. While nuclear DNA is normally affected, AFB_1 has been shown to bind covalently to liver mitochondrial DNA preferential to nuclear DNA (92). Cellular macromolecules other than nucleic acids are possible sites for aflatoxins. The site of the aflatoxin molecule responsible for mutagenicity is the C_2-C_3 double bond in the dihydrofurofuran moiety. Its reduction to the 2,3-dihydro (AFB_2) form reduces mutagenicity by 200–500-fold (see Stark, 116). Following binding to DNA, point mutations are the predominant genetic lesions induced by aflatoxins, although frameshift mutations are known to occur. The mutagenesis of AFB_1 has been shown to be potentiated twofold by BHA and BHT and much less by propyl gallate employing the Ames assay, but whether potentiation occurs in animal systems is unclear (108).

The LD_{50} of AFB_1 for rats by the oral route is 1.2 mg/kg, and 1.5–2.0 mg/kg for AFG_1 (17). The relative susceptibility of various animal species

Table 22-1. Aflatoxin levels in dent corn grown in Virginia, 1976–80 (112).

Total aflatoxin, ng/g	Collected from trucks by FGIS										Collected at harvest by SRS			
	1976		1977		1978		1979		1980		1978		1979	
	No. of samples	(%)	No. of samples	(%)	No. of samples	(%)	No. of samples	(%)	No. of samples	(%)	No. of samples	(%)	No. of samples	(%)
ND[a]	77	(63)	52	(51)	63	(64)	81	(71)	18	(18)	79	(88)	93	(79)
<20	13	(10)	17	(17)	10	(10)	13	(11)	20	(20)	2	(2)	9	(8)
20–100	21	(17)	18	(18)	21	(21)	8	(7)	32	(32)	5	(6)	7	(6)
101–500	9	(7)	10	(10)	5	(5)	10	(9)	26	(26)	4	(4)	7	(6)
501–1000	1	(1)	1	(1)	—	—	2	(2)	1	(1)	—	—	1	(1)
>1000	2	(2)	3	(3)	—	—	—	—	2	(2)	—	—	1	(1)
Total	123		101		99		114		99		90		117	
% Incidence	37		49		36		29		82		12		21	
% ≥ 20 ng/g	27		32		26		18		61		10		13	
% > 100 ng/g	10		14		5		11		29		4		7	
Av. level (ng/g), all samples	48		91		21		34		137		13		36	
Av. level (ng/g), pos. samples	130		187		58		118		167		110		176	

[a]ND = not detected

to aflatoxins is presented in Table 22-2. Young ducklings and young trout are among the most sensitive, followed by rats and other species. Most species of susceptible animals die within 3 days after administration of toxins and show gross liver damage, which, upon postmortem examination, reveals the aflatoxins to be hepatocarcinogens (140). The toxicity is higher for young animals and males than for older animals and females, and the toxic effects are enhanced by low protein or cirrhogenic diets.

Table 22-2. Comparative lethality of single doses of aflatoxin B_1 (140).

Animal	Age (or weight)	Sex	Route	LD_{50} mg/kg
Duckling	1 day	M	PO	0.37
	1 day	M	PO	0.56
Rat	1 day	M-F	PO	1.0
	21 days	M	PO	5.5
	21 days	F	PO	7.4
	100 g	M	PO	7.2
	100 g	M	IP	6.0
	150 g	F	PO	17.9
Hamster	30 days	M	PO	10.2
Guinea pig	Adult	M	IP	ca. 1
Rabbit	Weanling	M-F	IP	ca. 0.5
Dog	Adult	M-F	IP	ca. 1
	Adult	M-F	PO	ca. 0.5
Trout	100 g	M-F	PO	ca. 0.5

PO = oral
IP = intraperitoneal

Circumstantial evidence strongly suggests that aflatoxins are carcinogenic to humans. Among conditions believed to result from aflatoxins is the EFDV syndrome of Thailand, Reye's syndrome of Thailand and New Zealand (18), and an acute hepatoma condition in a Ugandan child. In the latter, a fatal case of acute hepatic disease revealed histological changes in the liver identical to those observed in monkeys treated with aflatoxins, and an aflatoxin etiology was strongly suggested by the findings (107). Two researchers who worked with purified aflatoxin developed colon carcinoma (33). These and other instances of aflatoxin involvement in human diseases have been reviewed more extensively by Busby and Wogan (17).

Alternaria toxins
Several species of *Alternaria* (including *A. citri*, *A. alternata*, *A. solani*, and *A. tenuissima*) produce toxic substances that have been found in apples,

tomatoes, blueberries, and the like. The toxins produced include alternariol, alternariol monomethyl ether, altenuene, tenuazonic acid, and altertoxin-I. On slices of apples, tomatoes, or crushed blueberries incubated for 21 days at 21°C, several *Alternaria* produced each of the toxins noted at levels up to 137 mg/100 g (117). In another study, tenuazonic acid was the main toxin produced in tomatoes, with levels as high as 13.9 mg/100 g; while on oranges and lemons, *A. citri* produced tenuazonic acid, alternariol, and alternariol monomethyl ether at a mean concentration of 1.15–2.66 mg/100 g (118). The fruits were incubated at room temperature for 21–28 days. For more detailed information, see review by King and Schade (69).

Citrinin

This mycotoxin is produced by *Penicillium citrinum, P. viridicatum,* and other fungi. It has been recovered from polished rice, moldy bread, country-

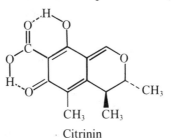

Citrinin

cured hams, wheat, oats, rye, and other similar products. Under long-wave UV light, it fluoresces lemon yellow. It is a known carcinogen. Of seven strains of *P. viridicatum* recovered from country-cured hams, all produced citrinin in potato dextrose broth and on country-cured hams in 14 days at 20°–30°C but not at 10°C (141). Growth was found to be poor at 10°C. While citrinin-producing organisms are found on cocoa and coffee beans, this mycotoxin as well as others is not found to the extent of growth. The apparent reason is the inhibition of citrinin in *P. citrinum* by caffeine (7). The inhibition of citrinin appears to be rather specific, since only a small decrease in growth of the organism occurs (9).

Ochratoxins

The ochratoxins consist of a group of at least seven structurally related secondary metabolites of which ochratoxin A (OA) is the best known and the most toxic. OB is dechlorinated OA and along with OC, it may not

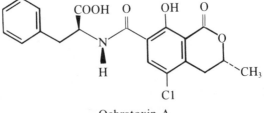

Ochratoxin A

occur naturally. OA is produced by a large number of storage fungi including *A. ochraceus, A. alliaceus, A. ostianus, A. mellus,* and other species of aspergilli. Among penicillia that produce OA are *P. viridicatum, P. cyclopium, P. variable,* and others.

OA is produced maximally at around 30°C and a_w 0.95 (24). The minimum a_w supporting OA production by *A. ochraceus* at 30°C in poultry feed is 0.85 (4). Its oral LD_{50} in rats is 20–22 mg/kg, and it is both hepatotoxic and neprotoxic.

This mycotoxin has been found in corn, dried beans, cocoa beans, soybeans, oats, barley, citrus fruits, Brazil nuts, moldy tobacco, country-cured hams, peanuts, coffee beans, and other similar products. Two strains of *A. ochraceus* isolated from country-cured hams produced OA and OB on rice, defatted peanut meal, and when inoculated into country-cured hams (38). Two-thirds of the toxin penetrated to a distance of 0.5 cm after 21 days, with the other one-third located in the mycelial mat. Of six strains of *P. viridicatum* recovered from country-cured hams, none produced ochratoxins. From a study of four chemical inhibitors of both growth and OA production by two OA producers at pH 4.5, the results were potassium sorbate > sodium propionate > methyl paraben > sodium bisulfite; while at a pH of 5.5, the most effective two were methyl paraben and potassium sorbate (122). Like most mycotoxins, OA is heat-stable. In one study, the highest rate of destruction achieved by cooking faba beans was 20%, and the investigators concluded that OA could not be destroyed by normal cooking procedures (36). Under UV light, OA fluoresces greenish, while OB emits blue fluorescence. It induces abnormal mitosis in monkey kidney cells.

Patulin

Patulin (clavicin, expansin) is produced by a large number of penicillia including *P. claviforme, P. expansum, P. patulum;* by some aspergilli (*A. clavatus, A. terreus,* and others); and by *Byssochlamys nivea* and *B. fulva.*

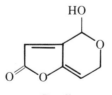

Patulin

Its biological properties are similar to those of penicillic acid. Some patulin-producing fungi can produce the compound below 2°C (3). This mycotoxin has been found in moldy bread, sausage, fruits (including bananas, pears, pineapples, grapes, and peaches), apple juice, cider, and other products. In apple juice, levels as high as 440 µg/liter have been found, and in cider, levels up to 45 ppm.

Minimum a_w for growth of *P. expansum* and *P. patulum* has been reported to be 0.83 and 0.81, respectively. In potato dextrose broth incubated at 12°C, patulin was produced after 10 days by *P. patulum* and *P. roqueforti* with the former organism producing up to 1,033 ppm (11). Patulin was produced in apple juice also at 12°C by *B. nivea* but the highest concentration was attained after 20 days at 21°C after a 9-day lag (99). The next highest amount was produced at 30°C, with much less at 37°C. These investigators confirmed that patulin production is favored at temperatures below the growth optimum, as was previously found by Sommer et al. (115). The latter investigators used *P. expansum* and found production over the range 5°–20°C, with only small amounts produced at 30°C. Atmospheres of CO_2 and N_2 reduced production compared to that in air. To inhibit production, SO_2 was found more effective than potassium sorbate or sodium benzoate (99).

The LD_{50} for patulin in rats by the subcutaneous route has been reported to be 15–25 mg/kg, and it induces subcutaneous sarcomas in some animals. Both patulin and penicillic acid bind to —SH and —NH_2 groups, forming covalently linked adducts that appear to abate their toxicities. Patulin causes chromosomal aberrations in animal and plant cells and is a carcinogen.

Penicillic acid

As noted above, this mycotoxin has biological properties similar to patulin. It is produced by a large number of fungi, including many penicillia (*P.*

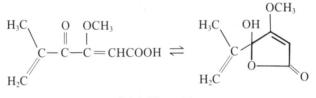

Penicillic acid

puberulum, for example), as well as members of the *A. ochraceus* group. One of the best producers is *P. cyclopium*. It has been found in corn, beans, and other field crops and has been produced experimentally on Swiss cheese. Its LD_{50} in mice by subcutaneous route is 100–300 mg/kg, and it is a proven carcinogen.

Of 346 penicillia cultures isolated from salami, about 10% produced penicillic acid in liquid culture media, but 5 that were inoculated into sausage failed to produce toxin after 70 days (25). In another study, some 183 molds were isolated from Swiss cheese; 87% were penicillia, 93% of which were able to grow at 5°C. Thirty-five percent of penicillia extracts were toxic to chick embryos, and from 5.5% of the toxic mixtures were recovered penicillic acid as well as patulin and aflatoxins (10). Penicillic acid was produced at 5°C in 6 weeks by 4 of 33 fungal strains.

Sterigmatocystin

These mycotoxins are structurally and biologically related to the aflatoxins and like the latter, they cause hepatocarcinogenic activity in animals. At least eight derivatives are known. Among the organisms that produce them are *Aspergillus versicolor, A. nidulans, A. rugulosus,* and others. The LD_{50} for rats by IP injection is 60–65 mg/kg. Under UV light, the toxin fluoresces dark brick-red. Although not often found in natural products, they have been found in wheat, oats, Dutch cheese, and coffee beans. While related to the aflatoxins, they are not as potent. They act by inhibiting DNA synthesis.

Zearalenone

There are at least five naturally occurring zearalenones, and they are produced by *Fusarium* spp., mainly *F. graminearum* (formerly *F. roseum,* = *Gibberella*

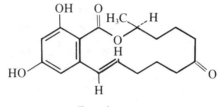

Zearalenone

zeae) and *F. tricinctum*. Associated with corn, these organisms invade field corn at the silking stage, especially during heavy rainfall. If the moisture levels remain high enough following harvesting, the fungi grow and produce toxin. Other crops, such as wheat, oats, barley, and sesame, may be affected in addition to corn.

The toxins fluoresce blue-green under long-wave UV and greenish under short-wave UV. They possess estrogenic properties, and promote estrus in mice and hyperestrogenism in swine. While they are nonmutagenic in the Ames assay, they produce a positive response in the *Bacillus subtilis* Rec assay (see Stark, 116).

VIRUSES

Much less is known about the incidence of viruses in foods than about bacteria and fungi, for several reasons. First, being obligate parasites, viruses do not grow on culture media as do bacteria and fungi. The usual methods for their cultivation consist of tissue culture and chick embryo techniques. Second, since viruses do not replicate in foods, their numbers may be expected to be low relative to bacteria, and extraction and concentration methods are necessary for their recovery. Although much research has been devoted to this methodology, it is difficult to effect more than about a 50% recovery of virus particles from products such as ground beef. Third, laboratory virological techniques are not practiced in many food microbiology laboratories. Finally, not all viruses of potential interest to food microbiologists can be cultured by existing methods (the Norwalk virus is one example).

In spite of these reasons, there does exist a growing body of information on viruses in foods, some of which is outlined and discussed below.

Since it has been demonstrated that any intestinal bacterial pathogen may under unsanitary conditions be found in foods, the same may be presumed for intestinal viruses, even though they may not proliferate in foods. Cliver et al. (27) have noted that virtually any food can serve as a vehicle for virus transmission, and they have stressed the importance of the anal-oral mode of transmission, especially for viral hepatitis of food origin. Just as nonintestinal bacteria of human origin are sometimes found in foods, the same may be true for viruses, but because of their tissue affinities, foods would serve as vehicles only for the intestinal or enteroviruses. These agents may be accumulated by some shellfish up to 900-fold (47). Viral gastroenteritis is believed to be second only to the common cold in frequency (29). A list of viruses that may be found in foods is presented in Table 22-3.

Table 22-3. Human intestinal viruses with high potential as food contaminants (79).

1. Picornaviruses
    Polioviruses 1-3
    Coxsackievirus A 1-24
    Coxsackievirus B 1-6
    ECHOvirus 1-34
    Enterovirus 68-71
    Probably hepatitis A
2. Reoviruses
    Reovirus 1-3
    Rotaviruses
3. Parvoviruses
    Human gastrointestinal viruses
4. Papovaviruses
    Human BK and JC viruses
5. Adenoviruses
    Human adenoviruses types 1-33

For further general information, see reviews by Gerba et al. (47, 49), Larkin (79), and Cliver et al. (27). Methods for recovery of viruses from foods have been outlined by Cliver et al. (28), Metcalf et al. (88), and Sobsey et al. (114).

Incidence in foods and the environment

The single most common food source of gastroenteritis-causing viruses is shellfish. While crustaceans do not concentrate viruses, molluscan shellfish do because they are filter feeders. When poliovirus 1 was added to waters, blue crabs were contaminated but they did not concentrate the virus (56). Shucked oysters artificially contaminated with 10^4 pfu (plaque forming units) of a poliovirus retained viruses during refrigeration for 30 to 90 days with a survival rate of 10–13% (34). It has been reported that the uptake of enteroviruses by oysters and clams is not likely when viruses in the water column are <0.01 pfu/ml (78). The recovery method employed by the latter authors was capable of detecting 1.5–2.0 pfu/shellfish.

Although the coliform index is of proven value as an indicator of intestinal bacterial pathogens in waters, it appears to be inadequate for enteroviruses, which are more resistant to adverse environmental conditions than bacterial pathogens (96). In a study of > 150 samples of recreational waters from the upper Texas Gulf, enteroviruses were detected 43% of the time when by coliform index the samples were judged acceptable, and 44% of the time when judged acceptable by fecal coliform standards (48). In the same study, enteroviruses were found 35% of the time in waters that met acceptable standards for shellfish harvesting, and the investigators concluded that the coliform standard for waters does not reflect the presence of viruses. From a study of hard-shell clams off the coast of North Carolina, enteric viruses were found in those from open and closed beds (136). (Closed waters are those not open to commercial shellfishing because of coliform counts.) From open beds, three of thirteen 100-g samples were positive for viruses, while all thirteen were negative for salmonellae, shigellae, or yersiniae. From closed waters, six of fifteen were positive for salmonellae, and all were negative for shigellae and yersiniae (136). The latter investigators found no correlation between enteric viruses and total coliforms or fecal coliforms in shellfish waters, or total coliforms, fecal coliforms, fecal streptococci, or APC in clams. Even though enteric viruses may be found in shellfish from open waters, < 1% of shellfish samples examined by the FDA contained viruses (79).

With respect to the capacity of certain viruses to persist in foods, it has been shown that enteroviruses persisted in ground beef up to 8 days at 23° or 24°C and were not affected by the growth of spoilage bacteria (58). In a study of fourteen vegetable samples for the existence of naturally occurring viruses, none were found, but coxsackievirus B5 inoculated onto vegetables did survive at 4°C for 5 days (72). In an earlier study, these investigators showed that coxsackievirus B5 showed no loss of activity when added to lettuce and stored at 4°C under moist conditions for 16 days. Several enteric viruses failed to survive on the surfaces of fruits, and no naturally occurring viruses were found in nine fruits examined (73). ECHOvirus 4 and poliovirus 1 were found in one each of seventeen samples of raw oysters by Fugate et al. (45), and poliovirus 3 was found in one of twenty-four samples of oysters. Of seven food-processing plants surveyed for human viruses, none was found in a vegetable-processing plant or in three that processed animal products (74). The latter investigators examined sixty samples of market foods but were unable to detect viruses in any. They concluded that viruses in the U.S. food supply are very low.

Destruction in foods

The survival of the hog cholera (HCV) and African swine fever (ASFV) viruses in processed meats was studied by McKercher et al. (86). From pigs infected with these viruses, partly cooked canned hams and dried pepperoni and salami sausages were prepared; while virus was not recovered from the partly cooked canned hams, they were recovered from hams after brining but not after heating. The ASFV retained viability in the two sausage products following addition of curing ingredients and starters but were negative after 30 days. HCV also survived the addition of curing ingredients and starter and retained viability 22 days later.

The effect of heating on destruction of the foot-and-mouth virus was evaluated by Blackwell et al. (6). When ground beef was contaminated with virus-infected lymph node tissue and processed to an internal temperature of 93.3°C, the virus was destroyed. However, in cattle lymph node tissue, virus survived for 15 but not 30 min at 90°C. The boiling of crabs was found sufficient to inactivate 99.9% of poliovirus 1, and a rotavirus and an ECHOvirus was destroyed within 8 min (56). A poliovirus was found to survive stewing, frying, baking, and steaming of oysters (34). In broiled hamburgers, enteric viruses could be recovered from eight of twenty-four patties cooked rare (to 60°C internally) if the patties were cooled immediately to 23°C (119). No viruses were detected if the patties were allowed to cool for 3 min at room temperature before testing.

Hepatitis A virus

There are more documented outbreaks of hepatitis A traced to foods than any other viral infection. Over 100 outbreaks have been recorded (27). The virus belongs to the family Picornaviridae, as do the polio, ECHO, and coxsackie viruses, and all have single-stranded RNA genomes. The incubation period for infectious hepatitis ranges from 15–45 days, and lifetime immunity usually occurs after an attack. The fecal-oral route is the mode of transmission, and raw or partially cooked shellfish from polluted waters is the most common vehicle food.

In the United States in 1973, 1974, and 1975, there were five, six, and three outbreaks respectively, with 425, 282, and 173 cases. The 1975 outbreaks were traced to salad, sandwiches, and glazed donuts served in restaurants. Several other outbreaks are noted in Table 22-4 along with summaries of outbreaks between 1945 and 1966. Of 66,747 reported cases of food poisoning in the United States from 1974–78, 2.5% were of viral etiology (27), with hepatitis A accounting for most.

Norwalk virus

This virus has a single-stranded RNA genome and its classification is uncertain, although it may be a calcivirus. The virus is believed to be responsible for about one-third of viral gastroenteritis in the United States (30). The Norwalk virus has not yet been propagated in any cell culture systems. It is infective

Table 22-4. Synopsis of some food-borne viral infections.

Virus	Vehicle foods/source/comments	Reference
Polio	10 outbreaks with 161 cases, 1914–49. Raw and pasteurized milk were the vehicles in 6 and 2 outbreaks, respectively.	(26)
Polio 3	Oysters from the Louisiana coast.	(45)
ECHO 9	Oysters from the New Hampshire coast.	(88)
ECHO 4	Oysters from the Texas Gulf waters.	(45)
Hepatitis A	Over 1,600 cases between 1945–66 from a variety of foods.	(26)
Hepatitis A	About 1,600 cases from oysters and clams, 1955–66.	(26)
Hepatitis A	A total of 3 outbreaks in 1975 (173 cases) from glazed donuts, sandwiches, and salad/sandwiches in New York, Oklahoma, and Oregon.	(126)
Hepatitis A	Eighteen of about 580 persons at risk contracted the disease from sandwiches in Pennsylvania.	(127)
Hepatitis A	Oysters.	(96)
Hepatitis A	From 3 outbreaks in New York state, there were 10 victims of about 238 at risk. Vehicle food was raw clams.	(129)
Hepatitis A	Fifty-six cases were contracted in a restaurant that employed a food handler with the disease in New Jersey.	(128)
Hepatitis A	Some 203 persons contracted the disease at a drive-in restaurant in Oklahoma in 1983.	(130)
Hepatitis A	From a salad bar–type restaurant in Texas, 123 cases occurred in 1983. A food handler was suspected.	(130)
Norwalk virus	Oysters. About 2,000 persons were infected in Australia in 1978.	(91)
Norwalk virus	Raw oysters were involved in this outbreak in Florida in 1980. Six of 13 persons at risk contracted the disease. The agent was identified by RIA.	(54)
Enteroviruses	Oysters from the Texas Gulf.	(53)
Coxsackie B-4	Oysters.	(88)

primarily for older children and adults, and about 67% of adults in the United States possess serum antibodies (29). Of seventy-four acute non-bacterial outbreaks reported to the U.S. CDC for 1976–80, 42% were attributed to the Norwalk virus (67). It is known to be associated with travelers' diarrhea, and polluted water is an obvious source.

The first documented food-borne outbreak occurred throughout Australia in 1978 and involved at least 2,000 persons (91). The vehicle food was oysters. The virus was found in 39% of fecal specimens examined by electron microscopy, and antibody responses were demonstrated in 75% of paired sera tested. The incubation periods ranged from 18–48 h, with most cases occurring in 34–38 h. The first symptom was nausea, usually accompanied by vomiting, nonbloody diarrhea, and abdominal cramps with symptoms lasting 2–3 days. Another food-borne outbreak of Norwalk virus occurred in Australia and involved five cases from two incidents traced to a single batch of bottled oysters (39). Symptoms occurred in 24–48 h, and all victims experienced diarrhea, while two also experienced nausea, abdominal cramps, and vomiting. From the stool of one victim, the Norwalk virus was detected by electron microscopy and an ECHO 5 virus was detected in cell culture. From the oysters, ECHO 8 and a reovirus were recovered. The oysters had an APC of 2.2×10^4/g and a fecal coliform count of 500/100 g. No salmonellae or *V. parahaemolyticus* were detected (39). The first documented food-borne outbreak in the United States occurred in January 1980 in northwest Florida, where six of thirteen persons who ate raw oysters contracted the illness (54). The etiologic agent was identified by use of a radioimmunoassay (RIA) method.

The Norwalk virus is more resistant to destruction by chlorine than other enteric viruses. In volunteers, 3.75 ppm chlorine in drinking water failed to inactivate this virus while poliovirus type 1 and human and simian rotaviruses were inactivated (68). Some Norwalk viruses remained infective at residual chlorine levels of 5 to 6 ppm. Hepatitis A viruses are not as resistant as Norwalk, but both are clearly more resistant to chlorine than the rotaviruses.

Rotaviruses

These viruses were first discovered in Australia in 1973 and first propagated in the laboratory in 1981. They belong to the family Reoviridae and contain double-stranded RNA. They are known to cause acute diarrhea in children, with the most susceptible age being 6 months to 1 year. The general lack of susceptibility of persons over 6 years of age is due apparently to the existence in the gut of secretory IgA (immune globulin A). The virus attacks the upper intestines, and symptoms generally occur within 1–3 days. Unlike bacterial pathogens, the minimal infectious dose is not known for these agents. The most common symptoms are watery diarrhea and vomiting. While adults are generally immune, diarrheal illness does occur—these pathogens are known to be associated with travelers' diarrhea (113) and they have been isolated from the stools of gastroenteritis patients (133). No food-borne outbreaks of rotavirus gastroenteritis have been documented.

Of forty-seven U.S. Peace Corps volunteers who traveled to Honduras, and sixty-six Panamanians who traveled to Mexico, 36 and 30%, respectively, showed about a fourfold increase in rotavirus antibody titers compared to pretravel specimens, thus associating rotavirus and travelers' diarrhea as noted above (109). The antibody found consisted of both IgG and IgA classes.

Levels of from 1 to 321 rotaviruses per fluorescent field (with a mean of 9.8) were found in Houston, Texas, sewage between May and September by use of indirect immunofluorescence (57). Generally higher levels were found during the winter and spring. While a seasonal pattern was observed for rotaviruses, such was not the case for total enteroviruses, which were found at levels of 7.5–800 pfu/l and were detected and enumerated by use of Buffalo green monkey cells (57). By use of an ELISA method, 100% of Australian aborigines and Europeans were found to contain rotavirus antibodies (49).

AEROMONAS HYDROPHILA, PLESIOMONAS SHIGELLOIDES, AND OTHER GRAM-NEGATIVE BACTERIA

The two named organisms are quite similar both taxonomically and in environmental habitats, and *P. shigelloides* was once placed in the genus *Aeromonas*. Both are associated with diarrhea in man, but whether either is a primary pathogen is not entirely clear.

A. hydrophila is an aquatic bacterium found more in salt waters than in fresh waters. It is a significant pathogen to fish, turtles, frogs, snails, and alligators, and a human pathogen especially in compromised hosts. It is a common member of the bacterial flora of pigs. Diarrhea, endocarditis, meningitis, soft tissue infections, and bacteremia are caused by *A. hydrophila* (32).

Virulent strains of *A. hydrophila* produce an enterotoxin and hemolytic cytotoxins (135). The enterotoxin has a molecular weight of 15,000–20,000, an isoelectric point of 4.0–5.7, and is heat labile. It is produced late in the logarithmic phase of growth, and the genes for its production are chromosomal (82). It can be demonstrated in the 18-h rabbit ileal loop assay, and it appears to be unrelated immunologically to either cholera toxin or the heat-labile enterotoxin of *E. coli*. Environmental isolates that initially do not produce the enterotoxin, or produce it in low quantities, produce higher quantities after a series of passages in rabbit intestinal loops (82).

In addition to enterotoxin, *A. hydrophila* produces other toxic substances including hemolysins, cytotoxin, and cytotonic toxins. The latter cause rounding and induce steroidogenesis in Y-1 adrenal cells. A cytotonic enterotoxin distinct from cytotoxin and hemolysin and giving positive responses in the rabbit ileal loop, suckling mouse, and CHO assays has been demonstrated (20).

A large number of studies have been conducted on *A. hydrophila* isolates from various sources. In one study, 66 of 96 (69%) isolates produced cytotoxins, while 32 (80%) of 40 isolates from diarrheal disease victims were toxigenic, with only 41% of nondiarrheal isolates being positive for cytotoxin production. Most enterotoxigenic strains are VP (Voges-Proskauer test) and hemolysin positive and arabinose negative (15), and produce positive responses in the suckling mouse, Y-1 adrenal cell, and rabbit ileal loop assays. In a study of 147 isolates from patients with diarrhea, 91% were enterotoxigenic, while only 70% of 94 environmental strains produced enterotoxin as assessed by the suckling mouse assay (16). All but 4 of the clinical isolates produced hemolysis of rabbit red blood cells. Of 116 isolates from the Chesapeake Bay, 71% were toxic by the Y-1 adrenal cell assay and toxicity correlated with lysine decarboxylase and VP reactions (66). In yet another study, 48 of 51 cultures from humans, animals, water, and sewage produced positive responses in rabbit ileal loop assays with 10^3 or more cells, and cell-free extracts from all were loop positive (2).

With regard to growth temperature and habitat, seven of thirteen strains displayed growth at 0–5°C, four of thirteen at 10°C, and one at a minimum of 15°C (100). The psychrotrophs had optimum growth between 15 and 20°C. The maximum growth temperature for some strains was 40–45 with optimum at 35°C (55). Regarding distribution, the organism was found in all but 12 of 147 lotic and lentic habitats (55). Four of those habitats that did not yield the organism were either hypersaline lakes or geothermal springs. Some waters contained up to 9,000/ml. An ecologic study of *A. hydrophila* in the Chesapeake Bay revealed numbers ranging from < 0.3/liter to 5 × 10^3/ml in the water column, and about 4.6 × 10^2/g of sediment (66). The presence of this organism correlated with total, aerobic, viable, and heterotrophic bacterial counts; and its presence was inversely related to dissolved O_2 and salinity, with the upper salt level being about 15%. Fewer were found during the winter than during the summer months.

Plesiomonas shigelloides is found in surface waters and soils, and has been recovered from fish, shellfish, other aquatic animals, as well as from terrestrial meat animals. It differs from *A. hydrophila* in having a G + C content of DNA of 51%, versus 58–62% for *A. hydrophila*. It has been isolated by many investigators from patients with diarrhea, and it is associated with other general infections in man. It has not been shown to produce an enterotoxin.

P. shigelloides was recovered by Zajc-Satler et al. (143) from the stools of 6 diarrheal patients. It was believed to be the etiologic agent, although salmonellae were recovered from 2 patients. Two outbreaks of acute diarrheal disease occurred in Osaka, Japan, in 1973 and 1974, and the only bacterial pathogen recovered from stools was *P. shigelloides*. In the 1973 outbreak, 978 of 2,141 persons became ill, with 88% complaining of diarrhea, 82% of abdominal pain, 22% of fever, and 13% of headaches (124). Symptoms lasted 2–3 days. Of 124 stools examined, 21 yielded *P. shigelloides* 017:H2. The same serovar was recovered from tap water. In the 1974 outbreak, 24 of 35 persons became ill with symptoms similar to those noted. *P. shigelloides* serovar 024:H5 was recovered from three of eight stools "virtually in pure culture" (124). The organism was recovered from 39% of 342 water and mud samples, as well as from fish, shellfish, and newts.

Growth of *P. shigelloides* has been observed at 10°C (100), and 59% of fifty-nine fish from Zaire waters contained the organism (134). In the latter study, the organism was found more in river fish than lake fish. It appears not to produce an enterotoxin since only four of twenty-nine isolates produced positive responses in rabbit ileal loops (102). Food-borne cases have not been documented, but the organism has been incriminated in at least two outbreaks (see 89).

Regarding other gram-negative bacteria, heat-stable enterotoxins have been shown for *Klebsiella pneumoniae* (70), and for *Enterobacter cloacae* (71). The STs were similar and resembled *E. coli* ST. Since LT and ST are plasmid-mediated toxins and since the Enterobacteriaceae are common inhabitants of the gastrointestinal tract, it is not unreasonable to expect that, through plasmid transfer, some non–*E. coli* organisms may acquire and express these toxin-synthesizing genes. The potential significance of some of these organisms in foods has been noted (125).

LISTERIOSIS

The genus *Listeria* consists of at least seven species of gram-positive non-sporing rods, only one of which is an animal pathogen—*L. monocytogenes.* This species was first recognized as an animal pathogen in 1926; since that time over fifty mammals, including man, have been shown to be susceptible to its infection. In addition to mammals, fowls, ticks, fish, and crustaceans are susceptible. *L. monocytogenes* is widely distributed in nature, where it tends to be associated more with vegetation than soils. It was isolated from 8.4 to 44% of samples taken from grain fields, pastures, mud, animal feces, wildlife feeding grounds, and related sources (137). Its survival in moist soils for 295 days and beyond has been demonstrated (138). It grows at 4°C and at least up to 38°C. It is destroyed at milk pasteurization temperatures.

In human adults, *L. monocytogenes* is known to cause meningitis, encephalitis, septicemia, endocarditis, abortion, and abscesses and local purulent lesions. In neonates (newborns), it is the third leading cause of bacterial meningitis after *E. coli* and *Streptococcus agalactiae* (106). The disease is characterized by an increase in the number of circulating monocytes, and the monocytosis-producing agent is associated with the cell wall. It is lipid in nature and is soluble in chloroform or ether. All virulent strains produce β-hemolysis on blood agar (horse or sheep cells) plates. To differentiate virulent from avirulent strains, serotyping is done. The serovars most frequently found in human infections are 1/2b and 4b, with the latter representing up to 80% of strains isolated in the United States and Canada.

Although sporadic cases of human listeriosis occur annually in the United States, the outbreak in 1985 in southern California has drawn new attention to this syndrome. There were at least eighty-six cases of listeriosis in this outbreak, which was traced to several brands of commercially produced Mexican-style cheeses (131). Fifty-eight of the cases were among mother-infant pairs, and sixteen were nonperinatal. There were twenty-nine deaths: eight neonates, thirteen stillbirths, and eight nonneonates (131). *L. mono-*

cytogenes serotype 4b was recovered from victims and from two varieties of cheeses. The bacteria entered the product by cross contamination of raw milk through faulty equipment.

Two documented outbreaks of human listeriosis traced to foods occurred before the one noted above. One occurred in 1981 in the Maritime Provinces of Canada and consisted of forty-one cases (seven adults and thirty-four perinatals). The probable source was coleslaw, and the serotype was 4b (106). An identical serotype was recovered from refrigerated coleslaw from the home of one victim. Another outbreak occurred during the summer of 1983 in Massachusetts and consisted of forty-nine cases with fourteen deaths (42). Forty-two of the forty-nine cases occurred in adults who had preexisting illnesses or conditions causing immunocompromise, and the other seven cases were among mother-infant pairs (42). Pasteurized milk was incriminated, although no isolates of *L. monocytogenes* were recovered. Of forty isolates tested from victims, thirty-two were serotype 4b. The milk came from a group of farms on which listeriosis in dairy cows was known to have occurred at the time of the outbreak.

The prevalence of *L. monocytogenes* in foods is not known, and the lack of studies in this regard probably results from the difficulty of isolating this organism from naturally contaminated samples. Typical of the recovery methods in use is an enrichment step with large jars containing brain-heart infusion or tryptose broth with incubations at 4°C for up to 3.5 months, accompanied by periodic platings onto selective media containing nalidixic acid, polymyxin, and/or acriflavin; or onto *Listeria* agar. To recover the organism from animal specimens, some investigators have found it necessary to incubate enrichments at 4°C for up to one year! The existence of this organism in the natural environment and its cause of disease among meat animals suggest that its presence in foods may not be sporadic, but more specific information must await the development of isolation/enumeration procedures that lend themselves to food analysis.

HISTAMINE-ASSOCIATED (SCOMBROID) POISONING

Illness contracted from eating scombroid fish or fish products containing high levels of histamine is often referred to as scombroid poisoning. Among the scombroid fishes are tuna, mackerel, bonito, and others. The histamine is produced by bacterial decarboxylation of the generally large quantities of histidine in the muscles of this group. Sufficient levels of histamine may be produced without the product being organoleptically unacceptable, with the result that scombroid poisoning may be contracted from both fresh and organoleptically spoiled fish. The history of this syndrome has been reviewed by Hudson and Brown (61), who questioned the etiologic role of histamine. This is discussed further below.

The bacteria most often associated with this syndrome are *Proteus* spp., especially *P. morganii*, of which all strains appear to produce histamine at levels up to 400 mg%. Among other bacteria known to produce histidine decarboxylase are *K. pneumoniae, Hafnia alvei, Citrobacter freundii, Clostridium perfringens, Enterobacter aerogenes, Vibrio alginolyticus,* and other *Proteus* spp. From room-temperature spoiled skipjack tuna, 31% of 470 bacterial isolates produced from 100 to 400 mg% of histamine in broth (94). The strong histamine formers were *P. morganii, Proteus* spp., and a *Klebsiella* sp., while weak formers included *H. alvei* and *Proteus* spp. Skipjack tuna spoiled in seawater at 38°C contained *C. perfringens* and *V. alginolyticus* among other histidine decarboxylase producers (142). From an outbreak of scombroid poisoning associated with tuna sashimi, *K. pneumoniae* was recovered and shown to produce 442 mg% histamine in a tuna fish infusion broth (120). This syndrome has been associated with foods other than scombroid fish, particularly cheeses including Swiss cheese, which in one case contained 187 mg% histamine, and the symptoms associated with the outbreak occurred in 30 min to 1 h after ingestion (121).

Histamine content of stored skipjack tuna can be estimated if incubation times and temperatures of storage are known. Frank et al. (45) found that 100 mg% formed in 46 h at 70°F; in 23 h at 90°F; and in 17 h at 100°F. A nomograph was constructed over the temperature range of 70°–100°F, underscoring the importance of low temperatures in preventing or delaying histamine formation.

Histamine production is favored by low pH, but it occurs more when products are stored above the refrigerator range. The lowest temperature for production of significant levels was found to be 30°C for *H. alvei, C. freundii,* and *E. coli;* 15°C for two strains of *P. morganii;* and 7°C for *K. pneumoniae* (5).

The syndrome is contracted by eating fresh or processed fish of the type noted; symptoms occur within minutes and for up to 3 h after ingestion of toxic food, with most cases occurring within 1 h. Typical symptoms consist of a flushing of the face and neck accompanied by a feeling of intense heat and general discomfort, and diarrhea. Subsequent facial and neck rashes are common. The flush is followed by an intense, throbbing headache tapering to a continuous dull ache. Other symptoms include dizziness, itching, faintness, burning of the mouth and throat, and the inability to swallow (see 61). The minimum level of histamine thought necessary to cause symptoms is 100 mg%. Large numbers of *P. morganii* in fish of the type incriminated in this syndrome and a level of histamine > 10 mg% is considered significant relative to product quality.

The first fifty incidents in Great Britain occurred between 1976 and 1979, with all but nineteen occurring in 1979. Canned and smoked mackerel was the most common vehicle, with bonita, sprats, and pilchards involved in one outbreak each. The most common symptom among the 196 cases was diarrhea (50).

Regarding etiology, Hudson and Brown (61) believe the evidence does not favor histamine per se as the agent responsible for the syndrome. They suggest a synergistic relationship involving histamine and other as yet un-identified agents such as other amines or factors that influence histamine absorption. This view is based on the inability of large oral doses of histamine or histamine-spiked fish to produce symptoms in volunteers. On the other hand, the suddenness of onset of symptoms is consistent with histamine reaction, and the association of the syndrome with scombroid fish containing high numbers of histidine decarboxylase-producing bacteria cannot be ignored. While the precise etiology may yet be in question, bacteria do play a significant if not indispensable role.

CESTODES

Of the flatworms that are parasitic for man, three are obtained by eating the flesh of pork, beef, and fish. The beef tapeworm, *Taenia saginata,* lives as an adult in the human intestine and in larval form in the muscles of bovines. The form of the organism found in bovine and porcine flesh is referred to as a **cysticercus**. They have been reported to occur in skeletal muscles, as well as in the muscles of the tongue, neck, jaws, heart, diaphragm, and esophagus, where they may remain viable for as long as a year (62). Upon the ingestion of cysticercus-containing beef by man, the scolex of the parasite evaginates and attaches to the intestine, where growth of the tapeworm begins. Development to adult stage requires from 8–10 weeks, at which time **proglottids** may appear in the feces.

The pork tapeworm, *T. solium,* has a life cycle similar to that of *T. saginata,* but unlike the latter, man can serve as the intermediate host as well. The larval stage of *T. solium* is *Cysticercus cellulosae.* Human in-festations from this form are referred to as **cysticercosis** (90). Autoinfection in man with eggs from the adult worm reaching the upper intestines is not unknown. Both the beef and pork tapeworms are distributed worldwide, with the incidence being highest where raw or improperly cooked pork is eaten. In the United States the pork tapeworm has been reported to be rarer than the beef tapeworm.

Prevention of tapeworm infections in man requires the rejection of tapeworm-infested meats—one of the primary functions of federal and local meat inspection laws. If present in meats, the cysts of these parasites can be destroyed by cooking to a temperature of at least 60°C (62). These forms are destroyed upon the freezing of meats to at least $-10°C$, with holding for 10–15 days, or by immersion in concentrated salt solutions for up to 3 weeks. The epidemiologic control of these infestations involves breaking the host cycle.

The third tapeworm that man acquires from food is the broad fish tapeworm, *Diphyllobothrium latum.* Man obtains this organism by ingesting fish such as pike, river perch, trout, and other fish that become infested upon the ingestion of **copepods**, in which the eggs of the parasite develop into **procercoid larva** measuring around 0.5 mm in length. The adult stage of this parasite develops in the intestinal tract of man, from which eggs are shed through feces. Once in water, the eggs hatch and the first larval stage is taken up by certain copepods, and the cycle continues. In addition to man, fish-eating mammals such as dogs and bears may become infested with this tapeworm. Like the other tapeworms, cooking for at least 10 min at 50°–55°C is adequate to destroy the larval forms of this organism. Freezing to at least −10°C has also been reported to destroy these forms.

Hydatidosis is an infestation in man caused by ingesting the larval stage (or hydatid cyst) of two species of cestodes, *Echinococcus granulosis* and *E. multilocularis*. Man becomes infested by these forms as a result of ingesting infective ova, which are shed by dogs and certain other carnivores. The natural host of the adult parasite is the dog. Problems arise only where dogs are allowed access to uncooked slaughterhouse viscera, or through the lack of proper hygiene of persons in close contact with dogs.

NEMATODES

Of the several species of nematodes pathogenic for man, the genus *Trichinella,* especially *T. spiralis,* is perhaps of greatest importance from the standpoint of foods. *T. spiralis* causes trichinosis in man. Unlike the tapeworm infestations discussed above, both larval and adult stages of *T. spiralis* are passed in the same host. Upon the ingestion of encysted larvae in muscle tissue, adults develop in the intestine, discharge larvae into the blood, and these larvae again settle in striated muscles and again become encysted. The cycle is repeated if another animal consumes these muscles containing encysted larvae. In addition to man and swine, other animals such as dogs, cats, and bears may be involved. Human trichinosis is contracted most often by eating infected pork. Between 10–15% of cases in the United States are contracted from wild game meats (see below).

This disease can be prevented by the thorough cooking of meats such as pork or bear meat. In a study on the heat destruction of trichina larvae in pork roasts, all roasts cooked to an internal temperature of 140°F or higher were subsequently found to be free of organisms (19). Larvae were found in all roasts cooked at 130°F or lower, and in some roasts cooked at 135°F. The USDA recommendation for pork products is that the product be checked with a thermometer after standing and if any part does not attain 76.7°C (170°F), the product should be cooked further (132).

Freezing will destroy the encysted forms, but freezing times and temperatures depend upon the thickness of the product (see Table 22-5). The lower the temperature of freezing, the more destructive it is to *T. spiralis,* as was demonstrated in the following study. Four selected temperatures were chosen for the freezing of infected ground pork that was stuffed into casings and packed into boxes. When frozen and stored at −17.8°C, the trichinae lost infectivity between 6 and 10 days; while at −12.2°C, infectivity was lost between 11 and 15 days (140). When frozen at −9.4°C, they

Table 22-5. Required period of freezing at temperatures indicated (75).

Temperature (°C)	Group 1 (days)	Group 2 (days)
−15	20	30
−23	10	20
−29	6	12

Group 1 = Less than 15.24 cm in depth.
Group 2 = More than 15.24 cm in depth.
From Sec. 18.10. Regulations Governing the Meat Inspection of the United States Department of Agriculture (9 CFR 18.10, 1960).
Copyright © 1983, Institute of Food Technologists.

remained infective up to 56 days, and for up to 71 days when frozen at −6.7°C (146). The destruction of trichina larvae by irradiation is discussed in Chapter 12.

The effect of curing and smoking on the viability of trichina in pork hams and shoulders was investigated by Gammon et al. (46). They employed the meat of hogs experimentally infested with *T. spiralis* as weanling pigs. After curing, the meat was hung for 30 days followed by smoking for approximately 24 h at 90°–100°F, with subsequent aging. Live trichinae were found in both hams and shoulders 3 weeks after smoking, but none could be detected after 4 weeks. The effect of NaCl concentration, a_w, and fermentation method on viability of *T. spiralis* in Genoa salami was evaluated by Childers et al. (21). Pork from experimentally infected pigs was used to prepare salami. The trichinae larvae were completely destroyed at day 30 and thereafter in salami made with 3.33% NaCl and given high-temperature (46.1°C) fermentation treatment, irrespective of product pH. No larvae were found in products made with 3.33% NaCl and given low-temperature fermentation after 30 days. In salami made with no salt, 25% of larvae were found at days 15–25, but none thereafter.

The efficacy of microwave ovens in destroying *T. spiralis* larvae has been investigated by several groups. In a homemaker-oriented study in which most trichina-infected pork roasts were cooked in microwave ovens by time rather than product temperature, Zimmerman and Beach (145) found that of fifty-one products (forty-eight roasts and three pork chops) cooked in six different ovens, nine remained infective. Six of the nine did not attain a mid-roast temperature of 76.7°C, while the other three exceeded this temperature at some point in the cooking cycle. Authors noted that the experimentally infected pork used in the study came from pigs infected with 250,000 *T. spiralis,* which produced around 1,000 trichina/g of tissue compared to about 1 trichina/g in naturally infected pigs. While the large number of trichinae/g may have been a factor in their survival at the cooking procedures employed, the inherent unevenness of cooking in microwave ovens is of concern. In another study, while trichinae larvae were not inactivated at 77° or 82°C in microwave ovens, cooking to an internal temperature of 77°C in a conventional convection oven, flat grill, charbroiler, or deep fat did inactivate the larvae (76). Further, infected larvae survived rapid cooking that involved thawing pork chops in an industrial microwave oven followed by cooking on a charbroiler to 71° or 77°C (75).

The cooking of pork in microwave ovens is clearly a matter of concern relative to the destruction of trichinae larvae, and two factors may explain the greater efficiency of convection ovens over microwave ovens. First, microwave cooking is rapid and herein may lie the problem. Oven heat has been shown to be more destructive to trichinae larvae in roasts when slow-cooked in conventional ovens at 200°F than when fast-cooked at 350°F (19). Second, a convection oven is more uniformly heated than some microwave ovens. This is minimized if the product is rotated in the latter type ovens, or if the oven is equipped with an automatic rotating device. Otherwise, uneven heating occurs, leading to undercooking of some parts of a roast while other parts may be overcooked. It has been shown that a set of criteria that leads to consistent doneness of pork products in microwave ovens will result in safe products (144).

Regarding incidence, there have been between 100 and 150 cases of trichinosis per year in the United States since 1966. For the 5-year period 1977–81, 686 cases with 4 deaths were reported (104). Pork was incriminated in 79% of the cases for the years 1975–81, with bear meat in 14%, and ground beef in 7%. Studies on pork in retail ground beef have revealed that from 3 to 38% of beef samples contained pork. The presence of pork in ground beef may be deliberate on the part of some stores or may result from using the same grinder for both products. The increasing consumption of ground beef along with the fact that it is often not well cooked offers the potential for a continuing increase in the incidence of trichinosis, unless the adulteration of beef with pork is reduced.

Trichinosis can be controlled by avoiding the feeding of infected meat scraps or wild game meats to swine and by preventing the consumption of infested tissues by other animals. The feeding of uncooked garbage to swine helps to perpetuate this disease. Where only cooked garbage is fed to pigs, the incidence of trichinosis has been shown to fall sharply.

A recently developed ELISA test for the detection of *Trichinella* antibodies in infested swine should prove to be of immense value in controlling this disease. While pork obtained directly from farms is sometimes the source of the disease to man, a significant number of cases occur from the consumption of federally inspected pork. The ELISA test will be of greatest value in eliminating the latter source.

TREMATODES

A large number of trematodes or flukes cause disease in man, but only one species will be mentioned here, *Clonorchis sinensis*. This organism, which is an important pathogen in Asia, carries out its adult stage in the bile ducts and gall bladder of mammals including man; its larval stage is found in some 40 species of fresh-water fish. Man contracts the organisms by eating one of the susceptible fish. Fish become reinfected when human feces or those of dogs, cats, hogs, and other mammals are deposited in water. Like the cestode infestations, the encysted larval forms in fish flesh can be destroyed by heating. A temperature of at least 50°C for 15 min has been recommended to destroy these forms. Unlike the cestode infestations, these forms have been reported to resist drying and salting (62).

PROTOZOA

The only protozoan disease of any known importance in food microbiology is **toxoplasmosis**, caused by *Toxoplasma gondii*. This organism has gained importance over the past 25 years, during which time it has been shown to cause a variety of diseases in man. The organism has been reported to multiply in many different cells of the human body. In chronic infections, it is found encysted in the brain and other tissues, especially the skeletal muscles. The cyst consists of many parasites packed together within a cyst wall. The disease is somewhat similar to rubella (German measles) in that during pregnancy it may cause serious damage to the fetus. In adults, symptoms consist of fever with rash, headache, muscle aches and pain, and swelling of the lymph nodes. The muscle pain, which is rather severe, may last up to a month or more. While the incidence is not known, estimates place the number of worldwide cases at about 1 million. The disease is known to be transmitted by congenital means, and evidence suggests the role of certain nematode ova in its transmission. Cats that are fed mice chronically infected with toxoplasmosis have been shown to excrete *T. gondii* in their feces for 7 days or longer, and up to 21–24 days after eating infected cat feces (35).

The importance of this disease to food microbiology is based upon the belief that infected meat serves as a source of human toxoplasmosis. Serologic studies show that a number of individuals have antibodies to the organism. In their study of 1,191 individuals, Feldman and Miller (41) found that 426 or 36% had toxoplasma antibodies, and 23% or 801 animals had antibodies to this organism. In another study of U.S. Army recruits ranging in age from 17 to 26, Feldman (40) found that 13% were positive for toxoplasma antibodies. Pork and mutton have been reported to show a high rate of infection by this organism, while infections of beef are rare although reported. The first demonstration of this organism in slaughter animals was made by Jacobs et al. (64), who showed that 24, 1.7, and 9.3% of diaphragm muscles of swine, beef cattle, and sheep, respectively, were infected with toxoplasma cysts. Jacobs (62) reports that it is similar to the *Trichinella* larvae with respect to heat, freezing, and irradiation destruction. The encysted parasites are scattered and too small to be seen with the unaided eye, thus making their detection by visual inspection all but impossible. For further information on this parasite relative to foods, see Jacobs (63).

A case of toxoplasmosis in a 7-month-old child who drank unpasteurized goat's milk has been recorded (98). Although *T. gondii* could not be isolated from milk, some of the goats had antibody titers as high as 1:512, and the child had titers as high as 1:16,384.

PARALYTIC SHELLFISH POISONING

This condition is contracted by eating oysters, mussels, or clams. Symptoms usually develop within 30 min after the ingestion of susceptible mollusks and death occurs within 3–12 h. The symptoms are characterized by paresthesia (tingling, numbness, or burning), which begins about the mouth, lips, and tongue and later spreads over the face, scalp, neck, and to the fingertips and toes. The disease is contracted from eating mussels, clams, or oysters that have fed upon certain dinoflagellates of which *Gonyaulax catenella* is representative of the U.S. Pacific Coast flora. Along the north Atlantic coast of the United States and over to northern Europe *G. tamarensis* is found, and its poison is more toxic than that of *G. catenella*. Along the coast of British Columbia *G. acatenella* is found. The toxin, which has been assigned an empirical formula of $C_{10}H_{17}N_7O_4 \cdot 2HCl$, exerts its effect in man through cardiovascular collapse and respiratory failure (139). A fatal dose of the poison for man, which may be obtained from a single serving of highly toxic shellfish, is 0.54–0.9 mg (103). The toxin acts by blocking the propagation of nerve impulses without depolarization (103). There is no known effective antidote. From 1793 to 1958, some 792 cases were recorded with 173 (22%) deaths (85). Paralytic shellfish or "mussel poisoning" in man has a mortality rate variously reported to range from 1 to 22%. The dinoflagellates from which mollusks and other organisms obtain the toxic principal of paralytic shellfish poisoning have a wide distribution in marine waters. Masses or blooms of toxic dinoflagellates give rise to "**red tide**" or other related conditions of seas. Prevention of this type of food hazard consists of avoiding seafood from waters laden with toxic dinoflagellates. The toxin can be reduced by heating above 100°C. Thorough cooking may effect a reduction of 70% of the poison in meat. A D 250°F value of 71.4 min in soft-shell clams has been reported (52).

CIGUATERA POISONING

According to Wills (139), ciguatera poisoning was first reported from the West Indies in 1555. It is contracted by man upon the ingestion of any one of approximately 300 species of fishes such as barracudas, groupers, sea basses, jacks, sharks, and eels (101). Almost all fishes involved in ciguatera poisoning are reef or shore species that become toxic by feeding upon herbivorous fishes, which in turn feed upon toxic algae or other toxophoric matter present in coralline reefs or from related areas. Russell has pointed out that the flesh of source fishes is less toxic than the viscera, with the liver being the most poisonous part of the fish. The toxin, which has been assigned an empirical formula of $C_{35}H_{65}NO_8$, acts by causing respiratory paralysis in man. The mortality rate has been reported to vary from 2 to 7%. Symptoms of ciguatera poisoning develop within 4 h following ingestion of the offending fish and consist of nausea, paresthesia about the mouth, tongue, throat, and sometimes over the face and distal parts of the fingers and toes. Weakness, abdominal pain, vomiting, diarrhea, and chills are often experienced (101).

OTHER POISONOUS FISHES

In addition to ciguatera poisoning, numerous other types of fish poisoning have been reported dating back to the Fifth Dynasty of the ancient Egyptians (101). Of these many types, only two others will be mentioned—**puffer fish** and **Moray eel poisoning**. Puffer fish (tetraodon, fugu) poisoning may be obtained from at least 50 of the approximately 100 species of puffer fish scattered among ten genera. The puffers are widely distributed and contain the most lethal of all fish poisons, having a mortality rate of 60–70% associated with their ingestion by man. The responsible toxin has been purified and assigned an empirical formula of $C_{11}H_{17}N_3O_8$. Except in Asia, these fishes are rarely used for food. Further information may be obtained from Kao (65) and Wills (139).

  Moray eel poisoning is caused by eating the Moray eel. This poisoning syndrome is much like ciguatera poisoning. It has a mortality rate of 10% and is restricted to those parts of the world where the Moray eel is eaten for food. Death is thought to be caused by paralysis of the diaphragm.

REFERENCES

1. Achmoody, J. B., and J. R. Chipley. 1978. The influence of metabolic inhibitors and incubation time on aflatoxin release from *Aspergillus flavus. Mycologia* 70:313–20.
2. Annapurna, E., and S. C. Sanyal. 1977. Enterotoxicity of *Aeromonas hydrophila. J. Med. Microbiol.* 10:317–23.
3. Ayres, J. C., J. O. Mundt, and W. E. Sandine. 1980. *Microbiology of foods.* 658–83. San Francisco: Freeman.
4. Bacon, C. W., J. G. Sweeney, J. D. Hobbins, and D. Burdick. 1973. Production of penicillic acid and ochratoxin A on poultry feed by *Aspergillus ochraceus:* Temperature and moisture requirements. *Appl. Microbiol.* 26:155–60.
5. Behling, A. R., and S. L. Taylor. 1982. Bacterial histamine production as a function of temperature and time of incubation. *J. Food Sci.* 47:1311–14, 1317.
6. Blackwell, J. H., D. Rickansrud, P. D. McKercher, and J. W. McVicar. 1982. Effect of thermal processing on the survival of foot-and-mouth disease virus in ground meat. *J. Food Sci.* 47:388–92.

7. Buchanan, R. L., M. A. Harry, and M. A. Gealt. 1983. Caffeine inhibition of sterigmatocystin, citrinin, and patulin production. *J. Food Sci.* 48:1226–28.
8. Buchanan, R. L., and D. F. Lewis. 1984. Caffeine inhibition of aflatoxin synthesis: Probable site of action. *Appl. Environ. Microbiol.* 47:1216–20.
9. Buchanan, R. L., G. Tice, and D. Marino. 1982. Caffeine inhibition of ochratoxin A production. *J. Food Sci.* 47:319–21.
10. Bullerman, L. B. 1976. Examination of Swiss cheese for incidence of mycotoxin producing molds. *J. Food Sci.* 41:26–28.
11. ———. 1984. Effects of potassium sorbate on growth and patulin production by *Penicillium patulum* and *Penicillium roqueforti*. *J. Food Protect.* 47:312–16.
12. ———. 1984. Formation and control of mycotoxins in food. *J. Food Protect.* 47:637–46.
13. Bullerman, L. B., P. A. Hartman, and J. C. Ayres. 1969. Aflatoxin production in meats. I. Stored meats. *Appl. Microbiol.* 18:714–17.
14. ———. 1969. Aflatoxin production in meats. II. Aged dry salamis and aged country cured hams. *Appl. Microbiol.* 18:718–22.
15. Burke, V., J. Robinson, H. M. Atkinson, and M. Gracey. 1982. Biochemical characteristics of enterotoxigenic *Aeromonas* spp. *J. Clin. Microbiol.* 15:48–52.
16. Burke, V., J. Robinson, M. Cooper, J. Beamon, K. Partridge, D. Peterson, and M. Gracey. 1984. Biotyping and virulence factors in clinical and environmental isolates of *Aeromonas* species. *Appl. Environ. Microbiol.* 47:1146–49.
17. Busby, W. F., Jr., and G. N. Wogan. 1979. Food-borne mycotoxins and alimentary mycotoxicoses. In *Food-borne infections and intoxications,* ed. H. Riemann and F. L. Bryan, 519–610. New York: Academic Press.
18. Butler, W. H. 1974. Aflatoxin. In *Mycotoxins,* ed. I. F. H. Purchase, 1–28. New York: Elsevier.
19. Carlin, A. F., C. Mott, D. Cash, and W. Zimmerman. 1969. Destruction of trichina larvae in cooked pork roasts. *J. Food Sci.* 34:210–12.
20. Chakraborty, T., M. A. Montenegro, S. C. Sanyal, R. Helmuth, E. Bulling, and K. N. Timmis. 1984. Cloning of enterotoxin gene from *Aeromonas hydrophila* provides conclusive evidence of production of a cytotonic enterotoxin. *Infect. Immun.* 46:435–41.
21. Childers, A. B., R. N. Terrell, T. M. Craig, T. J. Kayfus, and G. C. Smith. 1982. Effect of sodium chloride concentration, water activity, fermentation method and drying time on the viability of *Trichinella spiralis* in Genoa salami. *J. Food Protect.* 45:816–19.
22. Christensen, C. M. 1971. Mycotoxins. *CRC Crit. Rev. Environ. Cont.* 2:57–80.
23. Chu, F. S. 1984. Immunoassays for analysis of mycotoxins. *J. Food Protect.* 47:562–69.
24. ———. 1974. Studies on ochratoxins. *CRC Crit. Rev. Toxicol.* 2:499–524.
25. Ciegler, A., H.-J. Mintzlaff, D. Weisleder, and L. Leistner. 1972. Potential production and detoxification of penicillic acid in mold-fermented sausage (salami). *Appl. Microbiol.* 24:114–19.
26. Cliver, D. O. 1967. Food-associated viruses. *Hlth Lab. Sci.* 4:213–21.
27. Cliver, D. O., R. D. Ellender, and M. D. Sobsey. 1983. Methods for detecting viruses in foods: Background and general principles. *J. Food Protect.* 46:248–59.

28. ———. 1983. Methods to detect viruses in foods: Reading and interpretation of results. *J. Food Protect.* 46:345–57.
29. Cukor, G., and N. R. Blacklow. 1984. Human viral gastroenteritis. *Microbiol. Rev.* 48:157–79.
30. Cukor, G., N. A. Nowak, and N. R. Blacklow. 1982. Immunoglobulin M responses to the Norwalk virus of gastroenteritis. *Infect. Immun.* 37:463–68.
31. Davis, N. D., and U. L. Diener. 1978. Mycotoxins. In *Food and beverage mycology,* ed. L. R. Beuchat, 397–444. Westport, Conn.: AVI.
32. Davis, W. A., II, J. G. Kane, and V. F. Garagusi. 1978. Human *Aeromonas* infections. A review of the literature and a case report of endocarditis. *Med.* 57:267–77.
33. Deger, G. E. 1976. Aflatoxin—human colon carcinogenesis? *Ann. Intern. Med.* 85:204.
34. DiGirolamo, R., J. Liston, and J. R. Matches. 1970. Survival of virus in chilled, frozen, and processed oysters. *Appl. Microbiol.* 20:58–63.
35. Dubey, J. P., N. L. Miller, and J. K. Frenkel. 1970. Characterization of the new fecal form of *Toxoplasma gondii. J. Parasitol.* 56:447–56.
36. El-Banna, A. A., and P. M. Scott. 1984. Fate of mycotoxins during processing of foodstuffs. III. Ochratoxin A during cooking of faba beans (*Vicia faba*) and polished wheat. *J. Food Protect.* 47:189–92.
37. Enomoto, M., and M. Saito. 1972. Carcinogens produced by fungi. *Ann. Rev. Microbiol.* 26:279–312.
38. Escher, F. E., P. E. Koehler, and J. C. Ayres. 1973. Production of ochratoxins A and B on country cured ham. *Appl. Microbiol.* 26:27–30.
39. Eyles, M. J., G. R. Davey, and E. J. Huntley. 1981. Demonstration of viral contamination of oysters responsible for an outbreak of viral gastroenteritis. *J. Food Protect.* 44:294–96.
40. Feldman, H. A. 1963. A nationwide serum survey of United States military recruits, 1962. *Toxoplasma* antibodies. *Amer. J. Epidemiol.* 81:385–91.
41. Feldman, H. A., and L. T. Miller. 1956. Serological study of toxoplasmosis prevalence. *Amer. J. Hyg.* 64:320–35.
42. Fleming, D. W., S. L. Cochi, K. L. MacDonald, J. Brondum, P. S. Hayes, B. D. Plikaytis, M. B. Holmes, A. Audurier, C. V. Broome, and A. L. Reingold. 1985. Pasteurized milk as a vehicle of infection in an outbreak of listeriosis. *N. Eng. J. Med.* 312:404–7.
43. Frank, H. A. 1968. Diffusion of aflatoxins in foodstuffs. *J. Food Sci.* 33:98–100.
44. Frank, H. A., D. H. Yoshinaga, and I.-P. Wu. 1983. Nomograph for estimating histamine formation in skipjack tuna at elevated temperatures. *Mar. Fish. Rev.* 45:40–44.
45. Fugate, K. J., D. O. Cliver, and M. T. Hatch. 1975. Enteroviruses and potential bacterial indicators in Gulf Coast oysters. *J. Milk. Food Technol.* 38:100–104.
46. Gammon, D. L., J. D. Kemp, J. M. Edney, and W. Y. Varney. 1968. Salt, moisture and aging time effects on the viability of *Trichinella spiralis* in pork hams and shoulders. *J. Food Sci.* 33:417–19.
47. Gerba, C. P., and S. M. Goyal. 1978. Detection and occurrence of enteric viruses in shellfish: A review. *J. Food Protect.* 41:743–54.
48. Gerba, C. P., S. M. Goyal, R. L. LaBelle, I. Cech, and G. F. Bodgan. 1979. Failure of indicator bacteria to reflect the occurrence of enteroviruses in marine waters. *Amer. J. Pub. Hlth* 69:1116–19.

49. Gerba, C. P., J. B. Rose, and S. N. Singh. 1985. Waterborne gastroenteritis and viral hepatitis. *CRC Crit. Rev. Environ. Cont.* 15:213–36.
50. Ghose, L. H., R. D. Schnagl, and I. H. Holmes. 1978. Comparison of an enzyme-linked immunosorbent assay for quantitation of rotavirus antibodies with complement fixation in an epidemiological survey. *J. Clin. Microbiol.* 8:268–76.
51. Gilbert, R. J., G. Hobbs, G. K. Murray, J. G. Cruickshank, and S. E. J. Young. 1980. Scombrotoxic fish poisoning: Features of the first 50 incidents to be reported in Britain (1976–1979). *Brit. Med. J.* 281:71–72.
52. Gill, T. A., J. W. Thompson, and S. Gould. 1985. Thermal resistance of paralytic shellfish poison in soft-shell clams. *J. Food Protect.* 48:659–62.
53. Goyal, S. M., C. P. Gerba, and J. L. Melnick. 1979. Human enteroviruses in oysters and their overlying waters. *Appl. Environ. Microbiol.* 37:572–81.
54. Gunn, R. A., H. T. Janowski, S. Lieb, E. C. Prather, and H. B. Greenberg. 1982. Norwalk virus gastroenteritis following raw oyster consumption. *Amer. J. Epidemiol.* 115:348–51.
55. Hazen, T. C., C. B. Fliermans, R. P. Hirsch, and G. W. Esch. 1978. Prevalence and distribution of *Aeromonas hydrophila* in the United States. *Appl. Environ. Microbiol.* 36:731–38.
56. Hejkal, T. W., and C. P. Gerba. 1981. Uptake and survival of enteric viruses in the blue crab, *Callinectes sapidus*. *Appl. Environ. Microbiol.* 41:207–11.
57. Hejkal, T. W., E. M. Smith, and C. P. Gerba. 1984. Seasonal occurrence of rotavirus in sewage. *Appl. Environ. Microbiol.* 47:588–90.
58. Herrmann, J. E., and D. O. Cliver. 1973. Enterovirus persistence in sausage and ground beef. *J. Milk Food Technol.* 36:426–28.
59. Hesseltine, C. W. 1967. Aflatoxins and other mycotoxins. *Hlth Lab. Sci.* 4:222–28.
60. Holmquist, G. U., H. W. Walker, and H. M. Stahr. 1983. Influence of temperature, pH, water activity and antifungal agents on growth of *Aspergillus flavus* and *A. parasiticus*. *J. Food Sci.* 48:778–82.
61. Hudson, S. H., and W. D. Brown. 1978. Histamine (?) toxicity from fish products. *Adv. Food Res.* 24:113–54.
62. Jacobs, L. 1962. Parasites in food. In *Chemical and biological hazards in food*, ed. J. C. Ayres et al., 248–66. Ames: Iowa State Univ. Press.
63. ———. 1967. *Toxoplasma* and toxoplasmosis. *Adv. in Parasitol.* 5:1–45.
64. Jacobs, L., J. S. Remington, and M. L. Melton. 1960. A survey of meat samples from swine, cattle, and sheep for the presence of encysted *Toxoplasma*. *J. Parasitol.* 46:23–28.
65. Kao, C. Y. 1966. Tetrodotoxin, sanitoxin, and their significance in the study of excitation phenomena. *Pharmacol. Rev.* 18:997–1049.
66. Kaper, J. B., H. Lockman, R. R. Colwell, and S. W. Joseph. 1981. *Aeromonas hydrophila*: Ecology and toxigenicity on isolates from an estuary. *J. Appl. Bacteriol.* 50:359–77.
67. Kaplan, J. E., G. W. Gary, R. C. Baron, N. Singh, L. B. Schonberger, R. Feldman, and H. B. Greenberg. 1982. Epidemiology of Norwalk gastroenteritis and the role of Norwalk virus in outbreaks of acute nonbacterial gastroenteritis. *Ann. Intern. Med.* 96:756–61.
68. Keswick, B. H., T. K. Satterwhite, P. C. Johnson, H. L. DuPont, S. L. Secor, J. A. Bitsura, G. W. Gary, and J. C. Hoff. 1985. Inactivation of Norwalk virus in drinking water by chlorine. *Appl. Environ. Microbiol.* 50:261–64.

69. King, A. D., Jr., and J. E. Schade. 1984. *Alternaria* toxins and their importance in food. *J. Food Protect.* 47:886–901.
70. Klipstein, F. A., and R. F. Engert. 1976. Purification and properties of *Klebsiella pneumoniae* heat-stable enterotoxin. *Infect. Immun.* 13:373–81.
71. ———. 1976. Partial purification and properties of *Enterobacter cloacae* heat-stable enterotoxin. *Infect. Immun.* 13:1307–14.
72. Konowalchuk, J., and J. I. Speirs. 1975. Survival of enteric viruses on fresh vegetables. *J. Milk Food Technol.* 38:469–72.
73. ———. 1975. Survival of enteric viruses on fresh fruit. *J. Milk Food Technol.* 38:598–600.
74. Kostenbader, K. D., Jr., and D. O. Cliver. 1977. Quest for viruses associated with our food supply. *J. Food Sci.* 42:1253–57, 1268.
75. Kotula, A. W. 1983. Postslaughter control of *Trichinella spiralis*. *Food Technol.* 37(3):91–94.
76. Kotula, A. W., K. D. Murnell, L. Acosta-Stein, L. Lamb, and L. Douglass. 1983. Destruction of *Trichinella spiralis* during cooking. *J. Food. Sci.* 48:765–68.
77. Labuza, T. P. 1983. Regulation of mycotoxins in food. *J. Food Protect.* 46:260–65.
78. Landry, E. F., J. M. Vaughn, T. J. Vicale, and R. Mann. 1982. Inefficient accumulation of low levels of monodispersed and feces-associated poliovirus in oysters. *Appl. Environ. Microbiol.* 44:1362–69.
79. Larkin, E. P. 1981. Food contaminants—viruses. *J. Food Protect.* 44:320–25.
80. Leistner, L., and F. Tauchmann. 1979. Aflatoxinbildung in Rohwurst durch verschiedene *Aspergillus flavus*-Stämme und einer *Aspergillus parasiticus*-Stamm. *Fleischwirtschaft* 50:965–66.
81. Lie, J. L., and E. H. Marth. 1967. Formation of aflatoxin in Cheddar cheese by *Aspergillus flavus* and *Aspergillus parasiticus*. *J. Dairy Sci.* 50:1708–10.
82. Ljungh, Å., and T. Wadström. 1982. *Aeromonas* toxins. *Pharmac. Ther.* 15:339–54.
83. Maggon, K. K., S. K. Gupta, and T. A. Venkitasubramanian. 1977. Biosynthesis of aflatoxins. *Bacteriol. Rev.* 41:822–55.
84. Marth, E. H., and B. G. Calanog. 1976. Toxigenic fungi. In *Food microbiology: Public health and spoilage aspects,* ed. M. P. deFigueiredo and D. F. Splittstoesser, 210–56. Westport, Conn.: AVI.
85. McFarren, E. F., M. L. Shafer, J. E. Campbell, K. H. Lewis, E. T. Jensen, and E. J. Schantz. 1960. Public health significance of paralytic shellfish poison. *Adv. Food Res.* 10:135–79.
86. McKercher, P. D., W. R. Hess, and F. Hamdy. 1978. Residual viruses in pork products. *Appl. Environ. Microbiol.* 35:142–45.
87. Metcalf, T. G., E. Moulton, and D. Eckerson. 1980. Improved method and test strategy for recovery of enteric viruses from shellfish. *Appl. Environ. Microbiol.* 39:141–52.
88. Metcalf, T. G., and W. C. Stiles. 1965. The accumulation of enteric viruses by the oyster, *Crassostrea virginica*. *J. Infect. Dis.* 115:68–76.
89. Miller, M. L., and J. A. Koburger. 1985. *Plesiomonas shigelloides:* An opportunistic food and waterborne pathogen. *J. Food Protect.* 48:449–57.
90. Morgan, P. M. 1968. Meat animal parasites and the importance of an effective and standardized meat inspection system. In *The safety of foods.,* ed. H. D. Graham et al., ch. 16. Westport, Conn.: AVI.

91. Murphy, A. M., G. S. Grobmann, P. J. Christopher, W. A. Lopez, G. R. Davey, and R. H. Millsom. 1979. An Australia-wide outbreak of gastroenteritis from oysters caused by Norwalk virus. *Med. J. Austr.* 2:329–33.

92. Niranjan, B. G., N. K. Bhat, and N. G. Avadhani. 1982. Preferential attack of mitochondrial DNA by aflatoxin B₁ during hepatocarcinogenesis. *Science* 215:73–75.

93. Northolt, M. D., C. A. H. Verhulsdonk, P. S. S. Soentoro, and W. E. Paulsch. 1976. Effect of water activity and temperature on aflatoxin production by *Aspergillus parasiticus. J. Milk Food Technol.* 39:170–74.

94. Omura, Y., R. J. Price, and H. S. Olcott. 1978. Histamine-forming bacteria isolated from spoiled skipjack tuna and jack mackerel. *J. Food Sci.* 43:1779–81.

95. Park, K. Y., and L. B. Bullerman. 1983. Effect of cycling temperatures on aflatoxin production by *Aspergillus parasiticus* and *Aspergillus flavus* in rice and cheddar cheese. *J. Food Sci.* 48:889–96.

96. Portnoy, B. L., P. A. Mackowiak, C. T. Caraway, J. A. Walker, T. W. McKinley, and C. A. Klein. 1975. Oyster-associated hepatitis: Failure of shellfish certification programs to prevent outbreaks. *J. Amer. Med. Assoc.* 233:1065–68.

97. Potter, N. N. 1973. Viruses in foods. *J. Milk Food Technol.* 36:307–10.

98. Riemann, H. P., M. E. Meyer, J. H. Theis, G. Kelso, and D. E. Behymer. 1975. Toxoplasmosis in an infant fed unpasteurized goat milk. *J. Pediat.* 87:573–76.

99. Roland, J. O., and L. R. Beuchat. 1984. Biomass and patulin production by *Byssochlamys nivea* in apple juice as affected by sorbate, benzoate, SO₂ and temperature. *J. Food Sci.* 49:402–6.

100. Rouf, M. A., and M. M. Rigney. 1971. Growth temperatures and temperature characteristics of *Aeromonas. Appl. Microbiol.* 22:503–6.

101. Russell, F. E. 1968. Poisonous marine animals. In *The safety of foods,* ed. H. D. Graham et al., ch. 14. Westport, Conn.: AVI.

102. Sanyal, S. C., S. J. Singh, and P. C. Sen. 1975. Enteropathogenicity of *Aeromonas hydrophila* and *Plesiomonas shigelloides. J. Med. Microbiol.* 8:195–98.

103. Schantz, E. J. 1973. Some toxins occurring naturally in marine organisms. In *Microbial safety of fishery products,* ed. C. O. Chichester and H. D. Graham, 151–62. New York: Academic Press.

104. Schantz, P. M. 1983. Trichinosis in the United States—1947–1981. *Food Technol.* 37(3):83–86.

105. Schindler, A. F. 1977. Temperature limits for production of aflatoxin by twenty-five isolates of *Aspergillus flavus* and *Aspergillus parasiticus. J. Food Protec.* 40:39–40.

106. Schlech, W. F., III, P. M. Lavigne, R. A. Bortolussi, A. C. Allen, E. V. Haldane, A. J. Wort, A. W. Hightower, S. E. Johnson, S. H. King, E. S. Nicholls, and C. V. Broome. 1983. Epidemic listeriosis—evidence for transmission by food. *N. Eng. J. Med.* 308:203–6.

107. Serck-Hanssen, A. 1970. Aflatoxin-induced fatal hepatitis? *Arch. Environ. Hlth* 20:729–31.

108. Shelef, L. A., and B. Chin. 1980. Effect of phenolic antioxidants on the mutagenicity of aflatoxin B₁. *Appl. Environ. Microbiol.* 40:1039–43.

109. Sheridan, J. F., L. Aurelian, G. Barbour, M. Santosham, R. B. Sack, and R. W. Ryder. 1981. Traveler's diarrhea associated with rotavirus infection: Analysis of virus-specific immunoglobulin classes. *Infect. Immun.* 31:419–29.

110. Shih, C. N., and E. H. Marth. 1972. Experimental production of aflatoxin on brick cheese. *J. Milk Food Technol.* 35:585–87.
111. ———. 1974. Some cultural conditions that control biosynthesis of lipid and aflatoxin by *Aspergillus parasiticus*. *Appl. Microbiol.* 27:452–56.
112. Shotwell, O. L., and C. W. Hesseltine. 1983. Five-year study of mycotoxins in Virginia wheat and dent corn. *J. Assoc. Off. Anal. Chem.* 66:1466–69.
113. Smith, G. C., L. Aurelian, M. Santosham, and R. B. Sack. 1983. Rotavirus-associated traveler's diarrhea: Neutralizing antibody in asymptomatic infections. *Infect. Immun.* 41:829–33.
114. Sobsey, M. D., R. J. Carrick, and H. R. Jensen. 1978. Improved methods for detecting enteric viruses in oysters. *Appl. Environ. Microbiol.* 36:121–28.
115. Sommer, N. F., J. R. Buchanan, and R. J. Fortlage. 1974. Production of patulin by *Penicillium expansum*. *Appl. Microbiol.* 28:589–93.
116. Stark, A.-A. 1980. Mutagenicity and carcinogenicity of mycotoxins: DNA binding as a possible mode of action. *Ann. Rev. Microbiol.* 34:235–62.
117. Stinson, E. E., D. D. Bills, S. F. Osman, J. Siciliano, M. J. Ceponis, and E. G. Heisler. 1980. Mycotoxin production by *Alternaria* species grown on apples, tomatoes, and blueberries. *J. Agric. Food Chem.* 28:960–63.
118. Stinson, E. E., S. F. Osman, E. G. Beisler, J. Siciliano, and D. D. Bills. 1981. Mycotoxin production in whole tomatoes, apples, oranges, and lemons. *J. Agric. Food Chem.* 29:790–92.
119. Sullivan, R., R. M. Marnell, E. P. Larkin, and R. B. Read, Jr. 1975. Inactivation of poliovirus 1 and coxsackievirus B-2 in broiled hamburgers. *J. Milk Food Technol.* 38:473–75.
120. Taylor, S. L., L. S. Guthertz, M. Leatherwood, and E. R. Lieber. 1979. Histamine production by *Klebsiella pneumoniae* and an incident of scombroid fish poisoning. *Appl. Environ. Microbiol.* 37:274–78.
121. Taylor, S. L., T. J. Keefe, E. S. Windham, and J. F. Howell. 1982. Outbreak of histamine poisoning associated with consumption of Swiss cheese. *J. Food Protect.* 45:455–57.
122. Tong, C.-H., and F. A. Draughon. 1985. Inhibition by antimicrobial food additives of ochratoxin A production by *Aspergillus sulphureus* and *Penicillium viridicatum*. *Appl. Environ. Microbiol.* 49:1407–11.
123. Trenk, H. L., and P. A. Hartman. 1970. Effects of moisture content and temperature on aflatoxin production in corn. *Appl. Microbiol.* 19:781–84.
124. Tsukamoto, T., Y. Konoshita, T. Shimada, and R. Sakazaki. 1978. Two epidemics of diarrhoeal disease possibly caused by *Plesiomonas shigelloides*. *J. Hyg.* 80:275–80.
125. Twedt, R. M., and B. K. Boutin. 1979. Potential public health significance of non-*Escherichia coli* coliforms in food. *J. Food Protect.* 42:161–63.
126. United States Centers for Disease Control. 1976. Foodborne and waterborne disease outbreaks. *Annual Summary,* HEW Publ. No. (CDC) 76-8185.
127. ———. 1977. Foodborne outbreak of hepatitis A—Pennsylvania. *Morb. Mort. W. Rept.* 26:247–48.
128. ———. 1982. Outbreak of food-borne hepatitis A—New Jersey. *Morb. Mort. W. Rept.* 31:150–52.
129. ———. 1982. Enteric illness associated with raw clam consumption—New York. *Morb. Mort. W. Rept.* 31:449–51.
130. ———. 1983. Food-borne hepatitis A—Oklahoma, Texas. *Morb. Mort. W. Rept.* 32:652–54 and 59.

131. ———. 1985. Listeriosis outbreak associated with Mexican-style cheese—California. *Morb. Mort. W. Rept.* 34:357–59.
132. U.S.D.A. 1982. USDA advises cooking pork to 170 degrees Fahrenheit throughout. News release, U.S. Dept. of Agriculture, Washington, D.C.
133. Urasawa, S., T. Urasawa, and K. Teniguchi. 1982. Three human rotavirus serotypes demonstrated by plaque neutralization of isolated strains. *Infect. Immun.* 38:781–84.
134. Van Damme, L. R., and J. Vandepitte. 1980. Frequent isolation of *Edwardsiella tarda* and *Plesiomonas shigelloides* from healthy Zairese freshwater fish: A possible source of sporadic diarrhea in the tropics. *Appl. Environ. Microbiol.* 39:475–79.
135. Wadström, T., Å. Ljungh, and B. Wretlind. 1976. Enterotoxin, haemolysin and cytotoxic protein in *Aeromonas hydrophila* from human infections. *Acta Path. Microbiol. Scand. Sect. B* 84:112–14.
136. Wait, D. A., C. R. Hackney, R. J. Carrick, G. Lovelace, and M. D. Sobsey. 1983. Enteric bacterial and viral pathogens and indicator bacteria in hard shell clams. *J. Food Protect.* 46:493–96.
137. Weis, J., and H. P. R. Seeliger. 1975. Incidence of *Listeria monocytogenes* in nature. *Appl. Microbiol.* 30:29–32.
138. Welshimer, H. J. 1960. Survival of *Listeria monocytogenes* in soil. *J. Bacteriol.* 80:316–20.
139. Wills, J. H., Jr. 1966. Seafood toxins. In *Toxicants occurring naturally in foods*, 147–63. Pub. no. 1354. Washington, D.C.: National Academy of Science.
140. Wogan, G. N. 1966. Chemical nature and biological effects of the aflatoxins. *Bacteriol. Rev.* 30:460–70.
141. Wu, M. T., J. C. Ayres, and P. E. Koehler. 1974. Production of citrinin by *Penicillium viridicatum* on country-cured ham. *Appl. Microbiol.* 27:427–28.
142. Yoshinaga, D. R., and H. A. Frank. 1982. Histamine-producing bacteria in decomposing skipjack tuna (*Katsuwonus pelamis*). *Appl. Environ. Microbiol.* 44:447–52.
143. Zajc-Satler, J., A. Z. Dragav, and M. Kumelj. 1972. Morphological and biochemical studies of 6 strains of *Plesiomonas shigelloides* isolated from clinical sources. *Zbt. Baktr. Hyg. Abt. Orig. A.* 219:514–21.
144. Zimmermann, W. J. 1983. An approach to safe microwave cooking of pork roasts containing *Trichinella spiralis*. *J. Food Sci.* 48:1715–18 and 1722.
145. Zimmermann, W. J., and P. J. Beach. 1982. Efficacy of microwave cooking for devitalizing trichinae in pork roasts and chops. *J. Food Protect.* 45:405–9.
146. Zimmermann, W. J., D. G. Olsono, A. Sandoval, and R. E. Rust. 1985. Efficacy of freezing in eliminating infectivity of *Trichinella spiralis* in boxed pork products. *J. Food Protect.* 48:196–99.

VII

PSYCHROTROPHS,

THERMOPHILES AND

RADIATION-RESISTANT

MICROORGANISMS

23.

CHARACTERISTICS AND GROWTH OF PSYCHROTROPHIC MICROORGANISMS

In the older microbiological literature, the microorganisms that grow at refrigerator temperatures are most often referred to as psychrophiles. It has been suggested (42) that the term psychrophile be applied only to those organisms whose optimum and maximum temperatures of growth are about 15° and 20°C, respectively. Since the organisms that bring about the refrigerator spoilage of foods can grow at temperatures < 20°C, they are not psychrophiles. It was suggested in 1960 that the term psychrotroph (*psychros,* "cold"; and *trephein,* "to nourish upon" or "to develop") be applied to the organisms that grow at 5°C and below without regard to their optimum temperatures (10, 44). Based upon other findings and recommendations (30, 58), psychrotrophic microorganisms are those that can grow at temperatures between 0° and 7°C and produce visible colonies (or turbidity) within 7–10 days. Some authors suggest the use of optimum temperatures of growth to define psychrotrophs, and the temperature ranges suggested include 10°–20°, 12°–18°, 15°–25°, 18°–20°, and even 20°–40°C. Due to a lack of general agreement on optimum temperatures, psychrotrophs are most often defined based upon their ability to grow within the 0–7°C range. By the definitions noted, it is rather unlikely that psychrophiles exist in refrigerated meats, poultry, and vegetable products, although their existence in certain seafoods is more likely because of the known existence of true psychrophiles in oceanic environments.

Regarding the distribution of psychrotrophs among bacteria, the preponderant types are gram-negative rods, largely pseudomonads (13). The genera in which psychrotrophic strains are fairly common include *Pseudomonas, Flavobacterium, Alcaligenes, Alteromonas, Brochothrix,* and *Arthrobacter.*

Less frequently, but nevertheless regularly encountered, are psychrotrophic strains of *Yersinia, Escherichia, Aeromonas, Serratia, Proteus, Acinetobacter, Enterobacter, Chromobacterium, Vibrio, Clostridium, Citrobacter, Hafnia,* and *Bacillus* (13, 44, 69). Occasional reports of a less definitive nature on the isolation of psychrotrophic strains of *Corynebacterium, Lactobacillus,* and *Micrococcus* have been made. As far as low-temperature food preservation is concerned, the genus *Pseudomonas* represents the most important psychrotrophic bacteria by far. Psychrotrophic strains of yeasts have been reported from the genera *Candida, Cryptococcus, Rhodotorula,* and *Torulopsis,* with *Candida* strains being the more common.

Since psychrotrophic microorganisms are so important in the low-temperature preservation of foods, it is highly desirable to understand the basic mechanisms by which they are able to grow at such low temperatures. This information could lead to the preservation of foods by specifically altering or controlling the molecular mechanisms that allow for low-temperature growth and activity. While the basic mechanisms that underlie psychrotrophy are not yet well understood, most of what is known about this phenomenon can be grouped into three broad categories: (1) temperature-induced changes in the production of metabolic end products, (2) temperature effects on physiologic mechanisms, and (3) the low heat resistance of psychrotrophic microorganisms. Each of these is treated below.

TEMPERATURE-INDUCED CHANGES

There are at least four temperature-induced changes known to occur in psychrotrophic microorganisms that have been studied.

First, there is a greater increase in the proportion of unsaturated fatty acid residues in the lipids of psychrotrophs than in those of mesophiles. The usual lipid content of most bacteria is between 2 and 5%, most or all of which is in the cell membrane. Bacterial fats are glycerol esters of two types: neutral lipids in which all three or only one or two of the —OH groups of glycerol are esterified with long-chain fatty acids; and phospholipids in which one of the —OH groups is linked through a phosphodiester bond to choline, ethanolamine, glycerol, inositol, or serine, and the other two —OH groups are esterified with long-chain fatty acids (52). It has been shown by various investigators that most psychrotrophs synthesize neutral lipids and phospholipids containing an increased proportion of unsaturated fatty acids when grown at low temperatures, as compared with growth at higher temperatures. As much as 50% increase in content of unsaturated bonds in fatty acids from mesophilic and psychrotrophic *Candida* spp. grown at 10°C compared to 25°C was found by one group of investigators (35), with no effect on the phospholipid composition of the yeasts. Studies on a psychrotrophic *Candida* by other investigators (41) substantiated the findings of others that unsaturated fatty acids increased with decreasing growth temperatures (see Table 23-1). Linolenic acid increased at the expense of oleic acid at the lower temperatures.

Table 23-1. Effects of incubation temperature on the fatty acid composition of stationary cultures of *Candida utilis* (41).

Incubation temperature (°C)	Cell conc. (mg/ml)	Fatty acid composition[a]				
		16:0	*16:1*	*18:1*	*18:2*	*18:3*
30	2.0	18.9	4.6	39.1	34.3	2.1
20	2.0	20.3	11.4	31.6	27.7	6.1
10	2.0	27.4	20.6	20.7	17.6	10.7
5	1.7	19.2	15.9	18.2	16.3	27.3

[a]Values quoted are expressed as percentages of the total fatty acids. Fatty acids are designated x:y, where x is the number of carbon atoms and y is the number of double bonds per molecule.
Copyright © 1973, American Society for Microbiology.

The widespread occurrence of low-temperature-induced changes in fatty acid composition suggests that they are associated with physiological mechanisms of the cell. It is known that an increase in the degree of unsaturation of fatty acids in lipids leads to a decrease in lipid melting point. It has been suggested that increased synthesis of unsaturated fatty acids at low temperatures has the function of maintaining the lipid in a liquid and mobile state, thereby allowing membrane activity to continue to function. This concept is referred to as the **"lipid solidification"** theory and was first proposed by Gaughran (17) and Allen (3). It has been shown by Byrne and Chapman (8) that the melting point of fatty-acid side chains in lipids is more important than the entire lipid structure.

Although full support for the lipid solidification idea is wanting, there is circumstantial evidence available such as the phenomenon of **"cold shock,"** which is the dying off of many cells of mesophilic bacteria upon the sudden chilling of a suspension of viable cells grown at mesophilic temperatures. It has been shown for a large number of gram-negative bacteria including *E. coli* and is generally a property of gram-negative bacteria and not of gram positives. Cold shock has been shown to be accompanied by the release of certain low molecular weight cell constituents, an effect that presumably occurs by virtue of damage to the plasma membrane. According to Rose (52), cold shock seems to result from a sudden release of cell constituents from bacteria following the "freezing" of certain membrane lipids after sudden chilling, with consequent development of "holes" in the membrane. To support this hypothesis, Farrell and Rose (14) grew a mesophilic strain of *Ps. aeruginosa* at 30°C and showed that the cells were susceptible to cold shock while the same strain grown at 10°C was not susceptible. It has been proposed that the growth temperature range of an organism is dependent upon the ability of the organism to regulate its lipid fluidity within a given range (15).

Second, psychrotrophs display a greater synthesis of polysaccharides than do mesophiles. Well-known examples of this effect include the production of ropy milk and ropy dough, both of which are favored by low temperatures. The production of extracellular dextrans by *Leuconostoc* and *Pediococcus* spp. are known to be favored at temperatures below the growth optima of these organisms. The greater production of dextran at lower temperatures is due apparently to the fact that dextransucrase is very rapidly inactivated at temperatures in excess of 30°C (47). A temperature-sensitive dextransucrase synthesizing system has been shown also for a *Lactobacillus* sp. (9).

From a practical standpoint, increased polysaccharide synthesis at low temperatures manifests itself in the characteristic appearance of low-temperature spoiled meats. As discussed in Chapter 9, slime formation is characteristic of the bacterial spoilage of frankfurters, fresh poultry, and ground beef. The coalescence of surface colonies leads to the sliminess of such meats and no doubt contributes to the increased hydration capacity that accompanies low-temperature meat spoilage.

Third, pigment-producing microorganisms produce more pigment under psychrotrophic conditions than under mesophilic. This effect seems to be confined to those organisms that synthesize phenazine and carotenoid pigments. The best-documented example of this phenomenon involves pigment production by *Serratia marcescens*. The organism produces an abnormally heat-sensitive enzyme that catalyzes the coupling of a monopyrrole and bipyrrole precursor to give prodigiosin (the red pigment, 68). The increased production of pigments at suboptimum temperatures has been reported by others (64, 68). It is interesting that a very large number of marine psychrotrophs (and perhaps psychrophiles) are pigmented. This is true for bacteria as well as yeasts. On the other hand, none of the more commonly studied thermophiles are pigmented.

Finally, under psychrotrophic conditions, some organisms show a differential attack on certain metabolizable substrates. It has been reported that sugar fermentation at temperatures below 30°C gives rise to both acid and gas, while above 30°C only acid is produced (19). Similarly, others have found psychrotrophs that fermented glucose and other sugars with the formation of acid and gas at 20°C and lower, but produced only acid at higher temperatures (66). The latter was ascribed to a temperature-sensitive formic hydrogenase system. These investigators studied a similar effect and attributed the difference to a temperature-sensitive hydrogenase synthesizing system of the cell. Beef spoilage bacteria have been shown to liquefy gelatin and utilize water-soluble beef proteins more at 5°C than at 30°C (33), but whether this effect is due to temperature-sensitive enzymes is not clear.

THE EFFECT OF LOW TEMPERATURES ON MICROBIAL PHYSIOLOGIC MECHANISMS

Of the effects that low incubation temperatures have on the growth and activity of food-borne microorganisms, the five that have received the most attention are outlined below.

First, psychrotrophs have a slower metabolic rate than mesophiles. This is one of the best-known effects of low temperature on microorganisms, and as pointed out in Chapter 13, it is the rationale behind the low-temperature preservation of foods (see also Chapter 3). The effect of temperature on the generation time of a psychrotroph (a pseudomonad) and a mesophile (*E. coli*) is presented in Fig. 23-1, where the curve of the psychrotroph is shifted to a lower range by approximately 10°. An Arrhenius plot of these two organisms (Fig. 23-2) shows the general shape of the two curves to be similar (27). The striking difference between the two curves, however, is the difference in the slope of the linear region where the **temperature characteristic** (μ) of growth rate is about 28,000 cal/mole for the mesophile but only 18,000 cal/mole for the psychrotroph. It should be noted that an Arrhenius plot relates rate of enzyme reaction on a logarithmic scale to the reciprocal of the absolute temperature, while μ (and Q_{10}) are measures of the rate of decrease or rate of a given process with temperature. In general, as incubation temperatures are lowered, μ increases. Although the practice of employing μ values as a means of differentiating between mesophiles and psychrophiles is long standing, some investigators have questioned the appropriateness of this practice (23, 54). At least one investigator has found that μ values are about the same for mesophiles and psychrotrophs—about 12,000 cal/mole (54). In the more traditional sense, the deviation of the psychrotroph

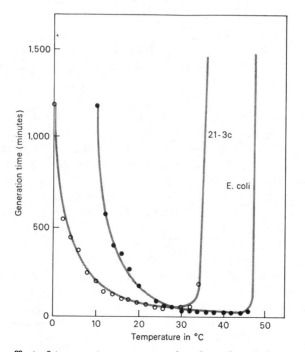

Fig. 23-1. The effect of temperature on generation time of a psychrotrophic pseudomonad (21-3c) and the mesophile *Escherichia coli*, K-12, (27).

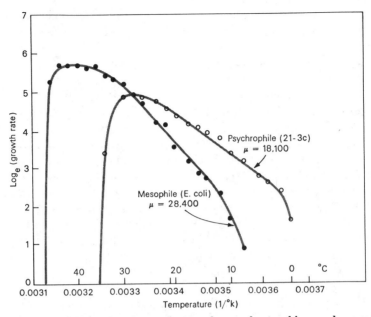

Fig. 23-2. An Arrhenius plot of growth rate of a psychrotrophic pseudomonad (21-3c) as compared with that of the mesophile *Escherichia coli*, K-12 (after Ingraham, 27, with corrected μ as suggested by Hanus and Morita, 23).

curve from linearity as seen in Fig. 23-2 is common for this group of organisms and, though rather difficult to interpret in kinetic terms, may be explained in part by the various parameters that affect growth. Other investigators have produced results that vary relative to the utility of μ and these have been reviewed and discussed (31, 42, 53).

The precise reasons why metabolic rates are slowed at low temperatures are not fully understood. As noted above, psychrotrophic growth decreases more slowly than that of mesophilic with decreasing temperatures. The temperature coefficients (Q_{10}) for various substrates such as acetate and glucose have been shown by several investigators to be lower for growing psychrotrophs than for mesophiles. The end products of mesophilic and psychrotrophic metabolism of glucose were shown to be the same, with the differences largely disappearing when the cells were broken (28). In other words, the temperature coefficients are about the same for psychrotrophs and mesophiles when cell-free extracts are employed.

As temperature is decreased, the rate of protein synthesis is known to decrease, and this occurs in the absence of changes in the amount of cellular DNA. One reason for this may be the increase in intramolecular hydrogen bonding that occurs at low temperatures, leading to increased folding of enzymes with losses in catalytic activity (36). On the other hand, the decrease in protein synthesis appears to be related to a decreased synthesis of individual enzymes at low growth temperatures. Although the precise mechanism of reduced protein synthesis is not yet well understood, it has been suggested that low temperatures affect the synthesis of a repressor protein (39), and that the repressor protein itself is thermolabile (63). Several investigators have suggested that low temperatures may influence the fidelity of the translation of mRNA during protein synthesis. For example, in studies with *E. coli,* it was shown that a leucine-starved auxotroph of this mesophile incorporated radioactive leucine into protein at 0°C (18). It was suggested that at this temperature all essential steps in protein synthesis apparently go on, and involve a wide variety of proteins. The rate of synthesis at 0°C was estimated to be about 350 times slower than at 37°C for this organism. It has been suggested that the cessation of RNA synthesis in general may be the controlling factor in determining low temperature growth (25), and the lack of polysome formation in *E. coli* when shifted to a temperature below its growth minimum has been demonstrated. The formation of polysomes is thus sensitive to low temperatures (at least in some organisms) and protein synthesis would be adversely affected (see 32).

Whatever the specific mechanism of lowered metabolic activity of microorganisms as growth temperature is decreased, psychrotrophs growing at low temperature have been shown to possess good enzymatic activity, since motility, endospore formation, and endospore germination all occur at 0°C (59). It has been shown that *Pseudomonas fragi,* among other organisms, produces lipases within 2–4 days at −7°C; within 7 days at −18°C; and within 3 weeks at −29°C (2). It has been noted that the minimum growth temperature may be determined by the structure of the enzymes and cell membrane as well as by enzyme synthesis (59). The lack of production of enzymes at high temperatures by psychrotrophs on the other hand is due apparently to the inactive nature of enzyme-synthesizing reactions rather than to enzyme inactivation (59), although the latter is known to occur (see below). With respect to individual groups of enzymes, it has been shown that yields of endocellular proteolytic enzymes are greater in *Pseudomonas fluorescens* grown at 10° than at either 20° or 35°C (49), while other investigators have shown that *P. fragi* preferentially produces lipase at low temperatures with none being produced at 30°C or higher (45, 46). *P. fluorescens* has been found to produce just as much lipase at 5°C as at 20°C, but only a slight amount was produced at 30°C (1). On the other hand, a proteolytic enzyme system of *P. fluorescens* showed more activity on egg white and hemoglobin at 25°C than at 15° and 5°C (26).

It has been suggested that there are preformed elements in microbial cells grown at any temperature that are selectively temperature sensitive (34). Microorganisms may cease to grow at a certain low temperature because of excessive sensitivity in one or several control mechanisms, the effectors of which cannot be supplied in the growth medium (29). According to the latter authors, the interaction between effector molecules and the corresponding allosteric proteins may be expected to be a strong function of temperature.

Second, psychrotrophs display a greater transport of solutes across the cell membrane than do mesophiles. Several investigators have shown that upon lowering the growth temperature of mesophiles within the psychrotrophic range, solute uptake is decreased. Studies by Baxter and Gibbons (4) indicate that minimum growth temperature of mesophiles is determined by the temperature at which transport permeases are inactivated. Farrell and Rose (13) offered three basic mechanisms by which low temperature could affect solute uptake: (a) inactivation of individual permease-proteins at low temperature as a result of low-temperature-induced conformational changes that have been shown to occur in some proteins; (b) changes in the molecular architecture of the cytoplasmic membrane that prevent permease action; and (c) a shortage of energy required for the active transport of solutes. Although the precise mechanisms of reduced uptake of solutes at low temperatures are not clear at this time, (b) above seems the most likely (13).

Psychrotrophs, on the other hand, appear to be more efficient in transporting solutes across their cytoplasmic membranes. As stated above, psychrotrophs tend to possess in their membrane lipids that enable the membrane to be more fluid. The greater mobility of the psychrotrophic membrane may be expected to facilitate membrane transport at low temperatures. In addition, the transport permeases of psychrotrophs are apparently more operative under these conditions than are those of mesophiles. Whatever the specific mechanism of increased transport might be, it has been demonstrated that psychrotrophs are more efficient than mesophiles in the uptake of solutes at low temperatures. Baxter and Gibbons (4) showed that a psychrotrophic *Candida* sp. incorporated glucosamine more rapidly than a mesophilic *Candida*. The psychrotroph transported glucosamine at 0°C, while scarcely any was transported by the mesophile at this temperature or even at 10°C.

Third, some microorganisms produce larger cells when growing under psychrotrophic conditions. Yeasts and molds have been reported to produce larger cell sizes when growing under psychrotrophic conditions than when growing under mesophilic. This observation has not been reported for bacteria. Employing *Candida utilis,* Rose (52) observed increase in cell size upon the lowering of growth temperature, and suggested that the increase in size was a consequence of the lowering of growth temperature, which might be correlated with increases in RNA and protein content of cells. Low-temperature-induced synthesis of additional RNA has been reported by several investigators (6, 24, 62). However, Frank et al. (16) found no increase in the amount of RNA at 2°C when *Pseudomonas* strain 92 cells were grown at 2° and 30°C at equivalent physiological stages. Also, no increases in cell size, protein content, or catalase activity could be found by these investigators.

Fourth, psychrotrophs are more efficient producers of flagella than mesophiles. It has been reported that flagella production is often favored at low temperatures though not at higher temperatures. Examples of this phenomenon include *E. coli, Bacillus inconstans, Salmonella paratyphi B,* and other organisms, including some psychrophiles. The opposite effect is also known for some psychrophiles.

Finally, psychrotrophs are more favorably affected by aeration than mesophiles. The effect of aeration on the generation time of *P. fluorescens* at temperatures from 4° to 32°C, employing three different carbon sources, is presented in Table 23-2. It may be noted that the greatest effect of aeration (shaking) occurred at 4° and 10°C, while at 32°C aerated cultures produced a longer generation time (48). The significance of this effect is not clear. In a study of facultatively anaerobic psychrotrophs under anaerobic conditions, the organisms were shown to grow more slowly, survive longer, die more rapidly at higher temperatures, and produce lower maximal cell yields under anaerobic conditions than under aerobic (65). It has been commonly observed that plate counts on many foods are higher with incubation at low temperatures than at temperatures of 30°C and above. The generally higher counts are due in part to the increased solubility, and consequently, availability of O_2 (55). The latter authors found that equally high cell yields can be obtained at both low and high incubation temperatures when O_2 is not limiting. This greater availability of O_2 in refrigerated foods undoubtedly exerts selectivity on the spoilage flora of such foods. The vast majority of psychrotrophic bacteria studied are aerobes or facultative anaerobes, and these are the types that are associated with the spoilage of foods stored at refrigerator temperatures. Relatively few anaerobic psychrotrophs have been isolated and studied. One of the first was *Clostridium putrefaciens* (40). Other investigators have isolated and studied psychrotrophic clostridia (51, 56).

Table 23-2. Effect of growth temperature, carbon source, and aeration on generation times (h) of *pseudomonas fluorescens* (48).

Growth medium[a]	Culture	Growth temp.					
		4°C	*10°C*	*15°C*	*20°C*	*25°C*	*32°C*
Glucose	Stationary	8.20	3.52	2.02	1.47	0.97	1.19
	Aerated	5.54	2.61	2.00	1.46	0.93	1.51
Citrate	Stationary	8.20	3.46	2.00	1.43	1.01	1.24
	Aerated	6.68	2.95	2.02	1.26	0.98	1.45
Casamino Acids	Stationary	7.55	3.06	1.78	1.36	1.12	0.95
	Aerated	4.17	2.57	1.56	1.12	0.87	1.10

[a]Basal salts + 0.02% yeast extract + the carbon source indicated.

NATURE OF THE LOW HEAT RESISTANCE OF PSYCHROTROPHS

It has been known for years that psychrotrophic microorganisms are generally unable to grow much above 30–35°C. Among the first to suggest reasons for this limitation of growth were Edwards and Rettger (11), who concluded that the maximum growth temperatures of bacteria may bear a definite relationship to the minimum temperatures of destruction of respiratory enzymes. The conclusion of these authors has been borne out by results from a large number of investigators over the past decade or so. It has been shown that many respiratory enzymes are inactivated at the temperatures of maximum growth of various psychrotrophic types (Table 23-3). Thus, the thermal sensitivity of certain enzymes of psychrotrophs is at least one of the factors that limit the growth of these organisms to low temperatures.

It has also been shown that when some psychrotrophs are subjected to temperatures above their growth maxima, cell death is accompanied by the leakage of various intracellular constituents (20, 22, 61). The leakage substances have been shown to consist of proteins, DNA, RNA, free amino acids, and lipid phosphorus. The latter was thought by Hagen et al. to represent phosphorus of the cytoplasmic membrane. While the specific reasons for the release of cell constituents is not fully understood, it would appear to involve rupture of the cell membrane. These events appear to follow those of enzyme inactivation noted above.

Whatever is the true mechanism of psychrotroph death at temperatures a few degrees above their growth maxima, their destruction at these relatively low temperatures is characteristic of this group of organisms. This is especially true of those that have optimum growth temperatures at and below 20°C. Reports by several investigators on psychrotrophs isolated and studied over the past two decades reveal that all are capable of growing at 0°C with growth optima at either 15 or between 20 and 25°C, and growth maxima between 20° and 35°C (20, 22, 37, 56, 57, 60). Included among these organisms

Table 23-3. Some heat-labile enzymes of psychrotrophic microorganisms (as reported by various investigators).

Enzyme	Organism	Temp. of maximum growth, °C	Temp. of enzyme inactivation, °C	Reference
Extracellular lipase[a]	*Ps. fragi*		30	(46)
α-oxoglutarate synthesizing enzymes and others	*Cryptococcus*	~28	30	(21)
Alcohol dehydrogenase	*Candida* sp.	<30		(4)
Formic hydrogenlyase	Psychrophile 82	35	45	(66)
Hydrogenase	Psychrophile 82	35	>20	(67)
Malic dehydrogenase	Marine *Vibrio*	30	30	(7)
Pyruvate dehydrogenase	*Candida* sp.	~20	25	(12)
Isocitrate dehydrogenase	*Arthrobacter* sp.	~35	37	(12)
Fermentative enzymes	*Candida* sp. P16	~25	35	(57)
Reduced NAD oxidase	Psychrophile 82	35	46	(50)
Cytochrome c reductase	Psychrophile 82	35	46	(50)
Lactic and glycerol dehydrogenase	Psychrophile 82	35	46	(50)
Pyruvate clastic enzymes	Psychrophile 82	35	46	(50)
Protein and RNA synthesizing	*Micrococcus cryophilus*	25	30	(38)

[a]Enzyme-forming system inactivated.

are gram-negative rods, gram-positive aerobic and anaerobic rods, spore-formers and nonsporeformers, gram-positive cocci, vibrios, and yeasts. One of these, *Vibrio fisheri* (*marinus*), was shown by Morita and Albright (43) to have an optimum growth temperature at 15°C and a generation time of 80.7 min at this temperature. In almost all cases, the growth maxima of these organisms were only 5–10 degrees above the growth optima. The Enterobacteriaceae isolated by Mossel and Zwart (44) appear to be exceptions to the foregoing. They were shown to be capable of growing at 6°C and were designated by these authors as psychrotrophs. They differed from the more classical psychrotrophs in being able to grow at 41–42°C.

REFERENCES

1. Alford, J. A., and L. E. Elliott. 1960. Lipolytic activity of microoganisms at low and intermediate temperatures. I. Action of *Pseudomonas fluorescens* on lard. *Food Res.* 25:296–303.
2. Alford, J. A., and D. A. Pierce. 1961. Lipolytic activity of microorganims at low and intermediate temperatures. III. Activity of microbial lipases at temperatures below 0°C. *J. Food Sci.* 26:518–24.
3. Allen, M. B. 1953. The thermophilic aerobic sporeforming bacteria. *Bacteriol. Rev.* 17:125–73.
4. Baxter, R. M., and N. E. Gibbons. 1962. Observations on the physiology of psychrophilism in a yeast. *Can. J. Microbiol.* 8:511–17.
5. Brown, A. D. 1957. Some general properties of a psychrophilic pseudomonad: The effect of temperature on some of these properties and the utilization of glucose by this organism and *Pseudomonas aeruginosa. J. Gen. Microbiol.* 17:640–48.
6. Brown, C. M., and A. H. Rose. 1969. Effects of temperature on composition and cell volume of *Candida utilis. J. Bacteriol.* 97:261–72.
7. Burton, S. D., and R. Y. Mortia. 1963. Denaturation and renaturation of malic dehydrogenase in a cell-free extract from a marine psychrophile. *J. Bacteriol.* 86:1019–24.
8. Byrne, P., and D. Chapman. 1964. Liquid crystalline nature of phospholipids. *Nature* 202:987–88.
9. Dunican, L. K., and H. W. Seeley. 1963. Temperature-sensitive dextransucrase synthesis by a lactobacillus. *J. Bacteriol.* 86:1079–83.
10. Eddy, B. P. 1960. The use and meaning of the term "psychrophilic." *J. Appl. Bacteriol.* 23:189–90.
11. Edwards, O. F., and L. F. Rettger. 1937. The relation of certain respiratory enzymes to the maximum growth temperatures of bacteria. *J. Bacteriol.* 34:489–515.
12. Evison, L. M., and A. H. Rose. 1965. A comparative study on the biochemical bases of the maximum temperatures for growth of three psychrophilic micro-organisms. *J. Gen. Microbiol.* 40:349–64.
13. Farrell, J., and A. Rose. 1967. Temperature effects on micro-organisms. *Ann. Rev. Microbiol.* 21:101–20.
14. ———. 1968. Cold shock in mesophilic and psychrophilic pseudomonads. *J. Gen. Microbiol.* 50:429–39.

15. Finne, G., and J. R. Matches. 1976. Spin-labeling studies on the lipids of psychrophilic, psychrotrophic, and mesophilic clostridia. *J. Bacteriol.* 125: 211–19.

16. Frank, H. A., A. Reid, L. M. Santo, N. A. Lum, and S. T. Sandler. 1972. Similarity in several properties of psychrophilic bacteria grown at low and moderate temperatures. *Appl. Microbiol.* 24:571–74.

17. Gaughran, E. R. L. 1947. The thermophilic micro-organisms. *Bacteriol. Rev.* 11:189–225.

18. Goldstein, A., D. B. Goldstein, and L. I. Lowney. 1964. Protein synthesis at 0°C in *Escherichia coli. J. Mol. Biol.* 9:213–35.

19. Greene, V. W., and J. J. Jezeski. 1954. The influence of temperature on the development of several psychrophilic bacteria of dairy origin. *Appl. Microbiol.* 2:110–17.

20. Hagen, P.-O., D. J. Kushner, and N. E. Gibbons. 1964. Temperature-induced death and lysis in a psychrophilic bacterium. *Can. J. Microbiol.* 10:813–22.

21. Hagen, P.-O., and A. H. Rose. 1962. Studies on the biochemical basis of low maximum temperature in a psychrophilic *Cryptococcus. J. Gen. Microbiol.* 27:89–99.

22. Haight, R. D., and R. Y. Morita. 1966. Thermally induced leakage from *Vibrio marinus,* an obligately psychrophilic marine bacterium. *J. Bacteriol.* 92:1388–93.

23. Hanus, F. J., and R. Y. Morita. 1968. Significance of the temperature characteristic of growth. *J. Bacteriol.* 95:736–37.

24. Harder, W., and H. Veldkamp. 1967. A continuous culture study of an obligately psychrophilic *Pseudomonas* species. *Arch. Mikrobiol.* 59:123–30.

25. ———. 1968. Physiology of an obligately psychrophilic marine *Pseudomonas* species. *J. Appl. Bacteriol.* 31:12–23.

26. Hurley, W. C., F. A. Gardner, and C. Vanderzant. 1963. Some characteristics of a proteolytic enzyme system of *Pseudomonas fluorescens. J. Food Sci.* 28: 47–54.

27. Ingraham, J. L. 1962. Newer concepts of psychrophilic bacteria. In *Proceedings, low temperature microbiology symposium—1961,* 41–62. Camden, N.J.: Campbell Soup Co.

28. Ingraham, J. L., and G. F. Bailey. 1959. Comparative study of effect of temperature on metabolism of psychrophilic and mesophilic bacteria. *J. Bacteriol.* 77:609–13.

29. Ingraham, J. L., and O. Maaløe. 1967. Cold-sensitive mutants and the minimum temperature of growth of bacteria. In *Molecular mechanisms of temperature adaptation,* ed. C. L. Prosser, Pub. no. 84, 297–309. Washington, D.C.: American Association for the Advancement of Science.

30. Ingraham, J. L., and J. L. Stokes. 1959. Psychrophilic bacteria. *Bacteriol. Rev.* 23:97–108.

31. Inniss, W. E. 1975. Interaction of temperature and psychrophilic microorganisms. *Ann. Rev. Microbiol.* 29:445–65.

32. Inniss, W. E., and J. L. Ingraham. 1978. Microbial life at low temperatures: Mechanisms and molecular aspects. In *Microbial life in extreme environments,* ed. D. J. Kushner, 73–104. New York: Academic Press.

33. Jay, J. M. 1967. Nature, characteristics, and proteolytic properties of beef spoilage bacteria at low and high temperatures. *Appl. Microbiol.* 15:943–44.

34. Jezeski, J. J., and R. H. Olsen. 1962. The activity of enzymes at low temperatures. In *Proceedings, low temperature microbiology symposium—1961*, 139–55. Camden, N.J.: Campbell Soup Co.

35. Kates, M., and R. M. Baxter. 1962. Lipid comparison of mesophilic and psychrophilic yeasts (*Candida* species) as influenced by environmental temperature. *Can. J. Biochem. Physiol.* 40:1213–27.

36. Kavanau, J. L. 1950. Enzyme kinetics and the rate of biological processes. *J. Gen. Physiol.* 34:193–209.

37. Larkin, J. M., and J. L. Stokes. 1966. Isolation of psychrophilic species of *Bacillus*. *J. Bacteriol.* 91:1667–71.

38. Malcolm, N. L. 1968. Synthesis of protein and ribonucleic acid in a psychrophile at normal and restrictive growth temperatures. *J. Bacteriol.* 95:1388–99.

39. Marr, A. G., J. L. Ingraham, and C. L. Squires. 1964. Effect of the temperature of growth of *Escherichia coli* on the formation of B-galactosidase. *J. Bacteriol.* 87:356–62.

40. McBryde, C. N. 1911. A bacteriological study of ham souring. Bull. no. 132. Beltsville, Md.: U.S. Bureau of Animal Industry.

41. McMurrough, I., and A. H. Rose. 1973. Effects of temperature variation on the fatty acid composition of a psychrophilic *Candida* species. *J. Bacteriol.* 114:451–52.

42. Morita, R. Y. 1975. Psychrophilic bacteria. *Bacteriol. Rev.* 39:144–67.

43. Morita, R. Y., and L. J. Albright. 1965. Cell yields of *Vibrio marinus*, an obligate psychrophile, at low temperatures. *Can. J. Microbiol.* 11:221–27.

44. Mossel, D. A. A., and H. Zwart. 1960. The rapid tentative recognition of psychrotrophic types among *Enterobacteriaceae* isolated from foods. *J. Appl. Bacteriol.* 23:185–188,

45. Nashif, S. A., and F. E. Nelson. 1953. The lipase of *Pseudomonas fragi*. I. Characterization of the enzyme. *J. Dairy Sci.* 36:459–70.

46. ———. 1953. The lipase of *Pseudomonas fragi*. II. Factors affecting lipase production. *J. Dairy Sci.* 36:471–80.

47. Neely, W. B. 1960. Dextran: Structure and synthesis. *Adv. Carbohy. Chem.* 15:341–69.

48. Olsen, R. H., and J. J. Jezeski. 1963. Some effects of carbon source, aeration, and temperature on growth of a psychrophilic strain of *Pseudomonas fluorescens*. *J. Bacteriol.* 86:429–33.

49. Peterson, A. C., and M. F. Gunderson. 1960. Some characteristics of proteolytic enzymes from *Pseudomonas fluorescens*. *Appl. Microbiol.* 8:98–104.

50. Purohit, K., and J. L. Stokes. 1967. Heat-labile enzymes in a psychrophilic bacterium. *J. Bacteriol.* 93:199–206.

51. Roberts, T. A., and G. Hobbs. 1968. Low temperature growth characteristics of Clostridia. *J. Appl. Bacteriol.* 31:75–88.

52. Rose, A. H. 1968. Physiology of microorganisms at low temperatures. *J. Appl. Bacteriol.* 31:1–11.

53. Rouf, M. A., and M. M. Rigney. 1971. Growth temperatures and temperature characteristics of *Aeromonas*. *Appl. Microbiol.* 22:503–6.

54. Shaw, M. K. 1967. Effect of abrupt temperature shift on the growth of mesophilic and psychrophilic yeasts. *J. Bacteriol.* 93:1332–36.

55. Sinclair, N. A., and J. L. Stokes. 1963. Role of oxygen in the high cell yields of psychrophiles and mesophiles at low temperatures. *J. Bacteriol.* 85:164–67.

56. ———. 1964. Isolation of obligately anaerobic psychrophilic bacteria. *J. Bacteriol.* 87:562–65.
57. ———. 1965. Obligately psychrophilic yeasts from the polar regions. *Can. J. Microbiol.* 11:259–69.
58. Stokes, J. L. 1963. General biology and nomenclature of psychrophilic microorganisms. In *Recent progress in microbiology, Symposium of the International Congress of Microbiologists, 8th,* Montreal, 1962, ed. N. E. Gibbons, 187–92. Toronto: Univ. of Toronto Press.
59. ———. 1967. Heat-sensitive enzymes and enzyme synthesis in psychrophilic microorganisms. In *Molecular mechanisms of temperature adaptation,* ed. C. L. Prosser, 311–23. Pub. no. 84. Washington, D.C.: American Association for the Advancement of Science.
60. Straka, R. P., and J. L. Stokes. 1960. Psychrophilic bacteria from Antarctica. *J. Bacteriol.* 80:622–25.
61. Strange, R. E., and M. Shon. 1964. Effect of thermal stress on the viability and ribonucleic acid of *Aerobacter aerogenes* in aqueous suspensions. *J. Gen. Microbiol.* 34:99–114.
62. Tempest, D. W., and J. R. Hunter. 1965. The influence of temperature and pH value on the macromolecular composition of magnesium-limited and glycerol-limited *Aerobacter aerogenes* growing in a chemostat. *J. Gen. Microbiol.* 41:267–73.
63. Udaka, S., and T. Horiuchi. 1965. Mutants of *Escherichia coli* having temperature sensitive regulatory mechanism in the formation of arginine biosynthetic enzymes. *Biochem. Biophys. Res. Commun.* 19:156–60.
64. Uffen, R. L., and E. Canale-Parola. 1966. Temperature-dependent pigment production by *Bacillus cereus* var. *alesti. Can. J. Microbiol.* 12:590–93.
65. Upadhyay, J., and J. L. Stokes. 1962. Anaerobic growth of psychrophilic bacteria. *J. Bacteriol.* 83:270–75.
66. ———. 1963. Temperature-sensitive formic hydrogenlyase in a psychrophilic bacterium. *J. Bacteriol.* 85:177–85.
67. ———. 1963. Temperature-sensitive hydrogenase and hydrogenase synthesis in a psychrophilic bacterium. *J. Bacteriol.* 86:992–98.
68. Williams, R. P., M. E. Goldschmidt, and C. L. Gott. 1965. Inhibition by temperature of the terminal step in biosynthesis of prodigiosin. *Biochem. Biophys. Res. Commun.* 19:177–81.
69. Witter, L. D. 1961. Psychrophilic bacteria—A review. *J. Dairy Sci.* 44:983–1015.

24.

CHARACTERISTICS AND GROWTH

OF THERMOPHILIC MICROORGANISMS

On the basis of growth temperature, thermophilic bacteria may be characterized as organisms with a minimum at 30°C, an optimum between 50°–60°C, and maximum growth temperatures between 70° and 80°C (36). Thermophilic algae, *Streptomyces,* and fungi have all been described (14, 15, 36). The thermophilic eubacteria, however, are by far the most important from the standpoint of food microbiology. The most important genera of eubacteria that contain thermophiles are *Bacillus* and *Clostridium,* with *Actinomycetes* and lactobacilli being of less importance in foods.

Thermophilic growth may be characterized as follows. The lag phase is short and sometimes difficult to measure. Spores germinate and grow rapidly. The logarithmic phase of growth is of short duration. Some thermophiles have been reported to have generation times as short as 10 min when growing at high temperatures. The rate of death or "die off" is rapid. Loss of viability or "autosterilization" below the thermophilic growth range is characteristic of organisms of this type. The growth curves of a bacterium at 55°, 37°, and 20°C are compared in Fig. 24-1.

The most basic question to be asked about this group of organisms is, why do they require such high temperatures for growth? As is the case with psychrophilic organisms, the precise mechanisms of growth of these forms is not too well understood, but research has greatly illuminated the underlying mechanisms and pointed to the following as being important to the welfare of these organisms.

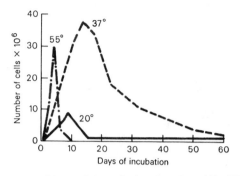

Fig. 24-1. Growth curves of bacterial strain incubated at 20°, 37°, and 55°C (37).

THERMOSTABILITY

Thermophilic microorganisms possess thermostable components that permit a high minimum growth temperature. The three groups of thermostable components that have been most studied are enzymes, ribosomes, and flagella.

THERMOSTABLE ENZYMES. The enzymes of thermophiles may be divided into three groups: (1) Those stable at the temperature of production but requiring slightly higher temperatures for inactivation. Examples of this type are malic dehydrogenase, ATPase, inorganic pyrophosphatase, aldolase, and certain peptidases. (2) Enzymes that are inactivated at the temperature of production in the absence of specific substrates. These include asparagine deamidase, catalase, pyruvic acid oxidase, isocitrate lyase, and certain membrane-bound enzymes. (3) Highly heat-resistant enzymes and proteins, among which are alpha-amylase, protease, glyceraldehyde-3-phosphate dehydrogenase, amino acid activating enzymes, flagella proteins, esterases, and thermolysin. The heat stability of some enzymes from mesophilic and thermophilic bacteria is presented in Table 24-1. It should be noted that the alpha-amylase of *B. stearothermophilus* remained completely active after being heated at 70°C for 24 h. However, in a study by Pfueller and Elliott (33) the alpha-amylase of *B. stearothermophilus* was found to have a molecular weight of approximately 53,000 instead of 15,500 and to be heat sensitive in the purified state, although it hydrolyzed starch at 70°C and above. These authors reported that the enzyme was protected from thermal denaturation above 55°C by metal ions, particularly Ca^{++}, and by protein. The alpha-amylase from a facultative thermophile grown at 37°C was shown by Campbell (11) to be heat-labile, whereas that produced at 55°C was heat stable. Studies on the heat-stable alpha-amylase by Campbell and Manning (10) and Manning and Campbell (27) showed that the enzyme contained a high content of proline, which resulted in a large negative optical rotation. It may be that since the enzyme exists in an unfolded configuration in the native state, heat has relatively little effect upon its denaturation. Koffler and Gale (24)

Table 24-1. Comparison of thermostability and other properties of enzymes from mesophilic and thermophilic bacteria (30).

Species	Enzyme	Heat stability[a] (%)		Half-cystine (mole/ mole of protein)	Molecular weight	Metal required for stability
B. subtilis	Subtilisin BPN'	45	(50°C, 30 min)	0	28,000	Yes
B. subtilis	Neutral protease	50	(60°C, 15 min)	0	44,700	Yes
Ps. aeruginosa	Alkaline protease	80	(60°C, 10 min)	0	48.400	Yes
Ps. aeruginosa	Elastase	86	(70°C, 10 min)	4.6	39,500	Yes
		10	(75°C, 10 min)			
Group A streptococci	Streptococcal protease	0	(70°C, 30 min)	1	32,000	
Cl. histolyticum	Collagenase	1.5	(50°C, 20 min)	0	90,000	
St. griseus	Pronase	60	(60°C, 10 min)			Yes
B. thermopro-teolyticus	Thermolysin	95	(60°C, 120 min)	0	42,700	Yes
		50	(80°C, 60 min)			
B. subtilis	α-Amylase	55	(65°C, 20 min)	0	50,000	Yes
B. stearother-mophilus	α-Amylase	100	(70°C, 24 h)	4	15,500	

[a]Activity remaining after heat treatment shown in parentheses.

studied the heat stability of cytoplasmic proteins isolated from four thermophiles and four mesophiles and showed that the enzymes from thermophiles were more heat-stable than similar preparations from mesophiles. These investigators further found that over 50% of the mesophile proteins coagulated when heated at 60°C and pH 6 for 8 min, while hardly any of the thermophile proteins coagulated under similar conditions. Ascione and Fresco (4) demonstrated the heat-resistance of amino acid activating enzymes from thermophilic algae. These authors found that enzymes from algae grown at 70°C withstood boiling for 5 min. The thermostable aldolase of *Thermus aquaticus* has been found to have optimal activity at 95°C with little activity < 60°C (18).

Just why thermophilic enzymes and proteins are so heat resistant is not clear at this time. Among the promising leads are findings that thermophiles have higher levels of hydrophobic amino acids in their proteins than do mesophiles. The amino acid sequences of clostridial ferredoxin from a mesophilic and a thermophilic strain reveal a somewhat higher number of charged amino acids in the thermophilic protein than in that of the mesophile. It is conceivable that this could lead to a tighter binding or a more hydrophobic character of the thermophile protein. A hydrophobic protein would presumably be more heat resistant than one with a more hydrophilic character. Thermophilic proteins have been found to undergo a type of conformational change around 55°–60°C, and this might have the effect of "melting" the hydrophobic cluster within the protein molecule. Many or most thermophilic proteins may owe their high stability to their binding of metal ions such as Mg^{2+}. This arrangement would make for a more tightly bound structure and consequently one more refractive to heat. The structural integrity of the membrane of *B. stearothermophilus* protoplasts has been shown to be affected by divalent cations (40). The proteins of thermophiles are similar in molecular weight, amino acid composition, allosteric effectors, subunit composition, and primary sequences to their mesophile counterparts. Caldoactive (extremely thermophilic) and obligate thermophiles synthesize macromolecules that have sufficient intrinsic molecular stability to withstand increased thermal stress (1). For more detailed information on thermophily, see (1) and (35).

THERMOSTABLE RIBOSOMES. It has been reported that ribosomes from thermophilic bacteria are more heat-stable than those from mesophiles such as *E. coli* (26). Furthermore, the thermal stability of ribosomes corresponds well with the maximal growth temperatures of organisms (Table 24-2). Heat-resistant ribosomal RNA has been reported (17), but not DNA (28). Friedman et al. (20) have reported that thermophile ribosomes have a higher melting point than their corresponding rRNA (ribosomal RNA), whereas this effect was not observed with *E. coli*. A study by Saunders and Campbell (34) of the amino acid composition of the ribosomes of *B. stearothermophilus* and *E. coli* showed them to be quite similar. As to why ribosomes of *B. stearothermophilus* are thermally stable, Saunders and Campbell suggested that they may reflect either an unusual packing arrangement of the protein to the RNA, or differences in the primary structure of the ribosomal proteins. Ansely et al. (2) were unable to find any unusual chemical features of *B. stearothermophilus* ribosomal proteins that could explain the thermal stability of the respective ribosomes, while Bassel and Campbell (5) found no significant differences in either the size or the arrangement of surface filaments of *B. stearothermophilus* and *E. coli* ribosomes.

The thermal stability of bacterial ribosomes is affected by rRNA base composition. For example, in a study of nineteen organisms, Pace and Campbell (32) found that with a few exceptions, the G-C (guanine-cytosine) content of rRNA molecules increased and the A-U (adenine-uracil) content decreased with increasing maximal growth temperature. An increased G-C content makes for a more stable structure through more extensive hydrogen bonding. The 16 S and 23 S fractions of thermophile rRNA have been shown by Saunders and Campbell to be more heat-stable than the corresponding components from *E. coli*. Friedman (19) has pointed out that the thermostability of ribosomes could be the limiting factor in determining the upper growth temperature for organisms. The thermal stability of soluble RNA from thermophiles and mesophiles appears to be the same (3, 21). The latter authors have shown that ribosomes and soluble components from *B. stearothermophilus* are capable of incorporating some amino acids at temperatures between 30° and 70°C.

Table 24-2. Ribosome melting and maximal growth temperatures of nineteen selected organisms (32).

Organism and strain no.	Max. growth temp. (°C)	Ribosome T_m (°C)
1. *V. marinus* (15381)	18	69
2. 7E-3	20	69
3. 1–1	28	74
4. *V. marinus* (15382)	30	71
5. 2–1	35	70
6. *D. desulfuricans* (cholinicus)[a]	40	73
7. *D. vulgaris* (8303)[a]	40	73
8. *E. coli* (B)	45	72
9. *E. coli* (Q13)	45	72
10. *S. itersonii* (SI–1)[b]	45	73
11. *B. megaterium* (Paris)	45	75
12. *B. subtilis* (SB–19)	50	74
13. *B. coagulans* (43P)	60	74
14. *D. nigrificans* (8351)[c]	60	75
15. Thermophile 194	73	78
16. *B. stearothermophilus* (T–107)	73	78
17. *B. stearothermophilus* (1503R)	73	79
18. Thermophile (Tecce)	73	79
19. *B. stearothermophilus* (10)	73	79

[a]*Desulfovibrio*
[b]*Spirillum*
[c]*Desulfotomaculum*

THERMOPHILIC FLAGELLA. In a series of studies by Koffler and his associates, the flagella of thermophilic bacteria were shown to be more thermostable than these structures from mesophilic organisms. Koffler (23) and Koffler et al. (25) have reported that flagella of thermophiles remain intact at temperatures as high as 70°C, while those of mesophiles are disintegrated at 50°C. Thermophilic flagella were also found by these investigators to be more resistant to urea and acetamide than mesophile flagella, suggesting that more effective hydrogen bonding occurs in those of thermophiles. Thermophile flagella were further shown to possess only one-half as many titratable or basic and acidic groups as did flagella from mesophiles. Other than having more overall intermolecular stability, the precise reasons why flagella from thermophiles are more heat-stable than those from mesophiles is not yet well understood.

In addition to the above thermostable components of thermophiles, other aspects of thermophilic microorganisms and thermophilic growth have received the attention of various investigators, and some of these are discussed below.

OTHER ASPECTS OF THERMOPHILIC MICROORGANISMS

NUTRIENT REQUIREMENTS. Thermophiles generally have a higher nutrient requirement than mesophiles when growing at thermophilic temperatures. This is illustrated for two strains of *B. coagulans* and three of *B. stearothermophilus* in Table 24-3. It may be noted that one strain of *B. coagulans* showed no differences in growth requirements regardless of incubation temperature. The obligately thermophilic strain of *B. stearothermophilus* showed one additional requirement as incubation temperature increased, while one temperature facultative strain of this organism showed additional requirements as the incubation temperature was lowered. Although this aspect of thermophilism has not received much study, changes in nutrient requirements as incubation temperature is raised may be due to a general lack of efficiency on the part of the metabolic complex. Certain enzyme systems might well be affected by the increased temperature of incubation as well as the overall process of enzyme synthesis.

Table 24-3. Effect of incubation temperature on the nutritional requirements of thermophilic bacteria (12).

Organism and strain no.	Nutritional requirements at		
	36°	45°	55°
B. coagulans 2 (F)[a]	his, thi, bio, fol	his, thi, bio, fol	his, thi, bio, fol
B. coagulans 1039 (F)	thi, bio, fol	thi, bio, fol	his, met, thi, nic, bio, fol
B. stearothermophilus 3690 (F)	met, leu, thi, nic, bio, fol	met, thi, bio, fol	met, thi, bio, fol
B. stearothermophilus 4259 (F)	bio, fol	met, his, nic, bio, fol	met, his, nic, bio, fol
B. stearothermophilus 1373b (O)[b]	no growth	glu, his, met, leu, bio	glu, his, met, leu, bio, rib

[a] F = facultative
[b] O = obligate thermophile
his = histidine; thi = thiamine; bio = biotin; fol = folic acid; met = methionine; leu = leucine; nic = nicotinic acid; glu = glutamic acid; rib = riboflavin

OXYGEN TENSION. Thermophilic growth is affected by oxygen tension. As the temperature of incubation is increased, the growth rate of microorganisms increases, thereby increasing the oxygen demand on the culture medium, while at the same time the solubility of oxygen is reduced. This is thought by some investigators to be one of the most important limiting factors of thermophilic growth in culture media. Downey (16) has shown that thermophilic growth is optimal at or near the oxygen concentration normally available in the mesophilic range of temperatures—143 to 240 μM. Although it is conceivable that thermophiles are capable of high-temperature growth due to their ability to consume and conserve oxygen at high temperatures, a capacity that mesophiles and psychrophiles lack, further data in support of this notion are wanting.

CELLULAR LIPIDS. The state of cellular lipids affects thermophilic growth. Since an increase in degree of unsaturation of cellular lipids is associated with psychrotrophic growth, it is reasonable to assume that a reverse effect occurs in the case of thermophilic growth. This idea finds support in the investigations of many authors. Gaughran (22) showed that mesophiles growing above their maximum range showed decreases in lipid content and more lipid saturation. According to this author, cells cannot grow at temperatures below the solidification point of their lipids. Marr and Ingraham (29) showed a progressive increase in saturated fatty acids and a corresponding decrease in unsaturated fatty acids in *E. coli* as the temperature of growth increased. The general decrease in the proportion of unsaturated fatty acids as growth temperatures increase has been found to occur in a large variety of animals and plants (13). Saturated fatty acids form stronger hydrophobic bonds than do unsaturated. Among the saturated fatty acids are branched chain acids. The preferential synthesis of branched heptadecanoic acid and the total elimination of unsaturated fatty acids by two thermophilic *Bacillus* spp. has been observed (39).

CELLULAR MEMBRANES. The nature of cellular membranes affects thermophilic growth. Brock (7) reported that the molecular mechanism of thermophilism is more likely to be related to the function and stability of cellular membranes than to the properties of specific macromolecules. This investigator pointed out that there is no evidence that organisms are killed by heat because of the inactivation of proteins or other macromolecules, a view that is widely held. According to Brock, an analysis of thermal-death curves of various microorganisms shows that this is a first-order process compatible with an effect of heat on some large structure such as the cell membrane, since a single hole in the membrane could result in leakage of cell constitutents and subsequent death. Brock has also pointed out that thermal killing due to the inactivation of heat-sensitive enzymes, or heat-sensitive ribosomes, of which there are many copies in the cell, should not result in simple first-order kinetics. The leakage of ultraviolet light–absorbing and other material from cells undergoing "cold shock" would tend to implicate the membrane in high-temperature death. Since most animals die when body temperatures reach between 40° and 45°C and most psychrophilic bacteria are killed at about this temperature range, the suggestion that lethal injury is due to the melting of lipid constituents of the cell or cell membrane is not only plausible; it has been supported by the findings of various investigators (35). The unit cell membrane is thought to consist of layers of lipid surrounded by layers of protein and to depend upon the lipid layers for its biological functions. The disruption of this structure would be expected to cause cell damage and perhaps death. In view of the changes in cellular lipid saturation noted above, the cell membrane appears to be critical to growth and survival at thermophilic temperatures.

EFFECT OF TEMPERATURE. Brock (7) has called attention to the fact that thermophiles apparently do not grow as fast at their optimum temperatures as one would predict or is commonly believed. Arrhenius plots of thermophile growth compared to *E. coli* over a range of temperatures indicated that, overall, the mesophilic types were more efficient. This author believes that thermophile enzymes are inherently less efficient than mesophiles because of thermal stability; that is, the thermophiles have had to discard growth efficiency in order to survive at all. Somewhat in support of the latter notion, Brock and Brock (8) have shown that the optimum temperature for glucose incorporation was similar to the environmental temperature for thermophilic bacteria occurring at various temperatures along a thermal gradient of a hot spring in Yellowstone Park. This is a good example of the adaptation of microorganisms to temperature changes. As was suggested for psychrotrophs in the previous chapter, the underlying control of thermophilism appears to be under genetic control.

GENETIC BASES. A significant discovery toward an understanding of the genetic bases of thermophilism was made by McDonald and Matney (31). These investigators effected the transformation of thermophilism in *B. subtilis* by growing cells of a strain that could not grow above 50°C in the presence of DNA extracted from one that could grow at 55°C. The more heat-sensitive strain was transformed at a frequency of 10^{-4}. These authors also noted that only 10–20% of the transformants retained the high-level streptomycin resistance of the recipient, which indicated that the genetic loci for streptomycin resistance and that for growth at 55°C were closely linked.

Although much has been learned about the basic mechanisms of thermophilism in microorganisms, the precise mechanisms underlying this high-temperature phenomenon still remain a mystery. The facultative thermophiles such as some *B. coagulans* strains present a picture as puzzling as the obligate thermophiles. The facultative thermophiles display both mesophilic and thermophilic types of metabolism. In their studies of these types from the genus *Bacillus,* which grew well at both 37 and 55°C, Bausum and

Matney (6) reported that the organisms appeared to shift from mesophilism to thermophilism between 44° and 52°C.

REFERENCES

1. Amelunxen, R. E., and A. L. Murdock. 1978. Microbial life at high temperatures: Mechanisms and molecular aspects. In *Microbial life in extreme environments,* ed. D. J. Kushner, 217–78. New York: Academic Press.
2. Ansley, S. B., L. L. Campbell, and P. S. Sypherd. 1969. Isolation and amino acid composition of ribosomal proteins from *Bacillus stearothermophilus. J. Bacteriol.* 98:568–72.
3. Arca, M., C. Calvori, L. Frontali, and G. Tecce. 1964. The enzymic synthesis of aminoacyl derivatives of soluble ribonucleic acid from *Bacillus stearothermophilus. Biochem. Biophys. Acta* 87:440–48.
4. Ascione, R., and J. R. Fresco. 1964. Heat-stable amino activating enzymes from a thermophile. *Fed. Proc.* 23:163.
5. Bassel, A., and L. L. Campbell. 1969. Surface structure of *Bacillus stearothermophilus* ribosomes. *J. Bacteriol.* 98:811–15.
6. Bausum, H. T., and T. S. Matney. 1965. Boundary between bacterial mesophilism and thermophilism. *J. Bacteriol.* 90:50–53.
7. Brock, T. D. 1967. Life at high temperatures. *Science* 158:1012–19.
8. Brock, T. D., and M. L. Brock. 1968. Relationship between environmental temperature of bacteria along a hot spring thermal gradient. *J. Appl. Bacteriol.* 31:54–58.
9. Campbell, L. L. 1955. Purification and properties of an alpha-amylase from facultative thermophilic bacteria. *Arch. Biochem. Biophys.* 54:154–61.
10. Campbell, L. L., and G. B. Manning. 1961. Thermostable α-amylase of *Bacillus stearothermophilus.* III. Amino acid composition. *J. Biol. Chem.* 236:2962–65.
11. Campbell, L. L., and B. Pace. 1968. Physiology of growth at high temperatures. *J. Appl. Bacteriol.* 31:24–35.
12. Campbell, L. L., and O. B. Williams. 1953. The effect of temperature on the nutritional requirements of facultative and obligate thermophilic bacteria. *J. Bacteriol.* 65:141–45.
13. Chapman, D. 1967. The effect of heat on membranes and membrane constituents. In *Thermobacteriology,* ed. A. H. Rose, 123–46. New York: Academic Press.
14. Cooney, D. G., and R. Emerson. 1964. *Thermophilic fungi.* San Francisco: Freeman.
15. Cross, T. 1968. Thermophilic actinomycetes. *J. Appl. Bacteriol.* 31:36–53.
16. Downey, R. J. 1966. Nitrate reductase and respiratory adaptation in *Bacillus stearothermophilus. J. Bacteriol.* 91:634–41.
17. Farrell, J., and A. Rose. 1967. Temperature effects on microorganisms. *Ann. Rev. Microbiol.* 21:101–20.
18. Freeze, R., and T. D. Brock. 1970. Thermostable aldolase from *Thermus aquaticus. J. Bacteriol.* 101:541–50.
19. Friedman, S. M. 1968. Protein-synthesizing machinery of thermophilic bacteria. *Bacteriol. Rev.* 32:27–38.
20. Friedman, S. M., R. Axel, and I. B. Weinstein. 1967. Stability of ribosomes and ribosomal ribonucleic acid from *Bacillus stearothermophilus. J. Bacteriol.* 93:1521–26.

21. Friedman, S. M., and I. B. Weinstein. 1966. Protein synthesis in a subcellular system from *Bacillus stearothermophilus*. *Biochim. Biophys. Acta* 114: 593–605.
22. Gaughran, E. R. L. 1947. The saturation of bacterial lipids as a function of temperature. *J. Bacteriol.* 53:506.
23. Koffler, H. 1957. Protoplasmic differences between mesophiles and thermophiles. *Bacteriol. Rev.* 21:227–40.
24. Koffler, H., and G. O. Gale. 1957. The relative thermostability of cytoplasmic proteins from thermophilic bacteria. *Arch. Biochem. Biophys.* 67:249–51.
25. Koffler, H., G. E. Mallett, and J. Adye. 1957. Molecular basis of biological stability to high temperatures. *Proc. Nat'l. Acad. Sci. U.S.* 43:464–77.
26. Mangiantini, M. T., G. Tecce, G. Toschi, and A. Trentalance. 1965. A study of ribosomes and of ribonucleic acid from a thermophilic organism. *Biochim. Biophys. Acta* 103:252–74.
27. Manning, G. B., and L. L. Campbell. 1961. Thermostable α-amylase of *Bacillus stearothermophilus*. I. Crystallization and some general properties. *J. Biol. Chem.* 236:2952–57.
28. Marmur, J. 1960. Thermal denaturation of deoxyribosenucleic acid isolated from a thermophile. *Biochim. Biophys. Acta* 38:342–43.
29. Marr, A. G., and J. L. Ingraham. 1962. Effect of temperature on the composition of fatty acids in *Escherichia coli*. *J. Bacteriol.* 84:1260–67.
30. Matsubara, H. 1967. Some properties of thermolysin. In *Molecular mechanisms of temperature adapation,* ed. C. L. Prosser, Pub. no. 84, 283–94. Washington, D.C.: American Association for the Advancement of Science.
31. McDonald, W. C., and T. S. Matney. 1963. Genetic transfer of the ability to grow at 55°C in *Bacillus subtilis*. *J. Bacteriol.* 85:218–20.
32. Pace, B., and L. L. Campbell. 1967. Correlation of maximal growth temperature and ribosome heat stability. *Proc. Nat'l. Acad. Sci. U.S.* 57:1110–16.
33. Pfueller, S. L., and W. H. Elliott. 1969. The extracellular α-amylase of *Bacillus stearothermophilus*. *J. Biol. Chem.* 244:48–54.
34. Saunders, G. F., and L. L. Campbell. 1966. Ribonucleic acid and ribosomes of *Bacillus stearothermophilus*. *J. Bacteriol.* 91:332–39.
35. Singleton, R., Jr., and R. E. Amelunxen. 1973. Proteins from thermophilic microorganisms. *Bacteriol. Rev.* 37:320–42.
36. Stokes, J. L. 1967. Heat-sensitive enzymes and enzyme synthesis in psychrophilic microorganisms. In *Molecular mechanisms of temperature adaptation,* ed. C. L. Prosser, Pub. no. 84, 311–23. Washington, D.C.: American Association for the Advancement of Science.
37. Tanner, F. W., and G. I. Wallace. 1925. Relation of temperature to the growth of thermophilic bacteria. *J. Bacteriol.* 10:421–37.
38. Tendler, M. D., and P. R. Burkholder. 1961. Studies on the thermophilic *Actinomycetes*. I. Methods of cultivation. *Appl. Microbiol.* 9:394–99.
39. Weerkamp, A., and W. Heinen. 1972. Effect of temperature on the fatty acid composition of the extreme thermophiles, *Bacillus caldolyticus* and *Bacillus caldotenax*. *J. Bacteriol.* 109:443–46.
40. Wisdom, C., and N. E. Welker. 1973. Membranes of *Bacillus stearothermophilus*. Factors affecting protoplast stability and thermostability of alkaline phosphatase and reduced nicotinamide adenine dinucleotide oxidase. *J. Bacteriol.* 114:1336–45.

25.

NATURE OF RADIATION RESISTANCE
IN MICROORGANISMS

Prior to 1956, the most radiation-resistant microorganisms known were the sporeforming bacteria, especially *Clostridium botulinum* strains. That spore-forming bacteria should be especially resistant to radiation is perhaps not surprising in view of their high levels of resistance to heat, drying, chemicals, and other environmental conditions. In 1956, Anderson et al. (2) announced the discovery of a nonsporeforming gram-positive coccus that was later named *Micrococcus radiodurans* (3). This organism was shown to be more radioresistant than any bacterium previously known. It has variously been referred to as a *Sarcina* sp. and as *M. rubens*. A second organism with a similar high degree of radioresistance was reported by Murray and Robinow (32). This isolate was later determined to be *M. radiodurans* and has been designated the Sark strain. A third radiation-resistant coccus was isolated by Davis et al. (8) from irradiated haddock and appears to be a variant of the original isolate by Anderson and co-workers, but differs from it in several ways as noted later in this chapter.

Another highly resistant coccus was isolated from Bombay duck (28). This isolate was subsequently named *Micrococcus radiophilus*. It differs from *M. radiodurans* in being smaller in size, salt tolerant, and more radioresistant. A hemolytic radiation-resistant micrococcus has been isolated from chicken (44). It is extremely resistant to drying and storage and exhibits its highest resistance during exponential growth unlike some *Moraxella-Acinetobacter* strains, which are most sensitive during this phase (40).

These organisms are more resistant to radiation than sporeformers. Following the early discoveries, a great deal of attention was focused on the mechanisms of their high level of resistance. There are several reasons for the relatively intense interest in these organisms. First, as food-borne organisms, an understanding of the mechanism of their radiation resistance would possibly enable one to prevent the accumulation of such types in foods through mutations or other means. This accumulation of radiation-resistant organisms would reduce the effectiveness of radiations when applied to foods, or encourage the radiation dosage to be increased to levels that would lead to undesirable changes in irradiated foods. Second, the doses of ionizing radiation necessary to effect sterilization of foods (4–5 Mrads) cause undesirable changes in some foods. The addition of suitable radiation-sensitizing agents would allow for a reduction in the dose of radiation and thereby lessen the development of undesirable changes. Research of this nature would also facilitate the search for or the design of radiation-sensitizing agents for food use. Third, since the advent of atomic energy, numerous unsuccessful attempts have been made to uncover a nontoxic radiation-protective agent for human and animal uses. It is not inconceivable that such compounds may be developed from a better understanding of the nature of the high level of radioresistance in microorganisms.

Presented below are characteristics of these highly radiation-resistant microorganisms in addition to other aspects of radiation resistance as determined by studies involving other organisms.

THE MICROBIOLOGY OF *M. RADIODURANS* STRAINS

MORPHOLOGIC AND CULTURAL CHARACTERISTICS. All three of the original isolates of this organism are gram-positive cocci that occur principally as tetrads and occasionally in pairs or singly. On culture media, they produce pigmented colonies that range from flesh-colored to pink to brownish-red and to bright red, depending upon age of culture and type of medium employed. The pigments have been identified as carotenoids (22). In the case of the original isolate of Anderson and co-workers, the pigment is water soluble while the strain isolated by Davis and co-workers is not. The temperature growth range for *M. radiodurans* was reported by Anderson et al. to be 5°–40°C with an optimum around 25°–30°C. No growth occurred at 45°C. Visible colonies are produced within 48 h at 35°C. Anderson and co-workers reported that *M. radiodurans* resembled *M. roseus* and *M. rubens* in its cultural and morphologic characteristics. The strain isolated by Davis and co-workers differs from the original isolate by its inability to reduce nitrate to nitrite, its inability to produce gelatinase, and its smaller cell size. Although facultative in nature, the growth of this organism was reported by Anderson and co-workers to be accelerated by aeration.

In regard to its nutrition, the R_1 strain (described below) grows on a synthetic medium containing four B vitamins, the amino acids glutamate and methionine, glucose as an energy source, and minerals. Methionine is the only essential amino acid, and it is incorporated rapidly into the cells

(33). Several mono saccharides are readily utilized by this organism, and it appears that they are oxidized through the tricarboxylic acid cycle.

RELATIVE RADIATION RESISTANCE. The R_1 strain of *M. radiodurans* was obtained by Anderson et al. from ground beef and pork that had been exposed to 2–3 Mrads of gamma radiation. A strain isolated from unirradiated meat was designated U_1. A variant of the R_1 strain having different colony form, pigmentation, and smaller cell size was designated R_4R, while the R_w strain is a variant of R_1 that possesses less color (10). Some of these strains survived 6 Mrads on agar slants. The isolate of Davis and co-workers was obtained from irradiated haddock and was shown to consist of smooth and rough strains and to be more resistant than the R_2 isolate of Anderson et al., especially in buffer (see Fig. 25-1). Of the smooth and rough strains of Davis and co-workers, the former was reported to be more resistant than the latter. The relative resistance of the four strains of *M. radiodurans* isolated by Anderson and co-workers is shown in Fig. 25-2. This organism has been shown to possess very high resistance to ultraviolet (UV) radiation as well as to ionizing radiations (13, 17).

The radiation survival of the R_1 strain was shown to be greater in raw beef and raw chicken than in raw fish and cooked beef (10). In beef, this strain was reduced by a factor of about 10^{-5} by 3 Mrads and by a factor of 10^{-9} by 4 Mrads, making it more resistant than sporeformers in this environment. The original isolate of R_1 required 4.8 Mrads for a 10^{-7} reduction.

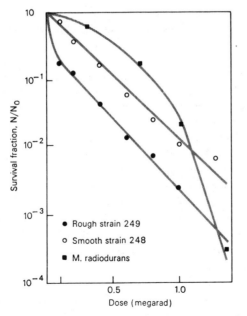

Fig. 25-1. Radiosurvival of washed cells of the rough and smooth variants of isolated coccus and *Micrococcus radiodurans* (8).

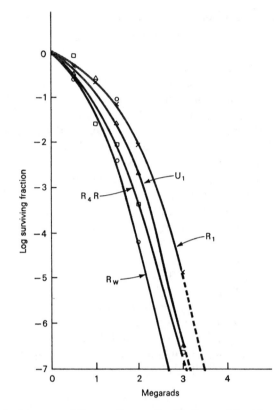

Fig. 25-2. Survival curves of four cultures of radiation-resistant bacteria in beef (10).

The effect of various treatments on the radiation-resistance of the R_1 strain was studied by Duggan et al. (11). These investigators found that freezing in raw puréed beef did not significantly affect its sensitivity. Irradiation of the organism in menstra at temperatures between 40° and 50°C reduced its resistance, as did pre-irradiation heat treatments. No effect upon radiation resistance was noted when the organism was irradiated at pH values of 5, 7, or 9 in buffer. The response of this organism under similar conditions was found to be essentially the same to UV light (13). The latter authors (11) also studied the effect of oxidizing and reducing conditions upon resistance to radiation of this organism in phosphate buffer (see Table 25-1). The flushing of buffer suspensions with nitrogen or O_2 had no significant effect on radiation sensitivity when compared to the control. Likewise, the presence of 100 ppm of H_2O_2 had no significant effect upon sensitivity. Treatment with cysteine rendered the cells less sensitive, while ascorbate increased their sensitivity. The effect of N-ethylmaleimide (NEM) and iodoacetic acid (IAA) upon resistance was investigated by Lee et al. (27). It was found that IAA reduced resistance but not NEM when tested at nontoxic levels.

Table 25-1.   Effects of oxidizing and reducing
conditions on resistance to radiation of *Micrococcus radiodurans* (table of means)[a] (11).

Condition	Log of surviving fraction[b]
Buffer, unmodified	−3.11542
Oxygen flushed	−3.89762
Nitrogen flushed	−2.29335
H_2O_2 (100 ppm)	−3.47710
Thioglycolate (0.01 M)	−1.98455
Cysteine (0.1 M)	−0.81880
Ascorbate (0.1 M)	−5.36050

[a]Determined by count reduction after exposure to 1 Mrad
of gamma radiation in 0.05 M phosphate buffer. LSD: P
= 0.05 (1.98116); P = 0.01 (2.61533).
[b]Averages of four replicates.

These compounds responded in the same way in the presence and absence
of O_2.

A study of Krabbenhoft et al. (25) showed that growth of *M. radiodurans*
in media supplemented with NZ-case caused significant reductions in radio
resistance. The LD_{50} for this organism was found to be 700 krads when
grown on plate count agar supplemented with DL-methionine, but was
approximately one-half as resistant when grown on this medium supplemented
with 0.5% NZ-case. Its resistance to UV light was shown also to change
in a similar manner. The growth patterns of the organism on the two media
were reported to be similar. In addition to being ten times more sensitive
on the NZ-case supplemented medium, the cultures were also less pigmented.

The carotenoid pigments of this organism have been shown to be radiosensitive and apparently do not contribute to its radioresistance (22). The
carotenoid pigments of *M. radiophilus* play no role in the resistance of this
organism to radiations (29).

Radiation death rate curves for *M. radiodurans* have been reported
by various authors to be sigmoidal in nature. This can be seen from
Fig. 25-1 for the R_1 strain. Some possible explanations for this are presented
later in this chapter.

THERMAL RESISTANCE. Studies by Duggan et al. (12) on strain R_1 showed
its thermal death rate to approximate exponential form, unlike its radiation
death rate. In beef, these investigators found this strain to possess a D_{140}
value of 0.75 and a z value of 10.65 min. Unlike its radiation resistance, it
is rather sensitive to heat and will not survive milk pasteurization temperatures.

DISTRIBUTION IN NATURE. The distribution of these organisms was studied
by Krabbenhoft et al. (24). These authors were able to isolate this organism

from ground beef, pork sausage, the hides of animals, and from creek water. All isolates were shown to possess high levels of radiation resistance. None could be found in soil, hay, or fecal matter samples. The finding by Davis and co-workers of this organism in haddock, along with the above, suggests its rather widespread distribution in nature. Its apparent low incidence may be due to its lack of competition with other members of its habitat (24).

SENSITIVITY TO OTHER INHIBITORS. The sensitivity of *M. radiodurans* to ten antibiotics was determined by Hawiger and Jeljaszewicz (17). These authors found this organism to be sensitive to all 10 antibiotics and thus to resemble *Staphylococcus aureus* in its sensitivity to these compounds. The sensitivity to ethylene oxide was investigated by Gammon et al. (15). In a mixture of ethylene oxide and Freon 12 at a level of 500 mg/liter, 50% R.H., and at a temperature of 130°F, the D value of *M. radiodurans* and several spore-formers on nonporous surfaces were found to be as follows:

M. radiodurans:	5.4 min
Bacillus globigii:	6.0 min
Clostridium sporogenes:	3.8 min
Bacillus stearothermophilus:	2.8 min

NATURE OF CELL WALLS. In general, the cell walls of this organism are atypical for gram-positive cocci. They have been shown to contain lipoproteins as well as mucopeptides in which L-ornithine is the principal diamino acid instead of the more usual diaminopimelic acid or lysine (45, 46). The cell-wall lipids consist of even- and odd-numbered straight-chain saturated and unsaturated fatty acids (23). Polysaccharides of the cell wall were found by Work and Griffiths to contain mannose and rhamnose but not heptose. One of the more unusual features of the cell wall of this organism is its thickness and the existence of at least three to four layers, with one of these layers consisting of hexagonally arranged subunits or holes (41, 42, 46). The existence of a multilayered cell wall in bacteria was previously unknown, and its significance relative to the radiation resistance of this organism is not clear at this time. Its cytoplasm and nuclear structures appear to be normal (42).

MECHANISM OF RADIATION RESISTANCE. Upon the irradiation of complex substrates such as foods, a series of chemical changes begins immediately, the extent and duration of which depend upon the dosage of radiation applied, nature of the substrate, and so forth (see Chapter 12). An overall view of some of the events that take place when radiation is applied may be seen from Fig. 25-3. When water-containing samples are irradiated, the radiolysis of water is one of the consequences that leads to the formation of free radicals, peroxides, and so on. Peroxides are produced by both UV and ionizing radiations (21). It has been known for some time that some of the damaging effects of irradiation can be minimized or halted by application

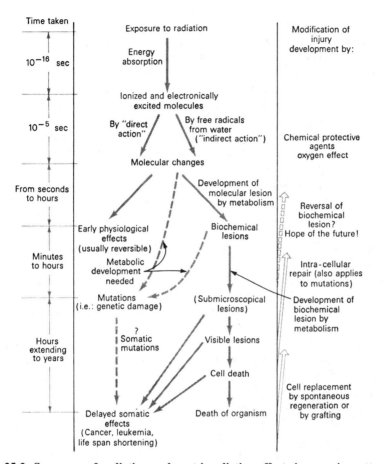

**Fig. 25-3. Summary of radiation and post-irradiation effects in organic matter.
(Bacq and Alexander, 4, reprinted with permission of the authors, *Fundamentals of
Radiobiology*, copyright © 1961 Pergamon Press)**

of certain radioprotective compounds prior to irradiation. The compounds
most effective for this purpose are those that contain —SH groups. According
to Serianni and Bruce (35), the most effective chemical radioprotective
agents found so far are amino-thiols of the following general formula:
R_2N—$(CH_3)_n$—SH, where R = H or NH_2 and *n* is not greater than 3. Since
such compounds possess the capacity of increasing radioresistance in microorganisms, most of the earlier work on the mechanism of radioresistance
in *M. radiodurans* sought to uncover and identify such compounds in this
organism.

An important finding in the studies on radioresistance in this organism
was that of Bruce (6), who was able to confer radioresistance upon *Escherichia
coli* by employing extracts from *M. radiodurans*. Similar extracts from
Sarcina lutea, a relatively radiosensitive organism, were shown to sensitize

rather than protect *E. coli*. This investigator stated that the amount of protective substance in the *M. radiodurans* extracts was sufficient to explain the high radioresistance of the organism. Additional information on this protective action was reported by Serianni and Bruce (35). These workers found that extracts from *M. radiodurans* cultured in chemically defined media had very little free —SH activity but, nevertheless, exhibited great protective activity. These authors further reported that extracts from stationary-phase cultures, while highly protective, were apparently devoid of components containing sulfur, thus suggesting that the radioprotective component or components in the extract were devoid of —SH groups. No evidence for the presence of either mercaptoethylamine (MEA) or mercaptopropylamine (MPA) was found. The protective substance was reported to be of low molecular weight and to be produced within the cell.

GENETIC ASPECTS OF RADIATION RESISTANCE. Kaplan and Zavarine (20) reported that the sensitivity of microorganisms to ionizing radiations appeared to increase with an increase in G-C content of their DNA. Various investigators have since studied DNA of *M. radiodurans* in an effort to determine its role in the high radioresistance of this organism. Setlow and Duggan (37) found that DNA from this organism was not resistant to UV light. These investigators postulated that the organism must have a very efficient DNA repair mechanism in view of its high UV resistance and the low resistance of its DNA. The $G + C/A + T$ ratio of *M. radiodurans* was found by these workers to be 1.6 compared to 1.0 for *E. coli*. Moseley and Schein (31) found a high G-C content (67%) in the DNA of this organism and noted that this level was the same as that for *Pseudomonas* spp., which represent some of the most radiosensitive bacteria. *M. radiodurans* does not fall into the pattern of those organisms studied by Kaplan and Zavarine (20), in which radiation resistance is correlated with a low G-C content. Moseley and Schein (31) found the quantity of DNA in *M. radiodurans* to be about the same as that for other bacteria.

In view of the evidence that the radioresistance of this organism is apparently not due to the radioresistance of its DNA, several investigators have postulated a rapid repair of DNA damage by this organism (30, 36, 37). Moseley and Laser have shown that ionizing radiation damage in *M. radiodurans* can be repaired enzymatically and is similar to that operative in UV dark repair. Setlow (36) found that the UV irradiation of DNA results in dimerization of adjacent thymines in the same polynucleotide chain, and that these dimers can be removed by the cell in the form of oligonucleotides. According to this author, formation of the dimers blocks DNA synthesis.

The resistance of this organism to X rays can be reduced by around 90% if irradiation of cells is carried out in the presence of 10^{-3} *M* iodoacetamide (9). These workers showed that this radiosensitizer must be present during irradiation to be effective. This compound was not effective against sensitive *E. coli* and *Ps. fluorescens*, but was effective against a radiation-resistant strain of *E. coli,* strain B/r. Moseley and Laser (30) have suggested that iodoacetamide acts by inhibiting the cell's DNA repair mechanism.

There is general agreement among researchers that the radiation resistance of these organisms is due in large part to their possession of an excellent pyrimidine dimer excision repair system. Studies with UV-sensitive mutants have shown that they possess reduced rates of pyrimidine dimer excision. A study by Lavin et al. (26) employing *M. radiophilus* revealed that this organism also possesses an efficient excision repair system, which these authors believe is at least in part responsible for its extreme radioresistance.

THE GENERAL NATURE OF RADIORESISTANCE IN OTHER MICROORGANISMS

ROLE OF RADIOPROTECTIVE COMPOUNDS. Much attention has been devoted to the finding of radioprotective substances in radioresistant microorganisms other than *M. radiodurans.* The finding by Raj et al. (33) that methionine was the only amino acid required by *M. radiodurans* in a synthetic medium provided some support for the notion of intracellular radioprotective substances, since this amino acid contains a sulfur group. More direct support came from studies on the radioresistance of bacterial endospores, structures known to be far more resistant than their corresponding vegetative cells. Romig and Wyss (34) studied UV resistance in sporulating *B. cereus* and found that UV resistance increased about 2 h before heat resistance as vegetative cells went into spore production. The same general phenomenon was shown by Vinter (43) for X-ray resistance. The latter author further demonstrated that spores contained higher levels of cysteine and cystine sulfur than corresponding vegetative cells. The increase in sulfur was found to coincide with an increase in radiation resistance of sporulating cells. Vinter (43) showed further that changes in cystine sulfur and radioresistance were distinct from the changes in Ca^{++} and dipicolinic acid. The cystine-rich substance occurred in spore coats and was not released into the medium upon spore germination as with dipicolinic acid. These studies were supported by the finding of Bott and Lundgren (5) that radioresistance in *B. cereus* increased with progressive spore maturity and that the spores had extremely low, or no, —SH content. Upon treating the spores of *B. cereus* with thioglycolic acid, which ruptured from 10–30% of the spore disulfide bonds to thiol groups, they were found to retain their resistance to gamma radiation and to heat (19).

ROLE OF CATALASE. Among the various characteristic features of *M. radiodurans* is the large quantity of catalase it produces (2). This property is apparently shared by all strains of this organism. The possible relationship of catalase production to radioresistance stems from the fact that H_2O_2 is known to be produced by both UV and ionizing radiations. This compound is, of course, destroyed by catalase. As noted above, the irradiation of *M. radiodurans* in the presence of H_2O_2 does not affect its resistance. In a study of the role of catalase and H_2O_2 on the radiation sensitivity of *Escherichia*, Engel and Adler (14) found that there was no positive correlation between catalase activity and sensitivity to ionizing radiation. It thus appears that the amount of catalase produced by an organism bears no relationship to its radioresistance. The clostridia are among the more radioresistant of all bacteria and yet they are anaerobic (produce no catalase).

ROLE OF OXYGEN. The sensitivity of microorganisms is generally higher when irradiation occurs in the presence of O_2. Conversely, the absence of O_2 during irradiation decreases sensitivity. Stapleton et al. (38) showed that O_2 removal protected *E. coli* against X-ray inactivation. These investigators suggested that O_2 removal by either cellular enzyme action or auto-oxidation could also serve to protect this organism against X rays. Cromroy and Adler (7) showed that *E. coli* could be protected by MEA more effectively than by O_2 removal. MEA apparently acts to remove this gas from the cell's environment.

GENETIC ASPECTS. Attempts to explain the sigmoidal death rate curve of *M. radiodurans* have caused various authors to suggest the role of packet or tetrad formation by this organism as well as the existence of a multinuclear state. The latter was found by Gunter and Kohn (16) to produce multihit curves when diploid yeasts were subjected to X rays. In a study of X-ray resistance in *E. coli*, Stapleton and Engel (39) found that the resistance was not due to ploidy. The irradiation of the more sensitive *Sarcina* sp. has shown that tetrad formation per se does not give rise to the multihit-type death rate curve.

Penicillin-resistant strains of the *Achromobacter-Alcaligenes* group have been reported to be more sensitive to radiation than their penicillin-sensitive variants (41). This finding may indicate some role of cell walls in radiation resistance, since penicillin affects these structures in bacterial cells. It also suggests that some property controlling penicillin resistance is at least partly responsible for radiation sensitivity (41). On the other hand, *M. radiodurans* is sensitive to penicillin (17).

Adler (1) has shown that radiation sensitivity in *E. coli* is under the control of genes that can be genetically transferred. In *E. coli* K12, one of these genes occupies a locus on the linkage map between those controlling the ability to utilize lactose and galactose. Both UV and ionizing radiation sensitivity are affected similarly. In studies with *B. subtilis*, Zamenhof et al. (47) were able to transform UV sensitivity to a more resistant strain. This feat has not been reported for *M. radiodurans*.

With respect to the role of DNA in the radiation resistance of microorganisms other than *M. radiodurans*. Hill and Simson (18) in their studies with a sensitive and a resistant strain of *E. coli* found no variation in the number of nuclei or in the content of DNA or RNA between these strains. DNA was shown by Zamenhof et al. (47) not to be responsible for radiation resistance in *B. subtilis*. Adler (1) has stated that genes affecting radiation sensitivity fall into two broad categories: (1) those that control systems concerned with repair of radiation-damaged DNA, and (2) those that seem to exert their influence on processes somewhat removed from DNA itself, such as cell division.

REFERENCES

1. Adler, H. I. 1966. The genetic control of radiation sensitivity in microorganisms. *Adv. Radiat. Biol.* 2:167–91.
2. Anderson, A. W., H. C. Nordan, R. F. Cain, G. Parrish, and D. Duggan. 1956. Studies on a radio-resistant *Micrococcus*. I. Isolation, morphology, cultural characteristics, and resistance to gamma radiation. *Food Technol.* 10:575–78.
3. Anderson, A. W., K. E. Rash, and P. R. Elliker. 1961. Taxonomy of a recently isolated radio-resistant *Micrococcus*. *Bacteriol. Proc.*, 56.
4. Bacq, Z. M., and P. Alexander. 1961. *Fundamentals of radiobiology*, 2d ed. Oxford: Pergamon.
5. Bott, K. F., and D. G. Lundgren. 1964. The relationship of sulfhydryl and disulfide constituents of *Bacillus cereus* to radioresistance. *Radiat. Res.* 21:195–211.
6. Bruce, A. K. 1964. Extraction of the radioresistant factor of *Microccus radiodurans*. *Radiat. Res.* 22:155–64.
7. Cromroy, H. L., and H. I. Adler. 1962. Influences of β-mercaptoethylamine and oxygen removal on the X-ray sensitivity of four strains of *Escherichia coil*. *J. Gen. Microbiol.* 28:431–35.
8. Davis, N. S., G. J. Silverman, and E. B. Masurovsky. 1963. Radiation-resistant, pigmented coccus isolated from haddock tissue. *J. Bacteriol.* 86:294–98.
9. Dean, C. J., and P. Alexander. 1962. Sensitization of radio-resistant bacteria to X-rays by iodoacetamide. *Nature* 196:1324–25.
10. Duggan, D. E., A. W. Anderson, and P. R. Elliker. 1963. Inactivation of the radiation-resistant spoilage bacterium *Microoccus radiodurans*. I. Radiation inactivation rates in three meat substrates and in buffer. *Appl. Microbiol.* 11:398–403.
11. ———. 1963. Inactivation of the radiation-resistant spoilage bacterium *Micrococcus radiodurans*. II. Radiation inactivation rates as influenced by menstrum temperature, preirradiation heat treatment, and certain reducing agents. *Appl. Microbiol.* 11:413–17.
12. ———. 1963. Inactivation rate studies on a radiation-resistant spoilage microorganism. III. Thermal inactivation rates in beef. *J. Food Sci.* 28:130–34.
13. Duggan, D. E., A. W. Anderson, P. R. Elliker, and R. F. Can. 1959. Ultraviolet exposure studies on a gamma radiation resistant *Micrococcus* isolated from food. *Food Res.* 24:376–82.
14. Engel, M. S., and H. I. Adler. 1961. Catalase activity, sensitivity to hydrogen peroxide, and radiation response in the genus *Escherichia*. *Radiat. Res.* 15:269–75.

15. Gammon, R. A., K. Kereluk, and R. S. Lloyd. 1968. Microbial resistance to ethylene oxide. *Bacteriol. Proc.,* 16.
16. Gunter, S. E., and H. T. Kohn. 1956. The effect of X-ray on the survival of bacteria and yeast. I. *J. Bacteriol.* 71:571–81.
17. Hawiger, J., and J. Jeljaszewicz. 1967. Antibiotic sensitivity of *Micrococcus radiodurans. Appl. Microbiol.* 15:304–6.
18. Hill, R. F., and E. Simson. 1961. A study of radiosensitive and radioresistant mutants of *Escherichia coli* strain B. *J. Gen. Microbiol.* 24:1–14.
19. Hitchins, A. D., W. L. King, and G. W. Gould. 1966. Role of disulphide bonds in the resistance of *Bacillus cereus* spores to gamma irradiation and heat. *J. Appl. Bacteriol.* 29:505–11.
20. Kaplan, H. S., and R. Zavarine. 1962. Correlation of bacterial radiosensitivity and DNA base composition. *Biochem. Biophys. Res. Commun.* 8:432–36.
21. Kelner, A., W. D. Bellamy, G. E. Stapleton, and M. R. Zelle. 1955. Symposium on radiation effects on cells and bacteria. *Bacteriol. Rev.* 19:22–44.
22. Kilburn, R. E., W. D. Bellamy, and S. A. Terni. 1958. Studies on a radiation-resistant pigmented *Sarcina* sp. *Radiat. Res.* 9:207–15.
23. Knivett, V. A., J. Cullen, and M. J. Jackson. 1965. Odd-numbered fatty acids in *Micrococcus radiodurans. Biochem. J.* 96:2–3C.
24. Krabbenhoft, K. L., A. W. Anderson, and P. R. Elliker. 1965. Ecology of *Micrococcus radiodurans. Appl. Microbiol.* 13:1030–37.
25. ———. 1967. Influence of culture media on the radiation resistance of *Micrococcus radiodurans. Appl. Microbiol.* 15:178–85.
26. Lavin, M. F., A. Jenkins, and C. Kidson. 1976. Repair of ultraviolet light-induced damage in *Micrococcus radiophilus,* an extremely resistant microorganism. *J. Bacteriol.* 126:587–92.
27. Lee, J. S., A. W. Anderson, and P. R. Elliker. 1963. The radiation-sensitizing effects of N-ethylmaleimide and iodoacetic acid on a radiation-resistant *Micrococcus. Radiat. Res.* 19:593–98.
28. Lewis, N. F. 1971. Studies on a radio-resistant coccus isolated from Bombay duck *(Harpodon nehereus). J. Gen. Microbiol.* 66:29–35.
29. Lewis, N. F., D. A. Madhavesh, and U. S. Kumta. 1974. Role of carotenoid pigments in radio-resistant micrococci. *Can. J. Microbiol.* 20:455–59.
30. Moseley, B. E. B., and H. Laser. 1965. Similarity of repair of ionizing and ultraviolet radiation damage in *Micrococcus radiodurans. Nature* 206:373–75.
31. Moseley, B. E. B., and A. H. Schein. 1964. Radiation resistance and deoxyribonucleic acid base composition in *Micrococcus radiodurans. Nature* 203:1298–99.
32. Murray, R. G. E., and C. F. Robinow. 1958. Cytological studies of a tetrad-forming coccus. In *International Congress of Microbiologists,* 7th, Stockholm, 427–28.
33. Raj, H. D., F. L. Duryee, A. M. Deeney, C. H. Wang, A. W. Anderson, and P. R. Elliker. 1960. Utilization of carbohydrates and amino acids by *Micrococcus radiodurans. Can. J. Microbiol.* 6:289–98.
34. Romig, W. R., and O. Wyss. 1957. Some effects of ultraviolet radiation on sporulating cultures of *Bacillus cereus. J. Bacteriol.* 74:386–91.
35. Serianni, R. W., and A. K. Bruce. 1968. Role of sulphur in radioprotective extracts of *Micrococcus radiodurans. Nature* 218:485–86.

36. Setlow, J. K. 1964. Physical changes and mutagenesis. *J. Cell. Comp. Physiol.* 64: Sup. 1, 51–68.
37. Setlow, J. K., and D. E. Duggan. 1964. The resistance of *Micrococcus radiodurans* to ultraviolet radiation. I. Ultraviolet-induced lesions in the cell's DNA. *Biochim. Biophys. Acta* 87:664–68.
38. Stapleton, G. E., D. Billen, and A. Hollaender. 1952. The role of enzymatic oxygen removal in chemical protection against X-ray inactivation of bacteria. *J. Bacteriol.* 63:805–11.
39. Stapleton, G. E., and M. S. Engel. 1960. Cultural conditions as determinants of sensitivity of *Escherichia coli* to damaging agents. *J. Bacteriol.* 80:544–51.
40. Tan, S.-T., and R. B. Maxcy. 1982. Inactivation and injury of a hemolytic radiation-resistant *Micrococcus* isolated from chicken meat. *J. Food Sci.* 47:1345–49, 1353.
41. Thornley, M. J. 1963. Radiation resistance among bacteria. *J. Appl. Bacteriol.* 26:334–45.
42. Thornley, M. J., R. W. Horne, and A. M. Glauert. 1965. The fine structure of *Micrococcus radiodurans. Arch. Mikrobiol.* 51:267–89.
43. Vinter, V. 1962. Spores of micro-organisms. IX. Gradual development of the resistant structure of bacterial endospores. *Folia Microbiol.* 7:115–20.
44. Welch, A. B., and R. B. Maxcy. 1979. Characteristics of some radiation-resistant hemolytic micrococci isolated from chicken. *J. Food Sci.* 44:673–75.
45. Work, E. 1964. Amino-acids of walls of *Micrococcus radiodurans. Nature* 201:1107–9.
46. Work, E., and H. Griffiths. 1968. Morphology and chemistry of cell walls of *Micrococcus radiodurans. J. Bacteriol.* 95:641–57.
47. Zamenhof, S., H. Bursztyn, T. K. R. Reddy, and P. J. Zamenhof. 1965. Genetic factors in radiation resistance of *Bacillus subtilis. J. Bacteriol.* 90:108–15.

APPENDIX

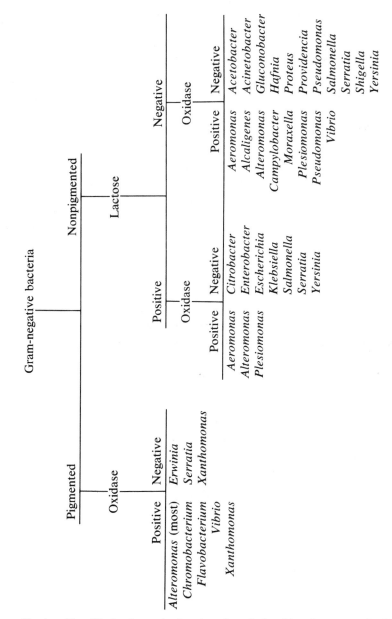

Appendix A. Simplified schematic showing the relationship of common food-borne genera of gram-negative bacteria to each other (for details, consult *Bergey's Manual*).

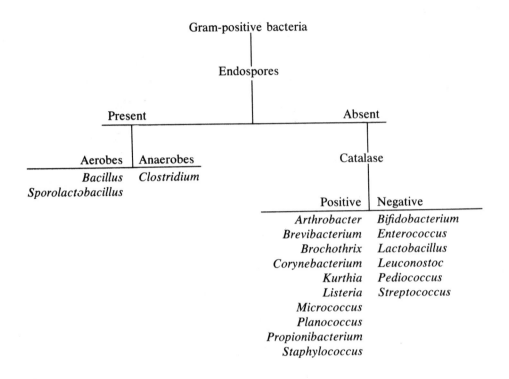

Appendix B. **Simplified schematic showing the relationship of common food-borne genera of gram-positive bacteria to each other (for details, see *Bergey's Manual*).**

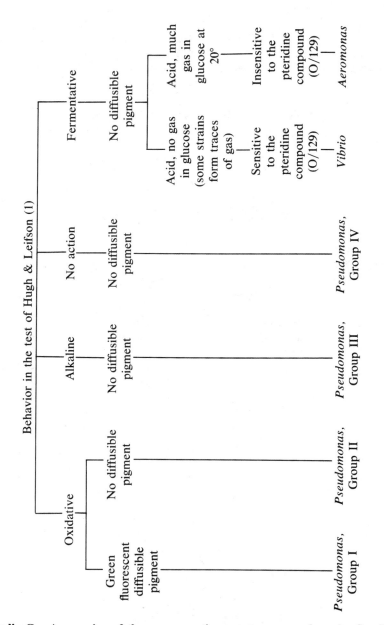

Appendix C. A grouping of the gram-negative asporogenous rods, polar-flagellate, oxidase positive, and not sensitive to 2.5 i.u. penicillin, on the results of four other tests (2). Copyright © 1953, Society for Applied Bacteriology

REFERENCES

1. Hugh, R., and E. Leifson. 1953. The taxonomic significance of fermentative versus oxidative metabolism of carbohydrates by various gram-negative bacteria. *J. Bacteriol.* 66:24–26.
2. Shewan, J. M., G. Hobbs, and W. Hodgkiss. 1960. A determinative scheme for the identification of certain genera of gram-negative bacteria, with special reference to the *Pseudomonadaceae. J. Appl. Bacteriol.* 23:379–90.

INDEX

Acetic acid, 221
  bacteria, 6, 18, 247
  preservative, 278
Acetobacter spp., 15, 16, *18*, 246–47, 264,
    373, 388–89, 390, 616
  aceti subsp. *aceti*, 18
  "overoxidizers," 365
  xylinum, 373
Acetoin, 221, 367
Acetomonas. See Gluconobacter
Acetylaldehyde, 388
Achiote, 277
Achromobacter spp., 18, 221, 230, 247, 612
  anaerobium. See Zymomonas anaerobia
Acidity. *See* pH
Acidophilus milk, 371, 375, 397
Acinetobacter spp., 15, 16, *18*, 22, 53, 68,
    70, 75, 85, 207, 210, 215, 217, 224–26,
    228–30, 239–40, 244, 309, 452, 501,
    580, 603, 616
  calcoaceticus, 18, 308, 392–93
  lwoffi, 18
Acriflavine, 438–39
Adenovirus, 308, 552
Adenylate cyclase, 481, 499, 506, 525
Adherance, 180–81, 183
Aerobacter. See Enterobacter
Aerobic plate count. *See* Standard plate
    count
Aerococcus spp., 364
Aeromonas spp., 15, *19*, 70, 207, 210, 215–
    16, 219, 239, 452, 579–80, 616, 618
  hydrophila, 19, 173, 181, 320, 392, 508,
    557–58
AFDOUS, 427
Aflatoxins, 8, *542–47*, 550

  allowable action level, 9, 545
  B_1, 130–31, 154, 542–45, 547
  B_2, 542, 545
  biosynthesis, 542
  in corn, 546
  destruction, 265, 544
  DNA binding, 545
  G_1, 542–45
  G_2, 542–43
  ELISA detection, 131, 154
  incidence in foods, 544–45
  L, 542
  LD_{50}, 545, 547
  LH_1, 542
  M_1, 130–31, 152, 154, 542–43
  M_2, 543
  mode of action, 545, 547
  organisms, 542
  p_1, 542
  production conditions, 541–42, 544–45
  Q_1, 542
  relative toxicities, 543, 547
  RIA detection, 130, 152
  structures, 542
  synthesis inhibition, 277, 544
African swine fever virus, 554
Agaricus spp., 392
  bisporus, 478
Agar droplet method, 103
Agar overlay method, 116
Agar sausage method, 108
Agar syringe method, 108
Aggregate-hemagglutination test, 131, 158
Air
  food-borne organisms in, 18
  sampling methods, 109–10

Air sampling methods, 109–10
Alarm water, 349
Alcaligenes spp., 15, 16, *19*, 49, 207, 210, 215, 217, 225, 239, 244, 392, 501, 612, 616
  eutrophus, 392–93
  faecalis, 35, 39
  viscolactis, 243
Alcohols, 230
Aldolase, 368
Ale, 30, 246, 372, *386*
Aleuriospores, 349
Allicin, 48, 277
Allyl isothiocyanate, 48
Altenuene, 548
Alternaria spp., 15, *25–26*, 200, 208, 224, 244, 248, *547–48*
  alternata, 547
  citri, 200, 349, 547–48
  solani, 547
  tenuis, 200
  tenuissima, 547
Alternariol, 548
Alteromonas spp., *19*, 49, 207, 219, 230, 579, 616
  nigrifaciens, 19
  putrefaciens, 19, 68, 77, 218, 220–21, 226, 230, 243
American Association of Medical Milk Commissions, 428
American Bottlers of Carbonated Beverages, 427–28
American Dry Milk Institute, 429
Ames assay, 545
Amino acid transport assay, 181
Amylase, 32
Anaerobes, 117
Analysis of foods, 98
Andersen sieve sampler, 110
Angular leaf spot, 197
Animal feeds, 17, 492
Animal hides, 17–18
Anthracnose, 199–201, 284
Anthrax, 19
Antibiotics, 7, 278–84
Anticancer effects, 399
Antimicrobial constituents, 48
Antioxidants, 274–75
Apiculate yeasts, 30
Appert, N., 5
Arizona spp., 147
Arrack, 372
Arrhenius law equation, 51, 584
Arthrobacter spp., 54, 207, 225, 579, 588, 617
Arthrospores, 25
Ascomycetes, 29
Ascorbate, 266
Ascospores, 25
Ascosporogenous yeasts, 30
Aseptic packaging, 287, 343–44
Aspergillus spp., 15, 16, *26–27*, 50, 208, 224, 244, 248, 371–72, 384, 392
  alliaceus, 549
  clavatus, 549
  echinulatus, 349
  flavus, 8, 307–8, 389, 542–45

  glaucus, 26–27, 349, 358, 371
  gracilis, 473
  mellus, 549
  nidulans, 551
  niger, 200, 276, 286–87, 358
  ochraceus, 389, 549–50
  oryzae, 363, 372, 383–84, 388
  ostianus, 549
  parasiticus, 115, 277, 542–45
  repens, 250
  rugulosus, 551
  soyae, 372, 383
  terreus, 549
  versicolor, 551
Asporogenous yeasts, 30
ATP
  assay for, 130, *141–43*
Autoagglutination, 525–26
Autosterilization, 593
Avidin, 239
a$_w$, *40–44*, 51
  aflatoxins, 543–44
  C. botulinum, 470, 472
  C. perfringens, 461
  defined, 40
  dried foods, 348–49, 351–52
  effects of low values, 41–44
  frozen foods, 322–23
  intermediate-moisture foods, 352
  minima for growth, 41
  patulin production, 549
  salmonellae, 496, 498
  S. aureus, 356, 440–46
    V. parahaemolyticus, 516–17
Azoreductase, 399

Bacillaceae, 19–20
Bacillus spp., 15, 16, 18, *19*, 20, 42–43, 50, 54, 80, 195, 207, 224, 232, 241–42, 245, 248, 331, 371, 381, 391–92, 452, 580, 593, 599–600, 617
  anthracis, 19
  cereus, 18, 19, 55, 83, 117, 131, 173, 179, 181, 275, 432, 474, *480–82*, 611
  coagulans, 8, 250–52, 286, 336, 474, 597–98, 600
  globigii, 608
  inconstans, 586
  larvae, 300
  licheniformis, 340, 475
  macerans, 250
  megaterium, 333, 597
  mesentericus, 242
  nigrificans, 248
  polymyxa, 244, 250
  pumilus, 308
  stearothermophilus, 8, 250, 252, 286, 342, 594–98, 608
  subtilis, 107, 242, 249, 283, 286, 336, 551, 595, 597, 600, 612–13
  thermoacidurans. See *B. coagulans*
  thermoproteolyticus, 595
  vulgatus, 249
Bacon, 72–73, 207, 224–25, 305, 449
Bacterial blight, of vegetables, 195–98
Bacteriocins, 462, 474
Bacteriophage typing, of *S. aureus*, 443

Bacteroidaceae, 246
Bacteroides spp., 17, 145, 207, 421
  fragilis, 47
Baker's barm, 390
Baker's yeast. *See also Saccharomyces*
  cerevisiae
  DNA content per cell, 135
  intracellular pH, 38
  single-cell protein, 394
Bakery products, 242
Bananas, 200, 284
Bay leaf, 277
Beans, 79
Becquerel (Bq), 299
Beef. *See* Meat
Beer, 3, 30, 243, 246, 372, 386
Benomyl, 284
Benzalkon-crystal violet, 142
Benzoate/benzoic acid, 6, 259–62, 355, 550
Betabacterium spp., 71, 365
  vermiforme, 373
Beta rays, 299
Bethesda-Ballerup group, 20
Bifidobacterium spp., 421, 617
  adolescentis, 421
  bifidum, 369
  breve, 421
Biken test, 158
Bile pigments, 223
Binding, assays for, 181
  C. perfringens enterotoxin, 181, 464
Binuburan, 372
Bioassays, 172–85
  animal systems, 178–80
  cell cultures, 180–85
  whole animals, 172–78
Biological acidity, 38
Biological activity, bioassays for, 180–81
Biological structures, of foods, 48–49
Biphenyl, 284–85
Birdseye, C., 7
Biscuit dough, 82, 241
Bitter rot. *See* Anthracnose
Black leg, of potatoes, 196
Black pepper, 277
Black rot, of vegetables, 197, 200
Black spot, of beef, 210
Blanching, 302–3, 321
Blue mold rot, of fruits, 200
Bologna, 72, 206, 222
Bone taint, of beef, 210
Bongkrek, 371, *385*
Bongkrekic acid, 385
Borden, G., 6
Botrytis spp., 15, 16, *27*, 197, 208, 224, 265,
  285, 318
  cinerea, 26, 200, 349
Bottom yeasts, 30
Botulinal toxins, 153–54, 156, 172, *475–76*
  A, 80, 173, 261, 320, 469, 472, 475–76
  B, 80, 153–54, 173, 320, 469, 472, 475–76
  C, 153–54, 469
  D, 154, 469
  E, 153–54, 173, 469, 473, 475
  F, 154, 173, 469, 475
  G, 153, 173, 469, 471–72
  LD$_{50}$, 475

mode of action, 475–76
Botulism, 20, *469–80*
  adult syndrome, 476–79
  history, 8, 469–70
  infant syndrome, 479–80
Bound water, 348, 350
Bourbon whiskey, 372, 387
Bouza beer, 372
Brandy, 387
"Bread molds," 29, 242
Breads, 242, 243, 373, 390–91
Breath hydrogen, 397
Breed count. *See* Direct microscopic count
Bremia spp., 200
Brettanomyces spp., 15, *30*, 246
Brevibacterium spp., 617
Brine bath, 6
Broccoli, 79, 195, 197
Brochothrix spp., 15, *20*, 49, 54–55, 69–71,
  207, 308, 579, 617
  thermosphacta, 20, 54–55, 69–72, 215,
  217, 219–21, 222, 309
Brown mold rot, of fruits, 200
Brucellosis, 409
Buffalo green monkey cell culture, 557
Buffering power, of foods, 38
Bulgarian buttermilk, 371, 375
Burong dalag, 371
Butter, 243–44, 370
Buttermilk, 370, 398
Butylated hydroxyanisole (BHA), 274–75,
  278, 545
Butylated hydroxytoluene (BHT), 274–75,
  545
t-Butylhydroxyquinoline (TBHQ), 274–75
Butyric anaerobes, 250–51
Byssochlamys spp., 15, *29*
  fulva, 8, 29, 250, 287, 549
  nivea, 549–50

Cabbage, 197
Caffeic acid, 48
Caffeine, effect on mycotoxin synthesis,
  544, 548
Cakes, 242, 243, 352, 532
Calcium content, of meat, 65
Calcium-dependent growth, 525–26
Calcivirus, 554
Caldoactive, 596
Camembert cheese, 375
Campylobacter spp., 8, 15, 18, *20*, 207,
  527, 616
  bioassay, 180
  coli, 508, 529
  distribution, 529–31
  enterotoxin, 9, 180, 531
  incidence in foods and specimens, 529–32
  intestinalis, 529
  jejuni, 67, 117, 173, 180, 508, *529–33*
Campylobacteriosis, 8, 81, *531–33*
Candida spp., 15, 17, *30–31*, 208, 210, 225,
  265, 318, 388, 580, 586, 588
  albicans, 276
  guilliermondii, 392
  intermedia, 393
  krusei, 371, 390

625

lipolytica, 210, 392–93
mycoderma, 246, 390
scottii, 349
tropicalis, 392–93
utilis, 349, 392–94, 581, 586
zeylanoides, 210, 349
Candy spoilage, 245
Canned foods
    antibiotics for, 283–84
    botulism from, 478–79
    hard swells, 252
    history, 5–7
    hydrogen swells, 252
    organisms in, 249–52
    soft swells, 252
    spoilage, 8, 249–53
Cantaloupe, 195
Caprylic acid, 260
Carbon dioxide, 7, 52–56, 69, 70, 71–72,
    226, 230, 550. *See also* Controlled at-
    mosphere storage; Vacuum packaging
Carcinogens, 268, 287, 399, 541, 547, 550
Carrier, 492
    salmonellosis, 502
Carrots, 195–96, 200–201
Carvacrol, 48
Carvone, 274, 276
Cassava, 384
Catalase, 113, 611–12
Catfish fillets, 76–77
Cauliflower, 79, 197
Celery, 195, 197
Cell culture systems, 180–85
    buffalo green monkey cells, 557
    CHO, 180, 184
    guinea pig intestine, 183
    HeLa, 180, 183–84
    human fetal intestine, 182
    human mucosal, 182
    Vero, 181, 184
    Y-1 adrenal, 181, 184–85
Cellulomonas spp., 392–93
    flavigena, 248
Cephalosporium spp., 15, 27, 384
Cephalothecium. *See Trichothecium*
Ceratocystis paradoxa, 200
Cereals, spoilage, 241
Cereolysin, of *B. cereus*, 480–81
Certified milk, 6, 81, 428
[137]Cesium, 299, 303
Cestodes, 562–63
Champagne, 387
Cheddar cheese, 79, 544
Cheeses, 80, 449, 561
    fermentation, 371, 375, 377
    hard, 375, 377
    history, 4, 7
    listeriosis from, 559–60
    semihard, 375, 377
    soft, 377
    spoilage, 244
    unripened, 377
Chilling temperatures, 318
Chinese hamster ovary (CHO) cell assay,
    158, 499, 521, 525, 531, 557
Chitterlings, 73
Chlamydospores, 25, 30

Chlorella spp., 392
    pyrenoidosa, 393
Chlorogenic acid, 48
Chlortetracycline. *See* Tetracycline
Chocolate, 245, 334, 496
Choleglobin, 223
Cholera, 8
Cholera toxin (CT), 51, 180–81, 499, 506,
    521, 557
Cholesterol, 398–99
Chromobacterium spp., 207, 580, 616
Chromogenic substrates, 145–46
    E. coli, 145
    Limulus test, 138
Chrysosporium spp., 210
Cider, 6, 372, 388–89
Ciguatera fish poisoning, 567–68
Cinnamon, 277
Cinnimic aldehyde, 48, 277
Citrinin, 548
Citrobacter spp., 16–17, *20*, 85, 147, 411,
    420, 580, 616
    freundii, 64, 561
Citrus canker, 197
Cladosporium spp., 15, *27*, 50, 208, 210,
    240, 244, 248, 318, 473
    herbarum, 26, 200
Clams, 76, 227, 231–32, 553
Claviceps purpurea, 4
Clonorchis sinensis, 565–66
Clostridium spp., 15–18, *20*, 42–43, 50, 64,
    67, 195, 207, 210, 224, 232, 245, 281,
    331, 338, 580, 593, 617
    acetobutylicum, 38–39
    bifermentans, 250
    botulinum 8, 55, 71–73, 78, 114, 116, 144,
        149, 154, 156, 172–73, 250, 261, 263–
        64, 266, 268–72, 281, 285–86, 300,
        305–6, 308, 311, 320, 336, 342–43,
        469–80, 603
    antagonists, 474–75
    a_w, 470, 473
    distribution, 469, 470–71
    D values, 306, *470*, *473–74*
    ecology of growth, 474–75
    effect of nitrite, 266–67
    growth temperatures, 320, *472–74*
    incidence in foods, 67, 72–73
    pH range, 342, 470, 472–73
    proteolytic/nonproteolytic strains, 470,
        472, 475
    recovery, 117
    vacuum-packaged meats, 71–72
    vacuum-packed fish, 474–75
    butyricum, 244, 250, 267, 279, 308
    difficile, 147, 479
    histolyticum, 595
    intestinal carriage, 421
    nigrificans, 250, 342
    pasteurianum, 244, 250
    perfringens, 8, 42, 55, 63–64, 67, 69–70,
        74, 77, 81, 83, 113–14, 149, 173, 175,
        178–79, 210, 267, 270–71, 301, 350,
        409, 426, 432, *459–67*, 474, 481, 561
    distribution, 460
    enterotoxin, 8, 131, 156, 181, 184, *462–*
        66

growth conditions, 460–62
incidence in foods, 74, 77
recovery, 117
sporulation, 463
toxins, 460
putrefaciens, 587
sordellii, 308
sporogenes, 55, 244–45, 267, 271, 281, 286, 308, 342, 474, 608
subterminale, 154
thermosaccharolyticum, 250–51, 342
tyrobutyricum, 267
Clove oil, 277
Coagulase, 117, 438
[60]Cobalt, 299, 303
Cocoa beans, 310, 371, 389–90, 498
Codex Alimentarius Commission, 310
Coffee beans, 372, *389*, 549, 551
Cold-boned meats, 67
Cold shock, 581, 600
Coleslaw, 81, 560
Coliforms, 21, 63–66, 74, 76, 81–83, 229, 242, 244, *410–14*, 421, 426, 521
  carbon dioxide effects, 53
  compared to enterococci, 420
  defined, 410–11
  detection by impedance, 129–30, 132
  fecal strains, 63, 74, 76, 130, *411–12*, 426, 503, 553, 556
  growth requirements, 412–13
  habitat and distribution, 413
  isolation, 98, 118
  meats, 62–64, 67, 69
  metabolic injury, 114
  poultry, 74
  radiometric detection, 144
  seafood, 76–77
  spices, 83
  standards for foods, 413–14, 433
  vegetables, 79, 418–19
  viruses in waters, 553
Colletotrichum spp., 15, *29*, 197
  coccodes, 200–201
  lindenmutheanum, 29
  musae, 200
Colonizing factor antigens, 503–4
Colony-blot method, 154
Colony forming unit (cfu). *See* Standard plate count
ColV plasmid, 505
Commercial sterility, 332
Common blight, of beans, 197
Conalbumin, 48, 239, 335
Condensed milk, 6
Conglutinin, 48
Conidia, 25
Conidiophores, 25, 286
Controlled atmosphere storage, 7, 52. *See also* Vacuum packaging
Copepods, 562
Corn, 195, 544–46, 551
Corn syrup, 479
Corynebacterium spp., 15, 16, *20–21*, 49, 54, 70, 85, 145, 207, 220, 225, 308, 384, 580, 617
  flaccumfaciens, 197
  manihot, 372, 384

  michiganense, 197
  sepedonicum, 197
Coryneform bacteria, 68, 75, 207, 217
Cottage cheese, 244, 430
p-Coumaric acid, 48
Country-cured hams, 371, *380*
  mycotoxigenic fungi, 544, 549
  ochratoxin production, 549
Coxiella burnetti, 331, 340
Coxsackie virus, 306–8, 552–53, 555
Crabmeat, 76, 227, 230–31, 414, 430, 519, 554
Cream, 243
Cream cheese, 80
Cream pies, 81–82
Critical control points, 425
Crown rot, 200
Crustacean shellfish, 227, 230–31, 430
Cryophilization. *See* Freeze-drying
Cryptococcaceae, 30, 32
Cryptococcus spp., 208, 580, 588
Cryptosporiopsis malicorticis, 200
Cucumbers, 197, 382
Cultured buttermilk, 370
Culture media, selective/differential
  APT broth, 221
  azide dextrose agar, 118
  Baird-Parker agar, 113, 119, 137
  BG agar, 119
  BGLB broth, 118
  bismuth sulfite agar, 120
  Campylobacter agar, 117
  cefsulodin-irgasan-novobiocin (CIN), 120
  Duncan-Strong, 463
  EC broth, 118, 146, 411
  EC-MUG, 146
  EE broth, 118
  Elliker's lactic agar, 118
  EMB agar, 118
  Endo-MUG, 146
  eugon agar, 118
  Glucose salt teepol broth (GSTB), 119
  Hektoen agar, 119
  KF agar, 118
  KG agar, 117
  lactose broth-MUG, 145
  lauryl tryptose broth-MUG, 145–46, 411
  Listeria agar, 560
  LSB-MUG, 146
  MacConkey agar, 118, 120, 413
  M broth, 150
  MRS agar, 118, 219
  MUG broth, 118
  PBAC agar, 118
  Pfizer selective enterococcus agar, 118
  phenol red-egg yolk polymyxin agar, 117
  PCA-monensin-KCl (PMK) agar, 118
  potato dextrose agar (PDA), 119, 549
  Preston medium, 117
  Sabouraud's agar, 543
  selenite cystine broth, 119, 150
  Skirrow agar, 117
  SPS agar, 117
  SS agar, 120
  TCBS sucrose agar, 119
  tetrathionate broth, 119, 150
  TPEY agar, 114

TSC egg-yolk agar, 117
TSN agar, 114
veal liver–egg yolk agar, 117
VRBA, 103, 112, 116, 118, 145
Wagatsuma's medium, 517
wort agar, 543
XLD agar, 119
XLPA, 114
YN-6 agar, 421
Culture methods, 97–102
Curie, 297
Cyanogenic glucoside, 384
Cyclic AMP, 481, 499, 506, 525, 531
Cysticercus cellulosae, 562
Cytophaga spp., 75, 207
Cytotoxicity, bioassays for, 181

Dairy mold, 28
Dairy products
  composition, 243
  fermented, 370–77
  freezer life, 318–19
  organisms in, 80–81
  pH range, 374
  standards for, 428
  spoilage, 242–44
Debaryomyces spp., 15, 30, *31*, 208, 318
  kloeckeri, 392
Decimal reduction time. *See* D values
Dehydrated foods, 84–85. *See also* Dried
  foods
Dehydroacetic acid, 287
Delicatessen foods, 81–83
Derivative toxin, of *C. botulinum*, 476
Desferal, 526
Desiccated foods. *See* Dried foods
Desulfotomaculum nigrificans, 597
Desulfovibrio spp., 45
  desulfuricans, 597
  vulgaris, 597
Dextran synthesis, 22, 245
Diacetate, sodium, 260, *286*
Diacetyl, 221, 274, 275–76, 365, 367, 396
Diarrheagenic toxin, 173, 480–81
Diethylpyrocarbonate, 262, *287*
Differential media. *See* Culture media
Dimethylamine, 269
Dimethyl disulfide, 220–21
Dimethyl sulfide, 220–21, 226
Dimethyl trisulfide, 220–21
Dinoflagellates, 567
Diphtheria, 409
Diphyllobothrium latum, 562–63
Dipicolinic acid, 338
Diplodia natalensis, 200
Diplodia strain, of onions, 202
Direct epifluorescent filter (DEFT), *101–2*,
  106
Direct microscopic count (DMC), 98, 101,
  105–6
Direct surface method, 108
Distilled spirits, 386–87
DNA binding of aflatoxins, 545
DNA-DNA hybridization, 139–40, 438
  basis, 139
  E. coli LT and ST probe, 140

lactic acid bacteria, 140, 367–68
rotaviruses, 140
salmonellae probe, 140
V. parahaemolyticus, 518
DNA homology of lactics, 366–68
DNAse. *See* Thermostable nuclease
Downey mildew, 200
Dried foods, 6, 84–85, 349, 417, 429, 498
  a_w, 348–49, 351
  effect on flora, 348–50
  intermediate-moisture, 351–59
  low-moisture, 346
  preparation, 346–48
  storage stability, 350–51
Drip, 322
Drosophila melanogaster, 201
Dry film plating method, 103
Dry rot, 200
Drying, effect on organisms, 348–50
Dust, food-borne organisms in, 18
D value, 339
  C. botulinum, 339–40, *470*, *473–74*
  C. jejuni, 529
  C. perfringens, 461
  chemicals, 286, 608
  D_r, 339
  heat, 333–34, *339–40*
  E. coli, 508
  irradiation, 301, 305–8
  other sporeformers, 342
  paralytic shellfish toxin, 567
  S. aureus, 444–46
  S. senftenberg, 334
  thermostable direct hemolysin, 518
  thermostable nuclease, 137, 445
  Y. enterocolitica, 523
Dye reduction tests, 98–99, *104–5*, 214
  methylene blue, 104
  resazurin, 99, 104–5, 214

Echinococcus granulosis, 563
ECHO viruses, 306, 552–54, 556
EDTA, 270, 274, 278
Edwardsiella spp., 420–21
EFDV syndrome, 547
Eggs
  diagram of hen's, 240
  D values in, 334, 497
  flora of, 239–40, 530
  powdered, 84
  rots, 240
  salmonellae in, 492–94, 497, 501
  spoilage, 7, *239–41*
  staphylococci in, 452
Eh. *See* Oxidation-reduction potential
Electrical impedance. *See* Impedance
Electrical stimulation, 68
Electroimmunodiffusion test, 131
Electron beams, 303–4
ELISA, 131, 152–55, 518, 557, 565
Embden-Meyerhof-Parnas pathway, 365
Emetic doses, of *S. aureus* enterotoxins,
  444
Emetic toxin, 173, 480–81
Endomyces vernalis, 349
Endomycopsis spp., 15, *32*
  fibuliger, 32, 392–93

Endospores, 262, 337–38, 342, 462–63
Endotoxins. *See* Lipopolysaccharides
Enrichment serology for salmonellae, 150–51
Entamoeba histolytica, 508
Enteral foods, 85
Enteritis necroticans, 459
Enterobacteriaceae, 20–21, 23–25, 64, 70, 73, 114, 118, 196, 218, 225, 266, 309, 420–21, 452, 507, 559, 589
Enterobacter spp., 15, 16, 17, *21*, 55, 73, 207, 215, 217, 219, 232, 240, 244, 248, 388, 420, 580, 616
 aerogenes, 21, 39, 42, 374, *410–12*, 561
 agglomerans, 64
 cloacae, 64, 85, 308, 508
 hafniae, 64
Enterobactin, 505
Enterochelin, 505
Enterococci, 62–63, 78, 118, 136, *414–20*, 553. *See also Enterococcus* spp.; *Streptococcus* spp.
Enterococcus spp., 415, 617. *See also Streptococcus*
 avium, *415*
 casseliflavus, 415
 durans, 415
 faecalis, 280, 415
 faecium, 415
 gallinarum, 415
 malodorus, 415
Enterotoxins, 24
 Aeromonas hydrophila, 173, 557
 Bacillus cereus, 131, *480–81*
 Campylobacter jejuni, 9, 173, *531*
 Clostridium perfringens, 131, 173, 177–79, 181, 184, *462–66*
 Enterobacter cloacae ST, 559
 Escherichia coli, 505–6
  LT, 131, 140, 152, 158, 173, 177, 180–81, 184, 504–7, 508, 531, 557
  ST, 140, 173, 175, *505*, 507–8, 525, 559
  ST_a, 130, 151–52, 173, 175, *505–6*
  ST_b, 151–52, 173, 175, *505–6*
 Klebsiella pneumoniae ST, 559
 Salmonella spp., 9, 499–500
 Staphylococcus aureus, 136, 156–57, 174, 437
  A, 130–31, 142, 151–53, 159, 177, 181, 356, 442–44, 446, 449, 451
  B, 130–31, 151, 153, 159, 174, 177, 181, 356, 438, 441–50
  C_1, 130–31, 151, 153, 159, 441–42, 444, 446–48, 450
  C_2, 130, 152, 442, 444
  C_3, 442–44
  D, 130–31, 151, 153, 159, 442–44, 446, 449, 451
  E, 130–31, 151, 153, 159, 181, 442–44, 447, 449–50
 Yersinia enterocolitica, 174–76, *525*
Enteroviruses, 552–53, 555
Entner-Doudoroff pathway, 369
Ent plasmids, 505
Enzyme-linked immunosorbent assay. *See* ELISA
Epifluorescent filters, 102

Equilibrium relative humidity (ERH), 358–59
Ergot poisoning, 4
ERV. *See* Extract-release volume
Erwinia spp., 15, 16, *21*, 194–97, 203–4, 616
 carotovora, 195–96
 carotovora var. *atroseptica*, 196
 carotovora var. *carotovora*, 196
 dissolvens, 372, 389
Erythemal activity assay, 173–74, *177*, 181
Erythorbate, 266, 270
Escherichia coli gastroenteritis, 8, 502–7
 enterotoxins, 131, 173, 180–81, 184, *505–6*
Escherichia spp., 15, 17, *21*, 64, 196, 207, 217, 232, 239, 248, 420, 580, 612, 616. *See also* Coliforms
 a_w minimum, 41
 coli, 21, 39, 63, 65, 68, 73–74, 76–77, 80, 82, 99, 113–14, 116, 147, 158, 173, 175, 179–81, 276, 308, 333, 374, *410–14*, 416, 418, 421, 426, 430, 433, 499, *502–8*, 515, 518, 521, 531, 557, 559, 581, 583, 585–86, 596–97, 599–600, 609–10, 612
  fecal and nonfecal, 411–13
  enteroinvasive (EIEC), 503, 505, 507
  enteropathogenic (EPEC), 503–4, 508
  enterotoxigenic (ETEC), 182, 503–5
  fluorogenic substrate, 145–46
  hemorrhagic colitis, 507–8
  travelers' diarrhea, 508
Ethanol, 221, 287
Ethylene oxide, 260, *285*
 D values, 286, 608
 synthesis, 313
Ethyl formate, 260
Ethyl vanillin, 274, 276
Eubacterium spp., 364
Eugenol, 48, 277
Eupenicillium. *See Penicillium*
Extract-release volume (ERV), 214–15, 219
Extrinsic parameters, 49–56

Facultative anaerobe, 45
False yeasts. *See* Asporogenous yeasts
FAO, 427
Fatty acids, antimicrobial activity, 277–78
Fecal coliforms, *411–12*, 433. *See also* Coliforms; *E. coli*
Fecal streptococci. *See* Enterococci
Felix-01 bacteriophage, 130, 146
Fermentation, 363–66, 386
Fermented foods, 369–91
 anticancer effects, 399
 effect on cholesterol, 398–99
 effect on lactose intolerance, 397–98
 lactic antagonism, 395–96
 pH, 374
 spoilage, 246–48
Ferredoxin, 271
Ferulic acid, 48
Film yeasts, 30
Fish and shellfish
 carbon dioxide–flushed packaging, 72
 composition, 227–28

fermented, 380–81
frozen, 6, 76–77
mechanically deboned, 66
organisms in, 75–77, 230
pH range, 37
spoilage, 227–32
vacuum-packaged, 474–75
Fish pastes, 380–81
Flat-sour organisms, 250–52, 301, 341, 427
Flavobacterium spp., 15, 16, *21*, 49, 54, 75, 207, 225, 229, 232, 240, 579, 616
proteus, 246
Flavoring agents, 275–76
Flipper, 252
Flour spoilage, 241
Flow cytometry, 134–35
Fluorescent antibody (FA), 98, 130, *147–50*
Fluorescein, 23
Fluorescein isothiocyanate, 147, 182
Fluorogenic substrates, 145–46
Flukes, liver. *See* trematodes
F_0. *See* F value
Food and Drug Administration (FDA), 9
Food, Drug & Cosmetic Act, U.S., 9
Food infection. *See* Food poisoning
Food intoxication. *See* Food poisoning
Food legislation, 9
Food poisoning, 8–9
Foot-and-mouth disease virus, 327
Formaldehyde, in wood smoke, 288
Frameshift mutations, 545
Frankfurters, 22, 72, 206, 222, 224, 426–27. *See also* Meats
Free water, 326
Freeze drying, 347–48
Freeze injury. *See* Metabolic injury
Freezer burn, 323
Freezer temperatures, 320
Freezing effects, 321–22
  C. perfringens, 462
  effect on organisms, 325–27
  fast (quick), 7, 321, 326
  freezing points of foods, 323
  slow, 321
  thawing, 328–29
French fries, 82
Fried-rice syndrome, 481–82
Frog legs, 430
Frozen foods, 63, 352, 417, 442
  coliforms, 419
  enterococci, 417–20
  freezing effects, 321–22
  ground meats, 63
  history, 6, 7
  meat pies, 84
  organisms in, 64, 76, 79, 419
  preparation for freezing, 320–21
  seafood, 75–77, 419, 433
  slow and fast freezing, 321–22, 326
  standards for, 414, 429, 433
  storage life, 318–19, 324–25
  thawing and refreezing, 328–29
  vegetables, 78–80, 417–19
Fruits, 6, 200, 203–4, 352
  composition, 203
  freezer life, 318–19, 324–25
  fungal spoilage inhibitors, 284–85

juice concentrates, 84, 352
mycotoxins, 548–49
pH range, 36
radiation preservation, 309–10
spoilage, 200, 203–4, 250–51
storage, 6
F_s. *See* F value
Fungi. *See* Molds; Yeasts
Fungi imperfecti, 25, 30
Fusarium spp., 15, 16, 27, *28*, 200, 208, 224, 248, 285, 384, 392, 551
  graminearum, 551
  roseum, 200, 551
  tricinctum, 551
F value, 340–41
  F_0, 267, 281, 340, 343

Gaffkya spp., 22
β-Galactosidase, 31, 288, 397
Game meats, 63
Gamma rays, 299, 303–4
Gangliosides, toxin binding, 475–76, 506, 517–18
Gari, 372, 384–85
Gas-impermeable packaging. *See* Vacuum packaging
Gastroenteritis, food-borne, 8
  A. hydrophila, 557–58
  B. cereus, 480–82
  campylobacteriosis, 531–33
  C. perfringens, 459–69
  E. coli, 502–7
  P. shigelloides, 558
  Salmonella, 489–502
  S. aureus, 437–53
  V. cholerae non-01, 520–21
  viral, 551–57
  V. parahaemolyticus, 518–20
  Y. enterocolitica, 527
Gelatin, 83–84
Gel diffusion methods, 131, 156–57. *See also* Toxins
Geotrichum spp., 15, *28*, 31, 208, 244, 318, 371–72, 385
  albidum, 28
  candidum, 106, 200–201, 244, 390
Gerbil lethality assay, 174
Giardia lamblia, 412, 508
Giardiasis, 412
Gibberella zeae, 551. *See also Fusarium*
Gloeosporium spp., 15, *28*, 197
  fructigenum, 28
  musarum, 200
  perennans, 200
Gluconobacter spp., 246, 616
  oxydans, 246
Glucose oxidase, 288
Glucuronidase test, 145–46, 399
Glutamate decarboxylase, 130, 146
Glycerol monoesters, 277
Goat's milk, 64, 567
Gonyaulax acatenella, 567
  catenella, 567
  tamarensis, 567
Gout, 395
Gram-negative bacteria, 616

LAL assay for, 137–39
recovery, 118
GRAS, 259
Gravy mixes, 84
Gray, 299
Gray mold rot, 27, 197, 200–201
Greening, of meats, 222
Green mold rot, of fruits, 200
Green salads, 81
Group D streptococci. *See* Enterococci
Guanylate cyclase, 506, 525
Guinea pig
cell assay, 181, 183
skin test, 173–74, 177

Haddock fillets, 75
Hafnia spp., 207, 219, 580, 616
alvei, 218, 220, 265, 561
Halibut pH, 37
Halobacterium spp., 207
Halo blight, of beans, 197
Haloduric organisms, 273
Halophile, 273
Hamburger meat, 63–64. *See also* Meats
Hams, 63, 305, 451–52, 554. *See also*
Meats
country-cured, 371, 380, 544, 549
spoilage, 224–25
Hansenula spp., 15, 30, *31*, 208
polymorpha, 392
Hazard analysis, critical control points
(HACCP), 425
Heat resistance, of microorganisms, 332–38
HeLa cell assay, 180, 183–84, 524, 526, 531
Helminthosporium spp., 15, *28*
satiuum, 28
Hemagglutination-inhibition test, 131, 156
Hemagglutination test, 156
Hemorrhagic colitis, 507–8
Henle 407 human intestine assay, 181, 526
Hennican, 354, 356–57
Hepatitis A virus, 554, 556
HEp-2 cell assay, 525–26
Herpes simplex virus, 308
Heterofermentative organisms, 22, 71, 219,
364–66, 388
Hexose isomerase, in lactics, 365
High-temperature vacuum drying, 345
Histamine, 560–62
Hog cholera virus, 554
Homofermentative organisms, 22, 70–71,
219, 364–66, 388
Homogenization, 99
Honey, 479
Hops, 386
Hormodendron spp., 240
Hot-boned meats, 66–67
Howard mold count, *106*, 429
Human fetal intestine assay, 181–82
Humectants in IMF, 354–56
Hydatidosis, food-borne, 563
Hydrogenase, 271
Hydrogen peroxide, 116, 222, 287, 288, 343
Hydrogen sulfide, 221, 226
Hydrogen swells, 252
p-Hydroxybenzoic acid. *See* Parabens
Hydroxycinnamic acid derivatives, 48

8-Hydroxyquinoline, 270
Hyperbaric CO_2, 52–55, 230
Hyperestrogenism, 551
Hypertonic, 273
Hypha, 25
Hypochlorite, 286
Hypotonic, 273

Ice cream, 81, 433, 493, 501
Idli, 373, 391
Ileal loop test. *See* Ligated loop
Imitation cheeses, 80, 449
Immunodiffusion methods, 156
Immunofluorescence assay, 185, 557
Impedance methodology, 128–30, 132–33
Impinger sampler, 110
IMViC formula, 411
Indicator organisms, 410–22
Indirect antimicrobials, 274–78
Infant botulism, 173, 479, 480
Infant mouse assay, 173. *See also* Suckling
mouse assay
Infant rabbit assay, 176
Infectious hepatitis, 554–56
Injury. *See* Metabolic injury
Insects, radiation destruction, 309–10
Intermediate-moisture foods (IMF), 351–59
International Commission on the Microbio-
logical Specification of Foods (ICMSF),
79, 427, 430
Intestinal hypolactemia, 397
Intestinal tract organisms, in foods, 17
Intrinsic parameters, 33–49
Invasiveness, assays for, 180–81, 183
Ionizing radiations, 297
Iron-dextran, 526
Irradiation of foods, 7, 9, 297–313, 498
Isoascorbate, 270
Isobutyric acid, 221
Isopropanol, 230
Isothymol. *See* Carvacrol
Isotonic, 273
Isovaleric acid, 221
Italian-type cured hams, 380
Izushi, 371, 478

Jellies, 243

Kaffir beef, 372
Kanagawa reaction, 182, 185, *517–18*
Katsuobushi, 371
Kauffmann-White scheme, 490–91
Kefir, 371, 375
Kenkey, 372
kGy, 299. *See also* Rad
Kidney meats. *See* Organ meats
Kimchi, 372
Kitten test for toxins, 174, *176–77*, 444
Klebsiella spp., 411, 420, 561, 616
pneumoniae, 64, 411, 508, 561
Kloeckera spp., 15, *31*
Kluyveromyces spp., 15, *31*
fragilis, 31, 392–94
lactis, 31
marxianus, 31
Koji, 363, 383

Kumiss, 371
Kurthia spp., 16, 207, 617

Lac genes, 369
Lactase, 31
Lactic acid, as preservative, 278
Lactic acid bacteria, 72, 241, 243, 247–48, 266, 272, 277, 309, *364–69*, 395
  CO_2-packaged meats, 54–55
  defined and characterized, 364
  fermentation, 371–72
  isolation, 118
  spoilage, 241
  vacuum-packaged meats, 69
Lactic antagonism, 72, *395–96*, 474
Lactobacillaceae, 20, *21–22*, 241, 452
Lactobacillus spp., 15–17, *21–22*, 49, 70–71, 207, 218, 222, 224, 230, 232, 241–42, 246, 250, 269, 308, 331, 364–65, 371, 373, 388, 580, 582, 617
  acidophilus, 365–66, 371, 375, 397, 398–99
  arabinosus, 309
  buchneri, 366–67, 377
  bulgaricus, 6, 365–68, 370–71, 374–75
  brevis, 249–51, 286, 366–67, 381, 390, 395
  casei, 309, 365–67, 395, 399
  cellobiosus, 70, 219, 367
  confusus, 366
  coprophilus, 366
  coryniformis, 366
  cucumeris, 247
  curvatus, 366
  delbrueckii, 366–68, 372, 383, 386–87, 391
  fermentum, 366–67, 391
  fructivorans, 249
  helveticus, 366–68
  hilgardii, 366
  intestinal flora, 421
  jensenii, 366
  jujurti, 366–68
  lactis, 366–68
  leichmannii, 366–67, 371
  lycopersici. See L. brevis
  plantarum, 247, 251, 287, 365–67, 369, 371–72, 377–78, 381–84, 395–96, 399
  salivarius, 366–67
  sanfrancisco, 366, 373, 391
  thermophilus, 22
  trichodes, 366
  viridescens, 222, 224, 367
  xylosus, 366
Lactoferrin, 48
Lactoperoxidase system, 48, 288
Lactose intolerance, 397–98
Lamb meats, 66, 69–71, 206
Lancefield's group D streptococci, 414–15, 421
Latex particle agglutination test, 131, 158
Lauric acid, 274
Lauricidin. *See* Monolaurin
Leakage, 322
Lebanon bologna, 371, 378
Leeuwenhoek, A., 5
Lenticel rot, 200
Lettuce, 195, 197

Leuconostoc spp., 15, 16, *22*, 73, 207, 222, 241–42, 364, 367–68, 388, 581–82, 617
  citrovorum, 370
  cremoris, 366
  dextranicum, 251, 366, 370
  lactis, 366
  mesenteroides, 7, 71, 219, 245, 250–51, 287, 366, 368, 371–73, 381–82, 390–91
  oenos, 247, 366
  paramesenteroides, 366
L-forms of bacteria, 118, 137, 197, 446, 462
Ligated loop techniques, 178–79
  mouse, 173, 179
  piglet, 173
  rabbit, 173–74, 178, 499, 503, 506, 517, 525, 557–59
  rat, 173
Limulus amoebocyte lysate test, 130, *137–39*
Linear accelerators, 303–4
Lipid solidification theory, 581
Lipopolysaccharides, 130, 137, 138
Liquid smoke, antimicrobial effects, 288
Listeria spp., 16, 559–60, 617
  monocytogenes, 320, 559–60
Liver, 68–69, 206–7, 325. *See also* Organ meats
  Campylobacter, 530–31
  composition, 217
  pH, 69, 217
  paste, 479
  pudding, 73
  S. aureus, 443
  sausage, 72
  spoilage, 216–18
Lobsters, 76–77, 227–28, 230–31
LT. *See* Enterotoxins
Luciferin-luciferase. *See* ATP assay
Luminometry. *See* ATP assay
Luncheon meats, 69–70, 72–73, 222, 224, 426–27. *See also* Meats
Lymph nodes, 62
Lyophilization. *See* Freeze drying
Lysis inhibition test, 158
Lysozyme, 48

Mace, 277
Machinery mold, 28
Macrophage assay, 181
Magon, 373
Maillard browning, 350–51, 358
Malo-lactic fermentation, 247
Malty flavor, 244
Margarine, 31, 243
Market diseases, of produce, 21
Mashing, 386
Mastitis, 24, 439
Maws, 73
Mayonnaise, 243, 248–49, 494, 496–97
Meat Inspection Law, 9
Meats, 4, 62–73
  campylobacters, 529–31
  coliforms, 63–64
  comminuted, 62–64, 309, 433
  composition, 206
  dark-firm dry, 215, 218, 220

electrical stimulation, 68
fermented, 377–81
food-poisoning outbreaks, 426
freezer life, 319, 324–25
greening, 218, 222
hot-boned, 66–67
hydration capacity, 214–16
irradiation of, 305, 309–10
mechanically deboned, 65–66
microorganisms in, 70, 207–8, 409, 493, 529
mycotoxins, 544
off-odor compounds, 211–12, 215, 220–22
organ/variety, 69–72
pH range, 37, 209
pigments, 223
pre- and postrigor effects, 209–10
processed, 207, 215, 218–25
rapid estimated SPC, 139
red, 62–68
restructured, 67
salmonellae in, 427, 493, 496
sliminess, 211–12, 222
spoilage, 206–16
staphylococcal gastroenteritis, 451–52
tenderization, 68
vacuum-packaged, 69–72, 218–20
viruses, 554
Melanconiales, 29
Membrane filter methods, 101–2
Menthol, 274, 276
Mesophile, 49
Mesophilic bacteria, 428
Metabisulfite, 264
Metabolic injury, of microorganisms, 110–17, 327
Metchnikoff, E., 6
Methanococcus spp., 38
Methionine incorporation, 146–47
2-Methoxy-3-isopropylpyrazine, 221
Methyl
    acetate, 221
    isobutyrate, 221
    mercaptan, 220–21, 226
    2-,3-methylbutanol, 221
    -2-methylbutyrate, 221
    sulfide, 221
Methylene blue reduction, 104
Methylococcus capsulatus, 393
Methylomonas spp., 392–93
    methanica, 393
4-Methylumbelliferyl-α-D-galactoside, 146
4-Methylumbelliferyl-β-D-glucuronide (MUG), 145–146, 508
Metmyoglobin, 223, 266
Mezcal, 373
Microaerophile, 415
Microbacterium spp., 20, 207, 242, 308, 364
Microbiological standards/criteria/guidelines, 423–30
    dairy products, 428–29
    in effect/proposed, 430–32, 433
    granulated sugar, 427–28
    liquid sugar, 428
    relative to food safety, 426–27
    sampling plans, 430–32
    starch and sugar, 427

tomato products, 429
Microcalorimetry, 130, 133, 134
Micrococcaceae, 22, 24
Micrococcus spp., 15–18, *22*, 49, 55, 68, 207, 224, 232, 240, 242, 388, 440, 580, 617
    aurantiacus, 379
    cryophilus, 588
    radiodurans, 286, 300, 308–9, 603–12
    radiophilus, 603, 607, 611
    roseus, 308, 604
    rubens, 603–4
Microorganisms
    factors influencing, 61–62
    sources of, for foods, 15–18
    types, in foods, 18–32
Microscope colony count method, 102
Microwaves, 300
    trichinellae destruction, 564–65
Milk, 5–8, 349, 442, 452, 493
    aflatoxins, 544–45
    certified, 6, 414, 428
    diseases from, 8, 80–81, 409, 527–28, 531–32, 559–60
    fermented, 371
    evaporated, 347
    pasteurized, 6, 9, 331, 500, 528
    raw, 80–81, 242, 495–96, 528
    spoilage, 7, *242–43*
    standards/criteria, 414, 428–30
    sweetened condensed, 250, 347
    UHT-treated, 382
Minimum radiation doses, 305
Miso, 372, 383–84
Modified atmosphere packaging. *See* Carbon dioxide; Controlled atmosphere storage
Moisture content. *See* a_w
Molar growth yields, 368–69
Molds, 16, 51, 224, 242, 244–46, 250, 260–61, 265, 379–80, 385, 550
    a_w minima, 41, 349
    common food-borne, 25–29
    beverage standards, 428
    fermented foods, 372
    incidence in foods, 83
    inhibitors, 260, 284–85
    intermediate-moisture foods, 358
    isolation, 119–20
    meats, 208, 210
    pH growth range, 34
Molluscan shellfish, 227, 231–32, 430, 552–53
Monascus spp., 208
    bisporus, 349
Moniliaceae, 30
Monilia spp., *28*, 208
    americana, 28
Monilinia fructicola, 200
Monkey emesis test, 173–74, *176*, 438
Monoclonal antibodies. *See* ELISA methods
Monolaurin, 274–75, 278
Moraxella spp., 15, 19, *22*, 53, 68, 70, 75, 207, 210, 217, 221, 226, 228–31, 308, 603, 616
    nonliquefaciens, 309

Moray eel poisoning, 568
Mortierella spp., 208
Moseley test, 129
Most probable numbers (MPN), 98, *103–4*
Mother of vinegar, 18
Mouse diarrhea assay, 176
Mouse lethality test, 172–75, 525
Mucor spp., 15, *28*, 208, 210, 224, 240, 244, 372
  spinosus, 349
MUG. *See* Methylumbelliferyl-glucuronide
Murine spleen cells assay, 181
Mushrooms, 478
Mussel poisoning. *See* Paralytic shellfish poisoning
Mutagen/mutagenesis tests, 541, 545
Mycelium, 25
Mycobacterium spp., 80, 392
  hominis, 331
Mycoderma spp., 15, *31*
  vini, 31
Mycoplasmas, recovery, 118
Mycotoxins, 154, *541–51*
  aflatoxins, 8, 542–47
  Alternaria toxins, 547–48
  assay methods, 154
  citrinin, 548
  ochratoxins, 548–49
  patulin, 549–50
  penicillic acid, 550
  production, 541–42
  sterigmatocystin, 550–51
  structures, 542, 548–51
  zearalenone, 551
Myoglobin, 223, 266

Nailhead spot, of tomatoes, 198
Natamycin, 280, 281–83
National Canners Association, 427
Neck rot, of bananas, 28
Necrotic enteritis, 459, 462
Neisseria spp., 207
Neisseriaceae, 22
Nematodes, 563–65
Neonatal mouse assay, 173
Nernst equation, 46
Neurospora spp., *28*, 208
  crassa, 43
  sitophila, 28, 242, 372, 385
Nisin, 7, 262, 280–82
Nitrate, 225. *See also* Nitrite
Nitric oxide, 266
  haemochromogen, 223
  myoglobin, 223
Nitrite, 70–71, 260, 265–72
  antibotulinal activity, 267–68
  antimicrobial spectrum, 266–67
  interaction with other chemicals, 267–70
  meat pigments, 266
  mode of action, 270–71
  nitrosamines, 268–69
  Perigo factor, 267
  with sorbate, 269–70
  summary of effects, 271–72
  vacuum-packaged meats, 70
Nitrocellulose colony-blot method, 154
Nitrogen atmospheres, 54–55, 71–72

Nitrogen cycle, 13–14
Nitroreductase, 399
Nitrosamines, 268–69, 307
Nitrosomyoglobin, 266
Nocardia spp., 80, 392–93
Nonagglutinating vibrios. *See V. cholerae* non-01
Nonane, 221
Nonenzymatic browning. *See* Maillard browning
Nordihydroguaiaretic acid, 274
Norwalk virus, diarrhea, 9, 508, 554–56
Nuclease. *See* Thermostable nuclease
Numerical taxonomy, 210
Nutmeats, 245–46
Nutmeg, 277
Nutrient content, and microbial growth, 47–48

Obesumbacterium. See Flavobacterium *proteus*
Ochratoxins, 131, 152, 154, *548–49*
Off-odors, of meats, 212, 215, 220–22
Ogi, 372, *384*
Oidia, 25
Oidium. See Geotrichum
Oils, 4
Olives, 248, 372, *382–83*
Onions, 195, 310
  rings and dips, 80, 82
ONPG, 130, 146, 490
Ontjom, 372, 385–86
Oo, 373
Oospora. See Geotrichum
  rot. *See* Sour rot
Oospores, 25
Orange juice concentrate, 129, 131, 144
Oregano, 277
Organ meats
  freezer life, 325
  organisms in, 64, 68–69, 207
  spoilage, 216–18
Osmoduric organisms, 273
Osmophiles, 8, 273
Osmophilic yeasts, 43, 349
Osmoregulators, 43
Ouchterlony. *See* Gel diffusion
Oudin. *See* Gel diffusion
Ovoflavoprotein, 241
Ovotransferrin, 241
Oxidation-reduction potential, 44–47, 461
Oxymyoglobin, 223, 266
Oxytetracycline. *See* Tetracycline
Oysters
  bacteria in, 76
  composition, 227
  ICMSF standards, 430
  spoilage, 231–32
  V. cholerae non-01, 520
  viruses in, 412, 552–53, 555–56
Ozone, 6, 52

Pacific hake, 229
Palm wine, 373, 388
Papovaviruses, 552
Parabens, *260–62*, 277, 549
Paracolons, 413

Paralytic shellfish poisoning, 152, 567
Paratyphoid, 23, 489
Parsley, 81
Parvoviruses, 552
Passive immune hemolysis test, 131, 158
Pasteur, L., 5
Pasteurization, 5, 331, 428–29
Patulin, 549–50
Peanut butter, 243
Peanuts, 545
Peas, 79, 195, 197
Pectinatus cerevisiiphilus, 246
Pediococcus spp., 15–16, *22–23*, 73, 364,
    581–82, 617
  acidilactici, 366, 378
  cerevisiae, 246, 366, 371–72, 378, 381–82,
    395–96
  pentosaceus, 366
Pemmican, 354
Penicillic acid, 550
Penicillium spp., 15–16, *28–29*, 208, 210,
    224, 240, 244, 248, 318, 371–72, 384,
    392, 550
  camemberti, 375
  chrysogenum, 39
  citrinum, 307–8, 389, 548
  claviforme, 549
  cyclopium, 389, 549–50
  digitatum, 200
  expansum, 389, 549–50
  patulum, 549
  puberulum, 550
  roqueforti, 375, 549
  variable, 549
  viridicatum, 548–49
2,3-Pentanedione, 276
1-Penten-3-01, 221
Peppers, 197
Peptococcus magnus, 47
Peptostreptococcus spp., 364
Perfringens food poisoning, 8, 20, 459, 462–
    69
Perigo factor, 267
Pet foods, intermediate-moisture, 351–52
Petrifilm. *See* Dry film
Peujeum, 372
pH, 33–40
  dairy products, 37
  effect of different acids, 39–40
  effect on organisms, 38–40
  freezing effects, 327
  fruits and vegetables, 36
  growth ranges, 34
  interaction with ingredients, 35
  meats, 37, 68, 209
  seafood, 37, 232
  vacuum-packaged meats, 69
Phagocytosis, 181
Phenylacetaldehyde, 274, 276
Phenylethyl alcohol, 230
Phoenix phenomenon, 112
Phoma spp., 248
Phomopsis citri, 200
Phosphoketolase, 365
Phospholipase, of *B. cereus*, 481
Phosphoroclastic system, 271
Photobacterium spp., 207

Phylctaena vagabunda, 200
Phytophora spp., 200–201
Pichia spp., 30, 265, 392
Pickles, 248, 372, *381*
Picornaviridae, 552, 554
Piglet assay, 173, 504–5
Pigments, of meats, 223
Pimaricin. *See* Natamycin
Pineapple black rot, 200
Pink mold rot, of vegetables, 200
Pinspot, of eggs, 240
Pizza dough, 241
Planococcus spp., 207, 617
Plaque forming unit (pfu), 552
Plasmapara viticole, 200
Plasmids, 559
  C. jejuni, 531
  E. coli, 183, 504–5
  lactic streptococci, 369
  salmonellae, 500
  S. aureus, 450
  V. parahaemolyticus, 518
  Y. enterocolitica, 525–26
Plasmolysis, 273
Plate count. *See* Standard plate count
Plesiomonas spp., 207, 616
  shigelloides, 557–59
Pod blight, of beans, 198
Poi, 372
Point mutations, 545
Poising capacity, 45
Poliovirus, 306, 552–54, 556
Polygalacturonase, 248
Polyhydric alcohol accumulation, 43
Polyphosphate, 262
Pork, 54–55, 63–64, 71, 206
  loins, 54–55
  modified atmosphere storage, 54–55
  salmonellae, 492–94
  spoilage, 209–12, 217
  trimmings, 63
  trichinosis, 563–65
  yersiniae, 527
Potatoes, 195, 197, 200, 310
Poultry Inspection Bill, 9
Poultry meats, 70, 73–75, 207–8, 442
  Campylobacter spp., 529–31
  effect of chilling, 73
  freezer life, 319
  off-odors, 221, 226
  New York–dressed, 225, 227
  radicidation, 307, 310
  salmonellae, 492, 496, 501
  sausage, 63
  slime, 226
  spoilage, 221, 225–27
  staphylococci, 443, 452
  volatiles, 221
Powdered eggs and milk, organisms in, 84
Prawns, 4
Preservation of foods, 257
  canning, 338
  chemicals, 259
  drying, 346
  freeze drying, 347
  freezing, 321
  intermediate-moisture, 351

irradiation, 297
refrigeration, 317
salting, 272
Preservative system, 278
Procercoid larva, 562
Processed meats. *See* Meats
Procholeragenoid, 499
Progenitor toxin, 475
Proglottids, 562
Propanol, 230
Propionaldehyde, 221
Propionibacterium spp., 242, 248, 364, 617
　shermanii, 375
Propionic acid/propionate, 260, 264, 365,
　376, 549
Propylene oxide, 260, *285*
Propyl gallate (PG), 274–75, 545
Protein A, 153, 438
Protein synthesis inhibition tests, 181, 500
Proteus spp., 15–17, *23*, 55, 196, 207, 224,
　230–32, 240, 242, 245, 420–21, 501,
　560–61, 580, 616
　intermedium, 240
　morganii, 219, 560–61
　vulgaris, 240, 262
Protomer, 505–6
Protopectinase, 196
Protozoa, 412, 566–67
Providencia spp., 616
Pseudomonadaceae, 23
Pseudomonas spp., 15–17, 19, *23*, 49, 54–
　55, 68–70, 75, 77, 114, 133, 197, 207,
　210, 217–18, 220–22, 225, 228–29, 231–
　32, 240, 242, 244, 327, 452, 501, 579–
　80, 583–84, 586, 610, 616, 618
　aeruginosa, 55, 85, 275, 308, 581, 595
　api, 197
　biotypes, 210
　cichorii, 197
　cocovenenans, 385
　fluorescens, 42, 55, 116, 210, 221, 240,
　　244, 275, 515, 585, 587, 610
　fragi, 55, 68, 210, 216, 221, 244, 329, 585,
　　588
　geniculata, 275
　glycinea, 197
　graveolens, 240
　lachrymans, 197
　maculicola, 197
　marginalis, 195
　mephitica, 244
　metabolic injury, 114, 116
　nigrificans, 244
　phaseolicola, 197
　pisi, 197
　putida, 210, 221
　putrefaciens. *See* Alteromonas
　Shewan's groups, 75, 216, 222, 225–27,
　　229, 618
　tomato, 197
Psychrophiles, 8, *317–18*, 579
Psychrotrophs, 7, 21, 23, 49, 53, 70, 80,
　318, *579–89*
　characterized, 579
　determination, 98, 118
　distribution, 80, 579–80
　heat lability, 587–89

physiology, 582–87
　temperature characteristic, 583
　temperature-induced changes, 580–82
Ptomaine poisoning, 8
Pufferfish poisoning, 568
Pullularia spp., 208
Pulque, 373
Pumpkin, 195
Putrefactive anaerobe, 64, 250–51, 342
Putridity, of butter. *See* Surface taint
Pyocyanin, 23
Pyruvate, 113

Q₁₀. *See* Temperature coefficient

Rabbit
　diarrhea model, 173, 176
　ileal loop tests, 173–74
　intestinal cells, 181
　skin test, 173–74, *177*
Rad, 299
Radappertization, 221, 304
　application, 305–7
　defined, 305
Radiation-resistant organisms, 603–4
　characteristics, 604–5
　destruction, 605–7
　genetics, 610–13
　mechanism, 608–10
Radiation, 299–313
　application, 303
　defined, 297, 299–300
　effect on foods, 311–13
　effect on microorganisms, 304–9
Radicidation, 304, 305, 307, 310
Radioimmunoassay (RIA), 130, 151–52, 556
Radiolysis products, 311–12, 608–10
Radiometry, 130, 133, *143–45*
Radioprotective compounds, 611
Radishes, 195
Radurization, 304, 305, 308–9
Rancidity, of butter, 243
Raoult's law of mole fractions, 352–53
Rate of destruction curve, 339
Rat hepatocyte assay, 181
Reading unit, 281
Red bread mold, 28
Redox potential. *See* Oxidation-reduction
　potential
Red tide, of oceans, 567
Refrigerator temperatures, 318
Relative humidity, 51–52, 351. *See also* aw;
　ERH
Reoviridae, 552, 556
Resazurin reduction, 99, 104–5, 214
Restructured meats, 67
Resuscitation. *See* Metabolic injury
Reverse passive hemagglutination test, 131,
　156
Reye's syndrome, 547
rh, 46
Rhesus monkey emesis assay, 173
Rhizoctonia spp., 200
Rhizopus spp., 15, *29*, 208, 210, 224, 241,
　244
　arrhizus, 383

oligosporus, 371–72, 383–85
oryzae, 372, 383
soft rot, 200–201
stolonifer, 29, 200–201, 242, 349
Rhodamine B, 147
Rhodopseudomonas spp., 392
Rhodotorula spp., 15–16, *31*, 208, 210, 225, 318, 392, 580
glutinis var. *glutinis*, 248
minuta var. *minuta*, 248
rubra, 248
Ribosomes, 596–97
Rickettsiae, 193
Rigor mortis, of meats, 37, 209
Ring rot, of potatoes, 197
Ristella spp., 421
RITARD assay model, 179–80
Rodac plate method, 107–8
Roentgen, 297
Roll tube method, 105
Root beer, 247
Ropiness
beer, 246
bread, 242, 264
milk, 243
Roquefort cheese, 375
Rosemary, 277
Rotavirus, 140, 508, 556–57
Rots, of eggs, 240
Rum fermentation, 387

Saccharomyces spp., 15–16, *31–32*, 208, 245–46, 249, 265, 372, 388, 392
bayanus, 265, 389
carlsbergensis, 30, 246, 265, 372, 386
cerevisiae, 30–31, 43, 47, 135, 246, 287, 369, 372–73, 386–87, 393
diastaticus, 246
ellipsoideus, 373, 387, 389
exiguus, 373, 391
intermedium, 373
marxianus, 389
sake, 373, 388
rosei, 369
rouxii, 43, 273, 349, 372, 383–84
Saccharomycopsis. See Endomycopsis
Sage, 277
Saké, 373, 388
Salad dressing, 243, 248–49
Salads, 81–83, 430, 554
Salami, 72–73, 275, 377
Salicylic acid, 262
Salinometer, 381–82
Salmonellae
antigenic structure, 491
antimicrobial resistant, 496
a_w, 496
classification, 490–91
distribution, 73–74, 491–96
D values, 497–98
ELISA detection, 131, 154–55
enrichment serology, 150–51
fluorescent antibody, 131, 147–50
growth requirements, 496–97
isolations, 74, 119, 501
meats, 63, 66–67, 427, 492
pH range, 496–97

radiation destruction, 498
radiometric detection, 131, 145
toxins, 499–500
Salmonella spp., 15, 17–18, *23*, 56, 72–73, 77, 83, 145, 174, 180–81, 196, 207, 240, 266, 275, 307, 326, 426, 430, 433, *489–502*, 521, 530, 553, 616
agona, 494–95
anatum, 74, 114, 326, 491–94
bareilly, 499
blockley, 74
choleraesuis, 490–91, 494, 496, 498
cubana, 492, 499
derby, 494
dublin, 492, 495–96, 500
eastbourne, 496, 499
enteritidis, 8, 85, 490–91, 494, 499–500
gallinarum, 326, 491
heidelberg, *494, 496, 498*
hirschfeldii, 490–91
infantis, 74, 494
johannesburg, 496
minnesota, 494
montevideo, 490–91, 494, 498
muenchen, 493, 500
newington, 326
newport, 490–91, 494, 496, 499–500
oranienburg, 43, 490–91, 494
paratyphi A, 489, 491
paratyphi B, 326, 489, 586
pullorum, 498, 499
reading, 74
saint-paul, 494, 496
sandiego, 74
schottmuelleri, 491
senftenberg, 114, 333–34, 336, 357, 492, 497–98
thompson, 494–95, 497
typhi, 326, 489, 491
typhimurium, 75, 114, 116, 139, 144, 155, 275, 326, 467, 491–92, 494, 496–97, 499–500
Salmonella surveillance program, 8
Salmonellosis
cases by year, 495
common serovars, 494
enterotoxin, 9, 180, 499–500
history, 8
incidence, 500–501
prevention and control, 501–2
syndrome, 498–99
vehicle foods, 81, 500–501
Salmon steaks, 76
Salt (NaCl), 40, *272–73*
Sampling plans
three-class, 431–32
two-class, 79, *430–32*
Sandwiches, 81–83
San Francisco sourdough bread, 373, 390–91
Sarcinae sickness, of beers, 246
Sarcina spp., 22, 603. *See also Micrococcus*
lutea, 609
Sassafras, antimicrobial effects, 277
Sauce and gravy mixes, 84
Sauerkraut, 247–48, 372, *381*

Sausages, 72–73, 305. *See also* Meats
  casings, 72
  dry, 377, 379–80
  fermented, 4, 377–80
  organisms in, 63, 72–73
  salmonellae, 493–94
  semidry, 377–78
  spoilage, 222, 224
  summer, 377
Scab, of tomatoes, 198
Scallops, 227, 232
Scarlet fever, 409
Scenedesmus spp., 392
  quadricauda, 393
Schewan's scheme, 618
Schizosaccharomyces spp., 15, *32*, 389
  pombe, 373
Sclerotinia spp., 28
  fructicola, 200
  sclerotiorum, 200
Sclerotium rolfsii, 55
Scombroid poisoning, 560–62
Scopulariopsis spp., 208
Scotch whiskey, 373, 387
Seafood, 53, *75–77*, 305, 308, 452, 561
  CO_2 storage, 53, 72
  freezer life, 319
  irradiation, 307–10
  organisms in, 76, 207–8, 414, 419, 471
  spoilage, 215–16, 227–32
  standards for, 430, 433
  V. alginolyticus, 522
  V. cholerae non-01, 520
  viruses, 552–56
  V. parahaemolyticus, 519
Selective media. *See* Culture media
Sereny test, 174, *177–78*, 525
Serovar, 23
Serratia spp., 15–16, *23–24*, 207, 215, 231,
    240, 388, 420–21, 580
  liquefaciens, 64, 70, 218, 265
  marcescens, 265
Serum resistance factor, 526
Shellfish. *See* Crustacean shellfish; Fish and
    shellfish; Molluscan shellfish; Specific
    types
Shellfish poisoning. *See* Paralytic shellfish
    poisoning
Shigalike toxin, 503
Shigella spp., 15, 17, *24*, 145, 181, 430,
    508, 530, 616
  dysenteriae, 503
  flexneri, 155
  isolation, 119
  shigellosis, 500
  sonnei, 114
Shrimp, 305
  organisms in, 76–77, 231, 519
  radurization, 305, 308, 310
  spoilage, 230–31
Side slime, 197
Single-cell protein, 391–95
Single radial immunodiffusion test, 131, 159
Skunklike odor, of butter, 244
Sliminess, 211, 212, 220, 226
Slimy brown rot, 200
Slimy curd, of cottage cheese, 244

Slow freezing, 321
Smoke, 4, 288
Smut, 200
Sodium benzoate, 259, 261, 262
Sodium chloride. *See* Salt
Sodium diacetate, 260, *286*
Sodium-α-phenylphenate, 284
Soft moist foods. *See* Intermediate-moisture
    foods
Soil, food-borne organisms in, 16
Sorbic acid/sorbate, 7, 85, 260, 262–64,
    277, 355–56, 388
  antimicrobial spectrum, 263, 549–50
  mode of action, 262–63
  with nitrite, 263, 269–70
Soul foods, 73
Soups, 84
Sour cream, 80, 370
Souring of meats, etiology, 210, 222
Sourness, of beers, 246
Sour pumpernickel, 373
Sour rot, 200–201
Soybean flour, 64–65
Soybeans, 373
Soy-extended ground meats, 64–65, 211
Soy sauce, 372, 383
Space foods, 83, 355
Spaghetti sauce mixes, 84
Spices
  antimicrobial properties, 274, 276–77
  organisms in, 64, 83
  radicidation, 310
  spoilage, 245
Spinach, 79, 195
Spiral plater, 100–101
Spirulina maxima, 392–94
Spirillum itersonii, 597
Spoilage. *See* Individual foods
Spoilage rate curve, 31
Sporangiophores, 25
Sporeformers, isolation, 119
Sporendonema. See Wallemia
Sporobolomyces spp., 208
Sporolactobacillus spp., 617
Sporotrichum spp., 15, *29*, 208, 210
Spray gun surface method, 109
Springer, 252
Squash, 79, 195
Squash seeds, 84–85
ST. *See* Enterotoxins
Standard plate count (SPC)
  dry film, 103
  factors affecting, 97–99
  homogenization of samples, 99
  pour plating, 98
  spiral plater, 100–101
  surface plating, 98
Standards. *See* Microbiological
    standards/criteria/guidelines
Staphylococcal gastroenteritis, 8, 437–53
  incidence, 439
  minimum toxin for, 451
  prevention, 453
  symptoms, 450–51
  vehicle foods, 4, 451–52
Staphylococcus spp., 15, 17–18, *24*, 56, 62,
    70, 224, 418–19, 437–52, 617

aureus, 24, 43, 65–67, 70, 72–74, 76–77, 81–83, 99, 110–14, 116, 144, 159, 266, 269, 275–76, 309, 352, 357, 395, 426, 430, 432–33, *438–52*, 608
  animal strains, 438–39
  antagonists of, 452
  a_w, 356–57, *440–41*, 446
  D and z values, 340
  detection methods, 130–31
  effect of salts, 440–41, 446
  enterotoxins, 131, 152–53, 157, 174, 356, *442–50*
  in foods, 63, 65–67, 69, 72–77, 439, 452
  growth requirements, 439–41
  habitat, 439
  isolations, 119
  pH range, 441
  thermostable nuclease, 135–37
  virulence factors, 438
 epidermidis, 24, 85, 358, 438, 440, 452
 hyicus subsp. *chromogens*, 438
 hyicus subsp. *hyicus*, 438
 intermedius, 438
 saphrophiticus, 438, 440
Starter cultures, 129, 370
Stem-end rot, of fruits, 200
Sterigmatocystin, 550–51
Sterilization, 331–32
Sticky film method for surfaces, 108–9
Stomacher, 99
Storage fungi, 549
Streptobacterium, 71, 365
Streptococcaceae, 22, 24
Streptococcus spp., 8, 15–17, *24*, 49, 68, 70, 83, 146, 207, 210, 217, 222, 224, 241, 308, 331, 364, 414, 617. *See also* Enterococci and *Enterococcus*
 agalactiae, 559
 bovis, 366, 414–16
 cremoris, 55, 140, 364, 366, 369–70
 diacetilactis, 140, 366, 369–70, 395
 durans, 414–16
 equinus, 414–16
 faecalis, 24, 50, 99, 115, 224, 269, 275, 300, 309, 335, 350, 368–69, 381, 391, 414–20
  subsp. *faecalis*, 414
  subsp. *liquefaciens*, 414
  subsp. *zymogenes*, 414
 faecium, 24, 115, 269, 300, 309, *414–16*
 fecal. *See* Enterococci
 group D, 73, 414–15
 lactis, 7, 24, 140, 243, 269, 280, 309, 366, 369–71, 375, 384, 395
  var. *maltigenes*, 244
  var. *taette*, 371
 sanguis, 369
 thermophilus, 140, 366, 370–71, 374–75, 395, 399
Streptomyces spp., 16, 49, 207, 593
 griseus, 595
 natalensis, 282
Sublethal heat injury, 114–15
Subtilin, 262, 280, 282, *283–84*
Suckling mouse assay, 173–76, 479, 499, 505–6, 525, 557–58. *See also* Infant mouse assay

Sucrose dicapyrlate, 278
Sucrose diesters, 277
Sufu, 372
Sugars
 criteria for, 427–28
 preservative properties, 272–73
 spoilage, 244–45
Sulfide spoilage organisms, 250–51, 427
Sulfite. *See* Sulfur dioxide
Sulfolobus, 38
Sulfur dioxide, 5, 260, 262, 264–65, 284, 347, 549
Sulphmyoglobin, 223
Summer sausage. *See* Sausage
Surface examination methods, 106–9
 agar syringe, 108
 direct surface, 108
 rodac plate, 107–8
 spray gun, 109
 sticky film, 108–9
 swab, 106–7
 swab-agar slant, 109
 ultrasonic devices, 109
Surface taint, of butter, 243
Swab/agar slant method, 109
Swab-rinse method, 106–7
Sweet acidophilus milk, 397–98
Sweetened condensed milk, 250, 347
Sweet potatoes, 195, 200
Swiss cheese fermentation, 375
Syrup, 479

Taenia saginata, 562. *See also* Trichinosis
 solium, 309, 562
Taette, 371
Takadiastase, 363
Talaromyces. *See Penicillium*
Talmadge-Aiken Act, 9
Tao-si, 372
Tapeworm. *See Taenia saginata*
Tarhana, 371
Tdh gene, 518
Teekwass, 373
Tempeh, 372, 383
Temperature characteristic, of psychrotrophs, 583
Temperature coefficient, *317*, 329, 584
Temperature cycling, 545
Temperature effects, on growth, 49–51
Temperature shock, 327
Tennecetin. *See* Natamycin
Tenuazonic acid, 548
Tequila. *See* Pulque
Tetracyclines
 mode of action, 283
 structures, 280
 use in foods, 211, 282–83
Tetrazolium reduction, for poultry spoilage, 226
Texturized soy protein, 64–65. *See also* Soy-extended meats
Thamnidium spp., 15, *29*, 50, 208, 210
Thawing effects, on foods, 321–22, 328–29
Thermal death point
 death time, 336, 338–39
 curve, 339, 341–42
 defined, 333, 338

Thermobacterium, 365, 368
Thermoduric organisms
  defined, 331
  isolation, 119
Thermophiles, 8, 337, 593–601
  defined, 49, 331, 593
  enzymes, 594–96
  flagella, 597–98
  isolation, 119
  membranes, 599–600
  nutrition, 598
  ribosomes, 596–97
Thermophilic anaerobes, 250–52
  canned foods, 250–52
  radiation resistance, 301
  spoilage, 250–52
  standards for, 427
Thermophilus milk, 398
Thermostable direct hemolysin, 174, *517–18*
Thermostable nuclease, 117, 130, *135–37*
  compared with coagulase, 136
  D values, 137, 445
  enterococci, 136
  methods, 136
  minimum cells for detectable levels, 136
  S. aureus, 135–36, 438
  S. epidermidis, 137
  S. hyicus, 438
  S. intermedius, 438
Thermus aquaticus, 595
Thibendazole, 284–85
Thielaviopsis paradoxa, 200
Thiocyanate, 288
3,3'-Thiodipropionic acid, 116
Thumba, 373
Thyme, 277
Thymol, 48
Tibi, 373
Tissue culture systems. *See* Cell culture
  systems
Titratable acidity, 278
Tofu, 82, 528
Tomato products, 195, 197, 200
  botulinal toxins, 472–73
  mold count methods, 106
  mold tolerances, 429–30
Top yeasts, 30, *363–64*, 386
Torula spp. *See Torulopsis* spp.
Torulopsis spp., 15–16, 18, *32*, 208, 210,
  225, 240, 244–45, 318, 371, 580
  candida, 276, 392
  globosa, 250
  glutinis, 248
  lactis-condensi, 250
  methanosorbosa, 392
  stellata, 251
Total counts, 97, 421–22. *See also* Standard
  plate counts
Tourne disease, 247
Toxic shock syndrome toxin, 442
Toxoflavin, 385
Toxoplasma gondii, 566–67
Toxoplasmosis, 566–67
Transposon, 505
Travelers' diarrhea, 158, 508, 555–56
Trematodes, 565–66

Trichinella spiralis, 563–65
  curing and smoking effects, 564
  drying effects, 350
  ELISA test for, 565
  fermented sausage, 378, 564
  freezing destruction, 327, 563–64
  heat destruction, 563, 565
  microwave cookery, 564–65
  radiation destruction, 310
  trichinosis, 563, 565
Trichoderma spp., 393
  viride, 393
Trichosporon spp., 15, *32,* 208
  cutaneum, 392
  pullulans, 32, 349
Trichothecium spp., 15–16, *29,* 30–31
  roseum, 29, 200
Trimethylamine (TMA), 221, 227–28, 230
Trimethylamine-N-oxide (TMAO), 228, 230
T-2 toxins, 28
Tuberculosis, 409
Tuna, 561
Tuna pot pies, 82, 84
Turbidity, of beers, 246
Turkey meat, 73–74, 206
Twelve-D, thermal process, 306, 342–43
Tylosin, 262, 280, 284
Typhoid fever, 8, 23, 409, 489

Ultra-high temperatures (UHT), 332
Ultrasonic device, 109
Ultraviolet light, 299, 605

Vacuum-packaged meats, 54–55, 63, *69–72,*
  73, 226
  effect on flora, 54–55
  history, 69
  spoilage, 218–20, 224
  volatiles, 220–22
V and W antigens, 526
Vanillin, 274, 276
Vapor pressure, of ice and water, 322
Variety meats. *See* Organ meats
Vegetables, 78–80, 449
  composition, 194–95
  freezer life, 318–19
  organisms in, 78–80, 411
  pH range, 36
  radiation preservation, 308–10
  spoilage, 193–203
  standards, 433
  viruses, 553
Verdohaem, 223
Vero cell culture assay, 181, 184, 500, 503
Vibrio spp., 15, 20, *24–25,* 207, 225, 515,
  518, 520–22, 527, 580, 588, 616, 618
  alginolyticus, 515, 522, 561
  cholerae 01 and E1 Tor biotype, 25, 140,
    151, 176, 180–81, 499, 503, 506, 515,
    520–21, 531
  cholerae non-01, 9, 18, 174, *520–21*
    bioassays, 174, 181, 521
    distribution, 521
    gastroenteritis, 9, 520
  fisheri, 589
  hollisae, 518

injury repair, 115
marinus, 589, 597
mimicus, 181
parahaemolyticus, 8, 18, 25, 77, 115–16,
    119, 174, 179–83, 185, 263, 275, 277,
    320, *515–20*, 533
    gastroenteritis, 8, *518–19*, 533
    growth conditions, 516–17
    minimum pH, 516–17
    occurrence, 515–16
    outbreaks, 519
    virulence properties, 174, 180–81, *517–
    18*
   species compared, *516*
   vulnificus, 174, 181, 515, *521–22*
Vibrionaceae, 24
Vibriosis. *See V. parahaemolyticus*
Vinegar, fermentation, 5, 18, 373
Viral gastroenteritis, 554–57
   hepatitis A, 554
   Norwalk, 554–56
   rotavirus, 556–57
   synopsis of outbreaks, 555
Viruses, 274, 412, *551–57*
   Adeno, 552
   African swine fever, 554
   coxsackie, 552
   destruction in foods, 554
   ECHO, 412, 552–53, 555
   entero, 552–53
   foot-and-mouth, 327
   hepatitis A, 554–56
   Herpes simplex, 308
   hog cholera, 554
   incidence in foods, 552–53
   Norwalk, 554–56
   persistence in foods, 553
   polio, 412, 552–53, 555
   radiation resistance, 301, 308
   recovery, 119
   reo, 552
   rotaviruses, 552, 554, 556–57
Visceral taint, of poultry, 227
Vitamins, 48, 312
Voges-Proskauer test, 21, 557–58
Volatiles, in spoiled meats, 220–22
Vomitoxin, 28

Wallemia spp., 208
Water
   campylobacteriosis, 532
   coliforms, 412, 414, 417
   open/closed, 553
   viruses, 553
Water activity. *See* a_w
Water-holding capacity, 214. *See also*
   Extract-release volume
Watermelons, 193
Water stress, 43
Watery soft rot, 200–201
Weaned piglet assay, 173
Whiskers, of beef, 29, 210
White rot, of onions, 202
White spot, of beef, 29, 210
Whole-animal assays, 172–78
Wholesome Meat Act, 9

Wild yeasts. *See* Asporogenous yeasts
Wiltshire bacon, 73, 224
Wines
   fermentation, 373, 387–88
   spoilage, 31, *246–47*
   turbidity, 31
   flowers, 31, 247
Wisconsin process, for bacon, 268
World Health Organization (WHO), 307,
   427
Wort, 386

Xanthomonas spp., 197, 248, 616
   campestris, 197
   citrus canker, 197
   phaseoli, 197
   vesicatoria, 197
Xeromyces bisporus. *See Monascus*
Xerophile, 41
Xerophilic molds, a_w, 41, 349
X-rays, 297, 303–4

Y-1 adrenal cell assay, 181, 184–85, 521,
   525, 531, 557–58
Y_{ATP}, 368–69
Yeasts, *30–32*, 50, 81, 204, 208, 217, 225,
   244, 250, 260, 265, 274, 287, 390, 392,
   474
   a_w minima for spoilage, 41, 349
   beverage standards, 428
   fermented foods, 372–73
   incidence in foods, 72, 83, 210
   isolation, 119–20
   metabolic injury, 115
   osmophiles, 349
   pH growth range, 34
Yersinia spp., 15, *25*, 207, 522, 580, 616
   biochemical characteristics, 522–23
   compared, 523
   enterocolitica, 8–9, 18, 25, 56, 67, 72,
     120, 174, 183, 218, 320, 508, *522–27*,
     533. *See also* Yersiniosis
     bioassays, 174–76, 180–81, 183
     biovars and serovars, 524, 528
     distribution, 523–24
     D value, 523
     growth requirements, 522–23
     incidence in foods, 526–27
     virulence factors, 525–26
   frederiksenii, 523–24
   intermedia, 523–24
   kristensenii, 523–24
   pestis, 25, 522, 525–26
   pseudotuberculosis, 526
Yersiniosis
   outbreaks, 8–9, 528
   prevention, 533
   symptoms, 527
   virulence factors, 525–26
Y_G, 368
Yogurt
   anticancer effects, 399
   antimicrobial properties, 374
   cholesterol, 398–99
   flora, 6, *374*
   lactose intolerance, 397–98

pH, 374
preparation, 370, 374
titratable acidity, 374

Zapatera spoilage, of olives, 248
Zearalenone, 28, 551
Z value
  C. botulinum, 342
  defined, 340

peroxidase, 342
S. aureus, 446
vitamins, 342
Zygorrhynchus spp., 208
Zygosaccharomyces spp., 32, 245
  bailii, 249, 265
Zygospores, 25
Zymomonas spp., 389
  anaerobia, 246
  mobilis, 369, 390